FLUID MECHANICS
AND
UNIT OPERATIONS

FLUID MECHANICS AND UNIT OPERATIONS

By

David S. Azbel
Nicholas P. Cheremisinoff

ANN ARBOR SCIENCE
THE BUTTERWORTH GROUP

Copyright © 1983 by Ann Arbor Science Publishers
230 Collingwood, P.O. Box 1425, Ann Arbor, Michigan 48106

Library of Congress Catalog Card Number 82-48638
ISBN 0-250-40541-5

Manufactured in the United States of America
All Rights Reserved

Butterworths, Ltd., Borough Green
Sevenoaks, Kent TN15 8PH, England

PREFACE

It is the intent of this book to provide fundamental instruction in the subject of applied fluid mechanics, relying heavily on the methodology and language of industry. The concept has evolved from the authors' teaching experiences and from consultations with academic and industrial colleagues who have concluded that the majority of chemical engineering curricula leave a void between theory and industrial practice. Often, the steps needed to relate the principles of fluid mechanics to the detailed design of a particular unit operation are only superficially examined in course work.

The term "unit operation" refers to industrial practices whose scientific principles of operation are the same, regardless of the specific industry to which they are applied. For example, the unit operation of extraction is employed in the paper, petroleum, pharmaceutical and chemical industries. Although the specific processes, materials and conditions may vary, the design principles for an extraction tower are identical. In the broadest sense, the design of unit operations draws on knowledge from three principal subjects: fluid mechanics, heat transfer and mass transfer. All three comprise the general subject of transport phenomena. Because it is not our intent to present an overview, but rather detailed instruction, we have elected to cover only those operations that demand a thorough understanding of fluid dynamics. Those operations requiring knowledge of heat and mass exchange principles are best covered in separate volumes.

The book is divided into four sections. The first, Chapters 1 through 4, reviews approaches to problem analysis, modeling and fluid transport properties. The material presented is applied throughout the balance of the book.

The second section covers the design of related practices, referred to as the *internal problems of hydrodynamics*. Extensive discussions are devoted to the dynamics of single-phase flows in Chapter 5, with emphasis on problem analysis. These principles and methodologies are applied to the operation of fluid transportation for both incompressible and compressible mediums in Chapters 6 and 7, respectively.

The third section, Chapters 8 through 11, is devoted to operations comprising the class of problems of *external hydrodynamics*. Chapter 8 provides background on heterogeneous systems (solid-liquid) and categorizes the specific unit operations of handling and separation. The operations covered are gravity sedimentation (Chapter 9), centrifugal separation techniques (Chapter 10) and mixing practices (Chapter 11).

The last section, Chapters 12 through 14, concerns the mixed problems of hydrodynamics. The practices of filtration are presented in Chapter 12. The final two chapters cover two-phase operations: gas-solid (Chapter 13) and gas-liquid relations (Chapter 14). Some of the methods outlined in the last section have not appeared in the engineering literature and are presented to stress new concepts and better understanding of multiphase operations.

Considerable thought has been given to the calculations and procedures presented. Each problem posed, along with selected solutions, is based on actual industrial encounters. Consequently, the book will be useful to the practicing engineer, as well as to the student.

David S. Azbel
Nicholas P. Cheremisinoff

ACKNOWLEDGMENTS

The authors express heartfelt gratitude to our numerous friends in industry who devoted time, effort and materials to this volume. Although too numerous to name here, their respective organizations are cited throughout the text. Thanks are also extended to the members of the chemical engineering faculty of the University of Missouri–Rolla for their suggestions, to Paul N. Cheremisinoff for his technical assistance, and to Ann Arbor Science Publishers.

Azbel   Cheremisinoff

David S. Azbel is Professor of Chemical Engineering at the University of Missouri–Rolla. He was previously Research Professor at the State University of New York at Stony Brook and Visiting Professor, University of Minnesota. From 1966 to 1972 he was Professor of Chemical Engineering at the All-Union Polytechnical Institute (Moscow, USSR); from 1961 to 1965 he was Chief of the Technology Department of the Organic and Synthetic Products Research Institute (Moscow, USSR). Previously he was project chief at the Chemical Engineering Research Institute and Petrochemical Research Institute (Irkutsk, Siberia, USSR). Dr. Azbel graduated from Moscow Institute of Chemical Engineering in 1939 and received his Doctor of Science degree from the Mendeleev Institute of Chemical Technology, Moscow, USSR. His engineering, design, research and teaching span the broadest range of chemical, petroleum and nuclear experience. Dr. Azbel is co-author of *Chemical and Process Equipment Design – Vessel Design and Selection*, published by Ann Arbor Science.

Nicholas P. Cheremisinoff heads the Reactor and Fluid Dynamics Modelling Group in the Chemical Engineering Technology Division of Exxon Research and Engineering Company, Florham Park, New Jersey, and was previously a research scientist for a paper company. He is involved in research and engineering on atomization, fluidized-bed reactor design, shale oil development, coal gasification, instrumentation, and equipment design and development. Dr. Cheremisinoff received his BS, MS and PhD degrees in Chemical Engineering from Clarkson College of Technology. He is author of numerous papers and books, including *Fluid Flow: Pumps, Pipes and Channels; Handbook of Fluids In Motion; Cooling Towers: Selection, Design and Practice; Chemical and Nonchemical Disinfection;* and *Gasohol for Energy Production*, published by Ann Arbor Science. He is a member of a number of professional and honor societies, including the American Institute of Chemical Engineers, Tau Beta Pi and Sigma Xi.

CONTENTS

CHAPTER 1

ELEMENTS OF FLUID MECHANICS AND
UNIT OPERATIONS

CONTENTS

INTRODUCTION AND SCOPE

Chemical engineering involves the application of the fundamental branches of science to the conception, design, implementation and operation of industrial processes for the manufacture of the products needed by civilization. It is, by nature, a profession based on practicality of solutions rather than on the platonic search for academic truth. The types of processes and

1

industries that require the skills of chemical engineers are so diversified that engineering curricula rarely specialize, except at advanced levels. Despite the diversification of industry, broad generalities enable engineering to proceed on the basis of fundamental principles. Furthermore, although the number of individual processes is significant, they may be categorized by a series of steps referred to as operations. These individual operations are based, in turn, on the same scientific principles. Regardless of the specific industry, operations overlap; for example, fluids are transported, feedstocks are combined or mixed, fluid-solid systems are separated, materials are dried. Hence, the concept of the unit operation is to approach all process steps with a unified and systematic approach.

Unit operations are applicable to both physical and chemical processes. Most often the physical processes are entangled in chemical synthesis. It is not the intention of this book to cover all unit operations. Instead, its scope is directed toward the hydrodynamics of unit operations. Unit operations aimed at the transfer of mass and heat between process constituents, as well as the thermokinetics in chemical synthesis, all depend on the governing principles of fluid mechanics. Without a thorough understanding of process hydrodynamics, the design of any unit operation becomes an art, specific to the particular process. It is this aspect that is largely responsible for our ability to treat unit operations as a unified subject.

On this premise, we divide unit operations into three broad categories based on the nature of the hydromechanics. Each category refers to a class of problems frequently encountered in industrial fluid mechanics.

1. those that constitute the *internal problems of hydrodynamics*, which include the unit operations of liquid pumping, as well as gas compression and transportation;
2. those that constitute the *external problems of hydrodynamics*, which include the unit operations of sedimentation, centrifugal separation and mixing; and
3. those problems that constitute the *mixed problems of hydrodynamics*, which include filtration, fluidized bed operations and two-phase gas-liquid operations.

UNITS OF MEASURE

First, we will define an engineering system of units that provides a standard format for measuring, interpreting and reporting phenomenological evidence, as well as a vehicle by which such information may be applied to benefit mankind. Historically, units of measure have been arbitrary; investigators working in a particular field would elect to devise convenient definitions. To standardize definitions of measure, the scientific and engineering communities have formulated a policy of an International System of Units (SI) through the International Organization of Standardization (ISO).

Transition to SI has advanced considerably during the past decade. The Metric Conversion Act of 1975 (PL 94-168) declared the coordination and planning of increasing use of the metric system (SI) in United States government policy. A memorandum by the Assistant Secretary of Commerce for Science and Technology in the *Federal Register* of October 26, 1977 interprets and modifies SI for the United States. The Act also provides for the establishment of the U.S. Metric Board to coordinate voluntary conversion. Despite this increased activity, complete changeover to SI has not yet materialized. Consequently, in this transition phase, practicing engineers, scientists and students must be familiar with several systems of units: English-British Units, cgs and SI. This book applies all three systems to problem-solving, with emphasis on SI.

As noted, SI is the abbreviation for the International System of Units (Le Systeme International d'Unités). The prelude to SI was the cgs (centimeter-gram-second) system of metric units. In contrast, SI is based on the meter, kilogram and second as the fundamental units.

The SI system utilizes three classes of units. The first comprises *base units*. By convention, these are dimensionally independent. The second includes *supplementary units*, which are used to measure plane and solid angles. *Derived units* are those formed by algebraic combinations of base units, supplementary units and other derived units. Specific names and symbols are assigned to the units in each class.

The advantage of SI over English units is the elimination of conversion factors within the system. That is, all derived combinations are in terms of unity. For example, the derived unit of power is the *watt*, which, in base units, is defined as 1 joule of work completed in 1 second of time.

There are seven *base units*, each considered dimensionally independent with specific definitions:

- meter (m) for length
- kilogram (kg) for mass
- second (s) for time
- ampere (A) for electric current
- kelvin (K) for temperature
- mole (mol) for the amount of a substance
- candela (cd) for luminous intensity

Table 1.1 provides the exact definitions of each of these units.

At present, there are only two *supplementary units* in the SI system. Both are purely geometric:

- radian (rad) for the unit of plane angle
- steradian (sr) for the unit of solid angle

The radian is the plane angle between two radii of a circle that cut off, on the circumference, an arc equal in length to the radius. The steradian is the solid angle that, having its apex in the center of a sphere, cuts out an area

Table 1.1 SI Base Units and Their Definitions
(courtesy of the American Petroleum Institute-API Pub. 2564)

Quantity	Name	Symbol	Definition
Length	Meter (or metre)	m	The meter is the length equal to 1 650 763.73 wavelengths in vacuum of the radiation corresponding to the transition between the levels $2p_{10}$ and $5d_5$ of the krypton-86 atom (Eleventh CGPM, 1960, Resolution 6).
Mass	Kilogram	kg	The kilogram is the unit of mass (not force); it is equal to the mass of the international prototype of the kilogram (Third CGPM, 1901, Resolution 3).
			This international prototype, made of platinum-iridium, is kept at the International Bureau of Weights and Measures. A copy of the international prototype is maintained by the national standards agency of each major country.
			The kilogram is the only base unit defined by an artifact and is the only base unit having a prefix.
Time	Second	s	The second is the duration of 9 192 631 770 periods of the radiation corresponding to the transition between the two hyperfine levels of the ground state of the cesium-133 atom (Thirteenth CGPM, 1967, Resolution 1).
Electric Current	Ampere	A	The ampere is that constant current that, if maintained in two straight parallel conductors of infinite length of negligible circular cross section and placed 1 meter apart in vacuum, would produce between these conductors a force equal to 2×10^{-7} newton per meter of length (CIPM, 1946, Resolution 2 approved by the Ninth CGPM, 1948).
Temperature	Kelvin	K	The kelvin, unit of thermodynamic temperature, is the fraction 1/273.16 of the thermodynamic temperature of the triple point of water (Thirteenth CGPM, 1967, Resolution 4).
			The unit kelvin and its symbol, K, are used to express an interval or difference of temperature (Thirteenth CGPM, 1967, Resolution 3).
			In addition to the thermodynamic temperature, Celsius temperature (formerly called Centigrade) is used widely. The degree Celsius (°C), a derived unit, is the unit for expressing Celsius temperatures and temperature intervals. Celsius temperature, t, is related to thermodynamic temperature, T, by the following equation:

$$t = T - T_0$$

where $T_0 = 273.16$ by definition (temperature
interval 1°C equals 1 K exactly)

Table 1.1, continued

Quantity	Name	Symbol	Definition
Amount of Substance	Mole	mol	The mole is the amount of substance of a system that contains as many elementary entities as there are atoms in 0.012 kilogram of carbon-12.
			When the mole is used, the elementary entities must be specified and may be atoms, molecules, ions, electrons, other particles or specified groups of such particles (Fourteenth CGPM, 1971, Resolution 3).
Luminous Intensity	Candela	cd	The candela is the luminous intensity, in a given direction, of a source that emits monochromatic radiation of frequency 540×10^{12} hertz and that has a radian intensity in that direction of 1/683 watt per steradian (Sixteenth CGPM, 1979).

of the surface of the sphere that equals that of a square, with sides of length equal to the radius of the sphere.

Derived units are expressed algebraically in terms of base units, having mathematical symbols for multiplication and division. A number of derived units have been given special names and assigned symbols. Examples of derived units and those given special names are noted in Table 1.2.

Several other units are used widely but are not part of SI. These include the minute, hour, day and year as units of time; degree, minute and second of arc (in addition to the radian); the metric ton (1000 kg); the liter (1 cubic decimeter); the nautical mile; and the knot. These may be used along with SI units.

Prefixes for Decimal Multiples and Submultiples of SI Units

These are listed in Table 1.3. The symbol of a prefix is combined with the unit to which it is directly attached, thus forming a new unit symbol that can be raised to a positive or negative power and can be combined with other unit symbols to form compound units.

Distinction between upper case and lower case symbols is important, as shown by the following examples:

- M = mega = 10^6
- m = milli = 10^{-3} (where m is a prefix)
- N = newton
- n = nano = 10^{-9}

Table 1.2 Examples of SI-Derived Units

Quantity	SI Unit			
	Name	Symbol	Expression in Terms of Other Units	Expression in Terms of SI Base Units
Frequency	Hertz	Hz		s^{-1}
Force	Newton	N		$m \cdot kg \cdot s^{-2}$
Pressure	Pascal	Pa	N/m^2	$m^{-1} \cdot kg \cdot s^{-2}$
Energy, Work, Quantity of Heat	Joule	J	N/m	$m^2 \cdot kg \cdot s^{-2}$
Power, Radiant Flux	Watt	W	J/s	$m^2 \cdot kg \cdot s^{-3}$
Electric Potential, Potential Differences, Electromotive Force	Volt	V	W/A	$m^2 \cdot kg \cdot s^{-3} \cdot A^{-1}$
Electric Resistance	Ohm	Ω	V/A	$m^2 \cdot kg \cdot s^{-3} \cdot A^{-2}$
Conductance	Siemens	S	A/V	$m^{-2} \cdot kg^{-1} \cdot s^{3} \cdot A^{2}$
Area	Square meter	m^2		
Volume	Cubic meter	m^3		
Speed, Velocity	Meter per second	m/s		
Acceleration	Meter per second squared	m/s^2		
Density	Kilogram per cubic meter	kg/m^3		
Concentration (of amount of substance)	Mole per cubic meter	mol/m^3		
Specific Volume	Cubic meter per kilogram	m^3/kg		
Luminance	Candela per square meter	cd/m^2		

Table 1.3 SI Prefixes for Forming Decimal Multiples and Submultiples

Prefix	Symbol	Factor by which Unit is Multiplied
tera	T	10^{12}
giga	G	10^9
mega	M	10^6
kilo	k	10^3
hecto	h	10^2
deca	da	10
deci	d	10^{-1}
centi	c	10^{-2}
milli	m	10^{-3}
micro	μ	10^{-6}
nano	n	10^{-9}
pico	p	10^{-12}
femto	f	10^{-15}
atto	a	10^{-18}

Caution must be exercised when a compound unit includes a unit symbol that is also a symbol for a prefix. As an example, the unit newton-meter should be written as N-m to avoid confusion with mN (millinewton).

Symbols can be used together with modifying subscripts and/or superscripts. Subscripts often are used to designate a place in space or time, or a constant or reference point. Superscripts can be used to designate a dimensionless form, reference or equilibrium value, or mathematical identification, such as an average value, derivative, tensor index, etc. Table 1.4 provides a partial list of commonly used symbols and their definitions by category.

Appendix C provides tables of units and conversion factors of SI, cgs and English units that have been prepared by the American Petroleum Institute (API), Washington, DC. Tables and conversion factors have been grouped into the following categories:

1. Space, Time
2. Mass, Amount of Substance
3. Heating Value, Entropy, Heat Capacity
4. Temperature, Pressure, Vacuum
5. Density, Specific Volume, Concentration, Dosage
6. Facility Throughput, Capacity
7. Flowrate
8. Energy, Work, Quantity of Heat, Power
9. Mechanics
10. Transport Properties
11. Electricity, Magnetism
12. Acoustics, Light, Radiation

Metric units recommended for general use are shown under the heading "API preferred metric unit."

Table 1.4 Commonly Used Symbols and Definitions

	Symbol	Unit or Definition
General Symbols		
Acceleration	a	m/s^2
of Gravity	g	m/s^2
Base of Natural Logarithms	e	
Coefficient	C	
Difference, finite	Δ	
Differential Operator	d	
Partial	δ	
Efficiency	η	
Energy, dimension of	E	J, N-m
Enthalpy	H	J
Entropy	S	J/K
Force	F	N
Function	ϕ, ψ, χ	
Gas Constant, universal	R	To distinguish, use R_O
Gibbs Free Energy	G, F	$G = H - TS$, J
Heat	Q	J
Helmholtz Free Energy	A	$A = U - TS$, J
Internal Energy	U	J
Mass, dimension of	m	kg
Mechanical Equivalent of Heat	J	Unity, dimensionless
Moment of Inertia	I	$(m)^4$
Newton's Law of Motion, conversion factor in	g_c	Unity, dimensionless
Number		
In general	N	
Of moles	n	
Pressure	p, P	Pa, bar
Quantity, in general	Q	
Ratio, in general	R	
Resistance	R	
Shear Stress	τ	Pa
Temperature		
Dimension of	θ	
Absolute	T	K (Kelvin)
In general	T, t	°C
Temperature Difference, logarithmic mean	$\bar{\theta}$	°C
Time		
Dimension of	T	s
In general	t, τ	s, hr
Work	W	J
Geometric Symbols		
Linear dimension		
Breadth	b	m
Diameter	D	m
Distance along Path	s, x	m
Height above Datum Plane	Z	m
Height Equivalent	H	m
Hydraulic Radius	r_H, R_H	m, m^2/m

Table 1.4, continued

	Symbol	Unit or Definition
Geometric Symbols		
Lateral Distance from Datum Plane	Y	m
Length, distance or dimension of	L	m
Longitudinal Distance from Datum Place	X	m
Mean Free Path	λ	m
Radius	r, R	m
Thickness		
In general	B	m
Of file	B_f	m
Wavelength	λ	m
Area		
In general	A	m^2
Cross section	S	m^2
Fraction-free Cross Section	σ	
Projected	A_p	m^2
Surface		
Per unit mass	A_w, s	m^2/kg
Per unit volume	A_s, a	m^2/m^3
Volume		
In general	V	m^3
Fraction voids	ϵ	
Humid volume	ν_H	m^3/kg dry air
Angle	α, θ, ϕ	
In x,y plane	α	
In y,z plane	ϕ	
In z,x plane	θ	
Solid angle	ω	
Other		
Particle-shape factor	ϕ_s	
Intensive Properties Symbols		
Absorptivity for Radiation	α	
Activity	a	
Activity Coefficient, molal basis	γ	
Coefficient of Expansion		
Linear	α	$m/(m\text{-}K)$
Volumetric	β	$m^3/(m^3\text{-}K)$
Compressibility Factor	z	$z = pV/RT$
Density	ρ	kg/m^3
Diffusivity		
Molecular, volumetric	D_v, δ	$m^3/(s\text{-}m)$, m^2/s
Thermal	α	$\alpha = k/C_p$, m^2/s
Emissivity Ratio for Radiation	e	
Enthalpy, per mole	H	J/kmol
Entropy, per mole	S	$J/(kmol\text{-}K)$
Fugacity	f	Pa, bar

Table 1.4, continued

	Symbol	Unit or Definition
Intensive Properties Symbols		
Gibbs Free Energy, per mole	G, F	J/kmol
Helmholtz Free Energy, per mole	A	J/kmol
Humid Heat	c_s	J/(kg dry air-K)
Internal Energy, per mole	U	J/kmol
Latent Heat, phase change	λ	J/kg
Molecular Weight	M	kg
Reflectivity for Radiation	ρ	
Specific Heat	c	J/(kg-K)
At constant pressure	c_p	J/(kg-K)
At constant volume	c_v	J/(kg-K)
Specific Heats, ratio of	γ	
Surface Tension	σ	N/m
Thermal Conductivity	k	$(J\text{-}m)/(s\text{-}m^2\text{-}K)$
Transmissivity of Radiation	τ	
Vapor Pressure	p*	Pa, bar
Viscosity		
Absolute or coefficient of	μ	Pa-s
Kinematic	ν	m^2/s
Volume, per mole	V	$m^3/kmol$
Symbols for Concentrations		
Absorption Factor	A	$A = L/K^*V$
Concentration (mass or moles per unit volume)	c	kg/m^3, $kmol/m^3$
Fraction		
Cumulative beyond a given size	ϕ	
By volume	x_v	
By weight	x_μ	
Humidity	H, Y_H	kg/kg dry air
At saturation	H_s, Y^*	kg/kg dry air
At wet-bulb temperature	H_w, Y_w	kg/kg dry air
At adiabatic saturation	H_a, Y_a	kg/kg dry air
Mass Concentration of Particles	c_p	kg/m^3
Moisture Content		
Total water to bone-dry stock	$X\phi^*$	kg/kg dry stock
Equilibrium water to bone-dry stock	X^*	kg/kg dry stock
Free water to bone-dry stock	X	kg/kg dry stock
Mole or Mass Fraction		
In heavy or extract phase	x	
In light or raffinate phase	y	
Mole or Mass Ratio		
In heavy or extract phase	X	
In light or raffinate phase	Y	
Number Concentration of Particles	n_p	$number/m^3$
Phase Equilibrium Ratio	K*	$K^* = y^*/x$
Relative Distribution of Two Components		
Between two phases in equilibrium	α	$\alpha = K_i^*/K_j^*$
Between successive stages	β	$\beta = (y_i/\nu_i)_x/(x_ix_i)_{n+1}$
Relative Humidity	H_R, R_H	
Slope of Equilibrium Curve	m	$m = dy^*/dx$
Stripping Factor	S	$S = K^*V/L$

Table 1.4, continued

	Symbol	Unit or Definition
Rate Symbols		
Quantity per Unit Time, in general	q	
Angular velocity	ω	
Feed rate	F	kg/s, kmol/s
Frequency	f, N_f	
Friction Velocity	u*	$u* = (\tau_w \rho)^{1/2}$, m/s
Heat Transfer Rate	q	J/s
Heavy or Extract Phase Rate	L	kg/s, kmol/s
Heavy or Extract Product Rate	B	kg/s, kmol/s
Light or Raffinate Phase Rate	V	kg/s, kmol/s
Light or Raffinate Product Rate	D	kg/s, kmol/s
Mass Rate of Flow	w	kg/s, kg/hr
Molal Rate of Transfer	N	kmol/s
Power	P	W
Velocity, in general	n	m/s
Revolutions per Unit Time	u	m/s
Longitudinal (x), component of	u	m/s
Lateral (y), component of	v	m/s
Normal (z), component of	w	m/s
Volumetric Rate of Flow	q	$m^3/s, m^3/hr$
Quantity per Unit Time, Unit Area		
Emissive power, total	W	W/m^2
Mass velocity, average	G	$G = w/S$, kg/(s-m^2)
Vapor or light phase	$G, \overline{G}$	kg/(s-m^2)
Liquid or heavy phase	$L, \overline{L}$	kg/(s-m^2)
Radiation, intensity of	I	W/m^2
Velocity		
Nominal, basis total cross section of packed vessel	v_s	m/s
Volumetric average	$V, \overline{V}$	m^3/(s-m^2), m/s
Quantity per Unit Time, Unit Volume		
Quantity reacted per unit time, reactor volume	N_R	kmol/(s-m^2)
Space Velocity, volumetric	Λ	m^3/(s-m^3)
Quantity per Unit Time, Unit Area, Unit Driving Force, in general	k	
Eddy Diffusivity	δ_E	m^2/s
Eddy Viscosity	ν_E	m^2/s
Eddy Thermal Diffusivity	α_E	m^2/s
Heat Transfer Coefficient		
Individual	h	$W/(m^2 \cdot K)$
Overall	U	$W/(m^2 \cdot K)$
Mass Transfer Coefficient		
Individual	k	kmol/(s-m^2) (driving force)
Gas film	k_G	To define driving force use subscript:
Liquid film	k_L	c for kmol/m^3
Overall	K	p for bar
Gas film basis	K_G	x for mole fraction
Liquid film basis	K_L	
Stefan-Boltzmann Constant	σ	5.6703×10^{-8} W/(m$^2 \cdot$K^4)

INDUSTRIAL STOICHIOMETRY

Fundamental Quantities of Matter

The properties of the three states of matter (solids, liquids, gases) are detected based on quantities related to our senses. These quantities include time, distance, mass, force and temperature. The first three are obvious. The concept of force is best introduced by way of Newton's second law of motion. This fundamental relation expresses the quantity force in terms of the product of mass and acceleration, or

$$F_0 = \frac{1}{g_c} ma \qquad (1.1)$$

This equation provides a relationship among the four fundamental quantities—time, distance, mass and force—which is valid regardless of the system of units employed. The quantity g_c is a conversion factor based on the appropriate unit of measure. For example, when time is measured in seconds, distance in centimeters and mass in grams, the unit of force is the dyne (i.e., the force that will cause a mass of 1 gram to accelerate 1 cm/s^2). Substituting these units into Equation 1.1 gives

$$1 \text{ dyne} = \frac{1}{g_c} (1 \text{ g})(1 \text{ cm/s}^2)$$

or

$$g_c = 1(\text{g-cm})/(\text{dyne-s}^2)$$

The dyne may be regarded merely as an abbreviation for the composite unit $g\text{-}cm/s^2$ and, hence, g_c is unity and dimensionless.

In the English engineering system, a pound force is the force exerted by gravity on a 1-lb mass of material under conditions in which the acceleration of gravity is 32.1740 ft/s^2. Hence, a 1-lb mass will acquire an acceleration of 32.1740 ft/s^2 by a force of 1 lb-force, or

$$1 \text{ lb}_f = \frac{1}{g_c} (1 \text{ lb}_m)(32.1740 \text{ ft/s}^2)$$

or

$$g_c = 32.1740 \; \frac{lb_m\text{-}ft}{lb_f\text{-}s^2}$$

Thus, in the English system, g_c is a dimensional constant having a numerical value equivalent to the standard of the acceleration of gravity, but with different units.

Temperature is a measure of the degree of hotness of a substance and is detected most commonly with a liquid-in-glass thermometer. By employing a uniform tube partially filled with an appropriate fluid such as mercury, alcohol or an oil, the property of thermal expansion of the fluid provides a measure when heated. That is, the degree of hotness is detected by measuring the length of the fluid in the column. The numerical values are again arbitrary in origin. In the centigrade or Celsius scale, the freezing point of water saturated with air at standard atmospheric pressure is defined as zero, and the boiling point of pure water at standard atmospheric pressure is assigned the value of 100. Other temperature scales are defined in terms of the Celsius scale. The Fahrenheit scale is defined as follows:

$$t°F = 1.8(t°C) + 32 \tag{1.2}$$

The freezing point on this scale is 32°F and the boiling point is 212°F.

Two other temperature scales of primary importance introduce the concept of an absolute lower limit of temperature. Accurate measurements establish this limit at −273.16°C, or −459.69°F. As this represents a lower limit, temperature scales are readjusted to assign zero to this limit. The *absolute scales* in use are the Kelvin scales (the size of the degree is equivalent to the centrigrade degree), in which all temperatures are 273.16 degrees higher. In the Rankine scale, the size of the degree is the same as that of the Fahrenheit degree, but all temperatures are 459.69 degrees higher.

Numerous other quantities enable the properties of matter to be characterized. Most are elementary and require no discussion. However, some of the thermodynamic quantities such as internal energy, work and enthalpy should be reviewed briefly before tackling engineering problems.

First, we note the quantities of volume and pressure. Volume, like mass, depends on the amount of material under examination. In contrast, specific volume is defined as the volume per unit mass or volume per mole of material, and is thus independent of the amount of material.

The pressure of a fluid is defined as the normal force exerted per unit area of surface ($p = F_0/F$, where F is area and F_0 force). Pressure is measured most frequently in terms of the height of a column of fluid under the influence of gravity. For example, a mercury manometer is used to obtain pressure measurements in millimeters of mercury. Such values are converted to force per unit area by multiplying the height of the column by the fluid's density. That is, the force exerted by gravity on the mercury column is

$$F_0 = \frac{1}{g_c} (m)(g) \qquad (1.3)$$

where g is the local acceleration of gravity and $m = (F)(h)(g)$. By eliminating mass, m, we obtain

$$\frac{F_0}{F} = p = (h)(\rho)\frac{g}{g_c} \qquad (1.4)$$

As the density of the manometric fluid is a function of temperature, the pressure corresponding to a given height of fluid column also depends on temperature. A common unit of pressure is the atmosphere, which corresponds to the force of gravity acting on air above the earth's surface. A standard atmosphere is equal to 14.696 $lb_f/in.^2$, or 29.921 in Hg at 0°C in a standard gravitational field. Note that in SI a *bar* is equivalent to 0.9869 atm or 10^6 $dyne/cm^2$. In many cases, measurements obtained represent a difference between a system under study and the pressure of the surrounding environment. These measurements are known as *gauge pressures*, which can be converted to *absolute* pressures by adding the barometric pressure.

Work is performed when a force acts through a distance. It is defined as the product of the force and the distance over which it is applied. If the force is constant, then

$$W = F_0 S \qquad (1.5A)$$

If it is variable, then

$$dW = F_0 dS \qquad (1.5B)$$

To determine the work for a finite process, Equation 1.5B must be integrated over appropriate limits. As an example, consider the compression or expansion of a gas by the movement of a piston. The distance over which the piston moves is equal to the volume change of the gas divided by the area of the piston, or

$$dW = (pF) d \frac{V}{F}$$

with constant F:

$$dW = pdV$$

or

$$W = \int_{v_1}^{v_2} pdV \qquad (1.6)$$

This is the general expression for work performed as a result of a finite compression or expansion process.

From Equation 1.5B a body moves through a differential distance, dS, over some differential time, $d\theta$. Denoting u as the body velocity, then

$$a = \frac{du}{d\theta}$$

whence

$$dW = \frac{m}{g_c} udu$$

Integrating over a finite change in velocity,

$$W = \frac{m}{g_c} \int_{u_1}^{u_2} udu$$

$$= \frac{mu_2^2}{2g_c} - \frac{mu_1^2}{2g_c} = \Delta \frac{mu^2}{2g_c} \qquad (1.7)$$

The term $mu^2/2g_c$ is termed the *kinetic energy* of the body. The work performed on a body in accelerating it from an initial velocity, u_1, to some final velocity, u_2, is equivalent to the change in the body's kinetic energy.

When a body of mass is elevated from an initial position, z_1, to a final level, z_2, an upward force that is equivalent at least to the weight of the body must be exerted on it. This force must move through the distance $z_2 = z_1$; as the weight of the body is the force of gravity on it, the minimum force required is given by Equation 1.3. That is, the minimum amount of work required to raise the body is

$$W = F_0(z_2 - z_1)$$

$$= mz_2 \frac{g}{g_c} - mz_1 \frac{g}{g_c} = \Delta \frac{mzg}{g_c} \qquad (1.8)$$

This is the work done on a body in elevating it (that is, in producing a change) or its *potential energy*. If the elevated body is permitted to fall freely, it will gain in kinetic energy what it loses in potential energy. Hence, the capacity for doing work remains unchanged:

$$\Delta KE = \Delta PE = 0 \qquad (1.9)$$

This is known as the principle of conservation of energy in mechanics and is a true statement provided that both kinetic and potential energy changes are equivalent to the work done in producing them. Work is essentially energy in transit. When work is performed and does not appear elsewhere in the system simultaneously, it is converted into another form of energy. In discussing specific unit operations, the body on or by which work is performed is referred to as the *system*. When work is performed, it is done by the surroundings on the system, or vice versa. Thus, energy is transferred from the surroundings to the system, or the reverse. During this transfer, the form of energy referred to as work exists. On the other hand, kinetic and potential energies reside with the system. They are measured with reference to the surroundings, and changes are independent of this reference or datum level.

Internal Energy and the First Law of Thermodynamics

The internal energy of a substance refers to the energy of the molecules comprising the material. Molecules are in perpetual motion and thus possess kinetic energies of translation, rotation and vibration. The addition of heat increases the molecular activity and, hence, causes an increase in the internal energy. Work performed on a substance has the same effect. In addition to kinetic energy, molecules also possess potential energy, which results from forces of attraction between them. Hence, internal energy is from within the system, whereas kinetic and potential energies are external forms.

As heat and internal energy are both forms of energy, the general law of conservation of energy (i.e., the first law of thermodynamics) may be stated in this way: *regardless of the form of energy, the total quantity of energy is constant, and when energy in one form vanishes, it appears simultaneously in other forms.* In relation to the system and surroundings, this is

$$\Delta(\text{energy of the system}) + \Delta(\text{energy of the surroundings}) = 0 \qquad (1.10)$$

The first term can be expanded to show energy changes in different forms:

$$\Delta(\text{energy of the system}) = \Delta U + \Delta KE + \Delta PE \qquad (1.11)$$

or

$$\Delta U + \Delta KE + \Delta PE = \pm Q \pm W \qquad (1.12)$$

where ΔU, ΔKE and ΔPE represent changes in internal, kinetic and potential energies of the system, respectively, and Q and W are heat and work, respectively. By convention, heat is regarded as positive when it is transferred to the system from the surroundings, whereas work is taken to be positive when transferred from the system to the surroundings.

In many cases, the system does not undergo a change in kinetic and potential energy, but only of internal energy. In these special cases,

$$dU = dQ - dW \qquad (1.13)$$

where the expression has been written to show differential changes.

From the standpoint of thermodynamics, there are two types of quantities: (1) those that depend on the path taken to achieve a final state, and (2) those that do not depend on the past history of the substance or on the path followed in reaching a given state. In the latter case, properties depend on the immediate conditions and are referred to as *point functions* or state functions.

Values of state functions may be represented as individual points on a graph. The differential of a state function represents an infinitesimal change in property. The integration of such differentials yields a finite difference between two values of the specific property (e.g., $\int_{u_1}^{u_2} dU = U_2 - U_1 = \Delta U$).

In contrast, work and heat are not state properties but are functions of the path followed. They are not represented by points on a graph but rather are denoted by areas. It should be noted also that a state function represents an instantaneous property of a particular system. In contrast, work and heat exist only when changes are caused in a system. This implies further that whenever heat is transferred or work performed, an increment of time occurs. The quantities of heat and work are critical in the design of equipment of unit operations, such as heat exchangers, distillation columns, evaporators, pumps, compressors, turbines, etc. State functions such as internal energy are properties of matter; however, once measured, they may be applied to calculating Q and W for any process through the laws of thermodynamics.

Enthalpy

Enthalpy is a thermodynamic function defined by the following relation:

$$dH = dU + d(pV) \tag{1.14}$$

where U is the internal energy of the system, p the absolute pressure and V the system volume. As U, p and V are all state functions, so is H. Equation 1.14 applies whenever a differential change to the system occurs. The equation may be expressed for any amount of material; however, it usually is expressed on a unit mass or mole basis. The importance of this thermodynamic concept to engineering problems will become apparent in the chapters to follow.

Heat Capacity and Specific Heat

Two additional thermodynamic terms of importance are a material's heat capacity and specific heat. The flow of heat is commonly thought of in terms of its effects on the substances that receive the energy. Heat capacity refers to the quantity of heat required to raise the temperature of a given mass by 1°. By convention, this quantity is based on either 1 mole or a unit mass of material. The mathematical relationship is

$$dQ = nCdT \tag{1.15A}$$

where n is the number of moles, C is the molal heat capacity, and dT is the temperature increment resulting from the quantity of heat, dQ. On a unit mass basis, this expression becomes

$$dQ = mCdT \tag{1.15B}$$

where m is mass, and C is per unit mass.

By rigorous definition, the term specific heat is the ratio of the heat capacity of a material to the heat capacity of an equal quantity of water. The specific heat of water is approximately 1 cal/(g)(°C) (the molal heat capacity of water is 18 cal/(g-mol)(°C)).

Let us assume a quantity of gas is contained in a vessel with rigid walls. On the addition of heat, the gas temperature rises, but the system volume is maintained constant. Then

$$dQ = C_v dT$$

or

$$Q = C_v \Delta T$$

As the system did not undergo a change in volume, no work was performed, and it is easy to show that the first law of thermodynamics reduces to

$$\Delta U = Q = C_v \Delta T \tag{1.16}$$

In contrast, assume the same operation is now performed as follows: a quantity of gas is retained in a cylinder by a frictionless piston at constant pressure, p. If heat is added, the gas will expand reversibly, and if the force on the piston is maintained constant, the process takes place at constant p. Then

$$Q = C_p \Delta T$$

As the gas expands, work is performed and the first law expression reduces to

$$\Delta H = C_p \Delta T = Q \tag{1.17}$$

The student should prove the definition. Thus, for a constant-pressure process carried out reversibly, the system's enthalpy change is equal to the heat added. For purposes of calculating changes in properties, the equation applying to the reversible constant-pressure process may be applied to actual or reversible processes that accomplish the same change in state (see Smith and Van Ness [1959]).

Properties of the Ideal Gas

Gases seldom, if ever, are handled at standard conditions (i.e., STP, standard temperature and pressure = 0°C, 1 atm). Consequently, it is necessary to estimate the molal volume at different temperatures and pressures. For a so-called ideal gas, a simple p-V-T relationship exists. An ideal gas is one characterized by the following:

1. The volume of individual molecules themselves is negligible in comparison to the total volume of the gas.
2. No forces exist between the molecules of the gas.

Applying these assumptions to the kinetic theory of gases (see Cheremisinoff [1981] or Bird et al. [1960] for derivation), the so-called *ideal gas law* is obtained:

$$pV = nRT \qquad (1.18A)$$

where   V = volume of n moles of gas
 p = absolute pressure
 T = absolute temperature
 R = universal gas law constant

For n = 1,

$$p\dot{v} = RT \qquad (1.18B)$$

where $\dot{v}$ is the gas volume per mole.

For most gases at pressures of a few atmospheres, the ideal gas law may be applied to engineering calculations. Various values of the constant R in different units are as follows:

$$
\begin{aligned}
R &= 1.987 \text{ Btu/(lb-mol)(°F) or cal/(g-mol)(°K)} \\
&= 0.730 \text{ (atm)(ft}^3\text{)/(lb-mol)(°R)} \\
&= 10.73 \text{ (psi)(ft}^3\text{)/(lb-mol)(°R)} \\
&= 1.314 \text{ (atm)(ft}^3\text{)/(lb-mol)(°R)} \\
&= 1545 \text{ (lb}_f\text{/ft}^2\text{)(ft}^3\text{)/(lb-mol)(°R) or ft-lb}_f\text{/(lb-mol)(°R)} \\
&= 82.06 \text{ (atm)(cm}^3\text{)/(g-mol)(°R)}
\end{aligned}
$$

The characteristics of ideal gases are best described in terms of the thermodynamics of commonly encountered situations. Therefore, we examine the behavior of 1 mole of ideal gas in a reversible nonflow process for five cases: isometric (constant-volume), isobaric (constant-pressure), isothermal (constant-temperature), adiabatic and polytropic.

In a *constant-volume (isometric) process*, Equation 1.16 is directly applicable to an ideal gas. In integral form, this expression is

$$\Delta U = Q = \int C_v dT$$

The equation states that the internal energy of an ideal gas is only a function of temperature (based on assumption (2) above). As U is independent of the specific volume at constant temperature, a plot of U versus $\mathring{v}$ will result in a straight line. For different temperatures, U has different values.

For a *constant-pressure (isobaric) process*, the first law expression developed earlier applies:

$$dH = dQ = C_p dT$$

or, for constant C_p,

$$\Delta H = Q = C_p \Delta T$$

As the energy of an ideal gas depends only on temperature, its enthalpy is also strictly a function of T. This is obvious from

$$dH = dU + d(p\mathring{v})$$

or

$$dH = dU + RdT \text{ (for an ideal gas)} \tag{1.19}$$

Equation 1.19 provides a relationship between C_p and C_v:

$$C_p dT = C_v dT + RT$$

whence

$$C_p = C_v + R \tag{1.20}$$

Note that because C_v is a function of temperature only, so is C_p.

In a *constant-temperature (isothermal) process*, the energy of an ideal gas cannot change. Hence,

$$\left.\begin{array}{l} dU = dQ - dW = 0 \\[2ex] Q = W \end{array}\right\}$$

(1.21)

or

Thus, for an ideal gas, we may write

$$Q = W = \int p d\dot{v} = \int RT \frac{d\dot{v}}{\dot{v}}$$

(1.22)

On integration at constant temperature, we obtain

$$\left.\begin{array}{l} Q = W = RT \, ln \, \dfrac{\dot{v}_2}{\dot{v}_1} \\[3ex] \quad\;\; = RT \, ln \, \dfrac{P_1}{P_2} \end{array}\right\}$$

(1.23)

because $\dot{v}_2/\dot{v}_1 = P_1/P_2$.

An *adiabatic process* is one in which there is no transfer of heat between system and surroundings, i.e., $dQ = 0$. For a reversible adiabatic process then,

$$dU = -dW = -p d\dot{v}$$

(1.24)

And because $C_v dT = -pd$ and $p = RT/\dot{v}$, we obtain

$$\frac{dT}{T} = -\frac{R}{C_v} \frac{d\dot{v}}{\dot{v}}$$

(1.25)

Denoting γ as the ratio of heat capacities, i.e., $\gamma = C_p/C_\dot{v} = (C_\dot{v} + R)/C_\dot{v} = 1 + R/C_\dot{v}$, Equation 1.25 may be rewritten as follows:

$$\frac{dT}{T} = -(\gamma - 1) \frac{d\dot{v}}{\dot{v}}$$

(1.26)

Integrating this expression gives

$$\left(\frac{T_2}{T_1}\right) = \left(\frac{\dot{\nu}_1}{\dot{\nu}_2}\right)^{\gamma-1} \tag{1.27A}$$

And through the ideal gas law relation, we also obtain

$$\left(\frac{T_2}{T_1}\right) = \left(\frac{P_2}{P_1}\right)^{(\gamma-1)/\gamma} \tag{1.27B}$$

$$\left(\frac{\dot{\nu}_1}{\dot{\nu}_2}\right)^{\gamma-1} = \left(\frac{P_2}{P_1}\right)^{(\gamma-1)/\gamma} \tag{1.27C}$$

or, finally,

$$P_1\dot{\nu}_1^{\gamma} = P_2\dot{\nu}_2^{\gamma} = P\dot{\nu}^{\gamma} = \text{constant} \tag{1.28}$$

Leaving the details of the derivation to the student, we write the expression for work of an adiabatic process $(-dW = dU = C_{\dot{\nu}}dT)$ as

$$W = \frac{P_1\dot{\nu}_1 - P_2\dot{\nu}_2}{\gamma - 1} \tag{1.29}$$

Or, rewriting this expression in terms of $\dot{\nu}_1$ only,

$$W = \frac{P_1\dot{\nu}_1}{\gamma - 1}\left[1 - \left(\frac{P_2}{P_1}\right)^{(\gamma-1)/\gamma}\right] = \frac{RT_1}{\gamma - 1}\left[1 - \left(\frac{P_2}{P_1}\right)^{(\gamma-1)/\gamma}\right] \tag{1.30}$$

Finally, a *polytropic process* is a general case in which no specific conditions other than reversibility are imposed. The above general equations for nonflow processes apply in this case to ideal gases:

$$dU = dQ - dW \ , \quad \Delta U = Q - W$$

$$dW = pd\dot{\nu} \ , \quad W = \int pd\dot{\nu}$$

$$dU = C_v dT \ , \quad \Delta U = \int C_v dT$$

$$dH = C_p dT \ , \quad \Delta H = \int C_p dT$$

and

$$dQ = C_v dT + pd\dot{v}$$

$$Q = \int C_v dT + \int pd\dot{v} \tag{1.31}$$

Work may be calculated directly from the integral $\int pd\dot{v}$.

Note that the expressions presented in this section were derived for reversible nonflow processes involving ideal gases. Those expressions, however, which relate state functions only, are applicable to ideal gases for both reversible and irreversible flow and nonflow processes because changes in state functions depend only on the initial and final states of the system. However, expressions for Q and W are specific to the cases considered.

BEHAVIOR OF FLUIDS

Unlike gases, which are nature's simplest substances, liquids are more complex in behavior and, hence, cannot be described by a simple thermodynamic relation. The complexity of liquid properties is due to their high density in comparison to the low molecular density of gases. The complex thermodynamic nature of liquids is beyond the scope of this text. For a discussion of the equations of states for liquids and transport properties, the reader is referred to the work of Sridhar [1983]. Rather, we shall describe briefly the physical behavior of fluids (liquids) when subjected to a force.

A *fluid* (gas or liquid) is any form of matter that deforms continuously under the action of a shearing stress. When this stress is removed deformation ceases, but the fluid does not return to its original configuration. This behavior is in contrast to that of a *solid*, which does not deform continuously under a shearing stress, but attains a definite equilibrium deformed state for a particular stress. On removal of the stress, the solid returns to its original configuration (assuming that the solid is being stressed below its yield value).

In practice, such an absolute definition is not applied to a fluid. There are many materials that exhibit both fluid-like and solid-like properties but are still considered to be fluids. For example, certain materials will not deform continuously under the action of any nonzero shearing stress until a certain "yield" stress is exceeded. Still other materials deform continuously, but when the stress is removed they partially recover their original configuration. As a compromise, we may consider a fluid to be a form of matter that exhibits continuous deformation under some range of shearing stress and may partially recover its original configuration when the stress is removed.

The deformation of a fluid under the action of a stress is the subject of the science of *rheology*. The study of the deformation and flow of materials is a complex subject that still is being developed. Extensive reviews of the subject have been done by Bird et al. [1977], Fredrickson [1964], Lodge [1964], Middleman [1968] and Schowalter [1978], among others.

Rheologists are concerned with the measurement of the deformation of a fluid with stress and the formulation of mathematical relations between deformation rates (or velocity gradients) and stress, τ. These relations are referred to as rheological models, or *constitutive equations*, which can be used in a momentum balance equation to compute for a given laminar flow situation, velocity profiles, volumetric flowrates, etc. While a large number of such equations have been formulated, their use often entails the evaluation of multiple parameters from data, which is difficult in practice. Furthermore, the complexity of many of these equations precludes their use for engineering design. Unfortunately, parameters evaluated from measurements under steady shear often cannot represent data from, say, oscillatory shear. As such, only very simple constitutive equations are applied in engineering work. Obviously, these simple rheological models are limited in their ability to represent all aspects of fluid behavior and are suitable only under restricted conditions.

Consider a fluid contained between two large parallel plates separated by a small gap, Y, as shown in Figure 1.1. The lower plate is stationary, while the upper plate is moved at a constant velocity, v, through the action of applied force, F_0. A thin layer of fluid adjacent to each plate will move at the same velocity as the plate (referred to as the "no slip" condition, which holds true for all except a few fluids). Molecules in the fluid layers between these two extremes will move at intermediate velocities. For example, a

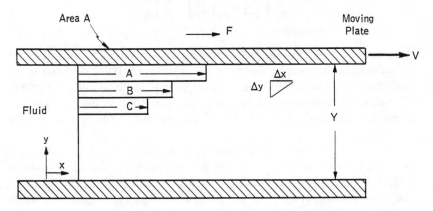

Figure 1.1   The deformation of a fluid.

fluid layer B immediately below A will experience a force in the x direction from layer A and a smaller retarding force from layer C. Layer B then will flow at a velocity lower than v. This progression continues to the layer adjacent to the lower plate, whose velocity is zero. Under steady-state conditions, the force F_0 required to produce motion becomes constant and is related to velocity as follows:

$$\frac{F_0}{A} = \psi\left(\frac{v}{Y}\right) \tag{1.32}$$

where A is the area of each plate, and ψ is a function of the fluid properties, temperature and pressure.

The ratio F_0/A is the shear stress on the fluid at the upper plate, and v/Y is the velocity gradient. On a local basis, that is, at any point within a fluid, the above expression may be generalized:

$$\tau_{yx} = \psi\left(\frac{dv_x}{dy}\right) \tag{1.33}$$

τ_{yx} is the local shear stress (or force per unit area) acting in the x direction on a plane perpendicular to the y axis; v_x is the velocity in the x direction; dv_x/dy is the local velocity gradient. Equation 1.33 represents the constitutive equation for the fluid in question. In graphic form it is referred to as the flow curve, or *rheogram* for the fluid.

It is instructive to interpret the velocity gradient as follows:

$$\frac{dv_x}{dy} = \frac{d}{dy}\left(\frac{dx}{dt}\right) = \frac{d}{dt}\left(\frac{dx}{dy}\right) \approx \frac{d}{dt}\left(\frac{\Delta x}{\Delta y}\right) \tag{1.34}$$

The term $\Delta x/\Delta y$ is the shear strain on the fluid, whence dv_x/dy is the rate of shear strain or, simply, the shear rate. Hence, for fluids the shear stress is a function of the rate of shear strain or, simply, the shear rate. In contrast, for solids the shear stress is a function of strain rather than the rate of strain. Shear rate often is called the rate of deformation and is denoted by the symbol $\dot{\gamma}$.

For a homogeneous fluid containing small molecules, the constitutive equation (1.33) is usually simple, but for a multiphase mixture, a solution or a liquid containing large molecules, complex relations result. In the first case, the shear stress-shear rate relation is linear through the origin. Such fluids are termed Newtonian. For these simple fluids, the internal structure is unaffected by the magnitude of the imposed shear rate. For complex fluids, an imposed shear results in changes in the internal structure of the fluid so that the constitutive relation becomes more complicated. For example, in a liquid containing large molecules (a molten polymer or a polymer solution), at low shear rates the molecules remain randomly coiled, much as they are in the fluid at rest. The fluid structure remains unchanged in this range, and the shear stress-shear rate relation is linear as for a Newtonian fluid. However, as the shear rate is increased, the randomly coiled molecules tend to line up in the flow direction, changing the structure of the fluid and, hence, the nature of the constitutive relationship. The progressive lining-up or disentangling of the molecules results in the fluid becoming less viscous, or "thinner," in its flow properties. Increasing the shear further would cause more and more molecules to line up, so that the stress-shear rate relation would remain nonlinear until a limiting shear is reached when all the molecules have lined up. Further shear rate increases would not result in structural change, and the constitutive relation would return to a Newtonian relation (albeit a different one from that for low shear rates). Thus, the progression is from Newtonian to non-Newtonian and back as the shear rate is increased. Structural changes also occur for suspensions of solid particles in a liquid or for liquid-liquid emulsions, resulting again in complex flow curves for such fluids.

Under the condition of steady flow as developed by the system in Figure 1.1, time-dependent properties and solid-like or elastic behavior cannot be detected. However, consider what would occur if the force moving the upper plate were removed. In most instances, the upper plate would continue to move, but with decreasing velocity until it and the fluid eventually come to rest. If the fluid were *viscoelastic* i.e., possessing both viscous and elastic properties, the plate and the fluid would first slow down as before; however, after coming to a stop, some motion in the negative x direction would occur as the fluid sought to recover its original configuration. Only partial recovery would be attained. Next consider what happens if the shear rate on a fluid is changed instantaneously (a situation difficult to achieve experimentally because in our system the inertia of the plate and the fluid itself will result in only a gradual change). For many fluids, any resulting changes in internal structure occur very rapidly, and a new stress level corresponding to the new shear rate is reached instantaneously following Equation 1.33. Such fluids are termed *time-independent*. For *time-dependent* fluids, structural changes are considerably slower, and the shear stress changes slowly until ultimately a steady value is reached corresponding to the new shear rate. Some fluids do not deform until a certain "yield" stress value is exceeded. In the experiment in Figure 1.1, F_0/A would have to exceed the yield stress for flow to occur for such a liquid.

Fluids may be divided into three classes based on their flow behavior:

time-independent purely viscous fluids, time-dependent purely viscous fluids and viscoelastic fluids. These are summarized in Table 1.5. Of these three classes, the time-independent purely viscous fluids are understood best. A great deal of work has been done on viscoelastic fluids, but few results of use in engineering design are available. Time-dependent fluids are less common and perhaps the least is known about them.

Time-Independent Fluids

Newtonian Fluids

As noted above, the simplest class of real fluids comprises the Newtonians, whose constitutive equation is given by

$$\tau = \mu\dot{\gamma} \tag{1.35A}$$

where, for convenience, we have dropped the yx subscripts on τ. Equation 1.35 is known as Newton's law of viscosity. Viscosity, μ, is a property of the fluid that is a function of temperature and pressure only (for a single-phase, single-component system). The viscosity of liquids decreases, whereas that of gases increases, with temperature. An *ideal* or perfect fluid is one whose viscosity is zero and, thus, can be sheared without the application of a shear stress. It is an artifact in that no such fluids exist, but finds use in the theory of potential flows.

The dimensions of μ are readily developed. The stress τ has dimensions ML^{-1} and $\dot{\gamma}$, t^{-1}. Thus, μ has dimensions $ML^{-1}t^{-1}$. In the cgs units system (M in g, L in cm, t in s), the viscosity has units g cm^{-1} s^{-1}, known as poise. For most Newtonian fluids this is a rather large unit, and the

Table 1.5 Classification of Fluid Behavior

Fluids			
Purely Viscous			
Time Independent		Time Dependent	
No Yield Stress	Yield Stress	Thixotropic	Viscoelastic
Newtonian Pseudoplastic Dilatant	Bingham Yield-Pseudoplastic Yield-Dilatant	Rheopectic	

centipoise (cp), which is 0.01 poise, is more convenient. For points of reference, the viscosity of water at 20°C is 1.00 cp and that of air is 0.0181 cp.

As noted, the flow curve or rheogram of a Newtonian fluid is a straight line through the origin in arithmetic coordinates. The slope of the line is the viscosity, so that the entire class of Newtonian fluids can be represented by a family of straight lines through the origin. It is often convenient to plot the rheogram using logarithmic coordinates as it allows a greater range of data and also permits easy comparison of Newtonian versus non-Newtonian behavior. On such coordinates, a Newtonian fluid has a rheogram that is a straight line of unit slope and whose intercept at a shear rate of unity is the viscosity. The plots in Figures 1.2(A) and 1.2(B) illustrate these alternative methods of representation.

Most low-molecular-weight liquids and solutions and all gases are Newtonian. Homogeneous slurries of small spherical particles in gases or liquids at low solids concentration are also frequently Newtonian. Thus, there is a large class of fluids for which the property of viscosity characterizes flow behavior.

A far larger class of fluids *does not* follow Newton's law of viscosity and, furthermore, may display other characteristics such as time-dependence, viscoelasticity and yield. A fluid with any of these characteristics is termed non-Newtonian. In general, there are no completely satisfactory constitutive equations for non-Newtonian fluids, in contrast to the situation for Newtonian fluids.

For non-Newtonian fluids, it is convenient to define an "apparent viscosity," η_a, as

$$\eta_a = \tau/\dot{\gamma} \qquad (1.35B)$$

The apparent viscosity is a function of $\dot{\gamma}$ for non-Newtonian materials and is analogous to the Newtonian viscosity, μ. However, whereas the Newtonian viscosity does not vary with shear rate, the non-Newtonian apparent viscosity does.

Pseudoplastic or Shear-Thinning Fluids

Pseudoplastic fluids often are referred to as shear-thinning because their apparent viscosity decreases with shear rate. That is, the rate of increase of shear stress for such fluids decreases with increased shear rate. Increased shear rapidly breaks down the internal structure within the fluid and does so reversibly, whereby no time-dependence is manifested. Examples of fluids that exhibit shear-thinning are polymer melts and solutions, mayonnaise, suspensions such as paint and paper pulp, and some dilute suspensions of

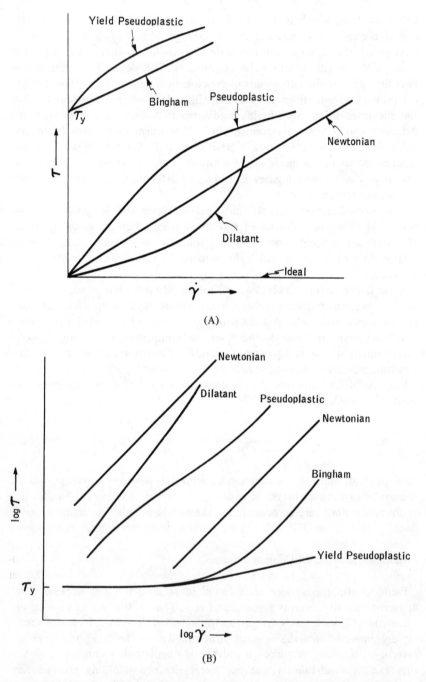

Figure 1.2   (A) Rheograms on arithmetic coordinates; (B) rheograms on logarithmic coordinates.

inert particles. Many of these fluids also exhibit other non-Newtonian characteristics such as viscoelasticity, in the case of polymer solutions and melts,
and time-dependence, in the case of paints. Thus, pseudoplasticity is but one
important characteristic of such a non-Newtonian fluid and does not
necessarily describe all its non-Newtonian features. Many pseudoplastics
are shear-thinning at intermediate shear rates and are Newtonian at low and
high shear rates. A mechanism for this phenomenon for fluids containing long
molecules such as polymer solutions was noted above. Rheograms for pseudoplastics are shown in Figure 1.2.

Many constitutive equations of varying complexity have been proposed
for these fluids. Because of its simplicity, a power law relation is used most
widely, even though it does not describe the Newtonian extremes found for
most pseudoplastics. The power law, or Ostwald-de Waele equation, is

$$\tau = K\dot{\gamma}^n \tag{1.36}$$

and is a two-parameter equation. Parameter K is called the *consistency
index* and n is the *flow behavior index*. Both are functions of temperature
and pressure, but K is more sensitive to temperature than n. Pressure dependence of these parameters has not been investigated.

Note that if $n = 1$, Equation 1.36 reduces to Newton's law of viscosity.
For $n < 1$, pseudoplastic or shear-thinning behavior is exhibited, whereas if
$n > 1$, dilatant or shear-thickening behavior is observed. The apparent viscosity for a power law fluid is

$$\eta_a \equiv \tau/\dot{\gamma} = K\dot{\gamma}^{n-1} \tag{1.37}$$

showing that if $n < 1$, η_a decreases with $\dot{\gamma}$. The value of n is the slope of the
rheogram on logarithmic coordinates.

A fit of the power law expression to experimental data is usually good
over several orders of magnitude in $\dot{\gamma}$. However, at low and high $\dot{\gamma}$, Newtonian
behavior often persists and agreement is poor.

Note that Equation 1.36 is valid only for $\dot{\gamma} > 0$. Depending on the coordinate system, $\dot{\gamma}$ may be negative over some portion of the flow field, in
which case the correct form of this equation is

$$\tau = K|\dot{\gamma}|^{n-1}\dot{\gamma} \tag{1.38}$$

In this book we assume $\dot{\gamma} > 0$, and "simple shear" contains only one nonzero
component of the velocity gradient. Extensions of Equation 1.37 to more
complicated flow fields and where $\dot{\gamma}$ may be negative are given by Bird

et al. [1960]. These authors place a negative sign in the constitutive equations because of their interpretation of τ as momentum flux rather than shear stress. This sign should be ignored to conform to the notation in this text.

Other equations have been proposed for pseudoplastics but have not been used as widely as the power law. Also, they are generally more complex, and some involve three or more parameters, rather than only two. The principal advantage of these more complex models is that they can predict shear-thinning behavior as well as the tendency to Newtonian behavior at one or both extremes of shear rate. Table 1.6 lists some of these models along with their main features. Further details are given by Fredrickson [1964], Skelland [1967] and Bird et al. [1960].

Dilatant or Shear-Thickening Fluids

Dilatant fluids (also called shear-thickening fluids) are those whose apparent viscosity increases with shear rate. The rate of increase of shear stress for such fluids increases with shear rate. From appropriate values of the parameters, equations developed for pseudoplastics can be applied to dilatant fluids. For example, the power law equation may be used with n greater than unity for dilatant fluids. Rheograms for these fluids are illustrated in Figure 1.2. Dilatancy is not as common as pseudoplasticity and generally is observed for fairly concentrated suspensions of irregular particles in liquids. A commonly accepted mechanism for dilatancy (refer to Metzner [1956]) is that at low shear rates the particles in such fluids are densely packed but still surrounded by liquid, which lubricates the motion of adjacent particles. At higher shear rates, the dense packing breaks up, progressively forcing liquid out of more and more of the interstices between particles. There is now insufficient liquid to lubricate the motion of such adjacent "dried out" particles, and the shear stress increases with shear rate at a higher rate than before. Many dilatant fluids also exhibit volumetric dilatancy, that is, an increase in volume with shear rate, as well as viscous dilatancy, the increase in apparent viscosity with shear rate. Examples of dilatant fluids are aqueous suspensions of titanium oxide, suspensions of starch and quicksand.

Plastic Fluids (Fluids with Yield)

Certain fluids do not deform continuously unless a limiting or "yield" stress is exceeded. These often are referred to as plastic fluids, or *Bingham fluids*, although the latter implies that they follow a particular constitutive equation. In the simplest case, such fluids behave in a manner identical to Newtonian fluids once the yield stress is exceeded, in that their rheograms are Newtonian rheograms shifted upward (Figure 1.2(B)). Pure Bingham behavior is rare in nature. In other cases, the flow curve is that of a

Table 1.6 Some Constitutive Equations for Pseudoplastic (or dilatant) Fluids

Model Name	Constitutive Equation	Apparent Viscosity	Limiting Newtonian Prediction		Remarks[a]
			Low Shear	High Shear	
Power Law (Ostwald-de Waele)	$\tau = K\dot\gamma^n$	$K\dot\gamma^{n-1}$	—	—	Two parameters: K, n. Newtonian limits not predicted.
Prandtl-Eyring	$\tau = A\,\sinh^{-1}(\dot\gamma/B)$	$A\dot\gamma\,\sinh^{-1}(\dot\gamma/B)$	A/B	—	Two parameters: A, B. Based on Eyring's kinetic theory of liquids.
Ellis	$\dot\gamma = (\phi_0 + \phi_1\tau^{\alpha-1})\tau$	$(\phi_0 + \phi_1\tau^{\alpha-1})^{-1}$	ϕ_0^{-1} for $\alpha > 1$ only	ϕ_0^{-1} for $\alpha < 1$ only	Three parameters: ϕ_0, ϕ_1, α. If $\phi_1 = 0$, gives Newtonian equation. If $\phi_0 = 0$, gives power law equation.
Reiner-Philippoff	$\dot\gamma = \left[\mu_\infty + \dfrac{\mu_0 - \mu_\infty}{1 + (\tau/\tau_S)^2}\right]^{-1}\tau$	$\mu_\infty + \dfrac{\mu_0 - \mu_\infty}{1 + (\tau/\tau_S)^2}$	μ_0	μ_∞	Three parameters: $\mu_0, \mu_\infty, \tau_S$. μ_0 and μ_∞ are the limiting viscosities.
Sisko	$\tau = a\dot\gamma + b\dot\gamma^c$	$a + b\dot\gamma^{c-1}$	a	—	Three parameters: a, b, c. Combination of Newtonian and power law.

[a] All these equations predict shear-dependent viscosity.

pseudoplastic shifted upward, and such fluids are termed yield pseudoplastics. Yield dilatant behavior also may be encountered. The commonly accepted mechanism for the behavior of plastic fluids is that the fluid at rest contains a structure sufficiently rigid to resist shear stresses smaller than the yield stress, τ_y. When this stress is exceeded, the structure collapses and deformation is continuous as for nonplastic fluids. Examples of fluids exhibiting plastic behavior are certain paints, suspensions of finely divided minerals such as chalk in water, and some asphalts. The yield values can be quite small for water suspensions and very large for materials such as asphalt. Plastic behavior is necessary in paints to prevent flow when applied in the form of a vertical film. Constitutive equations for such fluids are modifications of the Newtonian or the power law expressions.

Bingham Model. The Bingham equation is given by

$$\tau = \tau_y + \eta\dot{\gamma} \ , \quad \tau > \tau_y$$
$$\dot{\gamma} = 0 \ , \qquad \tau < \tau_y \tag{1.39}$$

where τ_y is the yield stress, and η is the "plastic viscosity" or "coefficient of rigidity" (i.e., the slope of the flow curve in rectangular coordinates). On logarithmic coordinates the curve is asymptotic to τ_y at low $\dot{\gamma}$ and approaches a slope of unity for high $\dot{\gamma}$ (Figure 1.2). Note that for $\tau < \tau_y$, $\dot{\gamma} = 0$, implying that there is no deformation in this region. The apparent viscosity for Bingham fluids is given by

$$\eta_a = \tau/\dot{\gamma} = \eta + \tau_y\dot{\gamma}^{-1} \tag{1.40}$$

showing that the apparent viscosity decreases with shear rate. For very high shear rates the effect of the yield stress becomes negligible, the apparent viscosity levels off to the plastic viscosity, η, and the fluid behaves Newtonian.

Yield-Power Law Model. This model is expressed as

$$\tau = \tau_y + K\dot{\gamma}^n \ , \quad \tau > \tau_y$$
$$\dot{\gamma} = 0 \ , \qquad \tau < \tau_y \tag{1.41}$$

and is simply a combination of the Bingham and power law equations. As before, pseudoplastic or dilatant behavior develops depending on whether n is less than, or greater than, unity. An example of yield-pseudoplastic liquids is a clay-water suspension. Yield-dilatant behavior is less common.

Time-Dependent Fluids

Fluids described thus far have the property that their flow behavior responds instantaneously to sudden changes in shear. For example if a pseudoplastic fluid were subjected to a sudden change in shear rate, the new stress given by its constitutive equation would be attained instantaneously. Such fluids are termed time-independent. Actually, a finite nonzero time is required for the new stress to be achieved, but this time is very short. There is a class of fluids termed time-dependent fluids for which the response time is appreciable. For such fluids, the sudden application of a change in shear rate results in the shear stress changing slowly with time until a new equilibrium shear stress is established corresponding to the changed shear rate. The postulated mechanism for time-dependence is that the time scale for structural changes within the fluid due to shear is large compared to the time scale of shear and the very short time scales for time-independent fluids. As the flow behavior depends on the fluid structure, the shear stress responds slowly to an imposed change in shear rate. Thus, the shear stress becomes a function of shear rate and time until steady conditions are attained. Time-dependency may be exhibited by fluids that otherwise may be termed pseudoplastic, dilatant or plastic. Time-dependent fluids generally are classified as either thixotropic or rheopectic.

Thixotropic fluids break down under shear. At a given shear rate, the shear stress slowly decreases until an equilibrium state is reached. Such fluids behave as time-dependent pseudoplastics. Thixotropic behavior is illustrated in Figure 1.3, which shows the change in apparent viscosity of the fluid when a higher shear rate, $\dot{\gamma}_2$, suddenly is imposed at time t_1. At times less than t_1, the fluid has been sheared at $\dot{\gamma}_1$ for a long time so that an equilibrium apparent viscosity, η_{a1}, is established. For a time-independent pseudoplastic, the apparent viscosity decays to η_{a2} immediately. For a thixotropic fluid, a slow decay over some measurable time $(t_2 - t_1)$ occurs until the equilibrium viscosity η_{a2} is reached. If the new shear rate is a *decrease* over the original value, $\dot{\gamma}_1$, the apparent viscosity will *increase* slowly with time. Such instantaneous changes in $\dot{\gamma}$ cannot be achieved in practice because of fluid and equipment inertia. A more practical approach to detecting thixotropy (or any time-dependent behavior) is to subject the fluid to a programmed change in shear rate with time, increasing from zero shear rate to some peak value and then decreasing back to zero. A thixotropic fluid under such a program would produce a hysteresis loop for τ versus $\dot{\gamma}$, as shown in Figure 1.4. For a pseudoplastic fluid no hysteresis is observed, and the same curve would be traced out for increasing and decreasing shear. Note

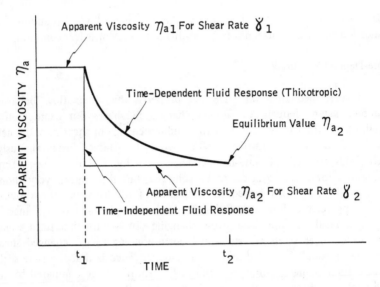

Figure 1.3   Response of time-dependent fluid to change in shear rate (thixotropic case).

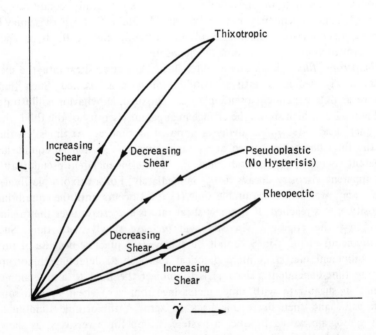

Figure 1.4   Hysteresis loops for time-dependent fluids. (Arrows show the chronology of the imposed shear rate.)

that the area within the loop depends both on the degree of thixotropy as well as on the time scale of change of the shear rate. If the shear rate is changed slowly enough, even a highly thixotropic fluid produces a curve with no hysteresis. Also to be noted is the position of the curve for increasing shear relative to that for decreasing shear, which follows from the fact that the fluid is basically pseudoplastic, with an apparent viscosity that decreases with increasing shear rate. Examples of thixotropic fluids are paints, ketchup and food materials, oil well drilling muds and some crude oils. Thixotropy is necessary in paint to allow it to flow easily after the large shear imposed by brushing and then to recover a more viscous character after a short period of standing. Govier and Aziz [1977] have summarized some of the rheological models proposed for thixotropic fluids. An extensive review of the subject is given by Bauer and Collins [1967].

For *rheopectic fluids*, shear stress at a constant shear rate increases slowly with time until an equilibrium value is reached. Such fluids behave as time-dependent dilatant fluids. Under the programmed change in shear rate described above, the stress-shear rate curve forms a hysteresis loop, but of a different shape than a thixotropic fluid (Figure 1.4). As before, the location and shape of the loop depends on the shear rate program, as well as on the degree of rheopexy. Rheopectic fluids are rare, examples being gypsum suspensions and bentonite clay suspensions.

Viscoelastic Fluids

For the fluids described, if the imposed shear stress is removed deformation ceases, but there is no tendency for the fluid to recover its original undeformed state. Certain fluids do have the property of partially recovering their original state after the stress is removed (i.e., they have memories). Such fluids thus have properties akin to elastic solids as well as viscous liquids, and are termed viscoelastic. Examples of viscoelastic liquids are molten polymers and polymer solutions, egg white, dough and bitumens.

The elastic property of such fluids leads to some interesting and unusual behavior. The classic example is the phenomenon of rod-climbing, or the so-called "Weissenberg effect," exhibited by these fluids. If a rotating cylinder or rod is immersed in a purely viscous liquid, the liquid surface is depressed near the rod because of centrifugal forces. With a viscoelastic fluid, on the other hand, liquid climbs up the rod because of normal stresses generated by its elastic properties. This can be observed during mixing of flour dough and in stirred polymerization reactors. Another phenomenon is the marked swelling in a jet of viscoelastic liquid issuing from a die. As a result, extrusion dies must be designed properly to produce the desired product cross section.

The work performed on a viscoelastic fluid (for example, by forcing it through a tube) is stored in the fluid as normal stresses, as opposed to all being dissipated into heat in the case of purely viscous liquids. This stored energy is released when the fluid emerges from the tube and results in a swelling of the emerging fluid jet. The normal stresses generated within the tube relax when the fluid emerges from the tube; such fluids are said to exhibit *stress relaxation*. Some liquid-liquid mixtures consisting of droplets of one liquid dispersed in the other also exhibit viscoelasticity. Elastic energy is stored when the spherical droplets are distorted by shear and is released through the action of interfacial tension when the shear is removed.

Elastic effects are important mainly during the storage or release of elastic energy. Hence, such effects are important in the entrance and exit sections of tubes and during flow accelerations and decelerations caused by changes in cross section, by imposed oscillations or by turbulence. For steady laminar flow in a tube or channel of constant cross section, elastic effects are not important except near the entrance and exit. On the other hand, for flow in fittings or in turbulent flow these effects may become important. Pronounced viscoelasticity is observed mainly in molten polymers and concentrated polymer solutions and generally is not considered important for pipeline flow of other non-Newtonians such as slurries.

The flow behavior of these fluids cannot be represented by a simple relation between shear stress and shear rate alone. Instead, it will depend on the recent history of these quantities, as well as on their current values. Constitutive equations for such fluids therefore must involve shear stress, shear rate and their time derivatives. It is clear also that the time derivatives involved must be those applicable to a particular quantity of fluid as it moves through the system and not those from the viewpoint of a stationary observer, for example. The latter has no direct impact on fluid behavior; the former represents time rates of change experienced by the fluid itself. A large number of constitutive equations of this general type have been proposed involving, among other features, various types of time derivatives. Unfortunately, their use in engineering calculations is not widespread. For illustrative purposes, the Oldroyd model is noted, which is applicable for low shear rates:

$$\tau + \lambda_1 \frac{d\tau}{dt} = \mu \left(\dot{\gamma} + \lambda_2 \frac{d\dot{\gamma}}{dt} \right) \qquad (1.42)$$

Here, λ_1 and λ_2 are relaxation times and μ is the viscosity. We first observe that if λ_1 and λ_2 are both zero, the equation reverts to that for a Newtonian fluid. If only λ_2 is zero, the equation reduces to that for a Maxwell fluid. The Maxwell fluid is an early and simple model for a viscoelastic fluid based on a mechanical analog, whereby the fluid is represented as a spring and dashpot in series, the spring representing the elastic part and the dashpot the viscous part. Both the Maxwell fluid and the fluid represented by Equation 1.42 show stress relaxation. If flow is stopped,

$$\dot{\gamma} = \frac{d\dot{\gamma}}{dt} = 0$$

the stress decays or relaxes as e^{-t/λ_1}. If stress is removed, the shear rate in a Maxwell fluid becomes zero immediately, while for the fluid of Equation 1.42 the shear rate decays as e^{-t/λ_2}. This simple three-constant model has been shown to represent the behavior of certain viscoelastic fluids at low shear rates.

TYPES OF OPERATIONS AND APPROACHES TO PROBLEM ANALYSIS

Industrial operations may be generalized under three broad categories: intermittent or batch, semibatch and continuous. An example of a batch operation is the production of paper pulp, in which wood chips, water and caustic cooking chemicals (called white liquor) are prepared in a reactor. The reactor vessel (called a digester) is charged with formulated proportions of these materials and cooked in the presence of high-pressure steam. After a required cooking time, the digester is blown (i.e., steam and vapors released), spent liquor is drained and the cooked pulp is sent on to the next stage of processing. The digester is then ready for a new charge or batch of materials. Still another example of an intermittent operation is the production of beer and other alcoholic beverages in batch fermenter vessels.

An example of a semibatch operation is the production of chlorophenols by chlorination of phenols. The chlorophenols of most commercial importance are 2,4-dichlorophenol, an intermediate in the manufacture of 2,4-dichlorophenoxy-acetic acid (2,4-D) and its derivatives, which are selected herbicides, and pentachlorophenol (PCP), used as a wood preservative due to its fungicidal properties. In this semibatch process, phenol is charged into two batch reactors—a primary reactor and a secondary scrubber-reactor. Chlorine is added only to the primary reactor. The offgas from the primary reactor, consisting of chlorine and hydrogen chloride, is sent to the scrubber-reactor, where sufficient phenol is charged to ensure complete reaction of the chlorine. The hydrogen chloride offgas from the scrubber-reactor is recovered by dissolving it in water in an absorption tower to produce commercial-grade hydrochloric acid. The primary reactor is a stirred-batch reactor. A period of 8-10 hours generally is required for the chlorination (flowsheets and other examples are given by Goldfarb et al. [1981]).

Continuous operations and processes are the most desirable both from operating and cost standpoints. The process flowsheet of one example, the production of acetaldehyde by liquid-phase ethylene oxidation, is shown in Figure 1.5. Acetaldehyde is used primarily as an intermediate for the production of other organic chemicals, the major derivatives being acetic acid, acetic anhydride, n-butanol and 2-ethylhexanol. Other products derived from acetaldehyde are pentaeythritol, trimethylol propane, pyridines, peracetic acid, crotonaldehyde, chloral, 1,3-butylene glycol, lactic acid, glyoxal and alkylamines. In Figure 1.5, ethylene gas (stream 2), oxygen (1) and a recycle gas stream (31) are fed continuously to a reactor containing an aqueous solution of palladium chloride and copper chloride. Exit stream (3) flows into a gas-liquid separator. The liquid stream from the separator (4) is split into two streams; stream (6) is recycled to the reactor and stream (5) is sent to a regenerator, where it is mixed with oxygen and steam to decompose copper oxalate and other organics prior to being returned to the reactor. The gas stream from the separator (11) contains product (acetaldehyde), which is sent to a quench scrubber. There it is cooled and scrubbed with water to condense the acetaldehyde and other condensables. Noncondensed vapors (12) consisting of unreacted ethylene, oxygen and various inerts are sent to a scrubber. Part of the stream (13) is purged to a flare to control the accumulation of inerts in the system, and the remainder (14) is recycled to the reactor vessel. Condensed stream (16) is split: stream (17) is recycled to the scrubber and stream (18) is heated and fed to a light ends distillation column, in which dissolved gases and low boiling material are removed overhead (stream 19). The bottoms (20) from the light ends distillation column are fed to the product recovery column, where acetaldehyde is removed as an overhead stream (22). A by-product stream (22) is removed from the middle of the column. Water contaminated with residual organics (29) is removed from the bottom of the column and sent to disposal.

As illustrated by the above examples, process operations can be complex and comprise a large number of different unit operations. Although the nature of a particular process essentially may be intermittent, continuous production often is approached by the combination of a number of unit operations simultaneously but in different stages. Most chemical reaction processes are in essence semicontinuous when viewed as a whole operation; however, in a strict sense, they may be classified as batch when one is dealing with individual unit operations.

As illustrated by the elementary flowsheet for acetaldehyde production (Figure 1.5), each plant involves knowledge of an overwhelming accumulation of information on quantities and compositions of raw materials, intermediates, waste products and by-products, which is essential to the production and accounting divisions and can be integrated into evaluating and improving an overall process. Quantitative bookkeeping of individual constituents is done by use of the generalized law of the conservation of mass. This is a

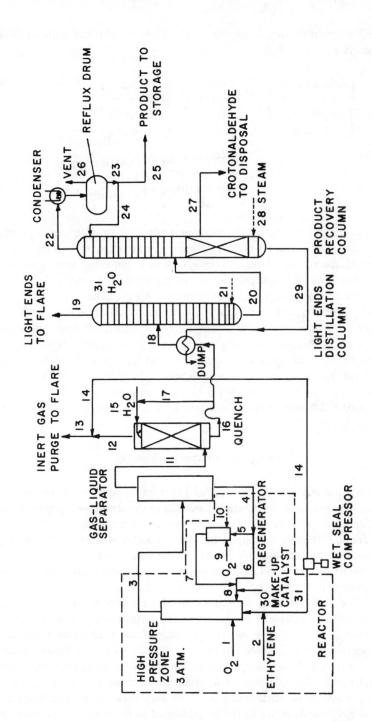

Figure 1.5   The continuous single-stage process for manufacturing acetaldehyde from ethylene.

general material balance applicable to all processes, with and without chemical reaction:

$$
\begin{Bmatrix} \text{Accumulation} \\ \text{of material} \\ \text{within the} \\ \text{system} \end{Bmatrix} = \begin{Bmatrix} \text{Input} \\ \text{through} \\ \text{system} \\ \text{boundaries} \end{Bmatrix} - \begin{Bmatrix} \text{Output} \\ \text{through} \\ \text{system} \\ \text{boundaries} \end{Bmatrix}
$$

$$
+ \begin{Bmatrix} \text{Material} \\ \text{accumulation} \\ \text{within the} \\ \text{system} \end{Bmatrix} - \begin{Bmatrix} \text{Consumption} \\ \text{within the} \\ \text{system} \end{Bmatrix} \tag{1.43}
$$

This relation may be applied to a single unit operation or to the entire process comprising many different unit operations. The same general balance may be applied to the energy relations of the system or to individual unit operations. In this form, another independent set of equations is formulated. The use of this principle is perhaps universal to engineering problems.

Most semicontinuous operations are repeated in an almost indefinite number of cycles, and both continuous and semicontinuous operations are performed over relatively long time periods. As such, the accumulation of materials and/or energy in a system or piece of equipment is limited. If the operations are performed over a sufficient time such that fluctuations of amounts are well known and negligible in comparison to input and output streams, the last two terms in Equation 1.43 may be neglected, whence we may state simply that

$$
(\text{Total input}) = (\text{output}) + (\text{losses}) \tag{1.44}
$$

This simple relation may be applied (1) to evaluating the performance of existing equipment and operations; and (2) in the design of equipment and planning new operations. This book is concerned primarily with the latter. Forthcoming chapters will apply this concept and some of the others briefly described in this introductory chapter to the design and proper selection of unit operations concerned with the physical combinations and separation of process streams.

We close this chapter with some general guidelines to problem analysis and solving. The success of a solution to any engineering problem is directly proportional to the degree of understanding of the mechanisms responsible for a particular phenomenon. For example, if our concern is with the production of a chemical feedstock, an understanding of the basic chemistry alone is insufficient. The environment in which the reaction is to take place is of equal concern. How feedstreams are to be combined and in what form in a commercial operation are questions addressed both by process kinetics and principles of unit operations. Emphasis in this text is on the hydrodynamics. Without this information the rational design of any process simply is not possible.

In the solution of problems the following steps will be found to be helpful:

1. Prepare a sketch of the phenomenon or process. This may be as detailed a drawing as needed to completely understand the requirements of the problem or the important mechanisms.
2. If a mathematical model is to be developed describing the phenomenon or process or an experimental unit is to be constructed, list all important mechanisms. From this list, decide which ones are the most important and which may be neglected in the analysis.
3. Select a basis for the design calculations. For example, if a piece of equipment is being scaled-up from a cold model unit, the basis to ensure proper operation might be residence time or an appropriate dimensionless group related to operating performance.
4. Tabulate all data and intermediate steps in design computations. In this day of computers, this last step is self-evident.

NOTATION

A area, m^2

a acceleration, m/s^2

C molal heat capacity, $cal/(g)(°C)$

F area, m^2

F_0 force, N

g local acceleration due to gravity, m/s^2

g_c conversion factor, $32.174 \ lb_m\text{-}ft/lb_f\text{-}s^2$

H enthalpy, J/mole

K consistency index

KE kinetic energy, J

L length, cm

M mass, kg

m mass, kg

n number of moles for flow behavior index

p pressure, Pa

PE potential energy, J

Q heat

R ideal gas law constant

S distance, m

T absolute temperature, °K or °R

t time, s, or temperature, °C or °F

U internal energy, J

u velocity, m/s

V volume, m^3

v velocity, m/s

W work, N-m

Z elevation, m

Greek Symbols

γ ratio of specific heats, c_p/c_v

$\dot{\gamma}$ shear rate (rate of deformation), s^{-1}

η_a apparent viscosity, poise

θ time, s

λ relaxation time, s

μ viscosity, poise

ρ density, kg/m^2

τ stress, Pa

τ_y yield stress, Pa

$\dot{v}$ molal volume, m^3/mol

REFERENCES

Bauer, W. H., and E. A. Collins (1967) In: *Rheology*, Vol. 4, F. R. Eirich, Ed. (New York: Academic Press Inc.), pp. 423-459.

Bird, R. B., R. C. Armstrong and O. Hassager (1977) *Dynamics of Polymeric Liquids*, Vols. I & II (New York: John Wiley & Sons, Inc.).

Bird, R. B., W. E. Stewart and E. N. Lightfoot (1960) *Transport Phenomena* (New York: John Wiley & Sons, Inc.).

Cheremisinoff, N. P. (1981) *Fluid Flow: Pumps, Pipes and Channels* (Ann Arbor, MI: Ann Arbor Science Publishers).

Federal Register (October 26, 1977) 42(206)56513-56514.

Fredrickson, A. G. (1964) *Principles and Applications of Rheology* (Englewood Cliffs, NJ: Prentice Hall, Inc.).

Goldfarb, A. S., G. R. Godgraben, E. C. Herrick, R. P. Ouellette and P. N. Cheremisinoff (1981) *Organic Chemicals Manufacturing Hazards* (Ann Arbor, MI: Ann Arbor Science Publishers).

Govier, G. W., and K. A. Aziz (1977) *The Flow of Complex Mixtures in Pipes* (New York: Krieger Publishing Co.).

Lodge, A. S. (1964) *Elastic Liquids* (New York: Academic Press, Inc.).

Metzner, A. B. (1956) In: *Advances in Chemical Engineering*, Vol. I, T. B Drew and J. W. Hoopes, Ed., pp. 79-150.

Middleman, S. (1968) *The Flow of High Polymers: Continuum and Molecular Rheology* (New York: Interscience Publishers).

Schowalter, W. T. (1978) *Mechanics of Non-Newtonian Fluids* (Oxford, England: Pergamon Press, Inc.).

Skelland, A. H. P. (1967) *Non-Newtonian Flow and Heat Transfer* (New York: John Wiley & Sons, Inc.).

Smith, J. M., and H. C. Van Ness (1959) *Introduction to Chemical Engineering Thermodynamics*, 2nd ed. (New York: McGraw-Hill Book Co.).

Sridhar, T. (1983) "Transport Properties of Liquids," in *Handbook of Fluids in Motion*, N. P. Cheremisinoff and R. Gupta, Eds. (Ann Arbor, MI: Ann Arbor Science Publishers), pp. 3-27.

CHAPTER 2

PRINCIPLES OF SIMILARITY, MODELING AND DIMENSIONAL ANALYSIS

CONTENTS

INTRODUCTION

Often, problems of a unique nature are encountered in industry. Therefore, many processes cannot be formulated rigorously in a purely mathematical sense. Physical and chemical phenomena may be too complex to be described from existing theories. To formulate mathematical relationships that characterize a process and are applicable to design, experimental investigations are conducted. From experimental observation and data, empirical equations can be developed. However, these correlations often are local in character and their range of applicability limited to the conditions over which studies were performed. Despite these shortcomings, such correlations have value and are used, perhaps too often, in designing and scaling up unit operations in chemical engineering.

The value of experiments can be enhanced greatly if results are generalized. In this manner, those parameters that best characterize the behavior of a process can be identified. By utilizing principles of *similarity theory*, experimental observations from small-scale tests may be applied with greater confidence to designing and predicting the performance of large-scale equipment.

Similarity theory is a method that provides a scientific approach to generalizing experimental results. It provides a means of evaluating important process parameters, and the technique can be used to limit the number of costly experiments needed to obtain unified equations. That is, it permits studies of the performance of the system (prototype) to be built on a small-scale replica (referred to as a "model").

The term *model* in the scientific literature does not always refer to a prototype's replica in which experiments can be conducted. It most generally refers to a cognitive or conceived physical mathematical description, i.e., a scheme that attempts to describe the most important variables of a process. In this chapter, however, the term will refer to physical and material approximations of the true system under investigation. Through physical modeling, the nature and behavior of a process may be investigated not only on a smaller scale than the conceptual design, but with different substances and at different temperatures, pressures, etc. In this manner, generalized dependencies can be defined mathematically. This allows experiments to be performed under less severe conditions than in the commercial case.

Similarity theory and physical modeling allow us to investigate phenomena more rapidly and economically, with a reasonable degree of reliability in making the transition from laboratory scale to commercial scale. Note, however, that they cannot provide complete theoretical understanding of the physicochemical phenomena in terms of fundamental equations. The theory and practice only provide integral solutions of the theoretical equations defining the process by generalizing experimental data. These data are valid only for a group of similar phenomena in the range of investigation and have the advantage of negating formal investigation of the theoretical equations.

PRINCIPLES OF SIMILARITY THEORY

Similarity theory is based on the principles of selecting a group of analogous phenomena from unrelated processes. For example, although the motion of fluids in the atmosphere and through piping are different systems, these flow phenomena are analogous because both represent the motion of viscous fluids under the influence of pressure gradients. Hence, the fluid motion for these different systems may be described by the unified Navier-Stokes equations and thus considered to be of the same class of phenomena. Similarly, the motion of incompressible and compressible viscous fluids through piping and equipment, although considerably different processes, comprise a class of similar phenomena. Phenomena are considered operationally similar when the ratios of certain parameters characterizing the system are compatible.

First, we shall consider the condition used most often to determine operational similarity, namely the *geometric* condition. For a model to be geometrically similar to the prototype, the ratio of the distances between any two common points in both systems must be constant. This ratio is referred to as the *geometric scale factor*. For example, the characteristic dimensions of a drum dryer are its diameter, D, and length, L. The geometric scale factor is thus $\frac{L}{D}$, and the model will be similar to the prototype when the following condition is fulfilled:

$$\Lambda = \frac{L_1}{D_1} = \frac{L_2}{D_2} \tag{2.1}$$

where L and D have the same units (m) and, thus, the ratio is dimensionless. Subscripts 1 and 2 refer to the model dryer and prototype, respectively. Thus, the model's diameter differs from the prototype by some constant scale factor. Similar ratios thus can be defined for any geometric configuration. Parameters of the same generic class are interchangeable; that is, those parameters that determine the similarity scale factors may be changed by similar values, which define the system geometry. Hence,

$$\frac{\ell'}{\ell''} = \frac{\ell'_1}{\ell''_1} = \frac{\ell'_2}{\ell''_2} = \frac{\ell'_1 - \ell'_2}{\ell''_1 - \ell''_2} = \frac{d\ell'}{d\ell''} = \Lambda_\ell \tag{2.2}$$

Geometric similarity between systems is a necessary condition for similarity of physical phenomena but is not sufficient in itself. For physical similarity to exist, all important parameters influencing the phenomena must be similar. These parameters often change as functions of time and space in each system. Hence, technological processes are similar only under conditions of mutual fulfillment of geometrical and time similarities over the fields of physical values, as well as similarities of initial and boundary conditions. Similarity among physical parameters such as velocity, acceleration, density, pressure, etc., may be defined in a manner analogous to geometric conditions:

$$\left.\begin{array}{l} \text{For velocity, } v_1 = \Lambda_v v_2 \\ \text{For acceleration, } a = \Lambda_a a_2 \\ \text{For density, } \rho_1 = \Lambda_\rho \rho_2 \\ \text{For pressure, } P_1 = \Lambda_p P_2 \\ \text{For time, } t_1 = \Lambda_t t_2 \end{array}\right\} \qquad (2.3)$$

The similarity scale factors Λ_ℓ, Λ_v, Λ_ρ, Λ_p . . . are constant for different compatible points of two similar systems but change depending on the size ratio of the prototype to the model.

Process similarity between the prototype and the model may be determined through the use of *characteristic parameters*. These parameters are expressed in the form of compatible ratios in the limit of each system and are illustrated in Figure 2.1. We shall denote a "dummy" characteristic parameter by i.

$$i = \frac{\ell_1'}{L'} = \frac{\ell_1''}{L''} = \text{inv.} \qquad (2.4)$$

where inv. denotes invariantly or "one and the same."

A parameter may be expressed in terms of relative units and, hence, an appropriate scale can be defined. For example, the diameter of the dryer may be selected as a scale rather than a length. For other systems, it may be appropriate to define this scale relative to common points in the model and prototype. Hence, from Equations 2.3 and 2.4 we write the following expressions in which the measuring scale is defined relative to the entrance of the system (i.e., in terms of the initial conditions v_0', v_0'', etc.):

$$\left.\begin{array}{l} \dfrac{t_1'}{T'} = \dfrac{t_1''}{T''} = i_t \\[4mm] \dfrac{v_1'}{v_0'} = \dfrac{v_1''}{v_0''} = i_v \\[4mm] \dfrac{\rho_1'}{\rho_0'} = \dfrac{\rho_1''}{\rho_0''} = i_\rho \\[4mm] \dfrac{P_1'}{P_0'} = \dfrac{P_1''}{P_0''} = i_p \end{array}\right\} \qquad (2.5)$$

The characteristic parameters, i_t, i_v, i_ρ, i_p . . . (referred to as simprexes), may not be equal to each other for different compatible points of similar systems, and are independent of the size ratio of the prototype and model.

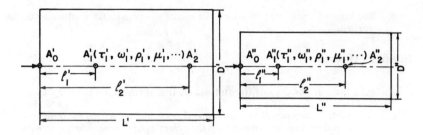

Figure 2.1   Scheme for the formulation of similarity conditions.

This means that in passing from one system to the other the characteristic parameters do not change. For example, assume a portion of fluid flows from a section I-I (a distance of 2 m from the inlet of a dryer) to the section II-II (a distance of 5 m from the inlet) for a total dryer length of $L = 10$ m. The characteristic parameter changes from $i_1 = \dfrac{2}{10} = 0.2$ to $i_2 = \dfrac{5}{10} = 0.5$; however, the scale factor remains unchanged if the sizes of the prototype and model are kept constant.

Let us assume the gas velocity in the dryer entrance to be $v_0' = 4$ m/s and, at the middle section of the dryer, $v' = 3$ m/s. Consequently, the characteristic parameter $i_v = \dfrac{v'}{v_0} = \dfrac{3}{4} = 0.75$. To provide similarity of the velocity field, it is necessary that $v' = v'' = 0.75$ for compatible points of the model. If the entrance velocity in the model $v_0'' = 3$ m/s, then at the middle of the model the velocity scale is

$$v'' = v_0'' \times v' = 3 \times 0.75 = 2.25 \text{ m/s}$$

METHODS OF SIMILARITY THEORY

The mathematical principles in similarity theory are simple and easily implemented. However, the student is cautioned against adopting a formal or stereotyped approach to avoid direct errors that can result if the physical concepts of the methods are not adopted. The application of specific techniques or theorems is determined by a volume of preliminary knowledge about the process under consideration. Application of the theory to the development of fully integrated equations is known as dimensional analysis. There are three basic theorems, each of which addresses a separate issue concerning the planning of model experiments. These issues are:

1. identification of important parameters to be measured in the experiments;
2. the final and most useful form in which the experimental results should appear; and

3.   the type of equipment to which the model's experimental results may be applied.

Newton's Theorem

Newton's theorem addresses the first issue. It is formulated in two ways:

1.   According to Newton, *the conditions necessary (and sufficient) for similarity of the phenomena are equality of the values of the dimensionless groups made up of the quantities given in the conditions.*
2.   According to Kirpichev [1953] (see also Gukhman [1965] and Sedov [1959]), *the conditions necessary (and sufficient) for similarity of the phenomena are equality of characteristic parameters (similarity indicators) to unity.*

We will illustrate the correctness of these formulations in the following example. Consider two similar systems that satisfy Newton's second law (the momentum equation), determining the relation between external forces and produced acceleration.

From Newton's law, the total force acting on a body is

$$f = m\frac{dv}{dt} \qquad (2.6A)$$

where f = force (N)
    m = mass (kg)
    v = velocity (m/s)
    t = time (s)

For two similar systems, Equation 2.6A may be written twice:

$$f_1 = m_1\frac{dv_1}{dt_1} \qquad (2.6B)$$

$$f_2 = m_2\frac{dv_2}{dt_2} \qquad (2.6C)$$

Physical values for the two systems differ only by a scale factor; therefore,

$$f_1 = \Lambda_f f_2 \;;\; m_1 = \Lambda_m m_2 \;;\; v_1 = \Lambda_v v_2 \;;\; t_1 = \Lambda_t t_2 \;;$$

$$dv_1 = \Lambda_v dv_2 \;;\; dt = \Lambda_t dt_2$$

Dividing Equation 2.6B by Equation 2.6C we obtain

$$\frac{f_1}{f_2} = \frac{m_1}{m_2}\frac{dv_1}{dv_2}\frac{dt_2}{dt_1} \quad \text{or} \quad \Lambda_f = \Lambda_m\frac{\Lambda_v}{\Lambda_t} \tag{2.7}$$

Consequently, we obtain

$$\frac{\Lambda_f\Lambda_t}{\Lambda_m\Lambda_v} = 1 = \tilde{j} \tag{2.8}$$

According to Kirpichev's statement, the nondimensional parameter $\tilde{j}$ (indicator of similar transformation) for two similar phenomena is equal to unity. Therefore, the selection of numerical values of scale factors is not arbitrary but rather subjected to the conditions of $\tilde{j} = 1$.

As follows from Equation 2.7

$$\frac{f_1 dt_1}{m_1 dv_1} = \frac{f_2 dt_2}{m_2 dv_2} \quad \text{and} \quad \frac{f_1 t_1}{m_1 v_1} = \frac{f_2 t_2}{m_2 v_2} \tag{2.9}$$

This last expression is a dimensionless group known as the Newton number:

$$Ne = \frac{ft}{mv} \tag{2.10}$$

or, taking into account that $t = \ell/v$,

$$Ne \equiv \frac{f\ell}{mv^2} \tag{2.11}$$

The symbol "$\equiv$" indicates that this is a definition and does not denote the Newton number to be a function of the f, t, m, v, i.e., $Ne \neq f\left(\frac{ft}{mv}\right)$. Thus, for a series (groups) of similar processes whose class is described by initial physical equations based on Newton's second law, the following equality is correct:

$$Ne_1 = Ne_2 = Ne_3 = \ldots = idem$$

Hence, if we consider similar processes of motion in a model and in a prototype, then

$$Ne_{model} = Ne_{prototype} \tag{2.12}$$

Equation 2.12 is the mathematical formulation of the first theorem of similarity. Thus, with properly planned experiments it is only necessary to measure those values that appear in the dimensionless groups of a process under evaluation. The dimensionless groups constitute generalized characteristics of a process consisting of dimensional physical values reflecting different characteristics of a phenomenon.

Because the dimensions entering these groups are reduced, they have a zero dimension and the numerical value of a dimensional group remains true whatever the system of units in which the various quantities are measured.

To derive and compute dimensionless groups properly, we must be sure that the initial equation or expression is dimensionally homogeneous, i.e., that the dimensions of all the terms in each group are consistent and cancel. The correctness of deriving dimensionless groups is checked by reducing the dimensions from which the group is formed. In the case considered,

$$Ne = \frac{ft}{m\nu} \left[\frac{N - s^2}{kg - m} = \frac{N}{N} \right]$$

i.e., the dimensions are reduced and, consequently, the dimensionless group is correct. The *dimensionless groups* are derived from the *dimensional* equation governing the physical process. The essential premise of deriving these groups is the availability of the equation governing the process, i.e., its mathematical description. Regardless how this equation is expressed—in algebraic or differential form—the dimensionless groups may be derived by the same technique from any homogeneous dimensional equation.

Dimensionless groups can be formed simply by dividing through the equation by any dimensional product that makes all the terms of the equation dimensionless. Thus, they can be formed directly from the basic equation of the system and *without its formal solution.*

In addition to the basic physical equation, the derived form of the dimensionless group has an intrinsic physical character. For example, the Newton number expresses the ratio of active to reactive forces and is therefore a measure of impulse, ft, and momentum, mv. In special cases of motion depending on concrete expressions for force, mass and velocity, the Newton number can take on another form, that is, it can be transformed into the Reynolds, Euler, Froude and/or Stokes numbers.

To derive these dimensionless numbers, it is necessary to substitute the acting forces into the expression of the Newton number accordingly. For the *Froude number,*

$$Ne_1 = \frac{[mg]\ell}{mv^2} = \frac{g\ell}{v^2} \; ; \quad Fr = \frac{v^2}{g\ell} \tag{2.13}$$

Fr is the ratio of inertia force on a fluid element to the gravity force.

For deriving the *Reynolds number* from the Newton number, we write
down the expression of the friction force in the flow of a viscous fluid as
it is determined by Newton's law:

$$f = \mu \ell^2 \frac{dv}{d\ell} \qquad (2.14)$$

Substituting f from Equation 2.14 into Equation 2.11 and denoting $m = \rho \ell^3$
(m is related to a unit volume), we obtain

$$Ne_2 = \frac{\left[\mu \ell^2 \frac{dv}{d\ell}\right]\ell}{[\rho \ell^3]v^2} \ ; \quad Ne_2 d\ell = \frac{\mu dv}{\rho v^2}$$

$$Ne_2 \int_0^\ell d\ell = \frac{\mu}{\rho} \int_0^{v_{max}} v^{-2}dv \ ; \quad Ne_2 = \frac{\mu}{\rho v \ell}$$

or

$$Re = \frac{dv\rho}{\mu} \qquad (2.15)$$

or

$$Re = \frac{dv\gamma}{\mu g} \qquad (2.15A)$$

The Reynolds number is interpreted as the ratio of inertia forces to the
viscous forces in the flow.

The *Euler number*, which characterizes the hydrodynamic processes
running under action of mechanical pressure, is the ratio of static pressure
drop, Δp, and dynamic head, ρv^2:

$$Ne_3 = \frac{[\Delta p \ell^2]\ell}{[\rho \ell^3]v^2} \ ; \quad Eu = \frac{\Delta p}{\rho v^2} \qquad (2.16)$$

The *Stokes number* is important in analyzing sedimentation processes.
Substituting the resistance force of the medium into the Newton number,

$$R = 3\pi d\mu v_1 \qquad (2.17)$$

$$Ne_4 = \frac{[3\pi d\mu v]\ell}{\left[\frac{\pi d^3}{6}(\rho_1 - \rho_2)\right]v_2^2} = \frac{\mu\ell}{d^2\rho_1 v^2} \qquad (2.18)$$

or

$$Stk = \frac{d^2\rho_1 v_2}{\mu\ell} \qquad (2.18A)$$

The Stokes number is the ratio of the resistance force of the medium to the buoyancy force of a particle. The intrinsic physical character of each dimensionless number (criterion of similarity) differs from *arbitrarily chosen* dimensionless complexes composed of random physical values.

From the equations governing the physical processes of different classes we obtain dimensionless numbers (criteria of similarity) of different processes: thermal, diffusion, hydrodynamic, and so on. The more complicated equations and systems of equations give at once *several dimensionless numbers*, describing different characteristics of a complicated process. The technique described above for deriving dimensionless numbers is not unique. Further discussions are given by Sedov [1959], Gukhman [1965], Ipsen [1960], Langhaar [1951] and Lighthill [1963]. Other methods are illustrated below.

Derivation of Dimensionless Groups from Process-Governing Equations

There are three basic methods for deriving dimensionless groups from a governing equation describing the process. Each is illustrated by considering the steady accelerated motion of a body:

$$w = w_0 + at \qquad (2.19A)$$

where w = velocity at time t (m/s)
w_0 = velocity at time $t = 0$ (m/s)
a = acceleration (m/s^2)
t = time from starting motion (s)

Method I:

This involves variable transformation through the use of scale factors used earlier in deriving the Newton number.
For the first phenomenon,

$$w_1 = w_{01} + a_1 t_1 \qquad (2.19B)$$

For the second phenomenon,

$$w_2 = w_{02} + a_2 t_2 \qquad (2.19C)$$

For similarity of two phenomena we have

$$\frac{w_1}{w_2} = C_w \; ; \quad w_1 = C_w w_2 \; ; \quad \frac{w_{01}}{w_{02}} = c_w \; ; \quad w_{01} = c_w w_{02}$$

$$\frac{a_1}{a_2} = C_a \; ; \quad a_1 = C_a a_2 \; ; \quad \frac{t_1}{t_2} = C_t \; ; \quad t_1 = C_t t_2$$

Substituting the new notations of w_1, w_{01}, a_1 and t_1 into the expression for the first phenomenon (Equation 2.19B),

$$C_w w_2 = c_w w_{02} + C_a C_t a_2 t_2 \qquad (2.20)$$

Equations 2.19B and 2.20 can coexist only under the condition of reduction of multiples formed from factors. This is equivalent to the condition of equality in pairs:

$$C_w = c_w \; ; \quad C_w = C_a C_t$$

The last condition gives two *characteristic parameters* (indicators of similarity):

$$\tilde{j}' = \frac{C_w}{c_w} = 1$$

$$\tilde{j}'' = \frac{C_a C_t}{C_w} = 1$$

In the expression for $\tilde{j}'$, C_w is related to *different* process velocities but it cannot be reduced.

Substituting the scale factors by the ratios of values, we obtain

$$\tilde{j}' = \frac{\dfrac{w_1}{w_2}}{\dfrac{w_{01}}{w_{02}}} = 1 \; ; \quad \frac{w_1}{w_{01}} = \frac{w_2}{w_{02}} = idem$$

Hence, the first dimensionless number, a *dimensionless velocity* (in this case a relative velocity) is

$$K_1 \equiv \frac{w}{w_0} \qquad (2.21)$$

In the same manner we obtain from the expression $\tilde{j}''$ the second dimensionless number:

$$\tilde{j}'' = \frac{\dfrac{a_1}{a_2} \times \dfrac{t_1}{t_2}}{\dfrac{w_{01}}{w_{02}}} = 1 \; ; \quad \frac{a_1 t_1}{w_{01}} = \frac{a_2 t_2}{w_{02}} = \text{idem}$$

Hence, the second dimensionless number of time similarity (i.e., the criterion of kinematic Homochronity) is

$$K_2 \equiv \frac{at}{w_0} \qquad (2.22)$$

A constant-scale value, w_0, is assumed in both expressions for the dimensionless numbers. According to the terms of the initial equation, we have to determine

$$\frac{w}{w_0} = f\left(\frac{at}{w_0}\right) \quad \text{or} \quad K_1 = f(K_2) \qquad (2.23)$$

In this simple case, the form of this function is understood in the second method of similarity transformation.

Method II:

This method consists of dividing all terms of the homogeneous equation by one of its terms serving as a scale (in this case by w_0):

$$\frac{w}{w_0} = 1 + \frac{at}{w_0}$$

Hence,

$$K_1 = \frac{w}{w_0} \; ; \quad K_2 = \frac{at}{w_0}$$

whence the form of the function is evident:

$$K_1 = 1 + K_2 \qquad\qquad (2.24)$$

The equations governing the process are usually much more complicated and the form of the function not so readily determined. However, for clarification of the combination of physical values entering into the dimensionless numbers, the method is applicable.

Method III:

This method consists of transition to new independent units of measurement of physical values. Assume we know a priori that there is an explicit relationship:

$$w = f(w_0, a, t) \qquad\qquad (2.25)$$

Here, the base units of measure are m and s, and the dependent units of measure are m/s (for velocity) and m/s^2 (for acceleration). Now we transfer to new independent units of measure, which are less in L and T times than the first ones. Then the numerical values of w, a and t will be changed to

$$w\frac{L}{T} = f\left(w_0\frac{L}{T}, a\frac{L}{T^2}, tT\right) \qquad\qquad (2.26)$$

Fluid motion is independent of the units chosen for measuring specific process characteristics. Therefore, the basic equation (2.25) has to retain its structure at different values of coefficients L and T. The numerical values of these coefficients are so chosen to provide expressions that are simple and convenient in application. In particular, we may choose L and T as

$$w_0\frac{L}{T} = 1 \; ; \quad a = \frac{L}{T^2} = 1 \; ; \quad tT = 1$$

Then,

$$T = \frac{1}{t} \; ; \; \frac{L}{T} = \frac{1}{w_0} \; ; \; \frac{L}{T^2} = \frac{1}{a} \; ; \; \frac{1}{w_0} = \frac{T}{a} = \frac{1}{at}$$

$$a\frac{L}{T^2} = a\frac{1}{w_0} \times \frac{1}{T} = \frac{at}{w_0}$$

Substituting the obtained values into Equation 2.26, we have

$$\frac{w}{w_0} = f\left(1, \frac{at}{w_0}, 1\right)$$

or

$$\frac{w}{w_0} = f\left(\frac{at}{w_0}\right)$$

or

$$K_1 = f(K_2)$$

This coincides with the result of transformation by the first method.

All these methods are applicable if one has the equation governing the physical process, i.e., the mathematical description of the process containing all necessary characteristic physical values.

However, as is more often the case, only the *qualitative* description of a process is known; the quantitative relationships among the different physical factors and their form are unknown. Then the theory of similarity helps to determine the *hypothetical* form and prioritizes the most important dimensionless numbers of the process. We then may use carefully planned experiments to validate this hypothesis. As noted earlier, the method used to predict the form of a dimensionless number without knowledge of the basic equation is called dimensional analysis. Its application is conjugated with Buckingham's "pi" theorem.

Dimensional Analysis

We shall illustrate this method on the same example by assuming that the governing equation of the process (Equation 2.19) is unknown. Hence, the only information that is known is

$$w = f(w_0, a, t)$$

Denoting the dimensions of the various parameters symbolically:

$$w\,[m/s] = LT^{-1}$$
$$w_0\,[m/s] = LT^{-1}$$
$$a\,[m/s^2] = LT^{-2}$$
$$t\,[s] = T$$

The dimensional equation describing the process may be assumed to have the form of a power function:

$$w = Cw_0^x a^y t^z \qquad (2.27)$$

where C = constant of the equation determined from experiments, and x, y, z are powers to be determined. Instead of Equation 2.27, we write the equation of dimensions in accordance with the assigned symbols:

$$LT^{-1} = C(LT^{-1})^x (LT^{-2})^y T^z$$

or

$$LT^{-1} = CL^{x+y} T^{-x-2y+z} \qquad (2.28)$$

The powers of the symbols on both sides of the expression must be equal:

$$1 = x + y$$
$$-1 = x - 2y + z$$

Thus, we have two equations with three unknowns, which may be solved by inspection. Assume $z = +1$; then $x = 0$ and $y = 1$. Consequently,

$$w = Cw_0^0 at = Cat$$

The dimensionless number $K_1 = \dfrac{w}{w_0}$ and w_0 result from the examination. Only one dimensionless number may be derived from the remaining physical values $-K_2 \equiv \dfrac{at}{w}$, with a *variable* scale w, which is undesirable in this case. Therefore, we will consider the second variant of Equation 2.27:

$$\frac{w}{w_0} = Ca^x t^y \tag{2.29}$$

For this equation, the formula of dimensions is

$$L^0 T^0 = C(LT^{-2})^x T^y$$

And comparing the powers we obtain

$$0 = x \ ; \quad 0 = -2x + y \ ; \quad y = 0$$

These conditions are possible at zero dimension of the right-hand side (RHS) of the equation, which is achieved by dividing through by w_0:

$$\left[\frac{LT^{-2}T}{LT^{-1}}\right] = [LT]^0$$

Therefore,

$$\frac{w}{w_0} = C\left(\frac{at}{w_0}\right) \ \text{ or } \ K_1 = f(K_2)$$

The form of this function may be determined experimentally.

As shown, the application of dimensional analysis involves the logical selection of values and may give a positive result only if the initial set of all factors governing the process is selected properly. However, it is not always possible for a process to be analyzed thoroughly without experiments. Therefore, we cannot exclude the possibility of missing a dimensionless number, especially from the numbers having the same name, such as $K_1 \equiv \dfrac{w}{w_0}$ (simplex number, differing from a complex number such as $K_2 \equiv \dfrac{at}{w_0}$ formed from different values). Therefore, dimensional analysis may be used in simple cases of generalization where there is not a large number of variables.

To restore the "lost" numbers, Buckingham's "pi" theorem may be employed with all methods of deriving dimensionless numbers. Further discussions of the above analysis are given by Klinkenberg [1955].

Buckingham Pi Theorem

The Buckingham pi theorem is a rule for determining the number of dimensionless groups that exist (called π's, in Buckingham's [1914] notation).

The theorem states that any dimensionally homogeneous equation connecting N physical values, the dimensions of which are expressed by n fundamental units, can be reduced to a functional relationship between π dimensionless numbers:

$$\pi = N - n \qquad (2.30)$$

The number of dimensionless groups (called simplexes) containing π numbers is equal to the number of pairs of the same values in the basic equation. Equation 2.30 states the rule for determining the number of dimensionless groups characteristic of the process. For our example, with the basic equation $w = w_0 + at$,

$$N = 4(w, w_0, a, t) \quad ; \quad n = 2(m, s)$$

Consequently, the total number of $\pi = N - n = 2 \left(\text{where } K_1 \equiv \dfrac{w}{w_0} ; K_2 \equiv \dfrac{at}{w_0} \right)$. Among them, one dimensionless number is a simplex $K_1 \equiv \dfrac{w}{w_0}$ because the basic equation contains one pair of the same values w and w_0.

Thus, using the first theorem of similarity and its associated methods we obtain a definite amount of dimensionless numbers, i.e., generalized process characteristics, for any basic physical equation. Among these could be simplex dimensionless numbers (so-called *parametric* dimensionless numbers of a geometric or physical nature, which are relative sizes of a system reflecting similarity) and complexes (called *numbers*, such as the Reynolds number, Froude number, Euler number, etc.). Any combination of the laws of similarity (dimensionless numbers) is also a criterion of similarity. The numbers may be divided, multiplied and/or raised to powers by each other to obtain new dimensionless groups. The transition from dimensional characteristics of a process to generalized characteristics decreases considerably the amount of variables. Further simplification may be achieved by combining dimensionless groups.

Lighthill [1963] notes that one of the most significant consequences of

Buckingham's theorem is the economy it makes possible in the number of variables that need to be included in an experimental investigation or theoretical analysis. The number of independent dimensionless products, whose values determine the properties of the system, is generally less than the number of different kinds of physical quantities in the system. The consequent reduction in the number of experimental conditions that need to be covered in a program of experiments represents an enormous simplification and economy, resulting in simplifications in the plotting or tabulation of results. Similar simplifications are achieved in analytical solutions when problems are formulated in terms of dimensionless products at the outset.

Federman-Buckingham's Theorem

Federman-Buckingham's theorem (see Gukhman [1965] and Kirpichev [1953] for details) constitutes the second theorem of similarity. It states that the quantitative results of experiments should be presented by equations expressing the relationship among nondimensional numbers. The dimensionless groups, K_1, containing the parameters of interest should be expressed as a function of other dimensionless numbers reflecting different sides of a process:

$$K_1 = f(K_2, K_3, K_4 \dots) \tag{2.31}$$

In the examples given above, the form of the function was known (see Equation 2.24). Generally, the form of this function is not known a priori and must be determined from experimental data. Unlike human beings, nature rigorously applies the laws of geometric progression and probability. Man attempts to approximate these laws through logarithmic relationships. The results of experimental investigations are approximated more readily in either form:

$$K_1 = CK_2^m K_3^n K_4^p \tag{2.32}$$

or exponential (for kinetic processes) forms:

$$K_1 = e^{-\tau/\theta} \quad \text{(damping process)} \tag{2.33}$$

$$K_1 = K_0(1 - e^{-t/\theta}) \quad \text{(growing process)} \tag{2.34}$$

where C, m, n and p are constants determined from the graphical analysis of experimental data.

K_0 is the initial value of the dimensionless number K_1 at time t = 0, and t denotes the time from the beginning of the process. θ is the time constant of the process that depends on the conditions of its realization and is expressed through the dimensionless numbers.

The second theorem of similarity is formally stated as follows: *The solution of any differential equation may be presented as a relationship among the dimensionless numbers obtained from this equation.* Analytical methods provide the initial description of phenomena in a form of complicated differential equations that determine interrelations among values in formulating a problem. However, their solution to design relationships usually is not achieved because of the complexity of the problem. Therefore, although a purely analytical investigation remains, which, as a rule, is the most desirable approach to problem solving, it often is not applied to engineering solutions. On the other hand, purely experimental approaches without knowledge of at least initial theoretical expressions often are doomed to failure. Blind or brute force experimentation often results in a tremendous volume of data, much of which is extraneous and unrelated to the desired solution.

Similarity theory brings to the *experimental* solution of the problem *physical laws* in the form of initial equations describing the process. The transition to generalized variables and modeling significantly facilitates and accelerates this solution.

Any dimensionless expression (despite the empirical method of its derivation in an explicit form) has a definite physical meaning because it reflects the laws of nature expressed by the initial system of physical equations. Modeling based on dimensional analysis produces only approximate solutions because only the most important parameters describing the phenomenon are included in the final expressions. Each dimensionless group in the generalized equation reflects some of the aspects of the process, and the total equation attempts to approximate the behavior of the process as a whole. The technique of treating experimental data in terms of power law expressions is illustrated in the following example.

Example 2.1

A process may be described by a relationship in which π = 2. Develop a relationship for the function $K_1 = f(K_2)$.

Solution

We may assume a general form of the relationship to be $K_1 = CK_2^n$, where C and n are the unknowns. Taking the logarithm of this expression, we obtain

$$\log K_1 = \log C + n \log K_2 \qquad (2.35)$$

This is the equation of a straight line on log-log coordinates. If we plot the experimental data and it can be correlated by a straight line, then the slope will determine the coefficient n, and C can be computed for any point on the line. If the data cannot be correlated by a straight line, then there is a portion of the phenomenon not accounted for in Equation 2.35 and further clarification of the system's physics is needed.

Kirpichev-Gukhman's Theorem

The first and second theorems do not provide information needed to establish which parameters should be known to satisfy similarity, neither do they provide a basis for designing the model and experiments.

This third theorem, along with the first two, serve as similarity indicators and address the issue concerning the field of application of similarity dimensionless equations. It should be obvious that when experimental results are represented in the form of a dimensionless function, its application may only be extended to a group of similar phenomena having common properties.

The dimensionless equation is correct only within the limits of maintaining similarity, which is characterized by definite intervals of changing dimensionless groups K_2, K_3, K_4 in the final expression. Beyond the investigated intervals of changing dimensionless groups the generalization loses its validity. Hence, an expression derived by the methods outlined in this chapter generally should never be extrapolated much beyond the range in which the important dimensionless groups have been studied experimentally.

This third theorem states that *phenomena are similar if they can be described by the same system of differential equations and have similar conditions of uniqueness.* Consequently, the phenomena are physically similar if they belong to the same class and enter the same group of phenomena, which differ only by a scale of physical values. The fulfillment of conditions of uniqueness in dimensionless treatment is as valid as the determination of uniqueness in analytical solutions of physical differential equations.

Differential equations describe a variety of phenomena of a given class that are based on common physical laws. For example, the class of phenomena of heat propagation is subordinated to the law of heat conduction. However, the general equation does not reflect specific indications of particular phenomena of a given class (for example, heating of a pellet in a catalyst layer). Therefore, many solutions correspond to the initial equation, that is, it is *many-valued*. The engineer is generally interested in a specific phenomenon within a given class, and most often under the observed operating conditions in a specific piece of equipment. Therefore, it is necessary *to select from a set* of possible solutions to an initial equation (or system of equations) *one* solution corresponding to the analyzed phenomenon.

Therefore, *additional conditions of uniqueness (boundary conditions)* are added to the problem to obtain a single-valued solution. The boundary conditions include the following:

1. information on geometric properties of a system (the shape and size of a piece of equipment and the working volume);
2. data on the physical properties of products and materials comprising the system to be investigated (conductivity of heat, heat capacity of equipment walls, viscosity, density of working media, etc.);
3. data on the condition of a system at its boundary (boundary or space boundary conditions) and on the system's interaction with the surrounding medium (intensity of heat or mass transfer, distribution of temperatures or concentrations on a surface, etc.); and
4. data on the system's condition in the initial and final moments of a process (i.e., time conditions).

The initial system of physical equations, together with boundary conditions, *uniquely* determine a specific phenomenon of a given class. The solution of such a system in the form of an equation relating to the basic parameters of a process contains information of practical importance. Unfortunately, the analytical solution of the equation system is not achieved because of its complexity. In this case, one has recourse to the method of generalized variables, i.e., to the theory of similarity.

It is evident that dimensionless numbers should be derived not only from a basic physical equation of a process, but also from equations of conditions of uniqueness. The similarity conditions of uniqueness on the boundary of a system and during a process will determine the physical similarity of processes for the entire system volume. The similarity of conditions of uniqueness will manifest *in coincidence of numerical values* of dimensionless numbers in similar processes.

PRINCIPLES AND METHODS OF MODELING UNIT OPERATIONS

There are two general approaches to modeling, namely, physical and mathematical. In *physical* modeling, the phenomenon occurring in the true or prototype system is approximated under laboratory conditions. For example, in a small model of a heat exchanger, the heat transfer process is reproduced between real working bodies, i.e., heating steam and product. In *mathematical* modeling, or modeling by analogy, the model reproduces a physically different phenomenon, described by the same kind of equations as the phenomenon in the prototype. In this approach, a property called *equation isomorphism* is involved. This property dictates that the same system of equations may be used to describe other phenomena in nature. The property is especially useful in describing fields of temperatures, concentrations, velocities, etc.

Physical and mathematical modeling are based on a unified method of generalized variables. Both methods solve the same problems but accomplish it differently. Each *differs* by specific requirements to initial data and some other properties at a considerable generality of the input information. Thus, for physical modeling it is sufficient to have physical equations in the most general form. In contrast, this is unacceptable for both analytical solutions and those achieved through the use of a mathematical model.

Mathematical modeling is only possible when the equations can be transformed conveniently to a useful solution form. Both methods permit differential equations to be solved; however, it often is easier to alter process parameters and analyze their influence on the performance in a prototype.

Regardless of the modeling approach, one must observe similarity of the following conditions of uniqueness in accordance with the third theorem of similarity:

1. *Geometric similarity* is necessary for physical and mathematical similarity of physical fields. This condition is eliminated in the mathematical modeling of systems with centered parameters.
2. *Time similarity* must be maintained in both physical and mathematical modeling; the time correspondance is called *homochronity*. The physical meaning of homochronity is the history of a process changing in time, including transient situations in which the similarity of physical values in a model and in a prototype occurs in *compatible* time moments from the beginning of the process.
3. *Similarity of physical values* should be maintained in all cases of modeling. In *mathematical* modeling, the scale factors have dimension because the model presentation of dimensional physical values of the prototype are values with other dimensions.
4. *Similarity of initial conditions* also is mandatory for both physical and mathematical modeling because the process being developed in time is determined by the *properties* of the process itself, as well as by *initial conditions*. In some cases, the system may be stable, while in others it is unstable, which is unacceptable.
5. *Similarity of boundary conditions* also is necessary for "outlining" the sphere in which the phenomenon proceeds. Other phenomena, independent of the main process under investigation, may occur on the geometric boundaries of the zone. These related phenomena could influence the overall process indirectly. For example, heat transfer from the surface of a body to the atmosphere is independent of heat conduction inside the body but has an influence on the rate of cooling. For boundary conditions, which change with time, it is necessary to maintain time similarity of corresponding variables, which enter into the boundary conditions.

The conditions of uniqueness determine *the scale* of variables and physical parameters of the process according to the requirements of the first theorem. Characteristic parameters should be chosen so that the similarity indicator is equal to unity.

The exact observance of conditions of uniqueness is difficult and may be achieved only in single cases. The fulfillment of uniqueness conditions is especially difficult in those cases in which different and interlinked processes proceed (for example, heat and mass transfer). The requirements of *exact* similarity of conditions of uniqueness would mean providing corresponding equality of all dimensionless groups, including those variables that are unknown. In a practical sense, this is not feasible; therefore, *total similarity* of a model and a prototype really is never achieved and one has recourse to an *approximate* modeling of the most important dimensionless numbers.

Short-cuts to physical modeling have, on occasion, been applied successfully to experimental investigations. The selection of a specific method is determined by an actual problem. Often, engineering solutions to problems are based on experimental studies of only those parameters or perhaps pieces of equipment that are most important. Common methods of approximate physical modeling are outlined below:

1. Experimental studies of an equipment's *element* sometimes can be made rather than simulating the entire unit. As an example, the investigation of heat transfer in a multitubular commercial exchanger could be simulated in a single pipe or small pipe bundle. With this type of approach, it is possible to model the prototype design in full scale (that is, a tube simply could be a vertical "extraction" of the commercial unit).

2. Analysis of the prototype is possible through local modeling of complex configurations of the full-scale unit. With this approach, different working portions of the prototype could be modeled separately. Data obtained could be summed graphically by constructing the volume field of measured values. The similarity is thus provided during the experiment but not over the entire volume of the apparatus. Instead, modeling is performed locally and over different time intervals.

3. Isothermal modeling of hydrodynamic processes can be performed by *averaging* physical values, depending on the temperature field in a prototype.

4. Modeling may be performed in the so-called *auto-modeling region* when one of the basic dimensionless numbers of the process becomes unimportant ("confluent") and its change does not influence the measured values. Then the condition of equality of this number in model and prototype may not be fulfilled.

5. The actual working media of the prototype sometimes can be substituted for with a "modeling" media. One example is substituting water for air in flow studies for furnaces. It is possible in some cases not to fulfill some conditions of uniqueness. In a number of chemical processes there is no need for geometric similarity of a system. For example, it is possible to create a chemical model of the sea in a glass of water by simulating only one dimensionless parameter—concentration. Geometric similarity of model and prototype is not always a sufficient condition of physical similarity. Consider

a group of wooden and iron spheres. All the spheres are geometrically similar, but the wooden spheres are not physically similar to the iron ones. It is evident that the more complicated the process, the more phenomena act in the model at the same time and, thus, the more difficult it is to construct an accurate mathematical description.

The confidence in similarity of phenomena is especially important when a prototype is to be scaled up from a small model. Often, a stepwise approach to design is taken, that is, "micromodels" on bench-scale ("models of models"), can be studied first to understand phenomena and to more carefully design larger-scale models. This approach often provides more confidence in scaling up equipment, especially if the conditions of similarity between prototype and model cannot be strictly maintained. This approach also has the advantage of identifying parameters that do not play dominant roles in the phenomenon and, therefore, may be neglected in experimental studies on larger models.

When changing model scales, often the operating conditions are changed. Typically, the smaller the model, the less accurate the prototype simulation. In transferring to larger models, the process becomes more complicated and, thus, the number of factors to be investigated increases.

Modeling usually can be performed in the following order:

1.   A mathematical description of the process is composed in terms of the physical equations and conditions of uniqueness.
2.   Dimensionless groups are identified along with the specific dimensionless number containing the design parameter of interest. The design dimensionless number is an implicit function of other dimensionless numbers, which are referred to as the *determining dimensionless groups*.
3.   From the condition of equality for dimensionless groups between model and prototype scale, factors are selected for each physical value.
4.   On the basis of the above information, a model is constructed whose working volume is geometrically similar to the prototype. The scale of the model is determined from considerations of the size and productivity of the prototype, providing the necessary velocities, rates, temperatures and other values of the working environment.
5.   Experiments should be planned such that dimensionless numbers in the model change over the same limits as in the prototype. By fulfilling this last requirement, the characterizing phenomenon studied in the model would be proportional to the prototype.

The method of dimensional analysis identifies dimensionless groups; however, their physical significance is not always evident. The use of differential equations allows one to physically interpret the derived dimensionless groups but still does not provide information on the fundamental mechanism of the process. This can be evaluated only from experimental observation. Table 2.1 summarizes the physical significance of common dimensionless groups.

Dimensional analysis and similarity theory shall be applied to selected

Table 2.1 Various Dimensionless Groups and Their Physical Significance

Dimensionless Group	Symbol	Definition	Significance, Ratio of
Reynolds Number	Re	$\dfrac{\rho v L}{\mu}$	Inertial force / Viscous force
		ρ = fluid density	
		v = fluid velocity	
		μ = fluid viscosity	
		L = characteristic dimension	
Froude Number	Fr	$\dfrac{v^2}{Lg}$	Inertial force / Gravitational force
Euler Number	Eu	$\dfrac{p}{\rho v^2}$	Pressure / 2 × Velocity head
		p = pressure	
Mach Number	Ma	$\dfrac{v}{v_c}$	Fluid velocity / Velocity of sound
Weber Number	We	$\dfrac{\rho L v^2}{\sigma}$	Inertial force / Surface tension force
		σ = surface tension	
Drag Coefficient	C_D	$\dfrac{(\rho - \rho')Lg}{\rho v^2}$	Gravitational force / Inertial force
		ρ = density of object	
		ρ' = density of surrounding fluid	
Fanning Friction Factor	f	$\dfrac{D}{L}\dfrac{\Delta P}{2\rho v^2}$	Wall shear stress / Velocity head
		D = pipe diameter	
		L = pipe length	
Nusselt Number (heat transfer)	Nu	$\dfrac{hL}{k}$	Total heat transfer / Conductive heat transfer
		h = heat transfer coefficient	
		k = thermal conductivity	
Prandtl Number	Pr	$\dfrac{C_p \mu}{k}$	Momentum diffusivity / Thermal diffusivity
		C_p = heat capacity	

Table 2.1, continued

Dimensionless Group	Symbol	Definition	Significance, Ratio of
Peclet Number (heat transfer)	Pe	$\dfrac{C_p \rho v L}{k} = RePr$	Bulk heat transport / Conductive heat transfer
Grashof Number	Gr	$\dfrac{g b^3 \rho^2 \beta \Delta T}{\mu^2}$	$Re \times \dfrac{\text{buoyancy force}}{\text{viscous force}}$
		β = coefficient of expansion	
		ΔT = temperature difference	
		b = height of surface	
Stanton Number	St	$\dfrac{h}{\rho v C_p} = NuRe^{-1}Pr^{-1}$	Heat transferred / Thermal capacity of fluid
J Factor for Heat Transfer	j_H	$\dfrac{h}{\rho v C_p}\left(\dfrac{C_p \mu}{k}\right)^{2/3}$	Proportional to $NuRe^{-1}Pr^{-1/3}$
Nusselt Number (mass transfer)	Nu	$\dfrac{k_c L}{\mathscr{D}}$	Total mass transfer / Diffusive mass transfer
		k_c = mass transfer coefficient	
		$\mathscr{D}$ = molecular diffusivity	
Schmidt Number	Sc	$\dfrac{\mu}{\rho \mathscr{D}}$	Momentum diffusivity / Molecular diffusivity
Peclet Number (mass transfer)	Pe	$\dfrac{Lv}{D} = ReSc$	Bulk mass transport / Diffusive mass transport
J Factor for Mass Transfer	j_D	$\dfrac{k_c}{v}\left(\dfrac{\mu}{\rho D}\right)^{2/3}$	Proportional to $NuRe^{-1}Sc^{-1/3}$

problems throughout this text where appropriate. The student should attempt the practice problems at the end of this chapter. Solutions to some of these exercises are given in Appendix A. Supplemental readings noted in the reference section of this chapter should be consulted for further illustrations.

PRACTICE PROBLEMS

Solutions are provided in Appendix A for those problems that have an asterisk (*) beside their numbers.

2.1 A liquid flows under the conditions of fully developed laminar flow through a vertical capillary tube 40 cm long by 10 mm in diameter. The pressure drop over the entire length of the tube is measured as a function of volumetric flowrate, and the data are tabulated in Table 2.2. Through principles of dimensional analysis, develop an expression that relates wall shear stress to the velocity group, 8w/D, i.e., $\tau_w = f(8w/D)$, where w is the average fluid velocity.

$\left(\text{Note: A force balance for the vertical tube is } \dfrac{\pi D^2}{4}\Delta P_f = \pi D L \tau_w,\right.$
$\left. \text{and Poiseuille's equation is } \Delta P = \dfrac{32\mu Lw}{g_c D^2}. \right)$

2.2 The pressure drop of an incompressible fluid flowing through a straight section of pipe has the following functional dependence:

$$\Delta P = f(L, \mu, \epsilon, \rho, D, w)$$

where   L = pipe length
μ = fluid viscosity
ϵ = pipe roughness
ρ = fluid density
D = pipe diameter
w = fluid velocity

Table 2.2 Data for Problem 2.1

Volumetric Flowrate (cm^3/s)	Frictional Pressure Drop (kg_f/cm^2)
0.79	362
1.58	512
3.95	810
5.53	960
7.11	1090
8.69	1232
10.27	1344
15.80	1648

Determine the number of dimensionless groups required to describe the phenomenon:

$$f(\Delta P, L, \mu, \epsilon, \rho, D, w) = 0$$

2.3 For Problem 2.2, determine the complete set of dimensionless groups.

*2.4 It is decided to model the mixing efficiency of a large unbaffled mixing tank by conducting experiments in a smaller, geometrically similar vessel. The model and prototype are illustrated in Figure 2.2. Determine the conditions under which the model study must be conducted to provide valid predictions for the prototype. Assume the operation is at steady state.

2.5 For the mixing tank study (Problem 2.4), should the same liquid for the prototype be used in the model studies to predict the proper vortex depth? If the answer is no and the prototype media is pure glycerine (at 90°F), select a suitable liquid for the model studies.

2.6 For Problem 2.4, if optimum mixing was achieved in the model at 300 rpm with a 0.25-hp motor, what rpm and size motor should be specified for the prototype? Assume the mixing efficiency is 85%.

2.7 A solid body is immersed in the main stream of a flowing fluid. The force of the fluid exerted on the body is known to be a function of velocity, w, the fluid density, ρ, the fluid viscosity, μ, and a characteristic length of the body, L. Determine the important dimensionless groups.

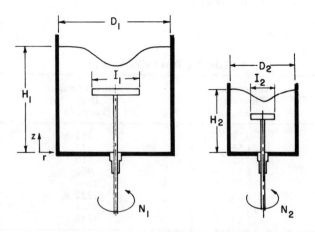

Figure 2.2 Mixing tanks for Problem 2.4.

2.8 The pressure distribution at the surface of a circular cylinder immersed in a nonviscous flow is described by the following equation:

$$P_\theta - P = \frac{\rho w^2}{2g_c}(1 - 4 \sin^2\theta)$$

where angle θ is measured from the forward point of the cylinder. Determine the dimensionless group that describes this system.

2.9 A spherical aerosol particle of mass, M, falls through the atmosphere under the influence of gravity. Neglecting interaction with the atmosphere, the particle's descent is a function of its density, ρ, surface tension, σ, diameter, d_p, and the acceleration of gravity, g. Perform a dimensional analysis of the system to identify the important dimensionless groups.

2.10 An atmospheric heat exchanger is being considered to cool brine water. The commercial prototype is to be designed to handle 500 gpm at an average temperature of 140°F. As no information exists on the unit's flow resistance, a scale model having a length ratio of 1:20 with the full-scale prototype will be studied in the laboratory. Determine the flowrate of nitrogen gas at 85°F and 1 atm pressure in the model unit that will satisfy similarity conditions with the prototype.

2.11 For Problem 2.10, if the pressure drop across the commercial unit is expected to be about 12 psig, what can be expected across the model unit for the flow conditions specified?

NOTATION

a	acceleration, m/s^2
C,c	dimensionless velocity variable
D,d	diameter, m
$\mathscr{D}$	molecular diffusivity, m^2/s
Fr	Froude number
f	force, N
g	gravitational acceleration, m/s^2
i	parameter scale factor
$\tilde{j}$	nondimensional parameter (transformation variable)
K	dimensionless velocity ratio, see Equation 2.21
L,ℓ	length, m
M,m	mass, kg
N	number of physical values
Ne	Newton number, see Equation 2.10

n   number of fundamental units
P   pressure, N/m^2
Re   Reynolds number
Stk   Stokes number, see Equation 2.18A
T,t   time, s
v,w   velocity, m/s

Greek Symbols

γ specific weight, N
θ time constant, or angle, in radians
Λ scale factor
μ viscosity, cp
ν kinematic viscosity, m^2/s
π number of functional dimensionless groups, refer to Equation 2.30
ρ density, kg/m^3

REFERENCES

Braines, Ya. M. (1961) "Podobie i Modelirovanie v. Khimicheskoy i Neftekhimicheskoy Technologi," *Gostoptekhizdat, Moscow.*

Buckingham, E. (1914) "On Physically Similar Systems: Illustration of the Use of Dimensional Equations," *Phys. Rev.* 4:345.

Focken, C. M. (1952) *Dimensional Methods and Their Application* (London: Arnold Press).

Gukhman, A. A. (1965) *Introduction to the Theory of Similarity* (New York: Academic Press, Inc.).

Ipsen, D. C. (1960) *Units, Dimensions, and Dimensionless Numbers* (New York: McGraw-Hill Book Co.).

Kirpichev, M. V. (1953) Teorya podobya, Moscow, Is-vo, *"AN USSR".*

Klinkenberg, A. (1955) "Dimensional Systems and Systems of Units in Physics with Special Reference to Chemical Engineering," *Chem. Eng. Sci.* 4(130):167.

Langhaar, H. L. (1951) *Dimensional Analysis and Theory of Models* (New York: John Wiley & Sons, Inc.).

Lighthill, M. J. (1963) in: *Laminar Boundary Layer*, L. Rosenhead, Ed., (Oxford, England: Clarendon Press).

Pankhurst, R. C. (1964) *Dimensional Analysis and Scale Factors* (New York: Van Nostrand Reinhold Co.).

Sedov, L. I. (1959) *Similarity and Dimensional Methods in Mechanics* (New York: Academic Press, Inc.).

Venikov, V. A., and A. V. Ivanov-Smolenski (1956) "Fisicheskoe Modelirovanie," *Gosenergoizdat, Moscow.*

CHAPTER 3

CLASSIFICATION OF HYDRAULIC PROCESSES

CONTENTS

INTRODUCTION

Most unit operations of chemical engineering involve the motion of fluids. These fluids may comprise liquids, gases, vapors or combinations of these. The manner in which fluids interact depends on the nature of the specific unit operation, with applications ranging from fluid transport to mixing, as well as separation of nonhomogeneous mixtures by precipitation, filtration, centrifugation, extraction, distillation, absorption, etc.

The rates of these processes are determined by the laws of hydro-mechanics and, hence, are called *hydromechanical processes*. The laws of hydromechanics are studied in hydraulics, which consists of *two parts:* hydrostatics and hydrodynamics.

Hydrostatics is concerned with the equilibrium state of fluids under stationary conditions, whereas *hydrodynamics* involves the laws of fluid motion. Classification of hydrodynamic processes may be established on the basis of flow patterns. On this basis, three groups of problems of hydrodynamics are defined in this volume—*internal, external* and *mixed*.

Internal hydrodynamic problems are related to fluid motion in pipes, channels and equipment. External problems of hydrodynamics cover the motion of bodies through fluid media and involve the analysis of fluid motion in relation to simple bodies (e.g., mechanical mixing, particle sedimentation in liquids or gases, etc.). Mixed problems of hydrodynamics involve the analysis of fluid motion caused by complex interaction with solid obstacles (e.g., liquid motion through a grain layer of a solid material or liquid flowing inside channels of complicated shapes while flowing simultaneously around solid particles.

It is essential to obtain a thorough understanding of the hydromechanical processes of any specific unit operation under consideration because hydrodynamic characteristics establish the rates of heat and mass transfer and of chemical reactions.

BASIC DEFINITIONS

Liquid and solid states of matter are the *condensed phases* of gases. The term "condensed phases" emphasizes the high concentration of solid and liquid molecules in comparison to the low density of gases. The volume per mole of gas is very large in comparison to liquids. At conditions of standard temperature and pressure (STP), any gas occupies 22,400 cm^3/g-mol. In contrast, most liquids occupy between 10 and 100 cm^3/g-mol. That is, the molar volume of a liquid is 500–1000 times smaller than in its gaseous state.

As the gas/liquid volume ratio is as large as 1000, from the kinetic theory of gases the ratio of distances between molecules in gas compared to in liquid state is $\sqrt[3]{1000}$, or 10. That is, the distance between gas molecules is ten times farther apart than in a liquid. For a liquid, the average distance between molecules is on the order of a molecular diameter. This large difference in molecular distance between gases and liquids accounts for the drastic difference in properties between these two states of matter. Intermolecular forces, called van der Waals forces, decrease dramatically with distances between molecules. For liquids, these forces are on the order of 10^6 times larger than for gases.

These large differences on the molecular scale cause liquids and gases to display widely different properties when subjected to the same environment. For example, the volume of a fluid as a function of temperature (pressure maintained constant) may be described by an expression of the following form:

$$V = V_0(1 + \kappa T) \tag{3.1}$$

where V_0 is the volume of the fluid at some temperature (usually $0°C$), and κ is the coefficient of thermal expansion. Gases follow a general dependency similar to Equation 3.1; however, κ is approximately constant for most gases. For liquids κ depends on the specific liquid.

This dependency of liquid volume on pressure may be expressed in terms of the coefficient of compressibility, β, as follows:

$$V_0 = \hat{V}_0[1 - \beta(P - 1)] \tag{3.2}$$

$\hat{V}_0$ is the volume of the fluid at STP. Coefficient β is observed to be constant over a wide range of pressures for a particular material but is different for each substance and for the solid and liquid states of the same material. Equation 3.2, an equation of state for liquids, states that volume decreases linearly with pressure. In its gaseous state, volume is observed to be inversely proportional to pressure. For a liquid, β is typically 10^{-6} atm^{-1}, whereas for gases it is significant. As an example, if water in its liquid state is subjected to a pressure change from 1 to 2 atm, Equation 3.2 predicts less than $10^{-3}\%$ reduction in volume. However, when this same pressure differential is applied to water as vapor, a volume reduction in excess of two occurs. Because of this property of insignificant volume changes over moderate pressure changes, liquids are referred to as *incompressible fluids*, while gases are *compressible*.

The terms compressible and incompressible are relative, however, because liquids can change appreciably if conditions are changed over wide limits. Gases also may behave as incompressible fluids if they are subjected to very small changes in pressure and temperature.

The term *liquid* is used in this text in a broad sense. Under the term "liquid," it is necessary to understand substances that possess fluidity. Unlike gases, liquids will not completely fill a volume of specified boundaries.

The general laws of motion describe both liquids and gases, provided gas velocities do not exceed the speed of sound. For liquids, the general laws of equilibrium and motion are expressed in terms of differential equations, where liquids are considered to be homogeneous continuous media. To derive the governing theorems of fluid mechanics, it is necessary to introduce the hypothesis of the *ideal* liquid, in contrast to *real* or *viscous* liquids. That is, for our initial discussions, a real fluid is one that is absolutely incompressible and does not undergo changes in density with variations of temperature. In addition, we assume the fluid is not viscous.

PHYSICAL PROPERTIES OF FLUIDS

Density and Specific Gravity

Fluids may be characterized by the following basic physical properties: density (or specific weight), viscosity and surface tension.

The mass of liquid per unit volume is called *density* and is denoted by the Greek symbol ρ:

$$\rho = \frac{m}{V} \left[\frac{kg}{m^3} \right] \tag{3.3A}$$

where   m = liquid mass (kg)
   V = liquid volume (m^3)

The weight per unit of fluid volume is called specific gravity:

$$\gamma = \frac{G}{V} \left[\frac{N}{m^3} \right] \tag{3.3B}$$

where   G = weight of the fluid (N)

Note that mass, m, equals G/g, where g is the acceleration due to gravity. Substituting Equation 3.2 into Equation 3.3 and denoting G/V by γ, we obtain

$$\gamma = \rho g \tag{3.4}$$

The density of gases is highly dependent on temperature and pressure. The relationship of temperature, pressure and density for gases is expressed by the *equation of state* for ideal gases:

$$PV = \frac{mRT}{M} \tag{3.5}$$

where   P = pressure (N/m^2)
   V = gas volume (m^3)
   m = mass of gas (kg)
   R = universal gas constant, R = 8314/k-mol ($^\circ$K)
   T = temperature ($^\circ$K)
   M = molecular weight of the gas

Equation 3.5 may be rearranged to solve for pressure:

$$P = \frac{m}{V} \frac{RT}{M} = \frac{\rho RT}{M} \qquad (3.6)$$

The volume per unit of mass of gas is called *specific volume:*

$$\dot{v} = \frac{V}{m} \left[\frac{m^3}{kg} \right] \qquad (3.7)$$

Specific volume is the inverse of density, i.e., $\dot{v} = 1/\rho$, and, consequently, Equation 3.5 may be rewritten as follows:

$$P\dot{v} = \frac{RT}{M} \qquad (3.8)$$

The following example illustrates the use of these definitions.

Example 3.1

Determine the density of gaseous ammonia at a pressure $P = 26$ atm gauge and a temperature of 16°C.

Solution

The absolute pressure is

$$P = 26 + 1 = 27 \text{ kg/cm}^2 = 265 \times 10^4 \text{ N/m}^2$$

The molecular weight of NH_3 (ammonia) is $M = 17$.
From Equation 3.6 the density of NH_3 is

$$\rho = \frac{PM}{RT} = \frac{(265 \times 10^4)(17)}{(8314)(273 + 16)} = 1.87 \text{ kg/m}^3$$

Viscosity

When a real fluid is set in motion, forces of internal friction arise acting in opposition to the direction of flow. This property, resisting motion, is called *viscosity.*

Consider a liquid flowing through a cylindrical tube. The fluid can be visualized as concentric rings or layers of fluid, as illustrated in Figure 3.1. If a certain layer has a velocity w, then the next layer has a velocity w + Δw. Experimental observations reveal that the velocity of the layers decreases from the axis to the tube wall, where the velocity is equal to zero. For

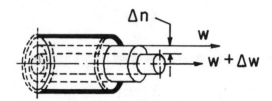

Figure 3.1   Conceptual interpretation of fluid flowing full through a tube.

displacement of one layer relative to the other, it is necessary to apply a force proportional to the surface contact area of the layers. The force, f, per unit of shearing plane, F, is called the *shear stress* and is denoted by the following:

$$\tau = \frac{f}{F} \tag{3.9}$$

Newton's law of viscosity states that the shear stress is proportional to the velocity gradient across the tube through a proportionality constant, μ:

$$\tau = \mu \frac{dw}{dn} \tag{3.10}$$

dw/dn is the velocity gradient normal from the tube wall (i.e., it is the relative velocity changing over unity of distance between layers in a direction perpendicular to the liquid flow).

The proportionality constant in Equation 3.10 depends on the physical properties of the fluid and is called the *coefficient of dynamic viscosity* or, simply, *viscosity*.

Substituting for τ from Equation 3.9, we obtain the dimensions of viscosity:

$$[\mu] = \left[\frac{f}{F \frac{dw}{dn}} \right] = \left[\frac{N}{m^2 \frac{m/s}{m}} \right] = \left[\frac{N \times s}{m^2} \right]$$

Note that $(N) = (kg\text{-}m/s^2)$, whence we obtain another form of viscosity:

$$[\mu] = \left[\frac{\frac{kg\text{-}m}{s^2} \times s}{m^2} \right] = \left[\frac{kg}{m \times s} \right]$$

In the cgs system, the unit of viscosity is the poise (ps). A poise refers to a force of 1 dyne that displaces liquid layers having a surface area of 1 cm^2, each situated at 1 cm from each other with a relative velocity of 1 cm/s. That is,

$$(ps) = \left[\frac{dyne \times s}{cm^2} \right] = \left[\frac{g \times cm}{s^2} \times \frac{s}{cm^2} \right] = \left[\frac{g}{cm \times s} \right]$$

The unit of viscosity equal to 0.01 ps is called a *centipoise* (cps). Typical values for different fluids as a function of temperature and pressure may be found in standard handbooks.

The ratio of viscosity, μ, to density, ρ, is the *coefficient of kinematic viscosity* or, simply, *kinematic viscosity*:

$$\nu = \frac{\mu}{\rho} \tag{3.11}$$

The dimensions of kinematic viscosity are

$$[\nu] = \left[\frac{N \times s/m^2}{kg/m^3} \right] = \left[\frac{\frac{kg \times m}{s^2} \times \frac{s}{m^2}}{\frac{kg}{m^3}} \right] = \left[\frac{m^2}{s} \right]$$

The following two examples illustrate the calculation of viscosity for a gas and a liquid and employ principles of physical chemistry. The reader who wishes to review these principles in more depth should consult the work of Cheremisinoff [1981] or Castellan [1971].

Example 3.2

Calculate the viscosity of sulfur dioxide gas at a temperature of 300°C and atmospheric pressure. The viscosity of SO_2 at 20° and 150°C is equal to 1.26×10^{-2} and 1.86×10^{-2} cp, respectively, and its critical temperature and pressure are as follows: $T_{cr} = 430°K$ at $P_{cr} = 77.7$ atm. Compare the calculated value with an experimentally determined value of 2.46×10^{-2} cp.

Solution

1. To calculate viscosity, the following equation is used:

$$\mu = 6.3 \times 10^{-4} \frac{M^{1/2} P_{cr}^{2/3}}{T_{cr}^{1/6}} \cdot \frac{T_{red}^{3/2}}{T_{red} + 0.8} \tag{3.12}$$

where $T_{red} = T/T_{cr}$

The subscript "red" refers to a *reduced* state:

$$M_{SO_2} = 64 \ ; \quad T_{red} = \frac{(300 + 273)}{430} = 1.335$$

Substituting these values into the above equation, we obtain the value of viscosity:

$$\mu = (6.3 \times 10^{-4}) \frac{(64)^{1/2} \times (77.7)^{2/3}}{(430)^{1/6}} \times \frac{(1.335)^{3/2}}{1.335 + 0.8} = 2.41 \times 10^{-2} \ cp$$

2. We now construct a graph of $y = T^{3/2}/\mu$ versus t using two known values of μ:

$$\text{at } t_1 = 20°C \ ; \quad y_1 = \frac{T^{3/2}}{\mu} = \frac{(239)^{3/2}}{1.26 \times 10^{-2}} = 3.98 \times 10^5$$

$$\text{at } t_2 = 150°C \ ; \quad y_2 = \frac{T^{3/2}}{\mu} = \frac{(423)^{3/2}}{1.86 \times 10^{-2}} = 4.68 \times 10^5$$

The plot is shown in Figure 3.2. From the plot in Figure 3.2, $y = 5.485 \times 10^5$ at 300°C. Therefore,

$$\mu = \frac{T^{3/2}}{y} = \frac{(573)^{3/2}}{5.485 \times 10^5} = 2.49 \times 10^{-2} \ cp$$

The same value of y may be obtained using the equation of a straight line passing the points t_1, y_1 and t_2, y_2.

3. We now determine the viscosity at $t = 300°C$, assuming we know μ_1 at $t = 150°C$. In this case, we use the following equation:

$$\mu = \mu_1 \left(\frac{T_{red}}{T_{red_1}} \right)^{3/2} \times \frac{T_{red_1} + 0.8}{T_{red} + 0.8} \tag{3.13}$$

$$\mu = 1.86 \times 10^{-2} \left(\frac{573}{423} \right)^{3/2} \times \frac{423/430 + 0.8}{573/430 + 0.8} = 2.44 \times 10^{-2} \ cp$$

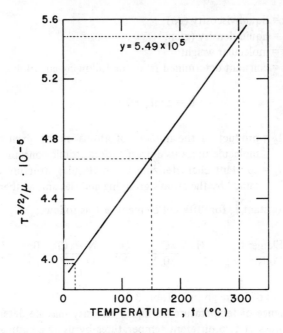

Figure 3.2   A plot of the function $y = T^{3/2}/\mu$ versus temperature, t. The plot is used in evaluating the viscosity of SO_2 gas by method 2 in Example 3.2.

Note that the errors associated in computing the viscosity of SO_2 by the different methods outlined above are small.

Example 3.3

Calculate the viscosity of acetic acid at t = 40°C. The density of the acid is 1.027 g/cm³ at 40°C. The viscosities of the acid at $t_1 = 20°$ and $t_2 = 100°C$ are $\mu_1 = 1.22$ cp and $\mu_2 = 0.46$ cp, respectively. Compare the computed values of μ with the experimental value of 0.9 cp.

Solution

1. Use the following equation:

$$\log(\log 10\mu) = K \frac{\rho}{M} - 2.9 \qquad (3.14)$$

where μ = liquid viscosity (cp)
          ρ = liquid density (g/cm^3)
          M = molecular weight
          K = constant determined from the following equation:

$$K = \Sigma m I_a + \Sigma I_c \qquad (3.15)$$

where $m I_a$ = product of the number of atoms, m, of given element in a
          molecule times its corresponding atomic constant, I_a
          I_c = constant characterizing the molecular structure and is deter-
          mined by the atom's grouping and ties among them

Values of constant I_a for different elements are as follows:

Element	H	C	O	N	Cl	Br	I
Constant I_a	2.7	50.2	29.7	37	60	79	110

Typical values of I_c are given in Table 3.1.

The influence of temperature on liquid viscosity may be determined from viscosity values at two different temperatures by using a comparison with a reference liquid. This can be stated as follows:

$$\frac{t_1 - t_2}{\theta_1 - \theta_2} = K \frac{t_1 - t}{\theta_1 - \theta} \qquad (3.16)$$

where t_1, t_2 = temperatures of the liquid
          θ_1, θ_2 = temperatures of the reference liquid at which its viscosity
          is equal to the viscosity of the liquid to be compared at
          t_1 and t_2

The molecular weight of acetic acid is 60.06. The constant, K, for acetic acid is

$$K = 2I_a(C) + 4I_a(H) + 2I_a(0) + I_c(COOH) = 2 \times 50.2 + 4 \times 2.7 + 2 \times 29.7$$
$$+ 7.9 = 162.7$$

Consequently,

$$\log(\log 10\mu) = 162.7 \frac{1.027}{60.06} - 2.9 = -0.118$$

Table 3.1 Values of Constant I_c

Compound, Grouping or Bond	Constant I_c	Compound, Grouping or Bond	Constant I_c
Double Bond	−15.5		
Five Ring C Atoms	−24.0	$CH_3-\overset{\parallel}{\underset{O}{C}}-R$	+5.0
Six Ring C Atoms	−21.0		
Substitution in Six-Ring Member In ortho- and para-position	+3.0	$-CH-CHCH_2X^a$	+4.0
In meta-position	+1.0		
$\underset{R}{\overset{R}{>}}CHCH\underset{R}{\overset{R}{<}}$	+8.0	$\underset{R}{\overset{R}{>}}CHX^a$	+6.0
$R-\overset{R}{\underset{R}{C}}-R$	+13.0	−OH −COO− −COOH −NO$_2$	+24.7 −19.6 −7.9 −6.4
$R-\overset{\parallel}{\underset{O}{C}}-H$	+10.0		

^{a}Electronegative group.

Hence,

$$\log(10\mu) = 0.762 \; ; \quad \mu = \frac{5.78}{10} = 0.578 \text{ cp}$$

2. Applying Equation 3.16 and using water as the reference liquid, where $\mu'_{H_2O} = 1.22$ cp and $\mu''_{H_2O} = 0.46$ cp at $\theta_1 = 12.5$ and $\theta_2 = 60.7°C$, respectively, we obtain the following:

$$\theta = \theta_1 + (t - t_1)\frac{\theta_2 - \theta_1}{t_2 - t_1} = 12.5 + (40 - 20)\left[\frac{60.7 - 12.5}{100 - 20}\right] = 24.55°C$$

At $\theta = 24.55°C$, $\mu_{H_2O} = 0.909$ cp. Therefore, the viscosity of acetic acid at 40°C is computed to be the same.

PRESSURE AND SURFACE TENSION

Liquid Pressure

A basic property of a static fluid is the pressure it exerts on the walls of its container. Pressure can be interpreted as a surface force exerted by a fluid against the walls and bottom of its container, as well as on a surface of a body submerged in the fluid medium. Liquid pressure exerted against a unit surface area is called *hydrostatic pressure,* or *pressure.* This is stated simply as follows:

$$P = \frac{p}{F} \qquad (3.17)$$

where p is the pressure force applied against surface area, F.

If a liquid is poured into a vessel, then the pressure force exerted against the vessel floor is equal to the weight of liquid in the vessel:

$$p = FH\rho g \qquad (3.18)$$

where   F = surface of the vessel's bottom
H = height of liquid column
ρ = liquid density
g = acceleration due to gravity

From Equation 3.18, the pressure exerted against the bottom of the vessel is

$$P = \frac{FH\rho g}{F} = H\rho g \qquad (3.19)$$

Equation 3.19 states that the liquid pressure exerted against the bottom of the vessel is equal to the weight of the column for a bottom area of unity (i.e., F = 1). If a pressure, P_0, is exerted over the upper liquid's exposed surface, then the hydrostatic pressure is

$$P = P_0 + H\rho g \qquad (3.20)$$

The pressure exerted against the vertical or sloped walls of a vessel is not constant over the vessel height. In fact, it should be considered as a limit of the ratio of pressure force, Δp, to an elemental area, ΔF, which is subjected to the force action at ΔF approaching zero:

$$P = \lim \left[\frac{\Delta p}{\Delta F} \right]_{\Delta F \to 0} \qquad (3.21)$$

This pressure is directed normal to the area of the walls. If this were not the case, the fluid would no longer be static.

Surface Tension

In a number of unit operations liquids are contacted with a gas or other immiscible liquids. The surface area of contact between such fluids approaches a minimum value due to the action of surface forces. For example, droplets suspended in a gas (vapor) or gas bubbles dispersed in a liquid medium approach a spherical geometry. This is explained by the fact that molecules at the fluid-fluid interface or close to it are subjected to attraction forces from molecules located inside the bulk fluid medium. Hence, there arises a pressure from within the bulk liquid that is perpendicular to the gas-liquid interface. The action of these forces is manifested by an attempt by the liquid to decrease its surface area. In developing a new surface it is necessary to consume some energy.

The work that is necessary for the formation of a unit of new surface is called *surface tension* and is denoted by the symbol σ. This work is measured in joules and is related to 1 m^2 of surface area.

The dimension of surface tension is

$$\sigma = \left[\frac{J}{m^2} \right] = \left[\frac{N \times m}{m^2} \right] = \left[\frac{N}{m} \right]$$

Surface tension also may be considered as a force acting on a unit length of the interface.

Further explanation of the material presented in this chapter is given by Bennett and Meyers [1964], Bird et al. [1960], Cheremisinoff [1981], Hougen et al. [1954], Perry and Chilton [1973] and Streeter [1971]. The student should solve some of the practice problems at the end of this chapter to strengthen understanding of this material.

PRACTICE PROBLEMS

3.1 Convert 50 psia to in. of mercury, ft of water, N/m^2 and Pascals (Pa).

3.2 Convert 500 mm Hg to psia.

3.3 Determine how many oz/ft^3 are in 1 g/l.

3.4 The density of a liquid is found to fit the following empirical correlation:

$$\rho = (A + Bt)\rho^{CP}$$

where ρ has units of g/cm^3, t is temperature, °C, and P is in N/m^2. Determine the units of constants A, B and C. The equation is dimensionally consistent.

3.5 The density of a certain liquid is 87 lb_m/ft^3. Determine the weight in kg for 8 m^3 of this liquid.

3.6 Determine how many g-mol of NO_2 gas exist in a 1-gal vessel at STP.

3.7 Compute the weight in pounds of:

   (a)   35 g-mol of nitrogen,
   (b)   10 g-mol of carbon dioxide, and
   (c)   100 g-mol of carbon monoxide.

3.8 A certain liquid has a specific gravity of 1.8. Compute its density in English units.

3.9 A 50/50 mixture (by volume) of liquid benzene and cyclohexane is prepared. Determine its molecular weight and specific gravity.

3.10 A gas mixture contained in an 8000-ft^3 vessel has the following composition by weight:

- CH_4 25%
- H_2S 15%
- CO_2 60%

Determine the average molecular weight of the gas.

3.11 Determine the average molecular weight of a gas mixture having the following volume % composition:

- H_2S 15%
- H_2O 15
- NO_2 40
- CO   10
- CO_2 10
- H_2 10

3.12 Determine the value for the universal gas constant in English units i.e., (psia) $(ft^3)/(°F)$ (lb-mol).

3.13 Compute the density of C_2H_4 at 95°F and 737 mm Hg in both English and cgs units.

3.14 Compare the density of He at 95°F and 750 mm Hg to air at STP.

3.15 A sealed vessel containing air has a volume of 50,000 ft^3. The composition of air is roughly 21% O_2 and 79% N_2. The vessel temperature is 85°F and it has a total pressure of 760 mm Hg.

   1.   Determine the partial volumes of oxygen and nitrogen in the vessel.
   2.   Determine the partial pressures of each constituent.

3.16 A Venezuelen crude oil of specific gravity 0.96 is stored in a 55-ft-deep open vessel. Determine the fluid pressure on the vessel floor.

3.17 A cylindrical storage tank (3 ft wide, 22 ft high) is three-quarters full with a 30 wt % nitric acid solution. Compute the pressure at the bottom of the tank.

3.18 The molal volume of a gas mixture of CH_4, C_2H_4 and H_2 is 315 cm^3 at 85 atm. Compute the temperature. (Hint: compute the pseudocritical ideal volume and a pseudoreduced ideal volume: $\hat{V}_{ci} = RT_c/P_c$.)

3.19 A sealed vessel is filled completely with liquid water at t = 18°C. If the temperature of the water is raised 15°C, determine the pressure in the vessel. The thermal coefficient of expansion for water is 2.1 X 10^{-4} deg^{-1} and β = 4.7 X 10^{-5} atm^{-1}.

3.20 Estimate the viscosity of CO_2 gas at 500°C and atmospheric pressure. Repeat the calculation for three other temperatures and prepare a plot of t versus viscosity.

3.21 Estimate the viscosity of H_2S gas at 1100°C and 12 atm pressure.

3.22 For Problem 3.20, prepare a plot of $T^{3/2}/\mu$ versus t and compare the predicted viscosity at 500°C to the value computed by the method used in solving 3.20.

3.23 (a) Compute the viscosity of acetic acid at t = 15°C. (b) Compute the viscosity of a 90% solution of sulfuric acid at the same temperature and compare. (c) Which is more viscous and what does this mean in terms of the fluidity of the materials?

NOTATION

F   area, m^2
f   force, N
G   weight of fluid, N
g   gravitational acceleration, m/s^2
H   height of fluid column, m

I_a, I_c atomic constants for liquid equation of state
K   constant defined by Equation 3.15
M   molecular weight
m   mass, kg, or number of atoms
n   normal distance, m
P   pressure, N/m^2
P_{cr} critical pressure, N/m^2
P_0 pressure exerted at liquid surface, N/m^2
R   universal gas law constant, $J/kg\text{-}mol\text{-}°K$
STP   standard temperature and pressure
T   absolute temperature, °K or °R
T_{red} reduced temperature, dimensionless
t   temperature, °C or °F
V   volume, m^3
V_0 volume at reference temperature t = 0°C, m^3
w   fluid velocity, m/s
y   function defined as $T^{3/2}/\mu$, $°K^{2/3}/poise$

Greek Symbols

β coefficient of compressibility, atm^{-1}
γ specific gravity (strictly defined as weight per unit volume, N/m^3) = ratio of mass of liquid to mass of equal volume of water at STP (for liquids) and = ratio of mass of gas to mass of equal volume of air at STP (for gases)
θ reference temperature, °C or °F
κ coefficient of thermal expansion, $m/m\text{-}°K$
μ viscosity, $poise = N\text{-}s/m^2$
ν kinematic viscosity, m^2/s
$\dot{\nu}$ specific volume, m^3/kg
ρ density, kg/m^3
σ surface tension, N/m
τ shear stress, N/m^2

REFERENCES

Bennett, C. O., and J. E. Myers (1964) *Momentum, Heat and Mass Transfer,* 2nd ed., (New York: McGraw-Hill Book Co.).
Bird, R. B., W. E. Stewart and E. N. Lightfoot (1960) *Transport Phenomena* (New York: John Wiley & Sons, Inc.).

Castellan, G. W. (1971) *Physical Chemistry,* 2nd ed. (Reading, MA: Addison-Wesley Publishing Co.).

Cheremisinoff, N. P. (1981) *Fluid Flow: Pumps, Pipes and Channels* (Ann Arbor, MI: Ann Arbor Science Publishers, Inc.).

Hougen, O. A., K. M. Watson and R. A. Ragutz (1954) *Chemical Process Principles,* Part I, 2nd ed. (New York: John Wiley & Sons, Inc.).

Perry, R. H., and C. H. Chilton, Eds. (1973) *Chemical Engineer's Handbook,* 5th ed. (New York: McGraw-Hill Book Co.).

Streeter, V. L. (1971) *Fluid Mechanics,* 5th ed. (New York: McGraw-Hill Book Co.).

Weast, R. C., Ed. (1968) *Handbook of Chemistry & Physics,* 49th ed., Cleveland, OH: The Chemical Rubber Co.).

CHAPTER 4

PRINCIPLES OF HYDROSTATICS

CONTENTS

INTRODUCTION

Hydrostatics is concerned with the conditions of equilibrium when the fluid is in a state of rest. That is, hydrostatics considers the *relative* state of rest of fluids; even if the fluid particles are in motion, they do not undergo displacement relative to each other. For example, a fluid contained in a vessel that is being transported via a truck is said to be in a state of relative rest with respect to any point of reference within the fluid volume. Another example is liquid inside a centrifuge rotating at constant angular speed. When fluids are totally at rest or stationary with respect to the surroundings, no internal frictional forces exist and the fluid is considered ideal.

At rest, a fluid's shape and volume do not change and, as in the case of a solid object, it displaces the surrounding environment by its own weight. A stationary fluid is subjected to gravity and pressure forces. For a fluid in a relative state of rest, inertia forces of transient motion are also at play. The relationship among the various forces that act against the fluid is determined by equilibrium conditions that can be expressed mathematically by Euler's differential equations.

EULER'S DIFFERENTIAL EQUATIONS

Consider a mass of static fluid in the form of an elementary parallelepiped of volume dV with edges dx, dy, dz, situated parallel to coordinate axes x, y and z. The system is shown in Figure 4.1. The gravity force acting on this volume is the product of its mass, dm, times acceleration due to gravity, g, i.e., gdm. The force of hydrostatic pressure acting against any side of the parallelepiped is equivalent to the product of pressure, p, over the side's area. To develop an expression for pressure, we must assume p to be a function of all three coordinates, $P = f(x,y,z)$.

According to basic principles of statics, at equilibrium the sum of the projections onto the coordinate axis of all forces acting on the elementary volume must be zero. If this were not the case, the fluid would be displaced. First we will consider forces acting in the z-direction. The force of gravity acting in the downward direction is parallel to the z-axis:

$$-gdm = -g\rho dV = \rho gdxdydz$$

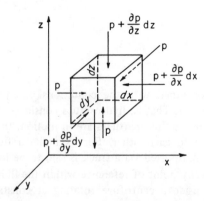

Figure 4.1   The system under consideration for developing Euler's equations.

The pressure force acting normal to the lower side of the parallelepiped of the z-axis is pdxdy. If the pressure change at a given point in the z-direction is $\partial P/\partial z$, then along edge dz it is $(\partial P/\partial z)$ dz.

The pressure acting on the opposite (upper) side is $[P + (\partial P/\partial z)$ dz$]$, and the projection of this pressure force onto the z-axis is

$$-\left(P + \frac{\partial P}{\partial z} \, dz\right) dxdy$$

Hence, the resulting pressure force on the z-axis is

$$Pdxdy - \left(p + \frac{\partial P}{\partial z} \, dz\right) dxdy = -\frac{\partial P}{\partial z} \, dzdxdy$$

and the sum of the forces projected onto the z-axis is equal to zero. That is,

$$-\rho gdxdydz - \frac{\partial P}{\partial z} \, dxdydz = 0 \qquad (4.1)$$

and because dxdydz = dV $\neq$ 0, we obtain

$$-\rho g - \frac{\partial P}{\partial z} = 0$$

The projection of gravity forces on the x-axis equals zero. Thus, we may write

$$\rho dydz - \left(P + \frac{\partial P}{\partial x} \, dx\right) dydz = 0$$

After simplification, we obtain

$$-\frac{\partial P}{\partial x} \, dxdydz = 0 \qquad (4.2)$$

or

$$-\frac{\partial P}{\partial x} = 0$$

And, correspondingly, for the y-axis,

$$-\frac{\partial P}{\partial y}\, dxdydz = 0 \qquad (4.3)$$

or

$$-\frac{\partial P}{\partial y} = 0$$

Thus, the equilibrium conditions of the elementary parallelepiped may be expressed by the following system of equations:

$$\left.\begin{array}{c} -\dfrac{\partial P}{\partial x} = 0 \\[2ex] -\dfrac{\partial P}{\partial y} = 0 \\[2ex] -\rho g - \dfrac{\partial P}{\partial z} = 0 \end{array}\right\} \qquad (4.4)$$

These expressions are known as Euler's differential equations.

THE BASIC EQUATION OF HYDROSTATICS

Equation 4.4 shows that the pressure of a static fluid varies vertically only. In other words, the pressure at any point on a horizontal plane of the fluid parallelepiped is the same. For the system of equations in Equation 4.4, the partial derivatives $\partial P/\partial x$ and $\partial P/\partial y$ are equal to zero. Consequently, the partial derivative $\partial P/\partial z$ may be substituted by dP/dz and, therefore,

$$-\rho g - \frac{dP}{dz} = 0$$

Hence,

$$-dP - \rho g dz = 0$$

or

$$dz + d\left(\frac{P}{\rho g}\right) = 0$$

For a homogeneous, incompressible fluid, density is constant. Therefore,

$$d\left(z + \frac{P}{\rho g}\right) = 0$$

After integration, we obtain

$$z + \frac{P}{\rho g} = \text{const.} \tag{4.5}$$

For any two horizontal planes 1 and 2, Equation 4.5 may be written in the following form:

$$z_1 + \frac{P_1}{\rho g} = z_2 + \frac{P_2}{\rho g} \tag{4.6}$$

Equation 4.6 expresses the condition of hydrostatic equilibrium.

For illustration, consider the system shown in Figure 4.2. Let us direct our attention to two liquid particles, one of which is point "A" located at a height z above some arbitrary reference plane, 0-0, and the other point "B" on the liquid surface at height z_0. P and P_0 are the pressures at points "A" and "B," respectively. From Equation 4.6 we may write the following:

$$z + \frac{P}{\rho g} = z_0 + \frac{P_0}{\rho g} \tag{4.7}$$

or

$$\frac{P - P_0}{\rho g} = z_0 - z \tag{4.8}$$

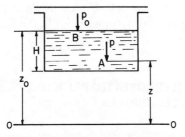

Figure 4.2 An example system for developing the basic hydrostatic equation.

Height z in Equation 4.6 is commonly referred to as the *leveling height* and has the following units:

$$\frac{P}{\rho g} = \frac{P}{\gamma} = \left[\frac{N\text{-}m^3}{m^2\text{-}N}\right] = [m]$$

The value $P/\rho g$ is referred to as the static or pressure head. As follows from Equation 4.6, for any point within the static liquid the sum of the leveling height and pressure head is constant.

The terms of the basic hydrostatic equation have relevance to the energy of the fluid. The term $P/\rho g$ may be expressed in units of $[(N\text{-}m)/N] = [J/N]$, which is the *specific energy* of the fluid, i.e., the energy per unit weight $[J/N$ or $(kg_f\text{-}m)/kg_f]$. The same definition may be used to describe the leveling height if we multiply and divide z by unity of weight.

Thus, leveling height, z, also referred to as the *geometric head*, characterizes the *specific potential energy of position* of a given point over a specified, but arbitrary, reference plane. The static head represents the *specific potential energy of pressure* at that point.

The sum of these two energies is equal to the total potential energy per unit weight of fluid. Therefore, the basic equation of hydrostatics is a specific case of the law of conservation of energy, which states simply that the specific potential energy at all points in a static fluid is constant. Equation 4.8 may be rewritten in the following form:

$$P + \rho g z = P_0 + \rho g z_0 \qquad (4.9)$$

or

$$P = P_0 + \rho g(z - z_0) \qquad (4.10)$$

Equation 4.10 is known as Pascal's law, which states that the pressure at any point in a static incompressible fluid is transmitted equally to all points within its volume. So, for example, if pressure P_0 at point z_0 is changed by some value, the pressure at any other point in the fluid also will change by an equivalent amount.

APPLICATION OF HYDROSTATIC PRINCIPLES TO MANOMETRIC TECHNIQUES

Efficient process operations rely heavily on accurate measurement and control of the amount of materials entering and leaving equipment. Many measuring and control devices indirectly monitor and regulate flow quantities through the equipment's pressure or some pressure difference. Through Pascal's equation, pressure or a pressure differential can be related to the amount of material present at any one time in a reactor or vessel. This is illustrated best by the *principle of communicating vessels*. Consider two vessels open to the atmosphere and connected to each other as shown by Figure 4.3(A). Both vessels contain the same liquid of density ρ. Note that the vessel on the right in Figure 4.3(A) is higher than the one on the left. The reference plane 0-0 passes through point A, which represents a position inside the liquid below each vessel. For point A below the left vessel, we have

$$P = P_{atm} + \rho g z_0'$$

For point A below the right vessel,

$$P = P_{atm} + \rho g z_0''$$

And as plane 0-0 passes through point A, we recognize that $z_0' = z_0'' = 0$:

$$\left. \begin{array}{c} P_{atm} + \rho g z_0' = P_{atm} + \rho g z_0'' \\ z_0' = z_0'' \end{array} \right\} \tag{4.11}$$

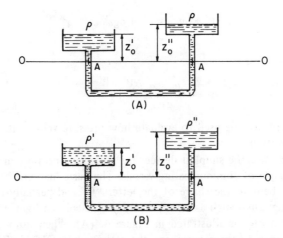

(A)

(B)

Figure 4.3 Equilibrium conditions established in communicating vessels: (A) represents the case of a homogeneous liquid system; (B) represents the case of a heterogeneous (immiscible) liquid system.

At equilibrium, the pressure is equivalent at any point in the liquid. Thus, in communicating vessels (either open or closed) under the same pressure and containing the same fluid, the liquid levels come to rest at the same height. This principle is used, for example, in measuring liquid levels in tanks by the use of water gauges.

Suppose we now drain the system and fill the left vessel with a liquid of density ρ' and the right with a liquid of density ρ'', as illustrated in Figure 4.3(B). The liquids are immiscible and, hence, we have a heterogeneous system. Using the same approach, we obtain the following expressions:

$$\rho' z_0' = \rho'' z_0''$$ (4.12)

or

$$\frac{z_0'}{z_0''} = \frac{\rho''}{\rho'}$$ (4.13)

Equation 4.13 states that the levels of the fluids are inversely proportional to their densities for the case of communicating vessels containing a heterogeneous fluid system.

A third situation worth considering is that in which both vessels are filled again with the same liquid of density ρ, but in which the pressures applied over each liquid surface are different. If the pressure is P' over the left vessel and P'' over the right vessel, then we may write the following:

$$P' + \rho g z_0' = P'' + \rho g z_0''$$ (4.14)

or

$$z_0'' - z_0' = \frac{P' - P''}{\rho g}$$

Equation 4.14 may be used to evaluate how pressure varies with a change in depth or elevation.

Manometers are the simplest devices for measuring pressure and are based on the principle of communicating vessels. The basic manometer consists of a glass tube bent in the shape of the letter "U" and partially filled with a standard gauge fluid such as water, mercury or a colored light oil. The basic U-tube manometer is illustrated in Figure 4.4(A). When both ends of the U-tube are open to the atmosphere, the pressure on each side is equivalent; thus, the column of liquid in each leg is balanced exactly. Hence, the liquid surfaces reach equilibrium at the same level.

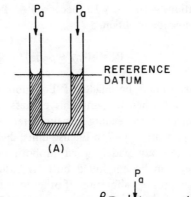

(A)

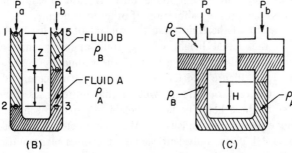

(B)   (C)

Figure 4.4 The basic U-tube manometer used in measuring pressure differences: (A) single fluid manometer open to the atmosphere; (B) a fluid manometer measuring the pressure differential; (C) the size of the manometer legs does not matter in measuring pressure differential; this particular design is a two-fluid manometer.

Let us now consider a case in which the manometer containing a gauge fluid of density $\rho_A(kg/m^3)$ is hooked up to a system containing a fluid of density ρ_B, in which the system fluid applies a greater pressure to one of the legs. The system is illustrated in Figure 4.4(B). We wish to develop an expression that relates the pressures at the top of each leg, i.e., between P_a and P_b. Starting with the left leg, the pressure at point 2, i.e., at the A-B interface is the following:

$$P_2 = P_a + (z + H)\rho_B g, \; N/m^2 \tag{4.15}$$

where H is the displacement of the heavier fluid in m. The pressure at point 3 must be equal to that at point 2 as per our previous discussion. The pressure at point 3 also is equal to the following:

$$P_3 = P_b + z\rho_B g + H\rho_A g \tag{4.16}$$

Combining Equations 4.16 and 4.15 (because $P_3 = P_2$) and rearranging terms, the following expression is obtained:

$$P_a - P_b = H(\rho_A - \rho_B)g \qquad (4.17)$$

Two observations should be made of Equation 4.17: (1) the elevation term, z, cancels out when deriving this expression, and hence, does not require measurement when reading a manometer; and (2) the expression is in SI units (to use Equation 4.17 in English units, divide the RHS by g_c).

Liquid manometers are widely used in both industrial and laboratory applications. They can be employed both as basic pressure measurement devices and as standards of calibration of other instruments.

Only the height of the fluid from the surface of one leg to the surface in the other is the actual height of the fluid opposing and balancing the applied pressure. This is so regardless of the geometry or size of the legs, as illustrated in Figure 4.4(C). Even if the manometer tubes are unsymmetrical, the only occurrence will be that more or less fluid will move from one leg to the other. The liquid height required to achieve equilibrium will depend only on the density of the manometer fluid and its vertical height. Table 4.1 gives equivalent values of various manometric fluids and demonstrates the versatility of the manometer. For example, when water is used as an indicating fluid, a 10-in. fluid height measures 0.360 psi, whereas the same measured height utilizing mercury corresponds to 4.892 psi. This represents a ratio of 13.57:1, which is the ratio of the specific gravities of the two fluids (refer back to Equation 4.13). Three types of pressure measurements can be made with manometers:

1. positive, or gauge, pressures, which are those greater than atmospheric;
2. negative pressures, or vacuums, which are pressures less than atmospheric; and
3. differential pressure, which is the difference between two pressures.

Connecting one leg of the U-tube to a source of gauge pressure causes the fluid in the connected leg to depress while the fluid rises in the vented leg. However, if the connection is made to a vacuum, the effect would be to reverse the fluid movement, causing it to rise in the connected leg and recede in the open leg. In obtaining differential pressures, both legs of the manometer are connected so that a pressure change between two points can be measured (as described in the derivation of Equation 4.17). The higher pressure depresses the fluid in one leg, while the lower pressure allows the fluid to rise in the other. The true differential is measured by the difference in height of the fluid in the two legs.

Manometers are available in configurations other than the U-tube to provide greater convenience and to meet different service requirements.

Table 4.1 Conversion Values for Common Manometer Fluids at 22°C

1 in. water	=	0.0360 lb/in.^2
		0.5760 oz/in.^2
		0.0737 in. mercury
		1.2131 in. red oil
1 ft water	=	0.4320 lb/in.^2
		6.9120 oz/in.^2
		0.8844 in. mercury
		62.208 lb/ft^2
		14.5572 in. red oil
1 in. mercury	=	0.4892 lb/in.^2
		7.8272 oz/in.^2
		13.5712 in. water
		1.1309 ft water
		16.4636 in. red oil
1 oz/in.^2	=	0.1228 in. mercury
		1.7336 in. water
		2.1034 in. red oil
1 lb/in.^2	=	2.0441 in. mercury
		27.7417 in. water
		2.3118 ft water
		33.6542 in. red oil

Figure 4.5 shows three other types of manometers. Figure 4.5(A) illustrates a well-type manometer. As shown, if one leg of the manometer is increased in area in comparison to the other, the volume of fluid displaced represents little change in the vertical height of the large-area leg compared to the change of height in the smaller-area leg. Thus, it becomes necessary to read only the scale adjacent to the single tube rather than two, as with the U-type.

The well-type design lends itself to use of direct reading scales that are graduated in appropriate units for the process variable involved. The higher-pressure source to be measured must remain connected to the well leg, while the lower pressure source must be connected to the top of the tube. In any measurement, therefore, the source of the higher pressure must be connected in such a manner to cause the manometer fluid to rise in the indicating tube.

The true pressure follows the same principles discussed previously, and the measurement obtained reflects the difference between the fluid surfaces. Obviously, there must be some decrease in the well level; however, this can be compensated for by spacing the scale graduations in the proper amount needed to reflect and correct for the well level drop.

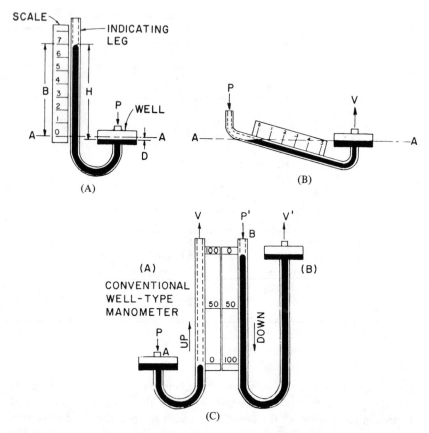

Figure 4.5   Alternative configurations of manometers: (A) a well-type manometer; (B) inclined-tube, well-type manometer for low-pressure measurements; (C) dual-tube manometer system for measuring high pressures.

A variety of applications require accurate measurement of low pressures, such as drafts and low differentials in air and gas installations. A common arrangement for these cases is the use of an inclined-tube well manometer (Figure 4.5(C)). This design has the advantage of providing an expanded scale. For example, 12 cm of scale length can represent 1 cm of vertical length. Using scale divisions of 0.01 inches of liquid height, an equivalent pressure of 0.000360 psi per division can be detected with water as the manometer fluid.

To measure relatively high pressures, a longer indicating fluid tube is required. Rather than using excessively long manometer tubes for high-pressure readings, a dual-tube manometer arrangement can be used, as shown in Figure 4.5(C). This arrangement provides for reading the full range of the instrument in only one-half the total vertical viewing distance. The system shown in Figure 4.5(C) consists of two separate manometers mounted on a single housing. Manometer (A) is a conventional well design having a zero scale at the bottom and graduated upward. Manometer (B) has a well and zero scale raised to the 100-inch level, with the scale graduated downward. Connecting a positive-pressure source at connections (A) and (B) causes the indicating fluid level to rise in tube (A) and fall in (B). Note, however, that these are essentially two separate manometers measuring the same pressure, with one indicating column rising and one falling. The fluid columns pass each other at the 50-inch mark and, as such, it becomes convenient to read the left scale on pressures from 0 to 50 inches progressing upscale and the scale on manometer (B) from 50 to 100 inches downscale. Only the lower 50 inches is required to read the entire range of 100 inches. Gauge pressure connections must be made at connections A and B, while vacuum connections must be made at V and V'.

Another system frequently used is a sealed tube or absolute manometer. The term "absolute pressure" originates from the fact that in a perfect vacuum the complete absence of any gas is referred to as "absolute" zero. For an absolute pressure manometer, the pressure measured is compared to the vacuum or absolute pressure in a sealed tube above the indicating fluid column. The most common type of sealed tube manometer is the conventional mercury barometer, which is used to measure atmospheric pressure. The barometer is a mercury-filled tube more than 76 cm high and immersed in a reservoir of mercury, which is exposed to the atmosphere. The mercury column is supported by the atmospheric or barometric pressure. A variety of processes, tests and calibrations are based on pressures near or below atmospheric conditions, and these are measured most conveniently with a sealed-tube manometer.

The principle of communicating vessels also may be applied to determining the height of hydraulic seals in different equipment. Figure 4.6 shows such an example in which an emulsion made up of two liquids of different densities continuously enters into vessel 1 through central tube 2, where it separates. The lighter liquid of density ρ' is skimmed off the fluid surface through nozzle 3, while the heavier fluid of density ρ'' is drained from below through the U-hydraulic seal 4.

The reference plane "0-0" is specified at the liquid-liquid interface. From Equation 4.13, the height of the hydraulic seal required will be

$$z_0'' = z_0' \frac{\rho'}{\rho''}$$

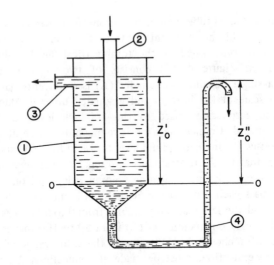

Figure 4.6   The use of a hydraulic seal in a continuous liquid separator.

We can assume that the pressures above the liquid surface inside the vessel and at the outlet of the seal are the same.

Figure 4.7 illustrates still another application of manometers. In this example, a manometer (number 3) is used to measure the liquid level in a tank as follows. Pressurized air is passed through tube 2 to the bottom of the tank. The air pressure is increased gradually until it overcomes the

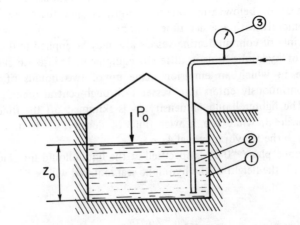

Figure 4.7   Application of pneumatic measurement of liquid level in a vessel.

resistance of the liquid column in the tank. At the point at which the static head is overcome, bubbling starts through the liquid and pressure P is maintained constant. From Equation 4.10, the value of the pressure required to overcome the liquid column resistance is

$$P = P_0 + \rho g z_0$$

whence we can solve for the liquid level in the tank:

$$z_0 = \frac{P - P_0}{\rho g} \tag{4.18}$$

where P_0 is the pressure at the liquid surface. From z_0 and the cross-sectional area of the vessel, one can compute the liquid inventory.

In addition to manometers, the mechanical Bourdon-type pressure gauge is employed extensively as a pressure-measuring device. Its principle of measurement is based on a coiled hollow tube housed by the gauge, which tends to straighten out when subjected to an internal pressure. The extent to which this straightening occurs is proportional to the pressure difference between the inside and outside of the gauge. The tube is connected to a point on a calibrated dial.

Further applications of manometers and various pressure-measuring devices used in industry and the laboratory are given by Cheremisinoff [1979,1981]. The following example problems illustrate the use of the principal equations derived in this subsection.

Example 4.1

A simple U-tube manometer is being used to measure the pressure drop of crude oil flowing through an orifice. As will be discussed later, this pressure drop can be related to the flowrate in the pipe. The manometric fluid is mercury (specific gravity = 13.6), and the specific gravity of the crude oil is 0.86. The reading obtained from the manometer is 18.7 cm. Compute the pressure differential.

Solution

This problem involves application of Equation 4.17:

$$P_a - P_b = H(\rho_A - \rho_B)g$$

where $$H = 18.7 \text{ cm} \times \frac{1 \text{ m}}{100 \text{ cm}} = 0.187 \text{ m} \ (0.61 \text{ ft})$$

$$\rho_A + 13.6 \times 1 \text{ g/cm}^3$$

$$\times \frac{\text{kg}}{10^3 \text{g}} \times \frac{10^6 \text{cm}^3}{\text{m}^3} = 13,600 \text{ kg/m}^3 \left(849 \frac{\text{lb}_m}{\text{ft}^3}\right)$$

$$\rho_A = 860 \text{ kg/m}^3 \ (53.7 \text{ lb}_m/\text{ft}^3)$$

In SI units, the pressure differential is

$$P_a - P_b = 0.187 \text{ m}[(13,600 - 860)\text{kg/m}^3](9.807 \text{ m/s}^2)$$

$$= 23,364 \frac{\text{kg}}{\text{m-s}^2} = 23,364 \text{ N/m}^2$$

In English units, the pressure differential is

$$P_a - P_b = 0.61 \text{ ft}[(849 - 53.7)\text{lb}_m/\text{ft}^3] \ \frac{\left(32.2 \frac{\text{ft}}{\text{s}^2}\right)}{\left(32.174 \frac{\text{ft-lb}_m}{\text{lb}_f\text{-s}^2}\right)} \times \frac{\text{ft}^2}{144 \text{ in.}^2}$$

$$= 3.37 \text{ psia}$$

Example 4.2

The system for the previous example is given in Figure 4.8. The pressure at point α is known to be P_α, N/m^2. Derive an expression for the pressure at point β.

Solution

The pressure at point 1 is

$$P_1 = P_\alpha + \gamma_A(H + z)$$

where γ_A is the specific weight of fluid A.

The pressure at point 2 is the same as at point 1 because they are at the same level ($P_1 = P_2$).

The pressure at point 3 is

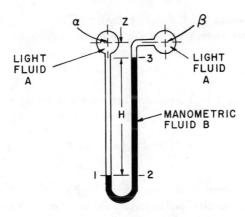

Figure 4.8   A differential manometer in operation for Example 4.2.

$$P_3 = P_\alpha + \gamma_A(z + H) - \gamma_B H$$

where γ_B is the specific weight of the manometric fluid.

The pressure at point β is, therefore,

$$P_\beta = P_3 - \gamma_A(z)$$

$$P_\beta = P_\alpha + \gamma_A(z + H) - \gamma_B(H) - \gamma_A(z)$$

or, simply,

$$P_\beta = P_\alpha + (\gamma_A - \gamma_B)H$$

We merely repeated our derivation of Equation 4.17. Hence, as stated earlier, the pressure differential $(P_\beta - P_\alpha)$ is a function of the *difference* between the specific weights of the two fluids.

Example 4.3

A two-fluid manometer (Figure 4.4(C)) is being used to measure a point pressure in an exhaust duct where gas is flowing. One leg of the manometer is hooked up to the duct and the other is open to the atmosphere. The atmospheric pressure is 750 mm Hg. The two manometric fluids are water ($\gamma_w = 1$) and a light oil ($\gamma_o = 0.86$). The diameter of each of the large reservoirs is 25.4 mm and that of the small ones is 3.0 mm. The manometer reading is 150 mm. Determine the point pressure in the duct.

Solution

Examining Figure 4.4(C), a pressure balance for the U-tube is made:

$$P_a - P_b = (H - H_0)\left(\rho_w - \rho_o + \frac{A}{A'}\rho_o\right)g$$

where H_0 = the manometer reading when $P_a = P_b$ (we can adjust the manometer reading to zero for no-flow in the duct)
   P_a = point pressure in the duct
   ρ_w = density of the heavier fluid (water) = 1000 kg/m^3
   ρ_o = density of the lighter fluid (oil) = 860 kg/m^3
   A' = cross-sectional area of each of the large reservoirs = $\frac{1}{4}\pi(25.4 \text{ mm})^2 \times 10^{-3}$ m/mm = 0.51 m^2
   A = cross-sectional area of each of the tubes forming the U = 7.06×10^{-3} m^2

Note that A/A' is relatively small and, hence, may be neglected. Also, because $H_0 = 0$, our equation reduces to Equation 4.17:

$$P_a - P_b = H(\rho_w - \rho_o)g$$

And solving for the pressure in the duct:

$P_a = P_b + H(\rho_w - \rho_o)g$

$P_a = (750 \text{ mm Hg})(1.3332 \times 10^2) + (0.150 \text{ m})[(1000 - 860) \text{ kg/m}^3]$
   $\times (9.8 \text{ m/s}^2)$

$P_a = 100.2 \text{ kN/m}^2$

PRESSURE FORCES ACTING ON SUBMERGED FLAT SURFACES

It is important to determine the pressure forces acting on the walls and floors of vessels when designing equipment. Excessive forces and pressure can warp and/or result in equipment failure. To evaluate such potential problems, an examination is needed of the physics of pressure forces. We shall begin the analysis by examining an elemental area of a vessel wall, $d\Omega$, submerged in the process fluid. By directing our attention to a small region, we can make some simplifying assumptions to the analysis, namely, that the surface under examination is flat and that the fluid pressure acting against it is distributed uniformly. The system under consideration is illustrated in Figure 4.9. We wish to estimate the magnitude of the resultant force on the area and determine the location of the center of pressure where the resultant force can be assumed to act.

Examining Figure 4.9, we take the coordinate axes locating the plane xoy on the fluid surface, with the oz axis in the downward direction. Then the pressure acting at the center of the elementary plane will be

$$P = \gamma z \qquad\qquad (4.19)$$

where z corresponds to the depth of the fluid from the free surface to the center of the plane. Hence, the resultant force on the area due to the fluid pressure is

$$dP = \gamma z d\Omega \qquad\qquad (4.20)$$

where dP is normal to the elemental plane.

We now expand the force dP in components dP_x, dP_y, dP_z parallel to their corresponding coordinate axes. So, for example,

$$dP_x = dP \cos\alpha = \gamma z d\Omega \cos\alpha$$

where α is the angle between the normal to the plane and the x-axis, and $d\Omega \cos\alpha$ is the projection of plane $d\Omega$ on the plane perpendicular to the ox-axis, i.e., on the plane zoy. We denote this projection as $d\Omega_{zy}$. Then,

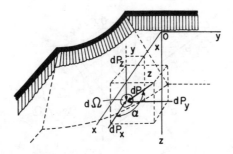

Figure 4.9   The coordinate system for analyzing the fluid pressure forces acting on an elemental area of a submerged vessel wall.

$$dP_x = \gamma z d\Omega_{zy} \tag{4.21}$$

In the same manner, the components dP_y and dP_z may be presented as follows:

$$dP_y = \gamma z d\Omega_{xz} \tag{4.22}$$

$$dP_z = \gamma z d\Omega_{yx} \tag{4.23}$$

The RHS of Equation 4.23 is the weight of an elementary fluid column having a cross section equivalent to the horizontal projection of a given flat element of a wall whose height is the depth of submergence of the element's center of gravity. Because the element $d\Omega$ is infinitesimal, the product $\gamma z d\Omega_{yx}$ may be equated to the weight of fluid column acting on the elemental flat wall.

Suppose we wish to determine the pressure acting against an inclined plane at some angle α, as shown in Figure 4.10. The ox-axis is directed along the free surface of the fluid to intersect with a wall and the oz-axis is vertically downward. To determine the total force, P, on the wall, we must sum the forces acting on the wall elements, i.e.,

$$P = \gamma \int_\Omega z d\Omega \tag{4.24}$$

Denoting ξ as the distance from the ox-axis to the center of gravity of $d\Omega$, we have

$$\xi = \frac{z}{\sin\alpha}$$

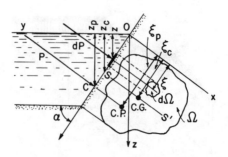

Figure 4.10   The coordinate system for analyzing the fluid pressure forces acting on an inclined vessel wall.

Accounting for this in Equation 4.20, we obtain

$$P = \gamma \sin\alpha \int_{\Omega} \xi d\Omega$$

The integral $\int_{\Omega} \xi d\Omega = S_x(\Omega)$ is the static moment S_x of the area Ω with respect to the ox-axis, which is equal to the product of the area and distance ξ_c from the center of gravity. Thus,

$$S_x(\Omega) = \Omega \xi_c$$

and

$$P = \gamma \Omega \xi_c \sin\alpha \qquad (4.25)$$

However, we must note the following:

$$P = \gamma \Omega z_c = P_c \Omega$$

where $z_c = \xi_c \sin\alpha$ is the submergence depth of the center of gravity and $P_c = \gamma z_c$ is the excessive pressure at the center of gravity of the wall. The component of pressure along the oz-axis may be expressed as follows:

$$P_z = \gamma \int_{\Omega} z d\Omega_{xy} \qquad (4.27)$$

Because $\int_{\Omega} z d\Omega_{xy}$ is the volume of the fluid over a flat wall, the vertical component of the total pressure is equivalent to the weight of this volume of liquid.

We will determine now the position of point C (the center of pressure) acting on the wall through which the line of action of total pressure, P, passes. The coordinate of the center of pressure (Figure 4.10) is ξ_p. From a principle of statics, the moment of the resultant of a force system is equal to the sum of the components of the forces. Taking momentums of forces with respect to the ox-axis, we obtain

$$\xi_p P = \int_{\Omega} \xi dP \qquad (4.28)$$

And, accounting for Equations 4.20 and 4.25, we obtain the following:

$$\xi_p \xi_c \gamma \Omega \sin\alpha = \int_{\Omega} \xi^2 \sin\alpha d\Omega$$

or

$$\xi_p \xi_c \Omega = \int_\Omega \xi^2 d\Omega$$

The value $\int_\Omega \xi^2 d\Omega = J_x$ is the moment of inertia of the wall area, Ω, about the ox-axis. J_x is basically the product of area Ω and the square of inertia radius ρ_x about the ox-axis:

$$\xi_p \xi_c \Omega = \rho_x^2 \Omega$$

Hence,

$$\xi_p = \frac{\rho_x^2}{\xi_c} \tag{4.29}$$

Trace the line SS′ through the center of gravity of the wall parallel to the ox-axis. In accordance with a known relationship between the momentum of inertia of an area about parallel axes, the moment of inertia, J_x, may be written in the following form:

$$J_x = \xi_c^2 \Omega + J_s \tag{4.30}$$

where J_s is the moment of inertia of the wetted wall area about the SS′-axis. Therefore,

$$\rho_x^2 \Omega = \xi_c^2 \Omega + \rho_s^2 S\Omega \tag{4.31}$$

where ρ_s is the radius of inertia of the wall about the SS′-axis. Thus, Equation 4.29 has the following form:

$$\xi_p \xi_c = \xi_c^2 + \rho_s^2 S \tag{4.32}$$

Hence,

$$\xi_p = \xi_c + \frac{\rho_s^2 S}{\xi_c} \tag{4.33}$$

Keeping in mind that all terms in Equation 4.33 are positive, we must conclude that the center of pressure for a flat inclined wall is always located below the center of gravity. The following example problems apply the above principles.

Example 4.4

An 18-ft-wide, 35-ft-high crude oil storage tank is vented to the atmosphere. It is believed that water accumulation at the tank bottom may have reduced the available storage volume seriously, as well as promoting corrosion of the tank shell. If the gauge pressure at the tank bottom is 13.9 psig, determine the volumes of oil and water in barrels. The specific gravity of the crude is 0.91.

Solution

As the tank is vented to the atmosphere, we can assume that the pressure over the oil surface is that of the atmosphere, i.e., $P_t = 14.696$ psia. From a pressure balance, the pressure at the bottom of the tank is

$$P_b = h_o \rho_o \frac{g}{g_c} + P_t + h_w \rho_w \frac{g}{g_c}$$

where ρ_o, ρ_w are the densities of oil and water, respectively, h_o is the height of the oil, and h_w is the height of the water layer in the tank. Assuming the tank is filled to capacity, the overall tank height is

$$H_T = h_o + h_w$$

or the oil height is

$$h_o = H_T - h_w = 35 - h_w$$

Substituting for h_o in our pressure balance and noting that the gauge pressure at the bottom is equal to the absolute pressure, P_b, minus the atmospheric pressure, $P_{gauge} = P_b - P_t$,

$$P_b - P_t = (H_T - h_w)\rho_o \frac{g}{g_c} + h_w \rho_w \frac{g}{g_c}$$

$$P_{gauge} = H_T \rho_o \frac{g}{g_c} + h_w \frac{g}{g_c}(\rho_w - \rho_o)$$

$$13.9 \; lb_f/in.^2 \times 144 \; in.^2/ft^2 = (35 \; ft)(0.91 \times 62.4 \; lb_m/ft^3)\left(1.0 \; \frac{lb_f}{lb_m}\right)$$

$$+ h_w \left(1.0 \; \frac{lb_f}{lb_m}\right)[(1.0 - 0.91) \times 62.4 \; lb_m/ft^3]$$

$$2001.6 = 1987.4 + 5.62 \; h_w$$

$$h_w = 2.5 \; ft$$

and the height of the oil is $h_o = 35 - 2.5 = 32.5$ ft. The cross-sectional area of the tank is $\frac{1}{4}\pi(18 \; ft)^2 = 254.5 \; ft^2$. The volume of oil is

$$V_o = 32.5 \; ft \times 254.5 \; ft^2 \times 7.48 \; \frac{gal}{ft^3} \times \frac{bbl}{42 \; gal} = 1473 \; bbl$$

and that of the water is

$$V_w = 2.5 \times 254.5 \times \frac{7.48}{42} = 113.3 \; bbl$$

Hence, about 7% of the tank volume has been lost to water accumulation.

Example 4.5

A reactor vessel is fitted with a 6-in.-diameter circular glass viewing port on an inclined wall ($\theta = 62°$). The system is illustrated in Figure 4.11. The centroid of the viewing port is at a depth of 3.5 ft from the liquid interface. The specific gravity of the liquid mixture is approximately 0.79.

1.   Evaluate the magnitude of the resultant fluid force on the viewing port.
2.   Determine the location of the center of pressure.

Solution

Part 1: The area of interest is the circular viewing port. For the simple area of a circle, the centroid exists at its center. From the problem description, we know the vertical depth from the fluid interface to the centroid of the viewing port, $d_c = 3.5$ ft. Note that ℓ_c and d_c in Figure 4.11 are related through angle θ as follows:

$$\sin\theta = d_c/\ell_c$$

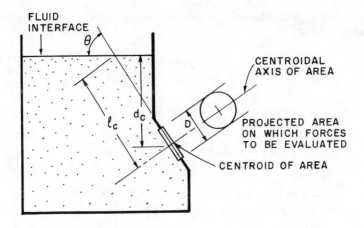

Figure 4.11 Defines system coordinates for Example 4.5.

Hence,

$$\ell_c = d_c/\sin\theta = 3.5 \text{ ft}/\sin 62^0 = 3.96 \text{ ft}$$

ℓ_c and d_c will be needed later on. The area of the viewing port is

$$A = \frac{1}{4} \pi (6/12)^2 = 0.20 \text{ ft}^2$$

The resultant force, F_R, acting on the port is

$$F_R = \gamma d_c A \frac{g}{g_c}$$

where γ is the specific weight of the liquid:

$$\gamma = 0.79 \times 62.4 \text{ lb}_m/\text{ft}^3 = 49.3 \text{ lb}_m/\text{ft}^3$$

Thus,

$$F_R = 49.3 \frac{\text{lb}}{\text{ft}^3} \times 3.5 \text{ ft} \times 0.20 \text{ ft}^2 = 34.5 \text{ lb}_f$$

Part 2: For a circle, the area moment of inertia is given by the following formula:

$$J = \frac{\pi D^4}{64}$$

or

$$J = \frac{\pi}{64} (6/12)^4 = 3.07 \times 10^{-3} \ ft^4$$

And we know that ℓ_c = 3.96 ft and A = 0.20 ft². The distance normal from the resultant force vector to the fluid interface is

$$x = \ell_c + \frac{J}{\ell_c A} = 3.96 + \frac{3.07 \times 10^{-3} \ ft^4}{(3.96 \ ft)(0.20 \ ft^2)}$$

$$x = 3.96 \ ft + 3.88 \times 10^{-3} \ ft = 3.964 \ ft$$

which means that the center of pressure will occur essentially at the centroid of the viewing port.

Example 4.6

The cross section of a vessel is in the form of a parabolic segment, as shown in Figure 4.12. Determine the resultant fluid pressure acting on the vessel's vertical front wall for a fluid depth H and fluid interface width 2a.

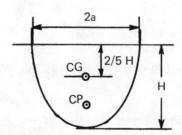

Figure 4.12   Parabolic cross section of a vessel for Example 4.6. CG and CP denote the center of gravity and the center of the pressure force, respectively.

Solution

The area of the parabolic segment is as follows:

$$\Omega = \frac{2}{3} aH$$

The distance from the center of gravity to the apex of a parabola is equal to ⅗ H; consequently,

$$z_c = H - \frac{3}{5} H = \frac{2}{5} H$$

The moment of inertia of the segment about boundary chord is

$$J_x(\Omega) = \frac{8H^2}{35} \Omega$$

Thus, the square of inertia radius is

$$\rho_x^2 = \frac{8H^2}{35}$$

and, from Equation 4.26, we obtain the following:

$$P = \gamma z_c \Omega = \gamma \frac{2}{5} H \frac{2}{3} aH = \frac{4}{15} \gamma aH^2$$

The coordinate of the center of pressure can be obtained from Equation 4.29. Hence, the answer is

$$z_p = \frac{8H^2}{35} \bigg/ \frac{2H}{5} = \frac{4}{7} H$$

Example 4.7

A conical vessel with no bottom is mounted on a horizontal surface. The dimensions of the vessel are given in Figure 4.13. The specific weight of the vessel shell is γ_m. Determine the liquid height that will cause the vessel to tear away from the horizontal surface.

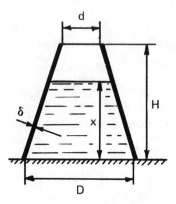

Figure 4.13   Conical vessel with no bottom mounted on a horizontal surface defining the system for Example 4.7.

Solution

The weight of the vessel is

$$G_v = \gamma_m \pi \delta \frac{D + d}{4} \sqrt{4H^2 + (D - d)^2}$$

The force that will tear the vessel away from the horizontal surface will result from the liquid pressure acting on one of the vessel's side walls. This force is directed vertically upward and is equal to the difference between the liquid weight in the cylinder and that in the cone, where

$$V_c = \frac{\pi D^2}{4} x$$

and

$$v_{cone} = \frac{\pi x}{12} \left[3D^2 - \frac{3D}{H} x (D - d) + \frac{x^2}{H^2} (D - d)^2 \right]$$

The tearing force is equal to the following:

$$G_\varrho = \gamma_\varrho \frac{\pi D^2}{12} x \left(1 - \frac{d}{D} \right) \left[\frac{3x}{H} - \frac{x^2}{H^2} \left(1 - \frac{d}{D} \right) \right]$$

Equating the above expression to the weight of the vessel, we obtain an equation for liquid level x raised to the third power:

$$\gamma_{\ell} \frac{\pi D^2}{12} \frac{x^2}{H} \left(1 - \frac{d}{D}\right) \left[3 - \frac{x}{H} \left(1 - \frac{d}{D}\right)\right] = \gamma_m \frac{\pi \delta (D + d)}{4} \sqrt{4H^2 + (D - d)^2}$$

This equation can have only one real root if

$$\left| \frac{\gamma_m}{\gamma_{\ell}} \frac{\delta(D^2 - d^2)\sqrt{4H^2 + (D - d)^2}}{D^3 H} \right| < \frac{4}{3}$$

HYDROSTATIC MACHINES

The design and operation of hydrostatic machines is based on the application of the basic hydrostatic equation. One example of a hydrostatic machine is illustrated in Figure 4.14. The system shown is a hydraulic press, which can be used for pressing and forming briquets of different materials. If a relatively small force is applied to piston 1, which moves in a cylinder of diameter d_1, a pressure, P, is established on the piston. In accordance with Pascal's law, the same pressure will be exerted on piston 2 in the large-diameter cylinder, d_2. The force acting on piston 1 is

$$P_1 = p \frac{\pi d_1^2}{4} \tag{4.34}$$

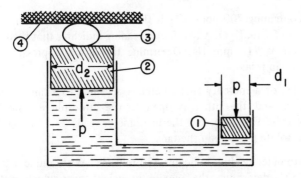

Figure 4.14   A scheme for a hydraulic press.

and the force acting on piston 2 is

$$P_2 = p \frac{\pi d_2^2}{4} \qquad (4.35)$$

As a result, the piston in the cylinder of greater diameter will transfer the force (which is many times greater than the force applied to the piston in cylinder of diameter d_1) to compressing material 3. The forces applied to compressing material 3 increase proportionally with the ratio of the diameters of the piston cylinders. Thus, by harnessing relatively small forces, it is possible to press material 3 located between piston 2 and the immovable plate 4.

Further discussions of hydrostatic machines and other principles described in this chapter may be found in the references listed. The reader should work through the practice problems at the end of this chapter to understand fully the computation methods outlined above.

PRACTICE PROBLEMS

4.1 The pressure gauge on a gas cylinder of CO_2 used to carbonate soda water bottles reads 67 psig. The barometric pressure is 749 mm Hg. Determine the absolute pressure in the tank in psia and N/m^2.

4.2 Convert 500 mm Hg to

   (a) ft of water,
   (b) psia,
   (c) N/m^2,
   (d) bars, and
   (e) Pascals.

4.3 Air containing 700 ppm SO_2 is flowing through an exhaust duct under a draft of 7 in. H_2O. A barometer reading indicates that the atmospheric pressure is 715 mm Hg. Determine the absolute pressure of the gas mixture in psia.

4.4 One leg of a mercury manometer is attached to a tank under vacuum pressure and the other is open to the atmosphere. The barometric pressure is 1.09 atm, and the manometer reads 23.2 in. Hg. Determine the absolute pressure in the tank.

4.5 A barometer reading indicates 670 mm Hg. At an ambient temperature of 25°C, determine the maximum height over which water can be siphoned. Repeat the calculation for siphoning kerosene (specific gravity = 0.82).

4.6 A steam condenser for a turbine has a pressure reading of 13.2 psig of vacuum. The barometric pressure is 795 mm Hg. Determine the absolute pressure in the condenser.

4.7 The pressure drop across an orifice plate used to measure gas flow in a pipe is obtained with a mercury-filled glass U-tube, as shown in Figure 4.15. Both legs of the U-tube are 5.80 mm in diameter (i.d.). Note that H_1 has a value of 5 in. The manometer reading indicates a pressure drop across the orifice of 97 mm Hg. Determine the values of H_2 and H_3, in inches.

4.8 The manometer described in Problem 4.7 is broken and is replaced with glass tubing that is 9.5 mm i.d. The manometer is filled with the same volume of mercury. Calculate H_2 and H_3 for the same pressure drop.

4.9 A Bourdon gauge is placed on one of the tanks in the communicating vessel system shown in Figure 4.16. The manometer reads 14 in. Hg, and the barometric pressure is 753 mm Hg. Determine the reading on the Bourdon gauge.

4.10 The interstage pressure on compressors usually is defined as the geometric mean of the absolute suction and discharge pressures. A two-stage air compressor takes suction at 2.5 psig and has a final discharge pressure of 137 psig. If the barometric pressure is 690 mm Hg, determine the interstage gauge pressure.

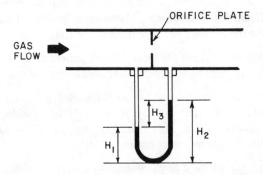

Figure 4.15 Mercury barometer used to measure gas flow through an orifice.

Figure 4.16   Bourdon gauge used in conjunction with a manometer.

4.11 A 4.6-m-diameter holding tank is filled with glycerine. The top of the tank is vented to the atmosphere, which has a pressure of 757 mm Hg. Determine the pressure exerted on the tank floor in both English and SI units.

4.12 A cylindrical holding tank is filled with gasoline to a depth of 30 ft. The gauge pressure at the top of the tank is 36 psig. Determine the head of the gasoline in ft, which corresponds to the absolute pressure at the tank bottom. Assume the density of gasoline to be 0.7.

4.13 A quantity of water is poured into the bottom of an open flask to a level of 5.1 cm. A sample of oil is then poured carefully over the top of the water so that the total level of fluid in the flask is 12 cm. The temperature of the surroundings and atmospheric pressure are 29°C and 745 mm Hg, respectively. The specific gravity of the oil is 0.82. Determine the pressure at the bottom of the flask.

4.14 Repeat Problem 4.13 if the second liquid poured in is a 45 wt% solution of acetic acid. Assume that the liquids are completely miscible.

4.15 A mercury-filled U-tube manometer is installed across an orifice meter. The liquid above the mercury is carbon tetrachloride, which has a specific gravity of 1.8. Determine the pressure differential across the manometer for a manometer reading of 10 in.

4.16 A 50-cm i.d. centrifuge bowl containing a layer of carbon tetrachloride is rotating at a speed of 10,000 rpm. If the pressure at the liquid surface is 28.9 in. Hg, determine the gauge pressure exerted on the walls of the centrifuge bowl.

4.17 A mercury-filled manometer is set up to measure the pressure difference between two vessels. The pressure in the tank containing more fluid is 9.3 psig. If the specific gravity of the liquid in the tanks is 2.3, determine the gauge pressure in the other tank. The manometer reading is 11.3 in. and atmospheric pressure is 29.93 in. Hg.

4.18 A two-fluid manometer is used to estimate the volume of nitrobenzene liquid contained in a holding tank vented to the atmosphere. One leg of the manometer is hooked up to the tank and the other is open to the atmosphere. The temperature of the nitrobenzene is 12°C. The manometric fluids are mercury (γ_m = 13.6) and water (γ_w = 1.0). The manometer is similar to the design shown in Figure 4.4(C), where the diameter of each of the large reservoirs is 31 mm and that of the legs is 4 mm. The manometer reading is 210 mm and the barometric pressure is 1.1 atm:

(a)   Determine the gauge pressure at the bottom.
(b)   If the inside diameter of the tank is 9 ft, determine the volume of nitrobenzene.

4.19 For the system described in Problem 4.18, one of the large reservoirs is broken accidentally and is replaced with a well having a diameter of 19 mm. The same volume of manometric fluids is put back into the system but the manometer reading is now 315 mm Hg. Repeat parts (a) and (b).

4.20 An inclined-tube, well-type manometer (Figure 4.5(B)) is used to measure a point pressure in a pipe in which gas is flowing. The leg of the manometer is hooked up to the pipe and the well is open to the atmosphere. The manometer fluid is a light oil (γ_o = 0.82). The diameter of the reservoir is 20 mm, the tube diameter is 6 mm and the manometer reading is 56 mm. If the tube is inclined 60° from the horizontal, what is the point pressure in the pipe?

4.21 A vessel is fitted with a rectangular access hatch that is 4.5 ft wide (W) by 3 ft high (H). The hatch is located on a vessel wall that is inclined at an angle of 50° from the horizontal. The centroid of the hatch is located at a depth of 6 ft from the liquid surface, and the specific gravity of the liquid is 1.5. Evaluate the magnitude of the resultant fluid force on the hatch area and determine the location of the center of pressure. For a rectangle, the area moment of inertia is $J = WH^3/12$.

4.22 A dam for a water reservoir consists of a slanted concrete wall with a trapezoidal sluice gate having the dimensions shown in Figure 4.17. The top of the sluice gate is located 17 ft below the surface of the water. Determine the magnitude of the resultant force and the location of the center of pressure.

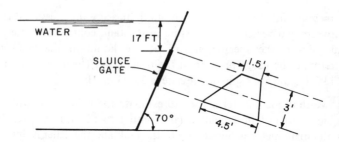

Figure 4.17   Trapezoidal sluice gate on a dam for Problem 4.22.

4.23 An oil-water separator is designed to operate with a residence time of 2 hours and at a constant total level of 8 ft. The volumetric loading is 790 gal/ft-hr (where loading = volumetric flow divided by the weir length).

   (a)   Assuming the separator to be cylindrical and the liquid mixture to be typically 15% by weight oil (γ_0 = 0.85), determine where the oil overflow leg should be placed.

   (b)   A mercury manometer is used to measure the gauge pressure at the separator's floor. If the manometer provides a reading of 6.3 psig, determine the position of the new oil-water interface.

4.24 The oil-water mixture described in Problem 4.23 is to be separated in a tubular centrifuge bowl. The centrifuge bowl has a 4-in. i.d. and rotates at 40,000 rpm. The free-fluid surface inside the centrifuge is 2 in. from the axis of rotation.

   (a)   Determine the radial distance from the rotational axis to the top of the oil overflow dam.

   (b)   Repeat part (a) for an equal volume mixture of oil and water.

4.25 A mercury manometer is hooked up to a water-piston system, as shown in Figure 4.18. The barometric pressure is 775 mm Hg. Determine the pressure at the piston that produces the manometer reading shown.

4.26 A nitrogen bubbling system as shown in Figure 4.7 is used to measure the glycerine inventory in a tank. The gauge pressure on the nitrogen line is 15 psig and the pressure at the top of the tank is 4 psig. Assuming ambient conditions, compute the inventory of glycerine in a 12-ft-diameter tank.

4.27 A 250-ft-diameter, 180-ft-high crude oil storage tank is filled to capacity. The tank is designed with a floating roof that normally has a blanket of gas between the roof and oil surface. This blanket occupies roughly 0.5% of the tank's overall height so that the vessel may be

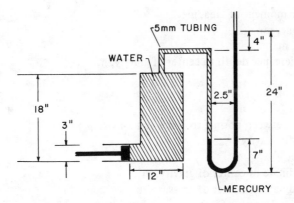

Figure 4.18   Piston manometer system described in Problem 4.25.

vented to the atmosphere. Water accumulation is always a problem with storage tanks. The gauge pressure at the tank bottom is 71.2 psig and the specific gravity of the crude about 0.91.

(a)   Determine the percentage of available storage volume that is lost to water accumulation.

(b)   If crude sells for $33/bbl, what is the worth of the storage volume lost to water accumulation?

NOTATION

A   cross-sectional area, m^2
CG   center of gravity
CP   center of pressure
D   diameter, m
d_c vertical depth from liquid surface, m
F_R resultant force, N
g   gravitational acceleration, m/s^2
g_c conversion factor, 32.174 ft-lb_m/lb_f-s^2
H   fluid displacement or height, m
h   height, m
J   moment of inertia, m^4
J_s moment of inertia of wetted surface, m^4
ℓ_c distance normal from center of gravity to fluid surface, m
P,p   pressure, N/m^2
RHS   right-hand side

S   static moment of area, m^3
V   volume, m^3
z   level, m
Z_c submergence depth to center of gravity, m

Greek Symbols

α angle between normal to plane and x-axis, °
γ specific gravity
θ angle, °
ξ_c coordinate of center of gravity, m
ξ_p coordinate of center of pressure, m
ρ_s radius of inertia of wall, m
ρ_x inertia radius, m
Ω area, m^2

REFERENCES

Cheremisinoff, N. P. (1979) *Applied Fluid Flow Measurement: Fundamentals and Technology* (New York: Marcel Dekker Inc.).

Cheremisinoff, N. P. (1981) Process Level Instrumentation and Control (New York: Marcel Dekker Inc.).

BIBLIOGRAPHY

Addison, H. (1949) *Hydraulic Measurements* (New York: John Wiley & Sons, Inc.).

Diederich, H., and W. Andrae (1930) *Experimental Mechanical Engineering*, Vol. I (New York: John Wiley & Sons, Inc.).

Perry, R. H., and C. H. Chilton, Eds. (1973) *Chemical Engineer's Handbook*, 5th ed. (New York: McGraw-Hill Book Co.).

Streeter, V. L. (1961) *Handbook of Fluid Dynamics* (New York: McGraw-Hill Book Co.).

CHAPTER 5

INTERNAL PROBLEMS OF HYDRODYNAMICS: SINGLE-FLUID FLOWS

CONTENTS

INTRODUCTION

An understanding of the basic physics governing flow dynamics is essential to developing governing equations, which we refer to as hydrodynamic laws. It is the proper application of these hydrodynamic laws that enables a multitude of unit operations to be designed and operated efficiently and safely throughout the process industries. Indeed, it is not possible to design rationally for heat and mass transfers without properly defining the hydrodynamics of the flow system.

The principles presented in this chapter are the foundation for the study of unit operations. Presented here are the governing equations describing the motion of single-phase fluids, which establish a basis for describing more complex multiphase flows encountered in many unit operations. Emphasis will be placed on the analysis of pipe flow for both compressible and incompressible systems. In addition, an analysis is included of the important system of film flow. The motion of single fluids through piping and equipment constitutes the first class of problems or system analysis encountered in unit operations, namely, the internal problems of hydrodynamics.

CHARACTERISTICS OF FLUIDS IN MOTION

Steady-State versus Transient Flows

Before addressing the complex behavior of fluids in motion and their subsequent effects on the transfers of heat and mass in various unit operations, an understanding is needed of the physics involved. Therefore, we will begin our discussions by describing certain characteristics common to all fluid behavior, regardless of the particular unit operation being performed. It is important, especially for later discussions, that the characteristics and definitions presented in this chapter be understood clearly. To begin, a distinction must be made between steady-state and unsteady-state, or transient flows.

For a process or operation described as being at steady state, the flowrates and properties of the fluid (e.g. temperature, pressure, composition, density and velocity) at each point within the system are constant with respect to time. These properties may vary from point to point within the system, but at any one location at any point in time they remain unchanged. Velocity, for example, may have different values at different points ($v_x = f(x,y,z)$), but at any one point it does not change with time, i.e., $\partial v_x/\partial t = 0$. Hence, for steady-state flows only special variations may exist.

In contrast, for systems that have *transient flows*, the properties that influence fluid motion *change with respect to time*. As an example, for a fluid flowing through a pipe in the x-direction, the velocity at any one location within the pipe is not only a function of the spacial coordinates x, y and z, but also of the time t, i.e., $v_x = f(x,y,z,t)$; consequently, $\partial v_x/\partial t \neq 0$. A specific example of a transient system is discharge through an orifice on a tank operating at variable liquid level. As the level in the tank decreases, so does the discharge velocity. Steady-state flow processes are typical in unit operations, whereas many transient flows are common to batch operations or are associated with startups and shutdowns. Transient flows also are associated with upset conditions in process operation flows. In continuous operations, transient conditions often result from flow regimes caused by phase changes or flow transitions at various points in the system.

To characterize an unsteady-state flow process with respect to time, we will examine the flow properties at the fixed points, i.e., we shall examine changes with time at constant spacial coordinates. For every moving fluid particle, the change of its parameters in time and space will not be described by a partial derivative but by a total derivative. This total derivative is called a *substantial derivative* or, more commonly, a *Lagrangian derivative*.

Let u be a dummy variable that denotes any property of the flow that changes with respect to time and space (e.g., temperature, pressure, density, concentration or any component of velocity w_x, w_y and w_z). Suppose we observe the flow and can measure the fluctuations of u in any moment at a fixed point in space (x,y,z). As stationary observers, we would note that the rate of change in u with time is constant. The rate of change of u may be described by a partial derivative, $\partial u/\partial t$, and, hence, the variation in u over a time interval, dt, is $\partial u/\partial t\, dt$. This value is the *local* variation of the given variable and, at steady state, is equal to zero.

If we now move along with the flow, then the measured value will be the sum of two components, that is, u is a resultant vector. Assume a fluid particle is displaced from point A(x,y,z) to point B ($(x + dx)$, $(y + dy)$, $(z + dz)$) over some time dt. As a result of this displacement, the corresponding path of projections dx, dy, dz becomes $\partial u/\partial x\, dx$, $\partial u/\partial y\, dy$ and $\partial u/\partial z\, dz$. Note that these quantities are not taken with respect to time. Thus, in the case of steady-state flow (where there are no local changes), u changes by a value resulting from the displacement from A to B:

$$du = \frac{\partial u}{\partial x} dx + \frac{\partial u}{\partial y} dy + \frac{\partial u}{\partial z} dz \qquad (5.1)$$

Equation 5.1 characterizes the *convective* change of parameter u.

For unsteady-state flow, where u = f(x,y,z,t), u changes over a time interval, dt, by a value $\partial u/\partial t$ dt. Consequently, the *total* change in u is *the sum of the local and convective variations*:

$$du = \frac{\partial u}{\partial t} dt + \frac{\partial u}{\partial x} dx + \frac{\partial u}{\partial y} dy + \frac{\partial u}{\partial z} dz \qquad (5.2A)$$

or

$$\frac{du}{dt} = \frac{\partial u}{\partial t} + \frac{\partial u}{\partial x} \frac{dx}{dt} + \frac{\partial u}{\partial y} \frac{dy}{dt} + \frac{\partial u}{\partial z} \frac{dz}{dt} \qquad (5.2B)$$

Assuming u to be fluid velocity, we note that the values $dx/dt = w_x$, $dy/dt = w_y$, $dz/dt = w_z$ are components on the corresponding coordinate axis and, hence, we obtain the following:

$$\frac{du}{dt} = \frac{\partial u}{\partial t} + \frac{\partial u}{\partial x} w_x + \frac{\partial u}{\partial y} w_y + \frac{\partial u}{\partial z} w_z \qquad (5.3A)$$

When $\partial u/\partial t = 0$ (steady state),

$$\frac{du}{dt} = \frac{\partial u}{\partial x} w_x + \frac{\partial u}{\partial y} w_y + \frac{\partial u}{\partial z} w_z \qquad (5.3B)$$

Equations 5.3A and 5.3B represent the substantial derivative of the given parameter, which characterizes changes in both time and space. We shall make the analysis more specific by examining a *continuous fluid flow* situation. Consider a flow in which the same amount of fluid enters and leaves a system, i.e., there is no accumulation or depletion of available fluid supply over some arbitrary time interval. To analyze this system, we must define the terms fluid mass velocity and linear velocity.

Fluid Mass and Linear Velocities

Let us first consider a very simple case of flow through a conduit of constant cross-sectional area. The amount of fluid flowing through a specified cross section is referred to as the *volumetric flowrate*, or *mass (weight) flowrate*, having units of m^3/s or kg/s, respectively.

From the mass rate W(kg/hr) and cross-sectional area $F(m^2)$, a *superficial mass velocity* can be computed. The superficial mass velocity is defined as the ratio of total mass flowrate divided by the total area of flow, i.e., in the case of a circular conduit, $F = \frac{1}{4}\pi D^2$,

$$G = \frac{W}{F}, \text{kg/m}^2\text{hr} \tag{5.4}$$

From the fluid density $\rho(kg/m^3)$ and mass rate W(kg/s), the volumetric flowrate is determined as follows:

$$V = \frac{W}{\rho}, \text{m}^3/\text{hr} \tag{5.5}$$

Volumetric flowrate is defined simply as the volume of fluid flowing through a given cross section per unit time. The ratio of volumetric rate to the cross section is the definition of the mean linear velocity of the flow:

$$\overline{W} = \frac{V}{F}, \text{m/hr} \tag{5.6}$$

The local linear fluid velocity, i.e., the velocity at a given point in the flow, depends strongly on whether the fluid is under the influence of solid boundaries. Fluid elements close to the pipe axis move faster than those in the region of the wall. Hence, to compute the volumetric flowrate, information on a local velocity (for example, along the pipe axis) is insufficient. We require information on how the velocity changes over the cross section. For a symmetrical velocity distribution about the pipe axis, w is a function of distance perpendicular from the axis to the pipe wall:

$$w = f(r) \tag{5.7}$$

Let us consider an elemental circular element of the flow having radius r and thickness dr, moving along a pipe of radius R. A front view of this system is illustrated in Figure 5.1. The perimeter of this elemental ring is $2\pi r$ and its area is

$$dF = 2\pi r dr$$

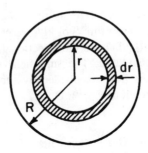

Figure 5.1   Front view of fluid flowing through a pipe.

If the linear velocity of the ring is w, i.e., the local velocity, then, following Equation 5.6, the local volumetric flow is

$$dV = w2\pi rdr$$

Hence, the volumetric rate over the total cross section is as follows:

$$V = 2\pi \int_0^R wrdr \tag{5.8}$$

Once the relationship between local linear velocity, w, and distance normal from the pipe axis to its walls is known, the integral in Equation 5.8 can be evaluated for the volumetric flowrate. The integral may be evaluated either exactly or graphically.

The average linear velocity may be evaluated similarly. For a circular flow section, $F = \pi r^2$ (using Equations 5.6 and 5.8), the following is obtained:

$$w = \int_0^1 wd\left(\frac{r}{R}\right)^2 \tag{5.9}$$

The local velocity, w, in Equation 5.8 can be rewritten in terms for the fraction of the pipe radius, r/R. Knowing w = f/(r), it is possible also to determine w = f(r/R) and w = f(r/R)^2. By integrating this relationship, graphically, for example, over the limits of r/R from 0 to 1, the average velocity of the flow is obtained. From Equations 5.4–5.6, several other useful relations may be written. The mass velocity can be expressed as the average velocity and specific weight of the fluid:

$$G = \overline{W}\gamma \qquad (5.10A)$$

The mass flowrate may be expressed as

$$W = \overline{W}\gamma F \qquad (5.10B)$$

Now we are ready to consider the continuous flow system described at the end of the previous subsection. According to our definition of continuous flow, the mass rate must be the same at any section within the system. That is,

$$W_1 = W_2 = W_3 = \ldots \qquad (5.11A)$$

Starting with Equation 5.4, the continuity of flow may be stated as follows:

$$G_1 F_1 = G_2 F_2 = G_3 F_3 = \ldots \qquad (5.11B)$$

And, from Equation 5.10B,

$$\overline{W}_1 \gamma_1 F_1 = \overline{W}_2 \gamma_2 F_2 = \overline{W}_3 \gamma_3 F_3 = \ldots \qquad (5.12)$$

Equations 5.11A,B and 5.12 represent the material balance for the flow system. Equation 5.12 has special relevance to compressible flows. In gas flow, pressure, temperature and specific gravity may vary at different cross sections and, hence, different linear velocities will result. Equation 5.12 permits us to establish the relationship between these velocities (when cross sections are changed) and their specific gravities. For liquid flow in a pipe of constant cross-sectional area, i.e., $F = F_1 = F_2 = F_3$, the balance equation is considerably simplified; and, from Equations 5.11B and 5.12, we obtain

$$G_1 = G_2 = G_3 = \ldots = \overline{W}_1 \rho_1 = \overline{W}_2 \rho_2 = \overline{W}_3 \rho_3 \qquad (5.13)$$

Equation 5.13 states that for any cross section of piping the mass rates of gas or liquid will be the same even if the temperatures and pressures undergo considerable changes. Applying Equation 5.12 to a liquid flow undergoing a small temperature change, and even at large pressure changes, we may assume that the specific density of the liquid is approximately constant. Consequently, the equation of continuity for liquid flow is

$$\overline{W}_1 F_1 = \overline{W}_2 F_2 = \overline{W}_3 F_3 \qquad (5.14A)$$

And, from Equation 5.6, the volumetric rate is

$$V_1 = V_2 = V_3 = \ldots \qquad (5.14B)$$

According to the above, the average linear velocities for liquid flow are inversely proportional to the cross-sectional areas of flow and, hence, the volumetric rates in different sections are the same. For piping of constant cross section, the linear velocities are therefore the same:

$$\overline{W}_1 = \overline{W}_2 = \overline{W}_3 = \ldots \qquad (5.15)$$

Equations 5.13, 5.14 and 5.15 may be used in practical calculations for liquids and, with some approximations, for gases in cases of small temperature and pressure changes, i.e., when the specific weight of the gas can be considered constant.

Residence Time

To determine *liquid residence time* in a system (e.g., in a pipe or piece of equipment), assume that the mass rate, W, does not change over the cross section and that there are no stagnant regions in the system. If dx is the distance along the direction of the flow and if the flow travels over dx in time dt, then the average linear velocity is simply

$$\overline{W} = \frac{dx}{dt} \qquad (5.16)$$

From Equation 5.10B, we have

$$W = \gamma F \frac{dx}{dt} \qquad (5.17)$$

where F is the cross-sectional area normal to the direction of flow. Note that the elemental volume of the pipe or piece of equipment is $dV_0 = Fdx$. When γ is constant, on integration of Equation 5.17 the residence time is obtained:

$$t = \frac{\gamma}{W} V_0 = \frac{V_0}{V} \qquad (5.18)$$

Hence, residence time depends only on the system volume, V_0, and the volumetric flow. Equation 5.18 states in simple terms that residence time is the ratio of system volume to the volumetric flowrate.

Further discussions of the principles presented in this subsection are given by Bird et al. [1960], Bennet and Myers [1964], Curle and Davies [1964], and Streeter and Wylie [1979]. The following two example problems apply some of the principles described; additional problems may be found at the end of this chapter.

Example 5.1

Oil (0.85 specific gravity) is flowing through two sections of piping that vary in cross section. The average velocity in section 1 is 1.2 m/s. The diameter of pipe section 1 is 8.9 cm and that of section 2 is 3.8 cm. Determine the following: (1) the velocity at smaller section 2; (2) the volumetric flowrate; (3) the mass flowrate; and (4) the mass velocity in each section:

Solution

1. The velocity at the smaller section is obtained from continuity Equation 5.14A:

$$\overline{W}_1 F_1 = \overline{W}_2 F_2$$

$$\overline{W}_2 = \overline{W}_1 \cdot \frac{F_1}{F_2}$$

$$F_1 = \frac{\pi}{4} D_1^2 = \frac{\pi}{4} (0.089 \text{ m})^2 = 6.221 \times 10^{-3} \text{ m}^2$$

$$F_2 = \frac{\pi}{4} D_2^2 = \frac{\pi}{4} (0.038 \text{ m})^2 = 1.134 \times 10^{-3} \text{ m}^2$$

Hence,

$$\overline{W}_2 = \left(1.2 \ \frac{m}{s}\right)\left(\frac{6.221 \times 10^{-3} \ m^2}{1.134 \times 10^{-3} \ m^2}\right) = 6.6 \ \frac{m}{s}$$

For the steady flow of a liquid, as the flow area is decreased the average velocity increases. This answer is independent of pressure and elevation changes.

2. To obtain the volumetric flowrate, we again use the principle of continuity, i.e., we can use the conditions either at section 1 or at section 2 to compute V (see Equation 5.14B):

$$V = F_1 \overline{W}_1 = (6.221 \times 10^{-3} \ m^2)(1.2 \ m/s) = 7.5 \times 10^{-3} \ \frac{m^3}{s}$$

or

$$V = F_2 \overline{W}_2 = (1.134 \times 10^{-3} \ m^2)(6.6 \ m/s) = 7.5 \times 10^{-3} \ \frac{m^3}{s}$$

3. The mass flowrate is simply:

$$W = V\gamma$$

$$= \left(7.5 \times 10^{-3} \ \frac{m^3}{s}\right)\left(850 \ \frac{kg}{m^3}\right) = 6.38 \ kg/s$$

4. The mass velocity is defined by Equation 5.10A:
In section 1,

$$G_1 = \overline{W}_1 \rho = \left(1.2 \ \frac{m}{s}\right)\left(850 \ \frac{kg}{m^3}\right) = 1020 \ \frac{kg}{m^2 \cdot s}$$

In section 2,

$$G_2 = \overline{W}_2 \rho = \left(6.6 \ \frac{m}{s}\right)\left(850 \ \frac{kg}{m^3}\right) = 5610 \ \frac{kg}{m^2 \cdot s}$$

We have invoked the assumption that the oil's specific gravity did not change as it passed from one section to the other.

Example 5.2

The exhaust portion of a ventilation system for a room consists of a rectangular withdrawal duct that is 12 in. X 4 in. The air is exhausted on the roof of the building through a circular stack with an inside diameter of 16 in. The mean temperature of the air in the room is 85°F at 14.7 psia, and the average velocity at the withdrawal duct is 1500 fpm. Neglecting temperature fluctuations in the exhaust ductwork and outside the building, determine the following:

1. the density of the air in the circular exhaust stack for a measured average velocity of 700 fpm; and
2. the mass flowrate of air exhausted.

Solution

1. The density of air at STP is approximately 0.0808 lb/ft^3. From the ideal gas law (Equation 3.7, Chapter 3), the density of air at 85°F is

$$\rho = \rho_0 \frac{T_0}{T} = \left(0.0808 \frac{lb}{ft^3}\right)\left(\frac{32 + 460°F}{85 + 460°F}\right) = 0.0729 \frac{lb}{ft^3}$$

According to the continuity equation for gas flow (Equation 5.12),

$$\overline{W}_1 \rho_1 F_1 = \overline{W}_2 \rho_2 F_2$$

$$\rho_2 = \rho_1 \left(\frac{F_1}{F_2}\right)\left(\frac{\overline{W}_1}{\overline{W}_2}\right)$$

$$F_1 = 12 \text{ in.} \times 4 \text{ in.} = 48 \text{ in}^2.$$

$$F_2 = \frac{\pi}{4}(16 \text{ in.})^2 = 201 \text{ in}^2.$$

$$\rho_2 = 0.0729 \frac{lb}{ft^3}\left(\frac{48 \text{ in}^2.}{201 \text{ in}^2.}\right)\left(\frac{1500 \text{ fpm}}{700 \text{ fpm}}\right) = 0.0373 \frac{lb}{ft^3}$$

2. The mass flowrate of air exhausted is

$$W = \gamma_1 F_1 \overline{W}_1 = \gamma_2 F_2 \overline{W}_2$$

$$W = 0.0729 \frac{lb}{ft^3} \times 48 \ in^2. \times 1500 \frac{ft}{min} \times \frac{ft^2}{144 \ in^2.} \times \frac{60 \ min}{hr}$$

$$W = 2187 \frac{lb}{hr}$$

REGIMES OF FLOW

Laminar versus Turbulent Flows

Two general regimes of flow describe the nature of fluid motion and the interactions that take place between fluid particles. These regimes are termed laminar and turbulent flows. The *laminar regime* occurs at relatively low fluid velocities. In this regime, the flow may be visualized as layers that slide over each other providing smooth flow patterns. No macroscopic mixing of fluid particles occurs in this regime. In the *turbulent regime*, fluid velocities are higher, and an unstable pattern within the bulk flow is observed in which eddies or small packets of fluid particles move at all angles to the axial line of normal total flow. However, a thin layer exists near the wall where the fluid motion is still laminar. This region is known as the *laminar boundary layer*.

Classical studies by Reynolds in 1883 showed that the transition from laminar to turbulent flow in tubes is a function of the fluid velocity, density and viscosity, and the tube diameter. The dependency of flow regime on these variables is summarized by a dimensionless group known as the *Reynolds number*:

$$Re = \frac{wD}{\nu} = \frac{wD\rho}{\mu} = \frac{WD}{\mu} \tag{5.19}$$

where ν = kinematic viscosity
        ρ = density
        μ = dynamic viscosity
        w = fluid velocity
        W = mass velocity
        D = tube diameter (or some characteristic size of the system through which flow occurs)

Equation 5.19 indicates that turbulent motion increases with tube diameter, fluid velocity and density, or with decreasing viscosity. The value of the Reynolds number corresponding to the transition from one regime to another is referred to as the *critical Reynolds number* (or, simply, the *critical value*).

For a straight circular pipe, when the Reynolds number is less than 2100 the flow is considered laminar. When the Reynolds number exceeds 4000 the flow is turbulent, except for some very special cases. The flow between these two values is referred to as the *transition region*, where the motion may be either laminar or turbulent.

Consider steady-state laminar flow of a fluid of constant density ρ through a tube of length L and radius R. In the analysis to follow, "end effects" shall be ignored, i.e., we will ignore the fact that at the tube's entrance and exit the flow will not be parallel to the tube axis. Visualize a portion of the flow as a cylindrical layer of length ℓ and radius r. The system is illustrated in Figure 5.2. The motion of the layer results from a difference in pressure forces P_1 and P_2 at both ends of the fluid cylinder:

$$P_1 - P_2 = (p_1 - p_2)\pi r^2 \tag{5.20}$$

where p_1 and p_2 are the hydrostatic pressures at sections 1 and 2 (Figure 5.2).

The motion of the fluid cylinder is resisted by a friction force

$$T = -\mu F \frac{dw_r}{dr} \tag{5.21}$$

where w_r = fluid velocity along the x-axis at a distance, r, from the tube centerline

F = $2\pi r\ell$ = the cylinder's outside surface area

μ = fluid viscosity

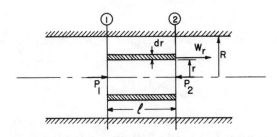

Figure 5.2   Plan view of fluid at laminar flow through a tube.

The minus sign in Equation 5.21 denotes that the fluid velocity decreases and radius r increases. That is, as r approaches the radius of the tube w_r becomes smaller, and at $r = R$, $w_r = 0$. The pressure gradient $P_1 - P_2$ represents the driving force for flow and must be sufficiently large to overcome frictional force T, i.e.,

$$(P_1 - P_2)\pi r^2 = -\mu 2\pi r \ell \frac{dw_r}{dr}$$

or

$$\frac{P_1 - P_2}{2\mu\ell} r\,dr = -dw_r \tag{5.22}$$

This differential equation may be integrated over the limits of $r = 0$ to R and $w = w_r$ to 0:

$$\int_r^R \frac{P_1 - P_2}{2\mu\ell} r\,dr = -\int_{w_r}^0 dw_r \tag{5.23}$$

Hence,

$$\frac{P_1 - P_2}{2\mu\ell}\left(\frac{R^2}{2} - \frac{r^2}{2}\right) = w_r$$

or

$$w_r = \frac{P_1 - P_2}{4\mu\ell}(R^2 - r^2) \tag{5.24A}$$

The maximum velocity occurs at $r = 0$ and, therefore, may be written as

$$w_{max} = \frac{P_1 - P_2}{4\mu\ell}R^2 \tag{5.24B}$$

Combining Equations 5.24A and 5.24B, we obtain

$$w_r = w_{max} \frac{R^2 - r^2}{R} = w_{max} \left(1 - \frac{r^2}{R^2}\right) \tag{5.25}$$

Equation 5.25 is known as *Stokes law for the parabolic velocity distribution* describing laminar flow in a pipe of constant cross section.

To evaluate the volumetric flowrate, examine Figure 5.2 in more detail. We will now consider only a small hollow cylinder of fluid having an inside radius r and outer radius (r + dr). The cross-sectional area of the annular fluid ring is dS = $2\pi r dr$ (examine both Figures 5.1 and 5.2). The volume rate passing through this section is

$$dV_{sec} = w_r dS = w_r 2\pi r dr \tag{5.26}$$

Combining Equations 5.26 and 5.24A, the following expression is derived:

$$dV_{sec} = \frac{P_1 - P_2}{4\mu\ell} \int_0^R (R^2 - r^2) 2\pi r dr$$

$$= \frac{P_1 - P_2}{4\mu\ell} \left(2\pi R^2 \int_0^R r dr - 2\pi \int_0^R r^3 dr\right) = \frac{P_1 - P_2}{8\mu\ell} \pi R^4 \tag{5.27A}$$

Denoting pipe diameter D = 2R and the pressure gradient $(p_1 - p_2) = \Delta P$, this expression can be rewritten as

$$V_{sec} = \frac{\pi D^4 \Delta P}{128 \, \mu\ell} \tag{5.27B}$$

Equations 5.27A and 5.27B express the volumetric rate for laminar flow in a straight circular pipe, and both are known as *Poiseuille's equation*.

The relationship between average velocity, $\overline{W}$, and maximum velocity, w_{max}, may be obtained by comparing the following expressions:

$$V_{sec} = wS = w\pi R^2 \quad \text{and} \quad \pi R^2 w = \frac{P_1 - P_2}{8\mu\ell} \pi R^4$$

Hence,

$$\overline{W} = \frac{P_1 - P_2}{8\mu\ell} R^2 \qquad (5.28)$$

Combining Equations 5.27B and 5.28, we obtain

$$\overline{W} = \frac{w_{max}}{2} \qquad (5.29)$$

Thus, the average velocity for laminar flow in a circular pipe is equal to half the maximum velocity at the pipe axis.

Similarly, the parabolic velocity distribution for pipe flow (Equation 5.25) may be rewritten as follows:

$$w_r = 2\overline{W}\left(1 - \frac{r^2}{R^2}\right) \qquad (5.30)$$

Further discussions are given by Knudsen and Katz [1958], and Curle and Davies [1968]. Before proceeding, let us apply some of the above expressions to problem solving.

Example 5.3

Water at 86°F is flowing through a 7.62-cm i.d. pipe at a velocity of 0.3 m/s. Calculate the Reynolds number in both English and SI units.

Solution

The density and viscosity of water at 86°F (30°C) are as follows:

- density, $\rho = 0.996$ (62.43 lb_m/ft^3) = 62.18 lb_m/ft^3
- viscosity, $\mu = 0.8007$ cp $\times$ (6.7197×10^{-4} lb_m/ft-s/cp) = 5.38×10^{-4} lb_m/ft-s

The Reynolds number is defined by Equation 5.19:

$$Re = \frac{wD\rho}{\mu}$$

Pipe diameter D = 7.62 cm $\times$ in./2.54 cm $\times$ ft/12 in. = 0.25 ft. Velocity in pipe $\overline{W} = 0.3$ m/s $\times$ 3.28 ft/m = 0.984 fps.

In English units,

$$Re = \frac{(0.984 \text{ ft/s})(0.25 \text{ ft})(62.18 \text{ lb}_m/\text{ft}^3)}{5.38 \times 10^{-4} \text{ lb}_m/\text{ft-s}}$$

$$= 28,440$$

In SI units,

$$Re = \frac{(0.3 \text{ m/s})(0.0762 \text{ m})(996 \text{ kg/m}^3)}{8.007 \times 10^{-4} \text{ kg/m-s}} = 28,440$$

Hence, the flow is turbulent.

Example 5.4

Consider the system shown in Figure 5.3. A liquid is flowing through a 20-m long, 25.4-mm i.d. tube that is inclined 15° from a horizontal datum plane. The pressure at point 1 is measured to be 27 psia and 30 psia at point 2. The viscosity and density of the liquid are 0.07 N-s/m^2 and 996 kg/m^3, respectively. Determine the following:

1. the direction of flow through the tube; and
2. the volumetric flowrate (assume the flow is laminar).

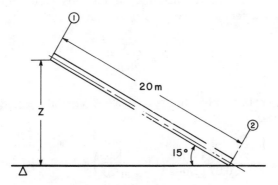

Figure 5.3   System under evaluation in Example 5.4.

Solution

1. To determine the direction of flow, we must determine the energy at points 1 and 2:
   At section 1,

$$P_1 = 27 \text{ psi} \times 6.985 = 186.2 \text{ kPa (or } 186{,}200 \text{ N/m}^2)$$

$$z = 20 \sin 15° = 5.18 \text{ m}$$

$$P_1 + \rho gz = 186{,}200 \text{ N/m}^2 + (996 \text{ kg/m}^3)(9.8 \text{ m/s}^2)(5.18 \text{ m})$$

$$= 236.8 \text{ kPa}$$

At section 2,

$$P_2 + \rho gz_0$$

$$= 30 \text{ psi} \times 6.985 + 0$$

$$= 206.9 \text{ kPa}$$

As the energy at point 1 is greater than at point 2, the flow is in the downward direction.

2. To obtain the volumetric flowrate, we first compute the differential pressure gradient per unit length of tube:

$$\frac{d}{d\ell}(P + \gamma z) = \frac{(236{,}800 - 206{,}900) \text{ N/m}^2}{20 \text{ m}} = 1495 \text{ N/m}^3$$

As the flow is laminar, we may use the Hagen-Poiseuille equation (Equation 5.27B) to determine the volumetric flowrate:

$$V = \frac{\pi D^4}{128 \, \mu} \frac{\Delta P}{\ell}$$

Pipe diameter D = 63.5 mm = 0.0635 m. Viscosity μ = 8.007 × 10^{-4} kg/m-s. Hence,

$$V = \frac{\pi (0.0254 \text{ m})^4}{128 \times 0.07 \text{ kg/m-s}} \left(1495 \frac{\text{kg}}{\text{m}^2\text{-s}^2}\right) = 2.182 \times 10^{-4} \text{ m}^3/\text{s}$$

As a check, we will compute the Reynolds number, Re = $D\overline{W}\rho/\mu$.

Area of pipe

$$F = \frac{\pi}{4} D^2 = \frac{\pi}{4} (0.0254 \text{ m})^2 = 5.067 \times 10^{-4} \text{ m}^2$$

Velocity

$$\overline{W} = \frac{V}{F} = \frac{2.182 \times 10^{-4} \text{ m}^3/s}{5.067 \times 10^{-4} \text{ m}^2} = 0.431 \text{ m/s}$$

$$Re = \frac{(0.0254 \text{ m})(0.431 \text{ m/s})(996 \text{ kg/m}^3)}{0.07 \text{ kg/m-s}} = 156$$

The flow is indeed laminar. If the Reynolds number had exceeded 2000, the Hagen-Poiseuille equation no longer would apply.

Turbulence

The phenomenon of turbulence is important to many branches of chemical engineering. Despite its far-reaching importance, no theoretically rigorous expressions have been developed to describe it. Often, in the development of turbulent flow expressions, so many assumptions are made that it is difficult to establish whether agreement with experimental observation results from the application of reasonable simplifications or from the fortuitous cancellation of errors.

In turbulent flow, chaotic motion of fluid particles causes a smoothing of velocity streamlines in the bulk flow. In pipe flow, this results in a velocity distribution that is very different from the parabolic profile obtained with laminar flows. A comparison of laminar and turbulent velocity profiles for pipe flow is shown in Figure 5.4. Note that the turbulent distribution does not show any significantly wider apex. Experiments have shown that unlike laminar flow, the average velocity is not equal to half the maximum. In fact, the average velocity is observed to be considerably greater and is a function of the Reynolds number, i.e., $w/w_{max} = f(Re)$. As a general rule, for turbulent Reynolds numbers up to 10,000 velocity $w \approx 0.8 \ w_{max}$ and, at $Re \geqslant 10^8$, $w \approx 0.9 \ w_{max}$.

Present understanding of turbulence does not enable velocity profiles to be computed a priori, as with laminar flow. The turbulent profile illustrated in Figure 5.4 does not represent an instantaneous distribution but rather one that is time-averaged. The instantaneous velocity at a point varies with

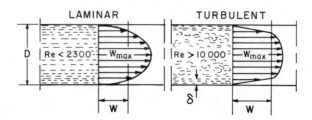

Figure 5.4   Velocity profiles for laminar and turbulent flows in a pipe.

time in both magnitude and direction. Therefore, it is an irregularly oscillat-ing function that is best illustrated by Figure 5.5. However, we can define a time-smoothed velocity, $\overline{W}_2$, in the x-direction by taking a time-average of w_x over some time interval, t. The time interval must be large with respect to the time of turbulent oscillation but small with respect to actual time changes:

$$\overline{W}_x = \frac{\int_0^\tau w_x dt}{t} \tag{5.31}$$

where w_x is the instantaneous velocity (a function of time t). w_x is the sum of the time-smoothed velocity, $\overline{W}_x$, and a velocity fluctuation, Δw:

$$w_x = \overline{W}_x \pm \Delta w \tag{5.32}$$

The value of the time-averaged velocity, $\overline{W}_x$, over a sufficiently large time interval is constant. As the frequency of the velocity fluctuation is extremely large, the time interval need be only a few seconds or even fractions of seconds. Consequently, we may consider a time-independent change of the velocities averaged over the cross section of piping, rather than the change of instantaneous values. That is, we shall consider local flow to be at quasi-stationary motion. Although the instantaneous fluctuations are small, they have a dramatic effect on the flow characteristics. For example, the trans-verse movement of a fluid particle from a faster-moving region to a slower-moving one has the effect of increasing the velocity in the slower region and thus acts as an equivalent shear stress. This turbulent shear stress may be hundreds of times greater than the laminar stress due to the sliding of one fluid layer over another.

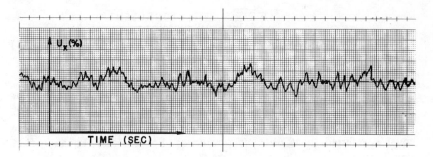

(A) OSCILLOGRAPH PRINTOUT FROM
HOT WIRE ANEMOMETER SHOWING VELOCITY
FLUCTUATIONS.

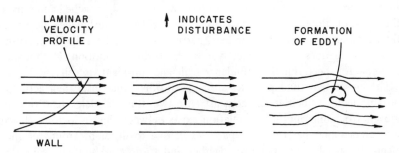

(B) DISTURBANCE INTRODUCED TO
FLUID CAUSING STREAMLINE INTERMIXING &
EDDY FORMATION.

Figure 5.5   Actual (instantaneous) and time-averaged velocities for turbulent flow.

To evaluate the importance of fluctuating velocities relative to their average, turbulence often is characterized by a parameter known as the *intensity of turbulence* (or intensity factor):

$$I_T = \frac{\overline{\Delta W}}{\overline{\overline{W}}}$$

(5.33)

where $\overline{\Delta W}$ is the root-average-square of fluctuating velocities, expressed as

$$I_T = \frac{\sqrt{\frac{1}{3}(\overline{\Delta W_x^2} + \overline{\Delta W_y^2} + \overline{\Delta W_z^2})}}{\overline{W}_x} \tag{5.34}$$

The intensity of turbulence *is an estimate of the fluctuations at a given point within the flow*. For turbulent flow in a piping, $I_T \approx 0.01-0.1$.

The special case in which the three average-square fluctuating velocities are equal is referred to as *isotropic turbulence*. For this case, Equation 5.34 simplifies to

$$I_T = \sqrt{\overline{\Delta W_x^2}} \tag{5.35}$$

In a practical sense, turbulent flows almost never reach isotropic conditions. In pipe flow, for example, the isotropic state is only approached and only at the axis of flow. Greater deviation from isotropic turbulence occurs in fluid regions removed from the axis, particularly near the wall. The intensity factor alone does not completely describe turbulent motion. Methods are available for specifying the turbulence scale and turbulent viscosity.

The closer two fluid particles are to each other, the closer their actual or instantaneous velocities. In contrast, for two fluid particles relatively removed from one another no relation exists between their velocities. Hence, we can assume that the fluid particles belong to different groups of fluid elements, which tend to move together and form some unified aggregates. These aggregates are called *eddies*. The size of eddies or the depth of their penetration before collapsing (which may be identified approximately by the distance between two close particles not belonging to the same eddy) depends on the rate of turbulence. This rate of turbulence is known as the *scale*. A correlation coefficient is conveniently defined as follows:

$$R(y) = \frac{\overline{\Delta W_{x_1} \Delta W_{x_2}}}{\sqrt{\overline{\Delta W_{x_1}^2}} \sqrt{\overline{\Delta W_{x_2}^2}}} \tag{5.36}$$

For flow in the x-direction, $\Delta \overline{W}_{x_1}$ and $\Delta \overline{W}_{x_2}$ are the fluctuating velocities occurring at the same time, t_1, at points 1 and 2, separated by distance y.

The turbulence scale is based on the correlation coefficient, $R(y)$, expressed as a function between two points:

$$L_T = \int_0^\infty R(y)\,dy \tag{5.37}$$

Note that the term "eddy" is not inclusive of turbulence. Eddy motion or currents also may exist in laminar flow because they are characterized simply by velocity differences over the flow area. The distinction between turbulent and laminar flows therefore is not denoted by eddy motion, but by the presence of chaotic fluctuations of velocities at different points throughout the flow area. These fluctuations produce particle displacements in a direction normal to the axis of flow.

To describe these chaotic fluctuations we need a parameter that relates the viscous forces of the fluid to its kinetic energy. In turbulent flow this parameter is the eddy viscosity. Let us consider two fluid particles flowing in the x-direction parallel to the pipe axis. Earlier, we defined y as the distance between these particles in the direction normal to the pipe axis. The particles' velocity components in the flow direction are $\overline{W}_{x_1}$ and $\overline{W}_{x_2}$, which differ from each other by $d\overline{W}_x$. Because of this difference, a shear stress results:

$$\tau_N = -\mu \frac{d\overline{W}_x}{dy} = -\rho\nu \frac{d\overline{W}_x}{dy} \qquad (5.38)$$

where μ and ν are the dynamic and kinematic viscosities, respectively. ρ is density, and subscript "N" denotes that we are dealing with a Newtonian fluid. Equation 5.38 is a general expression that applies only rigorously to laminar flow. In laminar flow, τ_N is the only stress existing between fluid layers located distance dy apart. In turbulent flow, however, fluid particles move relative to each other, not only in the longitudinal direction (i.e., together with the flow), but also in the transverse direction. This creates an additional shear stress, τ_T (subscript "T" refers to turbulent), which, by analogy with τ_N, may be expressed as follows:

$$\tau_T = -E_\nu \frac{d\overline{W}_x}{dy} \qquad (5.39)$$

Parameter E_ν in Equation 5.39 is called the *eddy viscosity* and is analogous to dynamic viscosity, μ. The quantity $\epsilon_m = E_\nu/\rho$ is the *eddy diffusivity of momentum*, which is analogous to kinematic viscosity. These two fluid properties (E_ν and ϵ_m) depend on the fluid velocity and geometry of the flow system. That is, they are functions of all factors that influence the nature of turbulence and the deviation of velocities. In particular, they are sensitive to location within the turbulent field, i.e., distance from the wall, etc. Thus, the total shear stress in a turbulent flow is the sum of viscous stresses and turbulent stresses:

$$\tau = \tau_N + \tau_T = -\rho(\nu + \nu_T)\frac{d\overline{W}_x}{dy} \qquad (5.40)$$

Examining the turbulent velocity profile again in Figure 5.4, we note that the time-averaged velocities comprise a relatively flat distribution in the bulk flow. It is only in the vicinity of the wall that velocities decrease rapidly and eventually become zero at the solid boundary. In the immediate vicinity of the wall, the fluid motion becomes less turbulent and more laminar. The solid boundary has the effect of dampening turbulent fluctuations in the transverse direction. Thus, turbulent motion never really exists in pure form; rather, it is accompanied by laminar motion.

Fluid motion is divided conditionally into two zones: (1) the central or *bulk flow*, where the motion is turbulent, and the *hydrodynamic boundary layer* (the zone near the wall where turbulent flow changes into laminar). Within the hydrodynamic boundary layer, a thin sublayer exists very close to the pipe wall. In this region viscous forces have a predominant influence on the fluid motion. The velocity gradient in the *laminar boundary sublayer* is very high. In turbulent flow the laminar sublayer is very thin (sometimes fractions of a millimeter in thickness), and decreases with increasing turbulence. Regardless of the thickness, the sublayer has a dramatic influence not only on the hydraulic resistance of fluid motion, but on the mechanisms of heat and mass transfer.

Between the bulk flow and the laminar sublayer exists a region known as the buffer zone. The laminar sublayer and buffer zone are termed the hydrodynamic boundary layer. The thickness of this layer is determined by shear stresses that exist between fluid particles caused by viscous forces and turbulent fluctuations. Hence, in this region the magnitudes of ν and ν_T become comparable. Further discussions are given by Knudsen and Katz [1958], Rohsenow and Hartnett [1973], and Perry [1950].

CONTINUITY EQUATION

We shall develop the general relationship among the velocities of flow through any system. Following the analysis presented in the previous chapter, Figure 5.6 shows a differential fluid element of volume dV = dxdydz. The velocity component along the x-axis is w_x. The input per unit time, dt, in the x-direction through the left face of the element (dS = dydz) is equal to

$$M_x = \rho w_x dydzdt \qquad (5.41)$$

where ρ = fluid density at the left face of the element

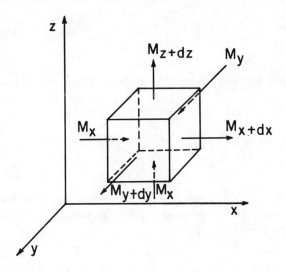

Figure 5.6   Fluid element showing mass transfer through defined boundaries.

The velocity and density at the parallel face (located at $x + dx$) are $(w_x + \partial(\rho w_x)dx/\partial x)$ and $(\rho + \partial\rho dx/\partial x)$, respectively. Hence, the output at the right face of the element in the same time interval, dt, is

$$M_{x+dx} = \left[\rho w_x + \frac{\partial(\rho w_x)}{\partial x} dx\right] dydzdt \qquad (5.42)$$

The mass differential in the element along the x-axis is

$$dM_x = M_x - M_{x+dx} = -\frac{\partial(\rho w_x)}{\partial x} dxdydzdt \qquad (5.43)$$

and the total mass differential in the elemental volume dV over time dt is

$$dM = -\left[\frac{\partial(\rho w_x)}{\partial x} + \frac{\partial(\rho w_y)}{\partial y} + \frac{\partial(\rho w_z)}{\partial z}\right] dxdydzdt \qquad (5.44)$$

For a mass differential to occur in the fluid element, density also must change with respect to time. Hence,

$$dM = \frac{\partial P}{\partial t} \, dxdydzdt \tag{5.45}$$

Equating these last two expressions through dM and rearranging terms, we obtain

$$\frac{\partial P}{\partial t} + \frac{\partial(\rho w_x)}{\partial x} + \frac{\partial(\rho w_y)}{\partial y} + \frac{\partial(\rho w_z)}{\partial z} = 0 \tag{5.46}$$

Equation 5.46 is the differential equation of continuity for unsteady flow of a compressible fluid. The equation may be written in another form by differentiating the product, ρw:

$$\frac{\partial \rho}{\partial t} + \frac{\partial \rho}{\partial x} w_x + \frac{\partial \rho}{\partial y} w_y + \frac{\partial \rho}{\partial z} w_z + \frac{\partial w_x}{\partial x} \rho + \frac{\partial w_y}{\partial y} \rho + \frac{\partial w_z}{\partial z} \rho = 0 \tag{5.47A}$$

or

$$\frac{1}{\rho} \frac{d\rho}{dt} + \frac{\partial w_x}{\partial x} + \frac{\partial w_y}{\partial y} + \frac{\partial w_z}{\partial z} = 0 \tag{5.47B}$$

where $\partial\rho/\partial t$ is a substantial derivative. In steady-state flows, density, ρ, does not change with time, i.e., $\partial\rho/\partial t = 0$, and Equation 5.46 simplifies to the following:

$$\frac{\partial(\rho w_x)}{\partial x} + \frac{\partial(\rho w_y)}{\partial y} + \frac{\partial(\rho w_z)}{\partial z} = 0 \tag{5.48}$$

For liquids, which are practically incompressible, and for gases under iso-thermal conditions and at velocities much less than velocity of sound, ρ = constant. Hence, Equation 5.48 is simply

$$\frac{\partial w_x}{\partial x} + \frac{\partial w_y}{\partial y} + \frac{\partial w_z}{\partial z} = 0 \tag{5.49}$$

Equation 5.49 is the differential continuity equation for the flow of incompressible fluids.

EQUATIONS OF MOTION FOR IDEAL FLUIDS

Following the analysis of Streeter and Wylie [1979], we shall consider steady flow of an ideal fluid having zero viscosity. As shown in Chapter 4, the projections of pressure and gravity forces onto the coordinate axes acting on the elemental volume $dV = dxdydz$ are the following:

For the x-axis,

$$-\frac{\partial P}{\partial x} dxdydz$$

For the y-axis,

$$-\frac{\partial P}{\partial y} dxdydz$$

For the z-axis,

$$-\left(\rho g + \frac{\partial P}{\partial z}\right) dxdydz$$

In accordance with the basic principle of dynamics, the sum of the projection of forces acting on the elemental volume in motion is equal to the product of mass and acceleration. Hence, for the x, y and z axis projections, we may write

$$\left.\begin{array}{c} \rho dxdydz \dfrac{dw_x}{dt} = -\dfrac{\partial P}{\partial x} dxdydz \\[2ex] \rho dxdydz \dfrac{dw_y}{dt} = -\dfrac{\partial P}{\partial y} dxdydz \\[2ex] \rho dxdydz \dfrac{dw_z}{dt} = -\dfrac{\partial P}{\partial z} dxdydz \end{array}\right\} \qquad (5.50)$$

And, after simplifying,

$$\left.\begin{array}{l} \rho\,\dfrac{dw_x}{dt} = -\dfrac{\partial P}{\partial x} \\[2em] \rho\,\dfrac{dw_y}{dt} = -\dfrac{\partial P}{\partial y} \\[2em] \rho\,\dfrac{dw_z}{dt} = -\rho g - \dfrac{\partial P}{\partial z} \end{array}\right\} \qquad (5.51)$$

The substantial derivatives of the velocity components are

$$\left.\begin{array}{l} \dfrac{dw_x}{dt} = \dfrac{\partial w_x}{\partial x}\,w_x + \dfrac{\partial w_x}{\partial y}\,w_y + \dfrac{\partial w_x}{\partial z}\,w_z \\[2em] \dfrac{dw_y}{dt} = \dfrac{\partial w_y}{\partial x}\,w_x + \dfrac{\partial w_y}{\partial y}\,w_y + \dfrac{\partial w_y}{\partial z}\,w_z \\[2em] \dfrac{dw_z}{dt} = \dfrac{\partial w_z}{\partial x}\,w_x + \dfrac{\partial w_z}{\partial y}\,w_y + \dfrac{\partial w_z}{\partial z}\,w_z \end{array}\right\} \qquad (5.52)$$

The above systems of Equations 5.51 and 5.52 are known as Euler's differential equations of motion for an ideal fluid under steady flow. In unsteady flow, velocity changes not only when fluid particles move from one point in space to another, but also with time. Therefore, the components of acceleration in Equation 5.51 must be expressed in the following form:

$$\left.\begin{array}{l} \dfrac{dw_x}{dt} = \dfrac{\partial w_x}{\partial t} + \dfrac{\partial w_x}{\partial x}\,w_x + \dfrac{\partial w_x}{\partial y}\,w_y + \dfrac{\partial w_x}{\partial z}\,w_z \\[2em] \dfrac{dw_y}{dt} = \dfrac{\partial w_y}{\partial t} + \dfrac{\partial w_y}{\partial x}\,w_x + \dfrac{\partial w_y}{\partial y}\,w_y + \dfrac{\partial w_y}{\partial z}\,w_z \\[2em] \dfrac{dw_z}{dt} = \dfrac{\partial w_z}{\partial t} + \dfrac{\partial w_z}{\partial x}\,w_x + \dfrac{\partial w_z}{\partial y}\,w_y + \dfrac{\partial w_z}{\partial z}\,w_z \end{array}\right\} \qquad (5.53)$$

The systems of Equations 5.51 and 5.53 represent Euler's differential equations of motion for unsteady flow.

DIFFERENTIAL EQUATIONS FOR VISCOUS FLUIDS

Navier-Stokes Equations

The motion of real or viscous fluids is influenced by frictional forces in addition to pressure and gravity. Frictional forces, τ, acting on a volume element $dV = dxdydz$ shown in Figure 5.7, result in shear stresses, τ, forming on the faces. For simplicity, first consider one-dimensional flow in the x-direction where the velocity projection, w_x, depends only on distance z above some horizontal reference plane. In this case, the shear stresses on the lower and upper faces of the volume element $dF = dxdy$ are τ and $\tau + (\partial\tau/\partial z)dz$, respectively. The partial derivative $\partial\tau/\partial z$ expresses the change in shear stress along the z-axis at points located on the lower face of the volume element, and $(\partial\tau/\partial z)dz$ is the change of shear stress along the total length of the element's edge, dz. The projection of the resultant shear forces in the x-direction is

$$\tau dxdy - \left(\tau + \frac{\partial\tau}{\partial z}dz\right)dxdy = -\frac{\partial\tau}{\partial z}dxdydz \qquad (5.54)$$

From Newton's law, $\tau = -\mu\ \partial w_x/\partial z$, we may rewrite this expression as follows:

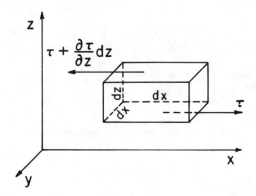

Figure 5.7   Forces acting on a volume element of moving fluid.

$$\mu \frac{\partial \left(\frac{\partial w_x}{\partial z}\right)}{\partial z} dxdydz = \mu \frac{\partial^2 w_x}{\partial z^2} dxdydz \tag{5.55}$$

For three-dimensional flow, component w_x changes along three coordinate axes, and the projection of the resultant shear forces in the x-direction is

$$\mu \left(\frac{\partial^2 w_x}{\partial z^2} + \frac{\partial^2 w_x}{\partial y^2} + \frac{\partial^2 w_z}{\partial x^2} \right) dxdydz$$

or

$$\frac{\partial^2 w_x}{\partial x^2} + \frac{\partial^2 w_x}{\partial y^2} + \frac{\partial^2 w_x}{\partial z^2} = \nabla^2 w_x$$

where ∇^2 is called the Laplacian operator and is defined as

$$\nabla^2 = \frac{\partial^2}{\partial x^2} + \frac{\partial^2}{\partial y^2} + \frac{\partial^2}{\partial z^2} \tag{5.56}$$

Using this notation, expressions can be written for the resultant projections of gravity, pressure and shear forces on respective coordinate axes:
    For the x-axis,

$$\left(-\frac{\partial P}{\partial x} + \mu \nabla^2 w_x \right) dxdydz$$

For the y-axis,

$$\left(-\frac{\partial P}{\partial x} + \mu \nabla^2 w_y \right) dxdydz$$

For the z-axis,

$$\left(-\rho g - \frac{\partial P}{\partial z} + \mu \nabla^2 w_z \right) dxdydz$$

Note that the gravity and pressure force projections are taken from Euler's equations.

From dynamics, the sum of the projection of forces equals the product of the fluid mass, $\rho dxdydz$, in the elemental volume and the projections of acceleration. Applying this principle and after simplification, we obtain the celebrated Navier-Stokes equations of motion for viscous incompressible fluids expressed in Cartesian coordinates:

$$\left. \begin{array}{l} \rho \dfrac{dw_x}{d\tau} = -\dfrac{\partial P}{\partial x} + \mu\nabla^2 w_x \\[2mm] \rho \dfrac{dw_y}{d\tau} = -\dfrac{\partial P}{\partial y} + \mu\nabla^2 w_y \\[2mm] \rho \dfrac{dw_z}{\partial \tau} = -\rho g - \dfrac{\partial P}{\partial z} + \mu\nabla^2 w_z \end{array} \right\} \qquad (5.57)$$

The Navier-Stokes equations for compressible fluids are as follows:

$$\left. \begin{array}{l} \rho \dfrac{dw_x}{d\tau} = -\dfrac{\partial P}{\partial x} + \mu\left(\nabla^2 w_x + \dfrac{1}{3}\dfrac{\partial \theta}{\partial x}\right) \\[3mm] \rho \dfrac{dw_y}{d\tau} = -\dfrac{\partial P}{\partial y} + \mu\left(\nabla^2 w_y + \dfrac{1}{3}\dfrac{\partial \theta}{\partial y}\right) \\[3mm] \rho \dfrac{dw_z}{\partial \tau} = -\rho g - \dfrac{\partial P}{\partial z} + \mu\left(\nabla^2 w_z + \dfrac{1}{3}\dfrac{\partial \theta}{\partial z}\right) \end{array} \right\} \qquad (5.58)$$

The partial derivatives $\partial\theta/\partial x$, $\partial\theta/\partial y$ and $\partial\theta/\partial z$ represent changes in velocities along the x-, y- and z-axes and are associated with compression and tension forces acting on the fluid.

The terms on the left-hand sides (LHS) of Equations 5.57 and 5.58 represent the rate of inertial forces for the moving fluid. Terms on the RHS describe three types of forces:

1. ρg denotes gravitational forces acting on the fluid volume element.
2. The partial derivatives $\partial P/\partial x$, $\partial P/\partial y$ and $\partial P/\partial z$ denote pressure forces.
3. The product of viscosity and the sum of the second derivatives of the velocity projections denote the viscous forces acting on the moving fluid element.

All terms in these two sets of equations have dimensions of force per unit volume. For ideal fluids, i.e., for $\mu = 0$, where there are no viscous forces, our expressions reduce to Euler's equations (Equations 5.52 and 5.53).

The Navier-Stokes equations represent a system of nonlinear partial expressions having no general solution. In fact, these equations are often of a higher order because viscosity can be a function of temperature, as well as velocity, for many flow systems. Two general approaches may be used in applying the Navier-Stokes equations to analyzing flow systems. The expressions may be reduced to specific solutions either by application of a series of simplifying assumptions or by transformation techniques based on the theory of similarity. The first approach is illustrated in the following example.

Example 5.5

Derive an expression for pressure drop for steady-state viscous flow in a horizontal tube of radius r. The flow is in one direction and is driven only by a constant-pressure differential. The derivation should be based on an incompressible fluid with constant viscosity.

Solution

Any model or expression developed from first principles will only have validity within the limitations of the constraints imposed and assumptions applied to its development. It is important that these assumptions be stated clearly along with the final expression to prevent improper application and, thus, erroneous results.

The constraints applied to our derivation are:

1. constant viscosity
2. incompressibility
3. one-directional flow
4. steady-state flow

For simplicity, we shall apply the assumption that the fluid is far from the tube's inlet and exit; hence, end effects will be ignored.

With the constraints and major assumptions established, we now must define a convenient coordinate system. As the flow is specified as being in one direction, the z-axis is assigned as the axis of symmetry of flow. Also, y shall denote the vertical direction and x the horizontal coordinate. The system under evaluation is represented by the sketch in Figure 5.8.

As w_x and w_y are zero (constraint 3), the continuity equation reduces to

$$\frac{\partial w_z}{\partial z} = 0$$

and, for steady-state flow (constraint 4),

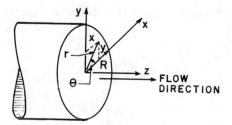

Figure 5.8   Flow system under evaluation in Example 5.5.

$$\frac{\partial w_z}{\partial t} = 0$$

Substituting these terms and assumptions into Equation 5.57 and noting the flow is horizontal, we obtain the following:

$$\frac{dP}{dz} = \mu \left(\frac{\partial^2 w_z}{\partial x^2} + \frac{\partial^2 w_z}{\partial y^2} \right)$$

Because of the geometry of this flow system, it is convenient to transform our starting expression into cylindrical coordinates:

$$z = z \quad y = r \sin\theta \quad \theta = \tan^{-1} y/x$$
$$x = r \cos\theta \quad r = \sqrt{x^2 + y^2}$$

Substituting these into our starting expression,

$$\frac{1}{\mu} \frac{dP}{dz} = \frac{\partial^2 w_z}{\partial r^2} + \frac{1}{r} \frac{\partial w_z}{\partial r} + \frac{1}{r^2} \frac{\partial^2 w_z}{\partial \theta^2}$$

As the z-axis has been assigned as the axis of symmetry of the flow,

$$\frac{\partial^2 w_z}{\partial \theta^2} = 0$$

As the flow is based on a constant pressure differential, $1/\mu\ dP/dz =$ constant C:

$$C = \frac{d^2 w_r}{dr^2} + \frac{1}{r}\frac{dw_r}{dr}$$

$$= \frac{1}{r}\frac{d}{dr}\left(r\frac{dw_r}{dr}\right)$$

The above expression now can be integrated twice to obtain an expression for the velocity profile. For the first integration, apply the boundary condition:

$$\frac{dw_z}{dr} = 0 \text{ at } r = 0$$

For the second integration, apply the boundary condition:

$$w_z = 0 \text{ at } r = R$$

where R is the tube radius. The resultant velocity profile is

$$w_z = \frac{1}{4\mu}\frac{dP}{dz}(r^2 - R^2)$$

This is the same as Equation 5.24A. Following the same development, we also can obtain an expression for the maximum velocity. $w_{z max}$ occurs at $r = 0$, Equation 5.24B. The above expression may be integrated over the tube cross section to obtain the average velocity:

$$\overline{W}_z = -\frac{R^2}{8\mu}\frac{dP}{dz}$$

And, finally, we can integrate this last expression over the limits of $z = 0$ for $p = p_1$ to $z = L$ for $p = p_2$, to obtain pressure drop:

$$p_1 - p_2 = \frac{8\mu\overline{W}_z L}{R^2}$$

This expression is the Hagen-Poiseuille equation derived earlier and, of course, is applicable only to laminar flow.

Additional illustrative examples are given by Bird et al. [1960], Curle and Davies [1968], Brodkey [1967], and Cheremisinoff [1979,1981a,b]. Problems at the end of this chapter provide additional instruction.

We will now use principles of similarity theory to transform the Navier-Stokes equations to forms suitable for flow analysis.

Transformation Techniques

Following the methods outlined by Sedov [1959] and Gukhman [1965], we will make certain transformations to the system of Equations 5.58, making use of the following characteristic parameters as measured units of the appropriate flow variables: ρ_0, P_0, μ_0, τ_0, w_0 and ℓ_0. This enables us to define the following scaling factors:

- $\rho/\rho_0 = P'$
- $P/P_0 = \mathcal{P}$
- $\mu/\mu_0 = M'$
- $\tau/\tau_0 = T'$
- $w/w_0 = \widetilde{W}$
- $w_x/w_{x_0} = W_x$
- $w_y/w_{y_0} = W_y$
- $w_z/w_{z_0} = W_z$
- $\ell/\ell_0 = x/x_0 = X'$
- $y/y_0 = Y'$
- $z/z_0 = Z'$

Rearranging Equation 5.58,

$$\rho\left[\frac{\partial w_x}{\partial t} + w_x\frac{\partial w_x}{\partial x} + w_y\frac{\partial w_y}{\partial y} + w_z\frac{\partial w_z}{\partial z}\right]$$

$$= \rho g - \frac{\partial P}{\partial x} + \mu\left(\nabla^2 w_x + \frac{1}{3}\frac{\partial\theta}{\partial x}\right) \qquad (5.59)$$

Then, substituting the above scaling factors for each variable, we obtain the following:

$$P\rho_0\left[\frac{\partial W_x}{\partial T'}\cdot\frac{w_0}{\tau_0} + W_x\frac{\partial W_x}{\partial x}\cdot\frac{w_0}{\ell_0} + w_0 W_y\frac{\partial W_y}{\partial y}\cdot\frac{w_0}{\ell_0} + w_0 W_z\frac{\partial W_z}{\partial z}\cdot\frac{w_0}{\ell_0}\right]$$

$$= P'\rho_0 g - \frac{\partial\mathcal{P}}{\partial x}\frac{P_0}{\ell_0} + \mu M'\left(\nabla^2 W_x\frac{w_0}{\ell_0^2} + \frac{1}{3}\frac{\partial\theta}{\partial x}\cdot\frac{w_0}{\ell_0^2}\right) \qquad (5.60)$$

where

$$\theta = \frac{\partial W_x}{\partial x} + \frac{\partial W_y}{\partial y} + \frac{\partial W_z}{\partial z} = \text{div } W$$

and

$$\nabla^2 W_x = \frac{\partial^2 W_x}{\partial x^2} + \frac{\partial^2 W_y}{\partial y^2} + \frac{\partial^2 W_z}{\partial z^2}$$

As ρ_0, w_0 and ℓ_0 are constants, they may be combined into a single coefficient:

$$\frac{\rho_0 W_0^2}{\ell_0}$$

To do this, w_0^2/ℓ_0 must be taken outside of the brackets on the LHS, while w_0/ℓ_0^2 is removed to the outside of the brackets on the last term in the RHS of Equation 5.60. Our expression then becomes

$$\left(\frac{\rho_0 w_0^2}{\ell_0}\right) P'\left[\frac{\partial W_x}{\partial T'}\left(\frac{\ell_0}{w_0 \tau_0}\right) + W_x \frac{\partial W_x}{\partial x} + W_y \frac{\partial W_x}{\partial y} + W_z \frac{\partial W_z}{\partial z}\right]$$

$$= (\rho_0 g) P' - \left(\frac{P_0}{\ell_0}\right)\frac{\partial \mathscr{P}}{\partial x} + \frac{\mu_0 w_0}{\ell_0^2} M'\left[\nabla^2 W_x + \frac{1}{3}\frac{\partial \theta}{\partial x}\right] \qquad (5.61)$$

Dividing both sides of the equation by $\rho_0 w_0^2/\ell_0$, we obtain

$$P\left[\frac{\partial W_x}{\partial T'}\left(\frac{\ell_0}{w_0 \tau_0}\right) + W_x \frac{\partial W_x}{\partial x} + W_y \frac{\partial W_x}{\partial y} + W_z \frac{\partial W_x}{\partial x}\right]$$

$$= \frac{g_0 \ell_0}{w_0^2} P' - \left(\frac{P_0}{\rho_0 w_0^2}\right)\frac{\partial \mathscr{P}}{\partial x} + \left(\frac{\mu_0}{\ell_0 w_0 \rho_0}\right) M'\left[\nabla^2 W_x + \frac{1}{3}\frac{\partial \theta}{\partial x}\right] \qquad (5.62A)$$

Equation 5.62A contains only similarity invariants (i.e., characteristic parameters), simplexes and complexes. These characteristic parameters now may be expressed in terms of recognizable dimensionless groups.

The following dimensionless groups are recognized in Equation 5.62(A):

$$\frac{w\tau}{\ell} = H_o = \text{Homochronity number}$$

$$\frac{w^2}{g\ell} = Fr = \text{Froude number}$$

$$\frac{\ell w\rho}{\mu} = Re = \text{Reynolds number}$$

$$\frac{P}{\rho w^2} = Eu = \text{Euler number}$$

Therefore, we may rewrite Equation 5.62A in the notation of identified dimensionless groups:

$$P\left[\frac{\partial w_x}{\partial T} \cdot \frac{1}{H_o} + w_x \frac{\partial w_x}{\partial x} + w_y \frac{\partial w_y}{\partial y} + w_z \frac{\partial w_z}{\partial z}\right]$$

$$= \frac{1}{Fr_0} P - Eu_0 \frac{\partial P}{\partial x} + \frac{1}{Re_0} M'\left[\nabla^2 W_x + \frac{1}{3}\frac{\partial \theta}{\partial x}\right] \qquad (5.62B)$$

Hence, in applying the Navier-Stokes equations to analyzing a flow system, the equations may be reduced to a workable solution by properly defining the physics of the system and, consequently, neglecting the least important dimensionless groups. For example, if gravity forces are unimportant (as in Example 5.5), then the Froude number may be neglected and the conditions of identity of a system of Navier-Stokes equations are determined strictly by the Homochronity, Reynolds and Euler numbers.

For dynamically possible flows at the condition of similarity of uniqueness, the identification of the Homochronity and Reynolds numbers is necessary to the existence of kinematic and dynamic similarity. It follows that the Euler number is not a premise but a consequence of similarity existence. Therefore, the Euler number should be considered as a nondetermining dimensionless group. Dimensionless groups of this type are unique functions of other determining groups. For this reason, the Navier-Stokes equations may be represented in the following dimensionless form:

$$Eu = \phi(H_o, Fr, Re) \qquad (5.63)$$

As noted earlier, the equations of motion alone are insufficient to describe a flow system. A second equation needed to characterize a flow system is the continuity equation (Equation 5.49). Continuity is dependent only on the Homochronity number:

$$\frac{1}{H_o}\frac{\partial P}{\partial T'} + \frac{\partial(\rho w_x)}{\partial x} + \frac{\partial(\rho w_y)}{\partial y} + \frac{\partial(\rho w_z)}{\partial z} = 0 \tag{5.64}$$

At steady-state, H_o is eliminated and similarity in the absence of viscous forces (i.e., the Froude number) is provided by identifying the Reynolds number. The Euler number, in this case, is a unique function of the Reynolds number whence Equation 5.63 reduces to

$$Eu = \phi(Re) \tag{5.65}$$

This last statement reflects only the kinematic and dynamic similarities of the flow. For total similarity it also is necessary to obtain geometric similarity. For flow in round pipes, the simplex of geometric similarity is the ratio of ℓ/d. Denoting this simplex number as $\widetilde{\Gamma} = \ell/d$, it is included in Equation 5.65 to give the following:

$$Eu = \phi(Re, \widetilde{\Gamma}) \tag{5.66}$$

THE TOTAL ENERGY BALANCE

Often the equations of motion and continuity are insufficient in defining a flow system. Another equation usually is required to solve for the flow system parameters of interest. For many flow systems this third equation is the energy equation. Consider a steady continuous-flow system comprising a pump and a heat exchanger, as shown in Figure 5.9. Because there is no accumulation of energy in this system, the overall energy balance is developed in the following manner.

The energy input to the system is the sum of the kinetic, E_{k_1}, potential, E_{p_1}, volumetric, E_{v_1}, and internal, E_{i_1}, energies at section 1; the heat, $\dot{Q}$, added through the exchanger; and the mechanical work, W', performed on the fluid by the pump. The energy output comprises kinetic, E_{k_2}, potential, E_{p_2}, volumetric, E_{v_2}, and internal, E_{i_2}, energies at section 2. Hence, we may equate the energy input to the system output to develop the energy balance for Figure 5.9.

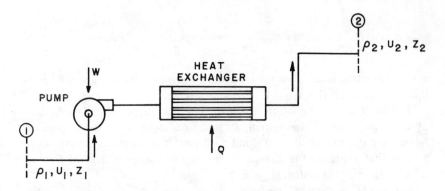

Figure 5.9   Flow system consisting of pump and heat exchanger.

$$E_{k_1} + E_{p_1} + E_{v_1} + E_1 + \dot{Q} + W' = E_{k_2} + E_{p_2} + E_{v_2} + E_2 \qquad (5.67)$$

For a basis of calculation, we will apply Equation 5.67 to a fluid mass of 1 kg of weight. (Note in the gravitational unit system that a kilogram is a unit of force.) Now consider each of the energy components in Equation 5.67.

Potential energy is the product of weight (1 kg) and the fluid's elevation above some specified datum plane. That is,

$$E_{p_1} = Z_1 \; ; \quad E_{p_2} = Z_2 \qquad (5.68)$$

where Z_1 and Z_2 are elevations at sections 1 and 2 relative to some reference plane.

The volumetric energy under pressure p is equivalent to the work expended to form volume v' at this pressure. The volumetric energies of 1 kg of fluid at the two sections are

$$E_{v_1} = P_1 v_1' \; ; \quad E_{v_2} = P_2 v_2' \qquad (5.69)$$

Kinetic energy is the product of mass and one-half the square of the linear velocity of the fluid. In this case, the weight of the fluid, i.e., the product of fluid mass and its acceleration due to gravity (g = 9.81 m/s^2), is equal to one kilogram-force. The mass of the fluid expressed in gravitational units is $1/g \times$ kg $\times$ s^2/m^4. If the fluid flows with velocity, w, then its kinetic energy is $w^2/2g$ and, for sections 1 and 2, we obtain

$$E_{k_1} = \frac{\overline{w}_1^2}{2g\alpha_1} \; ; \quad E_{k_2} = \frac{\overline{w}_2^2}{2g\alpha_2} \tag{5.70}$$

α is a correction coefficient for inaccuracies in measuring the average velocities. For turbulent flows, $\alpha = 1$.

The fourth term in Equation 5.67 is the internal energy, which is a thermodynamic property of the system. It is defined relative to some specified reference state, usually at $t = 0°C$ and $p = 1$ atm. It is convenient to introduce a value that expresses the change of the internal energy of 1 kg of fluid as it passes through the system $(E_2 - E_1)$.

Denoting W' as the work performed in pumping 1 kg of fluid and $\dot{Q}$ as the heat added in units of work, Equation 5.67 may be rewritten as follows:

$$P_1 \nu_1' + \frac{\overline{w}_1^2}{2g\alpha_1} + Z_1 + \dot{Q} + W' = P_2 \nu_2' + \frac{\overline{w}_2^2}{2g\alpha_2} + Z_2 + (E_2 - E_1) \tag{5.71}$$

Equation 5.71 is known as the law of conservation of energy and may be applied both to compressible and incompressible flows. Each term has units of length, m, or, more precisely, units of energy per unit weight (kg $\times$ m/kg = m).

For engineering calculations, the internal energy terms may be approximated by enthalpy. For this system, enthalpy must be expressed as follows:

$$i_1 = E_1 + P_1 \nu_1' \; ; \quad i_2 = E_2 + P_2 \nu_2' \tag{5.72}$$

Substituting Equation 5.72 into Equation 5.71, we obtain

$$\frac{\overline{w}_1^2}{2g\alpha_1} + Z_1 + \dot{Q} + W' = \frac{\overline{w}_2^2}{2g\alpha_2} + Z_2 + (i_2 - i_1) \tag{5.73}$$

For ideal gases, the change in enthalpy may be determined from the product of heat capacity at constant pressure and the system's temperature differential, $C_p(t_2 - t_1)$. The remainder of this section will concentrate on compressible flows.

Discharge of Gases

The discharge of gases through orifices or nozzles may be approximated as frictionless adiabatic flow. The reasons for this are as follows: (1) friction losses are minor because of the short distances traveled; and (2) heat transfer is negligible ($\dot{Q} = 0$) because the changes the gas undergoes are slow enough to keep velocity and temperature gradients small. Liepmann and Roshko [1957] provide evidence for these statements.

Applying this approximation, the energy balance expression (Equation 5.67) can be simplified for these flow systems. We shall consider the frequently encountered system of gas discharge from a tank. Due to the relatively low viscosities of gases, flows often are very turbulent and, hence, coefficient α in Equation 5.73 is unity. The height of flow before and after discharge is the same ($Z_1 = Z_2$), and no mechancial work is added or removed from such systems (i.e., $W' = 0$). Thus, Equation 5.73 may be rewritten as follows:

$$i_1 - i_2 = \frac{\overline{w}_2^2}{2g} - \frac{\overline{w}_1^2}{2g} \tag{5.74}$$

where i_1 and i_2 are the enthalpies of 1 kg of gas before and after the expansion, respectively. Thus, $\overline{w}_1$ and $\overline{w}_2$ are the average gas velocities before and after expansion.

As shown by Equation 5.74, the decrease in enthalpy is used to increase the kinetic energy of the system. The linear velocity in a tank, for example before discharge, is insignificant ($w_1 \approx 0$) compared to the velocity at the point of discharge. Therefore, from Equation 5.74,

$$w_2 = \sqrt{2g(i_1 - i_2)} \tag{5.75}$$

And noting that $i_1 - i_2 = C_p(T_1 - T_2)$,

$$w_2 = \sqrt{C_p(T_1 - T_2)2g} \tag{5.76}$$

For reversible adiabatic expansion of an ideal gas, Streeter and Wylie [1979] note the following expression from thermodynamics:

$$\frac{T_1}{T_2} = \left(\frac{P_1}{P_2}\right)^{\frac{\kappa-1}{\kappa}} \tag{5.77}$$

where κ is the specific heat ratio: $\kappa = C_p/C_v$ (C_p denotes heat capacity at constant pressure and C_v at constant volume).

From the ideal gas law ($P_1 \nu_1' = RT_1$) and the thermodynamic relationship $C_p - C_v = R$, an expression for the velocity of reversible adiabatic discharge of an ideal gas is

$$w_2 = \sqrt{2g \frac{\kappa}{\kappa - 1} (P_1 \nu_1') \left[1 - \left(\frac{P_2}{P_1}\right)^{\frac{\kappa-1}{\kappa}} \right]} \qquad (5.78)$$

The mass gas rate per unit discharge area may be expressed in terms of the discharge velocity:

$$G = w_2 \gamma_2 = \frac{w_2}{\nu_2'} \qquad (5.79)$$

where γ_2 is the specific weight of gas, i.e., the inverse of specific volume. Combining Equations 5.79 and 5.78 and including the expression for a reversible adiabatic fluid ($P_1 \nu_1^\kappa = P_2 \nu_2^\kappa$), the following equation is derived for mass rate in terms of the thermodynamic properties of the fluid:

$$G = \sqrt{2g \frac{\kappa}{\kappa - 1} \frac{P_1}{\nu_1'} \left[\left(\frac{P_2}{P_1}\right)^{2/\kappa} - \left(\frac{P_2}{P_1}\right)^{\frac{\kappa-1}{\kappa}} \right]} \qquad (5.80)$$

Equation 5.80 has a maximum with respect to P_2 or P_2/P_1. Differentiating the expression and setting the derivative equal to zero, the pressure at which the maximum flowrate occurs is obtained. This pressure is referred to as the "critical" pressure:

$$P_{cr} = P_1 \left(\frac{2}{\kappa + 1}\right)^{\frac{\kappa}{\kappa-1}} \qquad (5.81)$$

Because κ for gases does not change appreciably, it may be assumed that the critical pressure is in the range of $0.53\ P_1$ to $0.58\ P_1$, i.e., the critical pressure is approximately one-half the tank pressure.

Knowledge of the critical pressure is important for evaluating the efficiency of the flow process. If, for example, the pressure at the exit is higher than the critical value computed from Equation 5.81, then the flowrate in the orifice will not reach its maximum value. At complete expansion the gas velocity may be computed from Equation 5.78. If the exit pressure of the orifice is less than the critical value (because the maximum flowrate was exceeded), the amount of discharge must reach a maximum value, that is, the critical pressure will be achieved. Thus, further gas expansion will occur downstream of the orifice. The flow will expand, and, consequently, its head will decrease. Regardless, at $P_2 < P_{cr}$ the gas velocity will not correspond to pressure P_2, and Equation 5.78 should not be used. The discharge velocity will reach a lower value corresponding to P_{cr}. This critical velocity is determined by replacing P_2 in Equation 5.78 with P_{cr}. Taking into account

$$P_1 \nu_1'^\kappa = P_{cr} \nu_{cr}'^\kappa \tag{5.82}$$

the critical discharge velocity corresponding to the critical pressure is

$$w_{cr} = \sqrt{g\kappa P_{cr}\nu_{cr}'} \tag{5.83}$$

where critical volume, ν_{cr}', is obtained from Equation 5.82.

Equation 5.83 is an expression for the "sound velocity." Therefore, the maximum linear discharge velocity from a tank orifice is equal to the sound velocity, which is related to a corresponding temperature and pressure at the exit. This temperature may be computed from Equation 5.77 by replacing P_2 and T_2 by P_{cr} and T_{cr}, respectively:

$$T_{cr} = T_1\left(\frac{2}{\kappa + 1}\right) \tag{5.84}$$

At any point in the expanding flow the weight velocity is constant and flowrate changes according to Equation 5.80. Thus, the cross section of the gas flow

$$F = \frac{W}{G} \tag{5.85}$$

also will change. This is illustrated by the plot shown in Figure 5.10. Note that the velocity curve may be computed from Equation 5.78, the flowrate curve from Equation 5.80 and the section of flow from Equations 5.80 and 5.85.

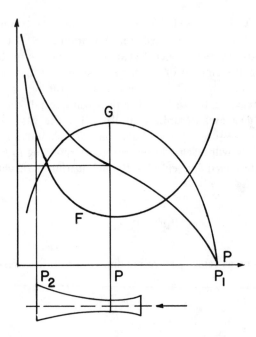

Figure 5.10   Plots of flow area and gas mass rate versus pressure for flow through a nozzle.

It is possible to design a nozzle with variable cross section for a specified weight flowrate using Equations 5.80 and 5.85. The critical pressure and velocity will be attained at the narrowest section of the nozzle. In the expanding section, pressure decreases and velocity increases and, according to Equation 5.78, the flow eventually reaches supersonic conditions. The use of a nozzle of variable cross section is logical when the counterpressure (downstream) is less than critical pressure, P_{cr}. If the downstream pressure exceeds P_{cr}, the effect is still the same because the orifice will have a diameter equal to the largest size of the nozzle's diameter.

Thus far we have considered thermodynamically reversible gas discharge through an orifice or nozzle, i.e., neglecting friction. However, friction can play an important role during discharge in some applications. Friction associated with the kinetic energy of the gas is converted into heat, thus increasing the enthalpy of the discharge. Therefore, enthalpy will be less than for the case of reversible flow at the same pressure and counterpressure. Furthermore, Equation 5.75 shows that the discharge gas velocity will decrease. To compute the discharge velocity for this case without considering counterpressure, the temperature of the discharging gas is needed. Knowing the exit temperature, T_2, and counterpressure, P_2, we find the enthalpy, i_2, and the discharge gas velocity is obtained from Equation 5.75. The following example illustrates the use of these equations.

Example 5.6

A reaction takes place in a nitrogen atmosphere in an autoclave at t = 130°C and p = 10 atm. The volume of the gas is 100 liters. After 10 minutes it was discovered that the pressure dropped to 9.7 atm because of leakage. The autoclave was found to be rated for pressure only to 6 atm. Determine how long it will take the vessel's pressure to drop to 6 atm and the rate at which nitrogen gas is escaping.

Solution

We will assume that the gas discharge takes place through a narrow slit and is adiabatic and reversible. First, determine the pressure in the autoclave. For nitrogen, $C_v = 5$ cal/mol and $C_p = C_v + R = 7$ cal/mol. Hence,

$$\kappa = C_p/C_v = 7/5 = 1.4$$

The critical pressure is

$$P_{cr} = P_1\left(\frac{2}{\kappa+1}\right)^{\frac{\kappa}{\kappa-1}} = P_1\left(\frac{2}{1.4+1}\right)^{\frac{1.4}{1.4-1}} = 0.53P_1$$

We note that the critical pressure always will be greater than the surroundings to which nitrogen is expanding. The gas discharge will occur at the sound velocity, which may be computed from Equation 5.83.

The absolute temperature in the autoclave is $T_1 = 273 + 130 = 403°K$, and, from Equation 5.84, we obtain

$$T_{cr} = 403\left(\frac{2}{1.4+1}\right) = 337°K$$

The specific volume of nitrogen at T_{cr} and P_{cr} is as follows (molecular weight of nitrogen is 28 and the volume of 1 kg-mol is 22.4 m^3):

$$v'_{cr} = \frac{22.4}{28} \times \frac{337}{273} \times \frac{10^4}{0.53 P_1} = \frac{18650}{P_1} \, m^3/kg$$

where $10^4 \, kg/m^2$ is the atmospheric pressure. Pressure P_1 is expressed in units of kg/m^2. As noted above, the discharge velocity through the slit may be computed from Equation 5.83:

$$w_{cr} = \sqrt{9.81 \times 1.4 \times 0.53 P_1 \times \frac{18650}{P_1}} = 368 \, m/s$$

The nitrogen weight rate at discharge is, therefore,

$$G = \frac{W_{cr}}{v'_{cr}} = \frac{368}{18650} P_1 = 0.0197 P_1 \, \frac{kg}{m^2 \text{-} s}$$

When the pressure is P_1, the specific volume of the nitrogen is as follows:

$$v' = \frac{22.4}{28} \times \frac{403}{273} \times \frac{10^4}{P} = \frac{11850}{P}$$

Because the volume of the autoclave is 100 ℓ ($0.1 \, m^3$), the weight of nitrogen is

$$W = \frac{0.1 P_1}{11850} = 8.42 \times 10^{-6} \, P$$

The nitrogen loss accompanied by the decrease dp is

$$dW = -8.42 \times 10^{-6} \, dp$$

The leakage occurs when the pressure decreases by some amount, dp, over a time interval, dt, through a slit having cross-sectional area, F. We can express the discharge by a rate expression:

$$dW = GFdt$$

or

$$-8.42 \times 10^{-6}\, dp = 0.0197\, pFdt$$

Integrating this expression over the pressure limits of 10 atm to P_1, the time over which the pressure drop occurs is obtained:

$$t = -\frac{4.27 \times 10^{-4}}{F} \int_{10}^{P_1} \frac{dP}{dt} = 9.48\, \frac{10^{-4}}{F} \log \frac{10}{P_1}$$

where P is expressed in units of absolute atmospheres and t in seconds.

From the problem statement we know that for t = 10 min = 600 s the pressure in the autoclave is 9.7 atm. From this we may compute the area of discharge, F, in mm^2:

$$F = \left(\frac{9.84 \times 10^{-4}}{600} \log \frac{10}{9.7}\right) 10^6 = 0.0213\, mm^2$$

Hence, the time it takes for the pressure to drop to 6 atm is

$$t = \frac{9.84 \times 10^{-4}}{0.0213 \times 10^{-6}} \log \frac{10}{6} = 10250\, s$$

Thus, the allowable operating time for the autoclave is about 3 hours. Calculation of the actual weight loss is left to the student.

Gas Flow Through Piping

The flow of gases in piping is more complex than that of liquids, mainly because of the dependency of specific weight on pressure changes. Because gases undergo different thermodynamic changes with pressure, specific gravity changes differently. First, we shall consider isothermal flow, i.e., flow at constant temperature. For this case we need to evaluate the amount of heat to be supplied to the flow to maintain a constant gas temperature. We note that the gas pressure decreases because of frictional resistances in the flow direction; hence, the gas specific volume will increase along with the linear velocity and kinetic energy of the gas.

Consider an ideal gas flowing through a section of horizontal piping of constant cross section. The energy balance (Equation 5.73) may be used to describe any two sections. If the flow is isothermal, then $i_1 = i_2$ and $Z_1 = Z_2$, and the energy balance simplifies to the following:

$$\frac{w_2^2}{2g} - \frac{w_1^2}{2g} = \dot{Q} \qquad (5.86)$$

Equation 5.86 implies that any increase in kinetic energy results from the introduction of heat from the outside. The flowrate can be expressed by

$$w_1 \gamma_1 = w_2 \gamma_2 \qquad (5.87)$$

and the ratio of these velocities for an isothermal system becomes

$$\frac{w_1}{w_2} = \frac{P_2}{P_1} \qquad (5.88)$$

Combining Equations 5.86 and 5.88 we obtain

$$\frac{w_1}{2g}\left[\left(\frac{P_1}{P_2}\right)^2 - 1\right] = \dot{Q} \qquad (5.89)$$

The above expression may be used to compute the amount of heat introduced to the system, provided we know the velocity at one section and the pressure ratio at the inlet and exit of the flow portion under evaluation.

To determine pressure drop, a differential form of the Bernoulli equation may be used (the Bernoulli equation is discussed in detail in the following section):

$$v' dp + \frac{dw^2}{2g} + dZ = 0 \qquad (5.90)$$

The change in head, dZ, may be obtained from the Darcy-Weisbach equation:

$$dZ = \lambda \frac{dL}{D} \times \frac{w^2}{2g} \tag{5.91}$$

where λ is a flow resistance coefficient.

For a small pressure drop, the specific gas volume may be assumed to be the average over the flow portion:

$$\nu'_{avg} = \frac{\nu'_1 + \nu'_2}{2} = \frac{RT}{2}\left(\frac{1}{P_1} + \frac{1}{P_2}\right) \tag{5.92A}$$

Similarly, average values for velocity and the resistance coefficient may be defined:

$$w_{avg} = \frac{w_1 + w_2}{2} \tag{5.92B}$$

$$\lambda_{avg} = \frac{\lambda_1 + \lambda_2}{2} \tag{5.92C}$$

Combining the Bernoulli and Darcy-Weisbach equations and including the definitions of Equations 5.92A–5.92C, we obtain the following expression for pressure drop:

$$\nu'_{avg}(P_1 - P_2) = \frac{w_2^2 - w_1^2}{2g} + \lambda_{avg}\frac{L}{D}\frac{w_{avg}^2}{2g} \tag{5.93}$$

The initial velocity, w_1, pressure, P_1 and temperature, T_1, usually are known. This information may be used to compute the specific volume, ν'_1, and resistance coefficient, λ_1. However, values downstream often are not known but may be solved for by trial and error using the following procedure: Assume a value for P_2 whence specific volume, ν'_2, velocity, w_2 (Equation 5.88) and resistance coefficient, λ_2, can be computed. Average values ν'_{avg}, w_{avg} and λ_{avg} now may be calculated. If Equation 5.93 is satisfied, then the initial guess for P_2 was correct. If the equation is not satisfied, then a new value for P_2 must be chosen and the computation repeated until convergence is achieved.

The above discussion was based on the assumption of small pressure drop. If, however, a large pressure drop is expected, the above expressions should be applied over a section of piping of length dL:

$$\nu' dp + \frac{wdw}{g} + \lambda \frac{dL}{D} \frac{w^2}{2g} = 0 \qquad (5.94)$$

where

$$\frac{wdw}{g} = d\left(\frac{w^2}{2g}\right)$$

Dividing Equation 5.94 by ν'^2 and noting that $G = W/\nu'$ and $P\nu' = RT$, we obtain the following expression after integration:

$$-\frac{1}{R} \int_{P_1}^{P_2} \frac{pdp}{T} = \frac{G^2}{g} \int_{\nu_1'}^{\nu_2'} \frac{d\nu'}{\nu'} + \frac{G^2}{2gD} \int_0^L \lambda dL \qquad (5.95)$$

The resistance coefficient, λ, for isothermal gas flow through a constant cross section must be constant because it is a function of the Reynolds number ($Re = DG/\mu g$), which is also constant for a specified flowrate.

On integration of Equation 5.95 we obtain

$$\frac{P_1^2 - P_2^2}{2RT} = \frac{G^2}{g} ln \frac{\nu_2'}{\nu_1'} + \frac{\lambda L G^2}{2gD} \qquad (5.96)$$

Equation 5.96 permits us to determine the pressure drop, $P_1 - P_2$, along a piping section of length L by a method of successive approximations. Let us assume that specific volume, ν_2', depends on unknown pressure, P_2. Denoting $\gamma = (P_1 + P_2)/2RT$ as a mean arithmetic value of ν_1' and ν_2', Equation 5.96 may be simplified to the following form:

$$P_1 - P_2 = \frac{G^2}{g\gamma} ln \frac{\nu_2'}{\nu_1'} + \frac{\lambda G^2 L}{2gD\gamma} \qquad (5.97)$$

This last expression is also applicable to nonisothermal flows. Now it is convenient to use the average specific volume from Equation 5.92A rather than the average specific weight in the Bernoulli equation:

$$P_1 - P_2 = \frac{G^2}{g}(v_1' - v_2') + \frac{\lambda G^2 L}{2gD\gamma} \qquad (5.98)$$

Solution of Equations 5.98, 5.97 and 5.96 can be obtained by the method of successive approximations.

We now will develop an expression for the maximum isothermal flowrate. Rearranging Equation 5.97 to solve for G,

$$G = \sqrt{\frac{(P_1^2 - P_2^2)g}{2RT\left(ln\frac{P_1}{P_2} - \frac{\lambda L}{D}\right)}} \qquad (5.99)$$

If P_2 changes, the flowrate maximizes at some critical value of pressure P_{cr}. The new critical pressure may be determined by differentiating Equation 5.99 with respect to P_2 and setting the derivative equal to zero:

$$ln\frac{P_1}{P_{cr}} + \frac{\lambda L}{D} = \frac{P_1^2 - P_{cr}^2}{2P_{cr}^2} \qquad (5.100)$$

The critical pressure from this expression is substituted into Equation 5.99 to obtain the maximum flowrate:

$$G_{max} = \sqrt{\frac{P_{cr}^2 g}{RT}} \qquad (5.101)$$

Noting that $v_{cr}' = RT/P_{cr}$, Equation 5.101 can be rearranged to give the maximum velocity for pipe flow:

$$w_{max} = \sqrt{gP_{cr}v_{cr}'} \qquad (5.102)$$

Thus, there is a limiting gas velocity in piping that corresponds to the critical pressure at the exit. If pressure falls below this value, the gas velocity will not increase.

Let us now consider another limiting case—that of frictionless adiabatic flow. That is, we will assume the pipe system to be perfectly thermally insulated. The energy balance equation simplifies to the following form:

$$i_1 - i_2 = \frac{w_2^2 - w_1^2}{2g} \tag{5.103}$$

or

$$C_p(T_2 - T_1) = \frac{w_1^2}{2g}\left[\left(\frac{T_2 P_1}{T_1 P_2}\right)^2 - 1\right] \tag{5.104}$$

This last expression is developed from Equation 5.103 using $i_1 - i_2 = C_p(T_2 - T_1)$ for perfect gases and $w_2/w_1 = T_2 P_1/T_1 P_2$ from $pv' = RT$ and $w_2/w_1 = v_2'/v_1'$. In practice, because of frictional resistances $P_1 > P_2$, and, consequently, a change in temperature does occur.

To develop an expression for pressure drop, the method of Lapple [1947] is used. The derivation is based on the Bernoulli and Darcy-Weisbach equations expressed in differential form:

$$v'dp + \frac{wdw}{g} + \lambda\frac{dL}{D}\times\frac{w^2}{2g} = 0 \tag{5.105}$$

denoting

$$v'dp = d(pv') - pv' \times \frac{dv'}{v'} \tag{5.106}$$

And, for the energy balance equation,

$$C_p dT = -d\left(\frac{w^2}{2g}\right) \tag{5.107}$$

where $dT = d(pv')/R$ and $R = C_p - C_v$.

Equation 5.107 may be rewritten as

$$d(pv') = -\left(\frac{\kappa - 1}{\kappa}\right)d\left(\frac{w^2}{2g}\right) \tag{5.108}$$

Substituting Equation 5.108 into Equation 5.106 and then into Equation 5.105, we obtain the following:

$$(1 + \kappa) \frac{dw}{w} - [2g\kappa P_1 \nu_1' + (\kappa - 1)w_1^2] \frac{dw}{w^3} + \frac{\kappa\lambda}{D} dL = 0 \qquad (5.109)$$

According to Equation 5.83, the velocity of sound C_1 at T_1 corresponding to P_1 and ν_1' is

$$C_1 = \sqrt{g\kappa P_1 \nu_1'} \qquad (5.110)$$

Hence, Equation 5.109 may be rewritten as

$$(1 + \kappa) \frac{dw}{w} - [2C_1^2 + (\kappa - 1)w_1^2] \frac{dw}{w^3} + \frac{\kappa\lambda}{D} dL = 0 \qquad (5.111)$$

On integration, we obtain

$$\lambda \frac{L}{D} = -\left(\frac{\kappa + 1}{\kappa}\right) ln \frac{w_1}{w_2} + \frac{1}{\kappa}\left(\frac{C_1^2}{w_1^2} + \frac{\kappa - 1}{2}\right)\left(1 - \frac{w_1^2}{w_2^2}\right) \qquad (5.112)$$

If the initial gas velocity, w_1, is known, w_2 may be computed from this expression for a specified pipe length, L. The value of λ does not change significantly and should be taken only as an average value when considering long lengths of piping. From w_2 and w_1, Equations 5.103 and 5.109 are combined to give

$$\frac{T_2}{T_1} = 1 + \frac{\kappa - 1}{2C_1^2} w_1^2 \left(1 - \frac{w_2^2}{w_1^2}\right) \qquad (5.113)$$

and, from the equation of continuity,

$$\frac{P_2}{P_1} = \frac{w_1}{w_2}\left[1 + \frac{\kappa - 1}{2C_1^2} w_1^2 \left(1 - \frac{w_2^2}{w_1^2}\right)\right] \qquad (5.114)$$

Analysis of the above equations for adiabatic gas flow in piping reveals that there is a maximum flowrate where the gas velocity at the exit reaches the velocity of sound. However, the adiabatic gas flow expressions provide essentially the same results as does the isothermal analysis. For very short pipes and high-pressure gradients, the adiabatic flowrate will be larger than the isothermal case; however, differences generally are no greater than 20%. Further discussions are given by Cambel and Jennings [1958], Shapiro [1958], Liepmann and Roshko [1957]. The following three examples illustrate applications of the flow analyses presented. Additional problems for the student are given at the end of this chapter.

Example 5.7

Air is flowing through a horizontal pipeline at a rate of 280 kg/hr. The pipe is 52.5 mm (2 in.) in diameter and 150 m in length. The exit pressure is atmospheric. Compute the pressure drop in the piping. The flow may be assumed isothermal at T = 20°C.

Solution

The cross section of the piping is

$$F = \frac{\pi \times 0.0525^2}{4} = 2.17 \times 10^{-3} \text{ m}^2$$

The mass flowrate is

$$W = \frac{280}{3600} = 7.76 \times 10^{-2} \text{ kg/s}$$

Hence, the specific weight flowrate is

$$G = \frac{W}{F} = \frac{7.76 \times 10^{-2}}{2.17 \times 10^{-3}} = 35.9 \frac{\text{kg}}{\text{m}^2 \times \text{s}}$$

The viscosity of air at 20°C is 0.02 cp, i.e., $\mu_g = 0.02 \times 10^{-3}$ kg/m-s. We now have enough information to compute the Reynolds number:

$$Re = \frac{GD}{\mu_g} = \frac{35.9 \times 0.0525}{0.02 \times 10^{-3}} = 94,200$$

The friction coefficient may be computed from the following turbulent correlation given by Perry [1950]:

$$\lambda = 0.0123 + \frac{0.7544}{Re^{0.38}} = 0.0123 + \frac{0.754}{94,200^{0.38}} = 0.022$$

The specific volume of the gas is

$$v'_2 = \frac{22.4}{29} \times \frac{(273 + 20)}{273} = 0.83 \ m^3/kg$$

We now can use Equation 5.98 to compute the upstream pressure:

$$P_1 = 10,333 + \frac{35.9^2}{9.81\gamma} \ ln \ \frac{0.83}{1} + \frac{0.022 \times 359^2}{2 \times 9.81 \times 0.0525\gamma}$$

or

$$P_1 = 10,333 + \frac{302}{\gamma} \ log \ \frac{0.83}{v'_1} + \frac{4120}{\gamma}$$

This expression may be solved by a trial and error solution, i.e., a method of successive approximations.

The gas constant R = 1.987 kcal/mol °K, and 1 kcal = 426.7 kg-m. Therefore,

$$R = \frac{1.987 \times 426.7}{2g} = 29.2 \ \frac{kg\text{-}m}{kg°K}$$

$$T = 273 + 20 = 293°K$$

$$RT = 29.2 \times 293 = 8550 \ m$$

$$v'_1 = \frac{RT}{P_1} = \frac{8550}{P_1}$$

The average specific volume is

$$\gamma = \frac{P_1 + P_2}{2RT} = \frac{P_1 + 10,333}{17,100}$$

Assuming a value for P_1 of 13,000 kg/m^2,

$$v_1' = \frac{8,550}{13,000} = 0.66 \text{ m}^3/\text{kg}$$

and the average specific weight is

$$\gamma = \frac{13,000 + 10,333}{17,100} = 1.36 \text{ kg/m}^3$$

Substituting these values into our equation,

$$P_1 = 10,333 + \frac{302}{1.36} \log \frac{0.83}{0.66} + \frac{4120}{1.36} = 13,380 \text{ kg/m}^2$$

This computed value is a little different from the assumed P_1 (13,000) and, hence, a new P_1 should be selected and the calculations repeated. To expedite computations, the equations were programmed in BASIC format with a desktop computer. The program is listed in Table 5.1, and tabulated values are given in Table 5.2. The program translated computations to a printer, which produced a plot of assumed P_1 versus computed P_1, as shown in Figure 5.11. The intersection of the curve generated by calculations with the bisectrix, i.e., the line of perfect agreement, provides the solution (P_1 = 13,350 kg/m^2). Hence, the pressure drop is

$$P_1 - P_2 = 13,350 - 10,333$$
$$= 3017 \text{ kg/m}^2 \text{ (or 0.3 atm)}$$

The student should prepare a separate program with a convergence routine to obtain the pressure drop. Specify a tolerance, i.e., the agreement between assumed and computed P_1, of 1%.

Example 5.8

Determine the upstream pressure for methane flowing at a rate of 1.2 kg/s in a pipe with a 130-mm i.d. and a 30-km length. Conditions at the exit are 2.5 atm and 20°C.

Table 5.1 Computer Program to Solve for Upstream Pressure in Example 5.7

```
10  DIM P1(15),N1(15),G(15),C(15
20  READ T,R,P2
30  P1(1)=8000
40  REM COMPUTES P1 VALUES
50  PRINT " ASMD.   CMPTD."
60  PRINT "  P1      P1      NU
65  ' GAMMA"
70  PRINT "---------------------

80  FOR I=1 TO 10
90  N1(I)=R*(T+273)/P1(I)
100 G(I)=(P1(I)+P2)/(2*R*(T+273)

110 H=302/G(I)*LOG(.83/N1(I))
120 C(I)=P2+H+4120/G(I)
130 PRINT USING 140 ; P1(I),C(I)
    ,N1(I),G(I)
140 IMAGE 000000.0000000,000.00
    D,DDDD.DD
150 PRINT
160 P1(I+1)=P1(I)+1000
170 P1(I)=P1(I)/1000
180 C(I)=C(I)/1000
190 NEXT I
200 DATA 20,29.2,10333
210 REM DRAWS PARITY PLOT
220 GCLEAR
230 SCALE 0,16,0,16
240 XAXIS 5,.5
250 YAXIS 5,.5
260 MOVE 5,5
270 DRAW 16,16

280 FOR I=1 TO 7
290 MOVE P1(I),C(I)
300 FOR I=2 TO 10
310 DRAW P1(I),C(I)
320 NEXT I
330 LDIR 0
340 FOR X=6 TO 16 STEP 2
350 MOVE X,4
360 LABEL VAL$(X)
370 NEXT X
380 LDIR 0
390 FOR Y=6 TO 16 STEP 2
400 MOVE 4,Y
410 LABEL VAL$(Y)
420 NEXT Y
430 LDIR 0
440 MOVE 6,2.5
450 LABEL "ASSUMED P1 (X1000)"
460 LDIR 0
470 MOVE 5,11
480 LABEL "COMPUTED"
490 MOVE 1.9,10
500 LABEL "P1"
510 MOVE 5,9
520 LABEL "(X1000)"
530 LDIR 0
540 MOVE 9,8
550 LABEL "LINE OF"
560 MOVE 9,7
570 LABEL "AGREEMENT"
580 COPY
590 END
```

Table 5.2 Tabulated Values Computed for Example 5.7

Assumed P_1	Computed P_1	v'	γ
8,000	14,107	1.069	1.07
9,000	13,943	0.951	1.13
10,000	13,792	0.856	1.19
11,000	13,653	0.778	1.25
12,000	13,525	0.713	1.31
13,000	13,406	0.658	1.36
14,000	13,295	0.611	1.42
15,000	13,192	0.570	1.48
16,000	13,096	0.535	1.54
17,000	13,007	0.503	1.60

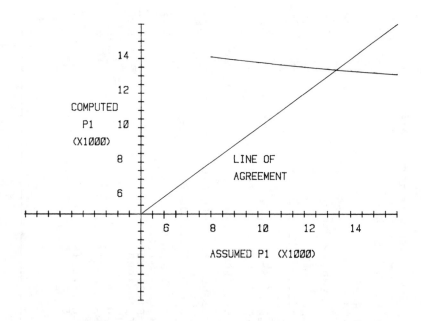

Figure 5.11   Plot of computed versus assumed values of P_1.

Solution

For isothermal flow of a real gas, the increase in internal energy equals zero and density is constant. Hence, the equation for a differential element of piping is as follows:

$$\frac{d(w^2)}{2} + \frac{dP}{\rho} + \delta\hat{F} = 0 \tag{i}$$

where $\hat{F}$ is the friction energy per unit mass.

The change in sign on $\hat{F}$ is associated with the inverse of terms containing infinitesimal differences. Velocity may be expressed as follows:

$$w = \frac{4G}{\pi d^2 \rho}$$

or

$$d(w^2) = \left(\frac{4G}{\pi d^2}\right) 2\nu' d\nu'$$

where ν = specific volume. Substituting this relationship into Equation i, we obtain

$$\left(\frac{4G}{\pi d^2}\right)^2 \frac{d\nu'}{\nu'} + \frac{dp}{\nu'} + \frac{\lambda}{2}\left(\frac{4G}{\pi d^2}\right)^2 \frac{d\ell}{d} = 0 \tag{ii}$$

To integrate ii, the relationship of pressure, specific volume and λ must be known.

For perfect gases at isothermal conditions,

$$P\nu' = \text{constant} = P_2\nu'_2$$

and

$$\lambda = \text{constant because } Re = \frac{wd}{\nu} = \frac{4G}{\pi d\mu} = \text{constant}$$

Substituting this expression in ii, we obtain the following:

$$\left(\frac{4G}{\pi d^2}\right)^2 \frac{dP}{P} + \frac{PdP}{P_2\nu'_2} + \frac{\lambda}{2}\left(\frac{4G}{\pi d^2}\right)^2 \frac{d\ell}{d} = 0$$

Integrating over the limits of 0 to ℓ, between P_1 and P_2, we obtain

$$\left(\frac{4G}{\pi d^2}\right)^2 \left(\ln\frac{P_1}{P_2} + \frac{\lambda}{2}\frac{\ell}{d}\right) = \frac{P_1^2 - P_2^2}{P_2 v_2'} \tag{iii}$$

To evaluate the friction factor, we first compute the Reynolds number:

$$Re = \frac{4G}{\pi d \mu} = \frac{4 \times 1.2}{\pi \times 0.13 \times 1.08 \times 10^{-5}} = 1.09 \times 10^6$$

Using a roughness coefficient, e = 0.1 mm, and relative roughness, e/d = 0.1/130 = 0.00077, we obtain from the Moody plot a value for λ:

$$\lambda = 0.018$$

The specific volume of methane at 2.5 atm, 20°C, is

$$v_2' = \frac{22.4}{16} \times \frac{1.033}{2.5} \times \frac{273 + 20}{273} = 0.62 \text{ m}^3/\text{kg}$$

Substituting these values into Equation iii,

$$\left(\frac{4 \times 1.2}{\pi \times 0.13^2}\right)^2 \left(\ln\frac{P_1}{2.5 \times 9.81 \times 10^4} + \frac{0.018 \times 30,000}{2 \times 0.13}\right)$$
$$= \frac{P_1^2 - (2.5 \times 9.81 \times 10^4)^2}{2.5 \times 9.81 \times 10^4 \times 0.62}$$

To simplify calculations, we will neglect the term $\ln P_1/P_2$, assuming it to be small in comparison to $\lambda/2 \times \ell/d$. Hence,

$$P_1 = \sqrt{(2.5 \times 9.81 \times 10^4)^2 + 2.5 \times 9.81 \times 10^4 \times 0.62\left(\frac{4 \times 1.2}{\pi \times 0.13^2}\right)^2 \frac{0.018 \times 30000}{2 \times 0.13}}$$

$$= 1.625 \times 10^6 \text{ N/m}^2 = 16.6 \text{ atm}$$

Checking the error obtained as a result of this assumption,

$$ln \frac{P_1}{P_2} = 1.89 \ll \frac{\lambda}{2} \times \frac{\ell}{d} = \frac{0.018 \times 30,000}{2 \times 0.13} = 2075$$

Hence, the calculation based on the initial pressure is sufficiently accurate.

Example 5.9

Determine the optimum pipe diameter for transporting 6000 Nm^3/hr of methane over a distance of 4 km. The efficiency of the gas blower is 0.5, and electrical costs are 50 ¢/kWh. The amortization cost of piping is $24/yr/m for 1-m i.d. pipe. Maintenance costs are $18/yr/m.

Solution

Assume that λ = 0.03 and local losses are 10% of the friction losses. The calculation will be based on a temperature of 30°C.

$$V_{sec} = \frac{6000 \times 303}{3600 \times 273} = 1.85 \text{ m}^3/\text{s}$$

$$w = \frac{V_{sec}}{0.785 \, d^2} = \frac{1.85}{0.785 \, d^2} = \frac{2.36}{d^2} \text{ m/s}$$

$$\Delta p = \Delta p_f + \Delta p_{\ell.\ell.} = 1.1 \, \Delta p_f$$

Hence, the pressure losses are

$$\Delta p = \frac{1.1 \, \lambda L w^2 \gamma}{d \times 2g} = \frac{1.1 \times 0.03 \times 4000 \times 2.36^2 \times 0.64}{d \times 2 \times 9.81 \times d^4}$$

$$= \frac{24}{d^5} \text{ kg/m}^2$$

The specific weight of methane is

$$\gamma = \frac{16 \times 273}{22.4 \times 303} = 0.64 \text{ kg/m}^3$$

We calculate the consumed power assuming that $\Delta p < 0.1$ atm (we will check it at the end of the design):

$$N = \frac{V_{sec}\Delta p}{102\eta} = \frac{1.85 \times 24}{102 \times 0.5 \times d^5} = \frac{0.87}{d^5} \, kW$$

One kW-yr costs $0.4 \times 24 \times 330 = \$3160$, assuming 330 working days. Thus, the cost of energy is

$$E = \frac{0.87 \times 3160}{d^5} = \frac{2750}{d^5} \, \$/yr$$

The amortization costs are

$$A = 24 \times L \times d = 24 \times 4000 \times d = 96{,}000 \, d \, \$/yr$$

Maintenance costs are

$$M = 18 \times L \times d = 72{,}000 \, d \, \$/yr$$

The total cost as a function of pipe diameter is

$$E + A + M = \frac{2750}{d^5} + 168{,}000 \, \$/yr$$

To determine the minimum costs (and, hence, the optimum diameter, assuming the Δp is acceptable), we differentiate this expression and equate it to zero:

$$\frac{\partial}{\partial d} (E + A + M) = -5 \times 2750 \, d^{-6} + 168{,}000 = 0$$

Hence,

$$d = 0.66 \, m$$

Finally, we check the pressure drop:

$$\Delta p = \frac{24}{d^5} = \frac{24}{0.66^5} = 193 \ kg/m^2 = 0.0193 \ atm$$

The pressure drop is less than the 0.1 atm assumed.

THE BERNOULLI EQUATION

General Application

In this subsection, a special case of the total energy balance is considered. Integration of Euler's equations of motion for steady flow leads to one of the most important and widely used expressions, namely, the Bernoulli equation. Multiplying both sides of the system of Equations 5.51 by dx, dy and dz, respectively, and then dividing by density, ρ, we obtain the following:

$$\frac{dx}{dt} dw_x = -\frac{1}{\rho} \frac{\partial P}{\partial x} dx$$

$$\frac{dy}{dt} dw_y = -\frac{1}{\rho} \frac{\partial P}{\partial y} dy$$

$$\frac{dz}{dt} dw_z = -gdz - \frac{1}{\rho} \frac{\partial P}{\partial z} dz$$

Summing these equations and noting that the derivatives dx/dt, dy/dt and dz/dt are velocity projections w_x, w_y and w_z on the corresponding coordinate axes, we obtain

$$w_x dw_x + w_y dw_y + w_z dw_z = -gdz - \frac{1}{\rho}\left(\frac{\partial P}{\partial x} dx + \frac{\partial P}{\partial y} dy + \frac{\partial P}{\partial z} dz\right) \quad (5.115)$$

The terms on the LHS are

$$w_x dw_x = d\left(\frac{w_x^2}{2}\right) ; \quad w_y dw_y = d\left(\frac{w_y^2}{2}\right) ; \quad w_z dw_z = d\left(\frac{w_z^2}{2}\right)$$

Hence, their sum may be written as

$$d\left(\frac{w_x^2}{2}\right) + d\left(\frac{w_y^2}{2}\right) + d\left(\frac{w_z^2}{2}\right) = d\left(\frac{w_x^2 + w_y^2 + w_z^2}{2}\right) = d\left(\frac{w^2}{2}\right) \qquad (5.116)$$

where $w = |\vec{W}|$ is the velocity vector with components along coordinate axes w_x, w_y and w_z.

The sum of the terms in parentheses on the RHS of Equation 5.115 is the total pressure differential, dp. Recall that the pressure at any one point does not change with time, i.e., only spacial coordinates are considered. Hence, we may write

$$d\left(\frac{w^2}{2}\right) = -\frac{dp}{\rho} - dz \qquad (5.117)$$

Dividing both sides by g and specifying incompressible homogeneous flow, i.e., ρ is constant,

$$d\left(\frac{w^2}{2g}\right) + \frac{dp}{\rho g} + dz = 0 \qquad (5.118)$$

The sum of the differential then is substituted by the differential of the sum

$$d\left(z + \frac{P}{\rho g} + \frac{w^2}{2g}\right) = 0$$

Hence,

$$z + \frac{P}{\rho g} + \frac{w^2}{2g} = \text{constant} \qquad (5.119)$$

This expression may be related to any two cross sections of a flow element in the following manner:

$$z_1 + \frac{P_1}{\rho g} + \frac{w_1^2}{2\alpha g} = z_2 + \frac{P_2}{\rho g} + \frac{w_2^2}{2\alpha g} \qquad (5.120)$$

Note that velocities and pressures are not equivalent at different points at any cross section (refer back to the velocity profiles shown in Figure 5.4).

Equation 5.120 is related not to the cross section as a whole, but to any pair of compatible points in these sections (for example, points located along the axis of the piping). To compare congruent values over total cross sections, it must be assumed that the terms in Equation 5.38 may be approximated by Equation 5.119, which represents the point form of the *Bernoulli equation for ideal fluids*.

It follows that the *hydrodynamic head* ($z + P/\rho g + w^2/2g$) is constant for all cross sections of ideal steady flow. The first two terms of the hydrodynamic head, z and $P/\rho g$, were introduced in Equation 4.7. z is the *leveling height* (also called the *geometric head*), which represents a specific potential energy at a given point within the cross section. $P/\rho g$ is the static or piezometric head characterizing the specific potential energy of pressure at a given point. Both terms may be expressed in units either of length or specific energy, i.e., energy per unit weight of fluid.

The third item, $w^2/2g$, also is expressed in units of length:

$$\left[\frac{w^2}{2g}\right] = \left[\frac{m^2 \times s^2}{s^2 \times m}\right] = [m]$$

or, after multiplying and dividing by unit weight, N, in units of energy:

$$\left[\frac{w^2}{2g}\right] = \left[\frac{N \times m}{N}\right] = \left[\frac{J}{N}\right]$$

$w^2/2g$ is called the *velocity or dynamic head*, which characterizes a specific kinetic energy at a given point within the cross section.

Thus, according to the Bernoulli equation, the sum of the static and dynamic heads and level do not change from one cross section of flow to another. This statement is inclusive only of steady-state flow of an ideal fluid. It further follows that the sum of the potential and kinetic energies is constant. This is illustrated further by the following discussions.

A conversion of energy occurs when the flow changes. This change in energy is reflected in the fluid velocity. Reducing a pipe's cross-sectional area causes part of the potential energy of pressure to be converted into kinetic energy and vice versa. When the flow area is increased, part of the kinetic energy is converted into potential; however, the total energy is unchanged. Hence, we may conclude that the amount of energy entering the initial pipe cross section is equal to that leaving the pipe. For this reason, Equation 5.120 is a special application of the principle of conservation of energy.

Multiplying Equation 5.120 by the specific weight $\gamma = \rho g$, we obtain

$$\rho g z_1 + P_1 + \frac{\rho w_1^2}{2} = \rho g z_2 + P_2 + \frac{\rho w_2^2}{2} \tag{5.121}$$

where each term expresses the specific energy of the flow per unit volume at a given point. For example,

$$[P] = \left[\frac{N}{m^2}\right] = \left[\frac{N \times m}{m^2 \times m}\right] = \left[\frac{J}{m^2}\right]$$

For a horizontal system, $z_1 = z_2$ and Equation 5.121 simplifies to

$$\frac{P_1}{\rho g} + \frac{w_1^2}{2g} = \frac{P_2}{\rho g} + \frac{w_2^2}{2g} \tag{5.122}$$

To illustrate application of the Bernoulli equation, consider the variable cross-sectional flow system illustrated in Figure 5.12. Points on the flow axis at sections 1 and 2 are at heights z_1 and z_2 above the datum, respectively. At each of these points, two piezometric tubes are inserted into the

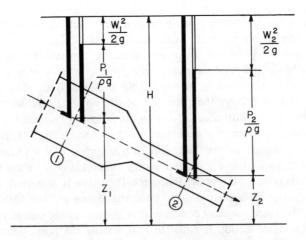

Figure 5.12   Ideal fluid flowing through a variable cross section located arbitrarily in space.

flow. One tube at each point has its end into the direction of flow. The levels of fluid in the straight vertical tubes settle at heights corresponding to the hydrostatic pressures at the points of their submergence, i.e., they provide a measure of the *static head* at corresponding points. The fluid height in the bent tubes is higher than in the straight ones because the measurement represents the *sum of the static and dynamic heads*. According to Equation 5.119, the levels in the bent tubes are the same because they are both referenced to the same datum plane and are a measure of the hydrodynamic head.

As the flow cross section at plane 2-2 is less than that of 1-1 from continuity, fluid velocity, w_2 (for a constant flowrate) must be greater than w_1. Hence, the kinetic energy at 2-2 is greater than 1-1 ($w_2^2/2g > w_1^2/2g$). Therefore, the difference between static and dynamic heads at plane 2-2 is greater than the difference at plane 1-1.

The Bernoulli equation states that the fluid level in the straight tube at 2-2 is less than the corresponding height in the straight tube in plane 1-1 and, by the same value, that the velocity head at 2-2 exceeds 1-1. This example illustrates the mutual conversion of potential energy into kinetic when the flow cross section is changed. The overall conclusion is that the sum of these energies in any cross section of the piping remains unchanged.

For real fluids, both shear and friction forces play important roles. These latter forces exert resistance to fluid motion. A portion of the flow energy must be devoted to overcome this *hydraulic resistance*. The total energy decreases continuously in the direction of flow as a portion is converted from potential energy into *lost energy* (the energy expended for friction). This conversion is irreversible and is lost in the form of heat dissipation to the surroundings. For the system just analyzed, this means that

$$z_1 + \frac{P_1}{\rho g} + \frac{w_1^2}{2g} > z_2 + \frac{P_2}{\rho g} + \frac{w_2^2}{2g}$$

Thus, for real (viscous) fluids, the levels in the bent tubes at planes 1-1 and 2-2 in Figure 5.12 are not the same. This difference in levels is attributed to energy losses in the fluid path from 1-1 to 2-2 and is referred to as the *lost head*, h_ϱ. The Bernoulli equation may be corrected for this frictional loss by adding the term h_ϱ to the RHS of Equation 5.120. Hence, the Bernoulli equation for real fluids is written as follows:

$$z_1 + \frac{P_1}{\rho g} + \frac{w_1^2}{2g} = z_2 + \frac{P_2}{\rho g} + \frac{w_2^2}{2g} + h_\varrho \qquad (5.123)$$

where the lost head, h_ϱ, characterizes the *specific energy spent for overcoming hydraulic resistances*.

Another form of this expression can be obtained by multiplying both sides by ρg:

$$\rho g z_1 + \frac{\rho w_1^2}{2\alpha} + P_1 = \rho g z_2 + P_2 + \frac{\rho w_2^2}{2\alpha} + \Delta P_\varrho \qquad (5.124)$$

ΔP_ϱ is the lost pressure drop, defined as

$$\Delta P_\varrho = \rho g h_\varrho \qquad (5.125)$$

The following examples apply to the above principles.

Example 5.9

Oil (specific gravity 0.89) is being pumped through a pipe system of constant diameter. The pressure just upstream of the pump is 25 kN/m² abs. and the pump's discharge pressure is 73 kN/m² abs. The discharge pipe of the pump is 8 m above the centerline of the inlet pipe. The pump supplies 180 J/kg of fluid flowing in the pipe. The flow through the pipe system was determined to be turbulent. Compute the frictional losses in the system.

Solution

We shall choose the centerline of the pipe inlet as the reference datum plane. Hence, $z_1 = 0$ and $z_2 = 8$ m. As the pipe diameters upstream and downstream of the pump are the same, $w_1 = w_2$.

This is a steady-state, incompressible flow problem involving application of the total energy equation (Equation 5.123). As the flow is turbulent, no correction is needed for the kinetic energy terms, i.e., $\alpha = 1.0$. Rewriting Equation 5.123 to the following to solve for friction losses:

$$\Sigma h_\varrho = -\hat{W}_s + \frac{1}{2\alpha}(w_1^2 - w_2^2) + g(z_1 - z_2) + \frac{P_1 - P_2}{\rho}$$

where $\hat{W}_s$ is the work performed by the pump, and Σh_ϱ represents the sum of the friction losses, i.e., head losses. Evaluating each term,

$$\frac{1}{2(1)} (w_1^2 - w_2^2) = 0$$

$$g(z_1 - z_2) = (9.8 \text{ m/s}^2)(0 - 8) \text{ [m]} = -78.4 \text{ J/kg}$$

$$\frac{P_1 - P_2}{\rho} = \frac{25 - 73 \text{ [kN/m}^2\text{]}}{890 \text{ kg/m}^3} \times 1000 = -53.9 \text{ J/kg}$$

Hence,

$$\Sigma h_\varrho = -(-180) + 0 + (-78.4) + (-53.9)$$
$$= 47.7 \text{ J/kg}$$

or

$$\Sigma h_\varrho = 47.7 \times 0.33485 = 15.96 \frac{\text{ft-lb}_f}{\text{lb}_m}$$

Example 5.10

A liquid is being pumped from an open tank to a height 37 ft above the initial level in the tank. The density of the liquid is 73 lb_m/ft^3 and the pumping rate is 35 gpm. The discharge line on the pump is 2.5 in. i.d. The total friction loss in the piping is 23 $\text{ft-lb}_f/\text{lb}_m$. The level in the tank is dropping at a rate of 0.25 fps, and the pump's rated efficiency is 59%. Assuming flow through the piping is turbulent, determine the horsepower of the pump.

Solution

The flow diagram for this system is given in Figure 5.13. The mechanical energy actually delivered to the fluid by the pump, i.e., the net mechanical work, is

$$\hat{W}_s = -\eta W_p$$

where η is the fractional efficiency of the pump, and W_p is the shaft work delivered to the pump.

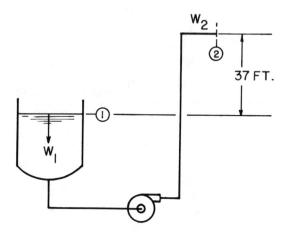

Figure 5.13   Flow system under evaluation in Example 5.10.

The volumetric flowrate is

$$V = \left(35 \ \frac{gal}{min}\right)\left(\frac{1 \ min}{60 \ s}\right)\left(\frac{ft^3}{7.481 \ gal}\right) = 0.07798 \ cfs$$

The fluid velocity in the tank is given, $W_1 = 0.25$ fps. The cross section of the discharge pipe is

$$F = \frac{1}{4} \pi (2.5/12)^2 = 0.0341 \ ft^2$$

Hence, the velocity downstream of the pump is

$$w_2 = \frac{0.07798 \ cfs}{0.0341 \ ft^2} = 2.288 \ fps$$

We will assume that the discharge is open to the atmosphere. Hence,

$$P_1 = P_2 = 1 \ atm$$

Therefore,

$$\frac{P_1}{\rho} - \frac{P_2}{\rho} = 0$$

And because $\alpha = 1.0$,

$$\frac{w_1^2}{2g_c} = \frac{(0.25 \text{ fps})^2}{2(32.174)} = 9.713 \times 10^{-4} \frac{\text{ft-lb}_f}{\text{lb}_m}$$

$$\frac{w_2^2}{2g_c} = \frac{(2.288 \text{ fps})^2}{2(32.174)} = 8.135 \times 10^{-2} \frac{\text{ft-lb}_f}{\text{lb}_m}$$

Assigning the initial level in the tank as the reference datum plane, we note that

$$z_1 = 0$$

and

$$z_2 \frac{g}{g_c} = (37.0 \text{ ft}) \left(\frac{32.2}{32.174} \right) = 37.0 \frac{\text{ft-lb}_f}{\text{lb}_m}$$

We now rearrange the mechanical energy equation to solve for mechanical work:

$$\hat{W}_s = \frac{g}{g_c}(z_1 - z_2) + \frac{1}{2g_c}(w_1^2 - w_2^2) + \frac{P_1 - P_2}{\rho} - \Sigma \hat{F}$$

$$\hat{W}_s = -37.0 + (9.713 \times 10^{-4} - 8.135 \times 10^{-2}) + 0 - 23$$

$$= -60.08 \frac{\text{ft-lb}_f}{\text{lb}_m}$$

Hence,

$$\hat{W}_p = -\frac{\hat{W}_s}{\eta} = -\frac{(-60.08)}{0.59} = 101.83 \frac{\text{ft-lb}_f}{\text{lb}_m}$$

Mass flowrate = $(0.07798 \text{ cfs})(73 \text{ lb}_m/\text{ft}^3) = 5.69 \text{ lb}_m/\text{s}$. The pump horsepower therefore is

$$\left(5.69 \; \frac{\text{lb}_m}{\text{s}}\right)\left(101.83 \; \frac{\text{ft-lb}_f}{\text{lb}_m}\right)\left(\frac{1 \text{ hp}}{550 \; \frac{\text{ft-lb}_f}{\text{s}}}\right) = 1.05 \text{ hp}$$

Additional examples are given by Bird et al. [1960], Cheremisinoff [1981a,b], Mott [1972] and Pai [1956].

Variable-Head Meters

The determination of head losses is an important practical problem connected with the calculation of the energy required for fluid displacement in pumps, compressors, etc., as well as for measuring flow quantities entering and leaving process equipment. Many flow problems in this category can be addressed through the Bernoulli equation. We will discuss some practical applications of the Bernoulli equation through the use of variable-head meters. Such devices are used extensively throughout industry to measure and control flows through equipment.

We shall limit our discussion to the three most widely used head meters, however, namely, pitot tubes, orifice and venturi meters. Extensive data on the design of these devices are given by the American Society of Mechanical Engineers (ASME Publications [1959]) and applications to industrial problems are given by Cheremisinoff [1979,1981a,b].

The first of these devices, the *pitot tube*, is used for measuring local fluid velocities. The device consists of a stainless steel tube with its inlet opening turned upstream into the flow. Therefore, the inlet receives the full impact of the flow against it. The impact is converted completely into pressure head $w^2/2g$, superimposed on the existing static pressure of the fluid. In principle, the pitot tube consists of both an impact tube and a piezometer tube (Figure 5.14A). Because an *impact tube* is used in connection with a *piezometer tube*, the static pressure may be subtracted from the total pressure measured, the difference of which is the velocity head. This has been illustrated in Figure 5.12. The pressure difference is measured conveniently with a differential U-tube, as shown in Figure 5.14A. Note that the U-tube should contain a liquid that does not mix with the working fluid and has higher density. Both the impact and piezometer tube are combined into a standard S-shaped pitot tube by including static pressure taps downstream of the impact tube's tip. The design is illustrated in Figure 5.14B.

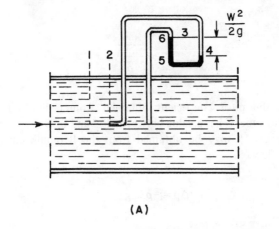

(A)

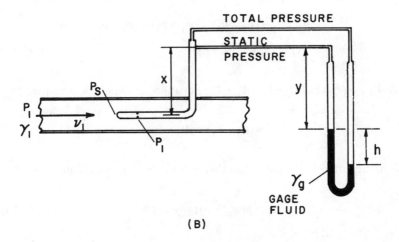

(B)

Figure 5.14   Pitot tube used to measure the maximum fluid velocity in a pipe.

The maximum flow velocity along a pipe axis may be computed from the measured pressure head, $w^2/2g$, using the Bernoulli equation. Consider, for example, an incompressible fluid flowing between points 1 and 2 in Figure 5.14A. As the velocity at 2 is zero, we may write

$$\frac{w_1^2}{2g} = \frac{P_2 - P_1}{\rho} \tag{5.126}$$

or

$$w = C \sqrt{\frac{2g\Delta p}{\rho}} \qquad (5.127)$$

where C is the pitot coefficient obtained from calibration. For a well-designed pitot tube, C has a value between 0.96 and 0.98.

The value of Δp may be obtained from the Bernoulli equation for the case of zero flow:

$$\Delta p = -\rho \Delta z g \qquad (5.128)$$

From Figure 5.14A, we see that

$$\Delta p = P_2 - P_1 = (P_2 - P_3) + (P_3 - P_4) + (P_4 - P_5)$$
$$+ (P_5 - P_6) + (P_6 - P_1) \qquad (5.129)$$

As $P_4 - P_5 = 0$ and $P_2 - P_3 = P_6 - P_1$, Equation 5.129 may be simplified to

$$\Delta p = (P_3 - P_4) + (P_5 - P_6) \qquad (5.130)$$

from Equation 5.128 and noting that $z_3 - z_4 = z_5 - z_6$, we obtain

$$\Delta p = g(\rho_m - \rho)(z_3 - z_4) \qquad (5.131)$$

Combining Equations 5.127 and 5.131,

$$w = C \sqrt{\frac{2g(\rho_m - \rho)\Delta h}{\rho}} \qquad (5.132)$$

where ρ_m is the density of the manometer liquid and $\Delta h = z_3 - z_4$.

To determine the average fluid velocity, measurements at successive points across the pipe cross section are needed. The velocity profile obtained from traversing the pipe then may be integrated over the pipe diameter to obtain an average value. Such a method for measuring velocity and flow rate is simple; however, in general it is not accurate due to the difficulties in positioning the instrument exactly along the pipe axis.

Further discussions on the pitot tube are given by Cheremisinoff [1979], the ASME Research Committee on Fluid Meters [1959], Stoll [1959], and Folsom [1956].

Example 5.11

For the pitot tube arrangement shown in Figure 5.14B, water is flowing through the pipe at 130°F. The manometer fluid is mercury (γ = 13.6). The manometer reading is 15 in. and the pitot tube coefficient is 0.97. Determine the point velocity of the water.

Solution

Equation 5.132 will be used:

$$w_1 = C\sqrt{2g(\rho_m - \rho)\Delta h/\rho}$$

$$\rho = 61.2\ lb_m/ft^3$$

$$\rho_m = (13.60)(62.4\ lb_m/ft^3) = 848\ lb_m/ft^3$$

$$\Delta h = (15\ in.)(1\ ft/12\ in.) = 1.25\ ft$$

The velocity in English units is

$$w_1 = 0.97\sqrt{\frac{(2)(32.2)(848 - 61.2)(1.25)}{61.2}}$$

$$= 31.2\ ft/s$$

An *orifice meter* consists of a thin plate mounted between two flanges, with an accurately drilled hole positioned concentric to the pipe axis. The flow measurement principle behind the orifice meter is based on the reduction of flow pressure accompanied by an increase in velocity, i.e., Bernoulli's principle. The reduction of the cross section of the flowing stream as it passes through the orifice increases the velocity head at the expense of pressure head. Figure 5.15 illustrates the operation of an orifice meter whereby a *manometer* or pressure gauges are used to measure the upstream and downstream pressures. By applying the Bernoulli equation, the discharge can be determined from the manometer's readings based on the known area of the orifice.

A *venturi meter*, illustrated in Figure 5.16, consists of a short length of straight tubing connected at either end of the pipe by conical sections. The

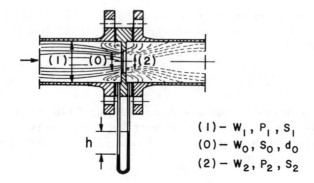

$$(1) - W_1, P_1, S_1$$
$$(0) - W_0, S_0, d_0$$
$$(2) - W_2, P_2, S_2$$

Figure 5.15   An orifice meter in operation.

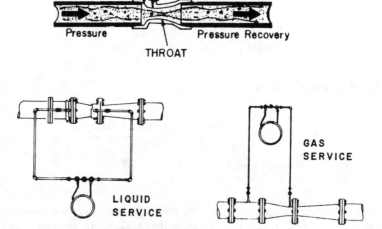

Figure 5.16   Flow measurement using a venturi meter.

measurement principle behind this instrument is based on the reduction of flow pressure accompanied by an increase in velocity of the venturi throat. The pressure drop experienced in the upstream cone section is used to measure the rate of flow through the venturi meter. On the discharge side of the meter the fluid velocity is decreased and the original pressure recovered. Because of its shape, pressure losses in a venturi meter are less than in an orifice meter. However, a venturi is large in comparison to an orifice meter, which can be mounted readily between flanges. (Detail design and performance data of venturis are given by Cheremisinoff [1979], ASME Committee on Fluid Meters [1959], Miner [1956], and Hooper [1950]).

Because of its size and expense, a smaller version of the venturi, called the *flow nozzle*, has been developed. A typical flow nozzle is illustrated in Figure 5.17. It often is used as the primary element for measuring liquid flows. In the design, the diverging exit of the venturi meter is omitted and the converging entrance altered to a more rounded configuration. In both the venturi meter and flow nozzle, the cross-sectional area of the compressed stream $(S_2 = \pi d_1^2/4)$ is equal to the cross-sectional area of the hole $S_0 = \pi d^2/4$. In the immediate discussions to follow, we shall derive formulas applicable to orifice meters, venturi meters and flow nozzles.

Let us apply the Bernoulli equation over two sections of horizontal pipe flow. The pressure drop between these two sections may be measured with a differential U-tube. Using the notation given in Figure 5.15, we obtain

$$\frac{P_1}{\rho g} + \frac{w_1^2}{2g} = \frac{P_2}{\rho g} + \frac{w_2^2}{2g}$$

Hence,

$$\frac{w_2^2}{2g} - \frac{w_1^2}{2g} = \frac{P_1 - P_2}{\rho g} = \Delta h \tag{5.133}$$

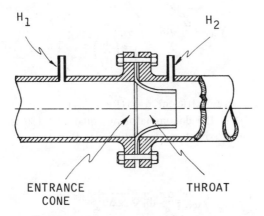

Figure 5.17   A typical flow nozzle.

where Δh is the pressure drop measured by the differential U-tube in m of working liquid column. Using the continuity equation, we determine the average velocity and fluid rate in the piping. w_1 denotes the average velocity in the pipe's large cross section and refers to the stream flow immediately after the orifice where pressure, p_2, is measured:

$$w_1 = w_2 \frac{S_2}{S_1} = w_2 \frac{d_2^2}{d_1^2} \qquad (5.134)$$

Substituting the value w_1 from Equation 5.134 into Equation 5.133, we obtain

$$\frac{w_2^2}{2g} - \frac{w_2^2}{2g}\left(\frac{d_2}{d_1}\right)^4 = \Delta h \qquad (5.135)$$

Hence,

$$w_2 = \sqrt{\frac{2g\Delta h}{1 - \left(\frac{d_2}{d_1}\right)^4}} \qquad (5.136)$$

The volume liquid rate, V_{sec}, in section S_0 of the hole in the orifice meter (and, consequently, in the piping) is

$$V_{sec} = \frac{\alpha\pi}{4} d_0^2 \sqrt{\frac{2g\Delta h}{1 - \left(\frac{d_0}{d_1}\right)^4}} \qquad (5.137)$$

where α is the discharge coefficient, $\alpha = f(Re\ d_0/d_1)$.

The term (d_2/d_1) in the denominator of Equation 5.137 is usually small. Hence, as a first approximation the volumetric flowrate may be computed from

$$V_{sec} = \frac{\alpha\pi}{4} d_0^2 \sqrt{2g\Delta h} \qquad (5.138)$$

Also, the mean velocity through the pipe may be approximated by

$$\overline{W} = \alpha \left(\frac{d_0}{d}\right)^2 \sqrt{2g\Delta h} \tag{5.139}$$

For compressible fluids, a correction coefficient must be applied to Equations 5.138 and 5.139. The following examples illustrate the use of the above formulas.

Example 5.12

A liquid of density 1237 kg/m^3 and viscosity 0.74 cp is flowing through a 19-cm i.d. pipe. A sharp-edged orifice having a diameter of 2.9 cm is installed in the pipeline. The measured pressure drop across the orifice is 112.7 kN/m^2. Calculate the volumetric flowrate and the average velocity of the liquid through the pipe.

Solution

Equation 5.137 is used:

$$V_{sec} = \frac{\alpha\pi}{4} d_0^2 \sqrt{\frac{2(P_1 - P_2)/\rho}{1 - (d_0/d_1)^4}}$$

$$P_1 - P_2 = 112.7 \text{ kN/m}^2 = 11.27 \times 10^4 \text{ N/m}^2$$

$$d_1 = 0.190 \text{ m} \qquad d_0 = 0.029 \text{ m} \qquad \frac{d_0}{d_1} = \frac{0.029}{0.190} = 0.153$$

Examining the discharge coefficient-Reynolds number plot (Figure 5.18), we note that for Re > 20,000, α is roughly the same regardless of the diameter ratio. Hence, we shall assume that $\alpha \simeq 0.61$:

$$V_{sec} = \frac{(0.61)\pi}{4} (0.029)^2 \sqrt{\frac{2(11.27 \times 10^4)/1237}{1 - (0.153)^4}}$$

$$V_{sec} = 0.00544 \text{ m}^3/\text{s (or 1.44 gpm)}$$

The average velocity of the liquid is

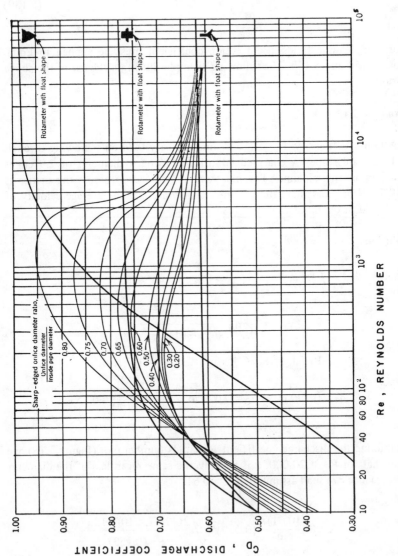

Figure 5.18 Plot of discharge coefficient versus Reynolds number for sharp-edged orifices and rotameters (from Brown et al. [1950]).

$$\overline{W} = \frac{0.00544 \text{ m}^3/\text{s}}{\frac{\pi}{4}(0.190)^2 \text{ m}^2} = 0.192 \text{ m/s (or } 0.63 \text{ fps)}$$

Re is calculated to determine whether it is greater than 20,000 for $\alpha = 0.61$. $\mu = 0.74 \times 10^{-3}$ kg/m-s $= 0.74 \times 10^{-3}$ Pa-s:

$$\text{Re} = \frac{d_1 W \rho}{\mu} = \frac{(0.190)(0.192)(1237)}{0.74 \times 10^{-3}} = 60,981$$

As the Reynolds number is greater than 20,000, a good value for the discharge coefficient was selected. The student should repeat the problem using Equation 5.139 and compare the results to the above.

Example 5.13

The venturi meter shown in Figure 5.16 is used to measure the flowrate in a pipe. The upstream diameter of the venturi is 4.5 in. (d_1) and the throat diameter $(d_0 = d_2)$ is 1.75 in. The pressure drop across the meter is 0.290 psi and the fluid is air $(\rho \approx 0.07 \text{ lb}_m/\text{ft}^3)$.

Solution

From continuity,

$$V = F_1 w_1 = F_2 w_2$$

where V is the volumetric discharge. Thus,

$$\frac{\pi}{4}\left(\frac{4.5}{12}\right)^2 w_1 = \frac{\pi}{4}\left(\frac{1.75}{12}\right)^2 w_2$$

or

$$V = 0.1104 \, w_1 = 0.0167 \, w_2$$

Applying the Bernoulli equation for $z_1 = z_2$,

$$P_1 - P_2 = 0.290 \times 144 = 41.76 \text{ lb}_f/\text{ft}^2$$

$$\frac{P_1 - P_2}{\rho} = \frac{w_2^2}{2g} - \frac{w_1^2}{2g}$$

$$\frac{41.76}{0.07} = \frac{V^2}{2g}\left[\left(\frac{1}{0.0167}\right)^2 - \left(\frac{1}{0.1104}\right)^2\right]$$

Solving for the volumetric discharge gives $V = 10.97$ cfs (or 2764 lb_m/hr of air flowing).

Efflux from Vessels and Pipes

We now direct our attention to another common application of Bernoulli's principle, that of efflux from a tank. Consider the system shown in Figure 5.19A, in which water is being discharged through a circular orifice located on the floor of a flat bottomed vessel open to the atmosphere. The tank has an inlet stream that maintains a constant liquid level, H. The reference plane 0-0 is designated to be below, and parallel to, plane 1-1 corresponding to the liquid surface, and plane 2-2 passes through the narrowest section of discharging jet. Hence, for an ideal liquid, Bernoulli's equation is

$$z_1 + \frac{P_1}{\rho g} + \frac{w_1^2}{2g} = z_2 + \frac{P_2}{\rho g} + \frac{w_2^2}{2g}$$

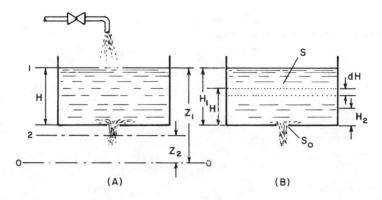

Figure 5.19   Efflux from a tank: (A) at constant liquid level in the tank; (B) at variable liquid level in the tank.

As this is an open vessel with constant level, $P_1 = P_2$ and $w_1 = 0$. Neglecting the small distance between the orifice plane at the vessel floor and plane 2-2, we may assume that $z_1 - z_2 = H$. Hence,

$$\frac{w_2^2}{2g} = H$$

and

$$w_2 = \sqrt{2gH}$$

For viscous fluids, a portion of the liquid head is lost to friction and to overcoming the resistance due to the abrupt jet constriction in the orifice. Therefore, the discharge or jet velocity of a viscous liquid is the following:

$$w_2 = \psi \sqrt{2gH}$$

where ψ is a correction factor called the velocity factor, which accounts for head losses incurred by discharging through the orifice. Values for ψ are less than 1. Because the jet cross-sectional area in the orifice, S_o, is larger than its narrowest section, S_2, the liquid velocity, w_o, in the orifice must be less than w_2. Then

$$w_o = \epsilon_j w_2 = \epsilon_j \psi \sqrt{2gH} = \alpha \sqrt{2gH} \tag{5.140}$$

where $\epsilon_j = S_2/S_o$ is the jet constriction coefficient. Coefficient α is the discharge coefficient defined as the product of the velocity coefficient, ψ, and the constriction coefficient, ϵ_j:

$$\alpha = \psi \epsilon_j \tag{5.141}$$

Values of discharge coefficient α may be found in handbooks such as Perry's [1950]. The coefficient is a function of the Reynolds number and is determined experimentally.

The volume liquid rate is equal to the velocity in the orifice, w_o, times the orifice cross-sectional area, S_o:

$$V_{sec} = \alpha S_o \sqrt{2gH} \tag{5.142}$$

Equation 5.142 shows that the rate of efflux through the orifice depends on the constant level in the tank and the orifice diameter but is independent of the vessel's shape. The expression also is valid for liquid discharging through an orifice in the tank's side wall. For the latter case, the level H must be measured from the centerline of the orifice.

For liquids in which viscosities are similar to that of water, $\alpha = 0.62$.

Example 5.14

Water is being discharged through a side wall nozzle in a large tank open to the atmosphere (Figure 5.20). The jet issues as a cylinder to the atmosphere and the water surface is located 10 m above the nozzle centerline. Determine the velocity of efflux from the nozzle.

Solution

This involves application of Bernoulli's principle:

$$z_1 + \frac{P_1}{\gamma_1} + \frac{w_1^2}{2g} = z_2 + \frac{P_2}{\gamma_2} + \frac{w_2^2}{2g}$$

where subscript 2 refers to a point immediately downstream of the nozzle (along its centerline) and 1 refers to the level of the tank. We may assume that the pressure along the centerline of the nozzle discharge is atmospheric. Hence, $P_1 - P_2 = 0$. And, as the elevation datum is at point 2, $z_2 = 0$ and $z_1 = H = 10$ m.

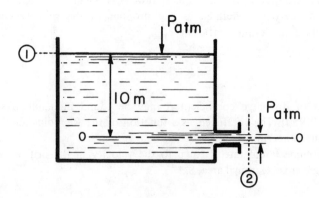

Figure 5.20   System for tank discharge problem (Example 5.14).

As this is a large vessel, we may further assume that the velocity of the reservoir is practically zero. Hence,

$$0 + 0 + H = \frac{w_2^2}{2g} + 0 + 0$$

or

$$w_2 = \sqrt{2gH} \quad \text{(derived earlier)}$$
$$w_2 = \sqrt{2 \times 9.806 \times 10} = 14.0 \text{ m/s}$$

Our solution has ignored frictional effects due to the nozzle and therefore is a maximum. It states that the velocity of efflux is equal to the velocity of free fall from the surface of the tank. This is known as **Torricelli's theorem**.

The student should continue with the problem by calculating w_2 for $\psi = 0.75$. For a fluid jet diameter of 55 mm, determine the volumetric discharge.

In short cylindrical nozzles, additional head losses in both the nozzle inlet and outlet make frictional corrections even more important (that is, ψ becomes smaller). However, the fluid entering a nozzle tends to expand after some constriction, eventually filling the entire nozzle cross section, i.e., it can be assumed that $\epsilon_j = 1$. The end result is that nozzles generally have a higher discharge coefficient than does a simple orifice (for water discharge, $\alpha \simeq 0.82$).

Let us now consider the second efflux problem shown in Figure 5.19B, that of tank discharge from a variable-level tank. Specifically, we wish to determine the discharge time through an orifice located on the floor of a thin-walled vessel. For a time interval, dt, the liquid level, H_1, is decreased to some height, H_2. According to Equation 5.142, the liquid volume discharged from the vessel is

$$dV = V_{sec}dt = \alpha S_0 \sqrt{2gH} \, dt \qquad (5.143A)$$

where S_0 is the cross-sectional area in the bottom of the vessel.

Over the same time interval, dt, the liquid level is lowered by the differential height, dH, and the differential volume lost for a vessel of constant cross-sectional area, S, is

$$dV = -SdH \qquad (5.143B)$$

From the principle of continuity, Equations 5.143A and 5.143B may be equated to each other:

$$\alpha S_o \sqrt{2gH} \, dt = -SdH$$

Hence,

$$dt = \frac{-SdH}{\alpha S_o \sqrt{2gH}} \qquad (5.144)$$

Assuming constant α and integrating over the limits of H_1 to H_2 for 0 to t, an expression for the time of discharge for a vessel of constant cross section is obtained:

$$\int_0^t dt = \int_{H_1}^{H_2} \frac{SdH}{\alpha S_o \sqrt{2gH}}$$

$$t = \frac{S}{\alpha S_o \sqrt{2g}} \int_{H_1}^{H_2} H^{-1/2} dH = \frac{2S}{\alpha S_o \sqrt{2g}} (\sqrt{H_1} - \sqrt{H_2}) \qquad (5.145)$$

If the vessel is emptied completely, i.e., $H_2 = 0$, the expression simplifies to the following:

$$t = \frac{2S\sqrt{H_1}}{\alpha S_o \sqrt{2g}} \qquad (5.146)$$

There are many operations in which vessels of variable cross section are utilized (for example, conical vessels, horizontal cisterns, etc.). To determine the discharge times from such vessels, the relationship between cross-sectional area and height must be known, i.e., $S = f(H)$. Consider the two systems illustrated in Figure 5.21: a funnel-shaped vessel and a horizontal cylindrical tank. The conical vessel with apex angle β has a cross-sectional area equal to the following:

$$S = \pi H^2 \tan^2 \beta/2 \qquad (5.147)$$

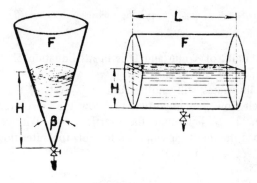

Figure 5.21   Liquid discharge from vessels of variable cross section.

The horizontal cylindrical vessel's cross section is

$$S = 2L \sqrt{HD - H^2} \qquad (5.148)$$

When these relationships are included in Equation 5.145, the discharge time between any two levels or for complete discharge can be computed for the two geometries.

Another common system encountered in plant operations is liquid discharge from partially filled pipes. Such a system is shown in Figure 5.22. The volumetric discharge rate depends on the pipe diameter, D, and a space factor defined by $K = H/D$. For water, the discharge rate may be obtained from the following empirical formula (refer to Folsom [1956] for details).

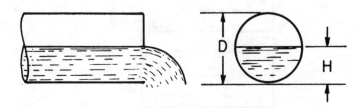

Figure 5.22   Liquid discharge from a partially filled pipe.

$$V = 2.54D^{2.56}K^{1.84} \qquad\qquad (5.149)$$

where V is in units of ℓ/min and D is in m. Equation 5.149 is valid for K = 0.2–0.6 and for D = 3–15 cm.

Figure 5.23 illustrates still another discharge system, that of flow over a weir. For weir flow it is important to know the crest over a weir for a given perimeter, L, to determine the overflow rate. Whitwell and Plumb [1939] developed the following empirical formula for water flow:

$$\frac{V}{L} = 304(H - 0.0032) \qquad\qquad (5.150)$$

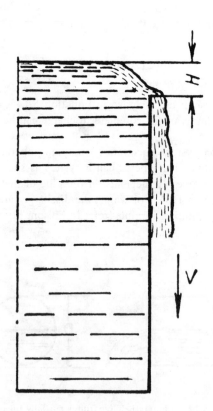

Figure 5.23   Flow over a weir.

where V has units of ℓ/s and L and H are in m. Equation 5.150 is applicable for $H < 0.01$ m. For $0.01 < H < 0.3$ m, the following formula is applicable:

$$\frac{V}{L} = 1670H^{1.455} \tag{5.151}$$

The above weir formulas were obtained with sharp-crested weirs. The shape of the weir edge (i.e., sharp or blunted) generally does not have a significant effect on the discharge rate.

Example 5.15

Determine the discharge time for a horizontal cylindrical cistern having flat ends 2 m in diameter and an overall length of 4 m. The cistern is filled with 12 tons of a liquid whose specific gravity is 1.1 and has a viscosity close to water. The diameter of the discharge hole is 27 mm.

Solution

The discharge time may be computed from Equation 5.145. We shall assume $\alpha = 0.62$. The cross-sectional area of the discharge opening is

$$S_o = \frac{3.14 \times 0.027^2}{4} = 5.7 \times 10^{-4}$$

Hence,

$$t = \frac{1}{0.62 \times 5.7 \times 10^{-4}\sqrt{2 \times 9.81}} \int_0^{H_1} SH^{-1/2}dH = 637 \int \frac{F}{\sqrt{H}}\,dH,\ s$$

where $F = L \times S$ (Figure 5.21).
We can write the following relationship for the tank radius:

$$R^2 = \left(\frac{S}{2}\right)^2 + (H - R)^2$$

where $R = D/2$.

Solving for the tank's cross section,

$$S = 2\sqrt{R^2 - (H - R)^2} = 2\sqrt{H(2R - H)} = 2\sqrt{H(D - H)}$$

Hence,

$$F = 2L\sqrt{H(D - H)} = 8\sqrt{H(2 - H)}$$

Substituting back into our expression,

$$t = 637 \int_0^{H_1} \frac{8H(2 - H)}{\sqrt{H}}\, dH = 5.1 \times 10^3 \int_0^{H_1} \sqrt{2 - H}\, dH$$

or

$$t = 9.62 \times 10^3 - 3.4 \times 10^3 (2 - H_1)^{3/2},\ s$$

H_1 is the initial liquid height, which is unknown. However,

$$V = \frac{12,000}{1,100} = 10.9\ m^3 \quad \text{and} \quad L = 4\ m$$

Therefore, $F_1 = 10.9/4 = 2.72\ m^2$, where F_1 is the initial vertical section of liquid.

The vertical section of the cistern is

$$F_2 = \pi D^2/4 = 3.14 \times 2^2/4 = 3.14\ m^2$$

Hence, the cistern vertical free section is

$$F_3 = 3.14 - 2.72 = 0.42\ m^2$$

For the ratio $F_3/D^2 = 0.105$, we find in Perry's *Chemical Engineer's Handbook* [1950] $H/D = 0.1915$. Therefore, $H = 0.1915 \times 2 = 0.383$ m, and the initial liquid depth in the cistern is

$$H_1 = D - H = 2 - 0.383 = 1.62 \text{ m}$$

Finally, discharge time, $t = 9620 - 3400(2 - 1.62)^{3/2} = 8824$ s, or

$$t = \frac{8824}{3600} = 2.5 \text{ hr}$$

HYDRAULIC RESISTANCES IN PIPE FLOW

Flow Through Pipes

A large chemical plant or refinery is analogous to the human body. Pumps and compressors perform functions similar to the heart, moving various chemicals and fluids to different reactors, which are the vital organs of the plant. Piping networks and ducts are the veins and arteries that transport the plant's processing fluids. The makeup of an entire plant can be almost as complex as the human body. As with most healthy humans, nature already has applied rigorous principles of hydrodynamics to properly sizing our hearts, to the energy consumption required by the heart and to the most appropriate layout of the arterial system.

In a similar manner, although perhaps less rigorously, we apply the same hydrodynamic principles to the distribution systems in plant operations. We direct our attention now to the determination of head losses or pressure drops. By determining such hydraulic resistances in piping distributions, proper selection and sizing of the plant's heart elements can be made, namely, the required energy ratings of pumps, compressors, fans, etc.

For the most general case, head losses are evaluated from friction and local resistances. *Frictional resistance* arises from the motion of real fluids through piping, and its value is influenced by the flow regime. Thus, in turbulent flow, resistances are characterized not only by viscosity but by the eddy viscosity, which depends on the hydrodynamic conditions. Such resistances result in additional energy losses.

Local resistances arise due to variations in fluid velocity caused by changes in flow area or direction. Among the many local resistances are friction losses caused by entrances and exits from piping, sudden contractions and expansions, losses in fittings, valves, etc. The loss of head, h_ϱ, therefore may be expressed by the sum of two terms:

$$h_\varrho = h_{fr} + h_{\varrho r} \tag{5.152}$$

where h_{fr} and $h_{\varrho r}$ are head losses due to friction and local resistances, respectively.

Friction losses for a laminar flow in a straight pipe may be determined analytically from the Poiseuille equation (Equation 5.27B).

For a horizontal pipe of constant cross section, i.e., $z_1 = z_2$ and $w_1 = w_2$, Bernoulli's equation is applied to determine the head loss due to friction:

$$\frac{P_1 - P_2}{\rho g} = \frac{\Delta P}{\rho g} = h_{fr}$$

Substituting $\Delta P = \rho g h_{fr}$ into Equation 5.27B and noting that V_{sec} may be estimated by the product of average velocity and the pipe's cross-sectional area, $\pi d^2/4$, we obtain

$$\overline{W} = \frac{\pi d^2}{4} = \frac{\pi d^4 \rho g h_{fr}}{128 \mu \ell}$$

where ℓ and d are the length and diameter of the pipe. Simplifying, and solving for the head loss,

$$h_{fr} = \frac{32 \overline{W} \mu \ell}{\rho g d^2}$$

Multiplying the numerator and denominator of the RHS of this expression by $2\overline{W}$, we obtain

$$h_{fr} = \frac{64 \mu}{\overline{W} d \rho} \frac{\ell}{d} \frac{\overline{W}^2}{2g}$$

or

$$h_{fr} = \frac{64}{Re} \frac{\ell}{d} \frac{\overline{W}^2}{2g} \tag{5.153}$$

Thus, the head losses due to friction are expressed in terms of the velocity head, $\overline{W}^2/2g$. The quantity 64/Re in Equation 5.153 is referred to as the *friction factor* for laminar flow, which we shall denote by λ.

Further, we denote ζ as the friction resistance coefficient defined as follows:

$$\zeta = \lambda \frac{\ell}{d} \tag{5.154}$$

Using this notation, Equation 5.153 is expressed in the following form:

$$h_{fr} = \zeta \frac{\overline{W}^2}{2g} = \lambda \frac{\ell}{d} \frac{\overline{W}^2}{2g} \tag{5.155}$$

And, for pressure losses ($\Delta P_{fr} = \rho g h_{fr}$),

$$\Delta P_{fr} = \lambda \frac{\ell}{d} \frac{\rho \overline{W}^2}{2} \tag{5.156}$$

Equation 5.155 has been found to agree well with experimental data for laminar flow, i.e., Re < 2100. For laminar flow, the friction factor is practically independent of pipe roughness.

Example 5.16

A small capillary tube with an inside diameter of 2.9 mm and length of 0.5 m is being used to measure the viscosity of a liquid having a density of 920 kg/m^3. The pressure drop reading across the capillary during flow is 0.08 m water (density = 998 kg/m^3). The volumetric flowrate through the capillary is 9.27×10^{-7} m^3/s. Ignoring end effects, determine the viscosity of the liquid.

Solution

First convert height, h, of 0.08 m of water to a pressure:

$$\Delta p = \rho g h_{fr}$$

$$= \left(998 \ \frac{kg}{m^3}\right)\left(9.80665 \ \frac{m}{s^2}\right)(0.08 \ m)$$

$$= 783 \ kg\text{-}m/s^2\text{-}m^2 = 783 \ N/m^2 \ (\text{or } 0.114 \ psia)$$

Assuming the flow to be laminar, we apply Poiseuille's equation (Equation 5.27B), solving for viscosity, μ:

$$\mu = \frac{\pi d^4 \Delta p_f}{128 \ell V_{sec}}$$

where
$\ell = 0.5$ m
$d = 2.9 \times 10^{-3}$ m
$V_{sec} = 9.27 \times 10^{-7}$ m^3/s
$\Delta p_f = 783$ N/m^2

Substituting in values,

$$\mu = \frac{\pi (2.9 \times 10^{-3}\ \text{m})^4 (783\ \text{N/m}^2)}{128(0.5\ \text{m})\left(9.27 \times 10^{-7}\ \dfrac{\text{m}^3}{\text{s}}\right)}$$

$$= 2.933 \times 10^{-3}\ \frac{\text{kg}}{\text{m-s}} = 2.933 \times 10^{-2}\ \text{p (or 2.93 cp)}$$

As it was assumed that the flow is laminar, the Reynolds number should be computed as a check:

$$\overline{W} = \frac{4V_{sec}}{\pi d^2} = \frac{4\left(9.27 \times 10^{-7}\ \dfrac{\text{m}^3}{\text{s}}\right)}{\pi (2.9 \times 10^{-3}\ \text{m})^2} = 0.1403\ \frac{\text{m}}{\text{s}}$$

$$\text{Re} = \frac{d\overline{W}\rho}{\mu} = \frac{(2.9 \times 10^{-3}\ \text{m})\left(0.1403\ \dfrac{\text{m}}{\text{s}}\right)\left(920\ \dfrac{\text{kg}}{\text{m}^3}\right)}{2.933 \times 10^{-3}\ \dfrac{\text{kg}}{\text{m-s}}} = 128$$

Thus, the flow is laminar as assumed.

Example 5.17

For the previous example problem, assume that the viscosity is known and the pressure drop, Δp_f, is to be predicted. Using the friction factor expression for laminar flow (Equation 5.156), compute Δp_f and compare.

Solution

The Reynolds number computed for Example 5.16 is Re = 128. Hence, the friction factor is

$$\lambda = \frac{64}{Re} = \frac{64}{128} = 0.50$$

Using Equation 5.156,

$$\Delta p_f = \lambda \left(\frac{\ell}{d}\right)\left(\frac{\rho \overline{W}^2}{2}\right) = 0.50 \left(\frac{0.5\ m}{2.9 \times 10^{-3}\ m}\right)\left(\frac{920\ \frac{kg}{m^3}}{2}\right)\left(0.1403\ \frac{m}{s}\right)^2$$

$$= 781\ \frac{kg}{m^2\text{-}s}$$

This agrees very well with the previous problem. Additional illustrative examples are given by Dodge [1968], King and Brater [1963], and King and Crocker [1967].

Equation 5.155 also may be used for conduits of noncircular cross sections, where diameter d is replaced by an equivalent diameter. The equivalent diameter, d_{eq}, is defined as 4 times the hydraulic radius, where the hydraulic radius is the cross-sectional area of flow divided by the wetted perimeter. In applying Equation 5.155, the laminar friction factor expression changes to the following:

$$\lambda = \frac{B}{Re} \qquad (5.157)$$

where B = 57 for a square cross section and B = 96 for an annular cross section. Equation 5.155 also may be applied to determining friction losses in the turbulent regime (see Hodson [1939], Rouse [1948] and Davies [1972]). However, in this case the friction factor relationship cannot be derived theoretically. Therefore, design equations for estimating λ in turbulent flow are obtained by generalized experimental data through the use of similarity theory. The generalized expression may be transformed into an exponential form:

$$Eu = A\,Re^m \left(\frac{\ell}{d}\right)^q \tag{5.158A}$$

From normalized data for fluid motion through smooth-walled piping in the Reynolds number range of 4000 to 100,000, the following constants in Equation 5.158 were evaluated:

$$A = 0.158 \ , \quad m = -0.25 \ , \quad q = 1$$

Hence, the design expression is

$$Eu = 0.158\,Re^{-0.25}\left(\frac{\ell}{d}\right) \tag{5.158B}$$

or

$$h_{fr} = 0.316\,\frac{\ell}{d}\,\frac{\overline{W}^2}{2g} \tag{5.158C}$$

Comparing Equations 5.158C and 5.155 reveals that the friction factor may be expressed in the following form:

$$\lambda = 0.316\,Re^{-0.25} \tag{5.159}$$

Hence, for laminar flow the frictional pressure drop is proportional to fluid velocity raised to the first power (Equation 5.153), whereas for the turbulent regime it is proportional to velocity raised to the 1.75 power.

The friction factor depends not only on the Reynolds number but also on the roughness of the pipe walls. Roughness is characterized by the effective height of the protrusions, e. Figure 5.24 shows a plot of friction factor versus Reynolds number for different piping materials. As noted earlier, the laminar friction factor shows no dependency on the condition of the pipe. However, a significant difference in the λ-Re relationship is seen for different material pipes in the turbulent regime. As shown by curve 2, rough pipes tend to lower λ, which increases frictional resistance. To normalize the effect of roughness, Moody [1944] developed the generalized friction factor plot shown in Figure 5.25. The ordinate on the right of Figure 5.25 is the inverse of relative roughness, where $\epsilon = e/d$.

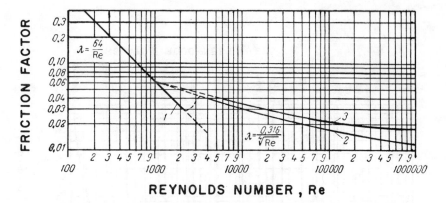

Figure 5.24   Friction factor vs Reynolds number chart: curve 1 = laminar flow for smooth and rough pipes; curve 2 = turbulent flow for smooth pipes made from copper, lead and glass; curve 3 = turbulent flow for rough pipes made from steel and cast iron.

In laminar flow through a pipe in which e/d is usually somewhat less than 0.01, the influence of wall roughness is insignificant because the fluid fills the spaces between the protrusions and the inner fluid layers slide smoothly over a pipe of effective diameter $d - 2e$. In some initial range of turblence, the wall roughness also can be neglected if it is smaller in height than the thickness of the viscous sublayer. In this case, the pipe is said to be hydraulically smooth; λ then can be calculated using Equation 5.159.

As the Reynolds number increases, the thickness of the viscous sublayer decreases. When the viscous sublayer becomes comparable to the effective height of wall protrusions, disturbances enter into the bulk flow, thus increasing turbulence and flow resistance. Under these conditions, the friction factor becomes more strongly dependent on roughness. Thus, λ and frictional pressure drop increase under the action of inertia forces due to the additional formation of vortexes about the protrusions.

From the above, it becomes apparent that as the Reynolds number increases, three regimes of flow or zones develop. The first is referred to as the zone of *smooth friction*, where $\lambda = f(Re)$. The second is the zone of *mixed friction*, where λ is both a function of Re and roughness. Finally, there is the *self-modeling* zone, where λ becomes practically independent of the Reynolds number and is primarily a function of wall roughness. The self-modeling zone also is called the *zone of quadratic resistance law* as $\lambda = f(\overline{W}^2)$.

The critical Reynolds numbers, $Re_{cr,1}$, where roughness begins to influence

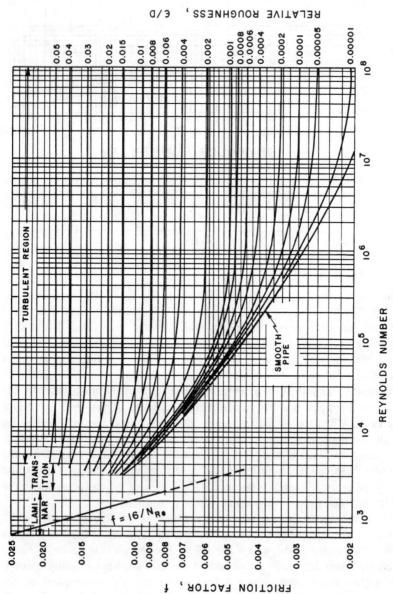

Figure 5.25   Reynolds number-friction factor chart as a function of relative roughness.

λ, as well as critical values, $Re_{cr,2}$, when λ is only a function of pipe roughness (refer to dashed line in Figure 5.25), depend on the *relative roughness*. These critical values have been determined experimentally to be the following:

$$Re_{cr,1} \simeq \frac{23}{\epsilon} \qquad (5.160)$$

and

$$Re_{cr,2} \simeq 220\epsilon^{-9/8} \qquad (5.161)$$

Hodson [1939] recommends the following correlation for all three zones of turbulent motion:

$$\frac{1}{\sqrt{\lambda}} = -2 \log \left[\frac{\epsilon}{3.7} + \left(\frac{6.81}{Re} \right)^{0.9} \right] \qquad (5.162)$$

For the zone of smooth friction, λ can be calculated either from Equation 5.159 or Equation 5.162. In the latter equation, the first term inside the parentheses is eliminated. This term reflects the influence of roughness, which, for this zone, is insignificant. Hence,

$$\frac{1}{\sqrt{\lambda}} = -2 \log \left(\frac{6.81}{Re} \right)^{0.9} = 1.8 \log Re - 1.5 \qquad (5.163)$$

For the self-modeling zone, where λ is independent of Re, Equation 5.163 has the following form:

$$\frac{1}{\sqrt{\lambda}} = 2 \log \frac{3.7}{\epsilon} \qquad (5.164)$$

Further clarification of the definition of friction factor is needed at this point. In this text we have elected to adopt the friction factor originally defined by Blasius and later used by Moody [1944] and others. The Blasius friction factor is defined as four times the Fanning friction factor (so, for laminar flow, $\lambda_F = 16/Re$). The chart given in Figure 5.25 is based on the

Blasius friction factor. Care must be exercised when performing calculations of frictional losses as some authors prefer to define a friction factor that is twice the size of the Fanning friction factor.

The following examples apply friction factors to estimate friction loss for steady flow in uniform circular pipes running full of liquid under isothermal conditions.

Example 5.18

Water is flowing full through a 50.8-mm (2.0-in.) i.d. wrought iron pipe at a rate of 2.7×10^{-3} m^3/s. The viscosity and density of the water under the process conditions are 1.09 cp and 992 kg/m^3, respectively. Determine the mechanical energy friction loss for a 73-m section of pipe.

Solution

The following information is known:

- $d = 0.0508$ m
- $\rho = 992$ kg/m^3
- $\ell = 73$ m
- $\mu = (1.09 \text{ cp})(1 \times 10^{-3}) = 1.09 \times 10^{-3}$ kg/m-s

The cross-sectional area of flow is

$$F = \frac{\pi}{4}(0.0508 \text{ m})^2 = 2.027 \times 10^{-3} \text{ m}^2$$

Hence, the liquid velocity is

$$w = \frac{2.7 \times 10^{-3} \text{ m}^3/\text{s}}{2.027 \times 10^{-3} \text{ m}^2} = 1.33 \text{ m/s}$$

The Reynolds number is computed as

$$Re = \frac{dw\rho}{\mu} = \frac{(0.0508)(1.33)(992)}{1.09 \times 10^{-3}} = 61,588$$

Hence, the flow is turbulent. For wrought iron pipe, a typical value for surface roughness is 0.00015 ft. Hence, the equivalent roughness is

$$\frac{e}{d} = \frac{0.00015 \text{ ft}}{0.0508 \text{ m}} \times \frac{\text{m}}{3.28 \text{ ft}} = 9.002 \times 10^{-4}$$

or

$$\left(\frac{e}{d}\right)^{-1} = \frac{1}{\epsilon} = 1111$$

From the friction factor chart, Figure 5.25, for $1/\epsilon = 1111$ and Re = 61,588, we obtain

$$\lambda = 0.023$$

The mechanical friction loss may be computed from Equation 5.156:

$$\hat{F} = \frac{\Delta p_{fr}}{\rho} = \lambda \frac{\ell}{d} \frac{w^2}{2}$$

$$= (0.023) \left(\frac{73 \text{ m}}{0.0508 \text{ m}}\right) \frac{(1.33 \text{ m/s})^2}{2}$$

$$= 29.23 \frac{\text{J}}{\text{kg}} \text{ or } 9.8 \frac{\text{ft-lb}_f}{\text{lb}_m}$$

Example 5.19

A liquid is flowing through a 250-mm i.d. commercial steel pipe (e = 2.5 mm) with a head loss of 5.3 m over 150 m of length. The kinematic viscosity of the process liquid is 2.3×10^{-6} m^2/s. Determine the volumetric flowrate.

Solution

We may use Equation 5.156 ($h_{fr} = \lambda(\ell/d)(w^2/2g)$) to solve for w:

$$e/d = \frac{2.5}{250} = 0.010$$

or

$$(e/d)^{-1} = \frac{1}{\epsilon} = 100$$

as λ is unknown, a trial and error solution for w must be made. An initial guess for λ is made. Assume $\lambda = 0.04$. Then,

$$5.3 \text{ m} = 0.04 \frac{150 \text{ m}}{0.25 \text{ m}} \frac{w^2}{2(9.806 \text{ m/s}^2)}$$

or

$$w = 2.08 \text{ m/s}$$

The Reynolds number computed from this velocity is

$$Re = \frac{wd}{\nu} = \frac{(2.08 \text{ m/s})(0.25 \text{ m})}{2.3 \times 10^{-6} \text{ m}^2/\text{s}} = 226,207$$

From the friction factor chart (Figure 5.25), this gives $\lambda = 0.037$. Interpolation between the assumed λ and the value obtained from Figure 5.25 gives a new $\lambda = 0.038$. Repeating the above calculations yields w = 2.14 m/s and a Reynolds number of 232,000. From Figure 5.25, Re of 232,000 reveals $\lambda \approx 0.038$.

Example 5.20

Oil is to be transferred at a rate of 28.9 m^3/hr through a horizontal wrought iron pipeline. The head of fluid available to overcome the friction loss is 7.3 m for a 530-m length of pipe. The density and viscosity of the liquid are 890 kg/m^3 and 3.8 cp, respectively. The roughness of the pipe to be used is 2.1×10^{-3} in. Determine the size of the pipe required to convey the oil.

Solution

The following information is given:
- $\rho = 890 \text{ kg/m}^3$
- $\mu = 3.8 \text{ cp} = 3.8 \times 10^{-3} \text{ kg/m-s}$

- $\ell = 530$ m
- $e = 2.1 \times 10^{-3}$ in. $= 5.33 \times 10^{-5}$ m
- $Q = 28.9$ m^3/hr $\times$ hr/3600 s $= 8.03 \times 10^{-3}$ m^3/s

And the friction loss, $\hat{F} = gh_{fr} = (9.80665 \text{ m/s}^2) (7.3 \text{ m}) = 71.59$ J/kg. The pipe diameter, d, is unknown, which means we also do not know the area of the pipe or the oil's velocity:

$$F = \frac{\pi d^2}{4}$$

$$w = \left(8.03 \times 10^{-3} \frac{\text{m}^3}{\text{s}}\right)\left(\frac{4}{\pi d^2 \text{m}^2}\right) = \frac{0.01022}{d^2} \text{ m/s}$$

Again we may apply Equation 5.126:

$$h_{fr}g = \lambda \frac{\ell}{d} \frac{w^2}{2}$$

However, a trial and error solution is required because λ and Re are functions of d and w.

For an initial guess, let d = 0.1 m. Then,

$$\frac{1}{\epsilon} = \frac{d}{e} = \frac{0.1 \text{ m}}{5.33 \times 10^{-5} \text{ m}} = 1876$$

$$w = \frac{0.01022}{(0.1)^2} = 1.02 \text{ m/s}$$

$$\text{Re} = \frac{dw\rho}{\mu} = \frac{(0.1)(1.02)(890)}{3.8 \times 10^{-3}} = 23{,}936$$

From Figure 5.25, for Re = 23,936 and $1/\epsilon = 1876$, we obtain a value for the friction factor, $\lambda = 0.026$. Hence,

$$h_{fr}g = (0.026)\left(\frac{530 \text{ m}}{0.1 \text{ m}}\right)\frac{(1.02 \text{ m/s})^2}{2} = 71.68 \text{ J/kg}$$

This agrees very well with the available head needed ($\hat{F} = 71.59$ J/kg). Hence, d = 0.1 m.

If the computed head did not agree, a second guess for d would have been necessary and the computations would have to have been repeated.

Alternative Solution

An alternative solution to Example 5.20 would be to program the equations using a convergence scheme. Equation 5.162 could be used to compute λ instead of using Figure 5.25. Such a computer program in BASIC language is listed in Table 5.3. Table 5.4 lists the results of several iterations performed by the program until convergence on the head loss is achieved. Note that the computer program predicts a required pipe diameter of 0.12 m. The difference between the two solutions is due to the inaccuracy in reading Figure 5.25.

The above friction factor expressions are based on fluids flowing under isothermal conditions. When the fluid is either heated or cooled through

Table 5.3 Computer Program for Solution to Example 5.20

```
10 DIM D(50)
20 REM PRGRM. TO CMPT. PIPE DIA

30 REM S=RHO/L=LENGTH/Q=FLOW/E=
   ROUGH/V=VISCOSIITY/F=FRICT./
   D=DIA./R=REYNOLDS/W=VELOCITY
   /A=AREA/G
40 READ S,L,Q,E,V,F,N
50 D(1)=.08
60 FOR I=1 TO N
70 A=.7854*D(I)^2
80 W=Q/A
90 E1=E/D(I)
100 R=D(I)*W*S/V
110 R1=E1/3.7+(6.81/R)^.9
120 G=-.5/LOG(R1)
130 F1=G*L*W^2/(2*D(I))
140 X=100*(F-F1)/F
150 IF X>-2 AND X<=2 THEN 180
160 D(I+1)=D(I)+.002
170 NEXT I
180 PRINT "AT ITERATIUN",I,"PIPE
    DIA. DETERMINED IN (M)=",D(
    I)
190 DATA 890,530,.00803,.0000533
    ,.0038,71.6,25
200 END



AT ITERATION           22
PIPE DIA. DETERMINED IN (M)=
   .122
```

Table 5.4 Computer Program Computations for Example 5.20

I	Re	Computed Head Loss (J/kg)	Difference in Head Loss (%)
1	29,932.4	582.6	−713.67
2	29,202.3	515.7	−620.21
3	28,507.0	457.8	−539.39
4	27,844.1	407.6	−469.27
5	27,211.2	363.9	−408.21
6	26,606.5	325.7	−354.89
7	26,028.1	292.2	−308.17
8	25,474.4	262.9	−267.12
9	24,943.6	237.0	−230.94
10	24,434.6	214.1	−198.99
11	23,945.9	193.8	−170.68
12	23,476.4	175.8	−145.54
13	23,024.9	159.8	−123.17
14	22,590.5	145.5	−103.21
15	22,172.1	132.7	−85.36
16	21,769.0	121.3	−69.37
17	21,380.3	111.0	−55.02
18	21,005.2	101.7	−42.11
19	20,643.0	93.4	−30.47
20	20,293.1	85.9	−19.97
21	19,954.9	79.1	−10.46
22	19,627.8	72.9	−1.86

AT ITERATION 22

PIPE DIA. DETERMINED IN (M) = 0.122

the pipe wall, viscosity changes over the flow cross section because of temperature gradients. Thus, the velocity field is modified and the λ varies, especially in the laminar region where the temperature gradients are higher than in turbulent bulk flow.

For all the friction factor expressions given (excluding Equation 5.164), correction factors must be applied. These corrections are based on the assumption that the fluid temperature can be represented by a mean bulk temperature (defined as the arithmetic average of the inlet and outlet temperature). The correction factors, denoted by ψ, are as follows:

$$
\text{For Re} > 2100: \ \psi = \begin{cases} \left(\dfrac{\mu}{\mu_w}\right)^{0.17} & \text{for heating} \\[2ex] \left(\dfrac{\mu}{\mu_w}\right)^{0.11} & \text{for cooling} \end{cases}
$$

$$\text{For Re} < 2100: \quad \psi = \begin{cases} \left(\dfrac{\mu}{\mu_w}\right)^{0.38} & \text{for heating} \\[2ex] \left(\dfrac{\mu}{\mu_w}\right)^{0.23} & \text{for cooling} \end{cases}$$

where μ = viscosity of fluid at the average bulk temperature
μ_w = viscosity at the temperature of the pipe wall

To obtain the friction factor corresponding to the fluid's average bulk temperature, divide the isothermal λ value by ψ.

Flow Through Varying Cross Sections

If the flow suddenly undergoes a change in cross-sectional area, additional head losses result. Figure 5.26 illustrates one example in which the flow experiences a sudden expansion. The head loss generated from this case is due to shocks as the flow from the smaller cross section impacts against a slower-moving fluid in the larger conduit.

The force of the rapid flow, pF_2, acts against the plane, F_2, in the larger section and in the opposite direction to the force from the slower-moving flow, p_1F_2. The momentum change, $d\phi$, with time, dt, is the following:

$$d\psi = (p_1F_2 - p_2F_2)dt \tag{5.165}$$

The fluid momentum over time, dt, is the product of mass and velocity. Hence, the change in momentum is

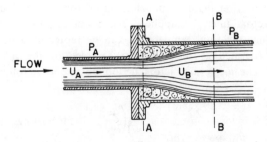

Figure 5.26   Flow through a sudden expansion.

$$(w_2 - w_1)dM = (w_2 - w_1) \frac{\gamma}{\rho} (F_2 w_2)dt \qquad (5.166)$$

where $F_2 W_2$ is the volumetric flowrate.

According to the second law of mechanics, Equations 5.165 and 5.166 may be equated:

$$\frac{p_1 - p_2}{\gamma} = \frac{w_2(w_2 - w_1)}{g} \qquad (5.167)$$

This expression shows that the total pressure change is due to resistance and changes in kinetic energy.

From Equation 5.167 it is determined that $p_2 > p_1$. Thus, despite resistances, the pressure in the larger pipe in Figure 5.26 will be greater than in the smaller one, resulting in a reversal of fluid motion, i.e., swirling occurs in the expansion's "corners." From the total pressure change, Equation 5.167 may be applied to calculate local resistances. For turbulent flow, the Bernoulli equation may be applied to the section preceding the expansion in the following form:

$$\frac{p_1}{\gamma} + \frac{w_1^2}{2g} = \frac{p_2}{\gamma} + \frac{w_2^2}{2g} + h_\varrho \qquad (5.168)$$

Substituting for $p_1 - p_2$ from Equation 5.167 into the above expression, we solve for resistance h_ϱ in units of height of liquid column:

$$h_\varrho = \frac{(w_1 - w_2)^2}{2g} \qquad (5.169)$$

Equation 5.169 is valid for any cross section of piping and also is applicable to liquid discharge from pipes into large vessels. For the latter case, $w_2 \ll w_1$, so resistance is equal to the kinetic energy of the flow in the piping, $h_\varrho = w^2/2g$. Experimental observations for turbulent flows of both compressible and incompressible fluids have verified Equation 5.169.

The effects of sudden contractions on flows are discussed by Mott [1972], Cheremisinoff [1979,1981a,b] and Simpson [1969]. Figure 5.27 illustrates flow through a sudden contraction. As shown, the stream cannot flow around sharp corners. The flow's cross section first contracts and then expands to fill the cross section of smaller piping. Because of these phenomena, energy

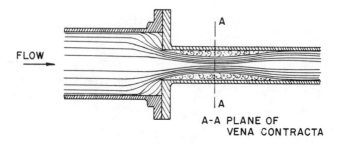

Figure 5.27   Flow through a sudden contraction.

losses or resistances arise, which may be estimated from the Weisbach equation:

$$h_\varrho = \left[0.04 + \left(\frac{1}{a} - 1\right)^2\right]\frac{w_2^2}{2g} = \psi\,\frac{w_2^2}{2g} \tag{5.170}$$

where a is the *contraction loss coefficient*, i.e., the ratio of the minimum flow cross section to the cross section of the smaller pipe, and W_2 is the average velocity in the smaller section.

The coefficients "a" and "ψ" depend on the ratio of the pipes' cross sections, F_2/F_1. Typical values are given in Table 5.5. Coefficient ψ also may be calculated from the following equation:

$$\psi = \frac{1.5\left(1 - \dfrac{F_2}{F_1}\right)}{3 - \dfrac{F_2}{F_1}} \tag{5.171}$$

Table 5.5 Coefficients of Contraction (for use in Equation 5.170)

F_1/F_2	0.01	0.1	0.2	0.4	0.6	0.8	1.0
a	0.6	0.61	0.62	0.65	0.7	0.77	1.0
ψ	0.5	0.46	0.42	0.33	0.23	0.13	0.0

These equations may be used to calculate the resistances encountered in liquid discharge from large vessels through orifices or for liquid discharges from vessels into piping. For the latter situation, F_2/F_1 in Equation 5.171 can be ignored. Recall that Equation 5.170 only expresses the head loss due to the change in flow cross section. The total head loss also accounts for the change in kinetic energy in accordance with the Bernoulli equation.

Example 5.21

Water is flowing at a rate of 1800 gal/hr through a distribution system. At one point in the system a 1-in. pipeline suddenly is expanded to 1.75 in. i.d.

1. Determine the energy loss that occurs due to the sudden enlargement.
2. Determine the difference between the pressure immediately ahead of the sudden enlargement and the pressure downstream from the enlargement.

Solution

Part 1

$$d_1 = 1 \text{ in.} = 0.0833 \text{ ft} \; ; \qquad F_1 = \frac{\pi}{4}(0.0833)^2 = 0.00545 \text{ ft}^2$$

$$d_2 = 1.75 \text{ in.} = 0.1458 \text{ ft} \; ; \qquad F_2 = \frac{\pi}{4}(0.1458)^2 = 0.01670 \text{ ft}^2$$

Hence,

$$w_1 = \frac{Q}{F_1} = \frac{1800 \text{ gph}}{0.00545 \text{ ft}^2} \times \frac{1 \text{ ft}^3/s}{449 \text{ gpm}} \times \frac{\text{hr}}{60 \text{ min}} = 12.3 \text{ fps}$$

$$w_2 = \frac{Q}{F_2} = \frac{1800}{0.01670} \times \frac{1}{449} \times \frac{1}{60} = 4.0 \text{ fps}$$

Using Equation 5.169,

$$h_\varrho = \frac{(w_1 - w_2)^2}{2g} = \frac{(12.3 - 4.0)^2 [\text{ft}^2/s^2]}{2(32.2 \text{ ft}/s^2)} = 1.07 \text{ ft}$$

This means that 1.07 ft-lb$_f$ of energy is dissipated from each lb$_m$ of water that flows through the sudden enlargement.

Part 2

To evaluate the pressure differential, we apply the Bernoulli equation in the following manner:

$$\frac{p_1}{\rho} + z_1 + \frac{w_1^2}{2g} - h_\ell = \frac{p_2}{\rho} + z_2 + \frac{w_2^2}{2g}$$

Solving for $p_1 - p_2$,

$$p_1 - p_2 = \rho[(z_2 - z_1) + (w_2^2 - w_1^2)/2g + h_\ell]$$

If we assume the expansion is horizontal, then $z_1 - z_2 = 0$. Hence,

$$p_1 - p_2 = 62.4 \frac{\text{lb}}{\text{ft}^3}\left(0 + \frac{(4.0)^2 - (12.3)^2}{(2)(32.2)} \text{ ft} + 1.07 \text{ ft}\right)$$

$$= 62.4\,(0 - 2.10 + 1.07)\text{lb/ft}^2$$

$$= -64.3 \frac{\text{lb}}{\text{ft}^2} \times \frac{1 \text{ ft}^2}{144 \text{ in}^2} = -0.447 \text{ lb}_f/\text{in}^2.$$

Hence, p_2 is greater than p_1 by 0.447 psi.

Example 5.22

By means of a momentum balance and mechanical energy balance, develop an expression for the loss that occurs for liquid flowing through the sudden expansion shown in Figure 5.26.

Solution

To write the momentum balance we select a control volume so as not to include the large pipe wall. The boundaries selected are defined by planes 0 and 2 in Figure 5.26. We may assume that the flow through plane 0 occurs only through an area F_1. The frictional drag force is neglected, and all the loss is assumed to be from the eddies within this volume. We note, therefore, that $p_0 = p_1$, $w_0 = w_1$ and $F_0 = F_2$. Making a momentum balance between planes 0 and 2,

$$p_0 F_2 - p_2 F_2 = Mw_2 - Mw_0$$

where the mass flowrates are

$$M = w_1 \rho F_1$$

And from continuity:

$$w_2 = (F_1/F_2)w_1$$

from continuity.

Substituting these terms into the momentum balance,

$$(p_1 - p_2)F_2 = w_1 \rho F_1 \left[\frac{F_1}{F_2} w_1 - w_1 \right]$$

Noting that

$$\frac{F_1}{F_2} = \frac{\frac{1}{4}\pi(d_1)^2}{\frac{1}{4}\pi(d_2)^2} = \left(\frac{d_1}{d_2}\right)^2$$

and rearranging, we obtain

$$\frac{p_2 - p_1}{\rho} = w_1^2 \left(\frac{d_1}{d_2}\right)^2 \left[1 - \left(\frac{d_1}{d_2}\right)^2 \right]$$

Applying the mechanical energy balance equation over planes 0 and 2,

$$\frac{w_1^2 - w_2^2}{2} - \Sigma \hat{F} = \frac{p_2 - p_1}{\rho}$$

Equating the last two expressions and solving for the friction loss, we obtain

$$\Sigma \hat{F} = \frac{w_1^2}{2} (1 - \beta)^2$$

where $\beta = (d_1/d_2)^2$.

The student should develop a derivation in a similar manner for flow through a sudden contraction (see Figure 5.27).

Similar losses occur when flows undergo gradual changes in cross section (Figure 5.28). For a smooth conical expansion, as shown in Figure 5.28A, where the cone's apex angle $\beta < 10°$, head losses may be estimated from the Eligner equation:

$$h_\varrho = \left(\frac{F_2}{F_1} - 1\right)^2 \sin\beta\left(\frac{w_1^2}{2g}\right) \tag{5.172}$$

For $\beta > 10°$, head losses may be estimated from the expression for a sudden expansion.

For a gradual expansion, as shown in Figure 5.28B, where β is in the range of 7 to 35°, head losses may be computed from the following empirical expression:

$$h_\varrho = 0.35\left(\log\frac{\beta}{2}\right)^{1.22}\frac{(w_1 - w_2)^2}{2g} \tag{5.173}$$

For gradual expansions with $\beta < 7°$, head losses are determined by integration of the Bernoulli and Darcy-Weisbach equations.

At $\beta > 40°$, head losses may become very high and even exceed those

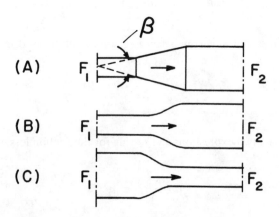

Figure 5.28   Flow in gradual expansions and contractions.

encountered in sudden expansions. Note also that the flow may be turbulent in the small cross section but laminar in the larger section.

For gradual contractions, as shown in Figure 5.28C, head losses are insignificant (especially for a smooth contraction surface and large Reynolds numbers). Equation 5.170 may be used, setting $h_\varrho = 0.05$, independent of the ratio F_2/F_1, provided that the flow is turbulent in the narrow cross section. If the flow in the contracted section is laminar (as in the case of discharge from a vessel), even when the resistance is small an unusual pressure decrease occurs that does not correspond to Poiseuille's law. This decrease occurs over a length equivalent to 0.065 ReD (the entrance region of a pipe). The pressure gradient, $p_0 - p_1$, for laminar flow at the entrance of a pipe of length L may be calculated from the data given in Table 5.6.

Example 5.23

Develop an expression for the average upstream velocity for liquid flow through a horizontal gradual contraction shown in Figure 5.28C. Assume the liquid to be of constant density. The expression should be written in terms of the pressure differential, $P_1 - P_2$, across the contraction.

Solution

Assign upstream conditions as $P_1(N/m^2)$, $F_1(m^2)$ and $w_1(m/s)$ and downstream conditions as $P_2(N/m^2)$, $F_2(m^2)$ and $w_2(m/s)$. As density is constant, $\rho_1 = \rho_2 = \rho$. And, from the mass balance continuity equation,

$$w_2 = \frac{w_1 F_1}{F_2}$$

where $F_1/F_2 = (d_1/d_2)^2$.

Table 5.6 Data for Estimating Pressure Losses[a]

$\frac{L}{D}$ Re	0.005	0.01	0.02	0.03	0.04	0.05	0.06
$\frac{p_0 - p_1}{\gamma \frac{w^2}{2g}}$	2.1	2.6	3.4	4.1	4.7	5.3	6.0

[a]D = pipe diameter; L = pipe length; γ = liquid specific gravity. (See Krylov [1947] for further details.)

As this is a horizontal flow system, $z_1 = z_2 = 0$, and the Bernoulli equation may be written as follows:

$$\frac{w_1^2}{2} + \frac{P_1}{\rho} = \frac{w_1^2(F_1/F_2)^2}{2} + \frac{P_2}{\rho}$$

or

$$\frac{\rho w_1^2}{2}[(F_1/F_2)^2 - 1] = P_1 - P_2$$

Hence,

$$w_1 = \sqrt{\frac{P_1 - P_2}{\rho}\,\frac{2}{(\beta^4 - 1)}}$$

in SI units, or

$$w_1 = \sqrt{\frac{P_1 - P_2}{\rho}\,\frac{2g_c}{(\beta^4 - 1)}}$$

in English units.

Example 5.24

Estimate the energy loss that occurs when 20 gpm of water flows from a 1-in. tube into a 2.9-in. tube through a gradual enlargement having an included angle of 30°.

Solution

$$Q = 20 \text{ gpm} \times \frac{ft^3}{7.48 \text{ gal}} \times \frac{\min}{60 \text{ s}} = 0.0446 \text{ ft}^3/\text{s}$$

$$F_1 = \frac{\pi}{4}\left(\frac{1}{12}\right)^2 = 0.00545 \text{ ft}^2 \;,\quad w_1 = \frac{0.0446}{0.00545} = 8.2 \text{ fps}$$

$$F_2 = \frac{\pi}{4}\left(\frac{2.9}{12}\right)^2 = 0.0459 \text{ ft}^2 \;;\quad w_2 = \frac{0.0446}{0.0459} = 0.97 \text{ fps}$$

Applying Equation 5.173,

$$h_\varrho = 0.35 \left(\log \frac{\beta}{2} \right)^{1.22} \frac{(w_1 - w_2)^2}{2g}$$

$$= 0.35 (\log 15)^{1.22} \frac{(8.2 - 0.97)^2}{2 \times 32.2} = 0.36 \text{ ft (or 0.16 psi)}$$

Alternative Solution

Another approach to estimating friction losses for any type of expansion or contraction is to apply Equation 5.169 with an appropriate discharge coefficient. That is,

$$h_\varrho = C_L \frac{(w_1 - w_2)^2}{2g}$$

Discharge coefficients for various cross-sectional changes are reported in the literature (see Streeter and Wylie [1970] or Perry [1950]). Figure 5.29 provides a plot of discharge coefficient versus diameter ratio for gradual enlargements.

Then, $d_2/d_1 = 2.9/1 = 2.9$ and, from Figure 5.29 (for $\beta = 30°$),

$$C_L \simeq 0.47$$

$$h_\varrho = 0.47 \frac{(8.2 - 0.97)^2}{2 \times 32.2} = 0.38 \text{ ft (0.17 psi)}$$

Note that often it is assumed for quick estimates that $w_1 \gg w_2$. Then,

$$h_\varrho = C_L \frac{w_1^2}{2g} = 0.47 \frac{(8.2)^2}{2 \times 32.2} = 0.49 \text{ ft (0.21 psi)}$$

The latter estimate actually is preferred when determining energy requirements for pumps (discussed in the next chapter). This provides a safety factor when selecting and sizing pump requirements.

Example 5.25

A large water holding tank is drained from a side nozzle 1.5-in. in diameter. Determine the energy loss as the water flow undergoes a sudden contraction. The water discharges at a rate of 27 gpm.

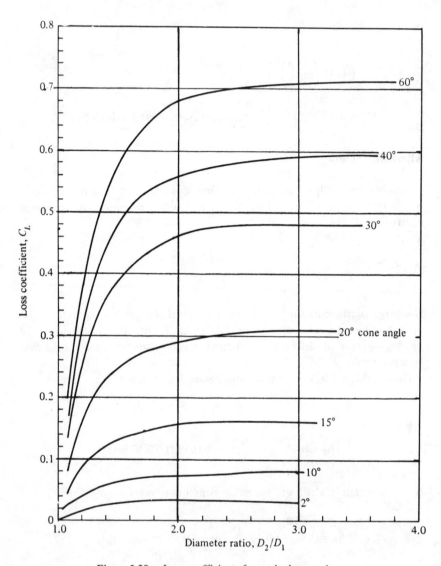

Figure 5.29   Loss coefficients for gradual expansions.

Solution

Energy losses may be estimated for the sudden contraction from the Weisbach equation (Equation 5.170):

$$h_\varrho = \left[0.04 + \left(\frac{1}{a} - 1\right)^2\right]\frac{w_2^2}{2g} = \psi\,\frac{w_2^2}{2g}$$

where $\quad \psi = 1.5(1 - F_2/F_1)/(3 - F_2/F_1) \simeq 0.5$ (as $F_1 \gg F_2$)

$Q = 27$ gpm $\times$ cfs$/449$ gpm $= 0.0602$ cfs

$F_2 = \pi/4\,(1.5/12)^2 = 0.0123$ ft^2; $w_2 = 0.0602/0.0123 = 4.91$ fps

and

$$h_\varrho = 0.5\,\frac{(4.91)^2}{2 \times 32.2} = 0.187 \text{ ft (or } 0.081 \text{ psi)}$$

Flow Through Piping Components

When fluids flow through fittings such as bends, elbows, etc., the flow direction is altered and the action of centrifugal force generates flow patterns that aggravate the bulk liquid motion. This phenomenon creates additional head losses that are best described for engineering calculations in terms of equivalent lengths of straight pipe of pipe diameters having the same head loss as the fittings. Beij [1938] first described such losses for fluid flow through 90° pipe bends. Head losses in fittings and valves may be expressed by a simple relationship of the following form:

$$h_\varrho = K_e\,\frac{w_1^2}{2} \tag{5.174}$$

where K_e is the head loss coefficient, which is a function of the ratio of the fittings' equivalent length to diameter ratio, L_e/D. Table 5.7 gives typical values of K_e for different valves and fittings. Note that w_1 is the velocity in the pipe immediately upstream of the fitting. Application of Equation 5.174 is limited exclusively to turbulent flows.

To evaluate total head loss for a piping system, equivalent lengths, L_e, of fittings and components are added to straight lengths of pipe in the system. That is, the total piping over which losses occur $= L_e$ + length of straight pipe. The Darcy equation then can be applied to obtain total head losses for a system:

$$\hat{F}_f = \frac{\Delta P_f}{\rho} = 4\lambda_F\,\frac{\Sigma L_e}{D}\,\frac{w^2}{2} \tag{5.175}$$

Table 5.7 Friction Loss Coefficients for Turbulent Flow Through Values and Fittings

Type Fitting or Valve	Head Loss Coefficient, K_e	L_e/D
Elbow, 45°	0.35	17
Elbow, 90°	0.75	35
Tee	1	50
Return Bend	1.5	75
Coupling	0.04	2
Union	0.04	2
Gate Valve		
Wide open	0.17	9
Half open	4.5	225
Globe Valve		
Wide open	6.0	300
Half open	9.5	475
Angle Valve		
Wide open	2.0	100
Check Valve		
Ball	70.0	3500
Swing	2.0	100
Water Meter, disk	7.0	350

where λ_F is the Fanning friction factor, i.e., one-fourth the value obtained from Figure 5.25.

For turbulent flow through copper bends, Konakov [1949] recommends the following formula:

$$\frac{L_e}{D} = 0.0202\, Xa^{1.10} Re^{0.032} \qquad (5.176)$$

where "a" is the bending angle in grades, and X is an empirical function of the ratio of bending radius R to pipe radius r. Values for X are given in Table 5.8.

Table 5.8 Values for Parameter X in Equation 5.176
(Data of Konakov [1949])

R/r	2	4	6	8	10	20	32
X	4	1.2	1.2	1.7	2.2	4.8	7.6

Values given in Table 5.8 indicate that L_e/D is, for practical purposes, independent of the Reynolds number under turbulent conditions, and that the minimum resistance occurs at $R/r = 5$. However, for laminar flow, equivalent length, L_e, is a function of Re. For 90° elbows, the data of Wilson [1922] (Table 5.9) are recommended.

Example 5.26

Determine the equivalent length of pipe (in feet) of a wide-open angle valve positioned in a 6-in. Schedule 40 pipe.

Solution

From Table 5.7, the equivalent length ratio L_e/D for a wide-open angle valve is 100. The inside diameter of 6-in. Schedule 40 pipe is 0.5054 ft. Hence,

$$L_e = (L_e/D)(D) = 100 \times 0.5054 = 50.5 \text{ ft}$$

Example 5.27

Estimate the pressure loss across a one-half-open gate valve located in a horizontal 6-in. Schedule 40 steel pipeline carrying 370 gpm of SAE 10W oil at 92°F.

Solution

To evaluate the pressure drop we shall apply the energy equation over a short section of piping containing the valve:

$$\frac{P_1}{\gamma} + z_1 + \frac{w_1^2}{2g} - h_\varrho = \frac{P_2}{\gamma} + z_2 + \frac{w_2^2}{2g}$$

Table 5.9 Equivalent Lengths of 90° Elbows for Laminar Flow
(Data of Wilson [1922])

Re	10	50	100	250	500	800	1000
L_e/D	2.5	3.5	6.0	12.0	22	27	30

where subscripts 1 and 2 refer to points immediately before and after the valve, respectively. Note that h_ϱ is the minor loss, which, for this system, is the valve alone.

Solving the energy equation for the pressure drop, and as this is a horizontal system, $Z_1 = Z_2 = 0$:

$$P_1 - P_2 = \gamma \left(0 + \frac{w_2^2 - w_1^2}{2g} + h_\varrho \right)$$

Now as the pipeline does not undergo a change in its cross section (except for the valve), we may assume that the flow quickly reaches steady state shortly after the valve and, hence, that $w_1 = w_2$. Therefore,

$$P_1 - P_2 = \rho g h_\varrho$$

The head loss can be determined from Equation 5.175:

$$h_\varrho = 4\lambda_F \frac{\Sigma L_e}{D} \frac{W^2}{2g} = \lambda \left(\frac{\Sigma L_e}{D} \right) \frac{W^2}{2g}$$

From Table 5.7, for a half-open gate valve, $L_e/D = 225$. The actual diameter of a 6-in. Schedule 40 pipe is 0.5054 ft. Hence,

$$F = \frac{\pi}{4}(0.5054)^2 = 0.201 \text{ ft}^2$$

and

$$w = \frac{370 \text{ gpm}}{0.201 \text{ ft}^2} \times \frac{1 \text{ ft}^3/\text{s}}{449 \text{ gpm}} = 4.1 \text{ fps}$$

The kinematic viscosity of the oil is $\nu = 1.7 \times 10^{-4}$ ft^2/s. Hence, the Reynolds number is

$$Re = \frac{wD}{\nu} = \frac{(4.1)(0.5054)}{1.7 \times 10^{-4}} = 12,190$$

The roughness e = 1.2×10^{-4} ft. Hence,

$$\epsilon = \frac{D}{e} = \frac{0.5054}{1.2 \times 10^{-4}} = 4212$$

From Figure 5.25, the friction factor λ is obtained; λ = 0.031 or λ_F = 7.75×10^{-3}. Hence,

$$h_\varrho = (0.031)(225)\frac{(4.1)^2}{(2)(32.2)} = 1.82 \text{ ft}$$

$$P_1 - P_2 = \gamma h_\varrho = (0.875)(62.4)\frac{lb}{ft^3} \times 1.82 \text{ ft} \times \frac{ft^2}{144 \text{ in}^2.} = 0.69 \text{ psi}$$

Hence, the pressure in the oil drops by 0.69 psi as it flows through the half-open valve. The student should repeat the problem for a fully open gate valve and a half-open globe valve. Compare results and comment.

Example 5.28

A large holding tank, 23 ft tall, is used to supply water to an irrigation ditch. The piping distribution for the system is shown in Figure 5.30. The farmer wants to maintain a flowrate of 150 gpm at the field but is not sure the tank is tall enough to maintain a water level to support that flow. Determine the level required in the tank to maintain a discharge rate of 150 gpm and whether the tank is tall enough. The pipe is all cast iron with a roughness e = 1.8×10^{-4} ft.

Solution

The mechanical energy balance is written over points 1 and 2:

$$z_1 \frac{g}{g_c} + \frac{w_1^2}{2\alpha g_c} + \left(\frac{P_1}{\rho_1} - \frac{P_2}{\rho_2}\right) - \hat{W}_s = z_2 \frac{g}{g_c} + \frac{w_2^2}{2\alpha g_c} + \Sigma\hat{F}$$

where $\rho = 62.4 \text{ lb/ft}^3$
$\mu = 0.35 \text{ cp} = 2.3 \times 10^{-4} \text{ lb}_m/\text{ft-s}$
$Q = 150 \text{ gpm} \times \text{cfs}/449 \text{ gpm} = 0.334 \text{ ft}^3/\text{s}$

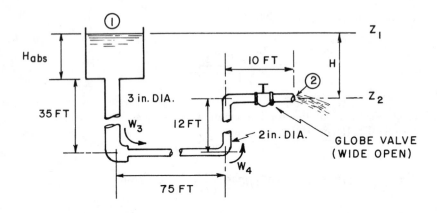

Figure 5.30   Water irrigation piping for Example 5.28.

For the 3-in.-diameter line,

$$F_3 = \frac{\pi}{4}\left(\frac{3}{12}\right)^2 = 0.0491 \text{ ft}^2 \;\; ; \;\; w_3 = \frac{0.334}{0.0491} = 6.80 \text{ fps}$$

For the 2-in.-diameter line,

$$F_4 = \frac{\pi}{4}\left(\frac{2}{12}\right)^2 = 0.0218 \text{ ft}^2 \;\; ; \;\; w_4 = \frac{0.334}{0.0218} = 15.31 \text{ fps}$$

Total friction losses, $\Sigma\hat{F}$, include:

1. contraction loss at tank exit,
2. friction in the 3-in. straight pipe,
3. friction loss in the 3-in. elbow,
4. sudden contraction from 3-in. to 2-in. pipe,
5. friction in the 2-in. straight pipe,
6. friction in 2-in. elbows, and
7. friction across the wide-open globe valve.

We will evaluate each loss individually:

1. **Contraction loss at the tank exit.** Equation 5.170 is applied where it is assumed that the tank's cross-sectional area is very large in comparison to the 3-in.-diameter exit. Hence,

$$\psi = 0.5$$

and

$$h_\varrho = \psi \, \frac{w_3^2}{2g} = 0.5 \, \frac{(6.80)^2}{2 \times 32.2} = 0.359 \text{ ft} \text{ or } 0.359 \text{ ft-lb}_f/\text{lb}_m$$

since $\Delta P/\rho = h_\varrho g/g_c$.

2. **Friction in the 3-in. pipe.** First compute the Reynolds number:

$$Re = \frac{d_3 w_3 \rho}{\mu} = \frac{\left(\dfrac{3}{12}\right)(6.80)(62.4)}{2.3 \times 10^{-4}} = 461{,}218$$

The flow is turbulent. Next, determine the relative roughness of the pipe:

$$\frac{d_3}{e} = \frac{3/12 \text{ ft}}{1.8 \times 10^{-4} \text{ ft}} = 1389$$

From Figure 5.25 the friction factor is determined to be $\lambda = 0.018$, or the Fanning friction factor is $\lambda_F = 4.5 \times 10^{-3}$. Applying Equation 5.175,

$$\hat{F}_f = 4\lambda_F \, \frac{\Sigma L_e}{D} \frac{w^2}{2} \cdot \frac{1}{g_c}$$

$$= 4(0.0045)\left(\frac{35 \text{ ft}}{0.25 \text{ ft}}\right) \frac{(6.80)^2 [\text{ft}^2/\text{s}^2]}{2} \times \frac{\text{lb}_f\text{-s}^2}{32.174 \text{ ft-lb}_m}$$

$$\hat{F}_f = 1.81 \text{ ft-lb}_f/\text{lb}_m$$

3. **Friction loss in the 3-in. elbows.** From Table 5.7 the head loss coefficient for a 90° elbow is $K_e = 0.75$. Applying Equation 5.174,

$$h_\varrho = K_e \, \frac{w_3^2}{2g_c} = (0.75) \, \frac{(6.80)^2}{2(32.174)}$$

$$= 0.539 \, \frac{\text{ft-lb}_f}{\text{lb}_m}$$

4. **Sudden contraction loss from 3-in. to 2-in. pipe.** Applying Equations 5.170 and 5.171, where

$$F_4/F_3 = \frac{0.0218}{0.0491} = 0.444$$

Hence,

$$\psi = \frac{1.5(1 - 0.444)}{3 - 0.444} = 0.326$$

$$h_\varrho = \psi \frac{w_4^2}{2g_c} = 0.326 \frac{(15.31)^2}{(2)(32.174)} = 1.187 \frac{\text{ft-lb}_f}{\text{lb}_m}$$

5. **Friction in the 2-in. straight pipe.**

$$Re = \frac{d_4 w_4 \rho}{\mu} = \frac{(2/12)(15.31)(62.4)}{2.3 \times 10^{-4}} = 692,300$$

and

$$\frac{d}{e} = \frac{0.167 \text{ ft}}{1.8 \times 10^{-4} \text{ ft}} = 928$$

From Figure 5.25, $\lambda = 0.020$ or $\lambda_F = 0.005$. Applying Equation 5.175, where the total length of 2-in. piping is $\Sigma L = 75 + 12 + 10 = 97$ ft,

$$\hat{F}_f = 4\lambda_F \frac{\Sigma L}{D} \frac{w_4^2}{2g_c} = 4(0.005) \frac{(97)(15.31)^2}{2(32.174)} = 7.07 \frac{\text{ft-lb}_f}{\text{lb}_m}$$

6. **Losses in the two 2-in. elbows.** From Table 5.7 $K_e = 0.75$. Applying Equation 5.174 again and noting there are two elbows,

$$h_\varrho = 2K_e \frac{w_4^2}{2g_c} = (2)(0.75) \frac{(15.31)^2}{2(32.174)} = 5.46 \frac{\text{ft-lb}_f}{\text{lb}_m}$$

7. **Friction loss across the globe valve.** From Table 5.7 $L_e/D = 300$ for a fully open globe valve. As determined from item 5, $\lambda_F = 0.005$. Note that $w_2 = w_4$. Hence,

$$h_\varrho = 4\lambda_F \frac{\Sigma L_e}{D} \frac{w_2^2}{2g_c} = 4(0.005)(300) \frac{(15.31)^2}{(2)(32.174)}$$

$$= 21.86 \frac{\text{ft-lb}_f}{\text{lb}_m}$$

Hence, the total friction loss for the system is the sum of items 1–7:

$$\Sigma \hat{F}_f = 0.359 + 1.81 + 0.539 + 1.187 + 7.07 + 5.46 + 21.86$$

$$= 38.29 \frac{\text{ft-lb}_f}{\text{lb}_m}$$

Assigning the reference datum at point 2, then $z_2 = 0$ and $z_1 = H$. If the tank is large, we may assume $w_1 = 0$. Also, $w_2 = w_4 = 15.31$ fps. And, as the flow is turbulent, $\alpha = 1.0$.

Both the tank and the discharge pipe are open to the atmosphere, thus $P_1 = P_2$. Furthermore, it is reasonable to assume $\rho_1 = \rho_2$, so

$$\frac{P_1}{\rho_1} - \frac{P_2}{\rho_2} = \frac{P_1 - P_2}{\rho} = 0$$

Finally, no work is performed on the fluid by a pump, so $\hat{W}_s = 0$. Substituting these values into the energy equation,

$$H \frac{g}{g_c} + 0 + 0 - 0 = 0 + \frac{(15.31)^2}{(2)(1)(32.174)} + 38.29$$

$$H = 41.93 \text{ ft}$$

Hence, the water level in the tank must be maintained at a height of 41.93 ft above the discharge outlet, or the absolute tank level should be

$$H_{abs} = H - (35 - 12) = 41.93 - (23) = 18.9 \text{ ft}$$

Hence, the tank will not overflow.

Example 5.29

Water at t = 20°C is being pumped from an open tank located 5 m below the pump to another open tank positioned 20 m above the pump. The pipe is 52.5 mm in diameter and the system is shown in Figure 5.31. The horizontal pipe section in which the pump is positioned is 100 m long, and the horizontal section above the elevated tank is 10 m long. A fully open globe valve is on the discharge line. The efficiency of the pump is 75%. Determine the pump's power.

Solution

The volumetric rate is

$$V_{sec} = \frac{5000}{1000 \times 3600} = 0.00139 \text{ m}^3/\text{s}$$

Pipe cross-sectional area is

$$F = \frac{3.14 \times 0.0525^2}{4} = 0.00216 \text{ m}^2$$

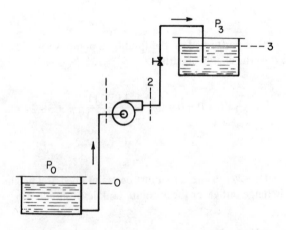

Figure 5.31   Pumping system for Example 5.29.

The superficial water velocity in the piping is

$$w = \frac{V}{F} = \frac{0.00139}{0.00216} = 0.64 \text{ m/s}$$

The kinetic energy of 1 kg of water is

$$\frac{w^2}{2g} = \frac{0.64^2}{2 \times 9.81} = 0.021 \text{ m}$$

$t = 20°C$ and $\mu \simeq 1$ cp, then $\mu g = 10^{-3}$ kg/m-s and the specific weight is $\gamma = 10^3$ kg/m^3. Hence, the Reynolds number is

$$Re = \frac{wd\gamma}{\mu g} = \frac{0.64 \times 0.0525 \times 10^3}{10^{-3}} = 33,600$$

As the flow is turbulent, we may use the following relationship to compute the friction factor:

$$\lambda = 0.0123 + \frac{0.7544}{Re^{0.38}}$$

$$\lambda = 0.0123 + \frac{0.7544}{33,600^{0.38}} = 0.0277$$

The actual length of piping from the bottom tank to the pump is 107 m. The piping has one 90° elbow, which has an L_e/D of 35 from Table 5.7. Consequently, the total equivalent piping length before the pump is

$$L = 107 + 35 \times 0.0525 = 108.8 \text{ m}$$

Applying the Darcy-Weisbach equation, the resistance before the pump is

$$Z_{0,1} = \lambda \frac{Lw^2}{2gd} = 0.027 \frac{108.8}{0.0525} \times 0.021$$

$$= 1.2 \text{ m}$$

The length of the piping after the pump is 30 m. This section has two elbows and a valve having an L_e/d = 300 (from Table 5.7). Hence, the equivalent length after the pump is

$$L = 30 + 0.0525 (2 \times 30 + 300) = 49 \text{ m}$$

The resistance after the pump will be

$$Z_{2,3} = 0.0277 \times \frac{49}{0.0525} \times 0.021 = 0.54 \text{ m}$$

The pressure difference before and after the pump is

$$\frac{P_0 - P_1}{\gamma} = (Z_3 - Z_0) + Z = 27.0 + 1.2 + 0.54$$

$$= 28.7 \text{ m}$$

because $P_3 = P_0 = 1$ atm, $Z = 0$ and $Z_3 = 27$ m. Consequently,

$$P_2 - P_1 = 28.7 \times 10^3 \text{ kg/m}^2$$

Hence, the pump's power is

$$N = \frac{(P_2 - P_1)V}{\eta} = \frac{(28.7 \times 10^3)0.00139}{0.75}$$

$$= 53.2 \frac{\text{kg-m}}{\text{s}}$$

or

$$N = \frac{53.2}{102} = 0.53 \text{ kW (or 0.71 hp)}$$

Coils, often used in a variety of process equipment, have critical Reynolds numbers, Re_{cr}, very different from those encountered in straight pipes. Here, the friction factor is a function of the ratio of coil diameter, D, to pipe diameter, d. Typical critical values for different D/d ratios are given in Table 5.10.

Table 5.10 reveals that the smaller the coil diameter, the higher the critical value of the Reynolds number and, consequently, the longer the flow remains in the laminar regime. That is, the critical Re of 2100 marking the transition between laminar and turbulent flows through pipes is greatly exceeded for coils. The resistance of laminar flow in coils may be estimated from the Darcy-Weisbach equation,

$$h_\varrho = -\frac{\Delta P}{\gamma} = \lambda \frac{L}{D} \frac{w^2}{2g}$$

using the following empirical friction factor expression developed by White [1929a,b]:

$$\lambda = C \frac{64}{Re} \qquad (5.177)$$

Coefficient C approaches unity with increasing coil radius of curvature. C may be computed from the following formula or from the values given in Table 5.11:

$$C = \frac{1}{1 - \left[1 - \left(\frac{11.6}{Re \sqrt{d/D}}\right)^{0.45}\right]^{1/0.45}} \qquad (5.178)$$

Table 5.10 Critical Reynolds Numbers for Coils

D/d	15.5	18.7	50	2050
Re_{cr}	7600	7100	6000	2270

Table 5.11 Resistances in Coils (values computed from Equation 5.178)

$Re \sqrt{d/D}$	10	50	100	250	400	600	1000	2000
C	1.0	1.2	1.5	2.0	2.5	3.0	4.5	5.0

For turbulent flow through coils of $D/d > 500$, resistances are comparable to those in straight pipes. At very high Reynolds numbers ($Re > 110,000$), Jeschke [1925] found that the resistance coefficient, λ, is independent of Re and could be expressed by the following:

$$\lambda = 0.0238 + 0.0891 \; d/D \qquad (5.179)$$

To estimate hydraulic resistances through coils, one may assume the coil to be a return bend and apply Equation 5.176 to obtain an L_e/D ratio.

Until now, the analyses presented have focused on evaluating flow resistances through piping of uniform or round cross sections. In practice, however, other configurations are encountered frequently. One example is an annular cross section in a heat exchanger, where the fluid must flow between two concentric pipes. The formulae and analyses presented are still valid for other configurations; however, the Reynolds number and, consequently, the friction factor, are redefined in terms of an equivalent diameter.

As previously noted, the *equivalent diameter* is defined as four times the hydraulic radius, r_h, where r_h is the ratio of the cross-sectional area of flow and the wetted perimeter, i.e., the perimeter of the channel contacting the fluid. That is,

$$d_{eq} = 4r_h \qquad (5.180A)$$

and

$$r_h = \frac{F}{p} \qquad (5.180B)$$

where   F = cross-sectional area of channel
         p = wetted perimeter of the channel

Table 5.12 gives specific formulas for the hydraulic radius for channels of various configurations.

For an elliptical cross section, coefficient k in Table 5.12 must be computed. k is a function of the ratio $\dot{s} = (a - b)/(a + b)$. Values are given in Table 5.13.

By redefining Re in terms of r_h, head losses for turbulent flows may be estimated from Equation 5.175 by virtually the same computation steps outlined in the example problems presented earlier. Unfortunately, this design method is not reliable for laminar flows.

Table 5.12 Hydraulic Radii for Different Channel Configurations

Cross Section	r_h
Circular pipe, diameter D	D/4
Annulus between two concentric pipes—D and d are the outside and inside diameters of the annulus	$\dfrac{D-d}{4}$
Rectangular duct with sides a and b	$\dfrac{ab}{2(a+b)}$
Square duct with a side a	a/4
Ellipse with axes a and b	$\dfrac{ab}{K(a+b)}$
Semicircle of diameter D	D/4
A shallow flat layer with depth h	h
Liquid film with thickness t on the vertical pipe of diameter D	$t - \dfrac{t^2}{D} \simeq t$

Table 5.13 Coefficients for Evaluating the Hydraulic Radius of an Ellipse

s	0.2	0.3	0.4	0.5	0.6	0.7	0.8	0.9	1.0
k	1.010	1.023	1.040	1.064	1.092	1.127	1.168	1.216	1.273

There are two approaches to evaluating resistances for laminar flows through irregular configurations. In the first method, the Reynolds number is evaluated from an equivalent diameter defined by Equations 5.180A and 5.180B, whence a friction factor may be determined from an expression of the following form:

$$\lambda = \frac{a}{Re} \tag{5.181}$$

The value of "a" depends on the flow geometry. Othmer [1945] has given values for different configurations. These values are given in Table 5.14, along with formulas for the equivalent diameter.

The second method for estimating losses in the laminar regime for non-circular cross sections is based on the derivation of Poiseuille-type equations.

Table 5.14 Equivalent Diameters and Values for Parameter "a" in Equation 5.18

Cross Section	h/b	D_{eq}	a
Circular with diameter D	–	D	64
Square with side h	–	h	57
Triangle with side h	–	0.58h	53
Ring with width b	–	2b	96
	0.7	1.17h	65
	0.5	1.30h	68
Ellipse with axes h and b	0.3	1.44h	73
(h = small axis)	0.2	1.50h	76
	0.1	1.55h	78
	1/∞	2h	96
	0.1	1.82h	85
Rectangle with sides h and b	0.2	1.67h	76
(h = small side)	0.25	1.60h	73
	0.33	1.50h	69
	0.50	1.33h	62

As an example, consider the flow configuration of an annular cross section. The derivation of Poiseuille's law, presented earlier, results in an expression for the resistance per unit length of channel:

$$-\frac{dP}{dL} = \frac{32\,\mu\overline{W}}{D^2 + d^2 - \dfrac{D^2 - d^2}{\ln D/d}} \qquad (5.182)$$

where $\overline{W}$ is the average superficial velocity. For a rectangular section of sides "a" and "b," Poiseuille's equation takes the following form:

$$-\frac{dP}{dL} = \frac{4\overline{W}\mu}{abn} \qquad (5.183)$$

n is a function of a/b; typical values are given in Table 5.15.

For an elliptical cross section, the head loss is expressed as

$$-\frac{dP}{dL} = \frac{4\overline{W}\mu(a^2 + b^2)}{a^2 b^2} \qquad (5.184)$$

where "a" and "b" are the semiaxes of an ellipse.

Table 5.15 Values of n in Equation 5.183 for a Rectangular Cross Section

a/b	0.1	0.2	0.3	0.4	0.5	0.6	0.7	0.8	0.9	1.0
n	0.03	0.06	0.08	0.10	0.11	0.133	0.136	0.138	0.139	0.140

Finally, let us consider two infinitely wide plates with a distance of 2b separation. Poiseuille's equation may be written as follows:

$$-\frac{dP}{dL} = \frac{3\mu \overline{W}}{b^2} \qquad (5.185)$$

The velocity distribution between the two parallel plates is given by the following equation:

$$w = \frac{3}{4} \frac{\Gamma(b^2 - y^2)}{b^3 \gamma} \qquad (5.186)$$

where Γ = weight rate per unit plate width
$\quad\quad\quad$ y = distance to the plane of symmetry (i.e., the plane in the middle between the plates)
$\quad\quad\quad$ γ = specific gravity

Velocity Dampening

For simple pipe flow we have observed the formation of a characteristic velocity distribution across the area of flow, with a maximum occurring at the pipe axis and zero at the pipe wall. The velocity gradient, especially in laminar flow, may have a relatively high value. In some industrial applications it is desirable to have almost a flat velocity profile, for example, in gas cleaning applications and sedimentation.

The dampening or smoothing of the velocity profiles may be accomplished by one of two methods described by Stokes [1946]. The first, shown in Figure 5.32A, is based on a continuous flow constriction; the second (Figure 5.32B) involves the use of a porous or a holed distributor. The advantage of the first method is that only a small additional resistance to the flow develops. Although the second method has a considerably higher head loss, costly fabrication and installation of the constriction are avoided. We shall analyze each approach from the principles already presented.

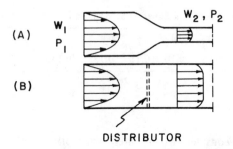

Figure 5.32 Two methods for dampening out velocity profiles.

Applying the Bernoulli equation to the first method (Figure 5.32A), flow elements moving along the axis may be expressed as follows:

$$w_{max_2}^2 - w_{max_1}^2 = \frac{2g}{\gamma}(P_1 - P_2) \tag{5.187}$$

where subscripts 1 and 2 refer to conditions before and after the constriction, respectively. If mean velocity $\overline{W}_1$ is known, it is possible to obtain a similar relationship in terms of average velocities:

$$\overline{W}_2'^2 - \overline{W}_1^2 = \frac{2g}{\gamma}(P_1 - P_2) \tag{5.188}$$

$\overline{W}_2'$ is not equivalent to the velocity in the constricted piping because the kinetic energy is determined from the average velocity, $\overline{W}^2/2g\alpha$, which contains a coefficient α accounting for the change in local velocities. The average velocity in the reduced section, $\overline{W}_2$, will be somewhat different from $\overline{W}_2'$ in Equation 5.188.

Equating Equations 5.187 and 5.188,

$$\overline{W}_2'^2(1 - \hat{\delta})^2 - \overline{W}_1^2 = w_{max_2}^2 - w_{max_1}^2 \tag{5.189A}$$

where

$$\hat{\delta} = \frac{\overline{W}_2 - \overline{W}_2'}{\overline{W}_2} \tag{5.189B}$$

From continuity, we note that

$$\overline{W}_1 F_1 = \overline{W}_2 F_2$$

where F_1 and F_2 are the cross-sectional areas of the large and constricted piping, respectively. Combining this with Equation 5.189A, the following relationship is obtained:

$$\frac{w_{max2}}{\overline{W}_2} = \sqrt{1 + \left(\frac{F_2}{F_1}\right)^2 \left[\left(\frac{w_{max1}}{\overline{W}_1}\right)^2 - 1\right] - 2\hat{\delta} + \hat{\delta}^2} \qquad (5.190)$$

Deviation of the ratio $w_{max2}/\overline{W}_2$ from unity indicates the rate of velocity dampening in the constricted piping. Parameter $\hat{\delta}$ in the expression is usually small and so may be neglected. Even for a very large velocity gradient the error incurred by neglecting $\hat{\delta}$ is less than 5% (the worst case being $w_{max1}/\overline{W}_1 = 2$). Hence, Equation 5.190 may be written as

$$\frac{w_{max2}}{\overline{W}_2} = \sqrt{1 + \left(\frac{F_2}{F_1}\right)^2 \left[\left(\frac{w_{max1}}{\overline{W}_1}\right)^2 - 1\right]} \qquad (5.191)$$

Equation 5.191 was applied to evaluating the most unfavorable velocity distribution, i.e., when $w_{max1}/\overline{W}_1 = 2$. The rate of dampening computed from Equation 5.191 for different F_2/F_1 ratios is given in Table 5.16.

Values in Table 5.16 indicate that the velocity distribution may be assumed dampened for F_2/F_1 less than about 0.11. Note, however, that as the fluid moves away from the entrance of the constriction, the fluid nearest the wall slows down again and the normal velocity distribution eventually is reestablished.

The other method of velocity dampening involves inserting a perforated

Table 5.16 Dampening Rates Computed from Equation 5.191

F_2/F_1	0	0.123	0.236	0.414	0.618	0.721	0.820	1.00
$\dfrac{w_{max2}}{\overline{W}_2}$	1.0	1.023	1.080	1.23	1.46	1.60	1.74	2.00

plate inside the piping (Figure 5.32B). The average velocities before and after the plate are equal to each other. The pressure drop of a turbulent liquid flow through the disk is proportional to the fluid's kinetic energy:

$$\frac{P_1 - P_2}{\gamma} = K \frac{\overline{W}^2}{2g} \tag{5.192}$$

The Bernoulli equation for liquid motion along the pipe axis is

$$\frac{P_1}{\gamma} + \frac{w_{max_1}^2}{2g} = \frac{P_2}{\gamma} + \frac{w_{max_2}^2}{2g} + K \frac{w_{max_2}^2}{2g} \tag{5.193A}$$

Evaluating $P_1 - P_2$ and substituting into Equation 5.192, we obtain

$$\frac{w_{max_2}}{\overline{W}} = \sqrt{\frac{(w_{max_1}/\overline{W})^2 + K}{1 + K}} \tag{5.193B}$$

This expression should be considered only as an approximation because the resistance coefficient, K, in Equation 5.192 is larger than the K value in 5.193A and 5.193B. In most cases, however, the difference between the two leads to small errors. At any rate, Equation 5.193A shows that at higher flow resistance (higher K) the velocity distribution is dampened more effectively, i.e., the ratio $w_{max_2}/\overline{W}$ is very close to unity.

FLOW NORMAL TO TUBE BANKS

Flow normal to a bank of tubes is encountered in a variety of industrial equipment, such as tubular heat exchangers and condensers. The magnitude of the flow resistance in such equipment is due to the contraction and expansion of the flow. This resistance also depends on the tube layout (e.g., staggered or unstaggered) relative to the flow direction, as illustrated in Figure 5.33. The critical Reynolds number for these flows has been determined as follows:

$$Re_{cr} = \frac{t'w_{max}\gamma}{\mu g} \simeq 40 \tag{5.194}$$

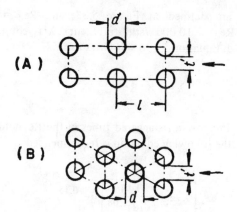

Figure 5.33   Flow normal to a bank of tubes. Flow resistance is a function of the tube layout: (A) an unstaggered layout; (B) a staggered layout.

where t' is the distance between tubes, and w_{max} is the maximum fluid velocity (in the narrow section).

For turbulent flow, i.e., Re > 40, flow resistance may be estimated from an expression of the following form:

$$h_r = \lambda N' \frac{w_{max}^2}{2g} \qquad (5.195)$$

where $h_r = \Delta P/\gamma$ = resistance of the liquid column
$\quad\quad$ N' = number of tube rows in the direction of flow
$\quad\quad$ λ = resistance coefficient depending on Re and the tube layout

The coefficient λ may be determined from the formulas reported by Jakob [1938].

For a single row of tube layout, λ is determined from the following:

$$\lambda = \left[0.175 + \frac{0.32 \frac{\ell}{d}}{(t'/d)^n} \right] Re_d \qquad (5.196)$$

where e, d, t' are defined in Figure 5.33, and $Re_d = t'w_{max}\gamma/\mu g = dw_{max}\gamma/\mu g$. At $Re_d \leqslant 10{,}000$ (still turbulent), λ is constant. Note that when $e = 2d$, $t' = d$, Equation 5.196 simplifies to

$$\lambda = 1.82 \, Re^{-0.15} \tag{5.197}$$

where $Re = Re_d$. For rows of staggered tubes in the Reynolds number range of 2000 to 4000, the following equation is recommended:

$$\lambda = \left[0.92 + \frac{0.44}{(t'/d)^{1.08}}\right] Re_d^{-0.15} \tag{5.198}$$

Or, for the case of $t' = d$,

$$\lambda = 1.40 \, Re^{-0.15} \tag{5.199}$$

Equations 5.196–5.199 give good estimates when the number of rows, N', exceeds 10. At $N' < 10$, actual head losses are somewhat higher than those predicted by applying these equations. At $N' = 4$, the error in applying these correlations is 7%; at $N' = 3$, 15%; and at $N' = 2$, up to 30%.

For flow around a single row of tubes ($N' = 1$), the data of Boucher [1947] should be used. Table 5.17 gives values of λ for different t'/d ratios at $Re = 10{,}000$ (recall that λ is constant at $Re \leqslant 10{,}000$ for a specific t'/d).

For laminar flow around tubes ($Re < 40$), the resistance may be calculated from an expression analogous to the Darcy equation (see Chilton [1938] for details):

$$h_r = \lambda \, \frac{\ell}{d_{eq}} \, N' \, \frac{w_{max}^2}{2g} \tag{5.200}$$

Table 5.17 Resistance Coefficients for Banks of Tubes

t'/d	0.2	0.5	1.0	1.5	2.0	5.0	10.0
λ	0.64	0.50	0.40	0.36	0.30	0.18	0.10

where $h_r = \Delta P/\gamma$ is the resistance of the liquid column
ℓ = distance between rows in the direction of flow
w_{max} = maximum velocity in the intertubular space
d_{eq} = equivalent diameter equal to four times hydraulic radius

Consider flow between two rows of tubes related to unity length of a tube, $\ell(t' + d) - \pi d^2/4$, where $\pi d^2/4$ is part of the volume occupied by corresponding segments of tubes. The surface of this tube is equal to πd; therefore, the equivalent diameter is

$$d_{eq} = 4 \frac{\ell(t' + d) - \dfrac{\pi d^2}{4}}{\pi d} = \frac{4\ell(t' + d)}{\pi d} - d \tag{5.201}$$

We then define an equivalent Reynolds number:

$$Re_{eq} = \frac{d_{eq} w_{max} \gamma}{\mu g} \tag{5.202}$$

For a row of staggered tubes, $Re_{eq} = 1 - 100$:

$$\lambda = \frac{106}{Re_{eq}} \tag{5.203}$$

Substituting this expression for λ into Equation 5.200, we obtain

$$h_r = \frac{53 \mu N' \ell w_{max}}{\gamma d_{eq}^2} \tag{5.204}$$

For a row of unstaggered tubes, Bergelin [1949] found that λ values for turbulent flow multiplied by 1.5 gave good estimates. Applications involving flow around tubes usually is accompanied by heat transfer. The tube surface has a temperature different from the average or bulk temperature of the fluid. The effect of wall temperature is considered in the method of Siedez [1936], in which λ_{is} corresponding to the temperature of the flow is estimated first and the true coefficient, λ, is determined from the following equation:

$$\lambda = \frac{\lambda_{is}}{a\left(\dfrac{\mu}{\mu_w}\right)^n}$$ (5.205)

where μ is the viscosity at the temperature of the flow and μ_w is the viscosity at the temperature of the wall surface (see also Chilton [1938] for details).

For laminar flow, $a = 1.1$ and $n = 0.25$; for turbulent flow, $a = 1.0$ and $n = 0.14$.

OPTIMUM PIPE DIAMETER

A very practical consideration in formulating plant layouts is the determination of optimum pipeline sizes for specified flow conditions and pump energy limitations. The optimum is defined as the pipe size that provides an acceptable resistance per unit length within specified economic constraints. In other words, it is the most economical pipe diameter for a given set of operating conditions. As illustrated in Example 5.9, the problem is somewhat complicated by the fact that the Darcy equation (Equation 5.156) cannot be applied in a straightforward manner. For a given volumetric flow, both the linear velocity, w, and Reynolds number (and, consequently, the friction factor) are functions of the pipe diameter.

Hence, the Darcy equation is an implicit function of the diameter, and direct determination of D, especially for turbulent flow, is often impossible. To address this problem, we begin by rewriting Darcy's equation in the following form:

$$-\frac{\Delta P}{\gamma} = \lambda \frac{8V^2 L}{D^5 \pi^2 g}$$ (5.206)

Also, the Reynolds number is denoted as

$$Re = \frac{4V\gamma}{\pi \mu g D}$$ (5.207)

Rearranging this last expression in terms of D and substituting into Equation 5.206, we obtain

$$\frac{128\, V^3 \gamma^4 \Delta P}{\pi^3 \mu^5 g^4 L} = \lambda\, Re^5$$ (5.208)

From a standard friction factor-Reynolds number chart such as in Figures 5.34 or 5.25, we may construct a plot of λRe^5 versus Re, where λRe^5 is equivalent to the LHS of Equation 5.208. From such a plot can be obtained a value for the Reynolds number corresponding to λRe^5. Once the Reynolds number is known, the pipe diameter may be computed (use Equation 5.207).

As is often the case, several pipe sizes will be determined acceptable in terms of head losses for a given flowrate; however, if we select a larger diameter, then construction costs will be high and, over a planned operating lifetime, maintenance and repair and amortization expenses, k_i, also will be high. At the same time, a large diameter means lower hydraulic resistance and, consequently, lower power consumption for transportation. Hence, production expenses, k_p, will be relatively low.

Conversely, if we select a smaller diameter, hydraulic resistance increases along with production costs, whereas amortization expenses decrease. The relationships of k_i (maintenance/amortization costs) and k_p (production costs) to pipe diameter are shown in Figure 5.35. The sum of k_i and k_p equals the total yearly expenses for a pipe network. This also is shown in Figure 5.35 as a plot of Σk versus diameter. As shown, the relationship has a minimum that corresponds to the optimum pipe diameter.

Determination of the optimum pipe diameter may be formulated into a rigorous design procedure, which we shall outline and illustrate below. Both series of expenses should be related to a unit length of piping and a basis of one year's operation. The cost per unit length of piping is directly proportional to the diameter and may be expressed by the following relationship:

$$C_1 = XD^n \tag{5.209}$$

From data supplied by manufacturers and contractors on the costs of piping and installation, coefficients "X" and "n" can be evaluated.

Pipe components such as fittings and valves also are included in the analysis. The cost of fittings depends on the corresponding pipe diameter and may be considered a fractional cost of the piping, i.e., $C_2 = jC$, where j is a coefficient denoting fractional cost. The total cost per unit length, C, of a piping network, including fittings and valves, is

$$C = (1 + j)XD^n \tag{5.210}$$

where $C = C_1 + C_2$.

By assigning an operating life for the system, e.g., 10-15 years, we may evaluate the portions of the total cost, C, attributed to amortization expenses

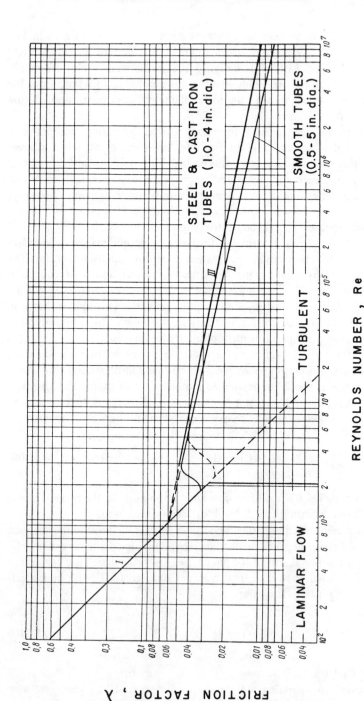

Figure 5.34   Expanded friction factor chart for steel and cast iron tubes and smooth tubes.

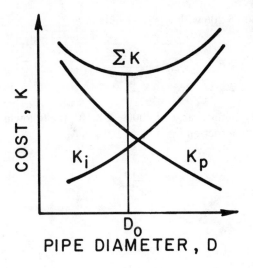

Figure 5.35   Relationship of operating costs and capital investment to pipe size.

(denote as a) and to maintenance costs (b), such as repairs, painting, insulation, etc. That is, the total yearly expenses for amortization and maintenance are

$$k_i = (a + b)(1 + j)XD^n \qquad (5.211)$$

Production or operating expenses depend on the number of hours of operation per year (Y), the hourly rate of energy needed for fluid transportation ($\tilde{N}$) and the cost of a unit of energy (C_e):

$$k_p = Y\tilde{N}C_e \qquad (5.212)$$

The power required for fluid transportation may be expressed as $V\Delta P/\eta$, where V = volumetric rate, ΔP = pressure drop per unit length of pipe and η = the mechanical pump efficiency. Hence,

$$k_p = \frac{W_g \Delta p k C_e}{Y\eta} \qquad (5.213)$$

where W_g is the weight rate.

From Equation 5.206 with $L = 1$, we obtain

$$\Delta p = \frac{8}{\pi^2} \frac{\lambda W^2}{g \gamma D^5} \tag{5.214}$$

For a Reynolds number range of $4000 - 2 \times 10^7$, the following equation may be used to obtain a friction factor:

$$\lambda = \frac{0.16}{Re^{0.16}} \tag{5.215}$$

where

$$Re = \frac{GD}{\mu g} \tag{5.216}$$

where

$$G = \frac{W}{\pi D^2 / 4}$$

Hence, the friction factor expression is

$$\lambda = \frac{0.16}{\left(\dfrac{W}{\pi D^2 / 4} \times \dfrac{D}{\mu g}\right)^{0.16}} = 0.154 \left(\frac{D \mu g}{W}\right)^{0.16} \tag{5.217}$$

Substituting λ from Equation 5.217 into Equation 5.214, then into Equation 5.213, we obtain, after some simplification,

$$k_p = 0.125 \frac{W^{2.84} \mu^{0.16} Y C_e}{D^{4.84} \gamma^2 \eta g^{0.84}} \tag{5.218}$$

Thus, Equations 5.218 and 5.211 provide estimates of production costs and amortization and maintenance expenses, respectively. The sum of these two is the total yearly cost for a piping system:

$$\Sigma k = (a + b)(j + 1)XD^n + 0.125 \frac{W^{2.84}\mu^{0.16}YC_e}{D^{4.84}\gamma^2\eta^2g^{0.84}} \qquad (5.219)$$

The optimum diameter is the minimum cost incurred for the pipe system. This minimum cost can be obtained by differentiating Equation 5.219 with respect to D and setting the derivative equal to zero:

$$\frac{d(\Sigma k)}{d(D)} = n(a + b)(j + 1)XD^{n-1} - \frac{0.605\,W^{2.84}\mu^{0.16}YC_e}{D^{5.84}\gamma^2\eta g^{0.84}} = 0$$

Thus, the optimum piping diameter is

$$D_o^{4.84} = \frac{0.605\,W^{2.84}\mu^{0.16}YC_e}{n(a + b)(j + 1)X\gamma^2\eta g^{0.84}} \qquad (5.220A)$$

For pipe diameters greater than ¾ in., exponent n in Equation 5.209 is, for practical purposes, unity. Hence, the final expression for the optimum pipe diameter is

$$D_o = 0.918 \left[\frac{YC_e}{(a + b)(j + 1)X\eta}\right]^{0.17} \frac{\mu^{0.027}}{g^{0.14}} \frac{W^{0.48}}{\gamma^{0.34}} \qquad (5.220B)$$

Examining this expression reveals that the dependency of D_o on viscosity is very small ($\mu^{0.027}$). For example, in the range of 0.02 cp (the viscosity of air at STP) to 30 cp (a typical oil), $\mu^{0.027}$ changes by only a few percentage points. Hence, $\mu^{0.027}$ can be assumed as constant. And denoting

$$k_1 = 0.918 \left[\frac{YC_e}{(a + b)(j + 1)X\eta}\right]^{0.17} \frac{\mu^{0.027}}{g^{0.14}} \qquad (5.221)$$

as a constant (actually assuming k_1 to be constant is a good approximation to within 10%), the optimum pipe diameter expression may be written simply as

$$D_o = k_1 \frac{W^{0.48}}{\gamma^{0.34}} \qquad (5.222)$$

The derivation of Equation 5.222 is based on the condition of turbulent flow. Hence, the D_o calculation should be checked by computing the Reynolds number and comparing against the criterion for turbulent pipe flow.

A similar design procedure is given by Normand [1948] for laminar flows. Following Normand, the derivation is the same through Equation 5.214. Departure from the above procedure comes from the use of the laminar friction factor, $\lambda = 64/Re$, or from definition:

$$\lambda = 16\pi D\mu g/W$$

Repeating the same derivation procedure, but using the laminar λ expression, total expenses are

$$\Sigma k = (a + b)(j + 1)XD^n + \frac{128}{\pi}\frac{W^2YC_e\mu}{\gamma^2D^4\eta}$$

Again, differentiating and setting the derivative equal to zero, we obtain

$$D_o^{4+n} = \frac{512}{\pi}\frac{W^2YC_e\mu}{\gamma^2\eta n(a + b)(j + 1)X} \tag{5.223}$$

For $D_o > \frac{3}{4}$ in., the exponent $n = 1$ and the expression simplifies to

$$D_o = 2.77\left[\frac{YC_e}{(a + b)(j + 1)X\eta}\right]^{0.2}\left(\frac{W^2\mu}{\gamma^2}\right)^{0.2} \tag{5.224}$$

Denoting

$$k_2 = 2.77\left[\frac{YC_e}{(a + b)(j + 1)X\eta}\right] \tag{5.225}$$

as a constant, we obtain the following expression for calculating the optimum pipe diameter for laminar flow:

$$D_o = k_2\left(\frac{W^2\mu}{\gamma^2}\right)^{0.2} \tag{5.226}$$

For laminar flow, $Re = GD_0/\mu g < 2100$. Or, substituting $W/(\pi D^2/4)$ for G,

$$Re = WD_0 \bigg/ \frac{\pi D_0^2}{4}\, \mu g < 2100 \tag{5.227}$$

Replacing D_0 in this expression with Equation 5.226, we obtain

$$W_g < 2.2 \times 10^5 (k_2 g)^{1.67}\, \frac{\mu^2}{\gamma^{0.67}} = W_{gcr} \tag{5.228}$$

This last expression evaluates the flow regime that will exist for the optimum pipe diameter. Hence, from information on the weight rate, W_g, viscosity, μ, specific weight, γ, and computed constant, k_2 (from Equation 5.225), the proper expression for calculating the optimum pipe diameter—Equation 5.222 or 5.226—can be selected. The following example problem illustrates the design procedure.

Example 5.30

Determine the optimum size pipe for transporting 5000 kg/hr of water. The minimum life of the pipeline may be assumed to be 10 years. Yearly maintenance and repair expenses are expected to be 5% of the initial piping costs. The cost of fittings and valves is roughly 10% of the cost of the pipe. The system is to be designed for 24-hour operation throughout the entire year. The pump efficiency, η, is 60%, and the cost of electrical power is $ 0.2/kWh. Table 5.18 provides quotations on costs per unit length of piping for different pipe diameters.

Solution

We will apply Equation 5.228 to evaluate the flow regime for the most economical or optimum pipe size and, depending on the results obtained, use either Equation 5.226 or 5.222 to compute D_0.

The cost per unit length of piping is described by Equation 5.209: $C_1 = XD^n$, where $n \simeq 1$. Plotting the data given in Table 5.18 and evaluating the slope (Figure 5.36), we find $X \simeq 150\ \$/m^2$.

Hence,

$$C_1 = 150 D(\$/m)$$

Table 5.18 Cost per Unit Length ($/m) of Different Pipe Diameters for Example 5.30

D (in.)	D (mm)	Price ($/m)
$3/4$	21.5	3.12
1	27.0	4.30
$1^1/4$	35.75	4.94
$1^1/2$	41.25	5.80
2	52.5	7.51
$2^1/2$	68	9.04
3	80.25	11.30
$3^1/2$	92.5	13.7
4	105	15.4
5	130	19.9
6	155.5	23.7

Electrical costs are

$$C_e = 0.02 \; \$/kWh = \frac{0.2}{3.67 \times 10^5} = 5.45 \times 10^{-7} \; \$/kg$$

Operating time is

$$Y = 365 \times 24 \times 3600 = 3.15 \times 10^7 \; s/yr$$

The amortization expenses are a = 0.10.
   Maintenance and repair costs are 5% of the pipe costs (or b = 0.05). Hence,

$$a + b = 0.10 + 0.05 = 0.15$$

The cost of all fittings is 10% of the capital costs for piping. Hence j = 0.1. From these values, the term common to Equations 5.221, 5.224 and 5.228 may be evaluated:

$$\frac{YC_e}{(a+b)(j+1)X\eta} = \frac{(3.15 \times 10^7)5.45 \times 10^{-7}}{0.15 \times 1.1 \times 150 \times 0.6} = 1.20$$

Using Equation 5.228, we evaluate the transition from laminar to turbulent flow:

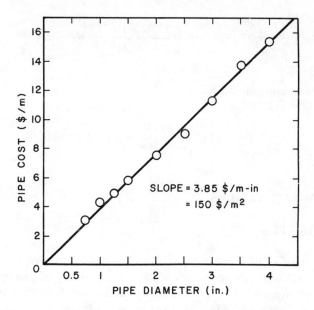

Figure 5.36   Plot of pipe costs versus diameter (prepared from data in Table 5.18).

$$\mu = \frac{1}{9.8 \times 10^3} = 1.02 \times 10^{-4} \frac{kg \cdot s}{m^2}$$

$$k_2 = 2.77(1.20)^{0.2} = 2.9$$

Hence, the critical weight velocity for flow through the optimum pipe size is

$$W_{g,cr} = 2.2 \times 10^5 (2.9 \times 9.81)^{1.67} \frac{(1.02 \times 10^{-4})}{1000^{0.67}}$$

$$= 7.35 \times 10^3 kg/s$$

For this problem, $W_g = 5000/3600 = 1.39$ kg/s. As $W_{g,cr} > W_g$, the flow is turbulent and Equation 5.222 should be used to compute D_0.

From Equation 5.221,

$$k_1 = 0.918(1.20)^{0.17}\frac{(1.02 \times 10^{-4})^{0.027}}{9.81^{0.14}} = 0.545$$

Finally, from Equation 5.222,

$$D_0 = 0.545\,\frac{1.39^{0.48}}{(1000)^{0.34}} = 6.1 \times 10^{-2}\,m$$

FLOW OF LIQUID FILMS

Vertical Film Flow

The flow of relatively thin liquid films in either a cocurrent or counter-current mode with gases or vapors is utilized in a variety of unit operations, such as absorption, distillation, humidification and a variety of cooling techniques. To a large extent the performance of equipment utilizing film flow depends on the flow regime, the film thickness and the fluid velocity.

Depending on the "film Reynolds number," three fundamental types of flow regimes are observed:

1. laminar liquid film flow with a smooth interface ($Re_f < 30$ to 50);
2. laminar flow with a rippled interface (30 to 50 $< Re_f < 100$ to 400); and
3. turbulent flow ($Re_f > 100$ to 400).

The film Reynolds number is defined as

$$Re_f = \frac{w d_{eq} \rho}{\mu} \tag{5.229}$$

where w = film's average velocity
           d_{eq} = equivalent diameter of the film

Re_f also may be expressed as follows:

$$Re_f = \frac{4\Gamma}{\mu} \tag{5.230}$$

where Γ is the mass rate of flow per unit width of wall.

This last definition allows evaluation of the film Reynolds number without the need for a separate measurement of the film thickness and average velocity. Expression 5.230 was derived based on the following reasoning. The equivalent diameter, d_{eq}, is defined as

$$d_{eq} = 4 \times \frac{\text{cross-sectional area of a film}}{\text{wetted perimeter of channel}} = \frac{4S}{p} = \frac{4p\delta}{p} = 4\delta \qquad (5.231)$$

where δ = film thickness

Substituting for d_{eq} from Equation 5.231 into Equation 5.229 yields

$$Re_f = \frac{4w\delta\rho}{\mu} \qquad (5.232)$$

Denoting

$$\Gamma = w\delta\rho$$

we obtain the modified expression (Equation 5.230) for the film Reynolds number.

Let us now consider laminar flow of a liquid film of constant thickness down a vertical wall. If the film is sufficiently thin, the flow may be assumed planar, even if it is flowing over a curved surface, e.g., flowing down the walls of a tube. Hence, the liquid interface is strictly parallel to the surface of the wall.

Figure 5.37 shows a portion of a liquid film having thickness, δ, and unit width flowing in the downward direction. For a Newtonian liquid, the shear stress at distance x from the wall is

$$\tau = -\mu \frac{dw}{dx} \qquad (5.233)$$

where w is now the local linear velocity.

Similarly, the shear stress at distance x + dx from the wall is

$$\tau + d\tau = -\mu \left[\frac{dw}{dx} + d\left(\frac{dw}{dx}\right) \right] \qquad (5.234)$$

The motion of the film of thickness dx is caused by the resultant of these two stresses:

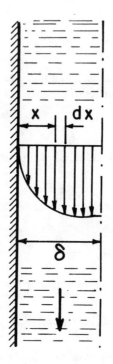

Figure 5.37   Liquid film flowing down a vertical wall. The downward arrows denote the velocity distribution.

$$d\tau = -\mu d\left(\frac{dw}{dx}\right) \tag{5.235}$$

This downward-directed stress corresponds to the ratio of the resultant shear forces acting on the film interface and is equal to the weight of the film, i.e., the product of specific weight, δ, and the liquid volume, $Ldx \cdot 1$, where L is the height of the wall. Hence, the resultant shear stress is

$$d\tau = \frac{Ldx\gamma}{L} = \gamma dx \tag{5.236}$$

Substituting for $d\tau$ from Equation 5.235 into the above expression, we obtain the following differential equation:

$$\frac{\gamma}{\mu} dx = -\left(\frac{dw}{dx}\right)$$

Integrating this expression once gives

$$\frac{\gamma}{\mu} x = -\frac{dw}{dx} + c_1 \qquad\qquad (5.237\text{A})$$

and integrating a second time gives

$$w = -\frac{x}{\mu} \cdot \frac{x^2}{2} + c_1 x + c_2 \qquad\qquad (5.237\text{B})$$

The constants of integration may be evaluated by making use of the following boundary conditions corresponding to the solid wall and the liquid-gas interface:

$$\left.\begin{array}{ll} \text{at } x = 0, & w = 0, \quad c_2 = 0 \\ \text{at } x = \delta, & dw/dx = 0 \end{array}\right\} \qquad\qquad (5.237\text{C})$$

Application of the boundary conditions yields

$$c_2 = 0$$

and

$$c_1 = \frac{w_{max}}{\delta} + \frac{\gamma}{\mu}\frac{\delta}{2}$$

Hence, an expression for the velocity distribution is

$$w = \frac{w_{max} x}{\delta} + \frac{\gamma}{2\mu}(\delta x - x^2) \qquad\qquad (5.238)$$

This expression is not used readily because it contains the maximum velocity, w_{max}. Therefore, we differentiate this expression and obtain

$$\frac{dw}{dx} = \frac{w_{max}}{\delta} + \frac{\gamma}{2\mu}(\delta - 2x) \tag{5.239}$$

Clearly, the velocity gradient at the interface is zero, i.e., at $x = \delta$, $dw/dx = 0$, because of the absence of shear stresses. Hence, Equation 5.239 reduces to an expression for the maximum velocity:

$$w_{max} = \frac{\delta^2 \gamma}{2\mu} \tag{5.240}$$

Substituting this expression for w_{max} into Equation 5.238 gives the velocity distribution for a thin liquid layer flowing down a vertical wall due to the influence of gravity force alone.

$$w = \frac{\gamma}{\mu}\left(\delta x - \frac{x^2}{2}\right) \tag{5.241}$$

This is our familiar parabolic expression, the profile of which is shown in Figure 5.37. The volumetric rate per unit width of film is obtained as follows:

$$V = \int_0^\delta w\,dx = \frac{\gamma}{\mu}\frac{\delta^3}{3} \tag{5.242}$$

The average film velocity is obtained by dividing V by the film thickness:

$$\overline{W} = \frac{\gamma\delta^2}{3\mu} \tag{5.243}$$

Comparison of Equations 5.243 and 5.240 reveals that the maximum velocity (on the surface of the layer) is 1.5 times higher than average velocity. Following Bird et al. [1960] the film thickness, δ, may be expressed in terms of the average velocity, the volumetric flowrate or the mass flowrate per unit width of the wall, $\Gamma = \rho\delta w$; hence,

$$\delta = \sqrt[3]{\frac{3\mu\overline{W}}{\rho g}} = \sqrt[3]{\frac{3\mu Q}{\rho g}} = \sqrt[3]{\frac{3\mu\Gamma}{\rho^2 g}} \tag{5.244}$$

Equation 5.243 accurately predicts the *film thickness* for laminar flow of low-viscosity liquids (<1 cp) up to a critical Reynolds number ($4\Gamma/\mu$) in the range of 1000 to 2000.

Higher-viscosity liquids (10–20 cp) have film thicknesses appreciably below the predictions of Equation 5.244. This is especially true at and beyond the inception of surface wave motion, as observed by Jackson [1955]. The inception of wave action has been observed to appear when the Froude number ($V^2/g\delta$) exceeds unity. From this, it follows that the Reynolds number at which waves start depends only on the kinematic viscosity of the vertical flowing liquid. This assumes that the surrounding gas is stagnant and of negligible density. Then $\Gamma/\rho' = 3(\mu'/\rho')$, which corresponds to a wave inception Reynolds number of $Re_f = 12$.

Early studies by Friedman and Miller [1941] showed the ratio w_{max}/w_{avg} to be 1.5 up to $Re_f = 25$, after which it increased to about 2.2 at $Re_f = 100$. Jackson found w_{max}/w_{avg} to be 1.5 up to $Fr_f = 1$ ($Re_f = 12$), increasing rapidly to 2.2 at $Fr = 9.0$ ($Re = 108$) and slowly decreasing to about 1.8 at $Fr_f = 200$ ($Re_f = 2400$).

Example 5.31

A liquid having a kinematic viscosity of 3×10^{-4} $m^2 \cdot s^{-1}$ and a density of 0.73×10^3 $kg \cdot m^{-3}$ is under gravity flow down a vertical wall. Determine the mass rate of flow for a measured film thickness of 3.2 mm.

Solution

An expression for the film thickness of a laminar flowing film under the influence of gravity is given by Equation 5.244. Rearranging this expression to solve for the mass flowrate per unit width of wall,

$$\Gamma = \frac{\delta^3 \rho g}{3\nu} = \frac{(3.2 \times 10^{-3})^3 (0.73 \times 10^3)(9.80)}{3(3 \times 10^{-4})} = 0.261 \text{ kg-m}^{-1} \cdot \text{s}^{-1}$$

To ascertain whether the flow is laminar, we compute the Reynolds number based on the calculated mass flowrate:

$$Re = \frac{4\Gamma}{\mu} = \frac{4\Gamma}{\rho\nu} = \frac{4(0.261)}{(0.73 \times 10^3)(3 \times 10^{-4})} = 4.8$$

As the Reynolds number is below the observed upper limit for laminar flow, the calculated value of Γ is correct.

Example 5.32

An experiment was devised to check the accuracy of the laminar film flow expressions presented above. The liquid film thickness was obtained experimentally by weighing an amount of water flowing down a wall 2 m high by 1.2 m wide. The measurements showed the average amount of liquid on the wall to be 395 g for a mass flow of 52 kg/hr. Compare the measured value to the predicted.

Solution

$$\Gamma = \frac{G}{L} = \frac{52}{3600 \times 1.2} = 0.012 \text{ kg/m-s}$$

The average film velocity is

$$\overline{W} = \frac{\Gamma}{\rho \delta}$$

and the Reynolds number is (the viscosity of water is 1 cp or 10^{-3} N-s/m^2 at 20°C)

$$Re_f = \frac{\overline{W}\delta}{\nu} = \frac{\Gamma}{\mu} = \frac{0.012}{10^{-3}} = 12$$

To evaluate the flow regime, compare the obtained Reynolds number ($Re_f = 12$) with the flow's critical value:

$$Re_{cr} = 2.4\left(\frac{\sigma^3}{g\rho^3\nu^4}\right)^{1/11} = 2.4\left[\frac{(7.26 \times 10^{-2})^3}{9.81 \times 1000^3(10^{-6})^4}\right]^{1/11} \simeq 22$$

As $Re < Re_{cr}$, the flow is laminar and the film thickness, δ, may be estimated from Equation 5.244:

$$\delta = \sqrt{\frac{3\mu\Gamma}{\rho^2 g}} = \sqrt{\frac{1.2 \times 10^{-2} \times 3 \times 10^{-3}}{9.81 \times 1000^2}} = 1.54 \times 10^{-4} \text{ m}$$

The amount of water retained on the wall when $\delta = 1.54 \times 10^{-4}$ m is

$$Q' = S\delta\rho = 1.2 \times 2 \times 1.54 \times 10^{-4} = 0.369 \text{ kg}$$

Comparing the experimental value of 0.395 kg with the prediction shows a relative error of about 7%.

Example 5.33

A system for contacting sulfuric acid with a process gas stream was proposed in the form of a vertical film flow system. A prototype was built to study the hydrodynamics. For this purpose water was used. For a water weight rate of 3.3 kg/s-m, the film thickness, δ, was measured to be 1.0 mm. Using dimensional analysis, determine under what conditions a 60% solution of sulfuric acid will flow with a film thickness of 1.5 mm. Table 5.19 provides data on viscosities (cp) and densities (g/cm^3) for the 60% H_2SO_4 solution at different temperatures. It may be assumed that the gas stream moves cocurrent with the downward liquid flow and at a much slower rate.

Solution

Film thickness, δ, is a function of the weight rate per unit width of a wall, Γ, viscosity, μ, specific gravity, γ, and acceleration due to gravity, g:

$$\delta = f(\Gamma, \mu, \gamma, g)$$

Table 5.19 Physical Properties Data for 60% H_2SO_4 Solution

t°, C	μ (cp)	ρ (g/cm^3)
0	10	1.52
10	8.0	1.51
20	6.5	1.50
30	5.5	1.49
40	4.5	1.48
50	3.7	1.47
60	3.2	1.47
70	2.7	1.46
80	2.3	1.45
90	1.9	1.44
100	1.6	1.43

Rewriting this function in series,

$$\delta = a\Gamma^a\mu^b\gamma^c g^d + a'\Gamma^{a'}\mu^{b'}\gamma^{c'}g^{d'} + \ldots$$

or

$$\Sigma a = \left(\frac{\Gamma^a\mu^b\gamma^c g^d}{\delta}\right) = 1$$

The units on each of the variables are

$$\Gamma \equiv \frac{kg}{m\text{-}s}$$

$$\gamma \equiv \frac{kg}{m^3}$$

$$\mu \equiv \frac{kg\text{-}s}{m^2}$$

$$g \equiv \frac{m}{s^2}$$

$$\delta \equiv m$$

The expression inside the brackets must be dimensionless. Therefore, we may write the following identity:

$$\left(\frac{kg}{m\text{-}s}\right)^a \left(\frac{kg\text{-}s}{m^2}\right)^b \left(\frac{kg}{m^3}\right)^c \left(\frac{m}{s^2}\right)^d m^{-1} = kg^0 m^0 s^0$$

or

$$kg^{a+b+c} m^{-a-2b-3c+d-1} s^{-a+b-2d} = kg^0 m^0 s^0$$

Comparing the corresponding exponents we obtain three equations:

$$a + b + c = 0$$

$$-a + b - 2d = 0$$

$$-a - 2b - 3c + d - 1 = 0$$

Thus, there are three equations with four unknowns. Solving in terms of one of the unknowns, we obtain

$$a = a$$

$$b = \frac{2}{3} - a$$

$$c = -\frac{2}{3}$$

$$d = \frac{1}{3} - a$$

The obtained results are substituted into the above series expression:

$$\Sigma a \Gamma^a \mu^{2/3-a} \gamma^{-2/3} g^{1/3-a} \delta^{-1} = 1$$

This equation then is transformed by exponents:

$$\Sigma a \left(\frac{\Gamma}{\mu g} \right)^a \left(\frac{\mu^2 g}{\gamma^2 \delta^3} \right)^{1/3} = 1$$

We further change the expression by the general function describing the given process:

$$\frac{\Gamma}{\mu g} = \Phi \left(\frac{\mu^2 g}{\gamma^2 \delta^3} \right)$$

Instead of $\mu g / \gamma$, we introduce the kinematic viscosity coefficient ν, then

$$\frac{\Gamma}{\mu g} = \Phi \left(\frac{\delta^3 g}{\nu^2} \right)$$

Assume

$$\delta = 0.001 \text{ m}$$
$$g = 9.81 \text{ m/s}^2$$
$$\mu = 1 \text{ cp (for water)}$$

Then,

$$\mu g = 10^{-3} \text{ kg/m-s}$$

and,

$$\nu = \frac{\mu g}{\gamma} = \frac{10^{-3}}{10^3} = 10^{-6} \text{ m}^2/\text{s}$$

Consequently,

$$\frac{\delta^3 g}{\nu^2} = \frac{(10^{-3})^3 \times 9.81}{(10^{-6})^2} = 9.81 \times 10^3$$

For the H_2SO_4 solution, this fraction will have the same numerical value, provided the proper kinematic viscosity is known. According to the problem statement, the film thickness should be 1.5 mm. That is, the kinematic viscosity coefficient should be

$$\nu = \sqrt{\frac{\delta^3 g}{9.81 \times 10^3}} = \sqrt{\frac{(1.5 \times 10^{-3})^3 \times 9.81}{9.81 \times 10^3}} = 1.84 \times 10^{-6} \text{ m}^2/\text{s}$$

From Table 5.19, values of the coefficient of kinematic viscosity are computed:

$T, °C$:	0	10	20	30	40	50	60	70	80	90	100
$\nu \times 10^{-6} \ (\text{m}^2/\text{s})$:	6.6	5.3	4.35	3.7	3.05	2.52	2.18	1.85	1.59	1.32	1.12.

We see that at $T = 70°C$ and, $\nu = 1.84 \times 10^{-6}$ m^2/s; therefore, at this temperature the dimensionless fraction on the right-hand side of the expression $\delta^3 g/\nu^2$ will be the same as that for H$_2$O. Only at this temperature is it possible for the acid solution to have the same velocity as the flowing water. That is,

$$\left(\frac{\Gamma}{\mu g}\right)_{H_2O} = \left(\frac{\Gamma}{\mu g}\right)_{H_2SO_4}$$

For water, $\Gamma = 3.3$ kg/m-s and $\mu g = 10^{-3}$ kg/m-s:

$$\frac{\Gamma}{\mu g} = \frac{3.3}{10^{-3}} = 3300$$

For temperature $T = 70°C$, the viscosity is $\mu = 2.7$ cp; therefore, $\mu g = 2.7 \times 10^{-3}$ kg/m-s and

$$\Gamma = 3300 \; \mu g = 3300 \times 2.7 \times 10^{-3} = 8.9 \text{ kg/m-s}$$

Hence, the acid must be fed to the system at a rate of 8.9 kg/m-s and at a temperature of 70°C.

Wavy Flow

In the above analyses it was implied that the gas stream flowed cocurrent to the falling liquid film and at a relatively low velocity. This enabled description of the hydrodynamics of the film strictly in terms of the liquid motion. Even for the case of countercurrent flow (liquid flowing vertically downward and gas upward), the film thickness will be independent of the gas stream velocity, provided the gas rate is low. If, however, the gas velocity is increased, the friction or shear force at the film interface, acting in the opposite direction to the liquid flow, becomes significant, and the film motion slows down.

At a certain gas velocity ($\sim$5–10 m/s), an equilibrium state arises between gravitational and friction forces. This condition leads to flooding, i.e., an accumulation of liquid in the column, eventually initiating entrainment and a dramatic increase in the hydraulic resistance. At still higher gas velocities, the film's flow direction becomes inverted and it begins to "creep up," that is, the liquid and gas move upward cocurrently. When this occurs, the hydraulic resistance is first observed to decrease comparatively to the flooding regime but then increases. At gas velocities greater than 15 m/s, the liquid is entrained. In contrast, at very high gas velocities in the downward cocurrent flow case, the gas greatly increases the liquid film's velocity while decreasing its thickness.

With this background, we will now extend the analysis of constant-thickness laminar film flow down a vertical wall to include interaction with the gas phase. The equation of motion for this system takes the form

$$\gamma' - \gamma'' + \mu' \frac{d^2 w'}{dy^2} = 0 \qquad (5.245A)$$

where γ', γ'' are the specific gravities of liquid and gas, respectively, and μ' is the liquid viscosity.

Boundary conditions for this differential equation are

$$y = 0 \ , \quad w' = 0$$

$$y = \delta \ , \quad \mu' \frac{dw'}{dy} = \pm \zeta'' \frac{\gamma'' w_2''^2}{2g} \qquad (5.245B)$$

where ($''$) refers to the gas phase and ζ is the resistance coefficient.

After integration and evaluation with the boundary conditions, we obtain

$$w' = \left(\pm \zeta \frac{\gamma'' w_r^2}{2g\mu'} + \frac{\gamma' - \gamma''}{\mu'} \right) y - \frac{\gamma' - \gamma''}{2\mu'} y^2 \qquad (5.246)$$

We shall assign the downward direction of flow as being positive. Hence, a plus sign is used on the second term in Equation 5.246 for the gas phase, i.e., the gas carries the film along in the same direction as the gravity force. To indicate upflow of the gas (the gas retards the downflow of the film) a minus sign is used. The velocity of the film on the phase boundary is

$$w_b' = \pm \zeta'' \frac{\gamma'' w_r''^2}{2g\mu'} \delta + \frac{\gamma' - \gamma''}{2\mu'} \delta^2 \qquad (5.247)$$

where w_r'' is the relative velocity of the gas and is equal to $w'' - w_b'$.

The average (flow) velocity of the liquid film is

$$\overline{W}' = \frac{1}{\delta} \int_0^\delta w' dy = \pm \zeta'' \frac{\gamma'' w_r''^2}{4g\mu'} \delta + \frac{\gamma' - \gamma''}{3\mu'} \delta^2 \qquad (5.248)$$

In upward gas flow, the film is carried along upward and the liquid velocity becomes negative at

$$\frac{3\zeta'' w_r''^2 \gamma''}{4g\delta(\gamma' - \gamma'')} > 1 \qquad (5.249)$$

This inequality represents the criterion for "flooding."

When 30 to $50 < \mathrm{Re_f} < 100$, a wavy regime is observed. That is, the interface is rippled and capillary waves appear at the interface that are occasioned by forces of gravity and viscosity acting on the film. The condition is illustrated in Figure 5.38.

Following the analysis of Kutateladze and Styrikovich [1958], we consider the limiting case that occurs when the gas velocity is close to zero. Hence, the tangential stresses on the free surface may be neglected. From Equations 5.246 and 5.248, the velocity profile is

$$w_x' = \frac{3\overline{W}_x}{\delta} \left(y - \frac{y^2}{2\delta} \right) \qquad (5.250)$$

During the wave motion, the average velocity along the cross section is a function of the x-coordinate and time t:

$$w_x' = \frac{3w_x'(x;t)}{\delta} \left(y - \frac{y^2}{2\delta} \right) \qquad (5.251)$$

Substituting this expression for w_x' into Equation 5.250 and integrating over the limits of $y = 0$ and $y = \delta$, we obtain the following:

$$\frac{\partial \overline{W}_x'}{\partial t} + \frac{9}{10} \overline{W}_x' \frac{\partial \overline{W}_x'}{\partial x} = \frac{6}{\rho'} \frac{d^3 \delta}{dx^3} - \frac{3\nu \overline{W}'}{\delta^2} + \frac{\gamma' - \gamma''}{\rho'} \qquad (5.252)$$

For waves of small amplitude, the film thickness may be expressed as a binomial function:

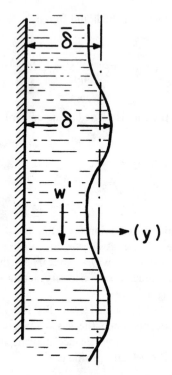

Figure 5.38   The formation of interfacial disturbances (waves) in film flow.

$$\delta = \bar{\delta} + \psi'\bar{\delta} \tag{5.253}$$

where $\bar{\delta}$ = the average thickness of the film
  ψ' = the deviation coefficient of the instantaneous value of the thickness from its average value

Capillary waves originating on the surface of a film are not dampened. Therefore, all the terms of Equation 5.252 are functions of the x-coordinate and the phase velocity, ν, i.e., they are functions of the argument $(x - \nu t)$. Then,

$$\left.\begin{aligned}
\frac{\partial \delta}{\partial t} &= -\bar{\delta}\nu \frac{\partial \varphi'}{\partial x} \\[2mm]
\frac{\partial \overline{W}'_x}{\partial t} &= -\nu \frac{\partial \overline{W}'_x}{\partial x}
\end{aligned}\right\} \tag{5.254}$$

Substituting these expressions into Equations 5.252 and 5.253, we have

$$\frac{9}{10}(\overline{W}'_x - v)\frac{\partial \overline{W}'_x}{\partial x} = \frac{6\overline{\delta}}{\rho'} \times \frac{d^3\varphi'}{dx^3} - \frac{3v'\overline{W}'_x}{\overline{\delta}^2(1+\varphi')^2} + \frac{\gamma' - \gamma''}{\rho'} \tag{5.255}$$

$$\frac{\partial}{\partial x}[(v - \overline{W}'_x)\overline{\delta}(1+\varphi')] = 0 \tag{5.256}$$

As follows from Equation 5.256,

$$\overline{\delta}(v - \overline{W}'_x)(1+\varphi') = const. = \overline{\delta}(v - w'_0) \tag{5.257}$$

where w'_0 is the average velocity in a cross section $\overline{\delta}$.

From this expression, the following relationship applies to waves at the film surface:

$$\overline{W}'_x = v - \frac{v - w'_0}{1 + \varphi'} \tag{5.258}$$

Expanding as a series,

$$\left.\begin{array}{l}\overline{W}'_x = w'_0 + (v - w'_0)(\varphi' - \varphi'^2 + \varphi'^3 \dots) \\[2mm] \dfrac{\partial \overline{W}'_x}{\partial x} = (v + w'_0)(1 - 2\varphi' + 3\varphi'^2 \dots)\dfrac{d\varphi'}{dx}\end{array}\right\} \tag{5.259}$$

Introducing Equation 5.259 into Equation 5.255 results in an expression for the dimensionless amplitude, φ. Small amplitudes ($\varphi' \ll 1$) allow us to consider only first approximations for w'_x and $\delta \overline{W}'_x/\delta x$.

The film's energy loss due to friction per unit area of wall surface is

$$-\frac{dE}{dt} \approx \mu' \int_0^\delta \left(\frac{\partial w'_x}{\partial y}\right)^2 dy = 3\mu'\frac{\overline{W}'^2_x}{\delta} \tag{5.260}$$

And, after averaging along the wavelength,

$$\left(\frac{d\overline{E}}{dt}\right)_{\lambda_0} = -\frac{3\mu'}{\lambda_0}\int_0^{\lambda_0}\frac{\overline{W}'^2_x}{\delta}dx \tag{5.261}$$

where λ_0 denotes the wavelength at any given instance.

The average work per unit area of wall due to gravity is

$$\bar{L} = \Gamma$$

In a steady-state process, the condition $|dE/dt| = \bar{L}$ must be satisfied, whence it follows, after transformation, that

$$\bar{\delta}^3 = \frac{3\nu'\Gamma}{g(\gamma' - \gamma'')} \Phi \tag{5.262}$$

where

$$\Phi = \frac{1}{\lambda_\varrho} \int_0^{\lambda_\varrho} \frac{\left(1 + \dfrac{\nu}{w_0'}\right)\varphi}{(1 + \varphi)^3} \, dx \tag{5.263}$$

Levich [1962] outlines the following relations as a second approximation:

$$\left.\begin{array}{c} \varphi = 0.21 \sin\left[k(x - \nu t)\right] \\[2mm] k = \left(0.9 \, \dfrac{g\dot{\Gamma}}{\gamma'\mu'}\right)^{1/2} \\[2mm] \nu = 2.4 w_0' \\[2mm] \Phi = 0.8 \end{array}\right\} \tag{5.264}$$

Investigations on wave profiles by Kapitsa and Zhetf (1948) showed good agreement between experiment and the above theory.

Turbulent Film Flow

This last subsection examines the system of turbulent film flow of constant thickness on a vertical wall. The analysis closely follows that of Kutateladze and Styrikovich [1958]. As a first approximation, turbulent flow may be divided into a laminar sublayer and a turbulent core. The generalized velocity profile then may be described by a system of two equations:

$$\tau_f \simeq \mu' \frac{dw'}{dy} \; ; \qquad \text{for } y < y_1 \Bigg\}$$

$$\tau_f \simeq \rho' \chi^2 y^2 \left(\frac{dw'}{dy}\right)^2 \; ; \quad \text{for } y > y_1 \Bigg\}$$

$$(5.265)$$

where y_1 is the thickness of the laminar sublayer, and parameter χ is a constant that characterizes the structure of turbulence. Kutateladze and Styrikovich evaluated χ and y_1 to have values 0.4 and 11.6, respectively. Note that Equations 5.265 (particularly the expression for the tangential stress due to turbulent friction) are only approximations.

From a steady flow balance, the tangential stresses in the liquid film may be expressed in terms of gravity force and the friction of the gas against the film interface:

$$\tau_f' = g(\rho' - \rho'')(\delta - y) \pm \zeta'' \frac{\rho' w_r'^2}{2} \tag{5.266}$$

At the wall, $y = 0$, and the tangential stresses become.

$$\tau_w = g(\rho' - \rho'')\delta \pm \zeta'' \frac{\rho' w_r'^2}{2} \tag{5.267}$$

At the interface, $y = \delta$, hence,

$$|\tau_i| = \zeta'' \frac{\rho' w_r'^2}{2} \tag{5.268}$$

Because the laminar sublayer is relatively thin, tangential stresses in this region may be considered practically constant and equal to τ. If the gas moves slowly, i.e., $w_r'' \simeq 0$, then we have the following expressions:

For the laminar sublayer,

$$g \frac{\gamma' - \gamma''}{\gamma'} \delta = \nu' \frac{dw'}{dy} \tag{5.269}$$

For the turbulent core,

$$g\left(\frac{\gamma' - \gamma''}{\gamma'}\right)(\delta - y) = \chi^2 y^2 \left(\frac{dw'}{dy}\right)^2 \qquad (5.270)$$

After integrating Equations 5.269 and 5.270 and applying appropriate boundary conditions, the following expression is obtained for the velocity profile across the entire film:

$$w' = w_1' + \frac{1}{\chi}\sqrt{g\left(1 - \frac{\gamma''}{\gamma'}\right)}\left[2\sqrt{\delta - y} - 2\sqrt{\delta - y_1}\right.$$

$$\left. + \sqrt{\delta}\; ln\; \frac{(\sqrt{\delta} - \sqrt{\delta - y})(\sqrt{\delta} + \sqrt{\delta - y_1})}{(\sqrt{\delta} + \sqrt{\delta - y})(\sqrt{\delta} - \sqrt{\delta - y_1})}\right]; \quad \text{for } y > y_1$$

$$(5.271)$$

In the vicinity of the solid wall, where $y \leqslant \delta$, it can be assumed that

$$\sqrt{1 - y/\delta} = 1 - \frac{y}{2\delta}$$

Hence, an approximate form of this expression is

$$w' \approx w_1' + \frac{1}{\chi}\sqrt{g\left(1 - \frac{\gamma''}{\gamma'}\right)}\left[\frac{y_1 - y}{\delta} + ln\; \frac{y\left(2 - \frac{y_1}{2\delta}\right)}{y_1\left(2 - \frac{y}{2\delta}\right)}\right]$$

$$\approx w_1' + \frac{1}{\chi}\sqrt{g\left(1 - \frac{\gamma''}{\gamma'}\right)}\;\delta\; ln\; y/y_1 \qquad (5.272)$$

This expression is the well-known logarithmic law of velocity distribution for turbulent flow in the vicinity of a solid wall. Note that in the derivation of this expression the boundary conditions on the free surface of the film were ignored. Nevertheless, Equation 5.272 does provide good predictions of velocities in various cross sections of the film.

The average velocity in the film is thus obtained from

$$\overline{W}' = \frac{1}{\delta}\left(\int_0^{y_1} w'\,dy + \int_{y_1}^{\delta} w'\,dy \right) \tag{5.273A}$$

And, substituting the value w' from Equation 5.272 and assuming that $y_1 \ll \delta$, we obtain

$$\overline{W}' \approx \frac{g\delta y_1}{\nu'}\left(1 - \frac{\gamma''}{\gamma'}\right) + \frac{1}{\chi}\sqrt{g\delta\left(1 - \frac{\gamma''}{\gamma'}\right)}\left(\ln\frac{\delta}{y_1} - 1\right) \tag{5.273B}$$

Finally, multiplying both sides of this expression by δ/ν' and replacing χ and y_1 with their respective values, we obtain

$$\text{Re} = \frac{G_1'}{g\mu'} \simeq \sqrt{\frac{g\delta^3}{\gamma'^2}\left(1 - \frac{\gamma''}{\gamma'}\right)}$$

$$\times \left\{ 11.6 + 2.5\left[\ln\left(\frac{1}{11.6}\sqrt{\frac{g\delta^3}{\nu'^2}\left(1 - \frac{\gamma''}{\gamma'}\right)} - 1\right)\right]\right\} \tag{5.274}$$

Thus, from information on the liquid film's physical properties and mass rate, G_1', the thickness of a turbulent film flowing down a vertical wall can be estimated. For further information on this subject, the reader is referred to the references.

PRACTICE PROBLEMS

Solutions are provided in Appendix A for those problems with an asterisk (*) next to their numbers.

5.1 Water at 110°F is flowing through a 1.5-in.-diameter pipe at an average velocity of 15 fps. The pipeline is expanded to a 2.25-in.-diameter pipe. Determine the following:
(a) the average velocity in the larger-diameter pipe;
(b) the volumetric flowrate;
(c) the weight flowrate; and
(d) the mass flowrate.

5.2 One section of an air distribution system consists of a round duct with a diameter of 19 in. The velocity of the air in this section is 700 fpm, and the temperature and pressure are 130°F and 1 atm, respectively. At another section the duct is rectangular, having dimensions of 11 in. by 15 in. The average velocity in the rectangular section is 1500 fpm. Determine the following:
(a) the density of the air in the rectangular section; and
(b) the weight flowrate of air in lb/hr.

5.3 A 25-mm i.d. straight tube, 18 m long, is used for transferring water at 110°F. The pressure at the inlet of the tube is 180 kPa and 290 kPa at the exit.
(a) If the bottom of the tube is inclined 25° from the horizontal, what is the direction of flow through the tube?
(b) Determine the volumetric flowrate in liter/s and gpm.
(c) Compute the Reynolds number for the flow.
(d) Determine at what angle of inclination the flow direction reverses.

5.4 Water at 298°K is flowing through a pipe having an inside diameter of 3.07 in. at a rate of 25 gpm. Compute the Reynolds number in both English and SI units.

5.5 A 30% solution of glycerine in water (by volume) at 85°F is flowing through a 7.62-cm i.d. pipe at an average velocity of 15.5 fps. Determine whether the flow is laminar or turbulent.

5.6 Water at 140°F is flowing through a 3-in. Schedule 40 steel pipe with a flowrate of 100 gpm. Determine whether the flow is laminar or turbulent.

5.7 SAE 10 oil at 70°F is flowing through a 3-in. Schedule 40 steel pipe. The specific gravity of the oil is 0.87. Determine the range of volumetric flows (in gpm) in which the flow would be in the transition region.

5.8 Determine the energy loss if SAE 10 oil at 60°F flows 150 ft through a 5-in.-diameter pipe at a volumetric rate of 60 gpm.

5.9 A capillary tube is used to obtain a measurement of the viscosity of a liquid. The length of the capillary tube is 700 mm. The mass flow of the liquid is 6.5×10^{-3} kg/s, and the recorded pressure drop across the entire length is 7.1×10^5 N/m^2. The kinematic viscosity and specific gravity of the fluid are 20×10^5 m^2/s and 0.91, respectively. Determine the diameter of the capillary. Assume that the capillary is oriented vertically.

*5.10 The shear stress of a pseudoplastic or power law fluid is given by the following relationship:

$$\tau = K \left(\frac{dw}{dy}\right)^{n-1} \frac{dw}{dy}$$

where K is the *consistency index* and n is the *flow behavior index*. Develop an expression for the velocity distribution of a power law fluid flowing through the cylindrical tube shown in Figure 5.39. Assume the flow to be laminar.

*5.11 Develop an expression for the velocity distribution for laminar flow of a power law fluid between two parallel plates. The system is illustrated in Figure 5.40.

5.12 An oil having a specific gravity of 1.3 is being pumped at a rate of 80 gpm from an open storage vessel of large cross-sectional area

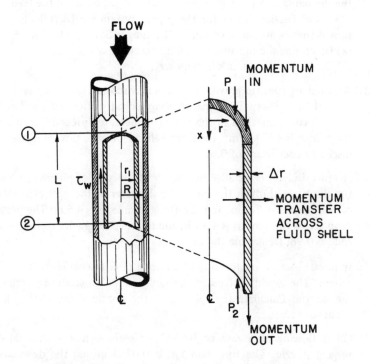

Figure 5.39   Flow through a vertical tube for Problem 5.10.

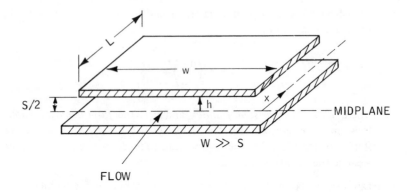

Figure 5.40 Flow between parallel plates for Problem 5.11.

through a 2.067-in. i.d. suction line. The pump discharges the oil through a 3.068-in. i.d. line to an open overhead tank. The end of the discharge line is 85 ft above the level of the liquid in the feed tank. The total friction losses for the piping system are 18 ft-lb$_f$/lb$_m$. The flow is turbulent and the pump's efficiency is 65% ($\eta = 0.65$):
(a) Determine the pressure the pump develops.
(b) Determine the pump's horsepower.

5.13 A liquid of specific gravity of 1.2 and viscosity of 1.3 cp is flowing through a 7.68-cm i.d. pipe. A sharp-edged orifice with a diameter of 0.75 cm is installed in the line. The measured pressure drop across the orifice is 187 kN/m^2. Determine the volumetric flowrate. Assume a discharge coefficient of 0.61.

5.14 A venturi meter is installed in a horizontal section of piping to measure the discharge through the pipe. The diameter of the upstream section of the venturi is 7.0 in. and the throat diameter is 4.5 in. The pressure drop across the venturi is 4.3 psi, and the specific gravity of the flowing fluid is 0.98. Determine the discharge.

5.15 A nozzle located on the side of an open tank discharges to the atmosphere. The liquid surface in the tank is 6 m above the centerline of the nozzle. Calculate the velocity in the nozzle if no friction losses are assumed.

5.16 Oil is flowing at a rate of 0.0573 m^3/s through a horizontal converging nozzle. The upstream i.d. is 0.0753 m and the downstream is 0.0386 m. Frictional losses may be considered negligible. The density of the oil is 889 kg/m^3. Compute the resultant force on the nozzle.

5.17 A liquid having a density of 950 kg/m^3 and a viscosity of 1.57×10^{-3} Pa-s is flowing through a small capillary with an inside diameter of 3.52×10^{-3} m. The velocity of the liquid is 0.185 m/s. Estimate the pressure drop over 0.7 m of length.

5.18 A liquid of viscosity 7.5 cp and density 750 kg/m^3 is flowing through a horizontal glass-lined pipe having a 1.75-in. i.d. Determine the mechanical energy friction loss in J/kg for a 15-m section of pipe.

5.19 A liquid of specific gravity 1.3 and viscosity 2.52 cp is to flow through a 500-m-long commercial stainless steel pipe at a rate of 250 gpm. A head of 8 m of fluid is available to overcome friction losses in the system. Write a computer program in BASIC or FORTRAN to compute the pipe diameter.

5.20 Water at 25°C is flowing through a 500-mm-diameter cast iron pipe. The roughness coefficient for the pipe is 4.2 mm. The head loss incurred over 700 m of pipe is 15 m. Determine the volumetric flow.

5.21 Compute the size of clean wrought iron pipe needed to transport 5200 gpm of a liquid having a kinematic viscosity of 0.0009 ft^2/s and specific gravity of 1.1. The head loss for the system is 105 ft-lb$_f$/lb$_m$.

5.22 Estimate the energy loss that occurs when 35 gpm of water flows through a sudden enlargement from a 1.5-in. tube to a 3.8-in. tube. Also, determine the difference between the pressure ahead of the sudden enlargement and the pressure downstream from the enlargement.

5.23 Estimate the energy loss that occurs when 50 gpm of water flows from a 1.75-in. i.d. pipe into a large vessel.

5.24 Estimate the energy loss that occurs when 30 gpm of oil (specific gravity 0.89) flows from a 3.5-in. tube into a 2.25 in. tube through a sudden contraction.

5.25 Estimate the energy loss that occurs when 20 gpm of oil (specific gravity 0.89) flows from a 1.5-in. tube into a 2.9-in. tube through a gradual enlargement, having an included cone angle of 28°.

5.26 Water at 75°F is being pumped at a mass rate of 800 kg/hr from a large storage vessel to an elevated receiving tank. A sketch of the pipe network and line dimensions are given in Figure 5.41.
    (a) Compute the total energy losses for the system (assume the valve to be fully open).

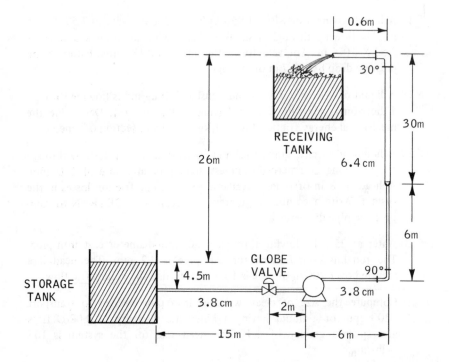

Figure 5.41 System diagram for Problem 5.26.

(b) Compute the horsepower required by the pump if the overall pump efficiency is 60% ($\eta = 0.60$).

(c) Compute the centerline velocity and the velocity at a distance of one-half the pipe radius for both pipe sizes.

5.27 In a chemical processing plant, toluene at 110°F is to be pumped to a point with a pressure of 95 psig. The pump is located at a point 30 ft below the discharge point. Approximately 900 ft of new Schedule 40, 2-in.-diameter steel pipe is used between the pump and the discharge point. For a volume flowrate of 35 gpm, compute the required pressure at the outlet of the pump.

5.28 Water at 120°F is flowing through the piping system shown in Figure 5.42 at a rate of 850 gpm. Determine the pressure drop between points 1 and 2.

5.29 The flow from a reservoir is controlled by a gate valve. The reservoir maintains a constant head of 22 ft. The system is shown in Figure 5.43.

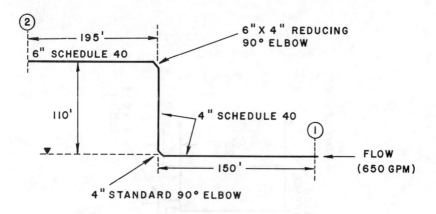

Figure 5.42   Piping distribution system for Problem 5.28.

Determine the maximum flowrate in gpm. (Hint: A trial and error solution is required.)

*5.30 Water is being pumped at a rate of 250 gpm to an elevation of 320 ft above the pump. The piping system is shown in Figure 5.44. All piping is 3-in. Schedule 40 steel pipe having a roughness coefficient of 0.092 in. Calculate the pump's brake horsepower for an efficiency of 73% ($\eta = 0.73$).

5.31 Carbon tetrachloride at 70°F is being pumped at a rate of 165 gpm from a holding tank open to the atmosphere. A gauge pressure on the discharge line reads 72 psig. The discharge line is 40 ft long, and the discharge point is 10 ft above the centerline of the pump. The pump is located adjacent to the holding tank and its efficiency is 65%. A globe valve is located immediately after the pump and there are four 90° standard elbows and one 45° elbow in the pipeline. The discharge pipe is 2-in. Schedule 40 stainless steel. Determine the head developed by the pump and the pump's horsepower. Assume the valve is normally 3/4 open.

5.32 Condensate water flows through the cylindrical copper coil shown in Figure 5.45 at a rate of 50 gpm. The coil tubing has an inside diameter of 0.567 in. Determine the pressure drop between points A and B. Note that there are 6.5 coils, i.e., twenty-six 90° bends in the coil.

5.33 Repeat Problem 5.32 for saturated steam at 190 psig flowing at a rate of 1800 lb$_m$/hr. In this problem, the coil tubing diameter is 1.56 in.

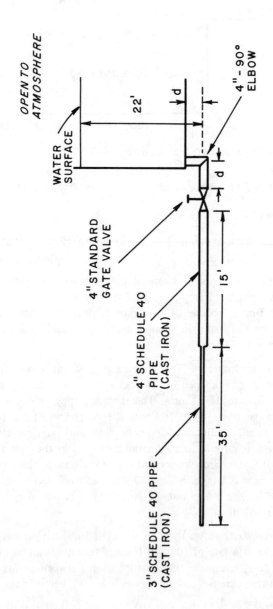

Figure 5.43   Piping arrangement for Problem 5.29.

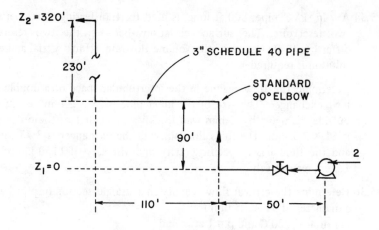

Figure 5.44   System diagram for Problem 5.30.

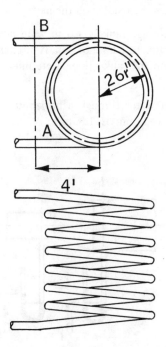

Figure 5.45   Condensate flow through a coil for Problem 5.32.

5.34 A 7-in. PVC pipe, 800 ft long, is used for transferring water between two reservoirs. The surface elevations between the two reservoirs differ by 75 ft. Calculate the volume flowrate. (Hint: A trial and error solution is required.)

5.35 Determine the flow regime in the intertubular space of a double-pipe heat exchanger. The size of the inner pipe is 25 × 2 mm i.e., 25 mm outside diameter by 2 mm wall thickness, and that of the outer tube is 51 × 2.5 mm. The mass flow through the exchanger is 3.73 ton/hr, and the fluid has a specific gravity and viscosity of 1150 kg/m^3 and 1.2 cp, respectively.

*5.36 Determine the critical flow velocity in a straight section of pipe having a diameter of 51 × 2.5 mm for
(a) air at t = 20°C and p = 1 atm; and
(b) oil with μ = 35 cp and γ = 0.963.

*5.37 A section of piping is contracted gradually from 200-mm i.d. to 100 mm (refer to Figure 5.46). Methane is flowing through the piping at a rate of 1700 m^3/hr at 30°C. A U-tube located on the larger piping shows an excess pressure of 40 mm of water. Neglecting resistances, determine the U-tube reading on the narrow section of the piping.

*5.38 A cylindrical storage tank is 1.0 m wide and 2.5 m high. The tank is filled with water to a level of 2 m. A discharge orifice, 2 cm in diameter, is located on the tank's floor. Determine the time it takes to drain the vessel.

*5.39 The velocity of water through a horizontal section of piping having a diameter of 152 mm is 1.3 m/s. A sharp-edged orifice meter is

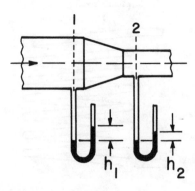

Figure 5.46   Gradual contraction for Problem 5.37.

installed in the piping as shown in Figure 5.47. The diameter of the orifice meter hole is 86.5 mm. Assuming the velocity coefficient to be $\psi = 0.97$ and jet constriction factor $\epsilon = 0.62$, determine:

(a) the rate coefficient; and

(b) the reading of the mercury differential manometer across the orifice meter.

*5.40 Determine the frictional pressure losses through the coil shown in Figure 5.48, where the water is flowing with a velocity of 1 m/s. The coil is made from a pipe 43 × 2.5 mm in diameter (43 mm outer diameter by 2.5 mm wall thickness); the coil diameter D = 1 m; and the number of coils is equal to 10. The average temperature of water is 30°C. The plot given in Figure 5.49 will be needed for the solution.

*5.41 A packed tower wet scrubber (Figure 5.50) is being used to scrub (absorb) NO_x (NO_2, NO) from an airstream using a caustic solution. The caustic solution is pumped from a supply tank located at the foot of the column to a set of spray nozzles located above the packed section of the column. The height, H, of the vertical line transferring the liquid is 10 m, and the piping is 102 × 3.75 mm in diameter. The solution is fed to the scrubber at a rate of 700 ℓ/min. The specific weight and viscosity of the solution are 1100 kg/m^3 and 1.1 cp,

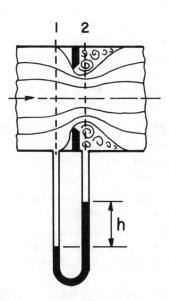

Figure 5.47   Orifice meter in pipe section for Problem 5.39.

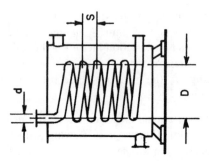

Figure 5.48   Cooling coil system for Problem 5.40.

respectively. The total pressure required for atomization is 0.5 atm. The piping has two valves, four 90° elbows and a total length of 25 m. Determine the pump's horsepower for an efficiency of $\eta = 0.6$.

*5.42 Mineral oil is pumped at a rate of 40,000 ℓ/hr through piping 108 mm I.D. X 4 mm wall thickness, to a tank located 20 m above the pump. The total length of piping is 430 m. Determine the necessary pump power for the following temperatures:
(a) at T = 15°C; and
(b) at T = 50°C.

*5.43 For Problem 5.42, the oil specific gravities are 0.96 and 0.89, and viscosities are 3430 cp and 187 cp for 15°C and 50°C, respectively. Determine whether it is economical to heat the oil to 50°C before pumping, if kilowatt-hour costs are 40¢ and 1 ton of heating steam (at p = 1 atm) costs $20. Assume a pump efficiency of $\eta = 0.5$.

5.44 For Problem 5.43, if steam heating costs rise at an annual rate of 5%/yr and electrical costs are expected to double at the end of 10 years, which option is more economical and at what point should a larger pump be installed?

*5.45 Liquid from montejus A is transferred by pressure to an open tank B, shown in Figure 5.51. The difference between the liquid levels in the two vessels is 8 m. Level variations in the montejus are negligible. The i.d. of the piping is d = 21 mm, and the resistance coefficient of the valve is 3.5. The liquid density is 1100 kg/m^3 and its viscosity is 2.5 cp.
(a) Determine the minimum pressure in the montejus required to transfer the liquid to tank B.

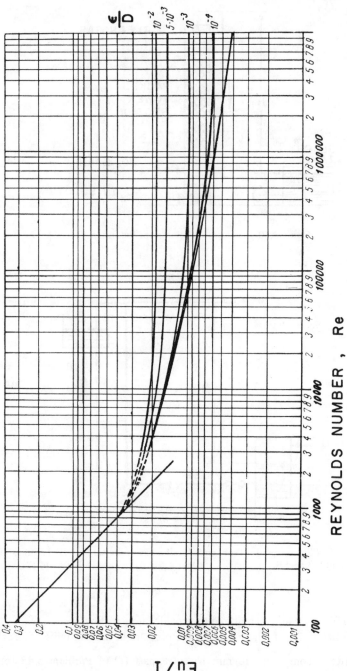

Figure 5.49   Plot of Euler number to weight rate ratio vs Reynolds number.

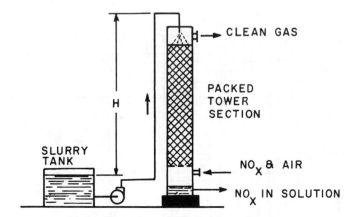

Figure 5.50   Absorption column for Problem 5.41.

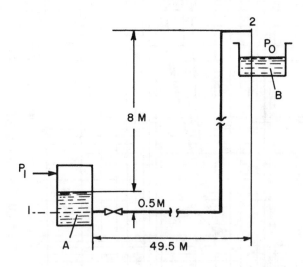

Figure 5.51   Flow system for Problem 5.45. The higher pressure in vessel A transfers the liquid to vessel B.

(b) What pressure is required to provide laminar flow?

(c) Determine the liquid rate if the pressure in the montejus is 2.2 atm.

5.46 Write a computer program to solve part (c) of Problem 5.45 for a pressure of 5 atm in the montejus.

*5.47 Determine the pressure loss for cooling water in a cooler having 40 coils with diameter D = 0.5 m. The i.d. of the coil tubing is d = 12 mm. The average temperature of cooling water T_1 = 20°C, and the average temperature of the tubing wall is T_2 = 40°C. The mass rate of cooling water is G = 0.1 kg/s.

*5.48 Determine the pressure losses for a single-pass heat exchanger (Figure 5.52). The exchanger is 3.5 m tall and has a shell diameter D = 600 mm. The exchanger contains 253 pipes, each having an inside diameter of 21 mm and an outer diameter of 25 mm. The pipes are arranged in a triangular fashion with a pitch t = 32 mm. The i.d. of the inlet and outlet nozzles are 140 mm each. There are 10 partitions in the intertubular space. The water solution circulating in the tubular space has a density of 1100 kg/m^3 and a viscosity of 1.2 cp. The solution's mass rate is G = 40 kg/s. The water rate in the inter-tubular space is G' = 50 kg/s at a temperature of 40°C.

*5.49 Liquid is discharged from rectangular tank 1 to tank 2 through a round orifice having an inside diameter of 2 cm. The system is shown in Figure 5.53. The initial difference in levels between the two tanks is H_0 = 2.0 m. The cross sections of the tanks are S_1 = 8 m^2 and S_2 = 6 m^2. Determine the time it takes for:
(a) the initial level in tank 1 to decrease to 1.0 m; and
(b) the levels in both tanks to become the same.

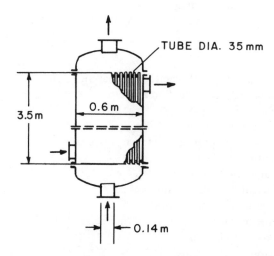

TUBE DIA. 35 mm

3.5 m

0.6 m

0.14 m

Figure 5.52   Single-pass heat exchanger for Problem 5.48.

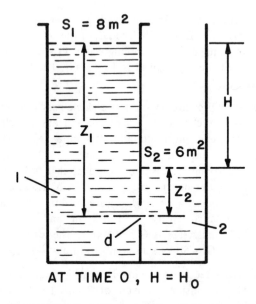

Figure 5.53 Tank discharge system for Problem 5.49.

5.50 For the packed tower described in Problem 5.41, it is determined that the liquid rate must be reduced to below 500 ℓ/min to avoid flooding the column. Determine by how much the pump is oversized for the operation specified.

NOTATION

a	contraction loss coefficient, see Equation 5.170
a	bending angle, see Equation 5.176
A	amortization costs, $/yr
A	coefficient in Equation 5.158A
a,b	fraction of costs for maintenance and repairs
a,b	sides of rectangle, m
C	discharge coefficient
c	coefficient defined by Equation 5.178
C_e	cost per unit of energy, $/kW
c_ϱ	cost per unit length of piping, $/m
c_1	velocity of sound, m/s
C_L	discharge coefficient
C_p, C_v	specific heat at constant pressure or constant volume, respectively, J/(kg-°K)

d_{eq}	equivalent diameter, m
D	diameter, m
D_o	optimum pipe diameter, m
e	roughness height, mm
E_ν	eddy viscosity, cp
E	energy costs, $/yr
E	energy term, J/kg
Eu	Euler number
F	area, m^2
$\hat{F}$	friction losses, N
Fr	Froude number
g	gravitational acceleration, m/s^2
G	superficial mass velocity, kg/m^2-hr
g_c	conversion factor, 32.174 ft-lb_m/(lb_f-s^2)
h_{fr}, h_{lr}	friction and local resistance losses, m
h_ϱ	lost head, m or J/kg
H	velocity head, m
H_o	Homochronity number
i	enthalpy, J/mol
I	turbulence intensity parameter
j	fractional cost
k	space factor, see Equation 5.149
k_1	cost parameter defined by Equation 5.221
k_2	cost parameter defined by Equation 5.225
k_e	head loss coefficient
k_i	yearly expenses for amortization and maintenance, $/yr
k_p	yearly production and operating expenses, $/yr
$\underline{\varrho}$	length, m
$\overline{L}$	average work per unit area due to gravity, J/m^2
L	length, m
L_e	equivalent length, m
L_T	turbulence scale, defined by Equation 5.37
M	mass rate, kg/s
M'	viscosity scaling factor, μ/μ_0
m	parameter in Equation 5.158A
n	exponent in Equation 5.209
n	function defined in Equation 5.183
N'	number of tube rows
N	pump power, hp
$\tilde{N}$	energy cost for fluid transport, $/hr
p	wetted perimeter, m
P'	density scaling factor, ρ/ρ_0
$\mathcal{P}$	pressure scaling factor, p/p_0
P_{cr}	critical pressure, Pa

P, p	pressure, Pa
q	exponent in Equation 5.158A
$\dot{Q}$	heat, J
Q	volumetric flow, cfs or m^3/s
r	pipe radius or distance from wall, m
r_h	hydraulic radius, m
R	tube radius, m
Re	Reynolds number
$R(y)$	correlation coefficient
$\dot{s}$	function defined in Table 5.13
S	surface area or hole area, m^2
t	time, s
t'	tube separation, mm
T	temperature, °C or °K
u	variable
V	volumetric flowrate, m^3/hr
v	velocity, m/s
$w_{x,y,z}$	velocity components, m/s
W	mass flowrate, kg/s
$\tilde{W}$	average velocity scaling factor
W'	mechanical work, J
$\overline{W}$	mean linear velocity, m/hr
W_g	weight rate defining flow regime for optimum pipe diameter, kg/s
$\hat{W}_s, \hat{W}_p$	shaft power or mechanical power, $ft\text{-}lb_f/lb_m$ or kJ
$W_{x,y,z}$	velocity scaling factors
x	parameter in Equation 5.209
X	empirical parameter in Equation 5.176
X', Y', Z'	distance scaling factors
y	distance from wall, m
Y	yearly operating time, hr/yr
Z, z	leveling height, m

Greek Symbols

α	correction coefficient for laminar or turbulent flow
β	diameter ratio or angle, °
Γ	mass rate per unit perimeter, kg/m-s
$\tilde{\Gamma}$	length to diameter ratio
γ	specific weight of fluid
∇^2	Laplacian operator, refer to Equation 5.56
δ	film thickness, mm
$\hat{\delta}$	velocity dampening parameter defined by Equation 5.189B
ϵ	relative roughness, e/D

ϵ_M	eddy diffusivity of momentum (E_ν/ρ), mm^2/s
ϵ_j	jet constriction coefficient
ζ	friction resistance coefficient, see Equation 5.154
η	pump efficiency
θ	velocity component for compressible fluid, m/s; also transformation to cylindrical coordinates
κ	ratio of specific heats, C_p/C_v
λ	friction factor
λ_{is}	friction factor at specified temperature
μ	viscosity, cp
ν	kinematic viscosity, mm^2/s; or specific volume as noted, m^3/kg
ν'	volume, m^3; or specific volume, m^3/kg
ρ	density, kg/m^3
σ	surface tension, N/m
τ	shear stress, lb$_f$/in.^{2} or Pa
τ_N, τ_T	viscous and turbulent stresses, respectively, lb$_f$/in.^{2} or Pa
T'	shear scaling factor, τ/τ_0
T	friction force, N
φ	velocity coefficient
ϕ	velocity factor
χ	constant defining turbulence structure, see Equation 5.265
ψ	correction coefficient for nonisothermal friction factors or Weisbach equation discharge coefficient (see Equation 5.171)
ψ'	deviation coefficient, see Equation 5.253

REFERENCES

Beij, K. H. (1938) "Pressure Losses of Fluid Flow in 90 Degree Pipe Bends," *J. Res. Nat. Bur. Standards* XXI.

Bennett, C. O., and J. E. Myers (1964) *Momentum, Heat and Mass Transfer* (New York: McGraw-Hill Book Co.).

Bergelin, O. P. (1949) *Chem. Eng.* 56(5):104-106.

Bird, R. B., W. E. Stewart and E. N. Lightfoot (1960) *Transport Phenomena*, (New York: John Wiley & Sons, Inc.).

Boucher, D. F. (1947) *Chem. Eng. Prog.* 44(10):527,601.

Brodkey, R. S. (1967) *The Phenomena of Fluid Motions* (Reading, MA: Addison-Wesley Publishing Co.).

Brown, G. et al. (1950) *Unit Operations* (New York: John Wiley & Sons, Inc.).

Cambel, A. B., and B. H. Jennings (1958) *Gas Dynamics* (New York: McGraw-Hill Book Co.).

Cheremisinoff, N. P. (1979) *Applied Fluid Flow Measurement: Fundamentals and Technology* (New York: Marcel Dekker, Inc.).

Cheremisinoff, N. P. (1981a) *Fluid Flow: Pumps, Pipes and Channels* (Ann Arbor, MI: Ann Arbor Science Publishers).

Cheremisinoff, N. P. (1981b) *Process Level Instrumentation and Control* (New York: Marcel Dekker, Inc.).

Chilton, T. H. (1938) *Trans. Am. Inst. Chem. Eng.* 29:161.

Cooper, C. M., T. D. Drew and W. H. McAdams (1934) *Ind. Eng. Chem.* 26:428-431.

Curle, N., and H. J. Davies (1968) *Modern Fluid Dynamics Vol. 1: Incompressible Flow* (New York: D. Van Nostrand Co.).

Davies, J. T. (1972) *Turbulence Phenomena* (New York: Academic Press Inc.).

Dodge, L. (1968) "How to Compute and Combine Fluid Flow Resistances in Components," *Hydraul. Pneum.* 21(9):118-121.

Folsom, R. G. (1956) *Trans. Am. Soc. Mech. Eng.* 78:1447-1460.

Friedman, S. G., and C. O. Miller (1941) *Ind. Eng. Chem.* 33:885-891.

Hodson, J. L. (1939) *Trans. ASME* 51:303.

Hooper, C. M. (1950) *Trans. Am. Soc. Mech. Eng.* 72:1009-1110.

Jackson, M. L. (1955) *Am. Inst. Chem. Eng. J.* 1:231-240.

Jakob, M. (1938) *Trans. ASME* 60:384.

Jeschke, J. (1925) *Tech. Mech. V. D. I.*, 69.

Kapitsa, L. P. (1948) "Volnovoye techenie Tonkikh Sloyev Vyazkoy Zhidkosti (Wave Flow of Thin Layers of Viscous Liquid), *Zhetf (J. Exp. Theoret. Phys.)*, No. 1.

King, H. W., and E. F. Brater (1963) *Handbook of Hydraulics* (New York: McGraw-Hill Book Co.).

King, R. C., and S. Crocker (1967) *Piping Handbook* (New York: McGraw-Hill Book Co.).

Knudsen, J. G., and K. D. L. Katz (1958) *Fluid Dynamics and Heat Transfer* (New York: McGraw-Hill Book Co.).

Kovakov, P. K. (1949) *Izv. an SSSR, OTN*, 7:1029.

Krylov, A. V. (1947) *Dan SSSR* 56(2):133.

Kutateladze, S. S., and M. A. Styrikovich (1958) *Hydraulics of Gas-Liquid Systems*, (Moscow: Wright Field Trans. F-TS-9814/V.).

Levich, V. G. (1962) *Physico-Chemical Hydrodynamics* (Englewood Cliffs, NJ: Prentice Hall, Inc.).

Liepmann, H. W., and A. Roshko (1957) *Elements of Gas Dynamics* (New York: John Wiley & Sons, Inc.).

Miner, I. O. (1956) *Trans. Am. Soc. Mech. Eng.* 78:475-479.

Moody, L. F. (1944) *Trans. ASME* 66:671.

Mott, R. L. (1972) *Applied Fluid Mechanics* (Columbus, OH: Charles E. Mezzill Publishing Co.).

Normand, C. E. (1948) *Ind. Eng. Chem.* 40(5):783.

Othmer, D. F. (1945) *Ind. Eng. Chem* 37(11):1112.

Pai, Shih-I (1956) *Viscous Flow Theory-Laminary Flow* (New York: D. Van Nostrand Co.).

Perry, J. H. Ed. (1950) *Chemical Engineer's Handbook*, 3rd ed. (New York: McGraw-Hill Book Co.).

Rohsenow, W. M., and J. P. Hartnett (1973) *Handbook of Heat Transfer* (New York: McGraw-Hill Book Co.).

Rouse, H. (1948) *Elementary Mechanics of Fluid* (New York: John Wiley & Sons, Inc.).

Shapiro, A. H. (1953) *The Dynamics and Thermodynamics of Compressible Fluid Flow* (New York: Roland Publishing Co.).

Siedez, E. N. (1936) *Ind. Eng. Chem.* 28(12):1429.

Simpson, L. L. (1969) "Process Piping: Functional Design," *Chem. Eng.* 76(8):167-181.

Stokes, R. L. (1946) *Ind. Eng. Chem.* 38(6):622.

Stoll, A. W. (1951) *Trans. Am. Soc. Mech. Eng.* 73:963-969.

Streeter, V. L., and E. B. Wylie (1979) *Fluid Mechanics* (New York: McGraw-Hill Book Co.).

White, S. M. (1929a) *Engineering* 128:69.

White, S. M. (1929b) *Proc. Roy. Soc. A.* 123.

Whitwell, J. C., and D. S. Plumb (1939) *Ind. Eng. Chem.* 31(4):451.

Wilson, R. E. (1922) *Ind. Eng. Chem.* 14(105).

BIBLIOGRAPHY

ASME Research Committee on Fluid Meters (1959) *Their Theory and Application*, (New York: The American Society of Mechanical Engineers).

Fallah, R., T. G. Hunter and A. W. Nash (1934) *J. Soc. Chem. Ind.* 53:369-379T.

Flowmeter Computation Handbook (1959) (New York: American Society of Mechanical Engineers).

Fluid Meters: Their Theory and Application, 5th ed. (1959) (New York: American Society of Mechanical Engineers).

Geankopolis, C. J. (1978) *Transport Processes and Unit Operations* (Boston, MA: Allyn and Bacon, Inc.).

Greve-Bull. (1928) *Purdue Univ.* 12(5):32.

Gukhman, A. A. (1965) *Introduction to the Theory of Similarity* (New York: Academic Press, Inc.).

Lapple, C. E. (1947) *Chem. Eng.* 56(2):96-104.

Sedov, L. E. (1959) *Similarity and Dimensional Methods in Mechanics* (New York: Academic Press, Inc.).

CHAPTER 6

FLUID TRANSPORT BY PUMPS

CONTENTS

INTRODUCTION

Transporting liquid and gaseous media to and from process equipment is integral to the chemical process and allied industries, and this transfer requires energy consumption. The amount of energy required depends on the height through which the fluid is raised, the length and diameter of the transporting conduits, the rate of flow and, of course, the fluid physical properties (in particular, viscosity and density). In some applications, external energy for transferring fluids is not required. For example, when liquid flows to a lower elevation under the influence of gravity, a partial transformation of the fluid's potential energy into kinetic energy occurs. When transporting fluids through horizontal conduits and especially to higher elevations within a system, mechanical devices such as pumps and compressors are employed.

Several methods are available for transporting fluids between process equipment:

1. centrifugal force inducing fluid motion;
2. volumetric displacement of fluids, either mechanically or with other fluids;
3. mechanical impulse;
4. transfer of momentum from another fluid;
5. electromagnetic forces; and
6. gravity-induced flow.

In this chapter, the first four methods are described; however, emphasis is placed on mechanical devices for transporting incompressible fluids, namely, pumps. Pumps, as well as their compressible fluid counterparts, fans and compressors, are hydraulic machines that transform the mechanical energy generated by a motor into the energy of moving fluids. It is the difference between the fluid's pressure in the hydraulic machine and that in the transferring lines that establishes the magnitude of displacement by gravity.

LIQUID TRANSPORTATION BY PUMPS

Classifications and Characteristics

The major types of pumps used in process plant applications are centrifugal, axial, regenerative turbine, reciprocating, metering and rotary. These classes are grouped under one of two categories: dynamic pumps or positive displacement pumps.

Dynamic pumps include centrifugal and axial types. They operate by developing a high liquid velocity, which is converted to pressure in a diffusing flow passage. These pumps generally are lower in efficiency than the positive displacement types. However, they do operate at relatively high speeds, thus providing high flowrates in relation to the physical size of the pump. Furthermore, they usually have significantly lower maintenance requirements than positive displacement pumps.

Positive displacement pumps operate by forcing a fixed volume of fluid from the inlet pressure section of the pump into the pump's discharge zone. This is performed intermittently with reciprocating pumps. In the case of rotary screw and gear pumps, the action is continuous. This category of pumps operates at lower rotating speeds than do dynamic pumps. Positive displacement pumps also tend to be physically larger than equal-capacity dynamic pumps.

Four characteristics describe all pumps:

1. **Capacity** $Q(m^3/s)$–the quantity of liquid discharged per unit time.

2. **Head** $H(m)$–the energy supplied to the liquid per unit weight, obtained by dividing the increase in pressure by the liquid specific weight. This specific energy is determined by the Bernoulli equation. Head may be defined as the height to which 1 kg of discharged liquid can be lifted by the energy supplied by a pump. Therefore, it does not depend on the specific weight $\gamma(kg_f/m^3)$ or density $\rho(kg/m^3)$ of liquid to be pumped.

3. **Power, N** $(kg_f\text{-}m/s)$–the energy consumed by a pump per unit time for supplying liquid energy in the form of pressure. Power is equal to the product of specific energy, H, and the mass flowrate, γQ:

$$N = \gamma QH = \rho g QH \qquad (6.1)$$

Effective power, N_e, is larger than N due to energy losses in a pump. Its relative value is evaluated by the pump efficiency, η_p:

$$N_e = \frac{N}{\eta_p} = \frac{\rho g QH}{\eta_p} \qquad (6.2)$$

4. **Overall efficiency**, η–the ratio of useful hydraulic work performed to the actual work input. This parameter characterizes the perfection of design and performance of a pump. The value of η reflects the relative power losses in the pump and is expressed by the following equation:

$$\eta = \eta_v \times \eta_h \times \eta_m \qquad (6.3)$$

where η_v is the volumetric efficiency defined as the ratio of liquid actually pumped to the liquid that theoretically should be discharged. In other words, it indicates the percentage of losses (or slip). In practice, slip should not exceed 5%. η_h is the hydraulic efficiency defined as the ratio of the actual head pumped to the theoretical head:

$$\eta_h = \frac{H}{H + \text{hydraulic losses}} \qquad (6.4)$$

Hydraulic losses are those head losses in the suction and discharge sections of a pump. In the suction end, these losses comprise

- Velocity head
- Entrance head
- Friction head in the suction line
- Losses in bends and losses in suction valves

Discharge line losses include

- Losses in the discharge valves
- Velocity head
- Friction in the discharge piping

η_m is the mechanical efficiency defining the relation between the indicated pump horsepower and the actual power input from the drive. It characterizes mechanical losses in the pump, e.g., in bearings, stuffing boxes, etc.

Overall efficiency η depends on the pump design and varies from 50% for small pumps to about 90% for large sizes. The power consumed by a motor (defined as the nominal power of a motor) N_m exceeds brake power by mechanical losses incurred in transmission and in the motor itself. These losses are accounted for in Equation 6.3 by including the efficiencies of a transmission η_{tr} and a motor η_m:

$$N_m = \frac{N_e}{\eta_m \times \eta_{tr}} = \frac{N}{\eta_e \times \eta_m \times \eta_{tr}} \tag{6.5}$$

The product η_e, η_m, η_{tr} is the total efficiency of a pump and may be defined as the ratio of hydraulic power to the motor's nominal power:

$$\eta = \frac{N}{N_m} = \eta_p \eta_{tr} \eta_{mot} \tag{6.6}$$

From Equations 6.3 and 6.6 therefore, the total efficiency of a pump may be expressed by the product of five values:

$$\eta = \eta_v \times \eta_h \times \eta_m \times \eta_{tr} \times \eta_{mot} \tag{6.7}$$

The actual power of a pump motor, N_A, should be based on the energy required to overcome the fluid's inertia at startup so as to avoid overloading the unit.

$$N_A = \beta N_m \tag{6.8}$$

Table 6.1 Typical Values of Coefficient β as a Function of Motor Power

N_A(kW)	Less than 1	1–5	5–50	Greater than 50
β	2–1.5	1.5–1.2	1.2–1.15	1.1

Coefficient β is determined from the size of the motor, of which typical values are given in Table 6.1.

Head and Suction Head

Head, H, characterizes the excessive energy (ℓ = gH) added to 1 kg of liquid in a pump. Figure 6.1 illustrates a simple pumping scheme for which we will write the Bernoulli equation.

At suction (sections 1-1 and 1′-1′),

$$\frac{p_1}{\rho g} + \frac{w_1^2}{2g} = H_s + \frac{w_s^2}{2g} + \frac{p_s}{\rho g} + h_{\varrho s} \qquad (6.9)$$

At discharge (sections 1′-1′ and 2-2),

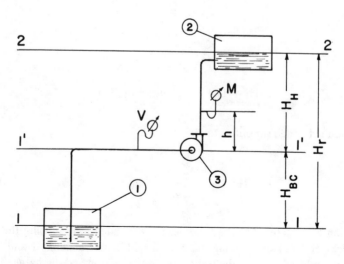

Figure 6.1   Scheme of pumping unit: (1) tank; (2) tank; (3) pump; (M) manometer; (V) vacuum gauge.

$$\frac{p_d}{\rho g} + \frac{w_d^2}{2g} = H_d + \frac{w_2^2}{2g} + \frac{p_2}{\rho g} + h_{\varrho d} \tag{6.10}$$

where w_1 and w_2 = liquid velocities in tanks 1 and 2, respectively
         w_s and w_d = liquid velocities in the suction and discharge pump
                nozzles
         $h_{\varrho s}$ and $h_{\varrho d}$ = head losses in the suction and discharge pipings

If velocities $w_1 = 0$ and $w_2 = 0$ are assumed, then the total dynamic head, H, of a pump is the difference between discharge head, H_d, and suction head in the nozzles:

$$H = \frac{p_d - p_s}{\rho g} \tag{6.11}$$

From Equations 6.9 and 6.10 we obtain

$$H = \frac{p_2 - p_1}{\rho g} + \frac{w_s^2 - w_d^2}{2g} + H_d + H_s + h_{\varrho d} + h_{\varrho s} \tag{6.12}$$

If the suction and discharge nozzles of a pump have the same diameter, then

$$w_d = w_s$$

and

$$H_t = H_d + H_s \quad ; \quad h_{\varrho d} + h_{\varrho s} = h_{\varrho t}$$

Equation 6.12 then simplifies to

$$H = H_T + \frac{p_2 - p_1}{\rho g} + h_{\varrho t} \tag{6.13}$$

Equation 6.13 states that the total head is expended in (1) lifting the liquid to height H_T; (2) overcoming the pressure difference in tanks 1 and 2; and (3) overcoming the hydraulic resistances in the suction and discharge pipings. If $p_1 = p_2$, Equation 6.13 becomes

$$H = H_T = h_{\varrho t} \qquad (6.14)$$

For a horizontal section of piping, $H_T = 0$,

$$H = \frac{p_2 - p_1}{\rho g} + h_{\varrho t} \qquad (6.15)$$

In the case in which $p_1 = p_2$ and $H_T = 0$,

$$H = h_{\varrho t} \qquad (6.16)$$

The total head of the operating pump may be evaluated from measurements obtained from the pressure p_m and vacuum gauges p_v.

Total suction head can be obtained from the measurement h_{sg} on the gauge on the pump's suction nozzle (corrected to the pump centerline and converted to feet of liquid), plus the barometer reading in feet of liquid and the velocity head, h_{vs}, (ft) at the point of gauge attachment:

$$H_s = H_{sg} + atm + H_{v.s} \qquad (6.17)$$

When the static pressure at the suction flange is less than atmospheric, the measurement obtained from a vacuum gauge replaces H_{sg} in Equation 6.17 and is assigned a negative sign.

Total discharge head, H_d, is obtained from a gauge, H_{dg}, at the pump's discharge flange (corrected to the pump centerline and converted to feet of liquid), plus the barometer reading and the velocity head, H_{vg}, at the point of gauge attachment:

$$H_d = H_{dg} + atm + H_{vg} \qquad (6.18)$$

Again, if the discharge gauge pressure is below atmospheric, the vacuum gauge measurement replaces H_{dg} with a negative sign. Before installation, it is possible to estimate the total discharge head from the static discharge head, H_{sd}, and discharge friction, H_{fd}, as follows:

$$H_d = H_{sd} + H_{fd} \qquad (6.19)$$

Static suction head, H_{ss}, is defined as the vertical distance (ft) between the free level of the source of supply and the pump centerline, plus the absolute pressure at this level (converted to feet of liquid).

Total static head, H_{ts}, is the difference between discharge and suction static heads.

The suction generated by a pump is derived from a pressure difference between the suction source, p_1, and the pump, p_s, or is due to the action of the head difference

$$\frac{p_1}{\rho g} - \frac{p_s}{\rho g}$$

Suction height can be determined as follows:

$$H_s = \frac{p_1}{\rho g} - \left(\frac{p_s}{\rho g} + \frac{w_s^2 - w_1^2}{2g} + h_{\varrho s} \right) \qquad (6.20A)$$

or

$$H_s = \frac{p_1}{\rho g} - \left(\frac{p_s}{\rho g} + \frac{w_s^2}{2g} + h_{\varrho s} \right) \qquad (6.20B)$$

because $w_1 = 0$. These expressions demonstrate that the suction head increases with p_1 and decreases with increasing w_s and $h_{\varrho s}$.

If liquid is pumped from an open tank, i.e., $p_1 = p_a$, where p_a corresponds to atmospheric, the suction pressure, p_s, must exceed pressure, p_t (the pressure of the saturated vapor of the liquid at the pumping temperature $(p_s > p_t)$), otherwise the fluid begins to boil. When the pumped liquid vaporizes, the suction head goes to zero at the limit and flow stops. Consequently,

$$H_s \leqslant \frac{p_a}{\rho g} - \left(\frac{p_t}{\rho g} + \frac{w_s^2}{2g} + h_{\varrho s} \right) \qquad (6.21)$$

This expression shows that the suction head is a function of atmospheric pressure, fluid velocity and density, temperature (and, correspondingly, the liquid's vapor pressure) and the hydraulic resistance of the suction piping. When pumping from an open tank, the suction head cannot exceed the head of pumping liquid, which corresponds to atmospheric pressure (the value of which depends on the height of the pump installation above a specified datum, normally sea level). Thus, for example, if water is pumped at $t = 20°C$, the suction head cannot exceed 10 m at sea level. If the same pumping system is used at an elevation of 2000 m, the suction head cannot be greater than 8.1 m, which corresponds to the atmospheric pressure in m of water column.

At temperatures approaching the boiling point of the liquid, the suction head becomes zero:

$$H_s = 0 \quad \text{at} \quad \frac{p_a}{\rho g} = \frac{p_1}{\rho g} + \frac{w_s^2}{2g} + h_{\varrho s}$$

In this situation, the pump must be installed below the suction line to provide a back liquid. This method is also used for pumping high-viscosity liquids. In addition to evaluating the friction head and local resistance losses, inertia losses (for piston pumps), H_i, and the effect of cavitation (for centrifugal pumps), h_k, must be accounted for in the overall suction head term.

Head losses due to overcoming inertia forces, H_i (in piston pumps), may be estimated by an expression that relates the pressure acting on the piston to the inertia force of a liquid column moving in the suction piping:

$$H_i = \frac{6}{5} \times \frac{\varrho}{g} \times \frac{f}{f_1} \times \frac{u^2}{r} \tag{6.22}$$

where $\quad \varrho$ = height of liquid column in the piping (for pumps having a gas chamber—the distance between pump centerline and the liquid level in the chamber)

g = acceleration due to gravity

f and f_1 = cross-sectional areas of the piston and piping, respectively

u = circumferential crank velocity

r = crank radius

Cavitation

Cavitation in centrifugal pumps arises from high velocities or when handling hot liquids under conditions of vaporization. It is a phenomenon caused by the formation and collapse of vapor cavities existing in a flowing liquid. Vapor cavities can form at any point in the fluid at which the local pressure approaches that of the liquid vapor pressure (at the operating temperature). At these positions, a portion of the liquid vaporizes to form bubbles or cavities of vapor. Low-pressure zones are generated in several ways: (1) by a local increase in velocity resulting in eddies or vortices near the boundary contours; (2) by rapid vibration of the boundary; (3) by separating or parting of the liquid due to "water hammer"; or (4) by an overall reduction in static pressure. Collapse of the bubbles initiates when they move into regions where the local pressure is higher than the vapor pressure. This often results in objectionable noise and vibration, as well as extensive erosion or pitting of the boundary materials in the immediate vicinity. Even more important, cavitation results in a decrease in pumping performance and efficiency. A dimensionless parameter called the cavitation number, σ_c, is used to correlate performance:

$$\sigma_c = \frac{p - p_\nu}{\rho w^2 / 2g_c} \qquad (6.23)$$

where   p = static pressure (absolute) in the undisturbed flow (lb_f/ft^2)
p_ν = liquid vapor pressure (abs) (lb_f/ft^2)
ρ = liquid density (lb/ft^3)
w = free-stream velocity of the liquid (fps)
g_c = dimensionless constant [32.17 (lb) (ft)/(lb_f) (s^2)]

The cavitation number's physical significance is the ratio of the net static pressure available to collapse a bubble to the dynamic pressure available to initiate bubble formation. The value of this dimensionless group at conditions of incipient cavitation, $\sigma_{c.i.}$, is dependent on the pump or equipment geometry. Typical values are reported by Perry and Chilton [1973], Karassik et al. [1976], Church [1944], and Cheremisinoff [1981].

A cavitation correction factor may be determined from the following correlation:

$$\widetilde{H}_c = 0.019 \frac{(Qn^2)^{2/3}}{H} \qquad (6.24)$$

where   Q = pump capacity (m^3/s)
n = number of revolutions (s^{-1})
H = fluid head on pump (m)

In practice, to avoid cavitation the suction head for pumping liquids with physical properties close to those of water should exceed the values in Table 6.2.

Table 6.2 Typical Suction Head Limits to Avoid Cavitation[a]

Temperature (°C)	10	20	30	40	50	60	65
Suction head, (m)	6	5	4	3	2	1	0

[a]Based on water flow.

CENTRIFUGAL PUMPS

Types and Applications

Centrifugal pumps are used extensively throughout the process industries because of their simplicity in design, low initial cost and maintenance, and their flexibility of application. This type of pump is available in a wide range of sizes; in capacities ranging from a few gpm up to 100,000 gpm and for discharge heads (pressures) from a few feet up to several thousand lb/in.^{2} Basically, a centrifugal pump consists of an impeller, which is a series of radial vanes of various shapes and curvatures rotating within a circular casing. Figure 6.2 illustrates the operation.

The liquid from suction piping (1) enters at the axis of a rotating impeller (2) into the pump chamber (3) and is thrown outwards by centrifugal action against the blades (4). The impeller's high speed of rotation causes the liquid to acquire kinetic energy. A pressure difference between the suction and discharge sides of the pump is produced by the conversion of kinetic energy of the liquid flow into pressure energy in the discharge piping (5). A reduction in pressure occurs at the entrance of the impeller, and the liquid is fed continuously into the pump from a supply tank. Without filling the pump chamber with liquid, the impeller cannot produce an adequate pressure difference, which is necessary for lifting liquid in the suction line.

Figure 6.3A shows a heavy-duty end-suction centrifugal pump used to pump large quantities of nonaggressive liquids. Pumps of this type are employed for general service applications in industry, irrigation and municipal water supplies. This design is generally all iron or bronze-fitted construction with replaceable shaft sleeves and wear rings. A separate cover is usually provided, and the casing has integrally cast support feet. Ball bearings are oil lubricated.

Most centrifugal pumps are not self-priming and, hence, cannot evacuate vapor from the suction line so that liquid can flow into the pump casing without external assistance. The impellers on centrifugal pumps are designed especially for efficient pumping and are not operated at high enough tip speeds to convert them into vapor compressors. The differential head that

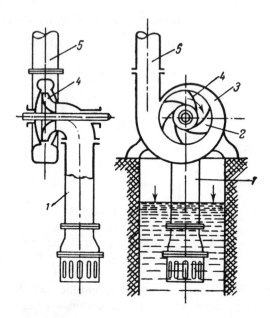

Figure 6.2   Operating scheme of a centrifugal pump: (1) suction piping; (2) impeller; (3) casing; (4) vanes; (5) delivery piping.

the pump impeller can deliver is the same on the vapor as on the liquid. However, the equivalent differential pressure rise capability is typically much lower with vapor. To prime a centrifugal pump, both the suction line and pump casing must be filled with liquid. When the suction source is at positive pressure or is positioned above the pump, priming is accomplished by opening the suction valve and venting the trapped vapor from a valve connection on the pump casing or discharge line. Liquid then flows into the suction line and pump casing to displace the escaping vapor.

Centrifugal pumps are used in a multitude of applications throughout the chemical industry. Designs may consist of an open impeller system mounted on an externally adjustable shaft for handling clear liquids, slurries or liquids with suspended solids. Also, they can be closed impellers for pumping clear liquids or light slurries. Figure 6.4 shows a cross section of one manufacturer's chemical process pump, and Figure 6.5 shows several of these units in operation. The specific features of this pump are given in Table 6.3

This particular design utilizes a dual volute. All pump volutes are designed to generate uniform radial thrust on the impeller shaft and bearings when operating at the best efficiency point on the pump curve (described in the next subsection). As a result, there is minimum radial thrust on the pump components. However, when a pump is not operating at the best efficiency point, the casing design no longer balances the hydraulic loads and radial thrust increases. A dual volute incorporates a flow splitter into the casing, which directs the liquid into two separate paths through the casing. The contour of the flow splitter follows the contour of the casing wall 180° opposite. Both are approximately equidistant from the center of the impeller; thus, the radial thrust loads acting on the impeller are balanced and greatly reduced.

Most chemical pumps are built with casings cast in high-cost alloys. Casings frequently are foot supported or bearing bracket supported rather than centerline supported. Chemical pumps are available in a wide range of operating conditions but most often are limited to low to moderate flows. Figure 6.6 shows still another chemical process pump for high-pressure, high-temperature service. This particular unit has hydraulic coverage to 9500 gpm, 700 ft of total dynamic head and 570 psig discharge pressure. A cross-sectional view of this pump is shown in Figure 6.6(B); its major features are listed in Table 6.4.

Many centrifugal pumps have single casings, that is, a single wall between the liquid under discharge pressure and the atmosphere. Double casings are used in horizontal, multistage, high-pressure pumps and in vertical pumps. In the former, a heavy barrel-shaped casing surrounds the stack of stage diaphragms. The stack of diaphragms comprises the inner casing, while the barrel forms the outer casing. This type of arrangement is used most often in boiler feed pumps.

Casings may be joined on the same plane as the shaft axis (called axially split) or perpendicular to the shaft (called radially split). Axially split horizontal pumps most commonly are referred to as "horizontally split." Figure 6.7 shows a single-stage, double-suction, horizontally split case centrifugal pump.

Radial split horizontal pumps are commonly called "vertically split." Radial joining is used on horizontal overhung pumps to allow ready removal of the rotor-and-bearing bracket assembly for maintenance. This design configuration also is employed on high-pressure multistage pumps because of structural problems associated with bolting together the halves of axially split casings exposed to high internal pressure.

The term *single-stage overhung* refers to the impeller mounting/support arrangement. The casings for these designs are supported at the centerline. Two shaft bearings are mounted close together in the same bearing bracket, with the impeller cantilevered or overhung beyond them. Normally, this type configuration utilizes top suction and discharge flanges; wearing rings both on the front and back of the impeller and casing; a single-suction closed impeller; and a single stuffing box fitted with a mechanical seal. These pumps

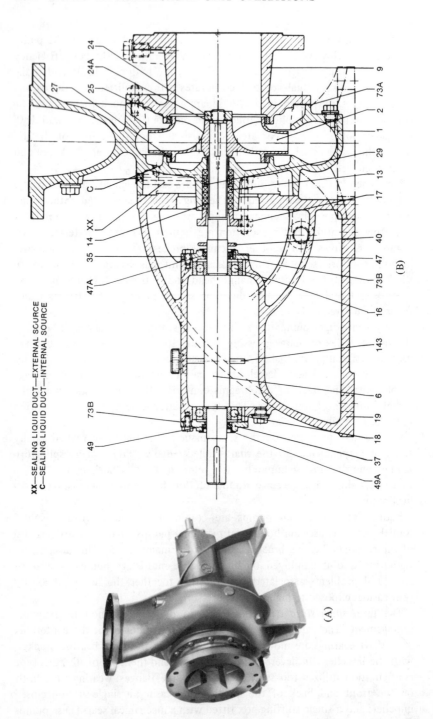

XX—SEALING LIQUID DUCT—EXTERNAL SOURCE
C—SEALING LIQUID DUCT—INTERNAL SOURCE

(A)

(B)

PART NO.	PART NAME	STD. MATERIAL	BRONZE FITTED
1	CASING	CAST IRON	CAST IRON
2	IMPELLER	CAST IRON	BRONZE
6	SHAFT	CARBON STEEL	CARBON STEEL
9	SUCTION COVER	CAST IRON	CAST IRON
13	PACKING	ASBESTOS	ASBESTOS
14	SHAFT SLEEVE	CAST IRON	BRONZE
16	BALL BEARING — INBOARD	STEEL	STEEL
17	GLAND	CAST IRON	CAST IRON
18	BALL BEARING — OUTBOARD	STEEL	STEEL
19	FRAME	CAST IRON	CAST IRON
24	IMPELLER NUT	STEEL	STEEL
24A	LOCKING PLATE	STEEL	STEEL
25	WEAR RING — SUC. COVER	CAST IRON	BRONZE
27	WEAR RING — STUFF. BOX	CAST IRON	BRONZE
29	LANTERN RING	CAST IRON	CAST IRON
32	IMPELLER KEY	STEEL	STEEL
35	BEARING COVER — INBOARD	CAST IRON	CAST IRON
37	BEARING COVER — OUTBOARD	CAST IRON	CAST IRON
40	DEFLECTOR	STEEL	STEEL
47	OIL SEAL — INBOARD	FELT	FELT
47A	CAP — INBOARD SEAL	STEEL	STEEL
49	OIL SEAL — OUTBOARD	FELT	FELT
49A	CAP — OUTBOARD SEAL	STEEL	STEEL
73A	GASKET — COVER	ASBESTOS	ASBESTOS
73B	GASKETS — BEAR. COVER	PAPER	PAPER
143	OIL GAUGE	STEEL	STEEL

CAST IRON PARTS ARE ASTM A-48-35.

(B)

NO.	CONNECTIONS PROVIDED USE	SIZE
I	PRIMING, VENTING	¾
III	DRAINING	¾
V	DISCHARGE GAUGE	½
VI	INLET GAUGE	½
VII	OIL DRAIN	⅜
VIII	OIL FILL, OIL GAUGE	⅜
XVI	LEAKAGE DRAIN	1
XX	SEALING LIQUID	¼

Figure 6.3 (A) A horizontal end suction centrifugal pump (top); (B) a cross-sectional drawing of the pump (bottom) (courtesy Carver Pump Company, 2415 Park Avenue, P.O. Box 389, Muscatine, IA).

Table 6.3 Features of the Chemical Process Pump Shown in Figure 6.4 (courtesy Ingersoll-Rand Company, 253 East Washington Ave., Washington, NJ)

Feature	Advantage	Benefit
1. Open impellers or closed impellers	Allows selection of proper impeller for individual needs. Open impellers handle slurries and particles. Closed impeller handles clear liquid and smaller particles at higher efficiencies.	Increased reliability; less maintenance downtime; lowest power costs for specific application.
2. Dual volute	a. Reduced radial thrust. b. Reduced shaft deflection at face of stuffing box.	Increased bearing life; less downtime; lower maintenance costs. Extended seal and packing life; less downtime; lower maintenance costs.
3. External axial impeller adjustment	Allows field setting of impeller clearances to compensate for wear, thus restoring high efficiencies without overhaul.	Less downtime; lower maintenance costs; reduced power costs.
4. Replaceable casing ring for closed impellers and casing shroud plate for open impellers	Provides for quick renewal of casing surface with inexpensive flat plate or ring. No need to replace a cast part (casing, stuffing box cover or cast wear plate).	Lower maintenance costs; increased availability of replacement parts.

5. Replaceable hook-type sleeve with O-ring seal between shaft sleeve and impeller.	Assures accurate seal setting. Protects shaft and impeller threads from contamination by pumped liquid.	Higher reliability; less downtime; lower maintenance costs.
6. Dry rabbet fit construction	Provides for accurate, positive alignment and reduced possibility of crevice corrosion.	Increased reliability; less downtime; lower maintenance costs.
7. Heavy-duty shaft system (thick shaft, optimum bearing spans, short impeller overhang)	Reduced shaft deflection.	Extends seal and packing life; less downtime and lower maintenance costs.
8. Three bearing cradles for all sizes	Maximum parts interchangeability, reduced spare parts stock levels.	Lower maintenance costs; increased parts availability.
9. Studded casing	Reduced possibility of stripping casing threads during assembly and disassembly.	Lower maintenance repair costs; less downtime.
10. Double row thrust bearings	Enables the bearing to carry high thrust loads at all suction pressures and operating conditions.	Increased bearing life; reduced downtime and lower maintenance costs.
11. Versatile stuffing box designed for packing and all seal types	Provides capability to change from packing to seals without changing the stuffing box cover. All seals (balanced, unbalanced, outside and tandem) can be interchanged without changing the stuffing box cover.	Greater interchangeability; lower repair costs.

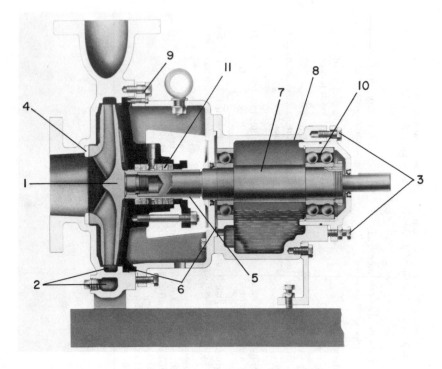

Figure 6.4   Cross section of a centrifugal pump used in handling process chemicals (courtesy Ingersoll-Rand Company, 253 East Washington Ave., Washington, NJ).

are well suited to high-temperature operations and can be used for handling flammable liquids.

A two-stage overhung pump is a modification of the single-stage process pump and is capable of higher head. Usually the stuffing box pressure is approximately halfway between suction and discharge pressures. Figure 6.8 shows a cross section of a two-stage, horizontally split pump.

Multistage centrifugal pumps generally are used for generating higher heads (pressures) than can be obtained by single-stage pumps. These pumps are available for pressures as high as 3000 lb/in.^{2} at capacities greater than 3000 gpm. The operation of this type pump is illustrated in Figure 6.9. As shown, these designs have (A) impellers in (B) one aggregate casing, which are located (C) in series on one shaft. Liquid discharged from the first impeller enters (D) through the offtake in the second impeller where it acquires additional energy from the second impeller through the offtake in the third impeller, and so on. Thus, multistage pumps may be thought of as several

Figure 6.5   Centrifugal chemical pumps being used to move harsh caustic through a chemical plant (courtesy Ingersoll-Rand Company, 253 East Washington Ave., Washington, NJ).

single-stage pumps on one shaft, with the flow in series. Hence, the total head developed is the head of one impeller multiplied by the number of impellers (usually designs do not exceed five impellers).

Further details and examples of multistage centrifugal pumps are shown in Figures 6.10–6.13. Figure 6.10 is a cross section of a multistage horizontally split pump. (Refer to Table 6.5 for an explanation of the features.) This design incorporates horizontally split channel rings and center bushings, so the rotor is easy to assemble, inspect and balance dynamically. Figure 6.11 shows the same pump used to produce a high-pressure stream of water to

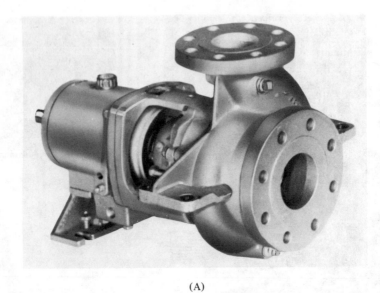

(A)

(B)

Figure 6.6　(A) High-pressure, high-temperature service chemical process pump; (B) cross-sectional view (refer to Table 6.4 for a listing of major features) (courtesy Carver Pump Company, 2415 Park Avenue, P.O. Box 389, Muscatine, IA).

Table 6.4 Features of the Chemical Process Pump Shown in Figure 6.6 (courtesy
Carver Pump Company, 2415 Park Ave., P.O. Box 389, Muscatine, IA)

1. **Pump casing**—back pull-out, self-venting top centerline discharge, vortex suppressing guide vane in suction nozzle, $\frac{1}{8}$ in. corrosion allowance, rugged integral casted centerline supports to allow thermal or pressure expansion without causing misalignment and shaft deflection.

2. **Positive alignment**—positively and permanently achieved by full-circle registered fit on all mating parts. All such fits are away from liquid being pumped, preventing crevice corrosion.

3. **Fully confined gaskets**—on wet and dry side of casing cover as well as between impeller nut, impeller and shaft sleeve provide safety against leakage.

4. **Enclosed impeller**—for high efficiency and low NPSH, with back vanes for axial thrust balancing and low stuffing box pressure, keeping erosive impurities out of shaft seal area, keyed to shaft for positive fastening. Positioned by acorn-type impeller nut with Heli-Coil lock insert, it cannot come loose under reserve rotation.

5. **Bearing frame**—five sizes for 27 models from $1\text{-}\frac{1}{2}$ in. to 12 in. discharge. Designed to carry maximum load with under 0.02 in. shaft deflection and at least 2-year bearing life.

6. **Built-in casing heating or cooling jacket**—increases the application scope without need of added costs for optional or extra parts. Special intensive cooling of stuffing box only is also available.

7. **Part interchangeability.**

8. **Stuffing box**—can be adjusted to particular application requirements. Replaceable shaft sleeve is provided with stuffing box packing or mechanical seal, providing maximum adaptability for various seal designs. Step design shaft sleeve simplifies accurate seal setting for long seal life.

process uranium at a mine. Multistage pumps are employed in a multitude of processing applications. Examples include hydrocarbon processing and refining, boiler feed operations, descaling operations, mine dewatering and hydraulic power recovery, in which excess plant energy is recovered to drive other equipment.

Figure 6.12 shows a high-pressure, horizontal shaft, vertically split multistage pump. The ring section stage casing construction provides a wide range of discharge pressures through the addition of individual stages. Its principal function is high-pressure boosting of clear, nonaggressive liquids.

Vertical pumps are another orientation used widely throughout the process industries. In this type, a vertical cylinder buried in the ground houses the pumping element. Suction liquid enters the outer cylinder, flows to the bottom and then up through the pumping element stages. The diaphragms of the stages in the pumping element comprise the inner casing. As *inline pumps*, the casings are designed to be bolted directly to the piping, much like a valve.

(A)

(B)

Figure 6.7   (A) Cross section of a single-stage, double-suction pump (top); (B) pumps in operation in a steel mill (bottom) (courtesy Ingersoll-Rand Co., 253 E. Washington Ave., Washington, NJ).

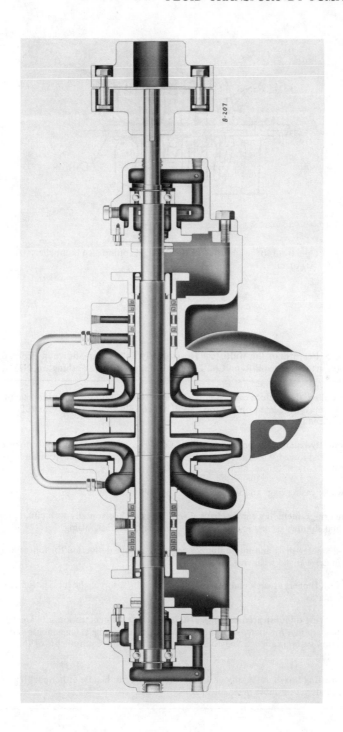

Figure 6.8   Cross section of a two-stage, horizontally split pump (courtesy Ingersoll-Rand Co., 253 E. Washington Ave., Washington, NJ).

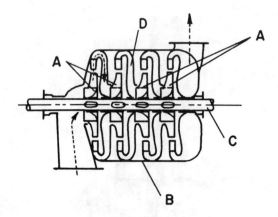

Figure 6.9   Operation of a multistage centrifugal pump: (A) impeller; (B) casing; (C) shaft; (D) offtake.

Table 6.5 Features of the Multistage Horizontally Split Pump Shown in Figure 6.10 (courtesy Ingersoll-Rand Co., 253 E. Washington Ave., Washington, NJ)

1. **Heavy-walled, dual-volute casing** allows for high working pressures and nozzle pipeloads. Capnuts are seated on the top half of the axially split casing to simplify assembly and disassembly.

2. **Dynamically balanced opposed impellers** of one-piece construction are shrunk on the shaft and keyed. Shaft is stepped by $^5/1000$-in. at each impeller fit to facilitate removal.

3. **Renewable casing rings and impeller rings** control interstage leakage.

4. **Channel rings, including center channel ring, are split horizontally** to facilitate replacement and simplify dynamic balancing of rotor without dismantling.

5. **Large-diameter shaft and minimum bearing span** reduce deflection for longer bearing, mechanical seal and wear ring life.

6. **Labyrinth flingers at each end** of the bearing housing protect the lubrication against contamination.

7. **Standard ring oil-lubricated bearings** assure complete oil penetration into the bearings without foaming, for increased bearing life. **Optional bearing arrangements and lubrication systems** allow a "customized" fit to meet the requirements of the application.

8. **Large stuffing boxes,** integrally cast with the casing, can handle either packing or single, tandem or double mechanical seals.

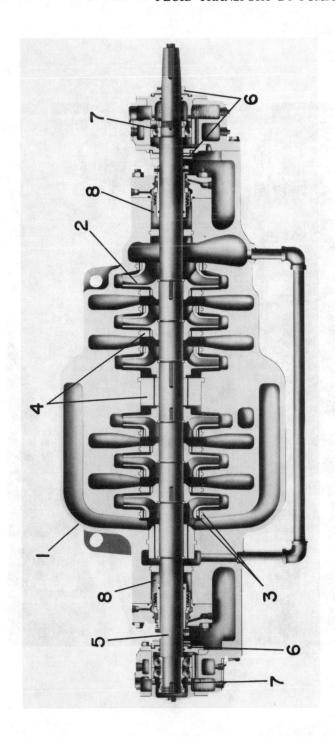

Figure 6.10   Cross section of a multistage, horizontally split pump (courtesy Ingersoll-Rand Co., 253 E. Washington Ave., Washington, NJ).

Figure 6.11   Multistage, horizontally split pumps used to produce a high-pressure stream of water to loosen uranium at a Colorado mine (courtesy Ingersoll-Rand Co., 253 E. Washington Ave., Washington, NJ).

Figure 6.12   High-pressure, horizontal-shaft, vertically split multistage centrifugal pump. The maximum capacity and head of this pump are 2200 gpm and 825 ft, respectively, with a maximum discharge pressure of 400 psi (courtesy Carver Pump Co., 2415 Park Ave., P.O. Box 389, Muscatine, IA).

Two basic configurations of inline pumps are coupled and close-coupled. Service life and maintenance requirements for both are about the same. Figure 6.13 shows the cross section of a vertical inline pump. The total head is the sum of the discharge pressure measured at the discharge nozzle above the floor plate, the velocity head at the same location and the vertical distance from the centerline of the pressure gauge to the liquid surface in the sump. These pumps generally are equipped with tail pipes, which allow pumping down below normal liquid level. The pump must be capable of operating under pumpdown conditions without allowing the throttle bushing to run dry.

Vertical multistage pumps can have 24 or more stages. High specific-speed impellers often are used. The first stage is usually at the bottom of the assembly, below grade. These pumps require a large number of close-running clearances and, thus, are sensitive to damage by solids ingestion and by dry or two-phase flow conditions. This type of pump is employed in a broad range of applications. Examples include (1) fossil power plants, where they are used for condensate service in large power generating plants; (2) nuclear power plants, where they are used in condensate and feedwater heater drain service; and (3) desalination, whose operating facilities require large-capacity pumping equipment that must perform with low net positive suction head

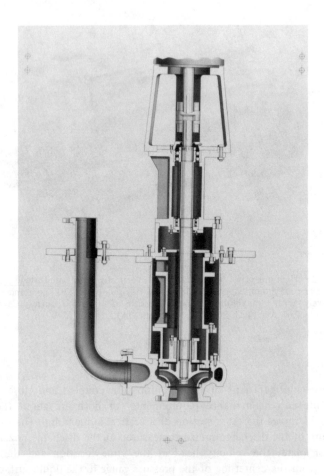

Figure 6.13   A top pullout vertical pump. This unit is suitable for continuous operation
in ambient temperatures up to 50°C (courtesy Carver Pump Co., 2415 Park Ave.,
P.O. Box 389, Muscatine, IA).

(NPSH) available. Figure 6.14 shows the details of a double-suction, single-
stage vertical pump. The features of this pump are explained in Table 6.6.

"Can" pumps are motor pump units with the rotating rotor and impeller
housed entirely within a pressure casing. This type design eliminates the need
for a stuffing box. The pumped fluid serves both as a lubricant for bearings
and as a coolant for the motor. Designs typically are limited to low-flow,
low-pressure and low-temperature services.

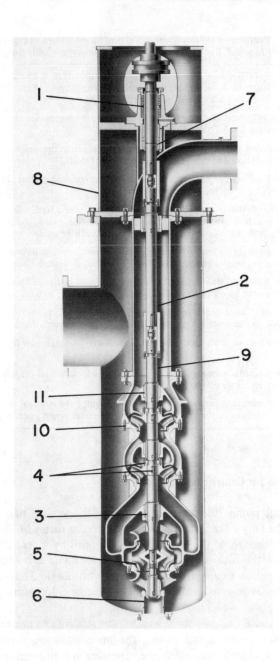

Figure 6.14   Cross section of a double-suction vertical pump (courtesy Ingersoll-Rand Co., 253 E. Washington Ave., Washington, NJ).

Table 6.6 Features of Double-Suction Vertical Pump Shown in Figure 6.14
(courtesy Ingersoll-Rand Co., 253 E. Washington Ave., Washington, NJ)

1. **Multi-element stuffing box** utilizes an advanced combination of bearing, serrated sleeve and throttle bushing, and a vent to suction. Vent vapors that collect in stuffing box increase packing and bearing life.

2. **Shaft surface** is ground after machining to assure shaft trueness for long bearing life and low vibration levels.

3. **Hardened stainless steel shaft sleeves** under carbon bearings prevent premature opening of clearances and premature replacement of shaft.

4. **Keys and lock collars** properly position impeller and transmit torque. This prevents impellers from breaking loose during operation.

5. **Double-suction first-stage impeller** reduces NPSH and meets Hydraulic Institute standards for suction-specific speed. This means a shorter, more rigid pump and reduces the possibility of cavitation over a wide operating range.

6. **Self-seating snubbers** on long settings eliminate column movement and prevent excessive wear on bearings during low-flow operations or system upset.

7. **Replaceable, hardened stainless steel shaft sleeve,** keyed and sealed with an o-ring, eliminates the need to replace the complete upper shaft due to wear.

8. **Discharge head's** fundamental frequency and reed critical frequency of motor are both analyzed to eliminate vibration problems caused by sympathetic resonance.

9. **Large shafts** with low shaft stress levels mean less shaft whip, longer bearing and ring life.

10. **Gaskets** in intermediate stages seal against interstage leakage and prevent premature failure of flange surface due to "wire drawing"

11. **Bearings** are available in carbon or bronze throughout, have inherent self-lubricating properties, last longer and are more durable in two-phase liquid operations.

Basic Equations for Centrifugal Machines

In centrifugal pumps the liquid flows along the surface of the impeller vanes while the tip of the vane moves relative to the casing of the pump. To develop an expression of the virtual head developed by a centrifugal pump, we shall assume the path followed by a volume of liquid as it passes through the pump in relation to a stationary impeller, with the fluid having the same relative velocity as an actual rotating impeller. Figure 6.15 defines the system under consideration.

Defining C_1 and C_2 as the vector sums of the relative and tangential velocities of the fluid entering and exiting the impeller, respectively, then the relative velocity components along the vanes are w_1 and w_2, and the tangential components (tangent to the circumference of rotation) are u_1 and u_2. Further, a reference datum is defined to be the surface of the impeller in Figure 6.15. We then may write an energy balance for the fluid passing through the impeller at $Z_1 = Z_2$:

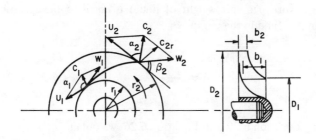

Figure 6.15 Defines system for the derivation of equations for centrifugal machines.

$$\frac{p_1}{\rho g} + \frac{w_1^2}{2g} = \frac{p_2}{\rho g} + \frac{w_2^2}{2g} \tag{6.25}$$

When the impeller rotates, the liquid obtains an additional energy, "A," which is derived from the work of centrifugal force along the path $r_2 - r_1$. Hence,

$$\frac{p_1}{\rho g} + \frac{w_1^2}{2g} = \frac{p_2}{\rho g} + \frac{w_2^2}{2g} - A \tag{6.26}$$

The centrifugal force, C, acting on the liquid particle of mass, m, is

$$C = m\omega^2 r = \frac{G}{g} \omega^2 r \tag{6.27}$$

where G = weight of fluid particle
ω = angular velocity
r = moving radius of particle rotation

The work, A_G, derived from the centrifugal force by displacement of this fluid particle along the path $r_2 - r_1$ is determined as follows:

$$A_G = \int_{r_1}^{r_2} \frac{G}{g} \omega^2 r\, dr = \frac{G\omega^2}{2g} (r_2^2 - r_1^2) = \frac{G}{g} \left(\frac{u_2^2 - u_1^2}{2} \right) \tag{6.28}$$

The specific work per unit weight of liquid is equal to the specific energy obtained by the fluid in the pump:

$$A = \frac{u_2^2 - u_1^2}{2g} \qquad (6.29)$$

Substituting Equation 6.29 into Equation 6.26, we obtain

$$\frac{p_2 - p_1}{\rho g} = \frac{w_1^2 - w_2^2}{2g} + \frac{u_2^2 - u_1^2}{2g} \qquad (6.30)$$

The heads of liquid at the inlet and outlet from the pump are

$$H_1 = \frac{p_1}{\rho g} + \frac{c_1^2}{2g} \;\; ; \;\; H_2 = \frac{p_2}{\rho g} + \frac{c_2^2}{2g} \qquad (6.31)$$

Hence, the head of the pump is equal to the difference of heads between the pump's inlet and its outlet:

$$H_T = H_1 - H_2 = \frac{p_2 - p_1}{\rho g} + \frac{c_2^2 - c_1^2}{2g} \qquad (6.32)$$

Substituting $\dfrac{p_2 - p_1}{\rho g}$ from Equation 6.30 into Equation 6.32, we obtain

$$H_T = \frac{w_1^2 - w_2^2}{2g} + \frac{u_2^2 - u_1^2}{2g} + \frac{c_2^2 - c_1^2}{2g} \qquad (6.33)$$

From the geometry of Figure 6.15 we have

$$w_1^2 = u_1^2 + c_1^2 - 2u_1 c_1 \cos \alpha_1 \;\; ; \;\; w_2^2 = u_2^2 + c_2^2 - 2u_2 c_2 \cos \alpha_2 \qquad (6.34)$$

Substituting this last set of expressions into Equation 6.33 results in an equation for the virtual head of a centrifugal pump:

$$H_T = \frac{u_2 c_2 \cos \alpha_2 - u_1 c_1 \cos \alpha_1}{g} \qquad (6.35)$$

This equation represents the theoretical maximum head that could be developed for a specified set of operating conditions. Note that the liquid entering the pump usually moves along the impeller in the radial direction. In this case, the angle between the absolute velocity value of the liquid entering the impeller and the tangential velocity is $\alpha_1 = 90°$, which corresponds to the liquid entering the impeller without any shock. Equation 6.35 then simplifies to the following:

$$H_T = \frac{u_2 c_2 \cos \alpha_2}{g} \qquad (6.36)$$

From Figure 6.15 we may write

$$c_2 \cos \alpha_2 = u_2 - w_2 \cos \beta_2$$

Hence,

$$H_T = \frac{u_2^2}{g} \left(1 - \frac{w_2}{u_2} \cos \beta_2 \right) \qquad (6.37)$$

From the width of the impeller, b, the length of the circumference, $2\pi r_2$, and the cross section of the flow leaving the impeller, $2\pi r_2 b$, the quantity of liquid being pumped is

$$V = 2\pi r^2 b w_2 \sin \beta_2 \qquad (6.38)$$

Hence,

$$w_2 \cos \beta_2 = \frac{V}{2\pi r_2 \tan \beta_2} \qquad (6.39)$$

Substituting Equation 6.39 into Equation 6.37, we obtain the following relationship between the head, H, and the volumetric flowrate through the pump, V:

$$H = \frac{u_2^2}{g} - \frac{V}{g(2\pi r_2 b)\tan\beta_2}$$

(6.40)

For a given speed of rotation there is a linear relation between the head developed and the rate of flow. This is illustrated for different vane configurations in Figure 6.16.

If the outlet vane angles are inclined backwards, β_2 is less than 90°; hence, $\tan\beta$ is positive and, therefore, the head decreases as the throughput increases (curve A, Figure 6.16). If β_2 is greater than 90°, i.e., the outlet vane is inclined forwards, the head increases at higher throughputs (curve B, Figure 6.16). Radial vanes provide a constant head (curve C, Figure 6.16). When the flow is zero, $V = 0$, then regardless of the vane angle our head expression is

$$H_T = \frac{u_2^2}{g}$$

(6.41)

Figure 6.16 shows that the maximum theoretical head achievable in a centrifugal pump is with vanes curved in the direction of rotation of the

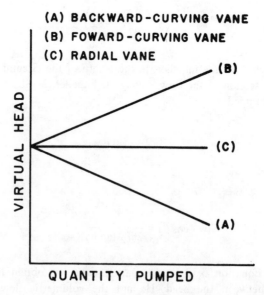

(A) BACKWARD-CURVING VANE
(B) FOWARD-CURVING VANE
(C) RADIAL VANE

Figure 6.16   Plot of virtual head versus flow capacity for different outlet vane angles.

impellers, whereas the minimum occurs with vanes curved in the opposite direction. However, pumps are fabricated with angles $\beta_2 < 90°$ because an increase in β increases hydraulic losses and decreases the hydraulic pump efficiency.

The actual head is always less than virtual head, for the following main reasons:

1. The fluid circulating in the spaces between the vanes forms eddies.
2. Frictional losses occurring in the suction port, the impeller and the discharge nozzle increase with pump speed and liquid viscosity.
3. Losses occur as the liquid is discharged from the impeller. The vane angles are correct only for the designed head and throughput. Deviation from these conditions causes an increase in the losses due to turbulence.
4. Leakage reduces the head developed, especially at low discharge rates.

The actual head is equal to

$$H = H_T \eta_h \epsilon \tag{6.42}$$

where η_h is the hydraulic efficiency (with typical values between 0.8 and 0.95), and ϵ is a coefficient that accounts for the number of vanes, $\epsilon = 0.6 - 0.8$.

Referring back to Figure 6.15, the throughput, Q, corresponds to the liquid discharge through the channels between the impeller vanes having widths b_1 and b_2:

$$Q = b_1(\pi D_1 - \delta z')C_{1,r} = b_2(\pi D_2 - \delta z')C_{2,r} \tag{6.43}$$

where
$$\delta = \text{vane thickness}$$
$$z' = \text{number of vanes}$$
b_1, b_2 = width of impeller at the internal and external circumferences, respectively
$C_{1,r}, C_{2,r}$ = radial components of absolute velocities at the inlet and outlet of the impeller ($C_{1,r} = C_1$)

The output and head of a centrifugal pump depend on the number of revolutions per unit of time made by the impeller. As follows from Equation 6.43, throughput is directly proportional to the radial component of the absolute velocity at the exit from the impeller, i.e., $Q \propto C_{2,r}$. If the number of revolutions is changed from n_1 to n_2 (thus changing from Q_1 to Q_2 correspondingly), the trajectories of the motion of liquid particles remain unaltered and the velocity parallelograms at any corresponding points will be geometrically similar. This is illustrated in Figure 6.17. Consequently,

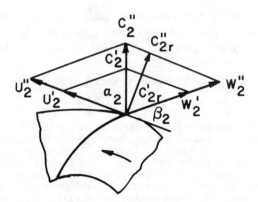

Figure 6.17   Shows that if the number of impeller revolutions is changed, similarity of the velocity parallelograms is maintained.

$$\frac{Q_1}{Q_2} = \frac{C'_{2r}}{C''_{2r}} = \frac{u'_2}{u''_2} = \frac{\pi D_2 n_1}{\pi D_2 n_2} = \frac{n_1}{n_2} \tag{6.44}$$

From Equation 6.37, the head is proportional to the square of circumferential velocity, i.e.,

$$\frac{H_1}{H_2} = \left(\frac{u'_2}{u''_2}\right)^2 = \left(\frac{n_1}{n_2}\right)^2 \tag{6.45}$$

The power developed by the pump is proportional to the product of volumetric flowrate, Q, and head, H. From Equations 6.2, 6.44 and 6.45 we then have

$$\frac{N_1}{N_2} = \left(\frac{n_1}{n_2}\right)^3 \tag{6.46}$$

Equations 6.44–6.46 are the *equations of proportionality* for centrifugal machines. These expressions state the following:

1. A change in the number of impeller revolutions (from n_1 to n_2) causes the pump throughput to change in a manner that is directly proportional.
2. The heads of the two systems are proportional to the number of revolutions raised to the squared power.
3. The powers developed by pumps are proportional to the number of revolutions to the third power.

In a practical sense, these relationships do not hold rigorously. The proportionality between pump parameters is not strictly maintained when the number of revolutions is changed by more than a factor of two.

The above theoretical treatment assumes the absence of friction, eddies, etc. In real pumps, all these complications exist and influence pump parameters. Now we shall examine commercial pump behavior, with attention given to backward-curved vanes. This type impeller configuration is used most extensively. Discussions on pump performance with forward-curved vanes have been given by Metzner [1961] and Cheremisinoff [1981].

The main parameters that reduce the virtual head of a centrifugal pump (using backward-curved vanes) to its developed head are summarized in Figure 6.18. The dashed line represents the "virtual head," and thus defines the theoretical characteristics of this type of impeller, $\beta_2 < 90°$. Physically, the virtual head line represents an infinite number of vanes and only infinitesimal amounts of liquid exist between them. In an actual pump, finite volumes of liquid exist between the vanes and inertia effects arise. Consequently, the portions of liquid near the impeller periphery have a greater velocity than inside the impeller. This creates circulation across the space between vanes. Hence, inertial effects cause the virtual head curve in Figure 6.18 to shift downward.

Now let us assume that the liquid velocity through the pump is increased but that the same number of impeller rotations is maintained. (This can be accomplished by increasing a valve opening ahead of the pump.) The head in this case will decrease because of an increase in friction resistance. These

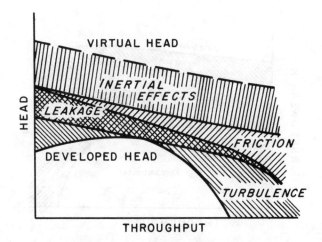

Figure 6.18   The head versus throughput plot summarizes the major factors that reduce virtual head.

losses are compensated for partially by a decrease in leakage, which at low throughputs is larger than at a high throughput. As shown in Figure 6.18, this counterbalancing effect is not that significant. The last factor noted in Figure 6.18 responsible for decreasing head is turbulence or eddy formation on vanes. All these factors result in the "developed head curve," which approaches zero on attaining a maximum throughput. Such a curve must be constructed from experimental data.

Returning for a moment to the ideal pump, a maximum volumetric throughput corresponding to zero head, H = 0, exists for a constant n. Conversely, a maximum head exists for the case of no flow. In real pumps, however, there is always some head at the maximum flow and some small flow at the pump's maximum head. The behavior of real pumps between these extremes is best explained by performance curves. Note that the product of the developed head (in units of pressure) and volumetric flowrate represents the power absorbed by the pumped fluid. Because the head approaches zero at the maximum flowrate, power first increases from zero, i.e., at V = 0, to a maximum and then decreases to zero at a maximum volumetric flowrate. This is illustrated by the pump horsepower versus throughput curve in Figure 6.19.

Power needed to drive the pump is the same as that required to overcome all the losses in the system and to supply the energy added to the liquid. The losses include frictional losses at the impeller, as well as turbulent losses; the disk friction (or energy required to rotate the impeller in the fluid); leakage from the periphery back to the eye of the impeller; and mechanical friction losses in various pump components, such as bearings, stuffing boxes and

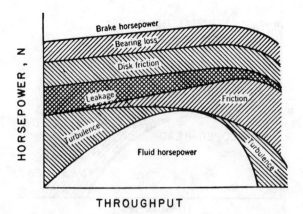

THROUGHPUT

Figure 6.19   Plot of pump horsepower vs throughput, summarizing major factors affecting pump power.

wearing rings. The sum of all these power consumption items produces the final brake horsepower curve shown in Figure 6.19. As shown, brake horsepower is required even when the volumetric flowrate is zero. With increasing flowrate, the brake horsepower increases even when the head is zero. In this case, the flowrate will be at a maximum. The brake horsepower will not drop to zero but will have a definite value.

An additional cross plot can be obtained from this last figure by dividing the fluid horsepower by the brake horsepower values (the definition of mechanical efficiency of a pump) and plotting efficiency versus throughput, as shown in Figure 6.20. All three performance curves—head (H-V), Figure 6.18; power (N-V), Figure 6.19; and efficiency (η-V), Figure 6.20—are combined conveniently into a single diagram (Figure 6.21 or Figure 6.22). Figures 6.21 and 6.22 are illustrative only. Manufacturers should be consulted for the specific performance data of a pump.

The plot given in Figure 6.21 contains several curves corresponding to different impeller diameters for a specific type of machine. Also shown are several lines of constant brake horsepower and constant efficiency, whose paths could be predicted from Figures 6.19 and 6.20. Such a diagram provides information on the characteristics of a pump for a definite head at a specified liquid flowrate. By specifying coordinates H,V, we can interpolate among the curves of brake horsepower, impeller diameter and efficiency to obtain all characteristic values of a pump under consideration.

It is important also to evaluate the influence of the number of impeller revolutions on pump performance. From Equation 6.40 we note that an increase in the number of rotations, n, is accompanied by an increase in head at a constant flowrate. This is illustrated graphically in Figure 6.23, showing the

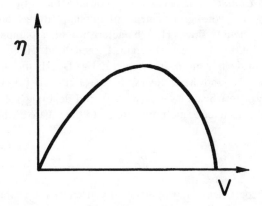

Figure 6.20   A typical curve of the mechanical efficiency of a pump.

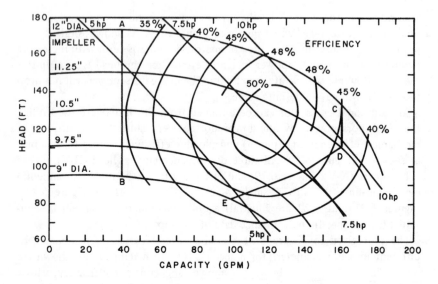

Figure 6.21   Total characteristics of a centrifugal pump.

influence of number of impeller rotations on head, H-V, brake horsepower, N-V, and efficiency (η-V).

Let us now consider the characteristic curves of a typical centrifugal pump for liquids of different viscosities. Higher viscosity translates to higher resistance to flow and, consequently, greater frictional losses. Hence, we may expect a decrease in the head, an increase in brake horsepower and a decrease in efficiency. Typical curves are shown in Figure 6.24.

In selecting a pump, it is necessary to consider the entire pump system's characteristics, i.e., the arrangement of piping, fittings and equipment through which liquids flow. The characteristics of a pumping system express the relationship between flowrate, Q, and head, H, needed for liquid displacement through a given arrangement. Head, H, is the sum of the geometric height, H_g, and head losses, h (see Equation 6.14). Taking into account that $V_{sec} = WS$ and $h_\varrho = \Sigma\zeta(w^2/2g)$, and denoting $V_{sec} = Q$, we determine that the head losses are proportional to the square of the flowrate:

$$h_\varrho = KQ^2 \tag{6.47}$$

where K is a coefficient of proportionality. Hence, the characteristics of a pump system may be expressed by the following parabolic expression:

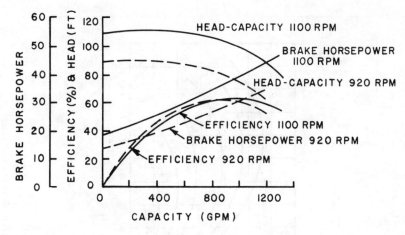

Figure 6.22 Head, efficiency and brake horsepower, plotted as a function of capacity for a centrifugal pump operating at two speeds with the same liquid.

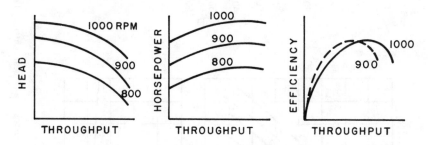

Figure 6.23 The influence of the number of impeller rotations on pump performance.

$$H = H_g + KQ^2 \qquad (6.48)$$

The characteristics of both the pump system and the pump can be represented on a common plot, as shown in Figure 6.25. Point A (the intersection of both characteristics curves) represents the *operating point* and corresponds to the maximum capacity of the pump, Q, while operating for a pump system. If a higher capacity is required, it is necessary either to increase the number of motor rotations or to change to a larger pump. Increased capacity can also be achieved by decreasing the hydraulic resistance of the pump system, h_Q. In this case, the operating point is displaced along the pump

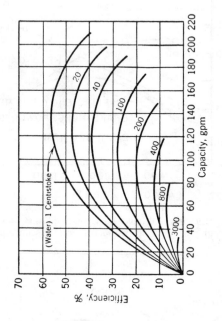

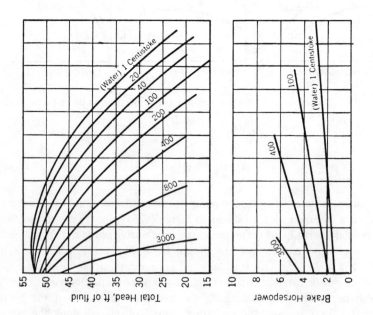

Figure 6.24   Characteristic curves of a typical centrifugal pump for liquids of different viscosities.

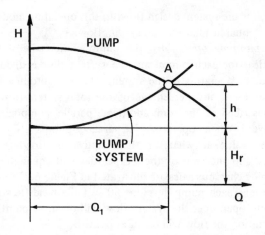

Figure 6.25 Pump and pump system characteristics represented on a single plot.

characteristics toward the right. Hence, a pump should be selected so that the operating point corresponds to a desired head and capacity.

In series pumping, the pump head component of total available head is the sum of the heads developed by each pump at any given flow. Each pump must be selected to operate satisfactorily at the system design flow. Pumps operating in series are referred to as "pressure additive." This principle is illustrated in Figure 6.26. With one pump operating, system flow will occur at point A. With both pumps operating, the system flow will occur at

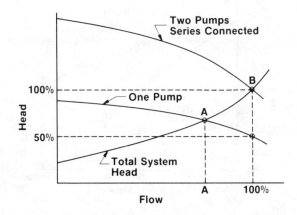

Figure 6.26 The effect of series pumping on the head curve.

point B, which is the system design flow. In this operating mode, both pumps are developing equal head at the full system flow.

In parallel pumping operations, the pump component of total available head is identical for each pump and the system flow is divided among the number of pumps operating in the system. The flows produced by individual pumps can represent any percentage of the total system flow. Where series pumping is described as "pressure additive," parallel pumping is described as "flow additive." When operating in parallel, pumps always develop an identical head value at whatever their equivalent flowrate for that developed head, and the sum of their capacities will equal the system flow. Parallel pumping characteristics are illustrated in Figure 6.27.

In this example, each pump develops 50% of the total flow at 100% head. With one pump operating, the system flow will occur at point A; with both pumps in operation the flow will occur at point B.

The high-speed coefficient (also referred to as the specific number of revolutions, n_s) is the number of revolutions of geometrically similar models of an impeller, which, at the same efficiency and capacity of $0.075 \ m^3/s$, has a head of 1 m. The high speed is a basic characteristic of a series of similar pumps having equal angles, α_2 and β_2, and coefficients, ϵ and η_h. The high-speed coefficient is expressed by the following relationship:

$$n_s = \frac{3.65 n \sqrt{Q}}{\sqrt[4]{H^3}} \qquad (6.49)$$

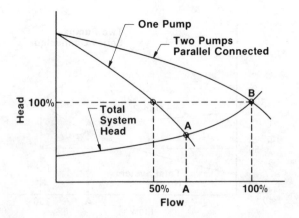

Figure 6.27   The effect of parallel pumping on the head curve.

Table 6.7 Range of Values for the High-Speed Pump Coefficient

Type Speed	n_S
Low	40–80
Normal	80–150
High Speed	150–300

where   n = the number of rotations (min^{-1})
Q = the maximum pump capacity (m^3/s)
H = the total pump head (m)

As follows from Equation 6.49, the high-speed coefficient, n_S, increases with increasing capacity and decreasing head. Therefore, low-speed impellers generally are used for obtaining higher heads at low capacities, and high-speed impellers are used for creating high capacities at low heads. The impellers are divided into three groups, depending on the value of the high-speed coefficient. Ranges of values are summarized in Table 6.7.

The student should work through practice problems 6.1–6.4, 6.10, 6.11, 6.16–6.18, 6.20 and 6.25 at the end of this chapter to strengthen understanding of principles presented thus far.

Additional discussions on centrifugal pumps are given by Perry [1973], Karassik [1976], Stepanoff [1965], Daugherty [1915], Church [1944], Cheremisinoff [1981], Hicks and Edwards [1971,1957], Brown [1950], and Karassik and Carter [1981].

POSITIVE DISPLACEMENT PUMPS

Positive displacement pumps operate on the principle of forcing a fixed volume of liquid from the inlet pressure zone into the discharge zone of the pump. The two basic types of positive displacement pumps are *reciprocating* and *rotary*. Whereas the total dynamic head developed by a centrifugal pump is determined uniquely by the speed at which the impeller rotates, a positive displacement pump ideally will produce whatever head is imposed on it by the restrictions to flow on the discharge side.

Reciprocating Pumps

Reciprocating pumps produce pulsating flow, develop high shutoff or stalling pressure, display constant capacity when motor driven and are subject to vapor binding at low NPSH conditions. There are three classes of reciprocating pumps: piston pumps, plunger pumps and diaphragm pumps.

The piston pump consists of a cylinder with a reciprocating piston connected to a rod that passes through a gland at the end of the cylinder. The basic design is illustrated in Figure 6.28, and the operating principle is outlined in Figure 6.29. The liquid suction and delivery in the pump are derived from the reciprocating motion of the piston (1) in the pump cylinder (2) of Figure 6.29. When the piston moves toward the right, a vacuum develops in the closed space between head (3) and the piston. Due to the pressure gradient between the suction tank and cylinder, the liquid is lifted through the suction piping and enters the cylinder via the suction valve (4). The delivery valve (5) is closed during the piston's motion to the right as it is subjected under liquid pressure from the suction piping. When the piston moves toward the left, the pressure generated in the cylinder closes one valve (4) and opens

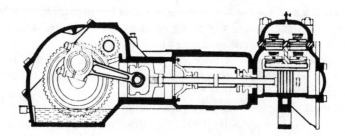

Figure 6.28   Basic design of the piston pump.

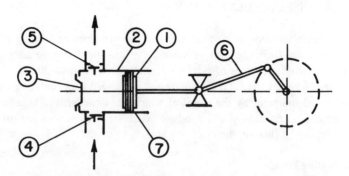

Figure 6.29   Operating principle for a horizontal, single-acting piston pump: (1) piston; (2) cylinder; (3) head of cylinder; (4) suction valve; (5) delivery valve; (6) crank and connecting rod assembly; (7) seal rings.

another (5). The liquid passing through the delivery valve enters into the discharge piping and then into the tank. Thus, both liquid suction and delivery by a single-acting piston pump vary with time because of the periodic motion of the piston. The volume delivery starts from zero at the instant the piston begins to move forward, reaches a maximum when it is fully accelerated at approximately the mid-point of its stroke, then gradually falls off to zero. There will be a short time interval during the return stroke when liquid fills the cylinder and the delivery remains at zero. This is illustrated by the plot shown in Figure 6.30.

The piston shown in Figure 6.29 is driven by a crankshaft assembly (6), which converts the rotary motion of the drive wheel to the back-and-forth linear movement of the piston rod. Piston pumps are classified as single acting and double acting. In a single-acting piston pump, a single revolution of the crankshaft provides suction and delivery; in a double-acting pump, two strokes of a piston effect delivery/suction. As noted in Figure 6.29, the main working component of a piston pump is the piston itself, consisting of disks and seal rings (7) housed inside a polished cylinder.

A *plunger pump* differs from a piston pump in that it has a plunger reciprocating through packing glands, causing the displacement of liquid from the cylinder. In this type pump, considerable radial clearance exists between the plunger and cylinder walls. Plunger pumps usually are thought of as single acting in the sense that only one end of the plunger is used for pumping liquid. Figure 6.31 provides details of the operating principle of a single-acting, horizontal plunger pump. As shown, a piston serves the role of the plunger (1), reciprocating in cylinder (2) and sealed with a packing gland (3). The internal cylinder surface of a plunger-type pump does not require as high a degree of finishing as does a piston pump. Also, leakage is readily minimized by tightening or replacing the packing gland. Plunger pumps are well suited to handling suspensions and viscous fluids because of their large radial clearances and ability to generate high pressures.

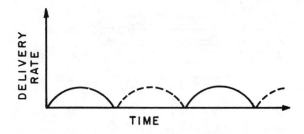

Figure 6.30   Liquid delivery rate from a simplex pump.

A more even delivery is achieved with piston and plunger pumps when operation is converted to double acting. The horizontal, double-acting plunger pump (Figure 6.32) may be considered to be an aggregate of two single-acting pumps having four valves (two suction and two delivery). When the plunger (1) moves toward the right of Figure 6.32, liquid enters into the cylinder (2) through the suction valve (3), while liquid passes through the delivery valve (6) into the discharge piping. During the reverse stroke of the plunger, suction occurs in the right side of the cylinder through the suction valve (4) and delivery takes place through the valve (5). Hence, with a double-acting pump, suction and delivery take place during each piston stroke. Consequently, the capacity of these pumps is greater and delivery is smoother than with single-acting designs.

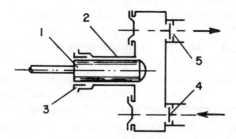

Figure 6.31 Operating principle behind a single-acting horizontal plunger pump: (1) plunger; (2) cylinder; (3) packing gland; (4) suction valve; (5) delivery valve.

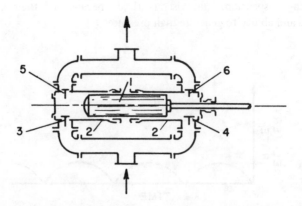

Figure 6.32 Operating principle behind a horizontal, double-acting plunger pump: (1) plunger; (2) cylinders; (3-4) suction valves; (5-6) delivery valves.

The operation of a three-plunger pump (triplex pump) is illustrated in Figure 6.33. These designs are widely used because of the pulsation-free flowrate they provide, as well as their smooth torque at maximum stroke frequencies, which can exceed 1500 strokes per minute (spm). The triplex pumps are single-acting tripled pumps having cranks located 120° from each other. The total triplex-pump delivery is the sum of the deliveries of the individual single-acting pumps. For one rotation of the crankshaft the liquid is suctioned and delivered three times. Figure 6.34 shows a cutaway view of a triplex plunger pump with an integrated gear drive. Figure 6.35 shows an actual triplex plunger pump.

Piston and plunger pumps may be actuated directly by steam-driven pistons or by rotating crankshafts through a cross head. The direct-acting steam pump consists of a steam cylinder end in line with a liquid cylinder with a straight rod connection between the steam and pump pistons or plunger.

Direct-acting steam pumps are available as simplex (one steam and liquid cylinder) and duplex (dual side by side) units. Duplex units are employed in

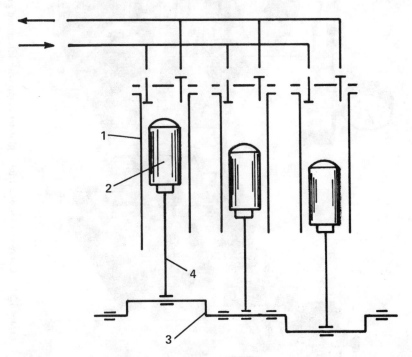

Figure 6.33   Operating principle of a triplex pump: (1) cylinders; (2) plungers; (3) crankshaft; (4) connecting rods.

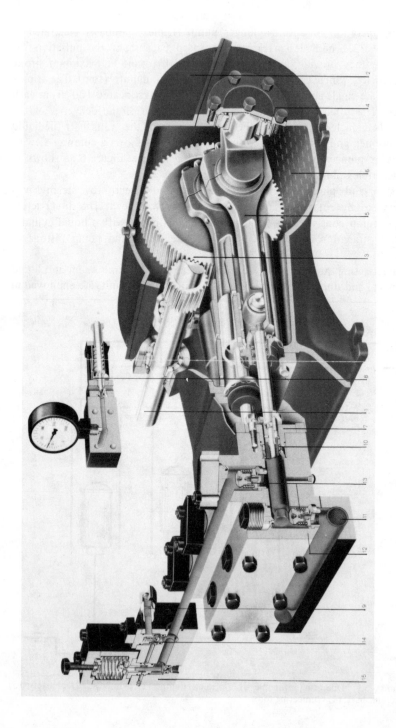

Figure 6.34   Cutaway view of a triplex plunger pump with integrated gear drive (courtesy American LEWA Inc., Natick, MA).

Figure 6.35   A triplex plunger pump with variable-speed drive (courtesy American LEWA Inc., Natick, MA).

large-capacity services and for reducing the flow pulsations below that of the simplex. Dual pumps are designed with an interconnecting steam valve linkage arrangement so that one side pumps when the other side reaches the end of its stroke. Steam pumps consist of a rod and piston design and are double acting, that is, each side pumps on every stroke. Consequently, a duplex pump will have four pumping strokes per cycle.

Power pumps convert rotary motion to low-speed reciprocating motion via speed-reduction gearing, a crankshaft, connecting rods and cross-heads. Plungers or pistons are driven by the cross-head drives. Rod and piston construction, similar to that in duplex double-acting steam pumps, is used by the liquid ends of the low-pressure, higher-capacity units. The higher-pressure units are normally single-acting plungers. This latter style generally employs three (triplex) plungers. Three or more plungers substantially reduce flow pulsations relative to simplex and even duplex pumps.

In general, the effective flowrate of reciprocating pumps decreases as viscosity increases because the speed must be reduced. High viscosity also leads to a reduction in pump efficiency. In contrast to centrifugal pumps, the differential pressure generated by reciprocating pumps is independent of fluid density. Rather, it is dependent entirely on the magnitude of force exerted on the piston.

Reciprocating pumps are used most often for sludge and slurry services, particularly where other pump types are inoperable or troublesome. Maintenance in such services tends to be high because of valve, cylinder, rod and packing wear.

The theoretical delivery of a piston pump is equal to the total volume swept out by the piston in the cylinder times the number of strokes of the piston per unit time. Thus, the theoretical delivery in a single-acting pump is

$$Q_{th} = F \times S \times n \qquad (6.50)$$

where   F = piston cross-sectional area
S = stroke length
n = number of crankshaft revolutions (or number of double strokes)

In the double-acting pump, there are two suctions and two deliveries per crankshaft revolution. When the piston moves to the right of Figure 6.32, the liquid volume sucked from the left side is equal to FS and that delivered from the right side is $(F - f)S$, where f is the rod cross-sectional area. When the piston moves to the left, the volume FS is delivered from the left side into the delivery piping; from the right side, the liquid volume $(F - f)S$ is sucked in from the suction line.

Consequently, for n rotations of the crankshaft the theoretical delivery of a double-acting pump will be

$$Q_{th} = FSn + (F - f)Sn = (2F - f)Sn \qquad (6.51)$$

The actual delivery may be less than the theoretical value because of leakage past the piston and valves or because of inertia in the valves. The actual delivery of the pump is

$$Q = Q_{th}\eta_v \qquad (6.52)$$

where η_v is the volumetric efficiency, defined as the ratio of the actual discharge to the swept volume.

The volumetric efficiency of large pumps is typically 0.97–0.99; for medium flowrates (Q = 20–300 m^3/hr), η_v = 0.9–0.95; and for small flowrates, η_v = 0.85–0.9.

In some cases, however, the actual delivery may be greater than the theoretical value because of fluid momentum in the delivery line and sluggishness in the delivery valve operation resulting in continued flow. Additional discussions of these pump types are given by Hicks and Edwards [1971], Kirk and Othmer [1963] and Hicks [1957].

Rotary Pumps

Rotary pumps combine the rotating movement of the working parts with the positive displacement of the fluid. There are a number of rotary pumps, but all operate on basically the same principle.

The rotating parts move in relation to the casing to create a space that first enlarges, drawing in the fluid through the suction line, sealing it, and then reducing the fluid volume and forcing it through the discharge port at a higher pressure. The rotating elements of the pump generate a reduced pressure in the suction line, thereby allowing the external pressure to force liquid into the pump. The capacity of a rotary pump is a function of its size and speed of rotation. This type of pumping equipment provides near constant deliveries in comparison to the fluctuating flows of reciprocating pumps. The main reason for selecting rotary pumps over centrifugals is to take advantage of their high-viscosity handling capability. In addition, rotaries are simple in design and efficient in handling flow conditions that generally are considered too low for economic application of centrifugals. Rotary pumps operate in moderate pressure ranges and have small to medium capacities.

Rotary pumps may be divided into five main types according to the character of the rotating parts: gears, screw, lobe, cam and vane. There is, however, considerable overlap among these types.

The simplest rotary pump is the *external gear pump,* illustrated in Figure 6.36. Two gear wheels (2) operate inside the casing (1), which provides

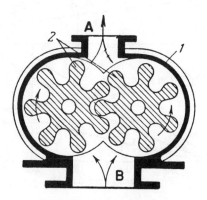

Figure 6.36   Operation of a gear-type rotary pump: (1) casing; (2) gear wheels.

a snug fit to effectively seal the spaces between each pair of adjacent teeth. One of the gear wheels is driven by the driver and the other rotates in mesh, as indicated by the arrows. As the spaces between the teeth of the gear wheel pass the suction opening "A," liquid is impounded between them, carried around the casing to discharge opening "B" and then delivered outside through the opening. The straight teeth in gear wheel pumps produce pulsations in the delivery with a frequency equivalent to the product of the number of teeth on both gear wheels and the speed of rotation. Pulsation can be eliminated by the use of gear wheels having *helical* teeth with a particular angle.

Gear pumps handle liquids of a very high viscosity but cannot be used with suspensions. Generally, spacings between gear teeth are too close to handle solids without suffering significant erosion. The rotating parts in the casing create a space that draws in the liquid from the suction line. As the parts rotate, the liquid is trapped between them and the pump casing, forcing the liquid out through the delivery side of the pump at the necessary higher pressure. Typical performance characteristics of external gear pumps are shown in Figure 6.37.

Internal gear pumps are of two principal types. Both kinds employ a modified spur gear, which rotates inside a larger gear rotating around an axis that is parallel to that of the spur gear and displaced somewhat to mesh snugly at one point on the periphery. Figure 6.38 shows one internal gear pump

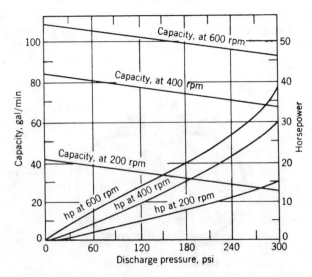

Figure 6.37   Typical performance characteristics of an external gear pump.

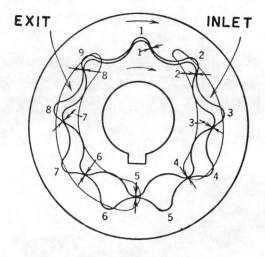

Figure 6.38 An internal gear pump with sliding seal.

design whose two gears maintain a continuous series of sliding seals with each tooth. This action forms pockets of fluid entirely between the gears. This type of pump can be equipped with any number of teeth in the spur gear (provided one more socket is included in the ring gear).

Screw pumps are modified helical gear pumps. They may have one, two or as many as three screws turning along the pump axis, with liquid flowing between the screw threads and the casing. Both single-rotor and multiple-rotor screw pumps are commercially available. Pumping action is accomplished by progressing cavities, which advance the fluid along the rotating screw from inlet to outlet. This axial flow pattern minimizes vibration, producing a rather smooth flow.

Figure 6.39 shows a single-rotor screw pump (also called a progressive cavity pump) for handling slurries with relatively large particles. This type of pump comprises a rotor that revolves within a stator, executing a compound movement. The rotor revolves about its axis while the axis itself travels in a circular path. Hence, the rotor is a true helical screw, and the stator consists of a double internal helical thread. With each complete revolution of the rotor, the eccentric movement allows the rotor to contact the entire surface of the stator. Voids between the rotor and stator contain entrapped fluid, which is moved continuously toward the pump outlet. This pumping action provides continuous flow at low, smooth and uniform rates. The action minimizes fracturing of particles as well as abrasion damage to the pump. Single-screw pumps are used extensively in the food processing

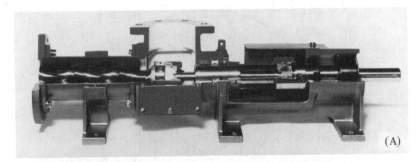

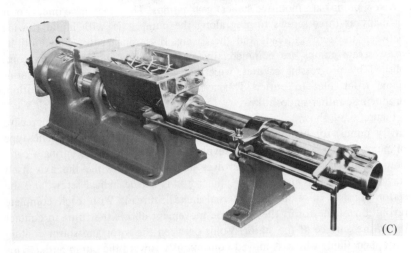

Figure 6.39 (A) cutaway view of single-screw, progressing cavity pump; (B) full view of progressing cavity pump; (C) pump with hopper, which permits full flow into the suction housing. An auger carries the product to the rotor-stator pumping elements. The design allows easy handling of high viscosity materials (courtesy Robbins & Myers, Inc., Fluids Handling Div., Springfield, OH).

and chemical industries for handling solid/liquid mixtures that are abrasive or require gentle handling of the solids.

Figure 6.40 shows a twin-screw pump consisting of two sets of screws that rotate and mesh in an accurately bored casing. Relatively tight operating clearances are maintained between the screws. This tight clearance usually is maintained by a pair of timing gears mounted on the shafts. The gears also transmit power from the drive shaft to the driven shaft. The body consists of a casing with two precision-machined bores, which house the rotating screws. Fluid passes from the inlet into the pumping chamber and then to the discharge zone.

Some screw pumps are operated without timing gears. In these units, a driver is used to turn the screws directly. The designs of twin-screw pump bodies and screw assembly arrangements are interrelated. Figure 6.41 illustrates typical body design.

The operation of a three-rotor-screw pump is shown in Figure 6.42. The center rotor (1) is the driving member, while the other two (2) are driven. Liquid enters at the end of the rotors, is sealed between the rotors and casing, and forced smoothly to the delivery end.

The mechanical displacement of liquid from inlet to outlet is generated by trapping a slug of fluid in the helical cavity (referred to as a "positive lock"), created by the meshing of the screws.

Two- or three-lobe pumps are similar in design to gear pumps (Figure 6.43), but gear wheels are replaced either by two or three lobes, which are driven separately through an external gearing mechanism. This arrangement makes it possible to avoid actual contact of the lobes with each other. Wear on the lobes and casing can be minimized by maintaining a small clearance between them. The clearances are a few thousandths of an inch, sufficient to reduce friction and wear but maintain minimum leakage from the delivery to the suction side. The characteristics of the lobe pumps are generally similar to those of the gear pumps.

Cam pumps consist of an eccentrically mounted circular rotor that sweeps a circle whose radius is the sum of the radius of the rotor and the eccentricity. Figure 6.44 shows such a pump utilizing a plunger valve. The circular cam is fixed rigidly to the shaft and is housed inside the rotor ring, which is free to rotate about the cam. The plunger slides freely through a slide pin to act as a discharge valve. The plunger may be replaced by any form of vane, which seals the suction line from the discharge line. In general, contacts between surfaces are almost free of friction and wear except for the cam inside the rotor.

As the cam rotates it expels liquid from the space ahead of it and sucks in liquid behind it. The characteristics again are similar to those of a gear pump.

Vane pumps (Figure 6.45) include a massive cylinder (1), which carries rectangular vanes in a series of slots arranged at intervals around the curved surface of the rotor located eccentric to the casing (2). The tip of the

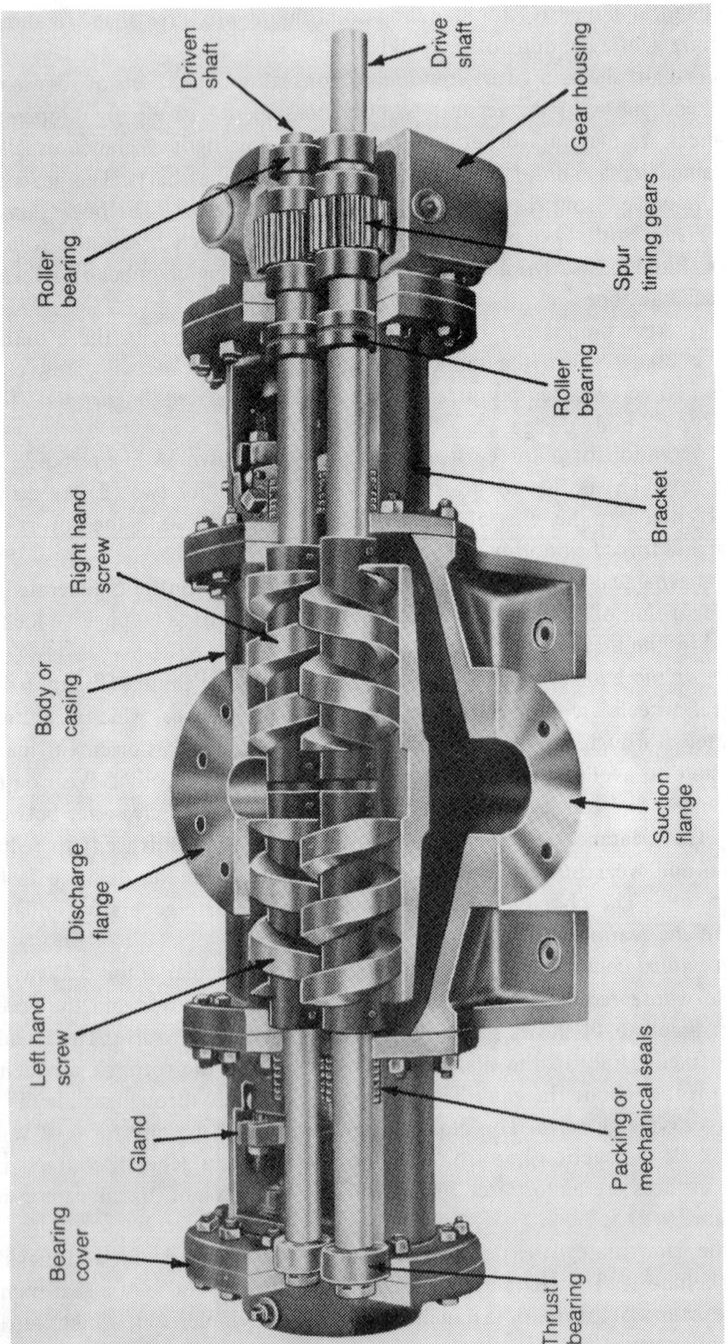

Figure 6.40 Cross section of a twin-screw pump, showing main features (courtesy Worthington Pump Inc., Mountainside, NJ).

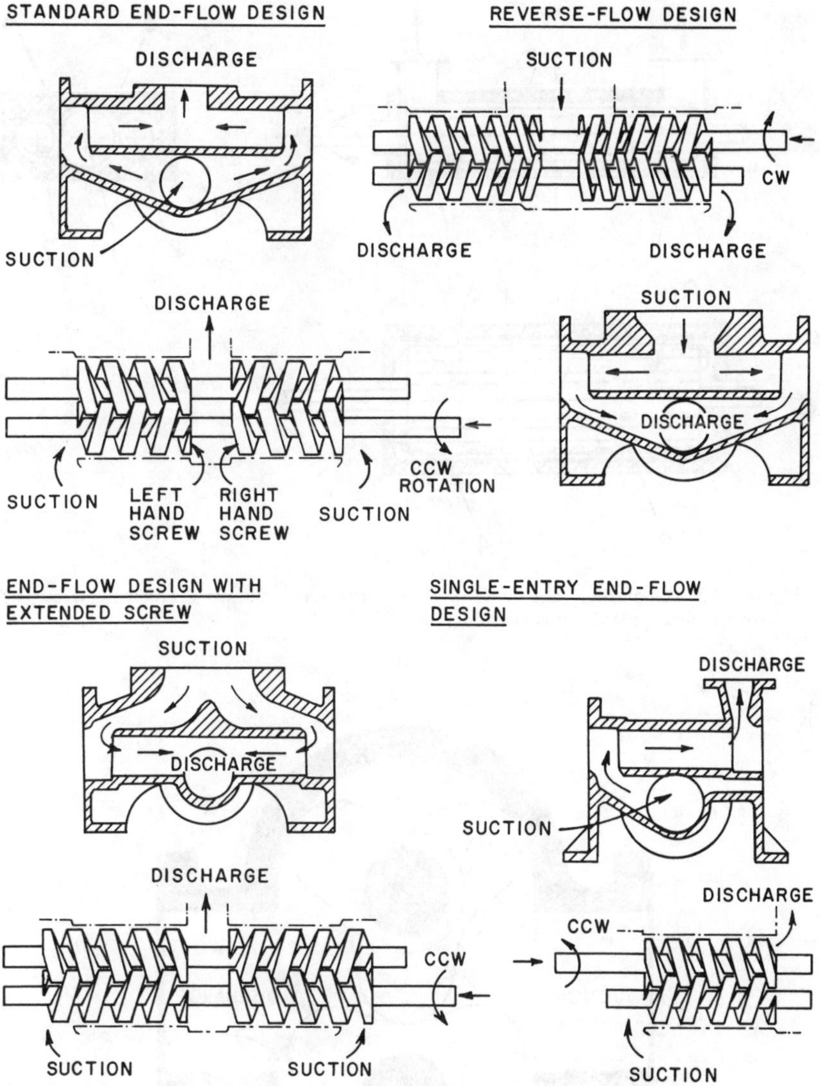

STANDARD END-FLOW DESIGN

REVERSE-FLOW DESIGN

Figure 6.41   Typical twin-screw pump bodies and flow patterns.

vanes (3) is thrown outward by centrifugal force. As the cylinder rotates, the space behind a vane enlarges when it moves from the suction nozzle (5) to the vertical axis of the pump, resulting in the formation of a vacuum in space (4), thus drawing in liquid. At this point in the cycle liquid is trapped

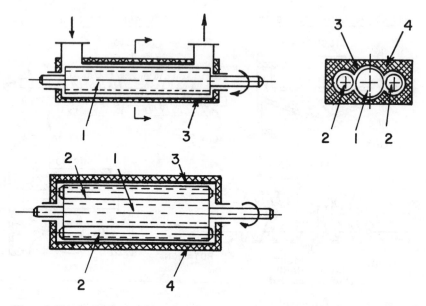

Figure 6.42 Operating scheme of a three-rotor screw pump: (1) driving screw; (2) driven screws; (3) sleeve; (4) casing.

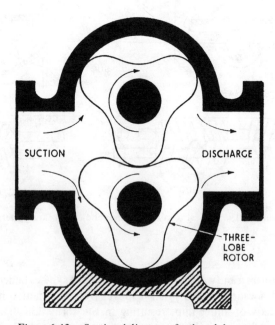

Figure 6.43 Sectional diagram of a three-lobe pump.

PLUNGER   SLIDE PIN

IN

OUT

ROTOR
RING

CAM

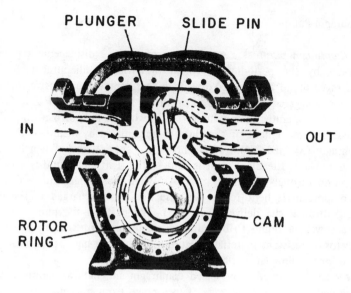

Figure 6.44   Cross-sectional view of a cam pump.

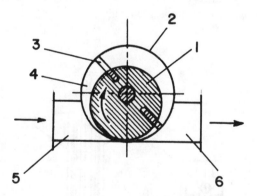

Figure 6.45   A sliding vane-pump: (1) rotor; (2) casing; (3) vanes; (4) working space;
(5) suction nozzle; (6) delivery nozzle.

between the vanes. When the vane moves from the vertical axis in the direction of rotation, the volume of the chamber (space 4) decreases, and the liquid eventually is forced out through the discharge nozzle (6). Because of wear, the vane dimensions change; however, this is compensated for somewhat until the seal is broken. At that time new vanes must be inserted.

Diaphragm Pumps

A *diaphragm pump* is a special type of positive displacement pump that operates by the periodic movement of a flexible diaphragm. It has the advantages of no stuffing boxes and high tolerance to abrasive slurries and chemically aggressive liquids. The operation is illustrated in Figure 6.46. The plunger (1) operates in a cylinder (2) in which a noncorrosive liquid is displaced. The movement of the fluid is transmitted by means of the flexible diaphragm (3) made from soft rubber or a special steel. When the plunger moves upward, the diaphragm is bent to the right and liquid is sucked into the pump through the globe valve (4). When the plunger moves downward, the diaphragm is bent to the left and liquid is discharged to the delivery piping through the delivery valve (5). The parts of the pump that are in contact with a liquid to be pumped are usually constructed of corrosion- and erosion-resistant materials. Figure 6.47 shows a high-pressure diaphragm pump for handling large volumes of process chemicals.

New and improved fully sealed diaphragm and bellows pumps have been developed to minimize leakage. Types range from micrometering pumps in the milliliter size to large diaphragm pumps having drives of several hundred kilowatts for high-pressure processes. These designs often achieve economic viability through reduction of maintenance and decontamination costs, and improvement of process reliability.

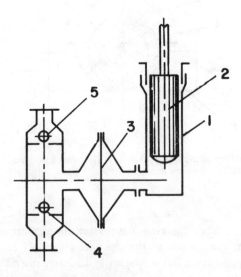

Figure 6.46   Operating scheme for a diaphragm pump: (1) cylinder; (2) plunger; (3) diaphragm; (4) suction valve; (5) delivery valve.

Figure 6.47   Diaphragm pump in horizontally opposed arrangement for high pressures up to 350 bar (courtesy of American LEWA Inc., Natick, MA).

Operating limits and the advantages of various constructions are described briefly in Table 6.8. Critical, toxic or abrasive media require the use of leakfree metering or production pumps, with a realistic upper size limit of about 300 kW.

A widely used construction utilizes a mechanical diaphragm drive (usually limited to less than 85 psi). The metering rate usually is limited to less than 200–500 charges/hr because the diaphragm loading becomes unfavorable with increasing diameter. The stroking rate also is restricted to less than 150 strokes/min because of acceleration shocks that occur at part-stroke settings with the widely used cam-and-spring-return or magnetic-drive units.

For a given membrane geometry, diaphragm life depends on stroke length, pressure, temperature and compatibility with the fluid being processed. The detailed design is empirical, based on extensive fatigue trials. Diaphragm life easily can reach 3000 hours under permissible maximum conditions and will considerably exceed this at lower pressures, shorter strokes, etc.

Diaphragm failure can be signaled by a float switch that responds to the presence of process fluid in the space behind the diaphragm. Another failure detection method involves two mechanically coupled diaphragms. A rise in pressure in the space between them indicates diaphragm failure and will cause simple sensors to trigger a warning signal.

Table 6.8 Operating Limits and Applications of Various Metering Pumps

Pump type	Limits	Application	Wetted Parts Made of
Low-pressure diaphragm pump with direct mechanical drive	<6-10 bar (85-140 psi), <500/h	Metering, pumping	PVC or austenitic stainless steel, elastomer diaphragm
Low-pressure bellows pump with direct mechanical drive	<5 bar (70 psi)	Metering, pumping	Glass; PTFE bellows
High-pressure microdiaphragm metering pump with hydraulic drive	<700 bar (9,940 psi), <10/h	Metering	Acid-resistant steel
Diaphragm pump with hydraulic compression of tubular member	<50 bar (710 psi)	Metering, pumping	Acid-resistant steel; elastomer diaphragm and tube
Diaphragm pump with hydraulic drive	<350 bar (5,000 psi) for PTFE diaphragm, <3,000 bar (42,600 psi) for metal diaphragm	Metering, pumping	PVC, PTFE, titanium, acid-resistant steel

Depending on the diaphragm geometry, the pumping characteristic is usually not quite linear. Metering accuracy is thus related to the tolerances of the operating conditions. Figure 6.48 shows the relationship between flow and pressure at various ratios of stroke length to delivery rate. The metering rate in relation to stroke frequency is linear.

Bellows-type metering pumps (Figure 6.49) are built almost exclusively of glass and polytetrafluoroethylene (PTFE) components. This limits their pressure level to less than 5 bar (70 psi) because of the glass parts. The output capacity is virtually unlimited because there is no problem in producing fatigue-free bellows with diameters of several hundred millimeters.

Bellows require more care than do diaphragms during forming and manufacture, but normally achieve lives of 5,000–10,000 hours at maximum load. Because of radial stiffness of the bellows, the metering characteristic curve is purely linear and only slightly pressure dependent.

As in the case of the diaphragm pump, bellows failure is monitored via level or pressure sensors in the chamber below the bellows. This chamber is sealed from the drive by an auxiliary packing.

Plunger displacement by means of hydraulic fluid offers the benefit of uniform diaphragm support. The loading of the diaphragm then becomes solely the result of elastic deformation and not of resistance to pressure

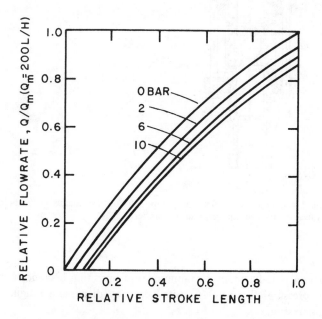

Figure 6.48   Diaphragm metering pump flow characteristics.

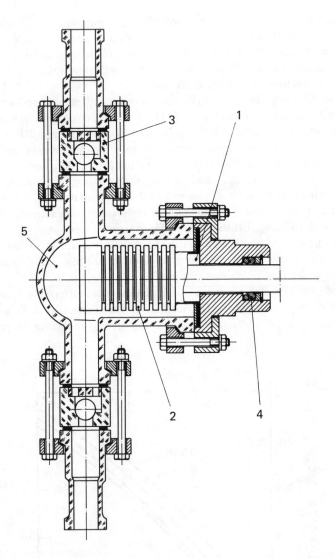

Figure 6.49   A bellows-type metering pump: (1) rupture chamber; (2) bellows; (3) check valve; (4) packed seal; (5) pump chamber.

forces. Strain is reduced to a minimum, and operation becomes possible at very high pressures. Pressure is limited only by diaphragm fatigue strength under compression.

Hydraulically driven pumps allow greater diaphragm displacements, but their operating pressure is restricted to about 5000 psi and 120°C maximum by the fatigue limit under compression for PTFE and by clamping effects.

The largest number of applications can be met by compact diaphragm metering pumps. Stroke-adjustable drives operate the piston via a connecting link, the piston sliding in a liner tight enough to provide a seal. The lubrication and hydraulic systems are conventional. Plunger movement displaces the diaphragm, which can deflect between two perforated support plates. Topping up the space between plunger and diaphragm (the hydraulic space) is controlled at negative pressure by a vacuum-replenishing valve when the diaphragm has reached the lower support plate.

The hydraulic system also contains a relief valve for protection against excess pressure. This sometimes can take the place of an external safety valve that otherwise would be in contact with the pumped fluid. The upper support plate, together with this relief valve, protects the diaphragm against excessive deflection under certain process conditions. An automatic vent valve in the highest point of the hydraulic system provides a degassed hydraulic medium and, thus, a faultless delivery characteristic. Compact diaphragm pumps with common oil systems generally require the use of sandwich diaphragms.

The pump is ideally suited to slurries of all types and cleaning-in-place in the food industry. Topping up the hydraulic system is controlled via a gate that is diaphragm position dependent, and by a relief-and-snifter valve. The continuous vent can be seen at the highest point.

For services beyond the practical limits of PTFE, *metal diaphragms* are used, made of dead-parallel cold-rolled sheet. Pumps with metal diaphragms are especially suited to:

1. pressures above 5000 psi and temperatures up to 200°C (in special cases, 400°C maximum);
2. microdiaphragm dosing pumps because of the low compressibility of diaphragm and clamping; and
3. applications where all-metal construction is mandatory, usually to ensure radiation resistance.

Because of the lower elasticity of metal diaphragms, the same stroke-displacement volumes require larger-diaphragm diameters and larger pump heads. For larger outputs, diaphragm pumps with metal diaphragms are more costly to manufacture than are pumps with nonmetallic diaphragms. In addition, because of the risk of notch formation and because of their lower deflection capability, thin metal diaphragms are not suitable for the unsupported type of diaphragm.

Important design considerations are an even flow distribution through the support perforations and the prevention of local adhesion. Because of diaphragm durability and adhesion avoidance, the dimensioning of the perforations and the profiling of the diaphragm support surface cannot be calculated mathematically. They depend on experience gained through successful construction and installation of high-pressure diaphragm pumps.

In micrometering pumps for high pressures, the displacement volumes and the diaphragm pump head dimensions are very small. Therefore, the design parameters have a different rank of importance than for high-output pumps. The pump cavity is designed for minimum dead volume and maximum rigidity. For improved degassing, oil circulation in the hydraulic system has proved to be a satisfactory concept. This is achieved by plunger displacement, aided by two nonreturn valves.

Hydraulic-linkage pipes (also referred to as "pendulum" or "remote-head" systems) serve to remove high temperatures and other troublesome and dangerous influences from the drive, making it feasible to extend the field of application of diaphragm pumps (e.g., to 400°C). A familiar application is in the radioactively heated zone of atomic fuel recovery plants, where extreme demands are made for safety. Figure 6.50 shows the major modules of these metering pumps.

Loss in suction pressure (largely influenced by the suction valve) is much the same for diaphragm and plunger pumps. In particular, types having

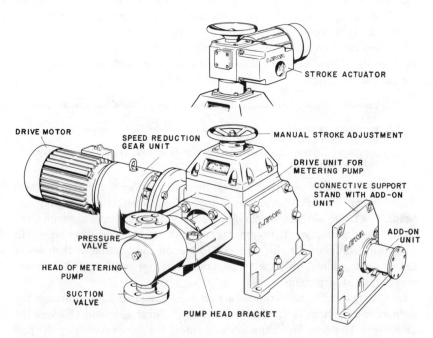

Figure 6.50   The modules of metering pumps (courtesy American LEWA Inc., Natick, MA).

unrestricted forward chambers experience no additional internal pressure losses on the process fluid side. However, at high stroking frequencies, in excess of 300 strokes/min, it is essential that internal flow patterns be optimized.

Some types of diaphragm pumps require minimum suction pressures of 7–42 psi. These include pumps with metal diaphragms for larger output and faster stroking rates. Metal diaphragms with their lower stiffness are more sensitive to cavitation than are the heavier PTFE diaphragms.

Cavitation can be minimized by proper layout of the pipe system and pressure conditions. Upsets in metering accuracy and performance owing to cavitation can be tremendous. Delayed compression, with cavitation present, leads to sizable pressure and loading shocks.

Because of elastic influences, the instantaneous delivery flow is not in phase with plunger displacement. Hence, the delivery often starts jerkily at the end of the compression phase (final plunger speed) with a corresponding shock wave as the result. This shock problem, which occurs in all high-pressure pumps, can be overcome by shock dampers that have been sized in line with pulsation theories. Further discussions are given by Vetter and Hering [1980].

Additional information on these pumps is given by Karassik [1976], Cheremisinoff [1981], Coulson and Richardson [1962], Brown [1950], and Foust et al. [1980].

PUMPS OF SPECIAL DESIGN

Thus far, we have described pumping operations utilizing mechanical components such as pistons, plungers, impellers, gears and screws. In addition to mechanically displaced pumping systems, specially designed pumps operate on the principle of liquid displacement by a secondary fluid. This class includes jet pumps, acid eggs, hydraulic rams and airlifts.

A *jet pump* (Figure 6.51) takes advantage of the momentum of a high-velocity secondary fluid stream (steam or water), referred to as the "working fluid," to impart momentum to the fluid to be pumped. In this operation, both streams are actually mixed. The working fluid (I) enters at a high velocity from the nozzle (1) through the mixing chamber (2) into the diffuser (3), thus entraining (because of surface friction) the liquid to be pumped (II). In the narrowest section of the diffuser (having a geometry similar to a venturi), the velocity of the mixture (working and pumped) reaches a maximum value, and the static pressure of the flow becomes minimum, according to Bernoulli's equation. The pressure drop in the mixing chamber and the diffuser provides the delivery of liquid (II) in the mixing chamber from the suction line. In the expanding section of the diffuser the flow velocity is decreased and the kinetic energy of the flow converted into potential pressure energy. Hence, liquid under pressure enters the delivery line.

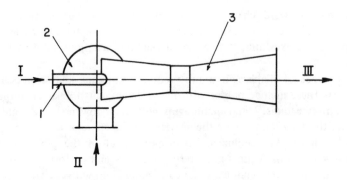

Figure 6.51 Operating principle of a jet pump.

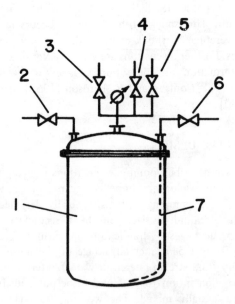

Figure 6.52 Operation of an acid-egg pump: (1) container; (2–6) valves; (7) outlet pipe.

The *acid egg pump* or *blowcase pump* (Figure 6.52) consists of a horizontal or vertical container (1) filled with the liquid to be pumped by means of gas pressure. In pumping liquids by kinetic, potential or pressure energy as a motive force, the egg operates by displacement of one fluid by another. As with the jet pump, this design has no moving parts and can be operated simply. Liquid enters the egg from the feed pipe through the open valve (2); normally the vent valve (3) is open if filling is done under atmospheric pressure. Liquid also may enter through the valve (4) if filling is to be done under vacuum. In displacing the liquid, valves (2-4) are closed and valve (6) on discharge pipe (7) and valve (5) on the line of compressed gas are opened. After the egg is discharged, some valves (5,6) are closed and one valve (3) is opened to connect the egg to the atmosphere. The acid egg is useful, despite its low efficiency (typically 10–20%), in cases in which conventional pumps are cost prohibitive in handling corrosive or erosion-causing liquids. Its low efficiency is due to its intermittent action and that at the end of each cycle compressed gas must be vented to the atmosphere without recovering work. Often these pumps require close manual operation or excessive instrumentation.

The *hydraulic ram* (Figure 6.53) utilizes the kinetic energy of a moving column of liquid (usually water) to raise a portion of the pumping liquid to a higher pressure or elevation. If the liquid flowing in the supply line, or fall pipe, is stopped abruptly by the closure of the escape check valves, the static pressure at the valve suddenly increases due to the stream's conversion of kinetic energy. When the delivery valve opens, some of the liquid flows into the delivery system. As the energy of the liquid in the supply line is absorbed, the static pressure at the base decreases. When the escape valve opens the delivery valve will close, thus causing the flow in the column to increase. This causes the motive fraction of the fluid to flow out to waste through the escape valve until the velocity becomes great enough to pick up and seal this valve. The cycle is repeated with a frequency as low as 15, or as high as 200 times per minute.

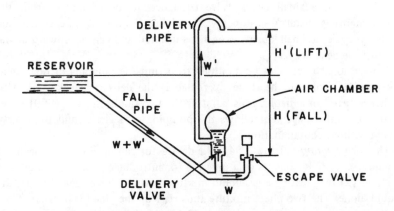

Figure 6.53   Operating principle of the hydraulic ram.

The efficiency of a properly designed hydraulic ram may be as high as 90%, where efficiency is defined as

$$\text{Efficiency } (\%) = \left[\frac{W'H'}{WH}\right] \times 100 \qquad (6.53)$$

where W' = mass of liquid delivered
         W = mass of motive liquid exhausted through the escape valve
         H = the fall (Figure 6.53)
         H' = the lift (Figure 6.53)

Thus, the essential components of this pump are the moving fluid column, the escape and delivery valves, and the air chamber. The efficiency and capacity strongly depend on the design of these elements.

The important factors of the moving fluid column are the mass of motive liquid and the friction of the flow of the motive liquid. These factors can be expressed as the ratio of the length of the fall pipe to the height of fall. For a vertical fall pipe an optimum ratio would contain too small a mass of liquid. For a relatively flat fall pipe, the ratio would have too great a friction loss. The optimum value thus varies with the lift.

Valve design for hydraulic rams depends on the weight and the length of the stroke. As a rule of thumb, efficiency varies inversely with the length of the stroke and the weight of the valve, whereas capacity varies directly with the weight of the valve and the length of the stroke. Decreasing the weight of the valve decreases the length of each cycle.

An air surge chamber is necessary to eliminate intermittent flow in the delivery line. In practice, the volume of the surge chamber is designed to be approximately the volume of the delivery pipe. To maintain a supply of air to the vessel, a small check valve (designed to open inward from the atmosphere) is installed just below the delivery valve. At the end of the delivery cycle, a small quantity of air is inspired and carried upward at the start of the next delivery.

Conventional rams may be modified to permit the pumping of one liquid by another. This can be done by replacing the delivery check valve with either a piston or a diaphragm separating the two fluids. Brown [1950] notes that such an arrangement allows the pumping of a clean fluid by a dirty liquid without contamination.

The *airlift pump* (Figure 6.54) is a device for raising liquid by means of compressed air. It consists of a pipe (1) for introducing compressed air and a mixer (2) for creating a gas-liquid mixture having a lower density than the liquid alone. The two-phase mixture thus rises up the pipe (3) because of the lower density and gas expansion. At the exit of the pipe (3) the mixture

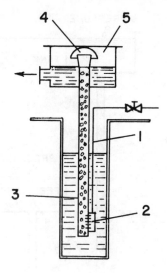

Figure 6.54 Operation of an airlift pump: (1) pipe for delivery of compressed air; (2) mixer.

flows around the baffle (4), and the liquid, after separation from air, enters the container (5).

For illustration, let us apply the airlift pump to transferring liquid from feed tank A to reservoir B at an elevation of h_r above the feed tank liquid surface in Figure 6.55. The net work performed on the liquid by a mass, "m," of air is Mgh_r, where M is the mass of liquid raised. Assuming that the air enters the lift line at pressure P, then the work done by the gas under isothermal expansion to atmospheric is

$$P_a v_a \, m \, ln \, P/P_a$$

where v_a = specific volume of air at atmospheric pressure.

From the above, we may write the efficiency of the pump as follows:

$$\eta = \frac{Mgh_r}{mP_a v_a \, ln \, (P/P_a)} \tag{6.54}$$

or rearranging this expression, the mass of air required to pump a unit mass of liquid is

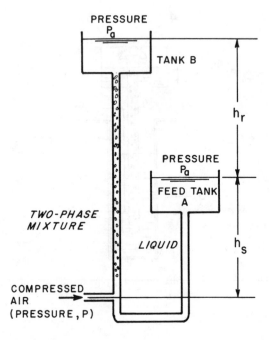

Figure 6.55   Example of a lift pump operation.

$$\frac{m}{M} = \frac{gh_r}{\eta P_a v_a \, ln \, (P/P_a)} \tag{6.55}$$

Coulson and Richardson [1962] note that when the density of the two-phase mixture in the riser is not significantly different than the feed liquid, the pump will deliver infinitely slowly at 100% efficiency, $\eta = 1$.

Under this condition, the pressure at the point of injection of the compressed air equals the sum of the atmospheric pressure and the pressure due to the column of liquid of height h_s, i.e., the vertical distance between the feed tank level and the air inlet point. Then the mass of air required to pump a unit of mass of liquid for a perfectly efficient pump is

$$\left(\frac{m}{M}\right)_{\eta=1} = \frac{h_r g}{P_a v_a \, ln \left(\dfrac{h_s + h_a}{h_a}\right)} \tag{6.56}$$

This expression is derived by noting that

$$P_a = h_a \rho g$$

and

$$P = (h_a + h_s)$$

where ρ is the liquid density. (Note that the derivation has ignored slip between the air and liquid.) Equation 6.56 represents the minimum air requirement as frictional losses have been ignored in the derivation.

The actual flow regime of the two-phase mixture in the riser depends on the design of the injection point and the size of the air bubbles. The size of the bubbles influences their velocity of rise relative to the liquid. With very small bubbles, the relative velocity is minimized; however, small bubbles are difficult to form at the rates needed for operation. Generally, bubble coalescence occurs near the injection point and, subsequently, in the upper portions of the riser. Usually coalescence is complete within a few pipe diameters of the injection point and, hence, the dispersion over most of the riser is not greatly different than by introducing large air bubbles. Very often, particularly in small lines (under 3 in. in diameter), a slug flow regime develops, as shown in Figure 6.56. The air slug is almost bullet shaped and occupies only the central core of the pipe cross section while a film of liquid flows downward at the walls. The gas slugs essentially push a volume of gas-entrained liquid up the pipe. As the liquid slug moves upward, it is continuously draining to the slug below. The pump functions properly only if the rate of drainage is less than the upward rate of transfer. Further discussion of this type of two-phase flow is given in Chapter 14.

Four sources lead to inefficient pumping with airlift pumps: (1) the injection point (footpiece), (2) at the pump discharge, (3) friction losses in the riser line, and (4) the air supply line.

The gas feed point is a critical part of the design of airlift pumps. The air normally is introduced to the riser through an arrangement of orifices. Two typical feed point arrangements, or footpieces, are illustrated in Figure 6.57. Energy losses occur because of the sudden expansion of the gas flow and as a result of friction at the orifice walls. In addition, the gas is accelerated as it passes through the footpiece (because the total volumetric gas flow in the lift line is increased by the addition of the air). To minimize slugging and accelerate the liquid as much as possible, the air should be injected to form relatively small bubbles over a considerable length of the pipe.

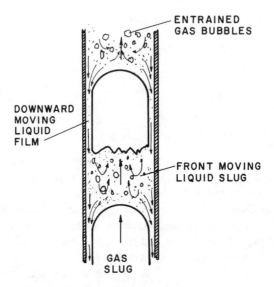

Figure 6.56   Slug flow in a vertical line.

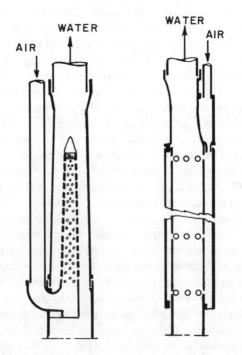

Figure 6.57   Two designs for gas injection into airlift pumps.

Losses from the discharge of the pump arise because the kinetic energy of the fluid cannot be converted effectively into pressure energy. This is attributed to the rapid fluctuations in the flow.

Friction losses in the riser generally are quite large due to the high degree of turbulence and unsteady slug flow. The relative velocity of the upward-moving air to the liquid promotes high turbulence, resulting in an increase in the flow of air required to maintain a desired fluid density in the riser. To minimize losses, an even feeding and distribution of air bubbles across the flow area is desirable.

Finally, minor losses in the air supply lines contribute to lower pump operating efficiency. Airlifts generally are more efficient than other pump types that employ compressed air directly, especially as the air is allowed to expand almost to atmospheric pressure when contacting the liquid. No valves are used other than those in the air supply lines and the nonreturn valve in the suction line. This type pump is well suited to handling liquids containing high concentrations of suspended solids. Both the pipe and the footpiece can be made of metal, stoneware or glass, according to the material to be handled and the duty required of the pump. They are used extensively in the oil industry and for circulating nitric acid in absorption equipment.

Their primary disadvantage is that the air must be introduced at a considerable depth below the liquid feed point to obtain a high efficiency. In addition, it often is difficult to predict accurately the total energy losses for such a system during design. Additional discussions of secondary fluid displacement pumps are given by Coulson and Richardson [1962], Foust et al. [1980], Datta [1948], Bergelin [1949], Simonin [1951], and Bonnington and King [1972].

PUMP SELECTION CRITERIA

Proper pump selection for an energy-efficient liquid transportation system demands understanding of the principles of mechanics and physics affecting the pumping system and the fluid. Pump efficiency is strongly dependent on the behavior of the pumping fluid. Principles of pump operations defining head and flow relationships must be understood clearly before the system performance can be evaluated accurately.

General performance characteristics of commercially available pumps can be obtained from the manufacturers. Selection is made on the basis of desired capacity and head, which are calculated in accordance with overall piping system layout. The electric motor for a pump is chosen from the brake horsepower determined from the equations given earlier. (Practice problems at the end of this chapter illustrate these concepts.)

The overall procedure in choosing a pump for a particular application is as follows:

1. Obtain information on the physical and chemical properties of the liquid at the intended operating conditions, i.e., specific gravity, viscosity, vapor pressure, corrosiveness, toxicity, etc.
2. Lay out the piping system on paper, defining major flow resistances in the system. Calculate total heads for the system.
3. Establish the capacity requirements in terms of a range. That is, define normal average capacity needs as well as system lows and peak flow required. If possible, estimate how long pumps will have to operate at peak loads.
4. Based on the above information, the class and type of pump can be selected. A more detailed specification can be made from examination of the manufacturer's literature.

Of the pumps described in this chapter, centrifugals are the most versatile and widely used throughout the chemical process industries.

The following advantages of centrifugal pumps should be remembered when comparing different pump classes for an application:

1. They are simple in construction and, as a general rule, less expensive than many positive displacement types. They are available in a wide range of materials.
2. They do not require valves for their operation.
3. They operate at high speeds (4000 rpm or higher) and, therefore, can be coupled directly to an electric motor. In general, higher speeds typically mean smaller pumps and motors for a given duty.
4. They give steady deliveries.
5. Depending on the application, maintenance costs are lower than for any other type of pump.
6. They are typically smaller than other pumps of equal capacity. Therefore, they can be made into a sealed unit with the driving motor and directly immersed in the suction tank.
7. Liquids having relatively high concentrations of suspended solids can be handled. At the same time, centrifugal pumps have several disadvantages, the primary ones being as follows:
   (a) Single-stage pumps cannot develop high pressures. Multistage pumps will develop greater heads but are much more expensive and cannot be readily constructed from corrosion-resistant materials without significantly higher costs due to their greater complexity. As a general rule, it is better to use very high speeds to reduce the number of stages required.
   (b) High-efficiency operation is usually only obtained over a limited range of conditions. This is especially true for turbine pumps.
   (c) The vast majority of centrifugal pumps commercially available are not self-priming.
   (d) A nonreturn valve must be installed in the delivery or suction line or the liquid will run back into the suction tank when the unit is not running.
   (e) Centrifugal pumps have problems handling highly viscous materials. They typically operate at greatly reduced efficiencies.

For pumping problems requiring relatively small capacities and high heads (e.g., 50–1000 or more atm) piston pumps are recommended. These pumps are well suited in these ranges to pumping liquids of high viscosity that are flammable and of an explosive nature (steam pumps). In addition, they make excellent metering pumps.

Screw pumps are best suited for handling high-viscosity liquids, fuels and petroleum products. These pumps are used for capacities up to 300 m^3/hr and pressures up to 175 atm at speeds of rotation up to 3000/min. The advantages of screw pumps are their high speed, compactness and quiet operation. Pump capacity is practically independent of pressure, and efficiency is rather high (in the range of 0.75 to 0.80). The field of application of single-screw pumps is restricted by capacity up to 3.6–7 m^3/hr and pressures up to 10–25 atm. Their cost and maintenance are similar to those of centrifugal pumps of low capacity operating under pressures up to 3–5 atm. Screw pumps are considerably more economical when their delivery pressures exceed 10 atm. Single-screw pumps are employed in handling dirty and aggressive liquids, solutions and high-viscosity polymer solutions.

Sliding-vane pumps are used for the displacement of clean liquids (without solid particles) at moderate capacities and heads.

Gear pumps are best suited to pumping viscous liquids without solid particles at low delivery rates (not higher than 5–6 m^3/min) and high pressures up to 100–150 atm.

Jet pumps are typically employed in operations in which pumping requirements are intermittent, where an inexpensive standby unit is desirable or corrosion is important. They are used for the displacement of low-viscosity clean liquids at low delivery rates up to 40 m^3/hr and relatively high heads (up to 250 m). Efficiencies are typically low (η = 20–50%).

Acid-egg pumps and airlifts are used in industries in which moving and friction parts are highly undesirable.

Further discussions of various pump types, operating limitations and applications are given by Tetlow [1950], Taylor [1950], Karassik and Carter [1981], Karassik [1976], Hicks and Edwards [1971], and Cheremisinoff [1981].

PRACTICE PROBLEMS

Solutions are provided in Appendix A for those problems with an asterisk (*) next to their numbers.

*6.1 A centrifugal pump is being used to pump water from a reservoir to an elevated tank at a rate of 12 m^3/min. The system is shown in Figure 6.58. A manometer located on the discharge side of the pump reads 3.8 atm, while a vacuum gauge on the suction line shows 21 cm of Hg. The distance between the manometer and vacuum gauge is

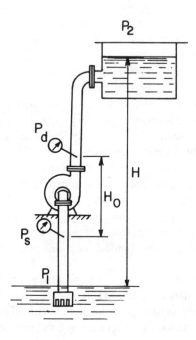

Figure 6.58   Pumping system for Problem 6.1.

410 mm. The diameter of the suction piping is 350 mm, while that of the discharge line is 300 mm. Determine the total head on the pump.

6.2 A horizontal pipeline carries oil (specific gravity (sg) of 0.87). Two pressure gauges along the pipe read 86.2 psig and 72.4 psig, respectively. Determine the energy loss between the two gauges.

6.3 A sump pump can deliver 1800 gph of water through a vertical lift of 17 ft. The pump's inlet is immediately below the water surface and the discharge is to the atmosphere through 1.5-in. Schedule 30 pipe. Determine the power delivered to the fluid by the pump.

6.4 The suction pressure at a pump inlet is 3 psi below atmospheric pressure. The discharge pressure at a point 4 m above the inlet is 55 psig. Both the inlet and discharge pipes are 3.5 in. Schedule 40. Calculate the power delivered by the pump for pumping water at a rate of 25 gpm.

6.5 A gear pump is sized for 0.93 hp to pump 11 gpm of oil (sg = 0.82) with a total head of 301 ft. Determine the mechanical efficiency of the pump.

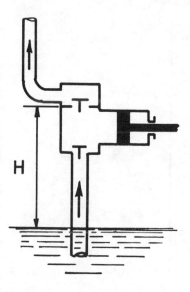

Figure 6.59   Piston pump scheme for Problem 6.6.

*6.6 A simplex piston pump (Figure 6.59) will operate at 150 rot/min to deliver water at a temperature of 60°C. Preliminary estimates show that total energy consumption for achieving the desired fluid velocity and overcoming inertia forces and hydraulic resistances in the suction line is 6.5 m H_2O. The hot water is being pumped to an elevated open vessel (p_a = 736 mm Hg). Determine the height over the water level supply tank at which the pump must be installed. The data in Tables 6.9–6.11 are needed.

*6.7 A double-action piston pump shown in Figure 6.60 delivers 22800 liter/hr of water at 65 rot/min. The piston diameter is 125 mm and the rod diameter is 35 mm. The crankshaft radius is 136 mm. Determine the pump delivery coefficient.

6.8 Repeat Problem 6.7 for delivery rates of 10,000 liter/hr and 35,000 liter/hr. Comment on the pump delivery coefficient.

*6.9 A single-acting piston pump (Figure 6.59) has a piston diameter of 130 mm and a stroke length of 200 mm. The pump is to deliver 430 liter/min of a liquid (sg = 0.93) from a tank open to the atmosphere into a vessel operating under a pressure of 3.2 atm gauge. The height over which the liquid is to be pumped is 19.5 m. Total head

Table 6.9 Atmospheric Pressure Data as a Function of Elevation

Height Above Sea Level (m)	−600	0	+100	200	300	400	500	600	700	800	900	1000	1500
P_a (m H_2O)	11.3	10.3	10.2	10.1	10.0	9.8	9.7	9.6	9.5	9.4	9.3	9.2	8.6

Table 6.10 Vapor Pressure Data for Water

Temperature (°C)	5	10	20	30	40	50	60	70	80	90	100
h_t (m H_2O)	0.09	0.12	0.24	0.43	0.75	1.25	2.02	3.17	4.82	7.14	10.33

Table 6.11 Recommended Suction Heights for Piston Pumps (m)

Number of Pump Rotations per Minute	Temperature of Water, °C						
	0	20	30	40	50	60	70
50	7	6.5	6.0	5.5	4.0	2.5	0
60	6.5	6.0	5.5	5.0	3.5	2.0	0
90	5.5	5.0	4.5	4.0	2.5	1.0	0
120	4.5	4.0	3.5	3.0	1.5	0.5	0
150	3.5	3.0	2.5	2.0	0.5	0	0
180	2.5	2.0	1.5	1.0	0.0	0	0

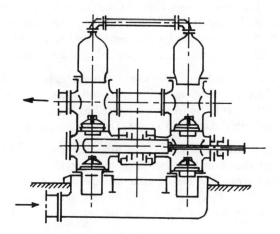

Figure 6.60   Double-acting piston pump for Problem 6.7.

losses in the suction and discharge lines are 1.7 m and 8.6 m, respectively. The pump delivery coefficient is 0.85, and the efficiencies of the pump and electric motor are 0.8 and 0.95, respectively. What is the number of rpm and the horsepower to be delivered by the pump?

6.10 A pump test on a centrifugal pump operating at 1200 rpm produced the data given in Table 6.12. The specific gravity of the liquid used in the tests was 1.12. Determine the efficiency of the pump at each capacity and prepare a plot of the pump characteristics.

*6.11 The same pump described in Problem 6.10 is to be used to deliver the liquid (sg = 1.12) at a rate of 115 m³/hr from a storage tank to a vessel

Table 6.12 Data from Centrifugal Pump Test in Problem 6.10

Q(ℓ/s)	0.0	10.8	21.2	29.8	40.4	51.1
H(m)	23.5	25.8	25.4	22.1	17.3	11.9
N(kW)	5.16	7.87	10.1	11.3	12.0	18.5

located at a height of 10.8 m. The pressure in the elevated vessel is 0.4 atm gauge and that in the storage tank is at atmospheric. The pipe diameter is 140 mm (wall thickness, 4.5 mm) and its design length (i.e., pipe length plus the equivalent length of the local resistance) is 140 m. The friction coefficient of the piping is 0.03 mm. Determine the necessary head of the pump. (Hint: the characteristic pump curve from Problem 6.10 must be determined first. This problem will require the use of the similarity pump equations to select a proper rpm.)

*6.12 Determine the delivery coefficient of a gear pump operating at 440 rpm. There are 12 teeth in the gear, and the tooth width is 42 mm. The tooth cross-sectional area is limited by the outer circumference of the neighboring gear to 960 mm^2. The pump capacity is 312 liter/min.

*6.13 Determine the theoretical vacuum that may be created by the working water jet shown in Chamber A of the jet pump in Figure 6.61. The discharge pressure is atmospheric and the jet velocity at that point is 2.7 m/s. The jet diameters in Sections I and II are 23 mm and 50 mm, respectively. Losses may be neglected.

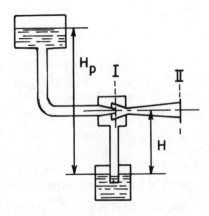

Figure 6.61   Jet pump for Problem 6.13.

*6.14 A water jet pump is used to lift a liquid (sg = 1.02) over a height of 4 m at a rate of 7.8 m^3/hr. The rate of the working fluid is 9.6 m^3/hr, and its head before the pump (H_p) is 22 m. Determine the efficiency of the pump.

*6.15 (a) For the airlift pump system illustrated in Figure 6.54, develop an expression for the minimum amount of compressed air needed for a unit mass of liquid from a consideration of the pressure distribution in the riser leg.

(b) If the liquid level in the feed tank lies on the same plane as the point of compressed air injection, will the pump work?

*6.16 Oil is to be pumped at a rate of 12,800 kg/hr through a horizontal pipeline that is 54 mm in diameter and 500 m long. Density and viscosity data at different temperatures are given in Table 6.13. As the oil is relatively viscous, it will have to be heated before pumping. Determine the temperature of the oil to be pumped by a unit having a rated power of 6.38 kW and an efficiency of 0.6. For a first approximation assume that there are no heat losses through the pipe walls and that the flow is isothermal.

6.17 Repeat Problem 6.16 for commercial-grade glycerine. The initial temperature of the glycerine is 50°F.

6.18 A volume of water (50 m^3) is to be pumped from an open tank to a vessel operating at a pressure of 1.5 atm and located at a height of 15 m. The inside diameter of the piping is 58 mm and the pipe length is 120 m. The equivalent height corresponding to the resistances is 350 pipe diameters.

(a) Determine the time required to deliver the water if a centrifugal pump with the characteristics given in Table 6.14 is used. Assume that the pump operates at constant speed.

Table 6.13 Viscosity and Density Data for Problem 6.16

Temperature, t (°C)	Viscosity, μ (N × s/m^2)	Density, ρ (kg/m^3)
20	0.140	850
30	0.100	840
40	0.075	830
50	0.055	830
60	0.042	820
70	0.032	820
80	0.025	810

Table 6.14 Characteristics of Centrifugal Pump in Problem 6.18

$V \times 10^3$ (m^3/s)	1	2	3	4	5	6
Head, H(m)	36.0	34.6	32.2	28.8	24.0	15.6

(b) Determine the pump's horsepower if its efficiency is 0.55. The friction coefficient, λ, is 0.03.

*6.19 Determine the maximum height relative to a feed tank at which a pump may be installed for delivering water at a rate of 2 kg/s at t = 60°C. The inside diameter of the suction piping is 38 mm and its length is 15 m. The sum of coefficients of local resistances is 3.5 m and the relative roughness, e/d = 0.003.

*6.20 Water at 36°C is delivered at a rate of 2.25 m^3/min through piping of 38 mm in diameter. The piping consists of a 150-m-long horizontal section and a 10-m-long vertical section. The total equivalent length of all valves in the system is 200 pipe diameters. The equivalent length of all bends and fittings is 60 pipe diameters. The system also has a heat exchanger with a pressure loss of 15000 N/m^2. Determine the pump horsepower if its efficiency is η = 0.6. The relative roughness of the piping is e/d = 0.005, and the viscosity of the water is 0.65 cp.

*6.21 Determine the maximum suction height for pumping water at t = 50°C through piping (25 mm in diameter and 10 m in length). The water capacity is 1.75 kg/s. The pipe has three 90° bends, and roughness is assumed to be equal to e = 0.01 mm.

6.22 Water is delivered from a tank to a condenser located at a height of 11 m. The piping diameter is 80 mm and its length is 200 m. The equivalent length of the local resistances corresponds to 100 pipe diameters. The condenser resistance coefficient is ξ = 10, and the friction factor is λ = 0.025. Determine the efficiency of the pump and water capacity if the horsepower of the pump is 1.8 kW. The pump characteristics for constant speed are given in Table 6.15.

6.23 A pump delivers a 30% solution of H_2SO_4. The pressure in the discharge line is 1.8 kg$_f$/cm^2. A gauge reading in the suction line is 29 mm Hg. The distance between the devices is 0.5 m. The suction and discharge pipes have the same diameter. What is the pump head?

6.24 A pump delivers liquid (ρ = 960 kg/m^3) from an open tank to an elevated vessel under pressure, p = 37 atm. The height of the raised tank

Table 6.15 Pump Characteristics for Problem 6.22

Capacity (m^3/hr)	Head (m)
10	25
14	23
18	20.3
20	16.4
21	11.8

is 16 m. The total resistance in the discharge line is 65.6 m. Determine the head of the pump.

6.25 Determine the efficiency of a pump whose capacity is 380 liter/min. The specific gravity of the liquid is 0.9 and the head is 30.8 m. The pump's power is 25 kW.

6.26 The capacity of a pump is 14 liter/s for a liquid whose specific gravity is 1.16. The total head on the pump is 58 m. The pump efficiency is 0.64; the efficiency of transmission is 0.97; and the electric motor efficiency is 0.95. Determine the size of motor needed for this operation.

*6.27 A piston pump is installed 200 m above sea level. The loss of suction height is 5.5 m H_2O. The geometric suction height is 3.6 m. At what water temperature is maximum suction possible?

6.28 (a) Determine the capacity of a differential piston pump. The diameters of the larger and smaller plungers are 340 mm and 240 mm, respectively. The plunger stroke is 480 mm, and the unit operates at 60 rpm. The delivery coefficient is 0.85.
(b) Determine the amount of liquid delivered by each side of the plunger.

6.29 A double-action piston-pump fills a tank (tank diameter = 3 m, tank height = 2.6 m) in 26.5 min. The plunger diameter is 180 mm, the diameter of the rod is 50 mm, the crankshaft radius is 145 mm and the unit operates at 50 rpm. Determine the pump delivery coefficient.

6.30 A centrifugal pump (1800 rpm) delivers 140 m^3/hr H_2O at a temperature of 30°C. The atmospheric pressure is 745 mm Hg. The head loss in the suction line is 4.2 m. Determine the theoretically permissible suction height.

6.31 A centrifugal pump delivers 280 liter/min H_2O with a head of 18 m.
(a) Determine whether this pump may be used for delivering 15 m^3/hr of liquid (sg = 1.06) through piping (70 mm i.d., 2.5 mm wall thickness) from an open tank to a vessel operating under a pressure of 0.3 atm gauge. The lift is 8.5 m. The design length of the piping is 124 m. The friction factor is $\lambda = 0.03$.
(b) Determine the motor horsepower if the pump efficiency is 0.55.

*6.32 A centrifugal pump for water delivery has the following specifications: Q = 56 m^3/hr, H = 42 m, N = 10.9 kW, rpm = 1140. Determine:
(a) pump efficiency,
(b) pump capacity,
(c) pump head, and
(d) pump horsepower at rpm = 1450, assuming the same efficiency.

6.33 A pump test resulted in the following information:

Q(ℓ/min):	0	100	200	300	400	500
H (m):		37.2	38.0	37.0	34.5	31.8	28.5

(a) What will be the pump capacity through 76-mm-diameter piping with 4-mm wall thickness and a design length of 355 m at a geometric height of 4.8 m? The friction factor is $\lambda = 0.03$, $\Delta p_{ad} = 0$.
(b) Plot the characteristics of the pump and pump-system, and find the working point.
(c) What will be the pump capacity if the height is 19 m?

6.34 Determine the capacity of a gear pump having the following characteristics: rpm is 650; number of teeth in the gear is 12; width of a tooth is 30 mm; area of tooth section is limited by the external circle of the neighboring gear to 7.85 cm^2; and the delivery coefficient is 0.7.

6.35 An injector pump is used to transfer 215 liter/min of a solution (sg = 1.06) out from a tank located in a basement. The lift is 3.8 m. The water pressure before the pump is 1.9 atm gauge, and the pump efficiency is $\eta = 0.15$. How much water will be consumed by the pump per hour?

*6.36 What is the pressure of air delivered to a montejus for lifting H_2SO_4 (sg = 1.78) to a height of 21.0 m? Neglect the hydraulic losses.

NOTATION

A	energy from centrifugal force, J
atm	atmospheric head or pressure, ft or atm
b	impeller width, m

C	centrifugal force, N, see Equation 6.26
C_1, C_2	vector sums of relative and tangential velocities, m/s
f, F	area, m^2
G	weight of fluid particle, kg
g	acceleration due to gravity, m/s^2
$\tilde{H}$	head, m
H_c	cavitation correction number, refer to Equation 6.24
H_d	discharge head, ft or m
H_{fd}	discharge friction head, ft or m
H_{ss}	static suction head, ft or m
H_{ts}	total static head, ft or m
h_ϱ	head loss, m
h_{vs}	velocity head, ft or m
K	proportionality coefficient in Equation 6.47
ℓ	height of liquid column, ft or m
m, M	mass, kg or lb
N	power, kW
N_e	effective power, kW
n	number of revolutions
n_s	specific number of revolutions
p	pressure, N/m^2
Q	flow capacity, m^3/s
r	crank radius, m
S	stroke length, m
t	temperature, °C
u	circumferential crank velocity, m/s
V	volume of liquid pumped, m^3
W	mass flow, lb/hr
w	average fluid velocity, m/s
z	height, m
Z'	number of vanes

Greek Symbols

α	angle, °
β	motor efficiency coefficient, see Equation 6.8
β_2	angle defined in Figure 6.15, °
γ	specific weight, kg_f/m^3
δ	vane thickness, mm
ϵ	coefficient accounting for number of vanes, see Equation 6.42
ξ	kinetic energy correction term
η_h	hydraulic efficiency
η_m	mechanical efficiency

η_p pump efficiency
η_{tr} transmission efficiency
η_v volumetric efficiency
μ viscosity, poise
ν specific volume, m^3/kg
ρ density, kg/m^3
σ_c cavitation number, refer to Equation 6.23
ω angular velocity, m/s

REFERENCES

Bergelin, O. P. (1949) "Flow of Gas-Liquid Mixtures," *Chem. Eng.* 56(5): 104.

Bonnington, S. T., and A. L. King (1972) "Jet Pumps and Ejectors," British Hydraulic Research Association: Fluid Engineering, Cranfield, England.

Brown, G. G. (1950) *Unit Operations* (New York: John Wiley & Sons, Inc.).

Cheremisinoff, N. P. (1981) *Fluid Flow: Pumps, Pipes and Channels* (Ann Arbor, MI: Ann Arbor Science Publishers).

Church, A. H. (1944) *Centrifugal Pumps and Blowers* (New York: John Wiley & Sons, Inc.).

Coulson, J. M., and J. F. Richardson (1962) *Chemical Engineering* (Elmsford, NY: Pergamon Press, Inc.).

Datta, R. L. (1948) "Studies for the Design of Gas Lift Pumps," *J. Imp. Coll. Chem. Eng. Soc.* 4(157).

Daugherty, R. L. (1915) *Centrifugal Pumps* (New York: McGraw-Hill Book Co.).

Foust, A. S., et al. (1980) *Principles of Unit Operations,* 2nd ed. (New York: John Wiley & Sons, Inc.).

Hicks, T. G. (1957) *Pump Selection and Application* (New York: McGraw-Hill Book Co.).

Hicks, T. G., and T. W. Edwards (1971) *Pump Application Engineering* (New York: McGraw-Hill Book Co.).

Karassik, I. J., and R. Carter (1981) *Centrifugal Pump Design and Selection,* R. P. Worthington Corp., Harrison, NJ.

Karassik, I. J., W. C. Krutzsch, W. H. Fraser and J. P. Messina (1976) *Pump Handbook* (New York: John Wiley & Sons, Inc.).

Kirk, R. E., and D. Othmer (1963) *Encyclopedia of Chemical Technology,* 2nd ed. (New York: John Wiley & Sons, Inc.).

Metzner, A. B. (1961) *Handbook of Fluid Dynamics* (New York: McGraw-Hill Book Co.).

Perry, R. H., and C. H. Chilton, Eds. (1973) *Chemical Engineers' Handbook,* 5th ed. (New York: McGraw-Hill Book Co.).

Reavell, E. A. (1936) "Some Aspects of Chemical Works Pumping in Acid Handling," *Proc. Chem. Eng.* 18:25.

Simonin, R. F. (1951) "Working of an Air Lift Water Pump," *Comp. Rend. Acad. Sci.* 233:465.

Stepanoff, A. J. (1965) *Pumps and Blowers* (New York: John Wiley & Sons, Inc.).

Swindin, N. (1924) *The Modern Theory and Practice of Pumping* (London: Ernest Benn Ltd.).

Taylor, I. (1950) "The Most Persistant Pumping Problems for the Chemical Plant Designer," *Chem. Eng. Prog.* 46 (637).

Tetlow, N. (1950) "Survey of Present-Day Pumping Practice in the Chemical Industry," *Trans. Inst. Chem. Eng.* 28(63).

Vetter, G., and L. Hering (1980) "Leakfree Pumps for the Chemical Process Industries," *Chem. Eng.*

CHAPTER 7

GAS COMPRESSION AND TRANSPORTATION

CONTENTS

INTRODUCTION

Numerous industrial operations are performed in the gaseous phase under pressure, which often leads to an increase in flowrates and a decrease in the volume of equipment needed for processing. Compression is used to displace gas through pipelines and process equipment, as well as to create vacuum conditions. Compressed gases are used in a variety of processes, such as mixing and atomization of liquids. Pressures utilized throughout the chemical process industries (CPI) vary over a tremendous range, typically from 10^{-3} to 10^8 N/m^2 (10^{-8} – 10^3 atm). Machines designed for the displacement

411

and compression of gases are called *compression machines,* or compressors. They increase the pressure of a gas, vapor, or their mixture, by reducing the fluid's specific volume either prior to or after its passage through various process equipment.

Compressors and fans are essentially pumps for gases. Although they differ in construction from liquid handling machines, the principles of operation are identical. Under normal operating pressures, the density of a gas is considerably less than that of a liquid, so that higher speeds of operation can be employed and lighter valves can be fitted onto delivery and suction lines. Because gases have very low viscosities there is a greater tendency for leakage. Hence, gas compressors are designed with smaller clearances between moving parts. Further differences in construction from pumps are necessitated by the decrease in the volume of a gas as it is compressed. As a large portion of the energy of compression is dissipated in the form of heat to a gas, the operation is accompanied by a considerable increase in temperature. This can limit the operation of the compressor unless suitable cooling can be effected. For this reason, gas compression often is done in stages, where cooling is performed between each stage. Gas not expelled from the cylinder at the end of a compression cycle, i.e., gas in the clearance volume, must be expanded again to the inlet pressure before a fresh charge can be admitted. Because continual compression and expansion of the residual gas are performed, efficiency is reduced because neither the compression nor the expansion can be carried out in a completely reversible manner. In the case of liquids, this factor has no effect on efficiency because the residual fluid does not undergo compression.

CLASSIFICATION OF COMPRESSION MACHINES

The ratio of the final pressure, P_2, established by a compression machine to the initial suction gas pressure, P_1, is called the *compression ratio.*

Depending on the value of the compression ratio, gas displacement machines usually are classified as *fans,* $P_2/P_1 < 1.1$ for displacement of large amounts of gas; *blowers,* $1.1 < P_2/P_1 < 3.0$, for moving gas at relatively high resistances in pipelines; *compressors,* $P_2/P_1 > 3.0$, for establishing high pressures; or *vacuum pumps,* for compressing gas that is below atmospheric pressure so that it can be discharged to the atmosphere.

The types of compression machines most often employed in process plant applications fall into the following classes: centrifugal, axial, reciprocating and rotary, depending on how the mechanical elements act on the gas. There are two categories of compression machines under which these classes are grouped: dynamic and positive displacement.

Dynamic machines use rotating vanes or impellers to impart velocity and pressure to the gas. They include centrifugal and axial designs. These machines operate by developing a high gas velocity and converting this velocity into pressure in the diffusing flow passage. In general, they tend to have a lower efficiency than positive displacement machines. However, they do operate at relatively high speeds to provide high flowrate in relation to the physical size of the machine. Also, dynamic machines usually have significantly lower maintenance requirements than positive displacement machines.

Positive displacement compressors consist of two types—reciprocating and rotary. Both confine successive volumes of gas within a closed space in which the pressure of gas is increased as the volume of the closed space decreases.

Reciprocating-type compressors have one or more cylinders, each fitted with a piston driven by a crankshaft through a connecting rod. Each cylinder is equipped with suction and delivery valves and means for cooling mechanical parts. Figure 7.1 illustrates both water-cooled and air-cooled compressor cylinders. Gas is introduced into the cylinder during the suction stroke. At the end of the stroke, the piston's motion is reversed, and gas is compressed and expelled during the delivery stroke. When only one end of the piston acts on the gas, the machine is referred to as a *single-acting compressor*. Machines in which compression is effected at both ends of the piston are called *double-acting machines*. They deliver about twice as much gas per cylinder per cycle as the single-acting machines. Figure 7.2 illustrates various frame arrangements for positive displacement piston compressors. *Single-stage compressors* compress the gas in each cylinder from the initial intake pressure to the final delivery pressure on each working stroke of the piston.

In two-stage compressors, the gas is compressed to an intermediate pressure in one cylinder while another cylinder is used to raise the pressure to the final delivery pressure. Machines utilizing two or more stages are called *multistage compressors.*

Vertical and horizontal compressors may be single-cylinder or multicylinder designs. Angle types are multicylinder, with one or more horizontal and vertical compressing elements. Single-frame (straight-line) machines are horizontal or vertical double-acting compressors with one or more cylinders in line with a single-frame having one crank throw and one connecting rod and cross head. The V- or Y-type compressor comprises a two-cylinder, vertical, double-acting machine with cylinders usually at a 45° angle with the vertical. These designs employ a single crank. Semiradial compressors are similar to the V or Y type, but have horizontal double-acting cylinders on each side. Duplex compressors have cylinders on two parallel frames attached by a common crankshaft. Duplex-tandem, steam-driven units employ steam cylinders in line with gas cylinders. Duplex four-cornered steam-driven compressors have one or more compressing cylinders on each end of the frame and one or more steam cylinders on the opposite end. Four-cornered motor-driven units have the motor on a shaft between compressor frames.

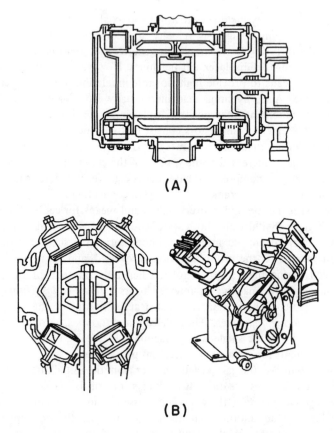

Figure 7.1   (A) Water-cooled compressor cylinder; (B) air-cooled compressor cylinder.

Reciprocating compressors are capable of gas capacities as high as 100,000 cfm, at pressures in excess of 35,000 psi. Special units with higher capacities or pressures can be custom made. Water generally is used as the coolant for cylinders, intercoolers and aftercoolers; however, other liquids, including refrigerants, are used.

The single-stage, horizontal, single-acting compressor consists simply of a cylinder with a reciprocating piston. Conceptual operation is illustrated in Figure 7.3(A). The cylinder is covered from one side with a head in which the suction valve (3) and the delivery valve (4) are located. The piston is connected directly to the rod (5) and crank (6) with a shaft having a flywheel (7). The piston movement from left to right effects gas discharging in the space between the head and piston. Because of the difference in pressure

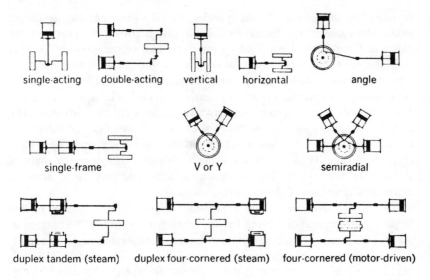

single-acting   double-acting   vertical   horizontal   angle

single-frame   V or Y   semiradial

duplex tandem (steam)   duplex four-cornered (steam)   four-cornered (motor-driven)

Figure 7.2   Various frame arrangements of positive displacement piston compressors.

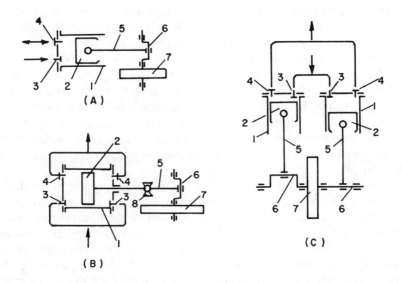

Figure 7.3   Diagrams of single-stage piston compressors: (A) one cylinder single acting; (B) one cylinder double acting; (C) two cylinder single acting. (1) cylinder; (2) piston; (3) suction valve; (4) delivery valve; (5) connecting rod; (6) crank; (7) flywheel; (8) cross head.

between the suction line and the cylinder, the valve (3) opens and, hence, gas enters the cylinder. On the reverse of the piston stroke, the suction valve closes and the gas in the cylinder compresses to a certain pressure. At that time, the valve (4) opens and the gas is delivered into the discharge piping.

In single-stage, double-acting compressors (Figure 7.3(B)), the gas is compressed in a cylinder (1) intermittently from both sides of the piston (2). For one double-stroke piston there are two suctions and two deliveries. The cylinder is fitted with two suction valves (3) and two delivery valves (4). Compressors employing double-acting pistons are more complicated; however, their capacity is two times higher than that of single-acting units of similar size and weight. This increased capacity also may be achieved in *multicylinder* single- and double-acting compressors.

The two-cylinder single-acting compressor (Figure 7.3(C)) consists of two single-acting compressors driven by one crankshaft, with cranks displaced at 90° or 180° from each other. Cylinders, and sometimes the compressor head, are equipped with waterjackets to remove the heat of compression. Although heat removal is not complete, it does decrease the energy expenses for compression significantly.

Vertical single-stage compressors have a number of advantages over horizontal designs. They are generally higher-speed units (for horizontal compressors n_s = 100–240 rpm; for vertical compressors n_s = 300–500 rpm or greater); consequently, their capacity is higher and they require less space. Cylinders and pistons undergo less wear. With horizontal cylinders, especially those of large diameters, local wear on the piston occurs due to the action of gravity force, which results in decreased piston velocity.

Multistage compression is employed when the required pressure ratio, P_2/P_1, is large and it is not practical to perform the entire compression in a single cylinder because of the high temperatures and the adverse effect of clearance volume on the efficiency. Mechanical construction also is difficult because single cylinders would require sufficient strength to withstand the final pressure but be of a large enough size to hold the gas at the initial pressure, P_1.

Staged compression is performed in *multistage* compressors in which the gas passes in series through a number of cylinders of gradually decreasing volume. Hence, the gas is compressed gradually to the final pressure. Multistage machines are provided with intercoolers between stages. Intercoolers are basically heat exchangers that remove the heat of compression from the gas and reduce the temperature to approximately the temperature at the intake. Cooling reduces the volume of gas entering the high-pressure cylinders, reduces the horsepower required for compression and, at high pressures, maintains the temperature within safe operating limits.

Stages typically are arranged in different cylinders, as shown in Figure 7.4(A,B,C), and in a single cylinder with a differential piston

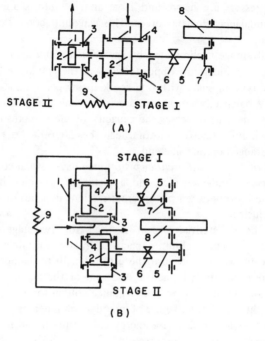

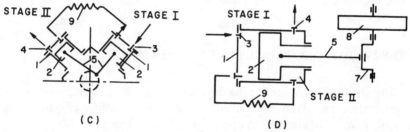

Figure 7.4   Schemes of positive displacement piston compressors with compression stages in separate cylinders: (A) single; (B) duplex; (C) V-type; (D) differential. (1) cylinder; (2) piston; (3) suction valve; (4) delivery valve; (5) connecting rod; (6) cross head; (7) crank; (8) flywheel; (9) intercooler.

(Figure 7.4(D)). V-type compressor cylinders are arranged at some angle (usually 45°) from the vertical.

Multistage compressors with stages arranged separately in several cylinders may have more than one cylinder in line (Figure 7.4(A)). They may also have duplex reciprocating compressors on one end of a frame (Figure 7.4(B)).

Such compressors are heavy and large and, therefore, are subject to considerable unbalanced inertia forces. This can limit compressor operation at high speeds.

This disadvantage is eliminated by *horizontally opposed reciprocating compressors* with cylinders on opposite sides of the crankcase. These employ a multithrow type of shaft with one crank throw per cylinder. Such units are dynamically balanced, which increases the crankshaft rotation speed 2–2.5 times. This, in turn, increases the capacity of the compressor. The system's weight, including electric motors, can be 50–60% less than that with cylinders arranged on one side end.

Compressors with a differential piston may have several compression stages formed by a cylinder surface and a piston of variable (differential) cross section. The ratio between piston cross sections depends on the compression ratio in each stage.

Usually the differential piston is used for two-stage compression in machines of low and medium capacities. In large machines, the differential piston may be wedged because the piston length to diameter ratio is small. The compression ratio (defined as the sum of the piston displacement and clearance volume divided by the clearance volume) in each stage is selected to provide the most effective use of cylinder volume, increase the compressor volumetric coefficient, decrease energy consumption and to establish the gas temperature at the end of compression.

THERMODYNAMIC PRINCIPLES OF COMPRESSION

The compression of real gases involves changes in pressure, volume and temperature only. Pressure-volume-temperature interrelations at pressures up to 10^6 N/m^2(~10 atm) can be expressed by the ideal gas law:

$$P\hat{V} = RT \tag{7.1}$$

where $\hat{V}$ is the molar volume of gas.

For pressures exceeding 10^6 N/m^2(p > 10 atm) van der Waals formula more accurately describes the pressure-volume-temperature relation:

$$\left(P + \frac{a}{v^2}\right)(v - b) = RT \tag{7.2}$$

where $\qquad$ P = gas pressure (N/m^2)
$\qquad\qquad\qquad$ v = specific volume of gas (m^3/kg)
$\qquad\qquad\qquad$ R = 8314/M = gas constant (J/kg-$^\circ$K)
$\qquad\qquad\qquad$ M = molecular weight (kg/kg-mole)
$\qquad\qquad\qquad$ T = temperature ($^\circ$K)
$\qquad$ "a" and "b" = empirical constants for a given gas

In the absence of data to establish "a" and "b," estimates may be made from the critical parameters of the gas, i.e., critical temperature, T_{cr}, and critical pressure, P_{cr}:

$$a = \frac{27R^2 T_{cr}^2}{64 P_{cr}} ; \quad b = \frac{RT_{cr}}{8P_{cr}} \tag{7.3}$$

The work expended for compression may be determined from Bernoulli's equation:

$$\tilde{\ell} + q = (i_2 - i_1) + \frac{w_2^2 - w_1^2}{2} \tag{7.4}$$

where $\qquad$ $\tilde{\ell}$ = work spent for compressing 1 kg gas (in a compressor)
$\qquad\qquad\qquad$ q = heat introduced per 1 kg gas
$\qquad$ i_2 and i_1 = enthalpies of gas before and after compression, respectively

Differences in gas velocities before and after compression usually are neglected, i.e., it is assumed that $w_1 = w_2$. Then Equation 7.4 takes the form

$$\tilde{\ell} + q = i_2 - i_1 \tag{7.5}$$

The work expended in the compressor, $\tilde{\ell}$, and the heat added, q, produce an increase in the enthalpy of the gas. The compression process may be either adiabatic or isothermal.

In adiabatic compression there is no input or output of heat from the compressor. That is, q = 0, and Equation 7.5 simplifies to

$$\tilde{\ell} = i_2 - i_1 \tag{7.6}$$

Work is converted into heat, thus raising the temperature of the gas. The net result is an increase in enthalpy. Hence, for adiabatic compression we can expect gas temperature to increase to high values.

Under isothermal conditions the gas is compressed at constant temperature. The internal energy of the gas remains unchanged, $U_1 = U_2$, and the following relationship applies, $P_1V_1 = P_2V_2$. Consequently, $U_1 + P_1V_1 = U_2 + P_2V_2$, or $i_1 = i_2$, i.e., the gas enthalpies at compression do not change. Equation 7.5 becomes the following for isothermal compression:

$$\tilde{\ell} + q = 0 \quad \text{or} \quad \tilde{\ell} = -q \tag{7.7}$$

The negative sign on q denotes the output of heat. Hence, all the work expended in compression is converted into heat and evacuated from the gas. Temperature, internal energy and the enthalpy of the gas therefore will not change. This means that the compressor must be cooled at isothermal conditions to evacuate an amount of heat equivalent to the work spent.

Expressing work in units of $kg_f - m$ and enthalpy in kcal/kg, Equation 7.5 becomes

$$A\tilde{\ell} + q = i_2 - i_1 \tag{7.8}$$

where $A = 1/427$, thermal equivalent of work (kcal/kg_f-m)

Correspondingly, for adiabatic compression Equation 7.6 is rewritten as

$$\tilde{\ell} = \frac{i_2 - i_1}{A} \tag{7.9}$$

And for isothermal compression, Equation 7.7 becomes

$$A\tilde{\ell} = -q \tag{7.10}$$

Compression may be illustrated graphically by a temperature-entropy diagram (T-S plot) (Figure 7.5). Entropy, S, is a thermodynamic function of state of a defined body or substance. It is increased on the addition of heat to the body. The higher the body's temperature, the smaller the degree of entropy increase.

For a reversible process, the increase in entropy due to the addition of heat, q(J/kg), is

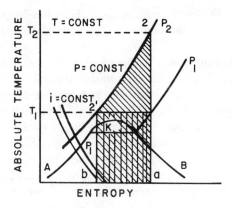

Figure 7.5   T-S diagram for gas compression.

$$\Delta S = \int_{T_1}^{T_2} \frac{\Delta q}{T} , \quad J/kg\text{-}°K \qquad (7.11)$$

Examining the T-S diagram in Figure 7.5, note that there are two types of lines—those corresponding to constant pressure (called isobars) and lines corresponding to constant temperatures (isotherms). The line AKB is a boundary curve. The region below the curve corresponds to wet vapor, and the branch AK corresponds to dry saturated vapor. Point K is a critical point. The region to the left of branch AK is the liquid state, and the region to the right of KB is superheated vapor. Because the processes of vaporization and condensation occur at constant pressures and temperatures, the isobars coincide with isotherms in the wet vapor region of the diagram. The condensation of a mixture of wet vapors occurs at variable temperatures and, hence, isobars do not coincide with isotherms in the wet vapor region.

Also shown in the T-S diagram are lines of constant enthalpies (i = constant). The enthalpies of ideal gases only depend on temperature, and for such gases the lines of constant enthalpy coincide with the isotherms. The enthalpy of a real gas is also a function of pressure and, hence, lines of constant i do not coincide with the isotherms.

Gas compression of a gas can be explained by Figure 7.5 as follows: For an adiabatic compression, q = 0, and from Equation 7.11 ΔS = 0, i.e., the process continues without a change in entropy. Consequently, adiabatic compression can be represented by the vertical line 1-2, where point 1 characterizes the state of the gas before compression. Point 1 is located at the intersection of isobar P_1 with isotherm T_1. Point 2 represents the condition of the gas after compression and is located on the isobar corresponding to pressure P_2.

An isothermal compression (temperature is constant) is represented by line 1-2 in Figure 7.5. Point $2'$ characterizes the state of the gas after compression and is located on isobar P_2. The amount of heat evacuated, q, (according to Equation 7.11), is $T\Delta S$, which is denoted by the cross-hatched rectangular area of rectangle a-1-$2'$-b. The height of this rectangular area is T_1, and the base represents the change in entropy, ΔS. We note that entropy decreases in this case, i.e., ΔS is negative. Therefore, the amount of heat also will be negative, that is, the process will be accompanied by an evacuation of heat. The same area a-1-$2'$-b expresses the work of isothermal compression in thermal units, whereas the area a-2-$2'$-b represents the work of adiabatic compression. Although adiabatic and isothermal compressions are ideal, in practice they may be approached fairly closely.

When real gases are compressed, their volumes, pressures and temperatures change. In addition, a portion of the heat is dissipated to the atmosphere. This is known as *polytropic* compression. The process of polytropic gas compression from pressure P_1 to P_2 is illustrated by the lines AC on the T-S diagram in Figure 7.6. The amount of heat generated in the compression of 1 kg of gas is equivalent to the specific work of polytropic compression, $\tilde{\ell}_{pol}$, and may be determined approximately from the diagram or analytically through the following relationship:

$$q_{pol} = \tilde{\ell}_{pol} \approx (S_A - S_C)\frac{T_A + T_C}{2} + C_p(T_C - T_A) \qquad (7.12)$$

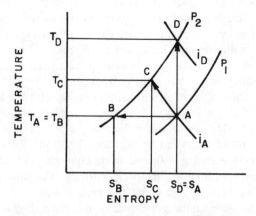

Figure 7.6   T-S diagram illustrating gas compression.

Let us examine a single-stage compression cycle in detail by considering the pressure-volume diagram shown in Figure 7.7. Gas enters at point A with pressure, P_1, and volume, V_1, and is compressed along the line AB. The delivery valve opens at point B when pressure P_2 is reached. At point C, the delivery valve closes and the piston begins its return stroke, allowing the pressure to decrease to P_1 along the line CM, opening the suction valve. Suction occurs along the line MA. The total work of compression from a pressure P_1 to a pressure P_2 may be defined as the product of the pressure difference and volume (represented by the shaded area ABCM). To evaluate the work performed along the compression line AB, let $d\tilde{\ell} = VdP$.

In an actual compression or expansion of a perfect gas, Boyle's law (PV = K, where K is a constant) is applicable, provided the process takes place at constant temperature. If a temperature change occurs, an exponent, n, is included on the volume:

$$P_1V_1^n = P_2V_2^n = K = (constant) \tag{7.13}$$

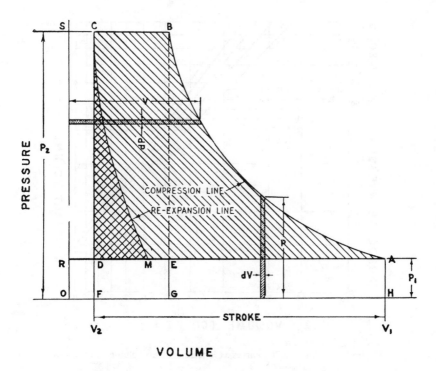

Figure 7.7   A single-stage compression cycle on a P-V diagram.

The subscripts refer to different conditions of the same weight of gas. In practice, most compression and expansion curves follow Equation 7.13, with the exponential term ranging between 1.0 and 1.5. The specific value of n depends on a number of factors, such as the peculiarities of the gas compressed, the specific heats of the gas, degree of cooling, the operating characteristics of the compressor cylinder, and/or the amount of ring leakage. Figure 7.8 shows different compression curves plotted on logarithmic coordinates, producing straight lines with slopes equal to the tangents of angles A,B,C, etc.

The isothermal case has a slope of 1.0 and is readily recognized as an equilateral hyperbola. The slopes of the other curves are usually greater than 1.0. When exponent n = 1.0, the pressure-volume change has taken place without any change in temperature. If n is greater than 1, the expansion is polytropic. In an adiabatic expansion or compression, the value of n is the ratio of the specific heats of the gas compressed, $n = \kappa = C_p/C_v$, where C_p is the specific heat at constant pressure, and C_v is the specific heat at constant volume.

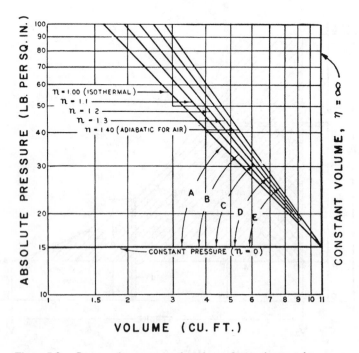

Figure 7.8   Compression curves as functions of gas volume and pressure.

From Equation 7.13, $P = KV^n$. By raising both sides of this equation to the $1/n$ power, we obtain

$$P_1^{1/n}V_1 = K^{1/n}$$

or

$$V = \left(\frac{K}{P}\right)^{1/n}$$

Substituting,

$$\tilde{\ell} = K^{1/n} \int_1^2 \frac{dP}{P} \tag{7.14}$$

Integration of Equation 7.14 produces two solutions, depending on whether the value of n is equal to unity. For isothermal compression (when $n = 1.0$), Equation 7.14 becomes

$$\tilde{\ell} = -K \, ln\left(\frac{P_2}{P_1}\right) = P_1V_1 \, ln\left(\frac{P_2}{P_1}\right) \tag{7.15}$$

This expression determines the work from zero volume to the line AB, including the area ABSR in Figure 7.7. Re-expansion takes place on the other side of the cylinder along the line CM, returning to the cycle the work (area CMRS) of opposite sign from the total area ABSR. Hence, the net work is the difference in areas, ABCM.

Isothermal compression is seldom achieved in practice; however, the ideal condition is instructive when comparing the performances of different compressors. For the second case, where $n \neq 1$, i.e., polytropic performance, Equation 7.14 again is integrated:

$$\tilde{\ell} = \frac{n}{n-1} K^{1/n} [P_2^{(n-1)/n} - P_1^{(n-1)/n}]$$

$$= -\frac{n}{n-1} P_1^{1/n}V_1 [P_2^{(n-1)/n} - P_1^{(n-1)/n}] \tag{7.16}$$

Multiplying the numerator and denominator on the right-hand side by $P_1^{(n-1)/n}$, we obtain

$$\tilde{\ell} = \frac{n}{n-1} P_1 V_1 \left[\left(\frac{P_2}{P_1} \right)^{(n-1)/n} - 1 \right] \tag{7.17}$$

Gill [1941] illustrates the comparative power requirements of a compressor for different values of n. In Gill's analysis, 10 ft³ of a gas (air) is compressed from 15 to 100 psia. The predictions of Equation 7.17 are shown plotted in Figure 7.9. In this plot, the work of isothermal compression is represented by the area ABGHA. The adiabatic value for air occurs at n = 1.4 and, hence, the work area is AFGHA. Comparing these two cases, then, a net savings in work denoted by area AFBA is achieved by isothermal compression over adiabatic compression. In reality, few compressors ever operate adiabatically due to radiation losses, condensation, etc. At the same time, it is impossible to maintain constant temperature and, hence, isothermal compression is also rare.

Figure 7.9 shows that the power coefficient becomes smaller as the compression cycle approaches the isothermal case. To approximate the isothermal

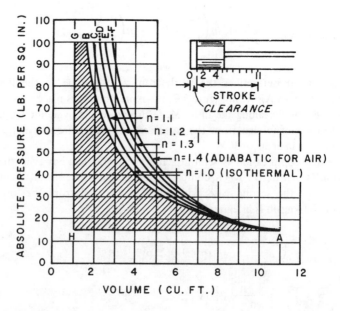

Figure 7.9   Comparative power requirements at different values of n, as presented by Gill [1941].

state as closely as possible, compressor cylinders are waterjacketed and inter-coolers are provided between stages in multistage compression.

The capacity of a piston compressor is defined as the actual amount of gas volume, V, handled by the machine per unit time at standard conditions. The theoretical capacity can be computed from Equations 6.50 and 6.51.

The actual capacity of a gas compressor is defined as the quantity of gas compressed and delivered, expressed in volume units per unit time at condi-tions of total temperature, total pressure and composition prevailing at the compressor inlet.

Capacity always is expressed in terms of the gas intake conditions, rather than in terms of STP.

$$V = \lambda_v Q_T = \lambda_v V_p \qquad (7.18)$$

where λ_v = delivery coefficient

For multicylinder compressors, the capacity computed from Equation 7.18 should be multiplied by the number of cylinders. The capacity of multistage compressors is determined by the capacity of the first stage. In this case, V_p in Equation 7.18 is the piston displacement volume of the first stage.

The delivery coefficient, λ_v, is the ratio of gas volume discharged in the delivery piping (but reduced to the suction conditions) to the displacement, V_p. The delivery coefficient accounts for all capacity losses, including those not shown on the indicator diagram. The losses detected on the diagram are those related to the gas expansion from the clearance and are denoted by coefficient λ_0. Additional losses are due to the loss of capacity as a result of gas leakage through piston rings, valves, stuffing boxes and gas expansion, when gas comes in contact with hot cylinder walls and mixes with hot gases in the clearance spaces. These losses are considered by a hermetic coefficient, λ_h, and a thermal coefficient, λ_T, correspondingly. Therefore, the delivery coefficient may be defined as a product of three coefficients:

$$\lambda_v = \lambda_0 \lambda_h \lambda_T \qquad (7.19)$$

For modern compressors, the values of these coefficients are $\lambda_h = 0.95 - 0.98$ and $\lambda_T = 0.9 - 0.98$. As noted earlier, the volume of gas taken in by the compressor, V_s, is less than the working volume of the cylinder, V_p. The ratio of the compressor capacity to its displacement is called the volumetric coefficient:

$$\lambda_0 = \frac{V_s}{V_p} \qquad (7.20)$$

We denote x as the ratio of the difference between total cylinder volume ($V_0 = V_p + \xi'V_p$) and the actual gas volume taken in ($V_s = \lambda_0 V_p$) to the piston displacement, V_p:

$$x = \frac{V_0 - V_s}{V_p} = \frac{V_p + \xi'V_p - \lambda_0 V_p}{V_p} = 1 + \xi' - \lambda_0 \qquad (7.21)$$

where ξ' is the clearance, defined as the volume remaining in the cylinder at the extreme position of the piston divided by the displacement of the cylinder. Hence, the relation between volumetric coefficient, λ_0, and clearance, ξ', for a given x is

$$\lambda_0 = 1 + \xi' - x \qquad (7.22)$$

Expansion of gas in the clearance space can be assumed to be polytropic, with an exponent, m_p, somewhat less than the exponent of the compression polytrope, m (for two atomic gases, m is equal to 1.2). Therefore,

$$P_2(\xi'V_p)^{m_p} = P_1(V_0 - V_s)^{m_p} = P_1(xV_p)^{m_p} \qquad (7.23)$$

where $xV_p = V_0 - V_s$ is the gas volume after expansion from pressure P_2 to P_1 in the clearance space (Figure 7.7).

From Equation 7.23 we obtain

$$x = \xi'\left(\frac{P_2}{P_1}\right)^{1/m_p} \qquad (7.24)$$

Substituting x into Equation 7.22 we obtain

$$\lambda_0 = 1 - \xi'\left[\left(\frac{P_2}{P_1}\right)^{1/m_p} - 1\right] \qquad (7.25)$$

Thus, the value of volumetric coefficient, λ_0, of the compressor depends on the value of relative clearance, ξ', the pressure ratio P_2/P_1 and the exponent of polytrope m_p of gas expansion. The capacity of a compressor will be higher at lower pressure ratios and clearances and with a greater exponent of expansion in this space. Equation 7.25 shows that the volumetric coefficient,

λ_0, decreases with an increased compression ratio and, at a certain value, may approach zero. The compression ratio $(P_2/P_1) = 0$ at a volumetric coefficient of zero is referred to as the *compression limit*.

At the limiting compression ratio the expanding gas in the clearance space occupies the total volume of the cylinder. The gas suction is terminated and the compressor capacity becomes equal to zero. On the indicator diagram shown in Figure 7.10, the curves of compression and expansion merge into a single line. Hence, the area of the indicator diagram and, consequently, the indicator horsepower at the compression limit, are zero.

The limit of pressure at polytropic gas expansion in the clearance is determined from Equation 7.25, taking into account that $\lambda_0 = 0$:

$$1 - \xi'\left[\left(\frac{P_2}{P_1}\right)_{\lambda_0=0}^{1/m_p} - 1\right] = 0$$

or

$$\left(\frac{P_1}{P_2}\right)_{\lambda_0=0} = \left(\frac{1}{\xi'} + 1\right)^{m_p} \tag{7.26}$$

Actually, the lower compression ratio is assumed to be the limiting case. It may be assumed that compressors with volumetric coefficients less than 0.7

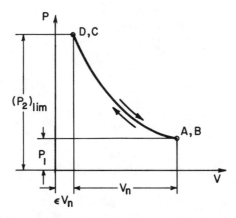

Figure 7.10   Indicator diagram of compressor operation at the compression limit.

are uneconomical. The corresponding *volumetric limit* of the pressure ratio $(P_2/P_1)_0$ is obtained from the following equation:

$$1 - \xi'\left[\left(\frac{P_2}{P_1}\right)^{1/m_p} - 1\right] = 0.7$$

or

$$\left(\frac{P_2}{P_1}\right)_0 = \left(\frac{0.3}{\xi'} + 1\right)^{m_p} \tag{7.27}$$

The gas temperature after compression in a single-stage compressor should not exceed 150–160°C. At higher temperatures lubrication is difficult due to carbonization of the oil. Furthermore, there is a risk of causing oil mist explosions in the cylinders when gases containing oxygen are being compressed. An expression for the limiting compression ratio due to the gas temperature rise during compression may be obtained by eliminating V between the following equations:

$$\frac{P_1 V_1}{T_1} = \frac{P_2 V_2}{T_2} = K \tag{7.28}$$

and

$$P_1 V_1^n = P_2 V_2^n = K \tag{7.29}$$

From Equation 7.28,

$$\frac{V_1}{V_2} = \frac{P_2 T_1}{P_1 T_2} \tag{7.30}$$

From Equation 7.29,

$$\frac{V_1^n}{V_2^n} = \frac{P_2}{P_1}$$

or

$$\frac{V_1}{V_2} = \left(\frac{P_2}{P_1}\right)^{1/n}$$

(7.31)

And equating,

$$\left(\frac{P_2}{P_1}\right)^{1/n} = \left(\frac{P_2}{P_1}\right)\left(\frac{T_1}{T_2}\right)$$

(7.32)

For an adiabatic compression,

$$\left(\frac{P_2}{P_1}\right)_T = \left(\frac{T_2}{T_1}\right)^{n/(n-1)}$$

(7.33)

There are some compressor designs in which the rate of compression is higher than that determined from Equation 7.33. In this case, the lubrication is carried out by water injection into the cylinder, which also provides for partial cooling of the gas. In practice, the compression ratio does not exceed a value of 5 (if the cylinder is cooled), with the exception of small compressors, in which the compression ratio may be increased up to 8.

Theoretical horsepower, N_T (wt), is defined as the horsepower required to compress the gas delivered by the compressor through the specified range of pressures. For a multistage compressor (with intercooling between stages), the *theoretical horsepower* assumes equal work in each stage and perfect cooling between stages. Theoretical horsepower may be calculated as a product of compressor capacity, $V\rho$(kg/s), and the specific work of compression, $\tilde{\ell}$(J/kg), as estimated from the above formulas (Equations 7.15 and 7.17):

$$N_T = V\rho\tilde{\ell}$$

(7.34)

where   V = volumetric capacity of compressor (m³/s)
        $\rho = 1/v$ = gas density (kg/m³)

If the volumetric capacity of the compressor and the gas density are reduced to conditions at suction, i.e., if $V = V_1$ and $\rho = \rho_1 = 1/v_1$, then considering Equations 7.15 and 7.17, and replacing n by κ to denote Cp/Cv, we obtain

$$N_{T,iso} = P_1 V_1 \, ln\left(\frac{P_2}{P_1}\right) \qquad (7.35)$$

$$N_{T,ad} = \frac{\kappa}{\kappa - 1} P_1 V_1 \left[\left(\frac{P_2}{P_1}\right)^{(\kappa-1)/\kappa} - 1\right] \qquad (7.36)$$

$$N_{T,pol} = \frac{m}{m - 1} P_1 V_1 \left[\left(\frac{P_2}{P_1}\right)^{(m-1)/m} - 1\right] \qquad (7.37)$$

Compressor efficiency cannot be evaluated from mechanical efficiency, which is the ratio of the horsepower imparted to the gas and the brake horsepower. Such an evaluation would suggest that the lowest efficiencies would require machines with intensive water cooling, as a considerable fraction of the energy of compression is absorbed in the form of heat by cooling water. However, it is well known that the desired increase of gas pressure is simply achieved with a minimum of energy spent in machines in which intensive cooling is provided. Therefore, the efficiencies of compression machines should be based on *the relative thermodynamic efficiency*. In this manner, a comparison can be made between a given machine and the most economical one of the same class.

Machines with water cooling are compared with a conditional machine that compresses gas isothermally. The ratio of the horsepower of an "isothermic" machine, $N_{T,iso}$, to the actual horsepower, N, of a machine equipped with heat removal is called the isothermal efficiency:

$$\eta_{iso} = \frac{N_{T,iso}}{N} \qquad (7.38)$$

Compressors operating without heat removal generate additional heat by friction between moving parts and gas, hydraulic resistances, etc. Gas compression in such machines is effected *polytropically* when heat is transferred from the gas during the compression and the exponent of the polytropic curve is $m > K$. To estimate the thermodynamic efficiencies of such machines, the conditional machine may be assumed to compress the gas adiabatically or *isentropically*.

An "isentropic" machine is the most economical system from this class of machines that work without heat removal. The ratio of compression horsepower of an isentropic machine, $N_{T,ad}$, to the horsepower, N, of a given compressor without heat removal is called the *isentropic (adiabatic)* efficiency:

$$\eta_{ad} = \frac{N_{T.ad}}{N} \qquad (7.39)$$

Brake horsepower or shaft horsepower, N_e, is the measured horsepower input to the compressor. It is equal to the *theoretical horsepower* defined as horsepower required to compress gas divided by *mechanical efficiency,* which characterizes horsepower losses for mechanical friction in the compressor:

$$N_e = \frac{N}{\eta_{mech}} \qquad (7.40)$$

Considering Equation 7.23

$$N_e = \frac{\eta_{T,iso}}{\eta_{iso} \times \eta_{mech}} \qquad (7.41)$$

The product of isothermic efficiency and mechanical efficiency is referred to as the total isothermal compressor efficiency:

$$\eta_{T,iso} = \eta_{iso} \times \eta_{mech} \qquad (7.42)$$

The horsepower of an electric motor, N_{em}, is greater than the compressor brake horsepower by the value of the horsepower losses in the transmission and the motor itself. These losses are estimated by the *efficiency of transmission,* η_{tr}, and *efficiency of the electric motor,* $\eta_{e.m.}$:

$$N_{em} = \frac{N_e}{\eta_{tr} \times \eta_{em}} \qquad (7.43)$$

The actual horsepower of an electric motor, N_{act}, usually is increased 10–15%, i.e.,

$$N_{act} = (1.1 \text{ to } 1.15)N_{e.m.} \qquad (7.44)$$

The value of an adiabatic efficiency, η_{ad}, is very close to unity (typically in the range 0.93 to 0.97). Isothermal efficiency is in the range of 0.64 to 0.78, and mechanical efficiency varies from 0.85 to 0.95.

OPERATIONAL CONTROL OF PISTON COMPRESSORS

The operational control of piston compressors is analyzed through the use of an indicator diagram similar to Figure 7.11. The diagram summarizes the relationship between pressure and gas volume at intake and delivery for one double piston stroke (i.e., for one rotation of the crankshaft). Point D corresponds to the extreme left piston position, which, in reality, never approaches the cylinder head.

In tracing the piston's movement through the cylinder, assume that the starting point for the cycle is represented by point D on Figure 7.11. The gas remaining in the cylinder in the clearance space now begins to expand. This step is represented by curve DA and is accompanied by an increase in volume and a decrease in the gas pressure. This process continues until pressure P_0 in the cylinder becomes somewhat less than pressure P_1 in the suction line. Because of this pressure difference $(P_1 - P_0)$ the suction valve opens and gas enters into the compressor at point A. Gas suction thus continues from point A to B on the diagram. The volume of gas being sucked in, V_s, is proportional to the line AB and may be expressed in terms of fractions of the cylinder working volume, $V_s = \lambda_0 V_p$, where λ_0 is the volumetric efficiency.

When the piston moves to the left suction ceases, and the gas compresses polytropically along curve BC to a pressure somewhat higher than delivery pressure, P_2. On reaching point C, the delivery valve opens. The discharge

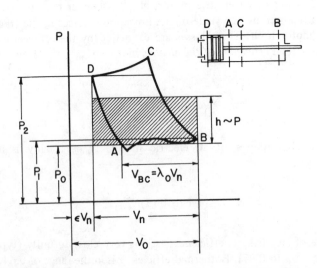

Figure 7.11   Indicator diagram showing operation of a single-stage piston compressor.

follows the path of curve CD and is proportional to the delivery volume. When the piston reaches point C it has compressed the gas to the pressure in the delivery line and must then push the compressed air out through delivery valves into the receiver. Due to the weight of the delivery valves and the tension of the springs holding them in their seats, the cylinder pressure in an actual system rises to a value slightly above that in the receiver just before the valves open. This explains why point C is slightly higher than point D. This pressure gradually drops to the receiver pressure at the end of the stroke.

When the indicator line follows the suction line, the inlet of the compressor is not restricted. Hence, the compressor receives a volume of gas at the suction pressure, represented by the travel of the piston from A to B. This then represents the actual suction capacity of the compressor.

The area enclosed by the indicator diagram is proportional to the useful work performed by the machine in compressing gas over a single shaft rotation per unit of piston area. This work corresponds to the *indicated horsepower* of the unit, which may be evaluated from the diagram through the following relation:

$$N_{ind} = \frac{F \times n_s \times f_{in}}{Sc_{sp}} \tag{7.45}$$

where   F = piston cross-sectional area
    n_s = number of piston double strokes or rotations of the crankshaft
    f_{in} = area of indicator diagram
    Sc_{sp} = scale of indicator spring

The scale of the indicator spring represents the number of pressure units corresponding to a unit length along the ordinate. Multiplying and dividing Equation 7.45 by the length of the stroke(s), we obtain

$$N_{ind} = \frac{FSn_s f_{in}}{S \times S_{sp}} = V_p P_{ind} \tag{7.46}$$

where V_p is FSn_s, i.e., the piston displacement for n_s rotations, $P_{ind} = f_{in}/(S \times S_{sp})$ is the ratio of the diagram area to the piston stroke, i.e., *the average indicator pressure.*

The average indicator pressure may be computed (based on an appropriate scale) as a height, h, of an equivalent rectangle constructed on the diagram, with the base equal to the diagram length because the volume of piston travel,

V_p, for a given compressor at constant F and n_s is proportional to the length of piston stroke, S. This rectangular area is the shaded region shown in Figure 7.11. The indicator horsepower for a double-acting compressor is determined for each side of the cylinder separately, and the values obtained then are summed.

Multistage compression is used for preparing gases at high pressures. If the required pressure ratio P_2/P_1 exceeds a value of 4 or 5, it is not practical to carry out the entire compression in a single cylinder (the temperature rise of the compressed gas will become excessive and thus adversely affect efficiency). As noted earlier, to minimize this lower efficiency, it is common practice to divide the compression into several stages, with intercooling between stages. In a multistage compressor, gas passes through a number of cylinders of gradually decreasing volume. This arrangement more closely approaches isothermal conditions, resulting in a considerable savings of power.

An indicator diagram again can be constructed to understand the operation of a multistage compressor. Such a diagram is shown in Figure 7.12. The area traced by ABCD denotes the work performed by compressing isentropically from pressure P_1 to P_2 in a single stage. The area traced by ABED denotes the work required to perform an isothermal compression. (We have, of course, ignored the effect of clearance volume.) If a multistage isentropic compression is performed, then the intermediate pressures between stages are P_{i1}, P_{i2}, P_{i3}, etc. To simplify our discussion, we shall assume that the gas can be cooled in an interstage cooler to its first stage intake temperature prior to entering each cylinder.

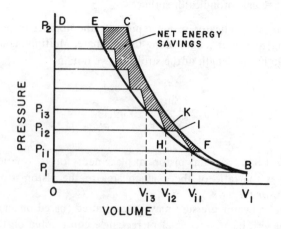

Figure 7.12 Operation of a multistage compression cycle with interstage cooling.

The compression cycle can be traced on Figure 7.12 as follows. The piston suction stroke of the first stage is represented by line AB, where gas volume V_1 is admitted at pressure P_1. The gas then is compressed isentropically to pressure P_{i1} (line BF) and then delivered from the first stage at this pressure (line FG). The suction stroke of the second stage is denoted by line GH, where the volume of the gas is reduced in the interstage cooler to V_{i1}. Note that V_{i1} is also the volume of the gas that would have been achieved as a result of an isothermal compression to pressure P_{i1}. In the second stage, the gas again undergoes isentropic compression, but from a pressure P_{i1} to P_{i2} (line HI). Line IJ traces the delivery stroke of the second stage. We then return to the isothermal compression curve (line BE) at point K (i.e., line JK denotes the suction stroke of the third stage).

The overall work performed on the gas is the intermediate area between that for a single-stage isothermal compression and that for an isentropic compression. That is, the net energy savings is denoted by the shaded region in Figure 7.12.

Following Coulson and Richardson [1962], the total work performed per cycle for an isentropic compression is

$$W' = P_1 V_1 \frac{\kappa}{\kappa - 1} \left\{ \left(\frac{P_{i1}}{P_1} \right)^{(\kappa-1)/\kappa} - 1 \right\} + P_{i1} V_{i1} \frac{\kappa}{\kappa - 1} \left\{ \left(\frac{P_{i2}}{P_{i1}} \right)^{(\kappa-1)/\kappa} - 1 \right\} + \ldots$$

$$(7.47)$$

For perfect interstage cooling, we note from the ideal gas law that

$$P_1 V_1 = P_{i1} V_{i1} = P_{i2} V_{i2} = \ldots$$

Hence, for a Z stage compressor the work performed is

$$W' = P_1 V_1 \frac{\kappa}{\kappa - 1} \left\{ \left(\frac{P_{i1}}{P_1} \right)^{(\kappa-1)/\kappa} + \left(\frac{P_{i2}}{P_{i1}} \right)^{(\kappa-1)/\kappa} + \ldots - Z \right\} \qquad (7.48)$$

The work performed by the compressor will be at a minimum when the partial derivative of W' with respect to the interstage pressures is zero, i.e., when $\partial W'/\partial P_{i1} = \partial W'/\partial P_{i2} = \partial W'/\partial P_{i3} = \ldots = 0$.

Hence, the minimum work for the first stage is

$$P_1 V_1 \frac{\kappa}{\kappa - 1} \left\{ \frac{\kappa - 1}{\kappa} \left(\frac{P_{i1}}{P_1} \right)^{(\kappa-1)/\kappa} P_{i1}^{-1} + \frac{1 - \kappa}{\kappa} \left(\frac{P_{i2}}{P_{i1}} \right)^{(\kappa-1)/\kappa} P_{i1}^{-1} \right\} = 0 \quad (7.49A)$$

or, on simplification,

$$\frac{P_{i1}}{P_1} = \frac{P_{i2}}{P_{i1}} \tag{7.49B}$$

This expression defines the optimum pressure, P_{i1}, for gas delivery from the first stage. In a similar manner, the optimum value for P_{i2} can be obtained (the derivation is left to the student). The net result is that the intermediate pressures should be arranged so that the compression ratio is the same in each cylinder (and, hence, the work done in each cylinder is the same).

The minimum work of compression in a Z-stage compressor becomes

$$P_1 V_1 \frac{\kappa}{\kappa - 1} \left\{ Z \left(\frac{P_2}{P_1}\right)^{(\kappa - 1)/Z\kappa} - Z \right\} = Z P_1 V_1 \frac{\kappa}{\kappa - 1} \left\{ \left(\frac{P_2}{P_1}\right)^{(\kappa - 1)/Z\kappa} - 1 \right\} \tag{7.50}$$

Defining the clearance volumes in the successive cylinders as C_1, C_2, $C_3, \ldots$, the theoretical volumetric efficiency for the first stage is

$$V_{cyl} = 1 + C_1 - C_1 \left(\frac{P_2}{P_1}\right)^{1/Z\kappa} \tag{7.51}$$

and the volume of gas admitted to the first cylinder is

$$V_{g1} = V_{s1} \left\{ 1 + C_1 - C_1 \left(\frac{P_2}{P_1}\right)^{1/Z\kappa} \right\} \tag{7.52}$$

The ratio of the volumes of gas admitted to successive cylinders is $(P_1/P_2)^{1/Z}$. This assumes that the same mass of gas is sent through each cylinder and that the interstage coolers are perfectly efficient. The volume of gas admitted to the second cylinder is

$$V_2 = V_{s2} \left\{ 1 + C_2 - C_2 \left(\frac{P_2}{P_1}\right)^{1/Z\kappa} \right\} = V_{s1} \left\{ 1 + C_1 - C_1 \left(\frac{P_2}{P_1}\right)^{1/Z\kappa} \right\} \left(\frac{P_1}{P_2}\right)^{1/Z} \tag{7.53}$$

or

$$\frac{V_{s1}}{V_{s2}} = \frac{1 + C_2 - C_2\left(\frac{P_2}{P_1}\right)^{1/Z\kappa}}{1 + C_1 - C_1\left(\frac{P_2}{P_1}\right)^{1/Z\kappa}}\left(\frac{P_2}{P_1}\right)^{1/Z} \qquad (7.54)$$

The swept volume of gas through each cylinder thus can be calculated from information on V_{s1} and the clearance volume $C_1, C_2, \ldots.$

The above equations are based on ideal gas behavior. When dealing with nonideal gases, the change caused by compression can be examined on a temperature-entropy or enthalpy-entropy diagram. Intermediate pressures $(P_{i1}, P_{i2}, P_{i3}, \ldots)$ can be chosen on the basis of the same enthalpy change in each cylinder. The following problem illustrates some of the above principles.

Example 7.1

Determine the effect of intercooling between stages for a two-stage air compressor having suction at atmospheric pressure and a delivery at 100 psia. The operating diagram for the compressor is shown in Figure 7.13.

Solution

The compression ratio is $100/14.73 = 6.79$. Assuming an initial temperature of $80°F$ and a value of $\kappa = 1.4$, we obtain the final temperature from Equation 7.32:

$$T_2 = T_1\left(\frac{P_2}{P_1}\right)^{(\kappa-1)/\kappa} = 473°F$$

The compression ratio can be split equally between the two stages such that equal work is done at each stage. Because the overall ratio of compression is 6.79, the value P_2/P_1 for each stage is $\sqrt{6.79} = 2.605$, and the intermediate pressure will be $2.605 \times 14.73 = 38.4$ psia. As a check, note that the second stage will have a compression ratio of $100/38.4 = 2.605$. This leads to the following equation:

$$P_2 = P_1\sqrt{\left(\frac{P_2}{P_1}\right)\left(\frac{P_3}{P_2}\right)} = P_1\sqrt{\frac{P_3}{P_1}} = \sqrt{P_1 P_3} = P_1\sqrt{R_t} \qquad (7.55)$$

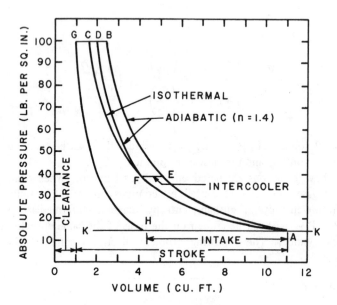

Figure 7.13   The effect of intercooling between two stages.

where P_1 = low-pressure intake
           P_2 = low-pressure discharge, intercooler and second-stage intake
           P_3 = high-pressure discharge
           R_t = overall ratio of compression

Equation 7.55 assumes that both high- and low-pressure cylinders have the same volumetric efficiency. Gas being compressed from 80°F and atmospheric pressure at point A in Figure 7.13 will obtain a temperature of 473°F when it reaches 100 psia at point B. Because the pressure after the first stage is 38.4 psia, the temperature will rise to only 250°F (point E). At this point the gas is cooled to 80°F (at F) and finally compressed to 100 lb abs along the line FD. The temperature at D also will be 250°F, as the compression ratio and initial temperature are the same as before. The area EFDB represents the savings in work for two stages over a single-stage compression. The line AC represents isothermal compression, beginning at point A. Figure 7.13 shows that the greater the number of stages used, the closer the compression line approaches the isothermal line AC.

For two-stage compression, temperature increase may be determined for each stage separately by Equation 7.32, taking into account that the temperature depends solely on the compression ratio and the value of exponent n.

Because the compression ratio of each stage is the square root of the overall ratio, then

$$T_2 = T_1 \sqrt{R_t^{(\kappa-1)/\kappa}} = T_1 R_t^{(\kappa-1)/2\kappa} \qquad (7.56)$$

This expression is valid in the case of perfect intercooling, i.e., the gas has to be cooled between stages so that the intake to the high-pressure cylinder is at the same temperature as at the low-pressure intake.

Cylinder sizes may be determined from Boyle's law:

$$\frac{P_2}{P_1} = \frac{V_1}{V_2} = \frac{V_{p1}\lambda_{v1}}{V_{p2}\lambda_{v2}} \qquad (7.57)$$

where V_p is the displacement of the cylinder, and λ_v is the volumetric coefficient.

Assuming the same displacement coefficient in both cylinders and taking into account that the displacements are directly proportional to the squares of the cylinder diameters, d, and that the compression ratio in the first stage, P_2/P_1, is equal to the square root of the overall compression ratio, we obtain

$$R_1 = \frac{V_{p1}}{V_{p1}} = \left(\frac{d_1}{d_2}\right)^2 \qquad (7.58)$$

Using the overall compression ratio,

$$\sqrt[4]{R_1} = \frac{d_1}{d_2} \qquad (7.59)$$

Equation 7.59 assumes that the stroke of both cylinders is the same.
From Equations 7.57 and 7.58, we obtain

$$R_1 = \left(\frac{d_1}{d_2}\right)^2 \left(\frac{\lambda_{v1}}{\lambda_{v2}}\right) \qquad (7.60)$$

where R_1 is the compression ratio of the low-pressure cylinder.

Substituting the expression for the volumetric coefficient from Equation 7.25 into Equation 7.60,

$$R_1 = \left(\frac{d_1}{d_2}\right)^2 \frac{1 - \xi_1[R_1^{1/m_p} - 1]}{1 - \xi_2[R_2^{1/m_p} - 1]} \tag{7.61}$$

This equation may be solved by trial and error.

Example 7.2

Continuing with the two-stage compressor of Example 7.1, assume that the low-pressure cylinder is 10 in. in diameter, with a 5% clearance. For equal work in each cylinder, the compression ratio in the first stage will be $\sqrt{6.79} = 2.605 = R_1 = R_2$.

The volumetric coefficient, from Equation 7.25, is

$$\lambda = 1 - \xi_1(R_1^{1/m_p} - 1) = 1.00 - 0.05(1.98 - 1.00) = 95.1\%$$

The approximate size of the high-pressure cylinder may be computed from Equation 7.59:

$$d_2^4 = \frac{d_1^4}{R_t}$$

$$d_2 = \sqrt[4]{\frac{10^4}{6.79}} = 6.2 \text{ in.}$$

Assume the following cylinders are available for use in this service:

- 6½-in. diameter, with 12¼% clearance
- 6¼-in. diameter, with 15% clearance
- 6-in. in diameter, with 15% clearance

Substituting values into Equation 7.61,

$$d_2^2 \lambda_{v_2} = \frac{10^2 \times 0.951}{2.605} = 36.5$$

Taking the compression ratio in the second stage as equal to 2.605, the values of the volumetric coefficient, λ_{v_2}, are computed as follows:

- 6½-in. cylinder, volumetric coefficient = 88%
- 6¼-in. cylinder, volumetric coefficient = 85.3%
- 6-in. cylinder, volumetric coefficient = 85.3%

The values of $d^2\lambda_v$ for these cylinders are: 6½-in. cylinder, 37.2; 6¼-in. cylinder, 33.3; 6-in. 30.7. We chose the 6½-in. cylinder. Because d_2 is now 6.5 in., instead of 6.2 in., as calculated, the compression ratio in the two cylinders will also have to change for Equation 7.61 to be satisfied. As the high-pressure cylinder is a little larger than calculated, its compression ratio will be a little less than the value of 2.605 required for equal work in both cylinders. Consequently, a little more work must be done by the low-pressure cylinder, and it will have a little higher compression ratio than computed earlier. After several trial values of R_1 and R_2 in Equation 7.61 (remembering that $R_1 \times R_2 = 6.79$), we find that $R_1 = 2.625$ and $R_2 = 2.585$ with the 6.5-in. cylinder in the second stage of compression. This yields a deviation of approximately 0.75% from the assumed value of 2.605. The intermediate pressure will be 2.625 × 14.73 = 38.7 psia, which is approximately 0.3 psi higher than the intercooler *pressure obtained from Equation 7.55.*

Two-stage energy computations usually are based on polytropic conditions because there would be little reason for two-staging if the isothermal condition were achievable. Two-stage compression may be assumed as two cylinders working separately on different ratios of compression. Hence, the energy required is the sum of the horsepowers for the two cylinders added together.

Transforming Equation 7.17 to describe the two-stage compression,

$$\tilde{\ell} = \frac{n}{n-1} P_1 V_1 (R^{(n-1)/n} - 1), \text{ft-lb} \qquad (7.62)$$

and defining the intake gas in terms of standard cubic feet at 14.73 psia, then

$$\tilde{\ell} = \frac{n}{n-1} 144 \times 14.73 V_1 [R^{(n-1)/n} - 1]$$

$$= \frac{n}{n-1} 2121 V_1 [R^{(n-1)/n} - 1], \quad \text{ft-lb} \qquad (7.63)$$

(Note that 14.73 V_1 is multiplied by 144 to convert to units of lb/ft^2.)

To express Equation 7.63 in terms of horsepower, V_1 is taken in standard cubic feet of gas handled per minute and divided by 33,000 ft-lb/min. (33,000 ft-lb/min is the equivalent of 1 hp).

$$HP = \frac{n}{n-1} \frac{2121}{33000} V_1 [R^{(n-1)/n} - 1]$$

$$= \frac{n}{n-1} 0.0643 V_1 [R^{(n-1)/n} - 1] \qquad (7.64A)$$

or

$$HP = \frac{0.0643nV_1}{n-1} [R_1^{(n-1)/n} + R_2^{(n-1)/n} - 2] \qquad (7.64B)$$

where V_1 is in cfm.

When the work is divided equally between two stages, $R_1 = R_2 = \sqrt{R_t}$, Equation 7.64B becomes

$$HP = \frac{0.1286nV_1}{n-1} [R_t^{(n-1)/2n} - 1] \qquad (7.65)$$

Equations 7.64B and 7.65 are based on double-acting compressors. Displacements of the two cylinders in two-stage compression are evaluated separately. Both capacity and the overall volumetric coefficient are based entirely on the low-pressure cylinder.

If the overall compression ratio exceeds 10, it is often recommended that another stage beyond two-stage compression be introduced to reduce the temperature rise.

For equal work in each stage of a *three-stage compression.*

$$R_1 = R_2 = R_3 = \sqrt[3]{R_t} \qquad (7.66)$$

Equation 7.66 is valid only for equal clearance in all cylinders and, consequently, equal volumetric coefficients. This is illustrated in Figure 7.14. Substituting for the pressures corresponding to R_1, R_2, etc., into Equation 7.66, the pressure in the first intercooler becomes

$$P_2 = P_1 \sqrt[3]{P_4/P_1}$$

or

$$P_2 = P_1 \sqrt[3]{R_t} \qquad (7.67)$$

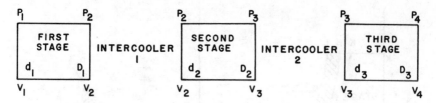

Figure 7.14 The operating principle for three-stage compression.

In a similar manner, the pressure in the second intercooler is

$$P_3 = P_1 \sqrt[3]{R_t^2} \tag{7.68}$$

The temperature increase in a three-stage compression is

$$T_2 = T_1 R_t^{(n-1)/3n} \tag{7.69}$$

where T_1 and T_2 are the initial and final absolute temperatures in each stage.
The power required in a three-stage compression is

$$HP = \frac{0.1929nV_1}{n-1} (R_t^{(n-1)/3n} - 1) \tag{7.70}$$

The indicator diagram for a three-stage compression (with equal compression ratios in each stage) is illustrated in Figure 7.15. In constructing these diagrams it is assumed that the gas is cooled to its initial temperature in the intercoolers and that the clearances for each stage, along with pressure losses, are zero. Line BC represents compression in the first stage from pressure P_1 to pressure P_2. The gas is then cooled along the line CE (isobar) to initial temperature, T_1. On entering cylinder II, it is compressed to pressure P_3 along the line EF, etc. The process of three-stage compression from pressure P_1 to pressure P_k, with cooling, is expressed by the broken line BCEFGHK.

The line BEGK corresponds to isothermal compression to the pressure P_k in a single-stage compressor, and line BCL represents polytropic compression in the same compressor. From the diagrams, one can see that multistage compression with interstage gas cooling is closer to the isothermal process and, consequently, demands less energy than a single-stage process over the same pressure limits.

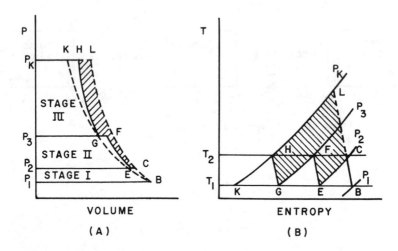

Figure 7.15 (A) Indicator and (B) entropy diagrams of multistage gas compression.

The area confined inside the indicator diagram and the area under the broken line BCEFGHK on the T-S diagram represent the work of multistage compression (based on an appropriate scale). The shaded area denotes the savings of power over a single stage. The closer the broken line of multistage compression is to the isotherm, the more stages that are required. However, the economical number of stages generally is limited to five or six. As the number of stages increases, so do capital and operating costs.

For extremely high compression ratios, *four-stage compression* often is used (Figure 7.16). In the case of equal work at each stage,

$$R_1 = R_2 = R_3 = R_4 = \sqrt[4]{R_t} \qquad (7.71)$$

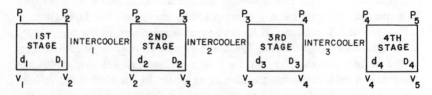

Figure 7.16 Diagram of four-stage compression.

By substituting in pressures for the compression ratios in Equation 7.71, an expression for the pressure in each intercooler can be obtained. For the first intercooler,

$$P_2 = \sqrt[4]{P_1^3 P_5} = P_1 \sqrt[4]{R_t} \qquad (7.72)$$

For pressure in the second intercooler,

$$P_3 = \sqrt{P_1 P_5} = P_1 \sqrt{R_t} \qquad (7.73)$$

And, for the pressure in the third intercooler,

$$P_4 = \sqrt[4]{P_1 P_5^3} = P_1 R_t^{3/4} \qquad (7.74)$$

The increase in temperature for four-stage compression is

$$T_2 = T_1 R_t^{(n-1)/4n} \qquad (7.75)$$

The power required for four-stage compression is

$$HP = \frac{0.2572 n V_1}{n - 1} (R_t^{(n-1)/4n} - 1) \qquad (7.76)$$

where V_1 is in cfm.

The theoretical work of compression becomes minimum when the compression ratios, R, are identical in each stage, i.e.,

$$R = \sqrt[Z]{\frac{P_k}{P_1}} \qquad (7.77)$$

where Z is the number of stages, P_k is the final pressure, and P_1 is the initial pressure. In practice, because of pressure losses between stages (in valves, intercoolers, etc.) the pressure ratio in each stage is somewhat higher than the theoretical prediction. To account for these losses, an empirical correction factor is used to adjust Equation 7.77:

$$R = \psi \sqrt[Z]{\frac{P_k}{P_1}} \tag{7.78}$$

where $\psi = 1.1 - 1.15$ is a coefficient accounting for pressure losses between stages.

The number of compression stages required may be determined from the following equation:

$$Z = \frac{\log P_k - \log P_1}{\log R - \log \psi} \tag{7.79}$$

To maintain the final gas temperature within allowable limits, a pressure ratio in the range of 2.5 to 3.5 may be assumed. This range generally provides effective use of the cylinder volume, increases the compressor volumetric coefficient and decreases the energy consumption.

Assuming equal compression ratios in each stage and ideal gas cooling in the intercoolers, the work at each stage will be the same. Then the theoretical work of a multistage compressor in adiabatic compression of 1 kg of gas is

$$\tilde{\ell}_{ad} = Z \frac{\kappa}{\kappa - 1} P_1 \nu_1 \left[\left(\frac{P_k}{P_1}\right)^{(\kappa-1)/Z\kappa} - 1 \right] \tag{7.80}$$

The limiting temperature at the end of compression is

$$T_k = T_1 \left(\frac{P_k}{P_1}\right)^{(\kappa-1)/Z\kappa} \tag{7.81}$$

The theoretical volumetric coefficient is

$$\lambda_0 = 1 - \xi \left[\left(\frac{P_k}{P_1}\right)^{1/Z\kappa} - 1 \right] \tag{7.82}$$

The theoretical compression work, temperature at the end of compression and volumetric coefficient at polytropic compression may be determined from Equations 7.80–7.82 by changing the exponent from an adiabatic to a polytropic process.

In practice, the work performed by each stage is not always equal because the polytropic exponents and the clearance spaces for each stage differ. In addition, the cooling in intercoolers is neither ideal or exactly the same.

MISCELLANEOUS COMPRESSOR DESIGNS

Rotary compressors generally are classified as sliding-vane type, two-impeller positive type, screw-type and liquid-piston type.

A *sliding-vane type compressor* (Figure 7.17), has a closed casing (1) in which the rotor (2) is located eccentric relative to the internal surface of the casing. Slits are provided on the rotor through which blades (3) move radially, affected by centrifugal force, which presses them tightly against the internal surface of the casing. Thus, the sickle-shaped working space between the rotor and the casing is divided by the blades into several nonequal chambers. Gas enters from the suction nozzle and fills the chambers. When the rotor turns to position B, suction ceases and the compression cycle now begins. When the chamber rotates to the right, its volume decreases and the gas inside is compressed. The compression cycle terminates when the chamber reaches position C, where it deposits the compressed volume of gas to the discharge piping.

In position D the gas is displaced completely from the working chamber. Clearance is provided between the rotor and the casing. From D to A, gas is expanded in the clearance space. The suction begins at point A, where the cycle begins again.

Rotary sliding-vane compressors are available at operating pressures up to 125 psig and in capacities up to 5300 cfm. Speeds vary from 3600 rpm for small-capacity units to 450 rpm in larger systems. Single-stage machines are

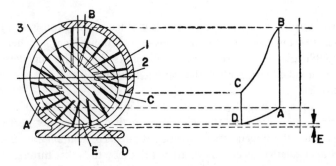

Figure 7.17   Operation of a sliding-vane compressor: (1) casing; (2) rotor; (3) sliding blades.

available at pressures up to 50 psig and vacuums to 29 in. Hg. Two-stage machines are capable of pressures up to 125 psig and vacuums to 29 in. Hg.

The capacity of a rotary sliding-vane compressor, $V_1(m^3/s)$, reduced to the suction conditions, may be calculated from the following equation:

$$V_1 = 2\ell en\,\lambda_v(\pi D - \delta\bar{z}) \qquad (7.83)$$

where ℓ = length of a blade (m)
   e = eccentricity of a rotor (m)
   n = rotor rotations per unit time
   D = inside diameter of the casing (m)
   δ = blade thickness (m)
   $\bar{z}$ = number of blades (typically 20 to 30)

The relative eccentricity, e/D, typically is 0.06–0.07.

The volumetric coefficient may be computed from

$$\lambda_v = 1 - K\,\frac{p_2}{p_1} \qquad (7.84)$$

where K = 0.05 for large machines (where capacity exceeds 0.5 m^3/s)
   K = 0.1 for small machines (where capacity is less than 0.5 m^3/s)

The brake horsepower of a rotary compressor is

$$N_e = \frac{V_1\rho_1\,ln\left(\dfrac{P_2}{P_1}\right)}{\eta_{iso}} \qquad (7.85)$$

A *two-impeller, positive-type compressor* (Figure 7.18) consists of a casing in which two impellers revolve in opposite directions. Each impeller consists of a double-lobe section symmetrical about its shaft. The impellers neither touch each other nor contact the casing. However, clearances are small to minimize leakage.

The rotation of impellers reduces the volume in which the gas is trapped, producing a pressure rise. The gas is drawn through the inlet port and discharged when the impellers pass the outlet port. To maintain the impellers at their proper relative speed, one shaft is driven by the other through a pair of gears.

The pitch diameter of these gears controls the size and capacity of the

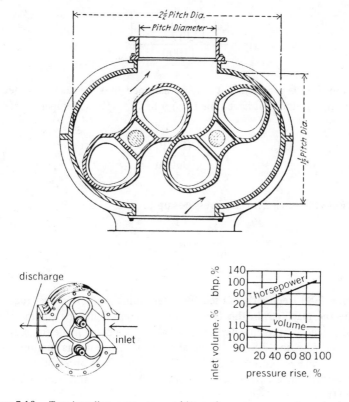

Figure 7.18   Two-impeller compressor and its performance curves at constant speed.

compressor. The radius of an impeller, or its half length, is designed to be three-quarters of the pitch diameter of the gears. The casing consists of semicylinders separated by a parallel section. The radius of the cylinders is equal to that of the impellers plus clearance. The width of the parallel section is equal to the pitch diameter of the gears plus the clearance.

The horsepower required at the shaft is proportional to the volume and pressure of the gas discharged. It is safe to assume that for each 100 ft^3/min/lb pressure of gas discharged, 5 hp is required. The following formulas sometimes are used in calculating the horsepower. Formulas 7.86–7.88 give the theoretical horsepower required. To determine the horsepower to drive the two-impeller, positive-type compressor, it is necessary to divide the results obtained by the efficiency of the machine. The usual efficiency is between 80 and 90%:

$$Hp = \frac{QP_1 \left[\left(\frac{P}{P_1} \right)^{1/3} - 1 \right]}{11000} \qquad (7.86)$$

Equation 7.86 is used when it may be assumed that the gas is compressed so quickly that it does not have time to cool to atmospheric temperature:

$$Hp = \frac{Q(P - P_1)}{33000} \qquad (7.87)$$

Equation 7.87 is used for low pressures up to 5 oz.

$$Hp = \frac{lb/in.^2 \times Q}{200} \qquad (7.88)$$

Equation 7.88 frequently is used for determining the horsepower required to operate the machine. In this last equation, Q represents the volume of gas in cfm displaced by the impellers, with no allowance being made for slippage. In Equations 7.86 and 7.87, P_1 is the absolute pressure in the suction line, and P is the discharge pressure, in lb/ft^2. These types of compressors are available for pressure differentials up to about 12 $lb/in.^2$ and capacities up to 15,000 ft^3/min.

A *screw-type rotary compressor* (Figure 7.19) can handle capacities up to about 25,000 ft^3/min at pressure ratios of 4:1 and higher. There is no

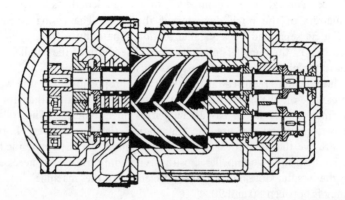

Figure 7.19   A screw-type rotary compressor.

unbalanced radial load on the rotor shafts. Hence, the magnitude of the pressure rise is not a limiting feature. Relatively small-diameter rotors are available to provide rotation speeds of several thousand rpm. In contrast to the straight-lobe rotary machine, the rotors are male and female whose rotation causes the axial progression of successive sealed cavities. Designs are staged with intercoolers when needed. High-speed operation usually necessitates the use of suction and discharge noise suppressors.

A *liquid-piston compressor* is shown in Figure 7.20. This design employs any low-viscosity liquid as the compressant to displace the gas handled. A round multiblock rotor forming a series of buckets rotates freely in an elliptical casing containing the liquid, which is carried around by the rotor and is caused to recede from its buckets at the wide point of the ellipse (major axis), permitting the buckets to fill with gas entering from the inlet ports. On reaching the narrow point of the ellipse (minor axis), it surges back into the buckets, compressing the gas and ejecting it through the discharge ports.

Liquid-piston compressors handle up to about 5000 cfm. Single-stage units can develop pressures to about 75 psi; multistage designs are used for higher pressures.

Additional discussions and applications are given by Stepanoff [1965], Kearton [1926], Rollins [1973] and the Compressed Air and Gas Institute [1954].

DYNAMIC-TYPE CENTRIFUGAL MACHINES

Fans, blowers and compressors are integral components in chemical and petrochemical plants. They are used for handling large volumes of gas over pressure rises ranging from 0.5 to several hundred psi. Typical applications include:

- Pneumatic Conveying
  In-plant bulk materials handling
  Bulk trailer loading and unloading
  Bulk railroad car unloading
  Barge and ship loading and unloading
- Waste Treatment
  Aeration
- Peak Shaving
  Air-gas mixing stations
- Petroleum and Chemical Process Industries
  Vapor recovery
  Gas circulation
- Compressor and Engine Supercharging
- Vacuum Filtration
- Food Processing
  Washing, cooling and drying
- Glass Manufacturing

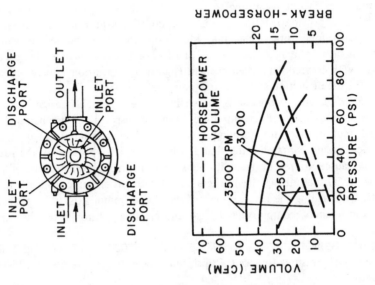

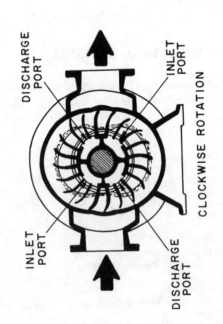

Figure 7.20   Liquid-piston rotary compressor and its constant-speed performance curves.

- Pulp and Paper-Industry
    Envelope production
    Printing presses
- Brewing Industry
- Aircraft and Missile Ground Support
    Cabin leakage testing
    Instrument cooling
    Jet engine starting
    Wind tunnels
    Flight and altitude simulation
- Packaging
    Bagging
- Plastic Molding
- Fuel Oil Atomization
    Large furnaces
- Wood Industry
    Chip and sawdust handling
    Impregnating equipment
- Foundries
    Conveying sand
    Core blowing
- Tunnel Pressurization
- Moving Film Processing
    Air knife
- Inflating Operations

Dynamic-type centrifugal machines for gas handling often resemble liquid centrifugal pumps. Both systems are based on the same operating principles. As noted previously, these gas-handling machines are classified according to compression ratios as *fans, blowers, compressors* or *vacuum pumps.* Gas is compressed by the dynamic action of rotating elements which imparts velocity and pressure to the fluid. With the thermodynamic principles of compression firmly in our grasp, we shall now discuss the detailed design and operation of gas-handling machinery.

Fans

Fans are classified by the direction of gas flow: radial flow and axial flow. In *centrifugal fans,* gas flow is fundamentally in a radial direction, whereas in *axial fans* flow is in an axial direction. *Centrifugal fans* are conditionally divided into three groups: *low* pressure ($P < 10^3$ N/m^2), *medium* pressure ($P = 10^3 - 3 \times 10^3$ N/m^2), and *high* pressure ($P = 3 \times 10^3 - 10^4$ N/m^2).

Centrifugal fans are available in three general impeller configurations: straight blade (or paddle wheel), forward-curved blades (or radial wheel), and backward-curved blades. The designs are shown in Figure 7.21.

The selection of a particular impeller configuration depends on the capacity requirements and the solids concentration in the gas stream. Figure 7.22 illustrates the operation of a backward-curved blade fan. As shown, in the volute fan casing (1), the drum (2) containing rows of blades rotates. The

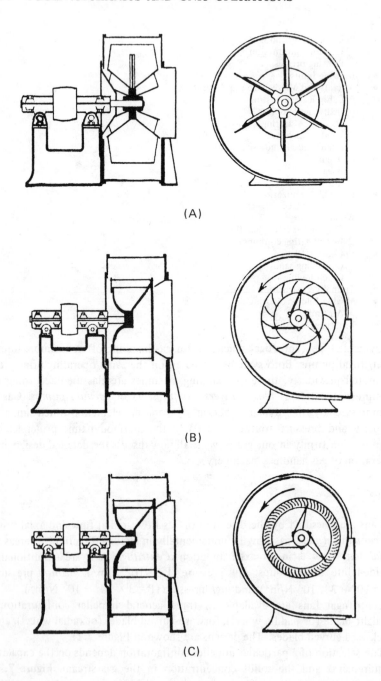

Figure 7.21   Three main impeller configurations for centrifugal fans: (A) paddle wheel; (B) forward curved; (C) backward curved.

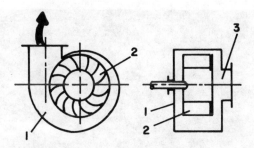

Figure 7.22   Operation of a low-pressure fan: (1) casing; (2) drum; (3) suction nozzle; (4) discharge nozzle.

blade width to length ratio depends on the required pressure generated and is typically lowest for high-pressure fans. For the system shown, gas enters along the fan axis through nozzle 3 and leaves through discharge nozzle 4. The shape and sizes of the fan casing, the drum (rotor), blades and nozzles are selected on the basis of minimum hydraulic losses. In general, low-pressure fans have backward-curved blades.

Heavy-duty industrial fans with radial flow blading are widely used for exhausting large volumes of dust-laden air or gas from industrial processes. Typical volume ranges for dust collecting and scrubber systems range between 25,000 to 250,000 cfm, with pressures up to 30 in. water. Figure 7.23 shows a large radial-tip blade unit.

In addition to handling process streams, fans are employed widely for ventilation and exhaust system operations. Figure 7.24 shows an airfoil blading used for mechanical-draft, high-velocity air conditioning systems and vehicular tunnel ventilation.

Characteristic curves for fans utilizing the three main impeller configurations are shown in Figure 7.25. In comparing the plots, note that the radial blade and forward-curved blade fans have medium efficiencies, whereas the backward-inclined blade fan has a higher efficiency.

In *axial-flow fans* the gas flow is parallel to the axis of rotation. These fans are manufactured in two general styles: *disk type* and *propeller type*. Disk-type fans have plain, or curved, blades (Figure 7.26). The particular unit shown is designed for exhausting a spray booth and can be mounted in horizontal or vertical stacks. An example of its application is shown in Figure 7.27.

Propeller-type axial fans are closer to an aeronautical design, usually consisting of two rows of blades. Figure 7.28 illustrates a two-stage axial-flow fan in which one row of blades rotates while the other is stationary. As

Figure 7.23   Radial flow blading design for a large industrial fan (courtesy Westing-house Electric Corp., Sturtevant Div., Hyde Park, MA).

the rotor blades revolve, they impart velocity and pressure to the gas. Both the rotor blades and the stator contribute almost equally to the gas pressure rise. Figure 7.29 shows a controllable-pitch axial fan. The controllable pitch allows high efficiencies to be maintained with frequently changing airflows. That is, the design permits changing the blade pitch angle while the fan is running.

Because of the low pressure exerted on the gas by fans, the efficiency of axial fans is significantly higher than that of centrifugal fans. At the same time, the head generated by axial fans is three to four times less than that of a centrifugal fan. Therefore, axial fans are used for transporting large volumes of gas against low resistances.

Design Principles for Centrifugal Fans

Visualize a fan impeller forming a rotating wheel of gas on which we assume there is no outlet. Now consider only a small prism of gas having a unit area

Figure 7.24   Airfoil blading design used for mechanical-draft ventilation applications (courtesy Westinghouse Electric Corp., Sturtevant Div., Hyde Park, MA).

bound between the limits r_1 and r on the system diagram in Figure 7.30. That is, let us direct our attention to the behavior of prism CB.

Following the analysis of Harris [1917], the centrifugal force acting on an elementary disk across the prism of thickness dx, is

$$dC' = \frac{Wu_x^2 dx}{gx} \tag{7.89}$$

where   W = weight of a cubic volume of gas
   u_x = the velocity of revolution at distance x from the center

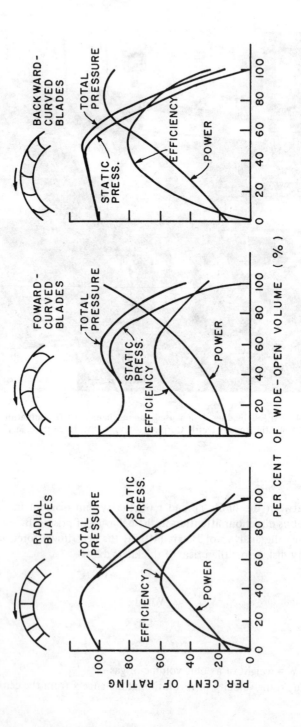

Figure 7.25   Fan performance curves for radial, forward-curved and backward-curved blades.

Figure 7.26   Propeller-type axial-flow fan used for spray booth or general laboratory hood exhaust (courtesy De Vilbiss Co., Div. Champion Spark Plug Co., Toledo, OH).

Because

$$u_x = \frac{x}{r}\, u \tag{7.90}$$

Then

$$dF = \frac{Wu^2}{gr^2}\, x\,dx$$

and the pressure on a unit area at the circumference of the disc is determined by the integral between the limits of $x = r$ and $x = r_1$:

$$C' = \frac{Wu^2}{2gr^2}\,(r^2 - r_1^2) = \frac{W}{2g}\,(u^2 - u_1^2) \tag{7.91}$$

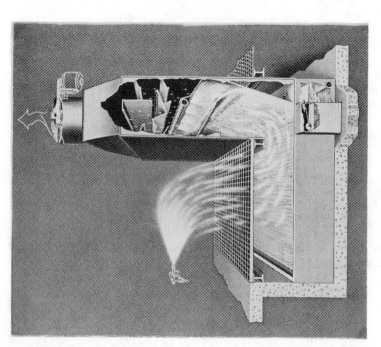

Figure 7.27 A water wash exhaust system and spray booth. The water wash provides scrubbing action of spray particles. In both designs, fans are located in vertical stacks (courtesy De Vilbiss Co., Div. Champion Spark Plug Co., Toledo, OH).

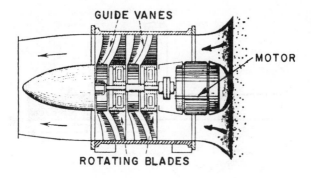

Figure 7.28   A two-stage axial-flow fan.

Figure 7.29   A controllable-pitch axial fan in operation (courtesy Westinghouse Electric Corp., Sturtevant Div., Hyde Park, MA).

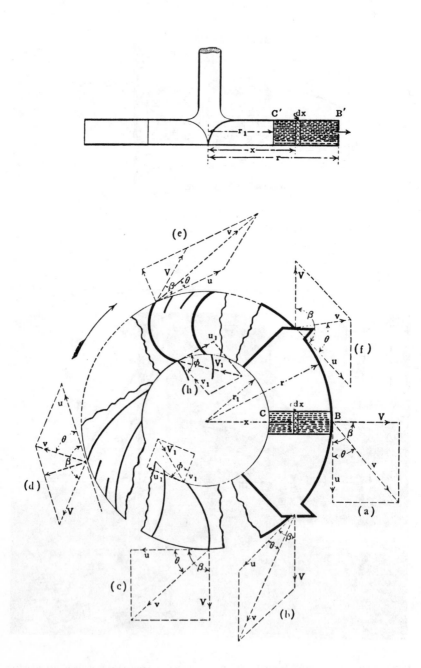

Figure 7.30   Design scheme for centrifugal fans.

As

$$u_1 = \frac{r_1}{r} u \qquad (7.92)$$

Hence, the pressure head against the walls of the wheel at the circumference is

$$h' = \frac{u^2 - u_1^2}{2g} \qquad (7.93)$$

If $r_1 = 0$, then $h' = u^2/2g$.

Note that h' does not include the velocity of rotation. If an opening is provided at the circumference (regardless of the direction and assuming the pressure outside and at the entrance are equivalent), the velocity of the discharge, relative to the rotating walls of the wheel, will be

$$v = \sqrt{2gh'} \quad \text{or} \quad v^2 = u^2 - u_1^2 \qquad (7.94)$$

Note that if $r_1 = 0$, then $v = u$.

The absolute velocity, $\acute{v}$, of discharge consists of two components: V and u:

$$\acute{v}^2 = u^2 + v^2 + 2uv \cos \beta = 2u^2 + 2uv \cos \beta \qquad (7.95)$$

when $r_1 = 0$.

The total head, H, in the departing gas is

$$H = \frac{\acute{v}^2}{2g} = \frac{u^2 + uv \cos \beta}{g} \qquad (7.96)$$

To describe the gas discharge, the initial velocity, v_1, at the entrance and the velocity in the final head within the wheel must be included in our expression. Hence, the total relative head at point B is

$$h' = \frac{v_1^2}{2g} + \frac{u^2 - u_1^2}{2g} \qquad (7.97)$$

And the velocity of the discharge, relative to the revolving components, is

$$v^2 = v_1^2 + u^2 - u_1^2 \qquad (7.98)$$

Assume that CB is a radial frictionless tube, open at both ends, and that the element of gas starts from a state, v_1, relative to the tube and moves outward, at constant pressure, from radius r_1 to r. The radial acceleration of the gas at distance x from the center is

$$\frac{u_x^2}{x} = \frac{dv_x}{dt} \qquad (7.99)$$

and

$$v_x = \frac{dx}{dt}$$

By eliminating dt, we obtain

$$v_x dv_x = \frac{u_x^2 dx}{x} \qquad (7.100)$$

Incorporating Equation 7.90 into Equation 7.100

$$V_x dV_x = \frac{u}{r^2} \, x dx \qquad (7.101)$$

Subscript x indicates the conditions at distance x from the center.
   Integrating Equation 7.101, we obtain, as before,

$$v^2 - v_1^2 = u^2 - u_1^2 \qquad (7.102)$$

We now must evaluate the effect of the gas entrance relative to the moving components of the wheel. Assume the gas to be at rest until influenced by the wheel. The total weight of gas, W, is accelerated to velocity $\dot{v}$, as illustrated in Figure 7.31. This causes the following force against the wheel:

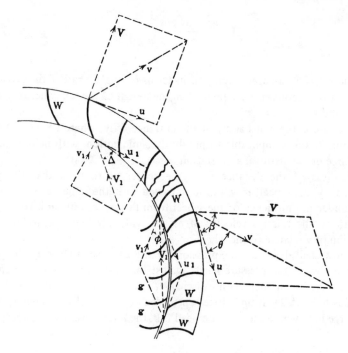

Figure 7.31 Acceleration of the entrance gas by the fan impeller.

$$C' = \frac{W\dot{v}}{g} \qquad (7.103)$$

The component of this velocity in the opposite direction to the rotation, u_1, is

$$\dot{v} \cos \theta = u - v \cos (180° - \beta) = u + v \cos \beta \qquad (7.104)$$

The work performed in overcoming this reaction is

$$u(u + v \cos \beta) \frac{W}{g} = (u^2 + uv \cos \beta) \frac{W}{g} \qquad (7.105)$$

Let H be the total head of the gas up to the point considered. Then the work performed is WH and, as all the head has been imparted by the wheel, we write the following:

$$H = \frac{u^2 + uv \cos \beta}{g} \qquad (7.106)$$

Equation 7.106 has been expressed in terms of angle β rather than α, as it is fixed. Also, β represents an element in the design of the fan, whereas α varies with u and v.

The above expressions apply at the entrance only, where $r - r_1 = 0$. Here, Equation 7.106 is applicable to purely dynamic action, with neither centrifugal force nor centrifugal acceleration.

Regardless of the distance from the center of rotation that the gas is engaged by the wheel, it will have imparted to it the same head as if it had been under the influence of the wheel from the center outward. Under these conditions, the relative velocity of escape will be $v = u_1$, regardless of the direction of v relative to the wheel.

We now shall develop an expression for the total head given to the gas by the wheel when the pressure heads against the outlet and inlet are P and P_1, respectively. When there is no discharge, the pressure developed *within the wheel* is $h' = u^2/2g$. If an orifice is provided at the wheel's periphery into the discharge line, where the pressure head is P, the relative velocity of escape is

$$v^2 = 2g \left(\frac{u^2}{2g} + P_1 - P \right) = u^2 + 2g(P_1 - P) \qquad (7.107)$$

and the total absolute head added to the gas by the fan will be

$$H = \frac{\dot{v}^2}{2g} + P - P_1 \qquad (7.108)$$

For suction at the inlet, P_1 is negative and, hence, a sign change is made to the last term in Equation 7.108. As

$$\dot{v}^2 = u^2 + v^2 + 2uv \cos \beta \qquad (7.109)$$

Equation 7.108 then becomes

$$H = \frac{u^2 + uv \cos \beta}{g} \qquad (7.110)$$

Equation 7.110 is applicable to both purely centrifugal and impulsive action, and is independent of entrance and discharge pressures.

Equations 7.106, 7.107 and 7.108 are theoretical relations and do not account for the effects of friction and/or imperfections of the design.

If $P_1 = P$, then $H = \dot{v}^2/2g$ and $v = u$. If $P_1 = P$ and $\beta = 90°$, then $\dot{v}^2 = 2u^2$. When $\beta = 90°$, $H = u^2/g$, regardless of the pressures. When β is less than $90°$, H is greater than u^2/g; and when β is greater than $90°$, H is less than u^2/g.

The pressure that a fan can produce depends on H. $P = WH$ when the whole energy is transformed into pressure head. Otherwise, in general,

$$P = W\left(H + P_1 - \frac{\dot{v}^2}{2g}\right) \qquad (7.111)$$

The work required (neglecting friction) is

$$WH = wQH \qquad (7.112)$$

where W and Q are total weight and total volume passed, respectively.

Note that even when v is zero, $H = u^2/g$. In this case, the gas revolving in the wheel has a pressure head $= u^2/2g$, with the total head being the sum. The work is zero because $W = 0$.

If the pressure head in the discharge duct is $u^2/2g$, there will be no discharge because the pressures inside and outside the wheel are balanced. As p decreases, v increases and, therefore, Q increases.

In designing a fan, the primary parameters to evaluate are H and Q. As seen from the above relations, these factors are interdependent (except when $\beta = 90°$) because for any completed fan Q is directly proportional to v.

To estimate the total volume of gas per unit time that can be handled by a fan, the following formulas may be used:

$$Q = 2\pi r b v \sin \beta = 2\pi r \sin \beta \sqrt{u^2 - 2g(P_1 - P)} \qquad (7.113A)$$

and

$$Q = 2\pi r_1 b_1 v_1 \sin \phi \qquad (7.113B)$$

where $v \sin \beta$ and $v_1 \sin \phi$ are radial components of the discharge and inlet velocities, respectively. If the designer has data on H and Q (or v) for a particular fan, it must show the same general dependency predicted by Equations 7.113A and 7.113B.

Angle β should be selected on the basis of the intended fan's service requirements and method of propulsion. Assuming constant u for β less than 90°, H increases with Q. When angle β is 90°, H is independent of Q. For β greater than 90°, H decreases as Q increases. In the case of constant H, u increases as β increases, and the embarrassing condition in the design is to apply a high rotative speed without getting excessive head. Hence, in such cases, angle β is chosen to exceed 90°.

Another advantage of backward-curved-vanes (β greater than 90°) is that the fan will not be overloaded when the head or resistance suddenly is reduced. If this were to occur, a sudden increase in the discharge would develop. This is illustrated by the head curves for different β values in Figure 7.32. In applications requiring constant pressure (head) with varying flow quantities (e.g., ventilating fans for buildings), the most rational design includes an adjustable discharge with radial vanes.

A very common design problem is proper fan selection for forcing gas through a long duct. The largest portion of the resistance to be overcome is the friction head, which varies with the square of velocity. Therefore, any increase in the flow quantity will be accompanied by a relatively greater resisting head. This situation is best met by selecting radial vanes at the discharge. Theoretically, $H = u^2/g$, and the flow quantity varies directly with

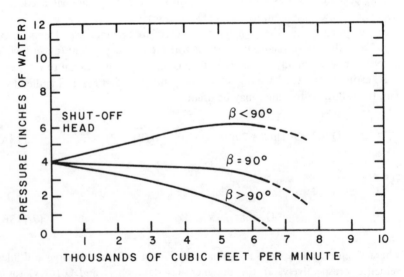

Figure 7.32   Characteristic curves for constant-speed fans with varying discharge.

u. In the long ducts, practically all the resistance is due to friction. For example, if the speed of the fan is doubled, the gas flow doubles, pressure increases by a factor of four, friction increases by a factor of four and the power required to operate the fan will be eight times higher.

For further discussion on fans, consult Stepanoff [1965], Rollins [1973] and Perry and Chilton [1973].

Blowers

The operation of a single-stage centrifugal blower is illustrated in Figure 7.33. The wheel (2), with blades similar to those in a centrifugal pump, rotates inside the casing (1). The wheel is inserted inside the diffuser (3), where the velocity generated is converted into pressure, partially in the wheel and partially in stationary diffusers following the wheel. The diffusers consist of two circular disks connected by curved blades having a slope opposite to that of the wheel. Gas enters the blower through the nozzle (4) and leaves the blower through the discharge nozzle (5). Figure 7.34 shows a single-stage blower.

The operating scheme for a multistage blower is illustrated in Figure 7.35. Contained in the casing (1) are several (usually 3 or 4) wheels (2). The gas passing through the first wheel enters into the guiding diffusers (3) and return channel (4), through which it is guided to the following wheel. The return channel (4) has fixed guiding diaphragms, which impart to the gas velocity and direction. Constant wheel diameters are used in multistage blowers; however, wheel widths decrease in the direction from the first to the last wheel in accordance with the change in gas volume at compression. Thus, gas compression is possible without changing the velocity of rotation or the form of the blades.

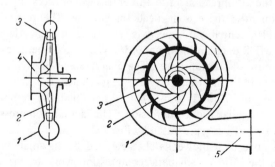

Figure 7.33   Scheme of a centrifugal single-stage blower: (1) casing; (2) guiding diffuser; (3) suction nozzle; (4) discharge nozzle.

Figure 7.34   Single-stage centrifugal blower, available in permanent-mold, precision-cast aluminum, fabricated steel or PVC housings (courtesy North American Mfg. Co., Cleveland, OH).

In general, the compression ratio in centrifugal blowers does not exceed 3–3.5. Consequently, there is no need for cooling. Theoretically, compression in noncooled centrifugal blowers is very close to adiabatic, as shown by line AB on the T-S diagram in Figure 7.36. The actual energy expended will be greater because of friction losses. The energy spent in overcoming the gas friction is converted almost entirely into heat. Hence, the final gas temperature (T_2' in Figure 7.36) is somewhat higher than the temperature, T_2, corresponding to adiabatic compression. The actual compression process from pressure P_1 to P_2 is represented by line AC.

The performance of a blower is characterized by its adiabatic efficiency, η_{ad}, which is the ratio of adiabatic compression work to the actual work expended:

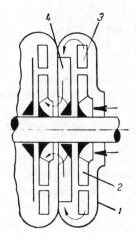

Figure 7.35 Operating scheme for a multistage blower: (1) casing; (2) wheel; (3) guiding diffusers; (4) return channel.

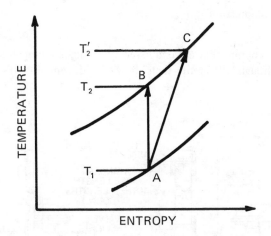

Figure 7.36 Entropy diagram of gas compression in a centrifugal blower.

$$\eta_{ad} = \frac{\tilde{\ell}_{ad}}{\tilde{\ell}} = \frac{C_p(T_2 - T_1)}{C_p(T_2' - T_1)} = \frac{T_2 - T_1}{T_2' - T_1} \qquad (7.114)$$

Temperature, T_2, may be calculated from the following equation:

$$T_2 = T_1 \left(\frac{P_2}{P_1}\right)^{(\kappa-1)/\kappa}$$ (7.115)

and temperature, T_2', is measured at the exit of the blower.

An additional blower design widely used in pneumatic bulk trailers is the rotary blower. Typical designs consist of helical four-flute gate rotors or two-lobe main rotors. The operating principle of a tractor-mounted unit is illustrated in Figure 7.37. The screw-type rotor is similar in design to a twin-screw pump. The meshing of the two screw-type rotors is synchronized by timing gears to provide controlled compression of air. Designs are axial flow, positive displacement types. These blowers are adaptable to various types of drives, such as electric motors (constant or variable speed), gasoline or diesel engines, or steam turbines. V-belt drive, direct connected or drive through a speed-regulating mechanism may be used. Figure 7.38 shows a rotary lobe-type, axial-flow, positive displacement blower ready for a horizontal installation.

Centrifugal Compressors

Centrifugal compressors are employed when higher compression ratios are needed. Although different than blowers, designs resemble multistage

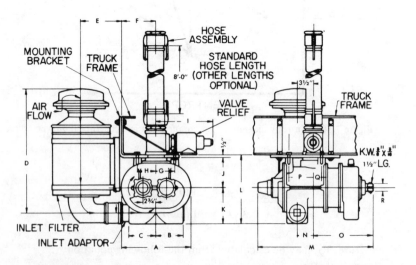

Figure 7.37   Operating features of a tractor-mounted rotary blower.

Figure 7.38 Small rotary blower for horizontal service installation (courtesy Gardner-Denver Co., Industrial Machinery Div., Quincy, IL).

blowers (similar to Figure 7.35). In contrast to blowers, to achieve greater discharge pressures in compressors the number of wheels must be increased, wheel sizes (including diameter) must be changed and the velocity of rotation must be raised. The peripheral velocities of the wheels can reach up to 240–270 m/s or more, with discharge pressures as high as 2.5–3.0×10^6 N/m^2 (25–30 atm).

Both smaller widths and wheel diameters are needed when adding stages of higher pressure. Centrifugal compressors do resemble blowers, however, in that the diffusers used for converting the gas kinetic energy into potential energy of pressure and the return channels are quite similar.

The working wheels of centrifugal compressors often are sectionalized and, thus, located in two or three casings. As noted in the previous section, high gas compression ratios produce significant increases in the gas temperature and, hence, the need for cooling. Gas cooling is effected either by introducing water in water-cooled diaphragms (located inside the compressor casing) or in external coolers. Cooling in external coolers is more effective and simplifies maintenance of heat transfer surfaces.

Figure 7.39 provides the entropy diagram of gas compression in a compressor with two external interstage coolers plus final cooling after the last stage. The diagram is interpreted by assuming that the gas is cooled (isobar) to the initial temperature of the gas, with zero pressure losses. The process is illustrated by the broken line ACDEFGH. The shaded area is equivalent to the savings in work obtained over compression without intermediate cooling.

Gas compression in centrifugal compressors is similar to that in blowers. As shown by Figure 7.39, after compression by the group of noncooled wheels (lines AC, DE and FG), the gas acquires temperature T_2', which is higher than T_2 at the end of adiabatic compression (point B). The degree at which the process approaches the isothermal case is characterized by the compressor's isothermal efficiency, i.e., the ratio of isothermal work to the work spent:

$$\eta_{iso} = \frac{\tilde{\ell}_{iso}}{\tilde{\ell}} \qquad (7.116)$$

Isothermal efficiency is in the range of 0.5 to 0.7.

The *axial-flow dynamic compressor* (Figure 7.40) meets the requirement of large capacity but at relatively low compression ratios (3.5–4). The characteristics of these machines are quite different from the centrifugal type.

The rotor blades impart velocity and pressure to the gas as the rotor turns,

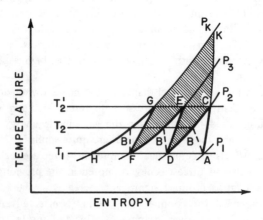

Figure 7.39   Entropy diagram for gas compression in a multistage centrifugal compressor.

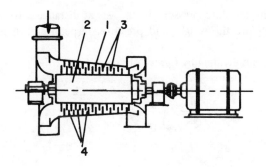

Figure 7.40   Cross section of a typical axial-flow dynamic compressor: (1) casing;
(2) rotor; (3) blades; (4) inlet guide blades.

with the velocity being converted to pressure in the stationary blades. Axial-type compressors are driven with steam or gas turbines with velocities of 5000–5500 rpm. They typically have 10–20 stages and operate without gas cooling.

In developing the governing equations for fans, we assumed gas density to be constant. This includes the case of centrifugal acceleration with a change in pressure and purely impulsive action. For blowers and compressors, the compression due to centrifugal force within the wheel is significant and therefore must be included in the performance equations.

In developing the equations, we shall assume isothermal conditions because the ratio of compression in each stage is low and intercooling can approximate this state. Referring back to Figure 7.30, assume that prism CB contains a compressible gas. The weight of a unit volume of gas, W_x, will depend on the distance, x, from the center and on the velocity of rotation.

The centrifugal force due to a unit area disk of radial thickness, dx, is

$$dP_x = \frac{W_x u_x^2}{gx}\, dx = \frac{W_x u^2}{gr^2}\, x dx \qquad (7.117)$$

because $u_x = \dfrac{x}{r}\, u$.

And

$$\frac{W_x}{W_1} = \frac{P_x}{P_1} \qquad (7.118)$$

where P_x is the absolute pressure in the gas at distance x from the center, and W_1 and P_1 are the weight and pressure, respectively, of the gas at the entrance.

Substituting and dividing by P_x, we obtain

$$\frac{dP_x}{P_x} = \frac{W_1 u^2}{P_1 g r^2} \, x dx \tag{7.119}$$

then,

$$\int_{P_1}^{P} \frac{dP_x}{P_x} = \frac{W_1 u^2}{P_1 g r^2} \int_{r_1}^{r} x dx \tag{7.120}$$

whence

$$ln \frac{P}{P_1} = \frac{W_1 u^2}{P_1 2 g r^2}(r^2 - r_1^2) = \frac{W_1}{P_1}\left(\frac{u^2 - u_1^2}{2g}\right) \tag{7.121}$$

because

$$r_1 = \frac{r}{u} u_1$$

Referring to Figures 7.31 and 7.41, we shall consider the more general and direct demonstration of Equation 7.110. The static pressure in the gas changes as it passes out of the rotating portion into the fixed outlet passage.

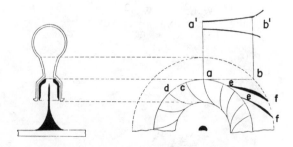

Figure 7.41   Gas discharge region of the fan.

It is this pressure drop that induces the relative discharge velocity, v. This difference in pressure (purely static) offers no resistance to the rotation of the wheel, as will be seen if we imagine the periphery of the wheel closed while rotating in a frictionless gas. The pressure in the frictionless gas must be normal to the periphery and, therefore, does not resist its rotation.

Then, in all cases (regardless of a change in pressure at the outlet) the resistance to rotation is due solely to the behavior of the departing jet. This behavior is in a direction opposite to that of the absolute velocity of discharge, $\dot{v}$ (Figure 7.30) and in the amount of $W(\dot{v}/g)$. However, the component opposed to rotation (that is, in the direction opposite to u) is $w(\dot{v}/g) \cos \theta$ and, as is apparent from the diagrams, $\dot{v} \cos \theta = u + v \cos \beta$. Therefore, the force opposed to rotation is $(W/g)(u + v \cos \beta)$. As the work done by the wheel equals force times distance, then

$$\text{Work} = \frac{W}{g} (u^2 + uv \cos \beta) \tag{7.122}$$

Evidently this is independent of the radial depth, $r - r_1$, of the vanes.

For a gas of uniform density (in low-pressure fans density changes are small), if the machine imparts a head, H, then work is WH, and

$$WH = \frac{W}{g} (u^2 + uv \cos \beta)$$

and

$$H = \frac{u^2 + uv \cos \beta}{g}$$

Let R_1 be the final ratio of compression when the gas has been brought to rest after one stage. Then work = $p_a v_a \, ln \, R_1$, where v_a is the volume of gas compressed. Then,

$$P_a v_a \, ln \, R_1 = \frac{W_a v_a}{g} (u^2 + uv \cos \beta)$$

and

$$ln \, R_1 = \frac{W_a}{P_a} \left(\frac{u^2 + uv \cos \beta}{g} \right) \tag{7.123}$$

This formula provides the ratio of compression produced by one stage in a centrifugal compressor.

If there are n stages, each contributing an additional ratio $(R_1, R_2, \ldots R_n)$, then

$$R_n = R_1^n$$

and

$$ln\ R_n = n\ ln\ R_1 \qquad (7.124)$$

The tremendous centrifugal force developed in these machines prompts most manufacturers to prefer the outer tips of the propeller blades to be radial $(\beta = 90°)$ to avoid cross bending.

In the case in which $\beta = 90°$, Equation 7.122 simplifies to:

$$ln\ R_1 = \frac{W_a u^2}{P_a g} \qquad (7.125)$$

Note that $W = p/53.35t$ and, if t is constant, $W_a/P_a g$ will be constant. To adopt the formula to common logarithms, divide the expression by 2.3026.

Obviously, perfect cooling cannot be accomplished. As an example, assume an average temperature of 580°R (120°F). Evaluating Equation 7.124,

$$\log R_1 = \left(\frac{1}{2.3026 \times 53.35 \times 580 \times 32.2}\right) u^2 = ku^2 \qquad (7.126)$$

$$\log k = 7.6398$$

Several practice problems are given to illustrate the above analysis.

Compressor Selection

Piston and centrifugal compressors are perhaps the compressors most often used in the CPI. Proper selection of a machine for a particular application is a complex process that requires detailed information on the advantages and disadvantages of each compressor type. Careful evaluation must be made of such factors as gas flowrate, inlet and discharge pressures, gas properties, power consumption, and investment and maintenance costs.

Turbocompressors and turboblowers are characterized by their compactness, simple design and steady delivery. Their important advantage is purity of gas delivered, that is, the gas is not contaminated with lubricant. The absence of inertia forces and high speed permits their installation on lighter foundations with direct connection to the drive (usually to gas or steam turbines) or to an electric motor through the gearbox by increasing the number of rotations.

The efficiencies of turbocompressors are lower than piston compressors. However, turbocompressors, with capacities of 6000 m^3/hr and higher, are often more economical because of lower capital and maintenance costs. Therefore, turbocompressors usually are applied in industries in which very large quantities of gas are handled (10,000–200,000 m^3/hr or more) at pressures up to 30 atm (average 10–12 atm). Modern multistage compressors develop pressures up to 300 atm.

In the range of lesser capacities (up to 10000 m^3/hr) and in the wide pressure range (up to 1000 atm), piston compressors are used almost exclusively. Rotary and screw compressors, having the advantages of turbocompressors, have higher efficiencies than centrifugal compressors and are used at capacities up to 6000 m^3/hr and at pressures up to 15 atm. Their disadvantages are the complexity of their fabrication and high maintenance costs.

Axial compressors are distinguished by their compactness and high efficiency; they are used at high deliveries (80,000 m^3/hr and higher) and at low pressures (up to 6 atm).

Further criteria for compressor selection are outlined by Church [1944], Foust et al. [1980], Kearton [1926], Gill [1941], James et al. [1971], Gibbs [1971], the Society of Automotive Engineers [1961] and Streeter and Wylie [1979].

PRACTICE PROBLEMS

Solutions are provided in Appendix A for those problems with an asterisk (*) next to their numbers.

*7.1 The fan shown in Figure 7.42 is being used to deliver nitrogen gas (sg = 1.2 kg/m^3) from a sphere to a countercurrent wetted wall absorption column. The excess pressure in the gas sphere is 60 mm H_2O and that in the reactor unit is 74 mm H_2O. Losses in the suction and discharge lines are 19 mm H_2O and 35 mm H_2O, respectively. The discharge velocity of the gas is 11.2 m/s. Determine the pressure developed by the fan.

7.2 For the fan system described in Problem 7.1, carbon dioxide is being delivered at a rate of 150 cfm. The absorption column is being fed

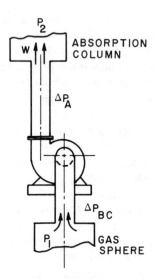

Figure 7.42   Fan used to deliver nitrogen gas (Problem 7.1).

with a 10% solution of sodium hydroxide at a rate of 50 gpm to a 76-cm i.d. column. The temperature of the operation is 65°F. Determine the fan pressure.

*7.3 The vacuum created in the suction line of a centrifugal fan is 15.8 mm H_2O. A manometer on the discharge side of the fan reads 20.7 mm H_2O for an air rate of 3700 m^3/hr. Both the suction and discharge duct sizes are identical, and the fan operates at 960 rpm. The fan consumes 0.77 kW of energy.

   (a) Determine the pressure developed by the fan.
   (b) Determine the fan's efficiency.
   (c) If the fan's speed is increased to 1150 rpm, what will be the new capacity?
   (d) Determine the fan's horsepower at the higher rpm.

*7.4 Tests were performed on a new centrifugal fan to assess its operating characteristics. Data obtained are summarized in Table 7.1. All tests were performed at a constant speed of 1440 rpm. The intended operation has the following pressure losses for the volumetric flow of 1350 m^3/hr:

   • $\Delta p_v = 8.7$ mm H_2O
   • $\Delta p_{fr} + \Delta p_{\varrho.r.} = 29.4$ mm H_2O

Table 7.1 Fan Performance Data for Problem 7.4

Q (m^3/hr)	100	350	700	1000	1600	2000
ΔP (kg/m^2)	45.8	43.2	44.0	43.5	39.5	32.2

The pressure differential (between discharge and suction) for the designed piping is $\Delta P_{a1} = P_2 - P_1 = 13$ mm H_2O. Determine the fan's capacity if it is operated at the same speed at which the tests were performed.

*7.5 Compare the theoretical work required for compressing 1 m^3 of air from 1 atm to:
(a) 1.1 atm.,
(b) 5 atm.
(c) Calculate the work spent from the thermodynamic formula for adiabatic compression as well as from the hydraulic formula (i.e., consider air to be incompressible).

*7.6 A single-stage piston compressor is used to deliver 460 m^3/hr of ammonia (NH_3) gas to a system. The gas is compressed from 2.5 to 12 atm and has an initial temperature of 10°C. The temperature-entropy diagram for ammonia is given in Figure 7.43. Determine the compressor's horsepower and the temperature of the compressed gas for a compressor efficiency of 0.7.

*7.7 A single-stage compressor is being considered to supply 80 kg/hr of air under a pressure of 4.5 atm gauge. The compressor's cylinder is 180 mm in diameter with a stroke length of 200 mm. The piston operates at 250 rpm with a 5% displacement clearance. The exponent of the polytrope is $\kappa = 1.25$. Evaluate whether this system is suitable for the intended service.

7.8 An adiabatic turbocompressor has radial blades at the exit of its 175-mm-diameter impeller. The unit compresses 0.8 kg/s of air at 104.7 kPa abs (t = 18°C) to 315.3 kPa abs. The entrance area is 73 cm^2 and the exit area is 42 cm^2. The compressor's efficiency is 0.72 and its mechanical efficiency is 0.89.
(a) Determine the rotational speed of the impeller.
(b) Determine the actual temperature of the air at discharge.

7.9 The speed of a hydraulic turbine is 310 rpm and the head at the site is 357 ft. The flowrate through the turbine is 650 cfs. The hydraulic

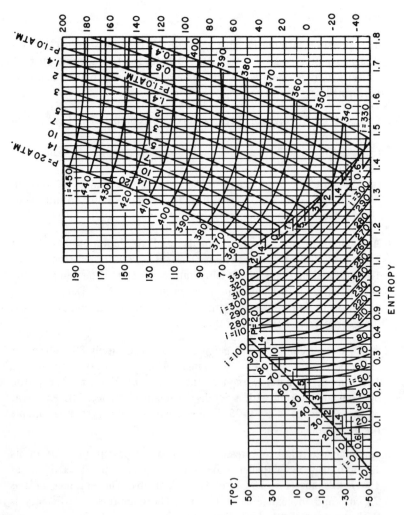

Figure 7.43   Temperature-entropy diagram for ammonia.

turbine efficiency is 0.89. Determine the specific speed (refer to previous chapter for help).

7.10 An axial-flow centrifugal compressor has straight radial blades and operates at 18,000 rpm. The tip diameter is 26 in. and the inlet stagnation temperature is 75°F.
   (a) Determine the work input to the rotor per pound of gas handled.
   (b) Determine the stage pressure ratio for isentropic flow. Assume $\kappa = 1.4$ and $C_p = 0.24$ Btu/lb$_m$ °R.

*7.11 The clearance in a single-stage compressor used for methane is 8.5% from the displacement. Considering gas expansion from the clearance volume to be adiabatic, determine at which limiting discharge pressure the capacity will be equal to zero. The suction pressure is atmospheric.

*7.12 Compare the temperatures at the end of compression, theoretical horsepowers and volumetric efficiencies for compressing air from 1 to 9 atm:
   (a) in a single-stage piston compressor; and
   (b) in a double-stage compressor with intermediate cooling, the initial temperature of the air and its temperature after the cooler being $t = 20°C$ and the clearance being 8% of the piston displacement.

*7.13 A compressor delivers 210 m^3/hr of methane at a pressure of 55 atm. The initial pressure is atmospheric, and the inlet temperature is 18°C. Determine
   (a) the number of stages and the pressure distribution by stages;
   (b) the compressor's horsepower, assuming an efficiency $\eta = 0.7$; and
   (c) the amount of cooling water required if the gas temperature rises 10°C.

*7.14 Determine the type of compressor required to deliver air to a classifier at a rate of 230 m^3/min under a pressure of 6 atm. Use the plot given in Figure 7.44 to make the selection.

*7.15 A vacuum of 0.9 atm is created by a piston vacuum pump. Assuming a polytropic compression ($\kappa = 1.25$), determine the theoretical work consumed under the following conditions:
   (a) at the moment when the vacuum is 0.1 atm (i.e., the residual pressure is 0.9 atm);
   (b) when the pressure in the delivery vessel is 0.3 atm; and
   (c) when the desired vacuum is achieved (i.e., the residual pressure in the vessel is 0.1 atm).

7.16 An axial flow fan is installed inside a laboratory walk-in hood. The exhaust duct on the hood is made from galvanized sheet metal and has

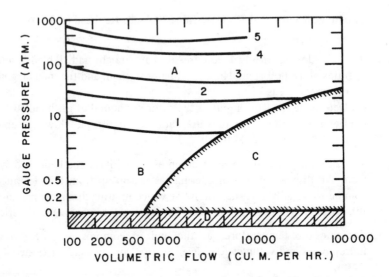

Figure 7.44   Performance curves for Problem 7.14: (A) Piston compressors: (1) single stage; (2) double stage; (3) three stage; (4) four stage; (5) five stage: (B) rotating compressors and blowers; (C) turbocompressors and turboblowers; (D) fans.

a diameter of 2.5 ft and a length of 385 ft. The fan rotates at 1320 rpm and moves air at a rate of 210 cfs (air density $\simeq 0.075$ lb_m/ft^3). Determine the following:

(a) the pressure drop in the duct, and

(b) the horsepower required to drive the fan.

7.17 A radial-flow blower operates at 970 rpm and moves 350 cfm of air at an average temperature of 120°F. The rotor has a width of 0.9 ft, an outer diameter of 1 ft and an inner diameter of 7 in. The blades are designed for radial inflow, i.e., $\alpha_1 = 90°$. Determine

(a) the blade angles;

(b) the required horsepower; and

(c) the head produced.

7.18 An air-compressor plant is being considered for operating a mine pump. The specifications for the operation are given in Table 7.2. Determine the following information to complete the design:

(a) the total pressure head against which the pump must work;

(b) the total work required in the air end of the pump; and

(c) the total work delivered to the water per minute.

Table 7.2 Plant Specifications for Problem 7.18

Water Volumetric Flow	2500 gpm
Net Water Lift	375 ft
Diameter of Water Pipe	12 in.
Length of Water Pipe	1033 ft
Length of Air Pipe	980 ft
Atmospheric Temperature	60°F
Elevation from Sea Level	6280 ft
Mechanical Efficiency of the Pump	0.9
Average Pump Piston Speed	250 fpm
Mechanical Efficiency of Air Compressor	0.82
Air Compressor Speed	95 rpm
Compressor Volumetric Efficiency	0.80
In Compression and Expansion, n =	1.3

7.19 For the system described in Problem 7.18, the compression is single stage to 95 lb gauge. The pump is direct acting without flywheels. No reheating has been specified, and it can be assumed that the air does not undergo expansion in the pump. Determine the following:
(a) the air pressure at the pump,
(b) the pressure loss in the air pipe,
(c) the ratios between the areas of air and water cylinders in the pump,
(d) the volume of compressed air in the pump,
(e) the diameter of the air pump, and
(f) the horsepower required in the steam end of the compressor.

7.20 Determine the number of revolutions for an isothermal compression with the following specifications: 18-in. stroke, 15-in. piston diameter, double-acting. The gas is air (initial temperature is 70°F), which is brought from atmospheric to 110 lb gauge in a cylindrical vessel that is 6 ft in diameter and 10 ft tall.

7.21 For the system described in Problem 7.20, determine the following:
(a) the compressor's horsepower if it is operated at 85 rpm,
(b) the horsepower if the compression is adiabatic,
(c) the volume of air handled per hour for an rpm of 100 and temperature of 85°F.

7.22 Determine the theoretical horsepower required to compress 20 lb/min of air from 14.5 psi (t = 25°C) to 110 psig:
(a) Assume adiabatic compression.
(b) Assume isothermal compression.

*7.23 Methane is to be compressed from atmospheric pressure to 3.5 atm at a rate of 1.5 kg/s.

(a) Determine the isothermic and adiabatic compressor efficiencies if the compressor horsepower is 430 kW and the mechanical efficiency is 0.91. The adiabatic exponent for methane is $\kappa = 1.31$.

(b) Determine the exponent for polytropic compression for a discharge temperature of 80°C. The temperature at suction is 20°C. The density of methane at STP is 0.717 kg/m^3.

*7.24 Air at 20°C and 1 atm is compressed to 110 atm at a rate of 0.5 kg/s.

(a) Determine the number of compression stages and the theoretical horsepower for compression if, after each stage, the air is cooled to 35°C. The compression is polytropic (n = 1.3).

(b) Compare the multistage compression with that of a single-stage in the isothermic and polytropic compression processes.

7.25 Air at t = 20°C is compressed from atmospheric pressure to 30 atm. Determine the number of compression stages, the intermediate pressures and the mechanical work for compressing 1 kg of air if the air is cooled between stages to the initial temperature of 20°C. The compression is adiabatic ($\kappa = 1.4$).

7.26 Determine the number of compression stages of a turbocompressor and its horsepower at a capacity of 1.5 kg/s air at t = 20°C. The air is compressed from 2 to 15 atm. The compression is adiabatic, and the total efficiency of the compressor is $\eta = 0.8$.

*7.27 Methane (at t = 20°C) is compressed in a multistage compressor from 1.5 atm to 90 atm. The system provides intermediate cooling to the initial temperature. Determine the percentage increase in horsepower needed if the gas is cooled to 40°C. The compression is adiabatic ($\kappa = 1.31$).

*7.28 A fan is used to ventilate a workroom with atmospheric air. The unit's capacity is 12,500 m^3/hr. Determine the weight capacity during winter (average temperature 15°C) and summer (30°C) service.

*7.29 Determine the pressure developed by a fan that delivers atmospheric air (t = 18°C) to a vessel operating at an excess pressure of 43 mm H$_2$O. Pressure losses in the ductwork are 28 mm H$_2$O, and the average air velocity through the duct is 11.5 m/s.

*7.30 Determine the horsepower of a motor for a fan that delivers 110 m^3/min of air at a head of 85 mm H$_2$O. The efficiency of the fan is 0.47.

*7.31 A centrifugal fan when operated at 960 rpm, delivers 3200 m^3/hr of air. The fan's horsepower is 0.8 kW and its pressure 44 mm H$_2$O.

(a) What will be the capacity, pressure and horsepower if the fan's speed is increased to 1250 rpm?

(b) What is the fan's efficiency at the higher speed?

*7.32 Determine the temperature of an airstream after it has been compressed adiabatically from 1 to 3.5 atm. The initial temperature of the air is $0°C$.

*7.33 For the previous problem, determine the work spent for compression of 1 kg of air.

*7.34 Determine the horsepower of a CO_2-piston compressor having a capacity of 5.6 m^3/hr at suction conditions. The gas is compressed from an initial pressure and temperature of 20 atm and $-15°C$, respectively, to a final pressure of 70 atm. The compressor efficiency is 0.65. The T-S diagram for CO_2 is given in Figure 7.45.

*7.35 For the compressor described in the previous problem, determine the volumetric efficiency for a clearance of 6% of the displacement. Use a value m = 1.4 for the exponent of the polytrope.

*7.36 A single-stage piston compressor has the following specifications: 250-mm piston diameter; 275-mm piston stroke; clearance 5.4% of displacement; and 300 rpm. Air is compressed from atmospheric conditions to 4 atm. The polytrope exponent for expansion is 10% less than that of the adiabatic case, and the total efficiency of the compressor is 0.72. Assume an ambient air temperature of $25°C$. Determine the capacity and horsepower of the unit.

*7.37 If the suction pressure of the single-stage piston compressor described in Problem 7.36 is increased to 0.4 atm gauge, what is the new capacity and horsepower. The final pressure remains unchanged (4 atm).

7.38 A single-stage piston compressor is used for delivering ethylene. The clearance is 7% of the piston displacement, and the gas expansion from the clearance may be considered adiabatic. Determine what discharge pressure will produce a volumetric efficiency of 0.2.

*7.39 The lubricating oil for a compressor may be used safely up to a temperature of $160°C$. Determine the maximum discharge pressure in a single-stage piston compressor for (a) air, and (b) ethane. The suction pressure is 1 atm, and the initial temperature of either gas is $25°C$. Assume the compression to be adiabatic.

*7.40 Determine the required number of stages of a piston compressor to compress nitrogen from 1 to 100 atm if the allowable discharge temperature should not exceed $140°C$. The compression is adiabatic and the initial temperature of the gas is $20°C$.

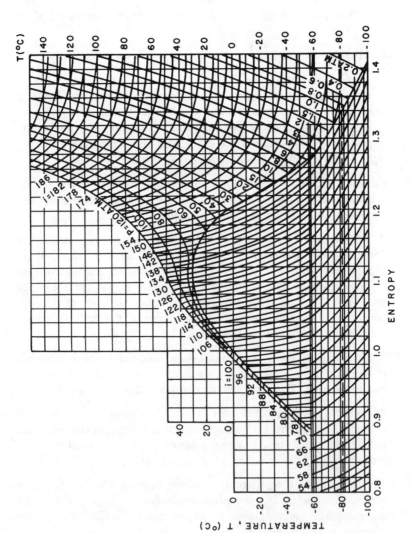

Figure 7.45   Temperature-entropy diagram for CO_2.

7.41 Determine the theoretical work for compressing hydrogen gas from 1.5 to 17 atm in both a single-stage and a two-stage compressor. The initial temperature of the gas is 20°C.

7.42 A small compressor is tested in the laboratory by delivering air to a balloon to a volume of 42.4 liters. In 10.5 minutes the pressure in the balloon is increased from 0 to 52 atm gauge and the temperature of the air increased from 17°C to 37°C. Determine the capacity of the compressor in m^3/hr.

7.43 Determine the horsepower and cooling water consumption required for two-stage and three-stage compression of ethylene from 1 to 18 atm. The efficiency of the piston compressor is 0.75 and its capacity is 625 m^3/hr. The initial temperature of the gas is 20°C, and the water temperature is raised by 13°C in the interstage coolers.

*7.44 Select the proper machine for compressing air for the following conditions:
(a) 4200-m^3/hr capacity, from atmospheric to 5.5 atm gauge; and
(b) 300-m^3/hr capacity, from atmospheric to 2 atm.
Capacities given are based on inlet conditions.

NOTATION

A	thermal equivalent of work, 1/427 kcal/kg_f-m
a,b	constants in van der Waal's formula, defined in Equation 7.3
$C_{1,2,3}$...	clearance volumes in successive stages, m^3
C_p	specific heat at constant pressure, J/kg-°K
C_v	specific heat at constant volume, J/kg-°K
C'	centrifugal force, N
D	inside diameter, m
d	cylinder diameter m
e	eccentricity of rotor, m
F	piston cross section, m^2
f_{in}	area of indicator diagram m^2
g	gravitational acceleration, m/s^2
H	total head, m
HP	required power of compressor, hp
h'	pressure head, m
i	enthalpy, J/kg or kcal/kg
K	Boyle's law constant, N-m
k	constant
ℓ	length of blade, m

$\tilde{\wp}$	work of compression, J/kg
M	molecular weight, kg/kg-mole
m	thermodynamic compression exponent
m_p	exponent for compression polytrope
N	theoretical horsepower, hp
n_s	number of piston double strokes or rpm
n	exponent in Boyle's law equation (Equation 7.13) in polytropic expansion it is C_p/C_v
P,p	pressure, N/m^2
P_{cr}	critical pressure, N/m^2
Q	volume capacity, cfm
Q_T	volumetric flow, m^3/s
q	heat, J/kg
R	gas constant, J/kg-mole-°K, or compression ratio
R_t	overall ratio of compression
S	entropy, J/kg °K
Sc_{sp}	scale of indicator spring
s	piston stroke, m
T	absolute temperature, °K
T_{cr}	critical temperature, °K
t	temperature, °C, °F
U	internal energy, N-m or J
u	velocity of revolution, m/s
V	volume, m^3
V	velocity at discharge from a fan blade, m/s
V_p	piston displacement volume, m^3/s
$\hat{V}$	molar volume, m^3/kg-mole
v	velocity, m/s
W	gas weight, kg; or mass rate, kg/s
W'	work performed per cycle, J
w	velocity, m/s
x	ratio of difference between total cylinder volume and actual gas volume to piston displacement
$\bar{z}$	number of blades
Z	number of stages

Greek Symbols

α	angle defined in Figure 7.31, °
β	angle defined in Figure 7.30, °
δ	thickness, m
ξ	fractional volume of clearance

η	efficiency
θ	angle defined in Figure 7.31
κ	ratio of specific heats
λ_h λ_T	hermetic and thermal coefficients, respectively
λ_v	delivery coefficient
ν	specific volume, m^3/kg
ν_a	volume of compressed gas, m^3
$\dot{\nu}$	absolute velocity of discharge, m/s
ϕ	angle defined in Figure 7.31, °
ψ	pressure loss coefficient

REFERENCES

Church, A. H. (1944) *Centrifugal Pumps and Blowers* (New York: John Wiley & Sons, Inc.).

Compressed Air and Gas Institute (1954) *Compressed Air Handbook* (New York: McGraw-Hill Book Co., Inc.).

Coulson, J. M., and J. F. Richardson (1962) *Chemical Engineering* (Elmsford, NY: Pergamon Press, Inc.).

Foust, A. S. et al. (1980) *Principles of Unit Operations,* 2nd ed. (New York: John Wiley & Sons, Inc.).

Gibbs, C. W., Ed. (1971) *Compressed Air and Gas Data* Ingersoll-Rand Co., NJ.

Gill, T. T. (1941) *Air and Gas Compressors* (New York: John Wiley & Sons, Inc.).

Harris, E. G. (1917) *Compressed Air* (New York: McGraw-Hill Book Co.).

James, E., A. John and W. Haberman (1971) *Introduction to Fluid Mechanics* (Englewood Cliffs, NJ: Prentice-Hall, Inc.).

Kearton, W. J. (1926) *Turboblowers and Compressors* (New York: Pitman Press).

Perry, J. H., and C. H. Chilton, Eds. (1973) *Chemical Engineer's Handbook,* 5th ed. (New York: McGraw-Hill Book Co.).

Rollins, J. P., Ed. (1973) *Compressed Air and Gas Handbook,* 4th ed. (New York: Compressed Air and Gas Institute).

Society of Automotive Engineers (1961) *Centrifugal Compressors,* Vol. 3, New York.

Stepanoff, A. J. (1965) *Pumps and Blowers* (New York: John Wiley & Sons, Inc.).

Streeter, V. L., and Z. Wylie (1979) *Fluid Mechanics* (New York: McGraw-Hill Book Co.).

CHAPTER 8

INTRODUCTION TO HETEROGENEOUS
SYSTEMS AND EXTERNAL PROBLEMS
OF HYDRODYNAMICS

CONTENTS

INTRODUCTION

Previous chapters have focused on standard unit operations for handling homogeneous fluid systems. Nature, however, tends toward complexity, so most process problems deal with mixed or heterogeneous systems. Industrial operations generate heterogeneous systems by one or more of the following general processes:

Mechanical. Operations that process various raw materials in their solid state or utilize materials in the processing of fluids, size reduction of particulates, mixing of solids and liquids are all examples of heterogeneous systems. The mixing of bulk materials with liquids results in suspensions. Drying bulk materials and mixing result in the formation of dust.

Thermal. An example of a heterogeneous system generated by a thermal process is fog, which is formed under natural conditions by the condensation of water vapor due to its cooling by air. Hence, fog is a gas-liquid system described as a suspension of water droplets in air. Many industrial operations apply intensive evaporation to different solutions producing solvent vapors. As these vapors travel through such process equipment as reactors, they normally flow at high velocities and entrain liquid particles, thus producing a system similar to fog.

Chemical. One example is sugar production, in which the cleaning of diffused liquors is accomplished by limewater and then by bubbling carbon dioxide through the fluid. This produces a suspension of saturated liquor, which consists of a sugar solution and solid particles of $CaCO_3$ formed by the chemical reaction between CaO and CO_2.

Diffusional. An example of a heterogeneous system generated by this process is crystal formation by the addition of a third substance to a suspension, which may react to form a precipitate or simply decrease the solubility of the precipitated material.

Biological. Milk is an example of a heterogeneous system produced by a biological process. It is essentially an emulsion comprising water and 2–5% fat globules in the size range of 1 to 10 μm.

Most heterogeneous systems are formed from a combination of the above processes. For example, drying often is accompanied by mechanical and pneumatic mixing of materials. This chapter will serve as a brief introduction to the characteristics of heterogeneous flow systems with an overview of the unit operations that involve their separations. This material will serve as a background to forthcoming discussions on external problems of hydrodynamics.

CLASSIFICATION OF HETEROGENEOUS SYSTEMS

The classification by state of aggregation and particle sizes provides a fundamental description of the properties of heterogeneous systems (HS). Each HS consists of two phases: dispersing (continuous) and dispersible (discrete). An interface separates these phases. It is the existence of an interface that distinguishes between a heterogeneous system and a simple solution.

Mixtures that have liquid discrete phases are referred to as liquid HS; those with gaseous discrete phases are called gaseous HS. Both classes are subdivided depending on the state of aggregation of the discrete phases. Liquid HS have solid particles as the discrete phases and are referred to as suspensions. When a liquid HS has another liquid (immiscible) or gas as the discrete phase, it generally is referred to as an emulsion or foam, respectively. The gaseous HS, with solids as the discrete phase, are called dusts, smokes and, with a liquid phase, fogs. Classification according to the state of aggregation is summarized in Figure 8.1. There is no total classification by particle size. Such a classification is used only for suspensions. Coarse suspensions comprise particles with sizes in excess of 100 μm. Fine suspensions comprise particles in the size range of 0.5 to 100 μm, with turbidity in the range of 0.5 to 0.1 μm. Colloidal solutions contain particles less than 0.1 μm.

Any classification by sizes is highly subjective because the discrete phase consists of particles of different sizes and shapes. Therefore, a system can be characterized by a conditionally assumed size. For example, in centrifugal processes the effective particle diameter is assumed to be the determining size (i.e., the characteristic particle size on which the process design is based). The term "particle size" is also highly subjective because particulates often are irregular in shape. A conventional definition for the size of a particle is its equivalent diameter, which is the diameter of a sphere occupying the same volume as that of the particle:

$$d_{eq} = \sqrt[3]{\frac{6V}{\pi}} = 1.24 \sqrt[3]{V} \qquad (8.1)$$

where V is the particle volume.

In the operation of centrifugation, the effective diameter is defined as the size range of particles whose concentration in the suspension is less than 17%. Based on the effective diameter, suspensions are classified according to particle size as follows:

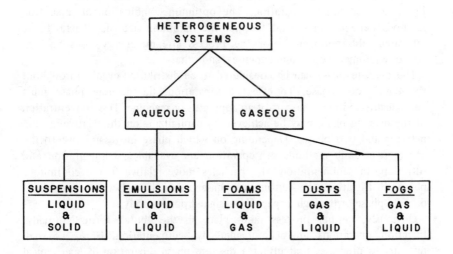

Figure 8.1   Summarizes the classification of heterogeneous systems.

1. Coarse suspensions at $d_{eq} > 1$ mm
2. Medium suspensions at $d_{eq} = 100 \, \mu m - 1$ mm
3. Fine suspensions at $d_{eq} = 5 - 100 \, \mu m$
4. Turbid suspensions at $d_{eq} = \leqslant 5 \, \mu m$
5. Colloidal suspensions at $d_{eq} = <1 \, \mu m$

This classification also may be applied as an approximate description for emulsions and dusts that are basically fine gas suspensions.

CLASSIFICATION OF SEPARATION METHODS

A large number of unit operations are designed to separate HS into their separate components for intermediate or final product processing. Examples for solid-liquid systems are crystallization, filtration, centrifugation and sedimentation. Solid particles can be separated from air and gases after drying to prevent pollution problems or to collect a valuable product, etc. The specific separation method is selected on the basis of the characteristics of the individual components of the system and the state of the phases (liquid, gaseous or solid).

In selecting a separation method, the physical and chemical properties of the medium (liquid and gas) should be considered, i.e., viscosity, density and chemical activity. Properties of the discrete phase to be examined include particle sizes, density, state of aggregation, etc. Generally, more than one method must be employed to achieve a desired separation. For example, a dust-laden gas stream may have its large particulates removed by cyclone separators and then a smaller size of particulates captured by a filter, i.e., secondary separation. The optimum combination of separation methods depends on several factors: the minimum size of particles to be separated; the temperature of the process stream; energy consumption; cost of equipment; operating expenses; and others.

The discrete phase can be considered to be distributed evenly throughout the continuous phase. To effect a separation, the discrete phase must be concentrated in some part of an apparatus or machine. This concentration of the discrete phase may be achieved by capitalizing on the relative motion between the two phases. Depending on which phase moves relative to the other, two basic methods of separation are available: sedimentation and filtration. In sedimentation, the particles move relative to the continuous phase. In filtration, the continuous phase passes through the concentrated discrete phase or through a specially designed porous body.

Obviously, neither process takes place in the "pure" form. Normally, filtration operations are accompanied by sedimentation. Here, we shall refer to all processes that effect a discrete phase separation by mechanical forces as *sedimentation*. The separation phenomenon achieved by passing the liquid or gas HS through a porous body, whereby particles are entrapped in the pores, is called *filtration*. The relative motion between the continuous and discrete phases may be influenced intentionally by forces other than gravity. The following force fields in addition to gravity are used in industry for the separation of HS: centrifugal, electrical and a field of surface forces of pressure in liquids and gases.

Gravitational and centrifugal forces are the primary mechanical forces that effect sedimentation. As will be discussed later, we also may include electrical force fields in this category. The condition that establishes separation under the influence of mechanical force fields is the difference between the densities of particles, ρ_p, and of the medium, ρ, i.e., the so-called "effective density":

$$\rho_{ef} = \rho_p - \rho \qquad (8.2)$$

The force field will influence the behavior of particles of equal size in both phases but at different intensities. Therefore, separation depends primarily on the relative motion between the discrete and continuous phases.

For sedimentation in an electrical field, the difference in the phase densities is not mandatory, although it occurs in a practical sense. Separation in the field of surface forces of pressure is evidently impossible because forces acting on the boundaries are equal but have opposite direction.

Filtration can be performed under the action of all these forces. Industrial applications rely primarily on gravity, centrifugal and pressure forces. In contrast to sedimentation, the difference in phase densities is not a criterion for separation.

Based on the above discussions we may classify the separation of HS into two general unit operations, as summarized in Figure 8.2.

MATERIAL BALANCES OF SEPARATIONS

The mechanical separation of phases can be represented diagrammatically as shown in Figure 8.3, where the areas of the rectangles illustrate (based on an appropriate scale) the volumes of substances at various stages of an operation. Consider a quantity of solid material AEFB having mass M_H, volume V_H, and density ρ_H. The solid is prepared in granular form and then mixed with a quantity of liquid ECDF of volume V_c, mass M_c and density ρ_c. If chemical reactions and other phenomena such as heating, foaming, etc., are ignored, then the mass and volume of the mixture ACDB will be equal to the mass and volume of the initial components.

The suspension next is passed through an apparatus or machine (A) that

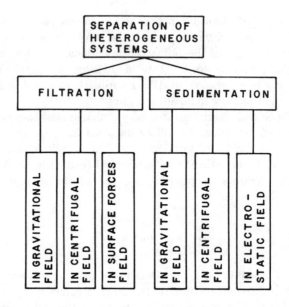

Figure 8.2   Classification of separation methods for heterogeneous systems.

divides the fluid system into two parts: sludge AMNB and filtrate (if filtration) or decanted liquid (if sedimentation) MCDN. The sums of the volumes and masses of the separation products will be equal to those of the initial suspension; however, the concentration of each product will be different. The concentration of the sludge, η_0, is higher than that of the initial suspension, and the concentration of the filtrate, η_f, is less than that of the suspension, i.e., $\eta_0 > \eta_f$.

Suppose that all the solid particles contained in the sludge and filtrate could be extracted and then compressed to the density of the solid lumps of the initial material. This would require combining the thickened sludge, having a volume of solid particles ARTB, with density ρ_H, mass $M_{H.0}$ and volume $V_{H.0}$, and the filtrate volume MQPN with $M_{H.\phi}$, $V_{H.\phi}$ and ρ_H. The same recombination can be performed for the liquid phase as well (RMNT plus QCDP). If the operation is 100% efficient, the masses and volumes of the solid particles in the sludge (ARTB and MQPN), as well as in the filtrate (RMNT and QCDP), will be equal correspondingly to those of AEFD and ECDF.

In the scheme described by Figure 8.3 (in the absence of chemical reactions), there are nine values represented by two unit measures—mass and volume. Hence, there are a total of 18 values for which 12 material balance equations can be written. These equations are written next to the

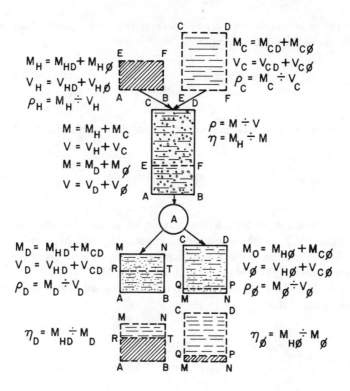

$$M_H = M_{HD} + M_{H\phi}$$
$$V_H = V_{HD} + V_{H\phi}$$
$$\rho_H = M_H \div V_H$$

$$M_C = M_{CD} + M_{C\phi}$$
$$V_C = V_{CD} + V_{C\phi}$$
$$\rho_C = M_C \div V_C$$

$$M = M_H + M_C$$
$$V = V_H + V_C$$
$$M = M_D + M_\phi$$
$$V = V_D + V_\phi$$

$$\rho = M \div V$$
$$\eta = M_H \div M$$

$$M_D = M_{HD} + M_{CD}$$
$$V_D = V_{HD} + V_{CD}$$
$$\rho_D = M_D \div V_D$$

$$M_O = M_{H\phi} + M_{C\phi}$$
$$V_\phi = V_{H\phi} + V_{C\phi}$$
$$\rho_\phi = M_\phi \div V_\phi$$

$$\eta_D = M_{HD} \div M_D$$

$$\eta_\phi = M_{H\phi} \div M_\phi$$

Figure 8.3   A process material balance.

appropriate stage in the scheme. To solve this matrix of equations, six more expressions are required. Additional expressions are often those that characterize the physical properties of the substances (densities, specific volumes) and the compositions of the mixtures.

Examples of mass concentration, η, and density, ρ, expressions also are given in Figure 8.3. In performing practical calculations, different variations of material balances usually are not necessary. Normally, only two or three equations are required to evaluate specific parameters of interest. The exact expressions required depend on the information known about the system and the parameters to be evaluated.

For filtration processes, we may assume that the concentration of solids in the filtrate, η_ϕ, is zero. This means that the entire solid phase of the suspension is transformed into a sludge. In this case (Figure 8.3) $M_H = M_{H.0}$ and $M_{H.d} = 0$.

We shall derive several relationships for the general operation of filtration. First, we define the parameter "m" to be the ratio of the mass (weight) of sludge to the mass (weight) of solids in the sludge phase. That is, m is the inverse of the mass concentration:

$$m = \frac{1}{\eta_0} = \frac{M_0}{M_{H.0}} \tag{8.3}$$

The volume of wet thickened sludge deposited by a unit filtrate volume is

$$x_0 = \frac{V_0}{V_\phi} \tag{8.4}$$

It is convenient to express the volume of sludge, V_0, and the volume of filtrate, V_ϕ, by their respective masses, M_0 and M_ϕ, with densities ρ_0 and ρ_ϕ. The mass of the filtrate can be expressed in terms of the mass of the initial mixture, M, and the mass of the sludge, M_0.

In this case, $M_H = M_{H.0} = M_\eta$ and, including Equation 8.3, we obtain the following expression:

$$x_0 = \frac{m\eta}{1 - m\eta} \times \frac{\rho_\phi}{\rho_0} \tag{8.5}$$

Substituting η_0 for M, Equation 8.5 transforms to

$$x_0 = \frac{1}{\dfrac{\eta_0}{\eta} - 1} \times \frac{\rho_\phi}{\rho_0} \tag{8.6}$$

If the densities of the medium (filtrate), ρ_ϕ, and the solid particles, ρ_H, are known, then the density of the sludge can be computed as follows:

$$\rho = \frac{1}{\dfrac{\eta_n}{\rho_a} + \dfrac{1 - \eta_n}{\rho_b}} \tag{8.7}$$

and Equation 8.6 can be rewritten as

$$x_0 = \frac{1 - \eta_0}{\dfrac{\eta_0}{\eta} - 1} \left(1 - \frac{\rho_0}{\rho_H} \right) \tag{8.8}$$

The amount of dry mass of the sludge, $M_{c,0}$, deposited per unit volume of filtrate is

$$x_{c,0} = \frac{M_{c,0}}{V_\phi} \tag{8.9}$$

It is evident that the mass of the dry sludge $M_{c,0} = M_{H.0} = M_H = M_\eta$; therefore, from Figure 8.3,

$$x_{c,0} = \frac{M\eta\rho_\phi}{M - \dfrac{M\eta}{\eta_0}} = \frac{\eta}{1 - \dfrac{\eta}{\eta_0}}\,\rho_\phi = \frac{1}{\dfrac{1}{\eta} - \dfrac{1}{\eta_0}}\,\rho_\phi \tag{8.10}$$

where M and η are the total mass of the heterogeneous system and its concentration, respectively. Transforming η_0 to m in Equation 8.10, we obtain

$$x_{c,0} = \frac{\eta}{1 - \eta m}\,\rho_\phi \tag{8.11}$$

Dividing Equation 8.5 by Equation 8.11 and including Equation 8.3, we obtain

$$\frac{x_0}{x_{c,0}} = \frac{m}{\rho_0} \tag{8.12}$$

where $m = 1/\eta_0$. Hence,

$$\frac{x_0}{x_{c,0}} = \frac{1}{\eta_0 \rho_0} \tag{8.13}$$

Equation 8.13 can be expressed in an alternate form by introducing the porosity of the sludge, ϵ. In filtrating a unit volume of filtrate, we obtain x_0 units of filtrate volume (Equation 8.4). The fraction of the total volume occupied by the liquid is ϵ and, hence, $1 - \epsilon$ represents the fractional volume occupied by solid particles. The total volume of solid particles is $(1 - \epsilon)x_0$ and their mass is $(1 - \epsilon)x_0\rho_H$. Consequently,

$$\frac{x_0}{x_{c,0}} = \frac{1}{(1 - \epsilon)\rho_H} \tag{8.14}$$

It is shown readily that the ratio $x_{c,0}/x_0$ is the density of dry sludge that occupies the same volume as the wet sludge presented in Figure 8.3 as rectangle AMNB. By dividing Equation 8.9 by Equation 8.4, we show that

$$\frac{x_{c,0}}{x_0} = \frac{M_{c,0}}{V_0} = \rho_{c,0} \tag{8.15}$$

SEPARATION EFFECT NUMBER

Separation operations are designed to decrease the concentration of one of the components in a heterogeneous system and to bring that concentration to a minimum level (or zero if possible), depending on the technological demands of the given plant. The purpose of these operations also may be to increase the concentration of one of the constituents. A parameter used to characterize a separation process quantitatively is the *separation effect number*, defined as the ratio of the mass of a given constituent extracted from the discrete phase, to its mass in the initial suspension. The separation effect number characterizes the rate of technical efficiency of a piece of equipment or process and is expressed by the following formula:

$$\eta_s = \frac{K_{H.0}}{K_{H.H.}} \tag{8.16}$$

where $K_{H.H}$ = mass of solids comprising the discrete phase, i.e., the mass of particles in the feed

$K_{H.0}$ = mass of solids separated, i.e., the solids mass transferred to sludge

The following examples illustrate these principles.

Example 8.1

The concentrations of a suspension, sludge and filtrate are $\eta = 4$, $\eta_0 = 30$ and $\eta_\phi = 0.05\%$, respectively. Determine the separation effect number for the operation.

Solution

Using the notation given in Figure 8.3, we write the following material balance expressions:

$$M = M_0 + M_\phi$$

$$M_H = M_{H.0} + M_{H.\phi} \quad ; \quad M_H = M_\eta$$

$$M_{H.0} = M_0 \eta_0 \quad ; \quad M_{H.\phi} = M_\phi \eta_\phi$$

Solving this system of equations with respect to Equation 8.16, the separation effect number is computed:

$$\eta_s = \frac{M_0 \eta_0}{M \eta} \times 100 = \frac{\eta_0}{\eta} \times \frac{\eta - \eta_\phi}{\eta_0 - \eta_\phi} \times 100 = \frac{30}{4} \times \frac{4 - 0.05}{30 - 0.05} \times 100 = 99.1\%$$

Example 8.2

The initial concentration of a suspension of total mass M = 123 kg is 30% calcium carbonate (solid). The relative density of the solid is (Δ_H) 2.65. After filtration, a sludge with a volumetric concentration of solid phase η_v = 50% and filtrate with density ρ_ϕ = 1030 kg/m^3 is obtained. Determine the densities of the initial suspension and separation products, as well as the amount of wet and dry sludge separated from 1 m^3 of filtrate.

Solution

Following the equations and notation of Figure 8.3, the mass of solid particles in the initial suspension is

$$M_H = M \frac{\eta_m}{100} = 123 \frac{30}{100} = 36.9 \text{ kg}$$

The density of solid particles in the initial suspension is

$$\rho_H = 1000 \Delta_H = 1000 \times 2.65 = 2650 \text{ kg/m}^3$$

The volume of these particles is

$$V_H = \frac{M_H}{\rho_H} = \frac{36.9}{2650} = 0.0139 \text{ m}^3$$

The mass of the liquid phase is

$$M_c = M - M_H = 123 - 36.9 = 86.1 \text{ kg}$$

The volume of the liquid phase is

$$V_c = \frac{M_c}{\rho_\phi} = \frac{86.1}{1030} = 0.0836 \text{ m}^3$$

The separation produces a pure filtrate and, hence, the mass of the solid phase in the initial suspension is converted to sludge:

$$M_{H.0} = M_H = 36.9 \text{ kg}$$

$$V_{\phi.0} = V_{H.0} = V_H = 0.0139 \text{ m}^3$$

Consequently, the total volume of the sludge is

$$V_0 = \frac{V_{H.0}}{50} \times 100 = 0.0278 \text{ m}^3$$

The volume of the dry sludge will be the same. The volume of filtrate is

$$V_\phi = V_c - V_{H.0} = 0.0836 - 0.0139 = 0.0697 \text{ m}^3$$

The calculated values are summarized in Table 8.1. The mass of liquid phase in the sludge is

$$M_{c,0} = V_{H.0}\rho_\phi = 0.0139 \times 1030 = 14.3 \text{ kg} = M_c - V_\phi\rho_\phi$$
$$= 86.1 - 0.0697 \times 1030 = 86.1 - 71.8 = 14.3 \text{ kg}$$

Table 8.1 Summary of Computed Values for Example 8.2

Phase	Sludge (kg)	Sludge (m^3)	Filtrate (kg)	Filtrate (m^3)	Suspension (kg)	Suspension (m^3)
Solid	36.9	0.0139	–	–	36.9	0.0139
Liquid	14.3	0.0139	71.8	0.0697	86.1	0.0836
Total	51.2	0.0278	71.8	0.0697	123	0.0975

The density of the suspension is

$$\rho = \frac{M}{V} = \frac{123}{0.0975} = 1263 \text{ kg/m}^3$$

The density of the sludge is

$$\rho_0 = \frac{M_0}{V_0} = \frac{51.2}{0.0278} = 1870 \text{ kg/m}^3$$

The density of the dry sludge is

$$\rho_{c,0} = \frac{M_{H.0}}{V_0} = \frac{36.9}{0.0278} = 1329 \text{ kg/m}^3$$

The specific amount of deposited sludge is

$$x_0 = \frac{V_0}{V_\phi} = \frac{0.0278}{0.0697} = 0.437 \text{ m}^3/\text{m}^3$$

The specific mass of deposited dry sludge is

$$x_{c,0} = \frac{M_{H.0}}{V_\phi} = \frac{36.9}{0.0697} = 579 \text{ kg/m}^3$$

The last set of values may be computed from Equations 8.5–8.15. For example, Equation 8.15 can be used to determine the amount of the dry sludge deposited per unit volume of filtrate:

$$x_{c,0} = x_0 \rho_{c,0} = 0.437 \times 1329 = 579 \text{ kg/m}^3$$

However, as a general rule, the use of material balances is recommended over Equation 8.15. Material balances, although sometimes cumbersome to solve, are more reliable. Their use provides a detailed description of the constituents' distributions through each stage of the separation process.

The student should solve the practice problems provided at the end of this chapter before proceeding to the more in-depth study of external hydrodynamic processes.

PRACTICE PROBLEMS

8.1 A small storage tank contains 13.5% (by weight) H_2SO_4. The remainder is distilled water. If 600 lb of 95% H_2SO_4 are added to the tank to make a final solution of 39.7% H_2SO_4, how many pounds of the total acid solution have been prepared?

8.2 Wet paper pulp containing 79% water is dried on a rotary-drum drier, which removes approximately 67% of the original moisture. Compute the composition of the dried pulp and the amount of water removed per pound of wet pulp.

8.3 After the pulp (Problem 8.2) is dried, it is sent through a series of presses, where another 10% of the remaining moisture is removed. Approximately 3% of the pulp is lost in the effluent water during this stage. Compute the final composition of the dried pulp and the amount of additional water removed during pressing per pound of wet pulp.

8.4 A vessel contains 25,000 lb of a saturated solution of $NaHCO_3$ at 65°C. Determine the temperature required to crystallize approximately 1200 lb of $NaHCO_3$ from this solution.

8.5 A small pilot plant generates water contaminated with about 12% oil. The oil is separated out by allowing the solution to stand for several hours in a tank and decanting off the oil layer. After 75 lb of oil are removed, it is found that the water contains only 2% oil. Calculate the total amount of oil in the process stream.

8.6 A radioactive salt tracer is used to obtain flow measurements through a large tubular reactor. The background salt concentration is 85 ppm. If 15 lb of the tracer are added to the feed stream uniformly over a 1-hour period and the analysis downstream of the reactor indicates a concentration of 2500 ppm salt, how many gallons of product are flowing per hour?

8.7 Determine how many lb of $MgSO_4 \cdot 7H_2O$ must be dissolved in 200 lb of H_2O to produce a saturated solution at 25°C. The solubility of magnesium sulfate at this temperature is 35.9 g/100 g H_2O.

8.8 A suspension containing 40% solid is fed at a rate of 370 kg/hr through a filter. The volumetric concentration of the solid phase downstream of the filter is 40% and the filtrate has a density of 985 kg/m^3.

a.   Determine the densities of the feed suspension and separation products.
b.   How efficient is the filter?

8.9 Calculate the separation effect number for a system in which the concentrations of the suspension, sludge and filtrate are 14, 42, and 0.3%, respectively.

8.10 We wish to measure the rate at which waste gases are emitted from a stack. The gases entering the stack contain 1.5% ammonia by weight. Pure NH_3 is fed to the bottom of the stack at a controlled rate of 2.9 lb/min. An ammonia monitor at the stack discharge indicates that the gas stream contains 2.4% NH_3 by weight. Determine the flowrate of gases in lb/min through the stack.

NOTATION

d_{eq} equivalent particle diameter, m
K_{HH} solid mass in discrete phase, kg
K_{HO} solid mass separated, kg
M   mass, kg
m   inverse of mass concentration, $1/\eta_0$
V   volume, m^3
x_0 mass concentration or mass of thickened sludge to filtrate volume, kg/m^3

Greek Symbols

Δ_H relative density
ϵ porosity or void fraction
η mass concentration, %
η_s separation effect number, refer to Equation 8.16
η_ϕ solid concentration in filtrate, %
ρ density, kg/m^3

PRINCIPLES AND PRACTICES OF PARTICLE SEDIMENTATION

CONTENTS

INTRODUCTION

The motion of solid particles in gas or liquid media is encountered all too frequently in chemical engineering applications. Unit operations directed at separation, handling, or processing of these heterogeneous systems include sedimentation of dust in chambers and cyclones, separation of suspensions in settlers, separation of liquid mixtures by settling and centrifuging, hydraulic and pneumatic transport, hydraulic and air classification, flotation, mixing by air, and others. Each of these operations involves the simultaneous flow of gas and solid, or liquid and solid phases.

The widespread and successful application of these hydrodynamic processes to a large number of industrial problems is based on our ability to take advantage of one or a combination of four primary forces: gravity, centrifugal, pressure, and electric. Gravity is the controlling force for separations achieved in settlers; centrifugal force is applied to cyclone separators, dryers and mixers; and pressure forces are employed in sprayers, pneumatic transport and filters. Electrical forces are employed in special techniques, such as precipitators.

During the motion of viscous flow over a stationary body or particle, certain resistances arise. To overcome these resistances or drag and to provide more uniform fluid motion, a certain amount of energy must be expended. The developed drag force and, consequently, the energy required to overcome it, depend largely on the flow regime and the geometry of the solid body. Laminar flow conditions prevail when the fluid medium flows at low velocities over small bodies or when the fluid has a relatively high viscosity. Flow around a single body is illustrated in Figure 9.1.

As shown in Figure 9.1(A), when the flow is laminar a well-defined boundary layer forms around the body and the fluid conforms to a streamline motion. The loss of energy in this situation is due primarily to friction drag. If the fluid's average velocity is increased sufficiently, the influence of inertia forces becomes more pronounced and the flow becomes turbulent.

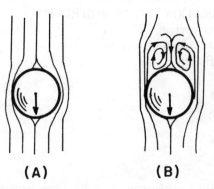

(A)   **(B)**

Figure 9.1   Flow around a solid particle: (A) laminar flow; (B) turbulent flow.

Under the action of inertia forces, the fluid adheres to the particle surface, forming only a very thin boundary layer and generating a turbulent wake, as shown in Figure 9.1(B). The pressure in the wake is significantly lower than that at the stagnation point on the leeward side of the particle. Hence, a net force, referred to as the *pressure drag*, acts in a direction opposite to that of the fluid's motion. Above a certain value of the Reynolds number, the role of pressure drag becomes significant and the friction drag can be ignored.

We shall begin discussions on the external problems of hydrodynamics by analyzing a dilute system that can be described as a low concentration of noninteracting solid particles carried along by a gas or liquid stream. In this system, the solid particles are far enough removed from one another to be treated as individual entities. That is, each particle individually contributes to the overall character of the flow.

SINGLE-PARTICLE MOTION

Pressure and Drag Force

We shall consider the dynamics of motion of a solid spherical particle immersed in a fluid independent of the nature of the forces responsible for its displacement. A moving particle immersed in a fluid (gas or liquid) experiences forces caused by the action of the fluid. These forces are the same regardless whether the particle is moving through the fluid or whether the fluid is moving over the particle's surface.

For our purposes, assume the fluid to be in motion with respect to a stationary sphere. The fluid shock acting against the sphere's surface produces an additional pressure, P. This pressure is responsible for a force, R (called the drag force) acting in the direction of fluid motion.

Consider an infinitesimal element of the sphere's surface, dF, having a slope, α, with respect to the normal of the direction of motion (Figure 9.2). The pressure resulting from the shock of the fluid against the element produces a force, $d\tau$, in the normal direction. This force is equal to the product of the surface area and the additional pressure, PdF_0. The component acting in the direction of flow, dR, is equal to $d\tau\cos\alpha$. Hence, the force, R, acting over the entire surface of the sphere will be

$$R = \int PdF_0\cos\alpha = \int PdF = PF \qquad (9.1)$$

where dF is the projection of dF_0 on the plane normal to the flow. The term F refers to a characteristic area of the particle, either the surface area or the maximum cross-sectional area perpendicular to the direction of flow.

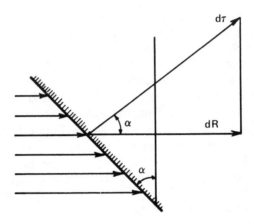

Figure 9.2   Infinitesimal element of sphere's surface inclined at an angle α from the direction of flow.

The pressure, P, i.e., the ratio of resistance force to surface area, R/F, depends on several factors, namely, the diameter of the sphere, d, its velocity, u, fluid density, ρ, and viscosity, μ:

$$P = f(d,u,\rho,\mu) \tag{9.2}$$

Using dimensional analysis, Equation 9.2 is transformed into a function of dimensionless groups:

$$Eu = \Phi(Re) \tag{9.3}$$

where the Euler number is defined as $P/u^2\rho$, and the Reynolds number is expressed in terms of the diameter of the sphere:

$$Re = \frac{ud\rho}{\mu} \tag{9.4}$$

Substituting for density with the ratio of specific gravity to the gravitational acceleration, an expression similar to the Darcy-Weisbach equation is obtained:

$$\frac{R}{F} = C_D \frac{u^2}{2g} \gamma \qquad (9.5)$$

where C_D is the drag coefficient, a dimensionless factor that is a function of the Reynolds number.

The relationship between C_D and Re for flow around a smooth sphere is shown in Figure 9.3. To facilitate machine calculations, the drag-coefficient curve may be approximated by the equations of three straight lines, each of which covers a definite Reynolds number range. These are noted by the dotted lines in Figure 9.3.

The equations for these lines and the range of the Reynolds numbers over which each applies are as follows:

For the laminar regime (referred to as the Stokes' law range),

$$C_D = \frac{24}{Re} \; ; \quad \text{for Re} < 2 \qquad (9.6)$$

For the intermediate (or transition) range,

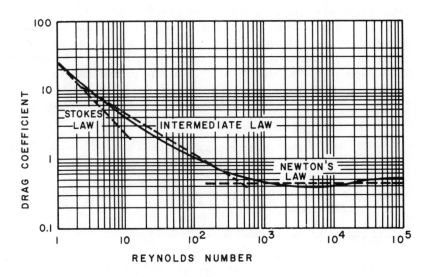

Figure 9.3   Drag coefficients for spheres.

$$C_D = \frac{18.5}{Re^{0.6}} \; ; \quad \text{for } 2 < Re < 500 \tag{9.7}$$

For Newton's law range,

$$C_D = 0.44 \; ; \quad \text{for } 500 < Re < 200,000 \tag{9.8}$$

Substituting the above drag coefficient expressions into Equation 9.5 shows that for the laminar regime the drag force is proportional to the velocity raised to the first power, i.e., $R \alpha w$; for the intermediate regime it is $R \alpha w^{1.4}$ and in the Newton's law range it is, $R \alpha w^2$.

Further discussions are given by Brodkey [1967], Hinze [1959], Heiss and Coull [1952], Pettyjohn and Christiansen [1948], and Hughes and Gilliland [1952].

Particle Settling by Gravity Force

If a particle at rest (with mass m and weight mg) begins to fall under the action of gravity, its velocity is increased initially over a period of time. The particle is subjected to the resistance of the fluid medium through which it descends. This resistance increases with particle velocity until the accelerating and resisting forces are equal. From this point, the solid particle continues to fall at a constant maximum velocity, referred to as the *terminal velocity*, u_t.

The force responsible for moving a spherical particle of diameter d can be expressed by the difference between its weight and the buoyant force acting on the particle. The buoyant force is proportional to the mass of fluid displaced by the particle, that is, as the particle falls through a surrounding medium it displaces a volume of fluid equivalent to its own weight:

$$w = \frac{\pi d^3}{6} g(\rho_p - \rho) \tag{9.9}$$

where ρ_p = density of the solid particle
         ρ = density of the fluid

From Equation 9.5, the resistance force exerted by the fluid is

$$R = C_D \frac{\pi d^2}{4} \frac{\rho u^2}{2} \tag{9.10}$$

The settling velocity of the particle is obtained when accelerating and resisting forces are equal:

$$\frac{\pi d^3 g}{6} (\rho_p - \rho) = C_D \frac{\pi d^2}{4} \frac{\rho u_s^2}{2}$$

Hence,

$$u_s = \sqrt{\frac{4gd(\rho_p - \rho)}{3C_D\rho}} \tag{9.11}$$

The drag coefficient, C_D, can be evaluated from Equations 9.6–9.8. For example, substituting Equation 9.6 into Equation 9.11 provides an expression for the settling velocity in the laminar regime:

$$u_s = \frac{d^2 g(\rho_p - \rho)}{18\mu} \tag{9.12}$$

where μ = viscosity of the fluid

The maximum particle size whose velocity follows Stokes' law can be established by substituting $\mu Re/d\rho$ for the settling velocity in Equation 9.12 and settling Re equal to 2, i.e., the limiting value for the laminar regime:

$$d_{max} = \sqrt[3]{\frac{36\mu^2}{\rho g(\rho_p - \rho)}} \simeq 1.56 \sqrt[3]{\frac{\mu^2}{\rho(\rho_p - \rho)}} \tag{9.13}$$

The minimum size of particles that do not follow Stokes' law occurs at $Re \simeq 10^{-4}$. The settling velocity in this region is less than that computed by Equation 9.12. In this case, U_s should be divided by the following correction factor:

$$K = 1 + A \frac{\lambda}{d} \tag{9.14}$$

where λ is the length of the mean free path of a fluid molecule. Constant A varies between 1.4 and 20 (A = 1.5 for air).

As an example, consider dust particles in the atmosphere. For this system $K = 1$, as evaluated from Equation 9.14, and those particles greater than 3 μm in size will settle out. For $d \simeq 0.1$ μm, dust particles will remain in suspension. In fact, particles in this size range undergo Brownian motion due to the bombardment of gas molecules, thus forming a very stable colloidal suspension.

In the intermediate Reynolds number range ($2 < Re < 500$), the settling velocity expression becomes, after substitution of Equation 9.7 for C_D into Equation 9.11,

$$u_s \simeq 0.78 \frac{d^{0.43}(\rho_p - \rho)^{0.715}}{\rho^{0.285}\mu^{0.43}} \tag{9.15}$$

And, similarly, for the Newton's law range,

$$u_s \simeq 5.46 \sqrt{\frac{d(\rho_p - \rho)}{\rho}} \tag{9.16}$$

Following Lyachshenko [1940], design formulas for the settling velocity are obtained by expressing u_s in terms of the Reynolds number in Equation 9.11 and raising both sides of the expression to the square power:

$$\frac{Re^2\mu^2}{d^2\rho^2} = \frac{4gd(\rho_p - \rho)}{3C_D\rho} \tag{9.17}$$

Hence,

$$Re^2C_D = \frac{4}{3}\frac{d^3\rho^2g}{\mu^2}\left(\frac{\rho_p - \rho}{\rho}\right) \tag{9.18}$$

The RHS of Equation 9.18 is the dimensionless Archimedes number:

$$Ar = \frac{d^3\rho^2g}{\mu^2}\frac{\rho_p - \rho}{\rho} \tag{9.19}$$

The desired settling velocity does not enter into the Archimedes number. It consists of values that usually are given or may be determined a-priori. Thus,

$$C_D Re^2 = \frac{4}{3} Ar \qquad (9.20)$$

Substituting for the critical (boundary) values of the Reynolds number corresponding to the transition from one range of settling to the other, appropriate values of the Archimedes number can be computed.

For Stokes' law range (Re < 2), on substituting Equation 9.6 for C_D into Equation 9.20, we obtain

$$\frac{24}{Re} Re^2 = \frac{4}{3} Ar$$

Hence,

$$Re = \frac{Ar}{18} \qquad (9.21)$$

The upper limiting or critical value of the Archimedes number for this range, i.e., at Re = 2, is

$$Ar_{cr.1} = 18 \times 2 = 36$$

Consequently, the existence of the laminar settling regime corresponds to the condition Ar < 36. For the intermediate range, where 2 < Re < 500, C_D is replaced by Equation 9.7 in Equation 9.20 to provide

$$Re^{1.4} = \frac{4}{3} \frac{Ar}{18.5}$$

or

$$Re = 0.152 Ar^{0.715} \qquad (9.22)$$

And for the critical value Re = 500, the limiting value of Ar for the intermediate range is

$$500 = 0.152 Ar_{cr.2}^{0.715}$$

or

$$Ar_{cr.2} = 83,000$$

Thus, the intermediate settling range corresponds to a change of the Ar number in the range $36 < Ar < 83,000$. For Newton's law range, where $Ar > 83,000$, the relationship between Re and Ar may be found by substituting $C_D = 0.44$ into Equation 9.20:

$$Re = 1.74 \sqrt{Ar} \qquad (9.23)$$

Evaluation of the Archimedes number establishes the settling range for particles. The Reynolds number then can be evaluated using one of the above equations (9.21, 9.24, 9.33), whence the settling velocity can be determined:

$$u_s = \frac{\mu Re}{d\rho} \qquad (9.24)$$

An interpolation formula valid for all three settling regimes also can be used:

$$Re = \frac{Ar}{18 + 0.575 \sqrt{Ar}} \qquad (9.25)$$

For low values of Ar, the second term in the denominator may be neglected, and Equation 9.25 reduces to Equation 9.21; at high Ar values, we may neglect the first term in the denominator and the expression simplifies to Equation 9.23, which corresponds to the Newton's law range.

The settling velocity of a nonspherical particle is less than that of a spherical one. A good approximation can be made by multiplying the settling velocity, u_s, of spherical particles by a correction factor, ψ, called the *sphericity factor*. The sphericity, or shape factor is defined as the area of a sphere divided by the area of the nonspherical particle having the same volume:

$$u_s' = \psi u_s \qquad (9.26)$$

The factor $\psi < 1$ must be determined experimentally for particles under study. Typical values are $\psi \simeq 0.77$ for particles of rounded shape; $\psi \simeq 0.66$ for particles of angular shape; $\psi \simeq 0.43$ for particles of a flaky geometry.

In designing a system based on the settling velocity of nonspherical particles, the linear size in the Reynolds number definition is taken to be the *equivalent diameter of a sphere, d,* which is equal to a sphere diameter having the same volume as the particle.

The above analysis applies only to the *free settling velocities* of single particles and does not account for particle-particle interactions. Hence, the application of these formulas only applies to very dilute systems. At high particle concentrations, mutual interference in the motion of particles exists, and the rate of settling is considerably less than that computed by the given expressions. In the latter case, the particle is settling through a suspension of particles in a fluid, rather than through a simple fluid medium. Descriptive information on particle settling in dense concentrations is given by Brown [1950], Coe and Clevenger [1916], and Comings [1940].

GRAVITY SEDIMENTATION

Thickeners and Clarifiers

A large number of processes in the chemical industries handle suspensions of solid particles in liquids. The application of filtration techniques for the separation of these heterogeneous systems is sometimes very costly. If, however, the discrete phase of the suspension largely contains settleable particles, the separation can be effected by the operation of *sedimentation*. Sedimentation involves the removal of suspended solid particles from a liquid stream by gravitational settling. This unit operation is divided into *thickening*, i.e., increasing the concentration of the feed stream, and *clarification*, removal of solids from a relatively dilute stream.

A thickener is a sedimentation machine that operates according to the principle of gravity settling. Compared to other types of liquid/solid separation devices, a thickener's principal advantages are:

- simplicity of design and economy of operation;
- its capacity to handle extremely large flow volumes; and
- versatility, as it can operate equally well as a concentrator or as a clarifier.

In a batch-operating mode, a thickener normally consists of a standard vessel filled with a suspension. After settling, the clear liquid is decanted and the sediment removed periodically.

The operation of a continuous thickener is also relatively simple. Figure 9.4 illustrates a cross-sectional view of a standard thickener. A

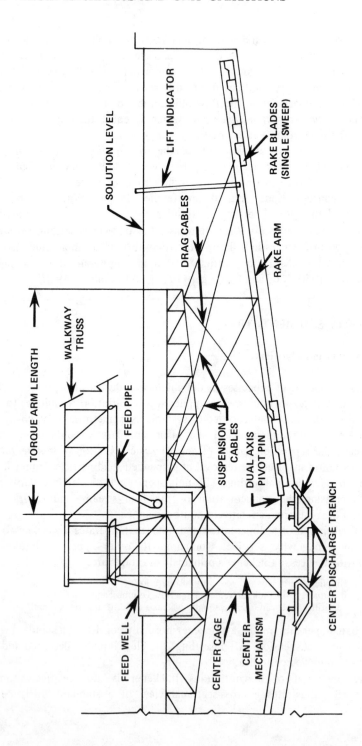

Figure 9.4   Cross-sectional view of a thickener (courtesy Dorr-Oliver Inc., Stamford, CT).

drive mechanism powers a rotating rake mechanism. Feed enters the apparatus through a feed well designed to dissipate the velocity and stabilize the density currents of the incoming stream. Separation occurs when the heavy particles settle to the bottom of the tank. Some processes add flocculants to the feed stream to enhance particle agglomeration to promote faster or more effective settling. The clarified liquid overflows the tank and is sent to the next stage of a process. The underflow solids are withdrawn from an underflow cone by gravity discharge or pumping. Table 9.1 provides further descriptions and typical size ranges of thickeners of different configurations. A listing of the major industries and typical applications served by thickeners is given in Table 9.2.

Thickeners can be operated in a countercurrent fashion. Applications are aimed at the recovery of soluble material from settleable solids by means of continuous countercurrent decantation (CCD). The basic scheme involves streams of liquid and thickened sludge moving countercurrently through a series of thickeners. The thickened stream of solids is depleted of soluble constituents as the solution becomes enriched. In each successive stage, a concentrated slurry is mixed with a solution containing fewer solubles than the liquor in the slurry and then is fed to the thickener. As the solids settle, they are removed and sent to the next stage. The overflow solution, which is richer in the soluble constituent, is sent to the preceding unit. Solids are charged to the system in the first-stage thickener, from which the final concentrated solution is withdrawn. Wash water or virgin solution is added to the last stage, and washed solids are removed in the underflow of this thickener.

The flow scheme for a three-stage CCD system is illustrated in Figure 9.5. The feed stream, F, is mixed with overflow O_2 (from thickener 2) before entering stage 1. The overflow of concentrated solution, O_1, is withdrawn from the first stage. The underflow from the first stage, U_1, is mixed with third-stage overflow, O_3, and fed to the second stage. Similarly, the second-stage underflow, U_2, is mixed with wash water and fed to thickener 3. The washed solids are removed from the third stage as the final underflow, U_3.

Continuous clarifiers handle a variety of process wastes, domestic sewage and other dilute suspensions. They resemble thickeners in that they are sedimentation tanks or basins whose sludge removal is controlled by a mechanical sludge-raking mechanism. They differ from thickeners in that the amount of solids and weight of thickened sludge are considerably lower. Figure 9.6 shows one of the various types of cylindrical clarifiers. In this type of sedimentation machine, the feed enters up through the hollow central column or shaft, referred to as a *siphon feed system*. The feed enters the central feed well through slots or ports located near the top of the hollow shaft. Siphon feed arrangements greatly reduces the feed stream

Table 9.1 Descriptions and Sizes of Different Configurations of

Type		Drive
Standard Bridge-Support Thickener (Type A)		Type A drive head
Standard Center-Pier Thickener (Type S)		Type S center mechanism
Caisson Center-Pier Thickener (superthickener)		Type S center mechanism
Traction Thickener		Electric motor
Hi-Rate Thickener		Type A or Type S drive unit
CableTorq Thickener		

Industrial Thickeners (courtesy Dorr-Oliver Inc., Stamford, CT)

Diameter	Operating Torque	Description
To 150 ft	To 750,000 ft-lb	"Bridge"—or "beam" or "truss"—spans diameter of tank and supports drive and rake mechanisms. Underflow is removed from discharge cone at bottom-center.
To 400 ft	To 2,400,000 ft-lb	Stationary center pier supports drive and rake mechanisms. Truss extending from center pier to tank periphery supports walkway, power lines and feed launder.
To 600 ft	To 4,000,000 ft-lb	Center pier has been enlarged to form a control and pumping station, as well as a support for the rake assembly. Underflow tunnel is eliminated as underflow is pumped *up* through the caisson.
To 400 ft	To 1,300,000 ft-lb	Stationary center pier partially supports rake mechanism and serves as pivot about which rake rotates. Power is supplied by electric motor mounted on two traction wheels running around the periphery of the tank.
To 140 ft	To 300,000 ft-lb for Type A to 600,000 ft-lb for Type S	Mechanically similar to Standard Type A or Type S units, except for special flocculating feed well and necessary support equipment. Designed to provide roughly 15× the throughput of the conventional machine of similar size.
		Although "CableTorq" is not, properly speaking, a *type* of thickener, it is *so important an engineering feature* that a thickener often is identified as a *CableTorq Thickener*. CableTorq is an exclusive Dorr-Oliver rake arm that has been designed to eliminate the problems caused by heavy sludge concentrations. A CableTorq rake arm is constructed with a dual-axis hinge that permits the arm to move with "universal joint" action whenever an overload is encountered. The rake arm automatically raises when heavy sludge occurs. Blades maintain their most efficient raking angle, and continuing raking action "works the solids down" until the blockage has been removed and the rake arm is back in normal position.

Table 9.2 Major Industries Served by Thickeners

Industry	Application
Alumina	Red mud and hydrate
Iron Ore	Hydroseparation, concentrate, tailings
Copper/Molybdenum Ore	Concentrate, tailings
Nickel/Cobalt	Concentrate, tailings
Steel	Mill wastes
Potash	Tailings
Phosphates	Slimes, phosphoric acid clarification
Flue Gas Desulfurization	Scrubber water
Coal	Refuse treatment
Cement	Acid-back residue
Uranium	Yellow cake
Zinc, Gold, Lead	Tailings
Sand	Slimes
Magnesia	Concentration
Chemical Processing	Waste treatment
Industrial Processing	Waste treatment

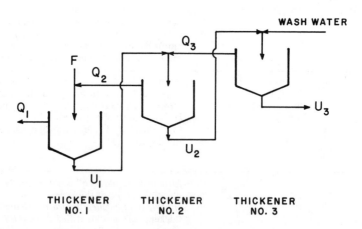

Figure 9.5   Flow scheme for a three-stage CCD system.

velocity as it enters the basin proper. This tends to minimize undesirable cross currents in the settling region of the vessel. Most cylindrical units are equipped with peripheral weirs; however, some designs include radial weirs to reduce the exit velocity and minimize weir loadings. The unit shown also is equipped with adjustable rotating overflow pipes.

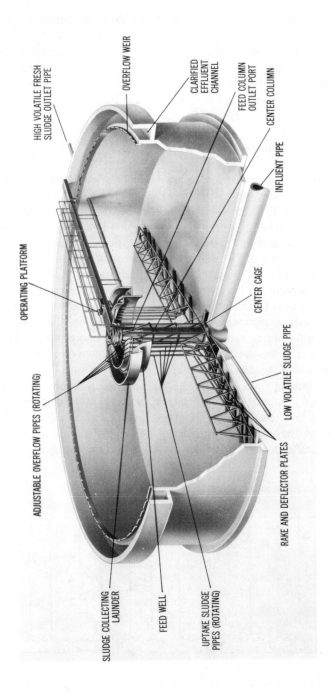

Figure 9.6   A cylindrical clarifier (courtesy Dorr-Oliver Inc., Stamford, CT).

Before attempting design problems for these separation machines, the mechanisms controlling sedimentation must be understood and defined in mathematical terms. Description of the type of sedimentation most frequently encountered in industrial applications goes beyond the analysis of single-particle settling presented in the previous subsection. The more detailed analysis below is based on careful experimental observations of the phenomenological process.

Principles of Sedimentation

To examine sedimentation in greater detail, let us examine the events occurring in a small-scale experiment conducted batchwise, as depicted by Figure 9.7(A). Particles in a narrow size range will settle with about the same velocity. When this occurs, a demarcation line is observed between the supernatant clear liquid (zone A) and the slurry (zone B) as the process continues. The velocity at which this demarcation line descends through the column indicates the progress of the sedimentation process. The particles near the bottom of the cylinder pile up, forming a concentrated sludge (zone D), whose height increases as the particles settle from zone B. As the upper interface approaches the sludge buildup on the bottom of the container, the slurry appears more uniform as a heavy sludge (zone D); the settling zone B disappears; and the process from then on consists only of the continuation of the slow compaction of the solids in zone D.

By measuring the interface height and solids concentrations in the dilute and concentrated suspensions, a graphic representation of the sedimentation rates can be prepared as shown in Figure 9.7(B). The plot shows the difference in interface height plotted against time, which is proportional to the rate of settling as well as to concentration.

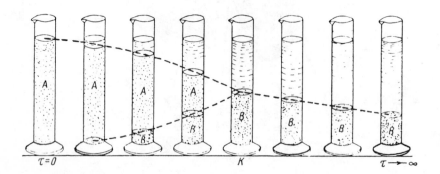

Figure 9.7(A)   Batch sedimentation occurring in a glass cylinder.

Examining these data in more detail as a plot of sediment height, Z, versus time, t, in Figure 9.8, we note that $Z \propto t$, meaning that the sedimentation rate is, and continues to be, constant. Then the sedimentation rate of the heavy sludge decreases with time, which corresponds to the curve on the graph after point K. Naturally, the higher the concentration of the initial suspension, the slower the sedimentation process.

Observations show that the solids concentration in the dilute phase is constant up to the point of complete disappearance of phase A. This is illustrated by the plot in Figure 9.9, and corresponds to a constant rate of sedimentation in the phase. Note, however, that the concentration in phase B changes with height Z and time t (as shown by Figure 9.8) and, hence, each curve in Figure 9.9 represents the distribution of concentrations at any given moment. The initial concentration is C_1, which remains in the dilute phase during the process. After a sufficient period of time, the concentration increases to C_2, but in zone D. Obviously, if the concentration of the feed suspension is too high, no dilute phase will exist, even during the initial period of sedimentation. Hence, there is no constant sedimentation rate. In this case, concentration, not height, will change with time only.

As follows from an earlier discussion on spheres falling through a fluid

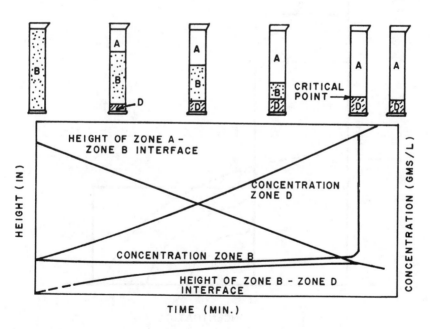

Figure 9.7(B)   Plots of interface height and solids concentration versus time for the batch sedimentation shown in (A).

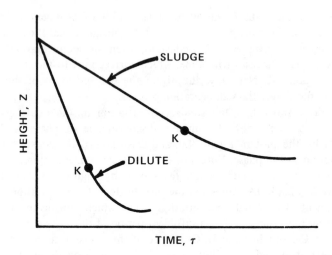

Figure 9.8   Graph describing the kinetics of sedimentation.

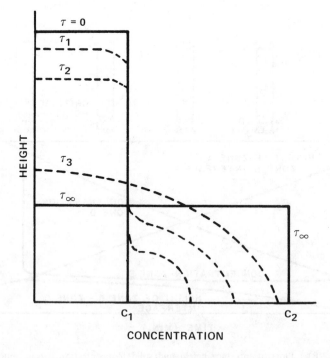

Figure 9.9   Plot of concentrations versus interface height. Denotes sedimentation rates.

medium, sedimentation is faster in liquids having low viscosities. Hence, sedimentation rates are higher at elevated temperatures. In addition to temperature, an increase in the process rate may be realized by increasing particle sizes through the use of coagulation or agglomeration. In the case of colloidal suspensions, this is achieved by the addition of an electrolyte. Instead of using the concentration of the initial suspension to describe the process, we introduce a void fraction for the suspension. The void fraction is the ratio of the liquid volume, V_f, filling the space among the particles, to the total volume which is the sum of the liquid volume and the actual volume of the solid particles, V_p:

$$\epsilon = \frac{V_f}{V_p + V_f} \tag{9.27}$$

As particles settle, forming a thickened zone, the void fraction, ϵ, decreases. At the total settling of the slurry, the void fraction is at a minimum, its value depending on the shape of the particles. For example, the minimum void fraction of spherical particles is $\epsilon_{min} = 0.215$; for small crystals, $\epsilon_{min} = 0.4$. For most systems, the void fraction of a thickened sludge is approximately $\epsilon_{min} \simeq 0.6$; however, values should be determined experimentally for the specific system.

As the sludge compacts, its void fraction and height, X, decrease. If the initial void fraction, ϵ_0, is known when the sludge height is X_0, an average void fraction, ϵ, can be estimated assuming that the height of the sludge decreases to X. For a vertical cylinder of cross-sectioned area F, the initial volume of the sediment is FX_0 and, hence, the volume of solid particles in the sediment is $(1 - \epsilon_0)X_0F$. Similarly, the volume of solid particles for voidage ϵ is $(1 - \epsilon)XF$. Consequently,

$$(1 - \epsilon_0)X_0F = (1 - \epsilon)XF \tag{9.28}$$

Hence,

$$\epsilon = \frac{X_0}{X}(1 - \epsilon_0) \tag{9.29}$$

For a unit volume of slurry, its weight, γ, is the sum of the weights of the solid particles, $\gamma_p(1 - \epsilon)$, and of the liquid, $\gamma_f\epsilon$, where γ_f = specific weight of liquid and γ_p = specific weight of particles:

$$\gamma = \gamma_p(1 - \epsilon) + \gamma_f \epsilon$$

or

$$\epsilon = \frac{\gamma_p - \gamma}{\gamma_p - \gamma_f} \tag{9.30}$$

Equation 9.30 can be used to compute the void fraction from experimentally determined values of the specific weights.

Let us now direct our attention to the sedimentation process in the zone of constant settling velocity, i.e., in the dilute phase. To simplify the analysis, we shall follow Steinour's [1944] suggestion and assume spherical particles of the same size. The process may be simplified further by viewing sedimentation as fixed particles in an upward-moving stream of viscous liquid, whose average velocity is u_f. Due to the viscosity of the liquid, a certain velocity gradient exists relative to the distance from the surface of spherical particle, du/dx. This velocity also depends on the average distance among particles, which is determined at any moment by a void fraction, ϵ, of the slurry and the particle diameter, d. The average velocity of the liquid may be presented in this case as

$$u_f = f\left(d, \epsilon, \frac{du}{dx}\right) \tag{9.31}$$

and rewritten on the basis of dimensional analysis in the following form:

$$u_f = K_1 \, d \, \frac{du}{dx} \, \Phi_1(\epsilon) \tag{9.32}$$

where K_1 = constant
$\Phi_1(\epsilon)$ = dimensionless function of the void fraction

The resistance to liquid flow around particles may be presented by an equation similar to the viscosity equation but with considering the void fraction. Recall that the shear stress is expressed by the ratio of the drag force, R, to the active surface, $K_2 \pi d^2$. The total sphere surface is πd^2 and K_2 is the coefficient accounting for that part of the surface responsible for resistance. Considering the influence of void fraction as a function $\Phi_2(\epsilon)$, we obtain

$$\frac{R}{K_2 \pi d^2} = \mu \, \frac{du}{dx} \, \Phi_2(\epsilon) \tag{9.33}$$

Dividing Equation 9.33 by Equation 9.32, we obtain

$$R = \frac{K_2}{K_1} \pi d\mu u_f \frac{\Phi_2(\epsilon)}{\Phi_1(\epsilon)}$$

(9.34)

For very dilute suspensions, in which the void fraction does not influence the sedimentation process, the function $\Phi_2(\epsilon)/\Phi_1(\epsilon) = \Phi(\epsilon)$ reduces to unity. It is known also that in dilute suspensions the sedimentation of small particles follows Stokes' law:

$$R_\infty = 3\pi\mu d u_f$$

(9.35)

Equating Equations 9.34 and 9.35, we find that (K_2/K_1) is equal to 3. Hence, the resistance of the liquid relative to a spherical particle in the sedimentation process is

$$R = \frac{3\pi\mu d u_f}{\Phi(\epsilon)}$$

(9.36)

This resistance is balanced by the gravity force acting on a particle:

$$W = \frac{\pi d^3}{6} (\gamma_p - \gamma)$$

(9.37)

where γ_p is the actual specific weight of a particle, and γ is the average specific weight of the sludge, which depends on void fraction ϵ. Using Equation 9.30, we replace $(\gamma_p - \gamma)$ with $(\gamma_p - \gamma_f)\epsilon$, where γ_f is the specific weight of the liquid:

$$W = \frac{\pi d^3}{6} (\gamma_p - \gamma_f)\epsilon$$

(9.38)

By comparing the gravity force acting on the particle (Equation 9.38) with the resistance to liquid flow (Equation 9.36), we obtain the average liquid velocity relative to the particles:

$$u_f = \frac{d^2(\gamma_p - \gamma_f)}{18\mu} \epsilon\Phi(\epsilon)$$

(9.39)

In practice, however, the liquid velocity relative to fixed particles, u_f, is not very useful. Instead, the velocity of settling relative to the walls of an apparatus, $u_f - u$, is of practical importance. The volume of the solid phase moving downward should be equal to that of liquid moving upward. This means that volume rates of these phases must be equal. Consider a column of slurry having a unit cross section and imagine the liquid and solid phases to have a well defined interface. The column of solid phase will have a base $1 - \epsilon$, and the liquid column phase will have a base ϵ. Hence, the volumetric rate of the solid column will be $(1 - \epsilon)u$, and that of the liquid column will be $(u_f - u)\epsilon$. Because these flowrates are equal to each other, we obtain

$$(1 - \epsilon)u = (u_f - u)\epsilon \tag{9.40}$$

Therefore, the settling velocity of the solid phase relative to the wall of an apparatus, depending on the average liquid velocity relative to the sludge with void fraction ϵ, will be

$$u = u_f \epsilon \tag{9.41}$$

Substituting this expression into Equation 9.39, we obtain the actual settling velocity:

$$u = \left[\frac{d^2(\gamma_p - \gamma_f)}{18\mu} \right] \epsilon^2 \Phi(\epsilon) \tag{9.42}$$

Note that the term in parentheses expresses the velocity of free falling, according to Stokes' law:

$$u_p = \frac{d^2(\gamma_p - \gamma_f)}{18\mu} \tag{9.43}$$

or

$$u = u_p \epsilon^2 \Phi(\epsilon) \tag{9.44}$$

For a very dilute suspension, i.e., $\epsilon = 1$ and $\Phi(\epsilon) = 1$, the settling velocity will be equal to the free-fall velocity.

As no valid theoretical expression for the function $\Phi(\epsilon)$ is available, we will use experimental data. Note that a unit volume of thickened sludge contains ϵ volume of liquid and $(1 - \epsilon)$ volume of solid phase, i.e., a unit volume of particles of sludge contains $\epsilon/(1 - \epsilon)$ volume of liquid. Denoting σ as the ratio of particle surface area to volume, we obtain the hydraulic radius as the ratio of this volume, $\epsilon/(1 - \epsilon)$, to the surface, σ, when both values are related to the same volume of particles:

$$r_h = \frac{\epsilon}{(1 - \epsilon)\sigma} \tag{9.45A}$$

For spherical particles, σ is equal to the ratio of the surface area, πd^2, to the volume $\pi d^3/6$, i.e., $\sigma = 6/d$. Hence,

$$r_h = d\frac{\epsilon}{(1 - \epsilon)6} \tag{9.45B}$$

For a specified void fraction, the diameter of the sphere is a measure of the distance between sludge particles in Equation 9.31. However, it is more practical to introduce the hydraulic radius, and instead of $\Phi_1(\epsilon)$ and $\Phi_2(\epsilon)$, according to Equation 9.44, we assume the following value:

$$\Phi(\epsilon) = \frac{\epsilon}{(1 - \epsilon)} \theta(\epsilon) \tag{9.46}$$

where $\theta(\epsilon)$ is the new experimental function of the void fraction. Hence, the settling velocity equation may be rewritten in the following form:

$$u = u_p \frac{\epsilon}{1 - \epsilon} \theta(\epsilon) \tag{9.47}$$

By representing the velocity in this manner, we can anticipate a small change in the function $\theta(\epsilon)$ because the influence of the flow pattern is, to a large extent, accounted for in the hydraulic radius. Experimental investigations by Steinour [1944] have shown that the function $\Phi(\epsilon)$ may be presented by the following empirical equation:

$$\Phi(\epsilon) = 10^{-1.82(1-\epsilon)} \tag{9.48}$$

Multiplying Equation 9.48 by $(1 - \epsilon)/\epsilon$, we obtain the function $\theta(\epsilon)$. For $\epsilon \leqslant 0.7$, i.e., for thickened sludges, this function is practically constant and equal to

$$\theta(\epsilon) = 0.123 \qquad (9.49)$$

Finally, the settling velocity of spherical particles is

$$u = u_p[\epsilon^2 \times 10^{-1.82(1-\epsilon)}] \qquad (9.50)$$

For more thickened sludges, Equation 9.47 takes the following form:

$$u = \left[0.123 \frac{\epsilon^3}{(1 - \epsilon)}\right] u_p \qquad (9.51)$$

Thus far, the analysis has been based on independently settling spherical particles. To relate to the design of the unit operations, we now must consider the kinetics of nonspherical particle settling and the sedimentation of flocculent particles. In contrast to single-particle settling, such systems form a certain structural unity similar to tissue. The sludge is compacted under the action of gravity force, i.e., the void fraction decreases and the liquid is squeezed out from the pore structure. The formation of a regular sediment from a flocculent may be achieved by the addition of electrolytes.

The general characteristic of normal settling of nonspherical particles (as well as flocculent ones) is that the sediment carries along with it a portion of the liquid by trapping it between particle cavities. This trapped volume of liquid flows downward with the sludge and is proportional to the volume of the sludge. That is, it can be expressed as $a(1 - \epsilon)$, where a is a coefficient and $(1 - \epsilon)$ is the volume of particles. Consequently, a portion of the liquid remains in a layer above the sludge, and a portion is carried along with the sludge corresponding to the modified void fraction:

$$\epsilon' = \epsilon - a(1 - \epsilon) \qquad (9.52)$$

This is the difference of the total relative liquid volume and liquid moving together with particles. Substituting ϵ' for ϵ in Equation 9.51, we obtain the settling velocity at $\epsilon \leqslant 0.7$:

$$u = 0.123(1 + a)^2 u_p \frac{\left(\epsilon - \dfrac{a}{1 + a}\right)^3}{1 - \epsilon} \qquad (9.53)$$

Denoting $a/(1 + a) = \beta$, Equation 9.53 is rewritten as

$$u = \frac{0.123 u_p (\epsilon - \beta)^3}{(1 - \beta)^2 (1 - \epsilon)} \qquad (9.54)$$

Similarly Equation 9.50 for slurries with nonspherical particles is

$$u = u_p \frac{(\epsilon - \beta)^2}{1 - \beta} 10^{-1.82(1-\epsilon)/(1-\beta)} \qquad (9.55)$$

Parameter β is equal to the ratio of the liquid volume entrained and the sum of the volumes of this liquid and particles. Values of β are determined experimentally from measured settling velocities. In general, the smaller the effective particle size, the more liquid is entrained by the same mass of solids phase. For example, particles of carborundum with $d = 12.2$ μm have $\beta = 0.268$; $d = 9.6$ μm, $\beta = 0.288$; and $d = 4.6$ μm, $\beta = 0.35$.

Application of Dimensional Analysis to Particle Settling

The forces acting on a single particle settling through a continuous fluid medium are gravity, $\overline{G}$, buoyant or Archimedes forces, $\overline{A}$, centrifugal field, $\overline{C}$, and electrical field, $\overline{Q}$, as summarized in Figure 9.10. Geometrically summing all the forces, we obtain Archimedes, $\overline{A}$, centrifugal, $\overline{C}$, and force of electrical field, $\overline{Q}$. Hence, for a state of equilibrium we may write:

$$\overline{P} = \overline{G} + \overline{A} + \overline{C} + \overline{Q} \qquad (9.56)$$

If force P is greater than zero, the particle will be in motion relative to the continuous phase with a certain velocity, w. At the beginning of the particle's motion, a resistance force develops in the continuous phase, $\overline{R}$, directed at the opposite side of the particle motion.

At low particle velocity (relative to the continuous phase), fluid layers

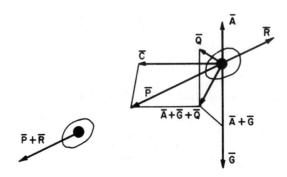

Figure 9.10   System of forces acting on a settling particle.

running against the particle are moved apart smoothly in front of it and then come together smoothly behind the particle (Figure 9.11A). The fluid layer does not intermix (a system analogous to laminar fluid flow in smoothly bent pipes). The particles of fluid nearest the solid surface will take the same time to pass the body as those at some distance away.

Because the liquid layers move at different velocities relative to each other, planes of slip exist between them and, from Newton's law, the forces of viscous friction arise. Consequently, the resistance force depends on the viscosity of the medium as is determined by the viscosity coefficient,

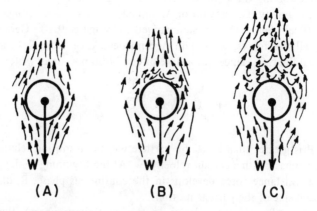

Figure 9.11   Flow around a particle by a fluid: (A) laminar regime; (B) transitional regime; (C) turbulent regime.

μ. At higher particle velocities (or higher medium velocities relative to the particle) the flow around the object is broken, forming swirling fluid patches (Figure 9.11B). The formation of vortices is influenced by the relative flow velocity, the shape of the particle and the smoothness of the object's surface. The higher the velocity, the more complicated the particle shape; and/or the greater the roughness, the more intense the vortex formation. Eventually, this leads to the generation of eddies along the downstream surface of the particle (Figure 9.11C).

The eddies are entrained by the flow and, at a certain distance from the particle, they disappear while being replaced by new eddies. Due to eddy formation and their breaking away from the particle, a low-pressure zone forms at the front of the particle. Hence, as described earlier, a pressure gradient is formed between the front and rear of the particle. This gradient is responsible primarily for the resistance to particle motion in the medium. The amount of this resistance depends on the energy expended toward eddy formation: the more intensive this formation, the greater the energy consumption and, hence, the greater the resistance force. The inertia forces generated by eddies play an important role. They are characterized by the mass and velocity of the fluid relative to the particle.

The total resistance is the sum of friction and eddy resistances. Both factors act simultaneously, but their contribution in the total resistance depends on the conditions of the flow in the vicinity of the particle. Hence, for the most general case the resistance force is a function of velocity, w, density, ρ, viscosity, μ, the linear size of a particle, ℓ, and its shape. Thus,

$$R = (w, \rho, \ell, \mu, \psi) \qquad (9.57)$$

Assuming this relationship as an exponential complex, we obtain

$$R = A' w^x \rho^u \ell^z \mu^\alpha \qquad (9.58)$$

where A' is dimensionless coefficient that includes the shape factor, ψ. Noting the dimensions of all parameters appearing in Equation 9.58,

$$[R] = LMT^{-2} \; ; \quad [w] = LT^{-1}$$
$$[\rho] = L^{-3}M \; ; \quad [\ell] = L$$
$$[\mu] = L^{-1}MT^{-1}$$

where L, M and T are the principal unit measures—length, mass and time.

Expressing Equation 9.58 in terms of its dimensions,

$$LMT^{-2} = (LT^{-1})^x(L^{-3}M)^y L^z(L^{-1}MT^{-1})^\alpha \qquad (9.59A)$$

or

$$LMT^{-2} = L^{x-3y+z-\alpha}M^{y+\alpha}T^{-x-\alpha} \qquad (9.59B)$$

For the dimensions on the LHS of this expression to satisfy the RHS, the exponents on the principal units of measure must be equal. Thus, we have the following system of three equations (corresponding to the number of values with independent dimensions):

$$\left.\begin{array}{r} x - 3y + z - \alpha = 1 \\ y + \alpha = 1 \\ -x - \alpha = -2 \end{array}\right\} \qquad (9.60)$$

Because this system of equations cannot be solved (there are fewer equations than variables), we express all exponents in terms of α:

$$x = 2 - \alpha \ ; \quad y = 1 - \alpha \ ; \quad z = 2 - \alpha$$

Equation 9.58 therefore may be written as

$$R = A'w^{2-\alpha}\rho^{1-\alpha}\ell^{2-\alpha}\mu^\alpha \qquad (9.61)$$

Equation 9.61 is a general expression that may be applied to the treatment of experimental data to evaluate exponent α. This, however, is a cumbersome approach that can be avoided by rewriting the equation in dimensionless form. Equation 9.57 shows that there are $n = 5$ dimensional values, and the number of values with independent measures is $m = 3$ (m, kg, s). Hence, the number of dimensionless groups according to the π-theorem is $\pi = 5 - 3 = 2$. As the particle moves through the fluid, one of the dimensionless complexes is obviously the Reynolds number:

$$Re = \frac{w\ell\rho}{\mu} \qquad (9.62)$$

Thus, Equation 9.61 may be rewritten as follows:

$$R = A' \left(\frac{w\ell\rho}{\mu} \right)^{-\alpha} \rho w^2 \ell^2 \qquad (9.63)$$

As one of two possible dimensionless numbers is now known, the second one can be obtained by dividing both sides of the equation through by the remaining values:

$$Eu = \frac{R}{\rho w^2 \ell^2} \qquad (9.64)$$

The result is a modified Euler number. The student should prove to himself that the pressure drop over the particle can be obtained by accounting for the projected area of the particle through particle size, ℓ, in the denominator. Thus, by application of dimensional analysis to the force balance expression, a relationship between the dimensionless complexes— Euler and Reynolds numbers—is obtained:

$$Eu = A'Re^{-\alpha} \qquad (9.65)$$

Coefficient A' and exponent α must be evaluated experimentally. Experiments have shown that A' and α are themselves functions of the Reynolds number.

Equation 9.63 shows that the resistance force increases with increasing velocity. If the force field (e.g., gravity) has the same potential at all points, a dynamic equilibrium between forces P and R develops shortly after the particle motion begins. As described earlier, at some distance from its start the particle falls at a constant velocity.

If the acting force depends on the particle location in space, in a centrifugal field, for example, it will move with uniformly variable speed until it travels outside of the boundary of the field's action or runs into an obstacle such as a vessel wall. We now shall define the relationship between particle motion and the acting factors.

Moving under the action of force P, with acceleration a_g at infinitesimal distance $\delta\ell$, the particle with mass m_p performs work $m_p a_g \delta\ell$. This work is spent on overcoming the resistance force and the displacement of fluid mass, m_c, in a volume equal to the volume of the particle, V, at the same distance, but in opposite direction, and with the same acceleration, a_g:

$$m_p a_g \delta\ell = m_c a_g \delta\ell + R\delta\ell \qquad (9.66)$$

Dividing through by $\delta\ell$ and expressing the masses of the particle and medium in terms of their volumes and densities, we obtain

$$V(\rho_p - \rho_c)a_g = R \tag{9.67}$$

Consider again the simple motion of a sphere. In this case, the equivalent diameter of a sphere, d_{eq}, is equal to its geometric diameter, d. Equating Equations 9.62 and 9.63 and replacing ℓ by d (and denoting the Euler number, Eu, by $\tilde{\psi}$), we obtain an expression for the resistance force:

$$R = \tilde{\psi}\rho_c w^2 d^2 \tag{9.68}$$

where

$$\tilde{\psi} = A'Re^{-\alpha} \tag{9.69}$$

In sedimentation operations, the Euler number is called the *resistance number*.

Multiplying and dividing the RHS of Equation 9.68 by $\pi/8$, we obtain

$$R = C_D F\rho_c \frac{w^2}{2} \tag{9.70}$$

where $C_D = (8/\pi)\tilde{\psi}$ = the resistance (or drag) coefficient
$F = \pi d^2/4$ = the cross-sectional area of the spherical particle

Equation 9.70 is Newton's resistance law. Substituting Equation 6.69 into the definition for C_D, we obtain

$$C_D = BRe^{-\alpha} \tag{9.71}$$

where

$$B = \frac{8A'}{\pi} \tag{9.72}$$

Because $\tilde{\psi} = f(Re)$, $C_D = f_1(Re)$.

The Reynolds number for a sphere is

$$Re = \frac{wd\rho_c}{\mu} = \frac{wd}{\nu}$$

(9.73)

Substituting the resistance force from Equation 9.70 into Equation 9.67 and expressing F and V in terms of d, the basic equation of sedimentation theory is obtained:

$$d^3 \frac{\rho_p - \rho_c}{\rho_c} a_g = \frac{3}{4} C_D w^2 d^2$$

(9.74)

or

$$d^3 \frac{\rho_f}{\rho_c} a_g = \frac{3}{4} C_D w^2 d^2$$

(9.75)

where $\rho_f = \rho_p - \rho_c$ is the effective density of the system.

In separation calculations for heterogeneous systems, the important parameter is settling velocity:

$$w = \sqrt{\frac{4}{3} \times \frac{d}{C_D} \frac{\rho_f}{\rho_c} a_g}$$

(9.76)

Application of the above formulas is difficult because the resistance coefficient, C_D, is a function of velocity and particle geometry. Lyachshenko [1940] proposed a generalized calculation procedure for settling under the influence of gravity. The method, however, also may be applied to settling due to the influence of any force field, provided the relationship between particle acceleration and the coordinates of field is defined. The procedure is based on expressing the basic equation of sedimentation (Equation 9.74) in terms of a relationship of criteria. For this, both sides of Equation 9.74 are divided by ν^2, and the RHS multiplied and divided by the acceleration due to gravity, g:

$$\frac{gd^3}{\nu^2} \frac{\rho_f}{\rho_c} \frac{a_g}{g} = \frac{3}{4} C_D \frac{w^2 d^2}{\nu^2}$$

(9.77)

The LHS of this expression contains a dimensionless group known as the Galileo number:

$$Ga = \frac{gd^3}{\nu^2} \tag{9.78}$$

Multiplying by a simplex composed of densities results in the Archimedes number:

$$Ar = Ga \frac{\rho_f}{\rho_c} \tag{9.79}$$

And introducing the ratio of accelerations,

$$K_s = \frac{a_g}{g} \tag{9.80}$$

The dimensionless parameter, K_s, indicates the relative strength of acceleration, a_g, with respect to the gravitational acceleration g. This is our familiar *separation number* described in the previous chapter. (Note that in centrifugal processes, K_s is called the *separation factor*.)

The LHS of Equation 9.77 contains a Reynolds number group raised to the second power and the drag coefficient. Hence, the equation may be written entirely in terms of dimensionless numbers:

$$ArK_s = \frac{3}{4} C_D Re^2 \tag{9.81}$$

The Archimedes number contains parameters that characterize the properties of the heterogeneous system and the criterion establishing the type of settling. The criterion of separation essentially establishes the separating capacity of a sedimentation machine. The product of these criteria is

$$S_1 = ArK_s \tag{9.82}$$

This product contains information on the properties of the suspension and characterizes the settling process as a whole. Substituting Equation 9.81 into 9.82 gives

$$S_1 = \frac{3}{4} C_D Re^2 \tag{9.83}$$

This expression represents the first form of the general dimensionless equation of sedimentation theory. As the desired value is the velocity of the particle, Equation 9.83 is solved for the Reynolds number:

$$Re = \sqrt{\frac{4}{3}\frac{S_1}{C_D}} \tag{9.84}$$

To determine the size of a particle having a velocity, w, in the gravitational field, both sides of Equation 9.83 are multiplied by the complex Re/ArK_s:

$$\frac{Re^3}{Ar}\frac{1}{K_s} = \frac{4}{3}\frac{Re}{C_D} \tag{9.85}$$

The dimensionless complex Re^3/Ar is the so-called *Lyachshenko number*:

$$Ly = \frac{Re^3}{Ar} = \frac{w^3}{g\nu}\frac{\rho_c}{\rho_f} \tag{9.86}$$

Denoting

$$S_2 = \frac{Ly}{K_s} \tag{9.87}$$

Equation 9.81 may be rewritten as

$$S_2 = \frac{4}{3}\frac{Re}{C_D} \tag{9.88}$$

This is the second form of the dimensionless equation for sedimentation. The Reynolds number also may be calculated from this equation:

$$Re = \frac{3}{4}C_D S_2 \tag{9.89}$$

As with S_1, the Reynolds number is the dependent variable and S_2 is the determining one.

For settling under the influence of gravity, we note that $a_g = g$, $K_s = 1$ and, hence, $S_1 = Ar$ and $S_2 = Ly$. Therefore, the general dimensionless equations for sedimentation (Equations 9.83 and 9.87) are applicable in any force field. They need be transformed only into the appropriate dimensionless groups describing the type of force field influencing the process. Again, for gravity settling,

$$Ar = \frac{3}{4} C_D Re^2 \tag{9.90}$$

and

$$Ly = \frac{4Re}{3C_D} \tag{9.91}$$

The dimensionless numbers of sedimentation, S_1 and S_2, as well as C_D and $\widetilde{\psi}$, are all functions of Re. The parameter $\widetilde{\psi}$ must be determined experimentally. Equation 9.69 can be written as a straight line when expressed in terms of its logarithms:

$$\log \widetilde{\psi} = \log A' - \alpha \log Re \tag{9.92}$$

Coefficient A' and exponent α can be evaluated readily from data on Re and $\widetilde{\psi}$. The dimensionless groups are presented on a single plot in Figure 9.12. The plot of the function $C_D = f_1(Re)$ is constructed from three separate sections. These sections of the curve correspond to the three regimes of flow. The *laminar regime* is expressed by a section of straight line having a slope $\beta = 135°$ with respect to the x-axis. This section corresponds to the critical Reynolds number, $Re'_{cr} \leqslant 0.2$. This means that the exponent α in Equation 9.69 is equal to 1. At this α value, the continuous-phase density term, ρ_c, in Equation 9.61 vanishes. Therefore, the inertia forces have an insignificant influence on the sedimentation process in this regime. Theoretically, their influence is equal to zero. In contrast, the forces of viscous friction are at a maximum. Evaluating the coefficient B in Equation 9.71 for $\alpha = 1$ results in a value of 24. Hence, we have derived the expression for the resistance coefficient of a sphere:

$$C_D = \frac{24}{Re}$$

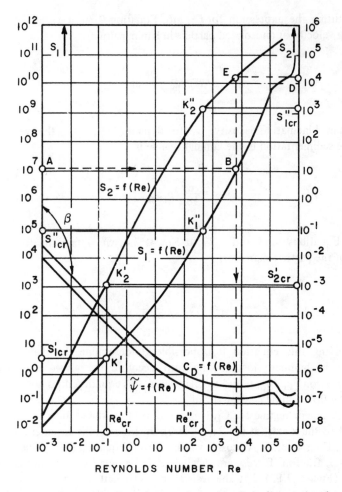

Figure 9.12   Plots of the dimensionless relationships for sedimentation theory.

The first critical values of the dimensionless sedimentation numbers, S_1 and S_2, are obtained by substituting for the critical Reynolds number value, $Re'_{cr} = 0.2$, into Equations 9.83 and 9.88:

$$S'_{1cr} \leqslant 3.6 \qquad\qquad (9.93)$$

$$S'_{2cr} \leqslant 0.0022 \qquad\qquad (9.94)$$

Substituting the expression for C_D into Equation 9.76, we again obtain the settling velocity of an isolated particle in laminar flow:

$$w = \frac{d^2}{18\nu} \frac{\rho_f}{\rho_c} a_g$$

Changing kinematic viscosity, ν, to dynamic viscosity, the velocity of particle sedimentation in the laminar regime is

$$w = \frac{d^2}{18\mu} \rho_f a_g \qquad (9.95)$$

From Equations 9.83 and 9.77 at $S'_{1cr} \leqslant 3.6$, we obtain the first critical value of the particle diameter:

$$d'_{cr} \leqslant 1.53 \sqrt[3]{\frac{\nu^2}{a_g} \frac{\rho_c}{\rho_f}} \qquad (9.96)$$

In applying this equation it is possible to determine the maximum size particle in laminar flow, taking into account the given conditions of sedimentation (ρ_c, ρ_p, μ and a_g). However, this equation does not determine what the flow regime is when $d > d'_{cr}$.

The turbulent regime for C_D is characterized by the section of line almost parallel to the x-axis (at the $Re''_{cr} \geqslant 500$). In this case, the exponent α is equal to zero. Consequently, viscosity vanishes from Equation 9.61. This indicates that the friction forces are negligible in comparison to inertia forces. Recall (Figure 9.12) that the resistance coefficient is nearly constant at a value of 0.44. Substituting for the critical Reynolds number, $Re'_{cr} \geqslant 500$, into Equations 9.83 and 9.88, the second critical values of the sedimentation numbers are obtained:

$$S''_{1cr} \geqslant 82,500 \qquad (9.97)$$

$$S''_{1cr} \geqslant 1,515 \qquad (9.98)$$

And substituting $C_D = 0.44$ into Equation 9.76,

$$w = 1.75 \sqrt{d \frac{\rho_f}{\rho_c} a_g} \qquad (9.99)$$

From Equations 9.83 and 9.77 at $S''_{1cr} \geqslant 82{,}500$, we obtain the second critical value of particle size:

$$d''_{cr} \geqslant 43.5 \sqrt{\frac{\nu^2}{a_g} \times \frac{\rho_c}{\rho_f}} \qquad (9.100)$$

Those particles with sizes $d > d''_{cr}$ at a given set of conditions (ν, ρ_c, ρ_p, and a_g) will settle only in the turbulent flow regime. For particles with sizes $d'_{cr} < d$, d''_{cr} will settle only when the flow around the object is in the transitional regime.

Recall that the *transitional zone* occurs in the Reynolds number range of 0.2 to 500. The sedimentation numbers corresponding to this zone are

$$3.6 \leqslant S_1 \leqslant 82{,}500$$

$$0.0022 \leqslant S_2 \leqslant 1{,}515$$

The slope of the curve in the transitional zone changes from 135 to 180°. It shows that the exponent in Equation 9.61 changes as follows:

$$0 \leqslant \alpha \leqslant 1$$

This means that the friction and inertia forces are commensurable in the process of sedimentation.

Several empirical formulas have been proposed for estimating the resistance coefficient in the transition zone. One such correlation is

$$C_D = \frac{13}{\sqrt{Re}} \qquad (9.101)$$

Introducing this to Equation 9.76 produces

$$w_s = 0.22d \sqrt[3]{\frac{(a_g \rho_f)^2}{\mu \rho_c}} \qquad (9.102)$$

When we consider many particles settling, the density of the fluid phase effectively becomes the bulk density of the slurry, i.e., the ratio of the total mass of fluid plus solids divided by the total volume. The viscosity of the slurry is considerably higher than that of the fluid alone because of the interference of boundary layers around interacting solid particles and the increase of form drag caused by particles. The viscosity of a slurry is often a function of the rate of shear of its previous history as it affects clustering of particles, and of the shape and roughness of the particles. Each of these factors contributes to a thicker boundary layer.

Experimental measurements of viscosity almost always are recommended when dealing with slurries and extrapolations should be made with caution. Most theoretically based expressions for liquid viscosity are not appropriate for practical calculations or require actual measurements to evaluate constants. For nonclustering particles, a reasonable correlation may be based on the ratio of the effective bulk viscosity, μ_B, to the viscosity of the liquid. This ratio is expressed as a function of the volume fraction of liquid x' in the slurry for a reasonable range of compositions:

$$\frac{\mu_B}{\mu} = \frac{10^{1.82(1-x)}}{x'} \tag{9.103}$$

A correction factor, R_c, incorporating both viscosity and density effects can be developed for a given slurry, which provides a more convenient expression based on the following equation:

$$u_t = \frac{(\rho_p - \rho)gd^2}{18\mu} \tag{9.104}$$

as

$$\nu_H = \frac{(\rho_p - \rho)gd^2}{18\mu} R_c \tag{9.105}$$

where ν_H is the terminal velocity in *hindered settling*.

Measurements of the effective viscosity as a function of composition may be fitted to Equation 9.103 or presented in graphic form as in Figure 9.13. The correction factor, R_c, also may be determined by accounting for the volume fraction, η_v, of particles through the Andress formula:

$$R_c = \frac{(1 - \eta_v)^2}{1 + 2.5\eta_v + 7.35\eta_v^2} \tag{9.106}$$

In summarizing sedimentation principles, we note that the particle settling velocity is the principal design parameter that establishes equipment

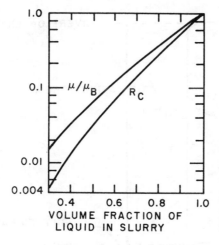

Figure 9.13   Settling factor for hindered settling.

sizes and allowable loadings for separating heterogeneous systems. However, design calculations are not straightforward because prediction of the settling velocity requires knowledge of the flow regime in the vicinity of the particles. Therefore, the following generalized design method is recommended.

From known values of d, ρ_p, ρ_c, μ and a_g, compute the first sedimentation dimensionless number (Equation 9.82). From the plot given in Figure 9.14, obtain the corresponding Reynolds number, Re, and evaluate the theoretical settling velocity. If the flow regime is laminar, the settling velocity may be calculated directly from Equation 9.95 and the regime checked by computing the Reynolds number for the flow around an individual particle. After determining w, determine the appropriate shape factor, ψ (either from literature values or measurements) and correction factor, R_c. The design settling velocity then will be

$$w_0 = R_c \psi w \tag{9.107}$$

Design Methodology

Sedimentation equipment is designed to perform two operations: to clarify the liquid overflow by removal of suspended solids and to thicken sludge or underflow by removal of liquid. It is the cross section of the apparatus that controls the time needed for settling a preselected size

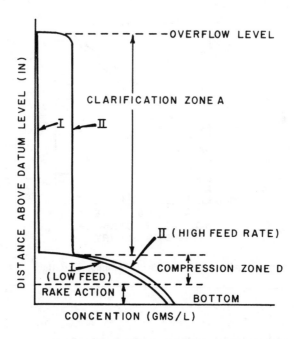

Figure 9.14   Plot of concentration versus height in a continuous sedimentation apparatus. Curves are for (I) a low feedrate and (II) a high feed rate.

range of particles out of the liquid for a given liquid feed rate and solids loading. The area also establishes the clarification capacity. The depth of the thickener establishes the time allowed for sedimentation (i.e., the solid's residence time) for a given feed rate and is important in determining the thickening capacity.

The *clarification capacity* is established by the settling velocity of the suspended solids. Sedimentation tests are almost always recommended when scaling up for large settler capacities. By the material balance method described in the previous chapter, the total amount of fluid is equal to the sum of the fluid in the clear overflow plus the fluid in the compacted sludge removed from the bottom of the thickener. The average vertical velocity of fluid at any height through the thickener is the volumetric rate passing upward at that level divided by the unit's cross section. Note that if the particle settling velocity is less than the upward fluid velocity, particles will be entrained out in the overflow, resulting in poor clarification. For those size particles whose settling velocity approximately equals that of the upward fluid velocity, particles remain in a balanced suspension, i.e., they neither rise nor fall, and the concentration of solids in the clarification zone increases. This eventually results in a reduction of the settling velocity until the point where particles are entrained out in the overflow. The thickener must be designed so that the settling velocity of particles is significantly greater than the upward fluid velocity, to minimize any increase in the solids concentration in the clarification zone.

Solids concentration varies over the thickener's height, and at the lower levels where the solution is dense, settling becomes retarded. In this region the upward fluid velocity can exceed the particle settling velocity irrespective of whether this condition exists in the upper zone or not. Figure 9.14 illustrates this situation, where curve II denotes a higher feed rate. A proper design must therefore be based on an evaluation of the settling rates at different concentrations as compared to the vertical velocity of the fluid.

If the feed rate exceeds the maximum of the design, particulates are unable to settle out of the normal clarification zone. Hence, there is an increase in the solids concentration, resulting in hindered settling. The result is a corresponding decrease in the sedimentation rate below that observed for the feed slurry. The feed rate corresponding to the condition of just failing to initiate hindered settling represents the limiting clarification capacity of the system. That is, it is the maximum feed rate at which the suspended solids can attain the compression zone. The proper cross-sectional area can be estimated from calculations for different concentrations and checked by batch sedimentation tests on slurries of increasing concentrations. Practice Problem 9.1 illustrates the need to check the thickener's calculated area against concentrations at various points in the vessel (including both the clarification and thickening zones).

Figure 9.15 shows the effect of varying the underflow rate on the

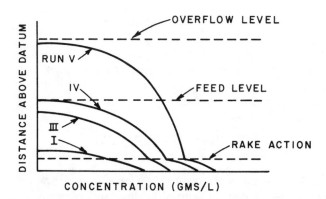

Figure 9.15 Plot of the effect of underflow rate (i.e., rate of sludge removal) on thickening capacity. The rate of sludge removal decreases from curve I to V.

thickening capacity. In this example, the depth of the thickening zone (compression zone) increases as the underflow rate decreases; hence, the underflow solids concentration increases, based on a constant rate of feed. The curves of concentration as a function of depth in the compression zone are essentially vertical displacements of each other and are similar to those observed in batch sedimentation. When the sludge rakes operate, they essentially break up a semirigid structure of concentrated sludge. Generally, this action extends to several inches above the rakes and contributes to a more concentrated underflow.

The required height of the compression zone may be estimated from experiments on batch sedimentation. The first batch test should be conducted with a slurry having an initial concentration equivalent to that of the top layer of the compression zone during the period of constant rate settling. This is referred to as the *critical concentration*. The time required for the sample slurry to pass from the critical concentration to the desired underflow concentration can be taken as the retention time for the solids in the continuous operation. The underlying assumption here is that the solids concentration at the bottom of the compression zone in the continuous thickener at any time is the same as the average concentration of the compression zone in the batch unit and at a time equal to the retention time of the solids in the continuous thickener. Hence, it is assumed that the concentration at the bottom of the thickener is an implicit function of the thickening time. The retention time is obtained from a batch test by observing the height of the compression zone as a function of time.

The slope of the compression curve is described by the following expression:

$$-\frac{dZ}{dt} = k(Z - Z_\infty) \tag{9.108}$$

where Z, Z_∞ are the heights of compression at times t and infinity, respectively, and k is a constant that depends on the specific sedimentation system. Integrating this expression gives

$$ln(Z - Z_\infty) = -kt + ln(Z_c - Z_\infty) \tag{9.109}$$

where Z_c is the height of the compression zone at its critical concentration. Equation 9.109 is that of a straight line and normally is plotted as $\log[(Z - Z_\infty)/(Z_0 - Z_\infty)]$ versus time, where Z_0 is the initial slurry concentration.

If batch tests are performed with an initial slurry concentration below that of the critical, the average concentration of the compression zone will exceed the critical value because it will consist of sludge layers compressed over varying time lengths. Roberts [1949] suggested a method for estimating the required time to pass from the critical solids content to any specified underflow concentration. The procedure is as follows:

1. Extrapolate the compression curve to the critical point or zero time.
2. Locate the time when the upper interface (between the supernatant liquid and slurry) is at height Z_0', halfway between the initial height, Z_0, and the extrapolated zero-time compression zone height, Z_0'. This time represents the period in which all the solids were at the critical dilution and went into compression. The retention time is computed as $t - t_c$, where t is the time when the solids reach the specified underflow concentration.

The procedure is illustrated in Figure 9.16. Brown [1950] recommends determination of the required volume for the compression zone to be based on estimates of the time each layer has been in compression. The volume for the compression zone is the sum of the volume occupied by the solids plus the volume of the entrapped fluid. This may be expressed as

$$V = \frac{Q(\Delta t)}{\rho_s} + \int_{t_c}^{t} \frac{m_\varrho Q}{m_s \rho_\varrho} \, dt \qquad (9.110)$$

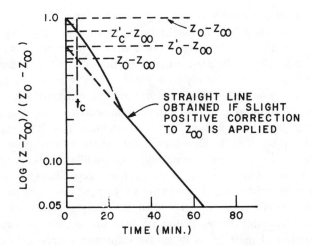

Figure 9.16   The extrapolation of sedimentation data to estimate the time for critical concentration.

where   Q = solids mass feed per unit time
         $\Delta t = t - t_c$ = retention time
         m_ϱ = mass of liquid in the compression zone
         m_s = mass of solids in the compression zone

Equation 9.110 is based on our earlier assumption that the time required to thicken the sludge is independent of the interface height of the compression zone. An approximate solution to this expression can be obtained if we assume m_ϱ/m_s to be constant, i.e., an average mass ratio in the thickening zone from top to bottom. Then,

$$V = Q\Delta t \left[\frac{1}{\rho_s} + \frac{1}{\rho_\varrho} \left(\frac{m_\varrho}{m_s} \right)_{avg} \right]$$

(9.111)

Coe and Clevenger [1916] suggest that more reliable results can be obtained by assuming average conditions over divided parts of the compression zone. That is, Equation 9.111 can be applied to divisions of the compression zone and the total volume obtained by the sum of these calculations.

Practice problems are given at the end of this chapter to strengthen the student's understanding of the principles covered in this chapter.

UNSTEADY PARTICLE MOTION IN FLUIDS

Transient Particle Settling

In this section we shall analyze two common cases of transient particle motions. The first relates to particle sedimentation in a fluid medium (just described), while the latter concerns particle injection into a moving fluid stream. The analysis of particle motion presented in the previous section was based on a steady-state phenomenon in which an equilibrium condition exists between the acting force and the resistance force. However, this state of equilibrium is not attained immediately. When a particle at rest is set into motion (i.e., it settles through the fluid medium) its velocity will increase at the same time as the resistance to the particle motion. Only after a certain time period will this equilibrium be achieved. The length of this time may be short or long, depending on the relative magnitudes of resistance and gravity forces. In most cases, this period is very short and therefore can be ignored in the analysis. There are, however, situations when this initial period of unsteady motion is important. A falling particle of mass, m, is under the influence of gravity force, G, and resistance force, R. The acceleration of the particle is du/dt; hence, the resulting force is

$$m \frac{du}{dt} = G - R \tag{9.112}$$

The weight of the particle in the liquid, according to Archimedes' law is

$$G = \frac{mg(\gamma_p - \gamma)}{\gamma_p} \tag{9.113}$$

The resistance force is

$$R = F\gamma C_D \frac{u^2}{2g} \tag{9.114}$$

where C_D = drag coefficient
   F = particle's projected area

Substituting Equations 9.113 and 9.114 into Equation 9.112, we obtain an expression for acceleration:

$$\frac{du}{dt} = g \frac{\gamma_p - \gamma}{\gamma_p} - \frac{F\gamma C_D u^2}{2gm} \tag{9.115}$$

Expressing the velocity, u, as a function of the Reynolds number,

$$u = Re \frac{\mu g}{d\gamma} \tag{9.116}$$

and substituting this value into Equation 9.115,

$$\frac{\mu F}{2md} dt = \frac{dRe}{\Phi - \lambda Re^2} \tag{9.117}$$

where

$$\Phi = \frac{2\gamma(\gamma_p - \gamma)md^2}{\gamma_p \mu^2 F} \tag{9.118}$$

To compute the area of the particle projection, F, use the equivalent diameter d (i.e., $F = \pi d^2/4$). As the mass of a particle is equal to the product of its density, γ_p/g, and volume, $\pi d^3/6$, Equation 9.118 simplifies to

$$\Phi = \frac{4}{3}\frac{\gamma(\gamma_p - \gamma)d^3}{\mu^2 g} \tag{9.119}$$

Comparing Equation 9.119 with Equation 9.18, we see that Φ is equal to $C_D Re^2$ when a dynamic equilibrium is reached between the gravity and drag forces:

$$\Phi = C_D Re_m^2 \tag{9.120}$$

where Re_m is the Reynolds number corresponding to the steady free-fall velocity. From a value of Φ, Re_m can be determined from Figure 9.17, which corresponds to steady falling of a spherical particle. From Re_m, the settling velocity, u_m, can be determined.

For nonspherical particles, the drag coefficient may be determined from the following empirical formula:

$$C_D = 5.31 - 4.88\psi \tag{9.121}$$

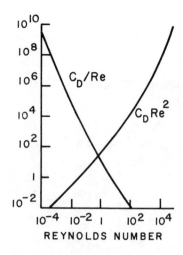

Figure 9.17   Plot for determining free-fall velocity.

Table 9.3 Values of Drag Coefficient C_D for Nonspherical Particles

ψ	Re				
	1.0	10	100	400	1000
0.670	28	6	2.2	2.0	2.0
0.806	27	5	1.3	1.0	1.1
0.846	27	4.5	1.2	0.9	1.0
0.946	27	4.5	1.1	0.8	0.8
1.000	26.5	4.1	1.07	0.6	0.46

where the particle shape factor ψ is related to a sphericity function a_s:

$$a_s = \frac{24}{0.843 \log \dfrac{\psi}{0.065}} \qquad (9.122)$$

Values of the drag coefficient for nonspherical particles are given in Table 9.3.

If the initial velocity of the particle is greater than zero, Equation 9.117 can be used to estimate the time it will take to achieve velocity u_m. Re_0 and Re first must be determined from Equation 9.116 (which corresponds to velocities u_0 and u) whence Equation 9.117 can be integrated over these limits:

$$\frac{\mu F}{2md} = \int_{Re_0}^{Re} \frac{dRe}{\Phi - C_D Re^2} \qquad (9.123)$$

Parameter Φ is constant and may be evaluated from Figure 9.17 (for spherical particles) or from Equations 9.121, 9.122 and Table 9.3 for nonspherical particles. Thus the integral of Equation 9.123 establishes a relationship between time and velocity during unstable falling. Knowing the relationship for u = f(t), we determine the distance of fall as

$$x = \int_0^t u\,dt \qquad (9.124)$$

An analytical solution to this expression can be obtained if the particle's motion is laminar, i.e. the particle follows Stokes' law. Then the drag coefficient may be defined as $C_D = a_s/Re$, and, on integration of Equation 9.123, we obtain

$$\frac{\mu F}{2dm} t = \frac{1}{a} \ln \left\{ \frac{\Phi - a_s Re_0}{\Phi - a_s Re} \right\} \tag{9.125}$$

From Equation 9.120, Φ may be replaced by $a_s Re_m$. Furthermore, the ratio of the Reynolds numbers is equivalent to the ratio of their respective velocities. Equation 9.125 therefore may be rearranged to solve for the time it takes for the particle velocity to increase from u_0 to u:

$$t = \frac{2dm}{\mu aF} \ln \left\{ \frac{u_m - u_0}{u_m - u} \right\} \tag{9.126}$$

For nonspherical particles, a_s may be evaluated from Equation 9.122. For spheres, a_s is equal to 24. Taking into account that $F = \pi d^2/4$ and $m = (\pi d^3/6)(\gamma_p/g)$, we obtain

$$t = \frac{d^2 \gamma_p}{18 \mu g} \ln \frac{u_m - u_0}{u_m - u} \tag{9.127}$$

The time may be estimated directly when the particle motion is turbulent. For example, when u_m corresponds to large Reynolds numbers, i.e., $Re > 1000$ for spheres, the initial velocity will be large enough that $u_0 > u_m$. In this case, the drag coefficient is a constant ($C_D = 0.44$), and Equation 9.123 may be integrated to give

$$\frac{\mu F}{2dm} t = \frac{1}{2\sqrt{2\Phi}} \ln \left[\frac{(\sqrt{\Phi} + Re\sqrt{C_D})(\sqrt{\Phi} - Re_0\sqrt{C_D})}{(\sqrt{\Phi} - Re\sqrt{C_D})(\sqrt{\Phi} + Re_0\sqrt{C_D})} \right] \tag{9.128}$$

Because $\sqrt{\Phi} = Re_m \sqrt{C_D}$ and the ratio of Reynolds numbers may be substituted for by the ratio of velocities, we obtain

$$\frac{\mu F}{2dm} t = \frac{1}{2\sqrt{\Phi C_D}} \ln \left[\frac{(u_m + u)(u_m - u_0)}{(u_m - u)(u_m + u_0)} \right] \tag{9.129}$$

In the case of upward turbulent motion of a particle, integration of Equation 9.123 at constant C_D, $Re \leqslant 0$ and $Re_0 \leqslant 0$ provides a somewhat different result:

$$\frac{\mu F}{2dm} t = \frac{1}{\sqrt{C_D \Phi}} \tan^{-1} \left[\frac{(u_0 - u)u_m}{(u_0 u + u_m^2)} \right] \qquad (9.130)$$

The relationship between the time of upward motion t and velocity can be obtained by accounting for appropriate sign changes on corresponding velocities, i.e., $u_m > 0$ and $u_0 < 0$ and $u < 0$.

Particle Injection into Flowing Streams

Particulates (solid or liquid) are injected into a moving fluid stream in a variety of process operations and equipment. Examples include atomization in furnaces, dryers, particle flows in pneumatic lift lines, solids transfer lines between fluidized beds, and others. We shall analyze the fluid dynamics of a simple system in which a particle is injected into an upward-moving gas stream. Due to the initial momentum of the particle and its resistance to flow, the particle velocity gradually decreases. For a first approximation, particles will be considered small and, hence, gravity forces can be neglected.

Consider a particle injected with velocity u_0 perpendicular to an upward-moving gas stream with velocity C_0. The system under consideration is illustrated in Figure 9.18. The particle will follow a path along some curve. The actual velocity, u, at any point in time is the sum of the vector components, u_y and u_x. The component u_x has a negative acceleration, $-du_x/dt$, due to resistance. The resistance force acting on a particle of mass, m, is

$$R = -m \frac{du_x}{dt} \qquad (9.131)$$

The resistance according to Stokes' law is $R = 3\pi \mu du_x$ and, assuming the particle to be spherical, we obtain

$$18 \frac{\mu g}{d^2 \gamma_p} u_x = - \frac{du_x}{dt} \qquad (9.132)$$

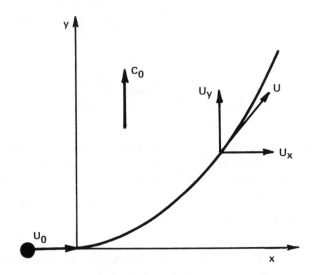

Figure 9.18   Particle injection into an upward-moving gas stream.

Integrating this expression over the limits $(u_0, 0)$ to (u_x, t),

$$t = \left(\frac{d^2 \gamma_p}{18\mu g}\right) \ln \frac{u_0}{u_x} \tag{9.133}$$

And, denoting

$$K' = \frac{18\mu g}{d^2 \gamma_p} \tag{9.134}$$

the relationship between u_x and t is

$$u_x = u_0 e^{-K't} \tag{9.135}$$

Hence, the distance traveled by the particle in the x-direction is

$$x = \int_0^t u_x dt = \int_0^t u_0 e^{-K't} dt = \frac{u_0}{K'}(1 - e^{-K't}) \tag{9.136}$$

or

$$x = \frac{u_0 - u_x}{K'}$$ (9.137)

At $t \to \infty$, $u_x \to 0$. Hence, the particle cannot move in the x-direction infinitely because at $u_x = 0$ the limiting distance is $x_\infty = u_0/K'$.

Thus, if a number of particles are distributed over the entire cross section of the flowstream, their initial velocity, u_0, should be greater than a certain minimum value.

At the initial point of injection, the velocity component, u_y, relative to the wall of a piece of equipment, is zero. Because the velocity of the gas is C_0, the velocity of the particle relative to the gas also is equal to C_0. Due to resistance, the particle velocity relative to the gas will decrease gradually and, at a certain moment when the particle reaches velocity, u_y (relative to the wall in the y-direction), the velocity of the particle relative to gas will be equal to $u_y - C_0$. Therefore, we may write an expression for the relative particle velocity that is analogous to Equation 9.132:

$$K'(u_y - C_0) = - \frac{d(u_y - C_0)}{dt}$$ (9.138)

As C_0 is constant relative to the wall, the integration of Equation 9.138 provides an expression for the upward velocity component:

$$u_y = C_0(1 - e^{-K't})$$ (9.139)

A second integration with respect to time over the intervals of 0 to t provides the path of the particle in the y-direction:

$$y = C_0 \left[t - \frac{1}{K'}(1 - e^{-kt}) \right]$$ (9.140)

Equation 9.140 can be used to compute the distance traveled by the particle in the direction of gas flow over time (or to compute the time needed for a particle to travel distance y in the direction of the flow). The resultant velocity of the particle is the vector sum of its components:

$$u = \sqrt{u_x^2 + u_y^2}$$ (9.141)

Substituting the expression for t (Equation 9.133) into Equation 9.140, an expression for the particle trajectory is obtained:

$$y = \frac{C_0}{K'} \left[ln\left(\frac{u_0}{u_0 - xK'} \right) - \frac{xK'}{u_0} \right] \tag{9.142}$$

As is often the case, particles may be injected into a piece of equipment, such as a furnace or a lift line, at some angle other than 90°. Consider when the particle is injected downward at angle β from the horizontal, as shown in Figure 9.19. Sell [1931] and Albrecht [1931] derived the equation of motion in the x-direction, i.e., perpendicular to the gas flow, for this case:

$$x = \frac{u_0 \cos\beta}{K'} (1 - e^{-K't}) \tag{9.143}$$

When $t \to \infty$,

$$x_\infty = \frac{u_0 \cos\beta}{K'} \tag{9.144}$$

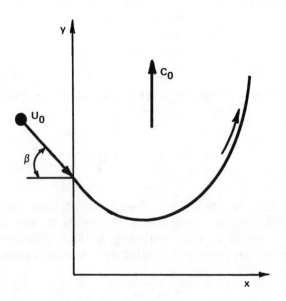

Figure 9.19   Particle injected into flowstream at an angle β from the horizontal.

The path in the y-direction is

$$y = \frac{C_0}{K'} [K't - (1 - e^{-K't})] + \frac{u_0 \sin\beta}{K'} (1 - e^{-K't}) \qquad (9.145)$$

Eliminating $1 - e^{-K't}$ from Equation 9.145, we obtain through Equation 9.144 an expression describing the particle trajectory. The particle velocity components can be obtained by differentiating Equations 9.144 and 9.145 with respect to time.

PARTICLE SEPARATION FROM GAS STREAMS

Sedimentation equipment for removing solid and/or liquid particles from gas streams is designed for the following reasons:

- Cleaning of ventilation air or fly ash removal for flue gases
- Product-quality improvement
- Recovery of valuable products
- Powdered-product collection

The forces utilized for separating particles from gas streams may be classified as (1) gravity settling, (2) inertial deposition, (3) flow-line interception, (4) diffusional deposition, (5) electrostatic deposition, (6) thermal precipitation, or (7) sonic agglomeration. We shall discuss gravitational and electrostatic deposition in the remainder of this chapter.

Gravitational Separation

Consider first *gravity settling*. This method is realized in a chamber in which the velocity of a gaseous suspension is reduced to enable particles (solid or liquid) to settle out by the action of gravity. To decrease the trajectory path of a particle and, consequently, the time of settling, horizontal plates are positioned within the chamber. These plates significantly improve the collection efficiency of the discrete phase. Figure 9.20 illustrates a typical multiplate settling chamber.

To establish the length of the chamber required to remove particles of a certain minimum size, or to determine the size particles removed by an existing chamber with a specified loading, the particle settling velocity must be evaluated. If the particulates in question follow Stokes' law and are approximately spherical, the familiar settling velocity expression developed earlier may be applied:

$$u = \frac{d^2(\gamma_p - \gamma)}{18\mu} \qquad (9.146)$$

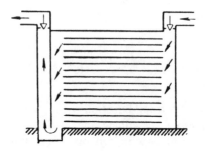

Figure 9.20   Multiplate settling chamber.

As $\gamma_p \gg \gamma$, Equation 9.146 simplifies to

$$u = \frac{d^2 \gamma_p}{18\mu} \tag{9.147}$$

For very large particles (typically above several hundred micrometers) the Allen equation may be used to estimate settling velocity:

$$U = \frac{0.153 d^{1.14} \gamma_p^{0.71}}{\left(\dfrac{\gamma}{g}\right)^{0.29} \mu^{0.43}} \tag{9.148}$$

If, for a given heterogeneous system, the settling velocity of the smallest particles that will separate out is u and the height of the chamber (or the distance between horizontal plates) is H, then the settling time will be H/u. The time required for a unit volume of gas to remain in the chamber, i.e., the gas residence time, must at least be equal to the settling time to allow the smallest particles to be removed. The linear gas velocity is equal to the volumetric flowrate, q, divided by the cross-sectional area of the chamber, F. For a chamber of length L, the residence time of the gas in the chamber is L(q/F). As this time must be equal to that of the smallest particle, we may equate the two times to give

$$L = \frac{Hq}{uF} \tag{9.149}$$

This expression shows that at a constant gas velocity, q/F, the length of the chamber, L, is proportional to its height, H. Obviously, the smaller the height, the smaller the chamber length required for a desired separation, which explains the advantage of a plate design (Figure 9.20) over a simple chamber.

Because the smallest particles will settle according to Stokes' law, we may substitute Equation 9.147 for u into Equation 9.149:

$$L = \frac{18Hq\mu}{Fd^2\gamma_p} \tag{9.150}$$

Thus, the length of a settling chamber is inversely proportional to the square of the particle diameter. For example, if it is desirable to separate out particles that are two times smaller than the selected size, then the length of the chamber must be increased by a factor of four. Equation 9.150 may also be used to determine the smallest particle diameter that can be removed by a chamber of specified dimensions.

The following example problem illustrates some of these design principles.

Example 9.1

Determine the dimensions of a simple settling chamber (Figure 9.21) required to remove 50-μm size particles under the following operating conditions:

- Gas capacity, $q = 2400$ m^3/hr
- Particle density, $\rho_p = 2400$ kg/m^3
- Gas temperature, $t = 20°C$
- Gas density, $\rho = 1.2$ kg/m^3
- Gas viscosity, $\mu = 1.8 \times 10^{-5}$ N-s/m^2

Solution

The settling regime for the particles must be determined first. Hence, the critical particle diameter is computed first:

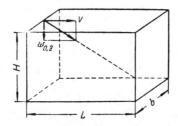

Figure 9.21 Simple settling chamber for Example 9.1.

$$d_{cr} = 2.62 \sqrt[3]{\frac{\mu^2}{g(\rho_p - \rho)\rho}}$$

$$d_{cr} = 2.62 \sqrt[3]{\frac{(1.8 \times 10^{-5})^2}{9.81(2400) - 1.2)1.2}} = 5.9 \times 10^{-5} \text{ m} = 59 \ \mu\text{m}$$

As the size range of particles to be removed is less than the critical diameter, we are confident that particles will follow Stokes' law. Hence, the settling velocity for a 50-μm particle is

$$w_0 = \frac{d^2}{18} \times \frac{\rho_p - \rho}{\mu} g = \frac{(50 \times 10^{-6})^2}{18} \frac{2400}{1.8 \times 10^{-5}} 9.81 = 0.182 \text{ m/s}$$

The dimensions of the chamber shown in Figure 9.21 are determined based on the criterion that the time required for the particle to settle out is proportional to the chamber length:

$$t = \frac{H}{w_0}$$

If the chamber volume is V_c, then the residence time is

$$t = \frac{3600 \, V_c}{q} = \frac{3600 \, V_c}{2400} = 1.5 \, V_c$$

Consequently, the maximum chamber height is

$$H = 1.5 \, w_0 V_c = 1.5 \times 0.182 \times V_c = 0.273 \, V_c, \text{ m}$$

The longitudinal cross section of the chamber is

$$F = \frac{V_c}{H} = \frac{V_c}{2.73 V_c} = 3.7 \text{ m}^2$$

To determine the specific dimensions, the chamber's cross section must be evaluated. This area is a function of the permissible gas velocity through the chamber. Assuming a gas velocity of w = 0.2 m/s, the cross-sectional area is

$$S = \frac{q}{3600w} = \frac{2400}{3600 \times 0.2} = 3.33 \text{ m}^2$$

As the two cross sections of the chamber are now known, by selecting one dimension, based on acceptable space or headroom availability, the other dimensions can be established. For example, let H = 1 m. Then

$$b = \frac{S}{H} = \frac{3.33}{1} = 3.33 \text{ m}$$

$$L = \frac{F}{b} = \frac{3.7}{3.33} = 1.11 \text{ m}$$

Practice problems 9.15–9.19 provide further exercises for mastering these concepts. Further examples and discussions are given by Green and Lane [1957], Drinker and Hatch [1954], Roberts [1939], and Cheremisinoff and Young [1977].

Electrostatic Precipitation

Particles typically less than 10 μm in size generally form a highly stable colloidal suspension in gas (called aerosols). These particles either will not settle out or will require an exceptionally long settling time. Recall that inertia forces are proportional to particle mass and, hence, in the case of aerosols these small forces lead to very low settling velocities. Particle separation from the gas can be achieved on a practical basis by passing the heterogeneous system between a pair of high-voltage electrodes that ionize the gas. The discharge electrode, responsible for ionizing the gas, consists of a small cross section that generates a high electrical field at its surface. The collecting electrode serves to precipitate the charged particles. Particles migrate according to an emf (electromotive force) gradient and are attracted to the appropriately charged electrodes, i.e., the negatively charged particles are attracted to the positive electrode and the positively charged particles to the negative electrode.

The direction of the velocity vector of the charged particles is determined by their sign and speed and, consequently, their kinetic energy, as determined by the field intensity. The electrical force field responsible for particle precipitation may exceed inertia forces by orders of magnitude.

There are several types of electrostatic precipitators. Normally, the flow section of a precipitation chamber (simply called a precipitator) consists of a bundle of vertical tubes arranged in parallel fashion (usually of round or hexagonal cross section), or as a packet of vertical parallel plates. Thin wires, approximately 2 mm in diameter, are suspended and stretched along the axes of the tubes, which are 150–300 mm in diameter. The same wires are stretched between the plate configurations. The wires and tubes are connected to a source of direct electrical current of high voltage (up to 90,000 V).

Cylindrical wires are connected to the negative terminal where a zone of ionized gas is generated. Gas molecules in this vicinity comprise both positively and negatively charged ions. The negative ions are repelled to the walls of the tubes and plates and fill the total volume of the precipitator. The dispersed particulates are contacted by the ions and entrained to the surface of the tubes and plates. Here, the particles lose their charge and descend downward along the electrode surfaces.

The ionized gas layer in the vicinity of the wires dissipates an energy field in the form of a thin layer called the corona, hence the name "wire corona-forming electrode." The plates and tubes in which the particles are precipitated are called collecting electrodes.

Depending on the shape of the electrodes, the electrical field may be either homogeneous or heterogeneous. A homogeneous electrical field is observed in the gaseous space between flat parallel electrodes, as shown in Figure 9.22(A). The lines of force uniformly fill the volume and are located in a parallel arrangement. A heterogeneous field is generated when the surface of one of the electrodes is reduced greatly by geometry changes. Examples are the tubular precipitator (Figure 9.22(B)) and the plate precipitator (Figure 9.22(C)). A thickening of the lines of force occurs around these electrodes, which corresponds to an increase in the field intensity. Hence, the field intensity, E_x(V/s-m), in the tube will be at a maximum when $x = r$, which follows from this theoretical formula:

$$E_x = \frac{\tilde{v}}{x \, ln \, \dfrac{r_0}{x}} \tag{9.151}$$

where $\tilde{v}$ = the potential difference between the electrodes, V
 x = the distance from the axis of the internal electrode (wire), m
 r_0 = the inside diameter of the tube, m
 r = the radius of the internal electrode, m

The field intensity at any point between parallel plates is

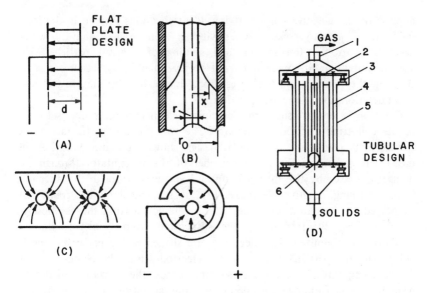

Figure 9.22   Different schemes for electrostatic precipitators: (1) duct outlet; (2) frame; (3) insulators; (4) corona-forming negative electrodes; (5) tubes; (6) duct for introducing suspension.

$$E_x = \frac{\tilde{\nu}}{d} , \quad V/m \tag{9.152}$$

As E_x increases to its critical value, E_{cr}, the gas gap, d, between parallel plates is "broken through" due to the fast (cumulative) ionization. In the space between the plates, a spark discharge is formed and the current greatly increases due to a shorting of electrodes. Such a process cannot be used for effective removal of particles.

The collision ionization in the heterogeneous field initiates before achieving E_x at the surface of the wire and is not propagated on the total volume of the gap. A corona discharge occurs in this case with uniform ion formation, that is, the negative ions are repelled to the collecting electrode. The critical field intensity, E_{cr}, corresponds to the critical potential difference, $\tilde{\nu}_{cr}$. As far as $\tilde{\nu}$ increases up to the spark potential, the current intensity is increased approximately by a square law (the electric current in gases does not follow Ohm's law) and, when $\tilde{\nu} > \tilde{\nu}_{spark}$, a spark discharge will also occur.

At the negative corona-forming electrode, $\tilde{\nu}_{cr}$ and $\tilde{\nu}_{spark}$ are higher. In addition, the mobility of the negative ions is greater; therefore, it usually is applied to the negative corona. It is evident that ions located in the force field are subject to greater forces with an increase in $\tilde{\nu}$ and E_x, which accelerates to the ion motion and settling of charged particles. The applied potential difference must not be less than $\tilde{\nu}_{spark}$ because at spark shorting the intensity on the wires drops drastically, resulting in coroning and a deterioration of the precipitation.

The primary design issue with a precipitator concerns the proper selection of the collecting electrode's geometry. The two primary configurations are tubular and plate. For the tubular design shown in Figure 9.22(D), the dust-laden gas enters the lower chamber of the precipitator through the duct (6), passes by the tubes (5) upward and exits through the duct (1). Corona-forming negative electrodes (4) are suspended on the frame (2), supported by insulators (3). The latter are removed from the gas flow to avoid fouling. By introducing the suspension from below, fouling of insulators is minimized. The electrodes are shaken with a percussion mechanism, and the particles fall off of the electrodes into the conical bottom of the precipitator. In the described precipitator, the ionization of the gas near the corona-forming electrode occurs simultaneously with the motion of the charged particles toward the collecting electrode.

The actual mechanical design of an electrostatic precipitator is simple; however, the phenomena occurring inside are extremely complicated. Because of this, a universal design procedure is unavailable. Hence, specific designs must be based on experimental data and approximate formulas. For general guidelines, the current intensity for a tubular electrode may be taken as $I = 0.3-0.5$ ma/m and, for plate-type precipitators, $I = 0.1-0.35$ ma/m. The field intensity is usually 4–450 kV/m (no more than 800 kV/m), with a working voltage of 35–70 kV.

For an electrical field length of 3–4 m, the allowable gas velocity is $w_g = 0.5-1.0$ m/s for plate-type precipitators. The gas residence time in a tubular precipitator depends on the particle settling velocity, w_s:

$$t_0 = \frac{r_0}{w_0} \qquad (9.153)$$

where w_0 is determined from Stokes' law incorporating the expression for the force of the electrical field.

The collection efficiency of a precipitator is a function of the time that the gas remains in the active field and is in the range of 90 to 99%.

For any given particle of uniform size and character, the collection efficiency, η_e, is related to the time, t, in seconds, that the gas remains in the active field:

$$\log(1 - \eta_e) = t \log \dot{K} = tEC' \tag{9.154}$$

where $\dot{K}$ = precipitation constant ($\dot{K} = 0.05 - 0.50$)
       E = voltage
       C' = constant

Early works by Loeb [1939] and Peek [1929] and more recent articles by Cheremisinoff and Cheremisinoff [1973] provide good introductory reading to this subject. Further information can be obtained from other discussions in the literature.

PRACTICE PROBLEMS

Solutions are provided in Appendix A to those problems with an asterisk (*) next to their numbers.

9.1   A waste stream from a pulp mill has an average concentration of 7.2 lb of water/lb of solids. A treatment plant to be designed will have a thickening stage that concentrates the stream to 1.8 lb of water/lb of solids with the production of a relatively clear overflow. Batch settling tests were conducted on different concentration slurries to ensure that the velocity of settling exceeds the upward flow of fluid at all concentrations normally encountered in the thickening of the specified feed. Tabulated results from these tests are given in Table 9.4. Prepare an additional column for this table showing the estimated minimum area required for clear overflow in units of 1 ton/day/ft^2 of solids feed. What is the minimum area to be used for design purposes?

Table 9.4 Batch Settling Test Results for Problem 9.1

Mass Ratio Fluid to Solids	Fluid Rising per Pound of Solids (lb/lb)	Calculated Fluid Rising (ft^3/hr)	Measured Rate of Settling (ft/hr)
7.2	5.9	7.3	2.9
6.3	4.1	4.8	1.5
4.3	3.2	3.9	1.1
3.4	2.6	3.1	0.85
2.5	1.7	2.6	0.59

*9.2 Determine the maximum size of quartz particles settling in water (t = 20°C) that can be described by Stokes' law. What is this particle's settling velocity? The specific weight of quartz is 2650 kg/m³.

*9.3 Determine the settling velocity of spherical quartz particles in water (d = 0.9 mm) using the dimensionless plot of the Lyachshenko and Reynolds numbers versus the Archimedes number in Figure 9.23.

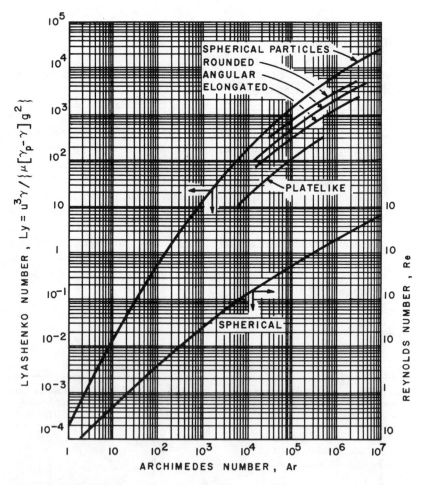

Figure 9.23   Dimensionless plot of Reynolds and Lyachshenko numbers versus Archimedes number for different shape particles.

The specific weight of the quartz is 2650 kg/m^3, and the temperature of the water is 20°C.

*9.4 Determine the maximum diameter of spherical chalk particles entrained by an upward-moving water stream with a velocity of 0.5 m/s. The liquid temperature is t = 10°C, and the specific weight of the chalk is γ_p = 2710 kg/m^3.

*9.5 Determine the settling velocity in the water (t = 20°C) for lead particles having an angular shape with d_{eq} = 1 mm. The specific weight of lead particles is 7560 kg/m^3.

*9.6 Calculate the sizes of elongated coal particles (ρ_{p_1} = 1400 kg/m^3) and plate-like particles of shale (ρ_{p_2} = 2200 kg/m^3) that have the same settling velocities of 0.1 m/s through water at 20°C.

*9.7 Determine the settling velocity of solid spherical particles if the particle diameter is d = 25 μm and particle density is ρ_p = 2750 kg/m^3. The density of the liquid phase is ρ = 1200 kg/m^3 and its viscosity is $\mu = 2.4 \times 10^{-3}$ N-s/m^2 (2.4 cp).

*9.8 Determine the velocity of hindered sedimentation of the suspension considered in Problem 9.2 if the concentration of solids in the feed is x = 30%, the density of suspension is ρ_s = 1440 kg/m^3, and the density of the solid phase is ρ_p = 2750 kg/m^3.

*9.9 Determine the capacity, cross-sectional area and diameter of a continuous sedimentation tank for liquid suspension clarification in the amount of Q_s = 20,000 kg/hr. The concentration of solids is x_1 = 50%, the settling velocity is u_0 = 0.5 m/hr, and the density of liquid phase is ρ = 1050 kg/m^3.

9.10 A large settling lagoon (approximately 0.5 ac in area and 75 ft deep) is used to separate a solid waste product whose particle density is roughly 1700 kg/m^3. The density of the dilute slurry is roughly 1300 kg/m^3 and its viscosity is 3.2 cp. The particles are spherical in nature with a 50 wt% size of 210 μm. If the lagoon is filled to 90% capacity with a solids concentration of 40%, how long will it take to achieve an 85% separation of sludge from the slurry? Ignore any evaporation losses.

9.11 The lagoon described in Problem 9.10 operates in the summer months at a mean temperature of 65°F. The mean ambient air temperature between the months of June and September is about 75°F. Assuming an average wind velocity of 5 m/s, determine the following:

(a) estimated water losses due to evaporation; and

(b) the concentration of the dilute slurry at the end of four months.

9.12 During the spring, the mean temperature over the bottom 3 ft of the lagoon described above is $10°C$, whereas the temperature at the surface is about $16°C$. Would you expect good separation of solids? Substantiate your conclusions.

9.13 Two primary settling basins are each 100 ft in diameter with an 8-ft side water depth. The tanks are equipped with single effluent weirs located on the peripheries. For a water flow of 10 mgd, calculate the overflow rate, gpd/ft^2, detention time, hr, and weir loading, gpd/ft. The overflow rate for a clarifier is defined as the surface settling rate, i.e., $q_0 = q/F$, q = volumetric flowrate and F = total surface area of basin.

9.14 A thickener handles 80,000 gpd of sludge, increasing the solids content from 2.0 to 8.0% with 85% solids recovery. Determine the quantity of thickened sludge generated per day.

*9.15 Determine the critical size of oil droplets settling in air at $t = 20°C$. The droplet density is $\rho_1 = 900$ kg/m^3, the air density is $\rho_2 = 1.2$ kg/m^3, and the air viscosity is $\mu = 1.8 \times 10^{-5}$ cp.

*9.16 For the conditions described in the previous problem, determine the settling velocity of those droplets that are 15 μm in diameter.

9.17 Determine the size of particulates that will be removed by a settling chamber operating under the following conditions:

- Gas capacity, $q = 3500$ m^3/hr
- Average particle diameter, $d = 70 \mu$m
- Particle density, $\rho_1 = 1800$ kg/m^3
- Air temperature, $t = 25°C$
- Air viscosity, $\mu = 1.8 \times 10^{-5}$ N-s/m^2

9.18 For the previous problem, the particle size range varies from 40 to 550 μm, with a 50 wt% particle size of 150 μm. Prepare a plot of particle removal efficiency versus particle size for the given operating conditions.

*9.19 Determine the required height between shelves of a multiplate settler for removing 8-μm sized pyrite dust particles. The rated capacity for the unit is 0.6 m^3/s at STP, and the overall dimensions of the unit are 4.1 m long by 2.8 m wide by 4.2 m high. The average temperature in the unit is $427°C$. The gas viscosity and density at process conditions are 0.034 cp and 0.5 kg/m^3, respectively. The specific weight of the dust is 4000 kg/m^3.

9.20 A spherical particle with a diameter of 25 μm and a density of 1.2 g/cm^3 is in free fall in air at standard conditions:
   (a) Compute the drag coefficient, the gravitational force and the terminal settling velocity of the particle.
   (b) Repeat the problem using a particle with a cubical geometrical shape.

9.21 The rate of particle agglomeration of aerosol particulates is observed to be approximated by a second-order rate expression:

$$-\frac{dN}{dt} = kN^2$$

where N is the number of particles per unit volume at time t, and k is the rate constant defined as

$$k = \frac{4kTC}{3\mu} \text{ in units of cm}^3/\text{s}$$

where   T = absolute temperature (°K)
   k = Boltzmann constant (1.38 $\times$ 10^{-16} g-cm^2/s^2-molecule-°K)
   μ = viscosity (p)
   C = a slip correction factor (C = 1 + 3.45 $\times$ 10^{-4}T/d)

   (a) Determine the agglomeration time to form a final concentration of dust particles of 10^8 particles/ft of air at STP from an initial concentration of 10^{11}.
   (b) Determine the final concentration of dust particles in the air after 4 hours if the initial concentration is 10^{20} particles/ft of air.

9.22 Determine the voltage required to generate a corona discharge in a cylindrical electrostatic collector using 10-gauge (AWG) discharge electrodes and a 3.5-in. i.d. collecting tube. Assume a smoothness factor of 0.95.

9.23 Determine the change in the required corona voltage if the collector in Problem 9.22 uses 12-gauge discharge electrode wire. Also, determine the collection efficiency of this unit on 0.2-μm particles in air if the gas flow is 10 fps.

NOTATION

A	constant in Equation 9.14
A$'$	dimensionless coefficient, see Equation 9.58
$\overline{A}$	Archimedes force, N
Ar	Archemedes number, Equation 9.19
a_s	sphericity function defined by Equation 9.122
a	coefficient accounting for liquid entrapped between sludge layers, see Equation 9.52
a_g	acceleration, m/s^2
B	parameter defined by Equation 9.72
b	dimension, m
C	concentration, kg/m^3
C$'$	constant
$\overline{C}$	centrifugal force, N
C_0	upward gas stream velocity, m/s
C_D	drag coefficient
d	particle diameter or distance, m
E	voltage, V
E_x	field intensity, V/s-m
Eu	Euler number
F	area, m^2
$\overline{G}$	average gravity force, N
Ga	Galileo number
H	chamber height, m
I	current density, ma/m
K	settling velocity correction factor, defined by Equation 9.14
K$'$	time constant defined by Equation 9.134
K_1, K_2	coefficients, refer to Equations 9.32 and 9.33
K_s	separation number
$\dot{K}$	precipitation constant
k	settling flux constant, Equation 9.108
L	length, m
Ly	Lyachshenko number
LHS	left hand side
ℓ	linear particle dimension, m
M	mass, kg
m	mass, kg$_f$
m_p	particle mass, kg
N	number of particles
P	pressure, N/m^2 or force, N
Q	mass feed rate, kg/s

Q_e	electric field force, N
R	drag or resistance force, N
R_c	physical properties correction factor for slurries
Re	Reynolds number
RHS	right hand side
r_h	hydraulic radius, see Equation 9.45, m
S	area, m^2
S_1	product of Archimedes number and separation number
S_2	sedimentation number described by Equation 9.88
T, t	time, s
u	velocity, m/s
u_t	terminal settling velocity, m/s
V	volume, m^3
W	total mass, kg
w	velocity, m/s^2
X	sludge height, m
x	distance, m
x'	liquid volume fraction
Z	height of compression zone, m

Greek Symbols

α	angle, °
β	ratio of liquid volume entrained to total sludge volume, or angle, °
γ	specific weight, kg_f/m^3
ϵ	void fraction
η_e	efficiency
η_v	volume fraction of particles in a slurry
$\theta(\epsilon)$	experimental function of void fraction
λ	length of mean free path, m
μ	viscosity, poise
ν	kinematic viscosity, m^2/s
ν_H	terminal velocity for hindered settling, m/s
$\tilde{\nu}$	potential difference, V
ρ	density, kg/m^3
σ	ratio of particle surface area to volume, m^{-1}
τ	force, N, or stress, N/m^2
Φ	parameter defined by Equation 9.118
ψ	sphericity factor
$\tilde{\psi}$	parameter defined by Equation 9.69

REFERENCES

Albrecht, F., (1931) *Phys. Zeit.* 32:48.

Bird, R. B., W. E. Stewart and E. N. Lightfoot (1960) *Transport Phenomena* (New York: John Wiley & Sons, Inc.).

Brodkey, R. S. (1967) *The Phenomena of Fluid Motions* (Reading, MA: Addison-Wesley Publishing Co.).

Brown, G. G. (1950) *Unit Operations* (New York: John Wiley & Sons, Inc.).

Cheremisinoff, N. P., and P. N. Cheremisinoff (1973) "Electrostatic Precipitators—Fundamentals and Applications," *Plant Eng.* 27:25.

Cheremisinoff, P. N., and R. A. Young (1977) *Air Pollution Control and Design Handbook, Part 1* (New York: Marcel Dekker Inc.).

Coe, H. S., and G. H. Clevenger (1916) *Trans. Am. Inst. Mining Met. Eng.* 55:356.

Comings, E. W. (1960) *Ind. Eng. Chem.* 32:663.

Drinker, P., and T. Hatch (1954) *Industrial Dust* (New York: McGraw-Hill Book Co.).

Green, H. L., and W. R. Lane (1957) *Particulate Clouds: Dusts, Smokes and Mists* (New York: Van Nostrand Reinhold Co.).

Heiss, J. F., and J. Coull (1952) *Chem. Eng. Prog.* 48:497.

Hinze, J. O. (1959) *Turbulence* (New York: McGraw-Hill Book Co.).

Hughes, R. R., and E. R. Gilliland (1952) *Chem. Eng. Prog.* 48:497.

Knudsen, J. G., and D. L. Katz (1958) *Fluid Dynamics and Heat Transfer* (New York: McGraw-Hill Book Co.).

Lapple, C. E., et al. (1951) *Fluid and Particle Mechanics* (Newark, DE: University of Delaware.

Loeb, L. D. (1939) *Fundamental Processes of Electrical Discharge in Gases* (New York: John Wiley & Sons, Inc.).

Lyachshenko, P. V. (1940) "Gravitatsionye Metody Obogachshenis," *Gostopisdat*, Moscow.

Peek, F. W. (1929) *Dielectric Phenomena in High-Voltage Engineering* (New York: McGraw-Hill Book Co.).

Perry, J. H., Ed. (1963) *Chemical Engineers Handbook*, 4th ed. (New York: McGraw-Hill Book Co.).

Pettyjohn, E. S., and S. Christiansen (1948) *Chem. Eng. Prog.* 44:175.

Roberts, E. J. (1949) "Thickening—Art or Science?" *Mining Eng.* 1:61.

Roberts, R. T. (1939) *Power* 83:345,392.

Sell, W. (1931) *Forsch. Geb. Ing. Wes.* 347.

Steinour, H. H. (1944) *Ind. Eng. Chem.* 36:618,640,901.

Zenz, F. A., and D. S. Othmer (1960) *Fluidization and Fluid—Particle Systems* (New York: Van Nostrand Reinhold Co.).

CONTENTS

INTRODUCTION

A second group of separation operations that falls within the external problems of hydrodynamics is based on the application of centrifugal force. The influence of a centrifugal field on particles in an accelerating fluid stream is analogous to the separation achieved in the presence of a gravitational field. However, unlike gravity, which is essentially constant, the strength of a centrifugal field can be varied through changes in rotational speed or equipment dimensions.

The equipment used to perform these unit operations includes cyclones for gas-solid separations; hydroclones for liquid-solid separations; and centrifuges for liquid-solid, liquid-liquid and gas-gas separations. In cyclone separators and hydroclones, the heterogeneous suspension is subject to centrifugal force by its own rotation with respect to a stationary apparatus boundary by introducing it through a tangential inlet. In centrifuges, the fluid undergoes rotation with respect to a revolving apparatus boundary. Principles of operation and design methodology are presented in this chapter.

CYCLONE SEPARATORS

A cyclone is a nonmechanical device that separates solid particulates from a relatively dry gas stream. Hydroclones are essentially the same device and effect solids separations from liquids. Both devices are the simplest and most economical separators (also called solids collectors or, simply, collectors). Their operations are identical, in which forces both of inertia and gravitation are capitalized on, and their primary advantages are high collection efficiency in certain applications, adaptability and economy in power. The main disadvantage lies in their limitation to high collection efficiency of large-sized particles only. In general, cyclones are not capable of high efficiencies when handling gas streams containing large concentrations of particulates less than 10 μm in size.

Cyclones generally are efficient handling devices for a wide range of particulate sizes. They can collect particles ranging in size from 10 to above 2000 μm, with inlet loadings from less than 1 gr/scfm to greater than 100 gr/scfm. There are many design variations of the basic cyclone configuration. Because of the cyclone's simplicity and lack of moving parts, a wide variety of construction materials can be used to cover relatively high operating temperatures of up to 2000°F.

Cyclones are employed in the following general applications:

1. collecting coarse dust particles;
2. handling high solids concentration gas streams between reactors such as Flexicokers (typically above 3 gr/scf);
3. for classifying particulate sizes;
4. in operations in which extremely high collection efficiency is not critical; and
5. as precleaning devices in line with high-efficiency collectors for fine particles.

Operating Principles

Figure 10.1 shows the primary design features of a cyclone separator. The unit consists of a cylindrical barrel section (1) fitted with a conical base (2). The solids-laden gas stream enters through the tangential inlet (3) while collected solids fall through the outlet duct (5) and the cleaned fluid discharges through the exit duct (4).

The flow pattern in a cyclone is quite complex and has been the subject of numerous investigations. Three main flow patterns prevail in all cyclones:

1. Descending spiral flow. This pattern carries the separated dust down the walls of the cyclone to a dust hopper.
2. Ascending spiral flow. This rotates in the same direction as the descending spiral, but the cleaned gas is carried from the cyclone or the dust receptacle to the gas outlet.
3. Radially inward flow. This feeds the gas from the descending to the ascending spiral.

The final solids separation occurs in the dust-collecting hopper located below the exit duct (5). In the duct leading to the hopper (called the dipleg) the total gas flow reverses direction and transforms to an ascending spiral flow.

The flow patterns are generated by the creation of a double vortex, which centrifuges the dust particles to the walls. The two distinct vortices present in a cyclone are (1) a large-diameter descending helical current in the body and cone; and (2) an ascending helix of smaller diameter extending up from the dust outlet section through the gas outlet.

Particles at the walls are transported into the collecting hopper, which is isolated from the influence of the spinning gases. The gas spirals downward and upward through the inside of the cyclone. On entering the cyclone, the gas undergoes a redistribution of its velocity so that the tangential velocity component exceeds the inlet gas velocity by several times. As the gas spins in a vortex in the cyclone body, the tangential velocities increase as the axis of the cyclone is approached at any horizontal plane. The tangential velocity at any radius appears to be relatively constant at all levels. However, tangential velocities at the extreme top of the cyclone are not included because the proximity of the cyclone cover slows the spin. When

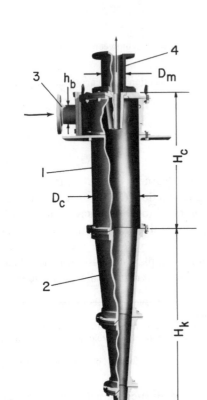
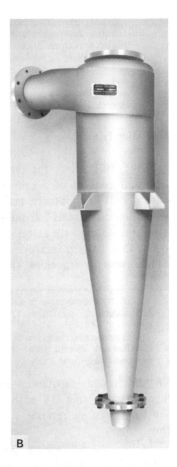

Figure 10.1   (A) Cross-sectional view of a cyclone separator suitable for gas or liquid services; (B) full view of a cyclone separator (courtesy Krebs Engineers, Menlo Park, CA).

the downward gas flow is smooth and unbroken, the dust particles flow spirally downward and pass the bottom dust outlet without reentrainment.

As depicted in Figure 10.2, the dust-laden gas enters the tangential inlet and swirls through several revolutions in the body and cone while dropping its dust load. The clean gas is emitted through the axial cylindrical gas outlet. Dust particles, which were dispersed uniformly in the entering gas stream, tend to concentrate in the layer of gas next to the cyclone wall under the influence of centrifugal force. The helical motion of the downward-moving main gas stream and the small quantity of gas through the cyclone's dust

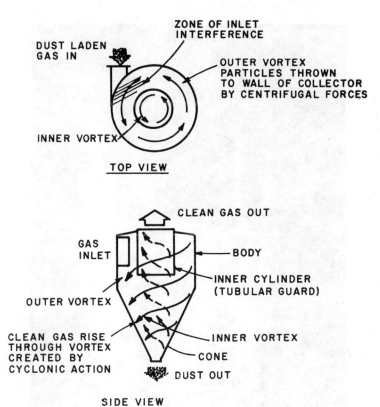

Figure 10.2   Details of internal flow patterns.

outlet assist the separated dust solids into the receiving bin. Ideally, the gas should be distributed to a cyclone in a radially thin layer so that the radial distance through which a particle must travel for separation is at a minimum. It is important that the inlet be exactly tangent to the body because separation is affected adversely by flow abnormalities. The gas discharge through the vertical exit duct maintains the upward helical pattern for some distance until pipe bends and/or flow restrictions dampen out the spiral. Figure 10.3 shows the exiting spiral in a duct immediately above the vertical discharge pipe. Smoke particles were injected into the Plexiglas® line at the cyclone exit to obtain this visual observation.

Three forces influence particle motion in a cyclone: (1) centrifugal, due to the particle rotation within the flow; (2) gravity force; and (3) Archimedes' force. The last two forces are small in comparison to centrifugal because the

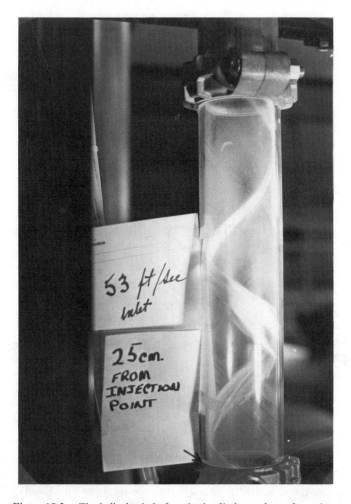

Figure 10.3   The helical spiral of gas in the discharge duct of a cyclone.

density of the medium is hundreds of times smaller than that of the solid particles.

In principle, the separation of finely dispersed solid particles from gases or liquids is governed by four factors: (1) the established centrifugal field; (2) the radial velocity pattern; (3) the residence time of the particles to be separated; and (4) the turbulence that develops.

Criteria for particle separation are based on the concept of "equilibrium orbits." This term refers to the regions of the flow comprising circular tracks having a radius such that the outwardly directed centrifugal force, acting on a particle in this orbit is in equilibrium with the inwardly directed drag according to Stokes' law, due to the radial flow.

In the case of a spherical particle, this equilibrium leads to

$$\frac{1}{6} \pi d^3 \rho_1 \frac{\omega^2 r_1^2}{r_2} = 3\pi\mu d \frac{dr}{d\tau} \tag{10.1}$$

where τ = residence of gas in the cyclone
μ = viscosity of gas
d = particle diameter
ρ_1 = gas density
ω = angular speed of gas
r_1 = outside radius of the exit pipe
r_2 = inside radius of the cyclone

Solving Equation 10.1 for τ, we obtain

$$\int_0^\tau d\tau = \frac{18\,\mu}{d^2 \omega^2 \rho_1} \int_{r_1}^{r_2} \frac{dr}{r} \tag{10.2}$$

or

$$\tau = \frac{18\,\mu}{d^2 \omega^2 \rho_1}\, ln\, \frac{r_2}{r_1} \tag{10.3}$$

For the separation of relatively large particles, i.e., for Re $>$ 2, the general resistance law is applicable:

$$\frac{\pi d^3}{6} \rho_1 \frac{\omega^2 r_1^2}{r_2} = \psi \rho_2 d^2 \left(\frac{dr}{d\tau}\right)^2 \tag{10.4}$$

or

$$d\tau = \sqrt{\frac{6\psi\rho_2}{\pi d \rho_1 \omega^2}} \times \frac{dr}{\sqrt{r}} \tag{10.5}$$

whence we obtain

$$\tau = 2 \sqrt{\frac{6\psi\rho_2}{\pi d \rho_1 \omega^2}}\, (\sqrt{r_2} - \sqrt{r_1}) \tag{10.6}$$

where ψ is the particle shape factor.

From the residence time of the gas in the cyclone, the working volume can be obtained:

$$V_c = \nu_{sec} \times \tau, \, m^3 \tag{10.7}$$

where ν_{sec} is the volumetric flowrate, m^3/s. The height of the cylindrical part of the cyclone is

$$H = \frac{V_c}{\pi(r_2^2 - r_1^2)}, \, m \tag{10.8}$$

If the residence time of the gas is specified, then the minimum particle size that will be collected can be computed:

$$d_{min} = \sqrt{\frac{18 \, \mu}{\omega^2 \rho_1 \tau} \, ln \, \frac{r_2}{r_1}} \tag{10.9}$$

Collection Efficiency

Figure 10.4 shows typical tangential and axial velocity distributions in a cyclone. The maximum values of the tangential velocities are observed close to the cyclone axis (but not on the axis), as shown in Figure 10.4(A). In tracing the tangential velocity component down the cyclone, the maximum shifts closer to the axis. This may be described by the following formula:

$$w = w_g \sqrt{\frac{4}{3} \frac{d}{C_D} \frac{\rho_{eff}}{\rho} \cdot \frac{1}{R}} \tag{10.10}$$

Equation 10.10 was derived from the settling velocity expression developed in the previous chapter:

$$w_s = \sqrt{\frac{4}{3} \times \frac{d}{C_D} \frac{\rho_{eff}}{\rho} a} \tag{10.11}$$

where the acceleration term, "a," is replaced by an expression for centrifugal acceleration:

$$a = \frac{w_g^2}{R} \tag{10.12}$$

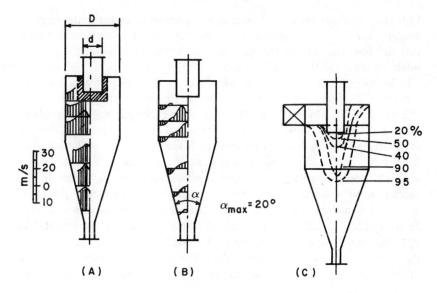

Figure 10.4 The variations of velocities in a cyclone: (A) tangential; (B) axial; (C) collection efficiency at different zones of a cyclone.

where   d = particle diameter
        C_D = drag coefficient
        ρ_{eff} = effective density ($\rho_{eff} = \rho_p - \rho$; ρ_p = particle density,
           ρ = fluid density)
        w_g = fluid rotational velocity
        R = radius of particle rotation

In contrast, the maximum axial velocities fall exactly on the centerline axis at the lower elevations of the cyclone, as shown in Figure 10.4(B). The axial components of velocity decrease the collection efficiency because particles can be entrained into the discharge pipe. Due to the counterbalancing effects of the velocity components, particle collection efficiency is not uniform in the different zones of cyclone, as shown by Figure 10.4(C).

Overall collection efficiency is defined as

$$\eta = \frac{c_1 - c_2}{c_1}$$

where c_1 = particle concentration in the entering gas stream
        c_2 = particle concentration in the exit gas

Efficiency depends on several factors, the primary ones being the physical properties of the heterogeneous system, i.e., gas and particle densities, particle sizes, viscosity of the medium; the linear dimensions of a cyclone; solids loadings; and the gas inlet velocity.

Collection efficiency is a major parameter in the selection and design of a cyclone. Cyclones may be designed for any required efficiency; however, efficiency is both a function of the energy expended and the available space for the unit. Thus, proper optimization of an acceptable efficiency at moderate pressure drop within reasonable space requirements is necessary. Efficiency can be improved markedly at the expense of pressure drop without tampering with space requirements. The major parameter in the prediction of collection efficiency is particle size. For each particular cyclone design there is a critical size particle at a given density on which the centrifugal and inward viscous forces are balanced, that is, at the equilibrium state the particle neither moves outward toward the walls nor inward toward the cyclone axis. Particles larger than this critical diameter (called the "theoretical cut," measured in microns) are collected, while all smaller particulates escape.

Efficiency is enhanced as the axis of the cyclone is approached until the edge of the core is reached. This occurs because the centrifugal force increases more rapidly than the inward drift up to the edge of the gas core. Therefore, particles rotate in orbits whose radii depend on the balance between the viscous forces due to the inward drift and outward centrifugal force. Particles then will be transferred to the outer orbits, where they pick up finer particles by collision or are captured by eddy currents. A knowledge of the theoretical cut has very little direct significance in the prediction of collection efficiency. It does, however, give a rough indication of cyclone performance.

The factors effecting particle settling depend on the relationship of forces acting on the particle in the flow:

$$\eta = \phi(C, G, R_D) \tag{10.13}$$

where

$$C = \frac{\pi d^3}{6} \rho_1 \frac{\omega^2}{r} \tag{10.14}$$

$$G = \frac{\pi d^3}{6} \rho_1 g \tag{10.15}$$

$$R_D = \psi' \rho_2 d^2 w_s^2 \tag{10.16}$$

and

$$\psi' = \phi(\text{Re}) \quad \text{or} \quad \psi' = \phi\left(\frac{dw\rho_2}{\mu}\right)$$

Consequently, the drag force is also a function of viscosity. Accounting for expressions 10.14–10.16, Equation 10.13 is rewritten as

$$\eta = \phi(\rho_1, d, \mu, \rho_2, w_s, r, g) \tag{10.17}$$

where w_s is the settling velocity of a particle.

Because it is difficult to measure the velocity of a particle in a cyclone itself, the velocity of the fluid is measured, typically at the inlet because it is most convenient. There is a definite proportionality between the fluid velocity and the actual velocity of the particle. Hence, w_s may be approximated by the velocity of the fluid if particles are sufficiently small. Note that the radius of rotation of a particle (or the gas-solid mixture itself) is a variable. It, too, may be approximated by a specific dimension of the cyclone, which is proportional to the other principal dimensions. This reference dimension, L, may be the cyclone radius, the diameter of the inlet duct or some other convenient size. Equation 10.17 therefore may be expressed in the following form:

$$\eta = \phi(\rho_1, d, \mu_1, \rho_2, w_g, L, g) \tag{10.18}$$

By means of dimensional analysis, the functionality of cyclone collection efficiency is deduced:

$$\eta = \phi\left(\text{Fr, Re, } \frac{\rho_2 L}{\rho_1 d}\right) \tag{10.19}$$

For the laminar regime of flow, i.e., Re $\leqslant$ 2, the inertia of the medium may be ignored and, hence,

$$\eta = \phi(\rho_1, d, \mu, w_g, L, g) \tag{10.20}$$

And, on the basis of dimensional analysis, we obtain

$$\eta = \phi(\text{Stk, Fr}) \tag{10.21}$$

where $\text{Stk} = \rho d^2 w_g / \mu L$ and $\text{Fr} = w_g^2 / gL$.

The most common method of estimating total collection efficiency is by plotting a fractional efficiency curve or grade efficiency curve. A fractional efficiency curve is a plot of particle size versus percent collection. Figures 10.5 and 10.6 show the specific features of such a curve and a typical plot based on experimental data, respectively.

Figure 10.5 shows that the cyclone has 0% efficiency for all particles smaller than the cut size and 100% for all larger sizes. Under actual operating conditions, however, a considerable amount of particulates smaller than the cut size are separated along with the coarser particles. This can be explained by collision between particulates or due to particle aggregation. Furthermore, a portion of the particles larger than the cut size escape collection. They are carried into the inner vortex by eddies or by collisions. Poor success has been achieved in predicting efficiencies with particle sizes less than the cut size. The degree of efficiency in this range depends largely on the properties of the particles.

It is apparent that the curve in Figure 10.5 may be used accurately in

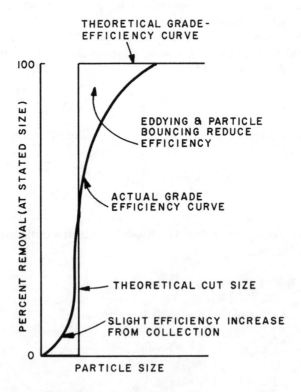

Figure 10.5   A typical grade efficiency curve.

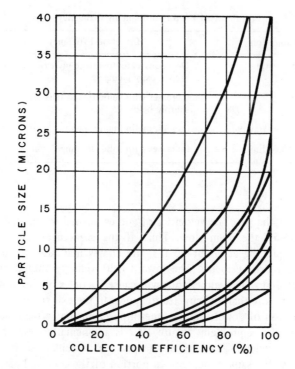

Figure 10.6   Typical collection efficiency curves. Data were obtained on a 12-in.-diameter cyclone using cracking catalyst.

predicting the total collection efficiency of a cyclone only if the particle size distribution of the suspension is known.

The specific operating parameters on which efficiency depends are (1) pressure drop, (2) particle size distribution, (3) inlet particle loading, (4) temperature of the inlet gas stream, and (5) specific gravities of gas and solids. Table 10.1 summarizes the relationships of these factors to efficiency. (Pressure drop is considered separately.)

Several operating characteristics contribute to lower collection efficiencies for particles exceeding the cut size. One important characteristic is that the distribution, or drift, is highly nonuniform. As noted by the contour plots shown earlier, at certain points in the cyclone local velocities may exceed the mean inlet velocity by a factor of 2 or 3. This is particularly pronounced at the ends of the cyclone where additional surfaces induce precession currents. A doubling or tripling of the drift velocity at any point results in particles 40% larger than the theoretical cut in reaching the exit.

Eddies in the vortex also are responsible for poor separation of coarse

Table 10.1 Factors Influencing Efficiency

Condition	Effect on Efficiency
Temperature	Decreases as temperature increases due to gas viscosity changes.
Velocity	Increases with velocity and falls off sharply below 25 fps.
Specific Gravity	Increases with higher specific gravity.
Inlet Loading	Increases with solids loading.

particles from the gas. The velocities of eddies normal to the main flow are generally 1/5 of the main flow. Thus, if the gases near the walls of the cyclone are spinning at about 50 fps, the eddies may add 10 fps or more to the inward drift velocity. The inward drift velocity should only be on the order of 1 fps. The eddies thus will cause particles of the magnitude of three times or more than the cut size to appear in the exit stream.

The generation of double eddies superimposed in the vertical plane on the main flow is also a problem. Figure 10.7 is a schematic view of the eddy currents in a cyclone. These assist the particle descent into the collection hopper but, at the same time, carry a portion of the collected particles back into the inner ascending vortex. Upswept particles have an opportunity to be separated while being transported to the clean gas exit because the gas spins very rapidly in the inner vortex. However, a large portion of particles does escape collection in this manner.

Base pickup is a major cause of excessive emissions. Some cyclone systems are designed with a predisengaging hopper, called a dustpot, before the final collecting hopper. Dustpots are designed to provide disengagement space for extracting particles from the return gas; however, when they are undersized additional vertical flow is induced and particulates become reentrained.

One approach to minimizing particle reentrainment from the base and cyclone walls is to irrigate the vessel walls. A 5–15% purge from the bottom of the cyclone also reduces base pickup. However, the added expense usually outweighs any additional increase in efficiency.

Improved separation can be achieved by increasing the effective particle settling velocity. Referring to Equation 10.11, we note that this can be accomplished for a given heterogeneous system of known physical properties and cyclone capacity by increasing the peripheral gas velocity, w_g, by decreasing the radius of the cyclone, or both. Note, however, that an increase in the gas flow leads to a drastic increase in the cyclone's hydraulic resistance.

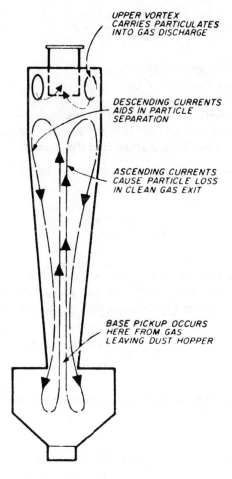

UPPER VORTEX
CARRIES PARTICULATES
INTO GAS DISCHARGE

DESCENDING CURRENTS
AIDS IN PARTICLE
SEPARATION

ASCENDING CURRENTS
CAUSE PARTICLE LOSS
IN CLEAN GAS EXIT

BASE PICKUP OCCURS
HERE FROM GAS
LEAVING DUST HOPPER

Figure 10.7   Eddying in a cyclone.

Pressure Drop

Pressure drop is one of the most important factors affecting efficiency and design.

Total pressure drop across a cyclone consists of separate losses in (1) the inlet pipe, (2) the vortex, and (3) the cyclone exit duct.

The pressure drop across a cyclone may be determined from Bernoulli's equation:

$$\frac{\Delta P}{\gamma} = \frac{w_0^2}{2g} - \frac{w_1^2}{2g} + Z \qquad (10.22)$$

where ΔP = total pressure drop
γ = specific gravity of gas
w_0 = superficial gas velocity at the inlet duct
w_1 = superficial gas velocity at the discharge duct
Z = total cyclone resistance consisting of the resistance inside the cyclone, Z_0, and at the discharge duct, Z_1

The gas resistance in the discharge duct may be obtained from Darcy's equation:

$$Z_1 = \lambda \frac{L_1}{d_1} \frac{w_1^2}{2g} \qquad (10.23)$$

As a first approximation, the resistance to gas flow in a cyclone will be proportional to the kinetic energy of the gas. As the gas velocity in the cyclone is practically equal to the initial velocity, w_0, we have

$$Z_0 = K_0 \frac{w_0^2}{2g} \qquad (10.24)$$

The resistance coefficient, K_0, for cyclones with a rectangular inlet duct (h – height, b – width) is

$$K_0 = 16 \frac{bh}{d_1^2} \qquad (10.25)$$

As the ratio w_0/w_1 is equal to the ratio of cross sections F_1/F_0, Equation 10.22 may be rewritten as follows:

$$\frac{\Delta P}{\gamma} = \frac{w_0^2}{2g} \left[Z_0 + Z_1 - 1 + \left(\frac{F_0}{F_1} \right)^2 \right] \qquad (10.26)$$

The section F_0 is equal to bh, and the section of the discharge duct is

$$F_1 = \frac{\pi d_1^2}{4}$$

Taking this, along with Equation 10.25, into account, the pressure drop expression becomes

$$\frac{\Delta P}{\gamma} = \frac{w_0^2}{2g} \left[\frac{16bh}{d_1^2} + \frac{4bh}{\pi d_1^2} \left(1 + \lambda \frac{L_1}{d_1} \right) - 1 \right]$$

(10.27)

The resistance coefficient, λ, is determined from the well-known expressions for turbulent motion in pipes.

From the data of Sheppard and Lapple [1939], flow resistance in a cyclone decreases with increasing particle concentration. The pressure drop in a cyclone at a given gas velocity and particle concentration, c, may be described by the following simple relation:

$$\frac{\Delta P_0 - \Delta P}{\Delta P} = \dot{a} \sqrt{c}$$

(10.28)

where ΔP_0 = the pressure drop at the same cyclone capacity for a gas flow without solids

$\dot{a}$ = a constant specific for a given cyclone

The volumetric flow rates of a clean gas, V_0, and gas suspension, V, under the action of the same pressure gradient are related by the following expression:

$$\frac{V - V_0}{V_0} = b \sqrt{c}$$

(10.29)

Excessive pressure losses can reduce collection efficiency greatly. Unfortunately, little can be done to correct for pressure drop. The spinning gases at the exit from a cyclone retain a great deal of kinetic energy (Figure 10.3). Attempts have been made to arrest this by trying to recover all, or part, of this lost kinetic energy. Some methods employed include (1) a tangential exit pipe, (2) a conical divergent pipe, (3) vanes in the exit pipe, and (4) disks inside and outside the pipe.

Common cyclones, whose bodies are three to five times the diameter of the inlet duct, are useful where large gas handling capacities and moderate separating efficiencies are required. Particle sizes in excess of 50 μm generally experience good separation in these large-diameter cyclones.

Operating Schemes and Design Principles

To achieve high separation efficiencies for particulates above 30 μm in size, small-diameter cyclones generally are employed. Diameters rarely exceed

4 ft and, consequently, gas handling capacities are limited to single units. Higher loading capacities generally are achieved by clustering a multiple of cyclones, as shown in Figure 10.8.

A battery of cyclones comprising miniature units (barrel diameters typically 150–250 mm) is referred to as a multiclone arrangement. A typical multiclone system is illustrated in Figure 10.9. The individual cyclones housed inside the retaining vessel (1) are similar to the unit shown in Figure 10.10. The cyclone units in Figure 10.9 are supported between two grids (2). The gas enters the battery through the duct (4) and is distributed among all the cyclone elements. On passing through the cyclones, the gas then enters the general chamber and the discharge duct. Particles are collected in the lower cone, part of the hopper, and then evacuated from it.

The theory in the development of small-diameter, long-cone cyclones for the collection of finely divided particles is that the heavy particle entering the

Figure 10.8   A cyclone cluster arrangement. All cyclones are fed from a common inlet manifold (courtesy Krebs Engineers, Menlo Park, CA).

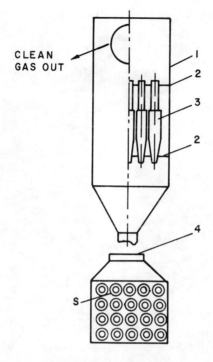

Figure 10.9   Schematic of a multiclone. Shown is a battery of cyclones: (1) hopper; (2) grid; (3) element of a battery cyclone; (4) nozzle for entering gas suspension.

cyclone reaches the walls with comparatively small angular movement. The lighter particles entering at the same place must travel through a much greater angle to reach the cyclone wall. Hence, the smaller the particle size, the greater the angle of rotation. Also, the greater the number of convolutions of the separating vortex, the smaller the particles that can be separated. High collection efficiencies are obtained from cyclones whose body diameters are less than 1 ft. For the highest separation, interiors must be well polished, with all eddy-forming projections eliminated. Efficiencies of more than 99% have been achieved in miniature cyclones handling average particle sizes of 5 μm. The capacities for larger units of similar design are proportional to the square of the diameter. Again, because of the limited capacity of a single unit, miniature cyclones can be installed in clusters and fed from the same duct. Greater pressure and more power are necessary to operate smaller combined units. The small-diameter cyclones tend to increase the gas stream's tangential velocity and centrifugal action and, because of their small size, the danger of choking is always imminent.

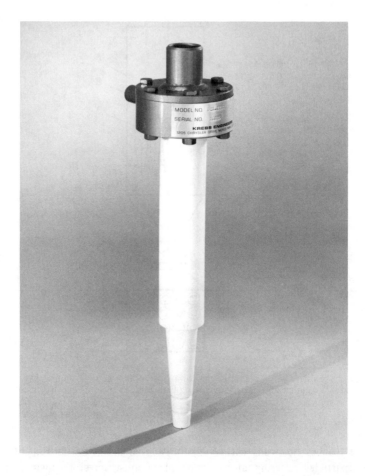

Figure 10.10 Cyclones are available in virtually all size ranges. Shown is a typical unit manufactured of nominal 0.5- and 1-in. diameters for separation of solids in the 5 to 15 μm size range (courtesy Krebs Engineers, Menlo Park, CA).

It is often more advantageous to employ several cyclones in parallel or series arrangements. Figure 10.11 shows two cyclones operated in series on a lime mud reburning system for lime recovery. In the process, lime mud filter cake is fed into a paddle mixer along with dry recycled fines and quench water. The mixture then goes to a cage mill disintegrator where precooled calciner stack gas at 1000°F dries and disintegrates the moist solids. The resultant fine carbonate is conveyed by the exhaust gas to the cyclone separators (in series). Solids collected are split, a portion being

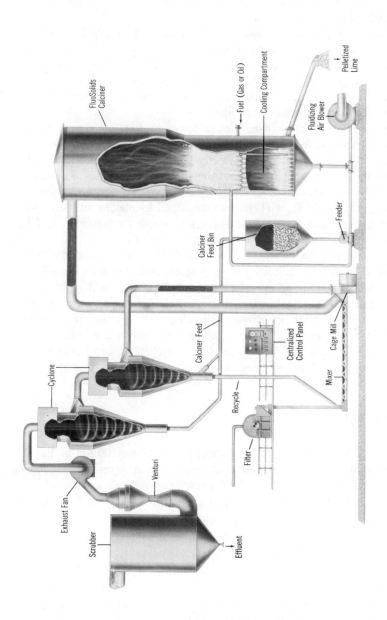

Figure 10.11   The Dorr-Oliver FluoSolids lime mud reburning system employing primary and secondary cyclone arrangements (courtesy Dorr-Oliver, Inc., Stamford, CT).

recycled to the mixture and the remainder to the calciner feed bin. Groups of cyclones, under some conditions and applications, can handle larger quantities of gas and more efficiently.

Parallel arrangements of cyclones can present difficulties, however, if proper design is not employed. Collection efficiency of a group in parallel could be less than that of one individual cyclone handling a comparable quantity of gas volume because of a difference in the pressures in various cyclone particle outlets, which causes a circulation of gas within the receptacle from one cyclone to another.

The particular cyclone exhibiting a lower pressure at the outlet will discharge more gas from the hopper as it enters through other cyclones. The effect is that the ascending gas velocity increases without a corresponding increase in radial velocity, thus decreasing the overall efficiency. Fortunately, this can be overcome by minimizing the difference in pressure between the various particle discharges. This is accomplished by making all the cyclones that feed a common hopper dimensionally identical. Furthermore, operating and loading conditions in each unit should be identical.

Groups of cyclones in series are employed when it is desirable either to operate at a higher efficiency than is possible with one stage of collection alone or to handle higher gas volumes. A setup such as this may prove more economical than a single high-efficiency cyclone that may have to handle a heavy concentration of abrasive particles. This involves the use of a two-stage arrangement, that is, a primary and secondary collector (Figure 10.11). In some applications, a tertiary cyclone is used. Generally, the tertiary unit is smaller because it is required to handle a smaller solids loading.

Example 10.1

Size a cyclone separator for removing particles above 100 μm in diameter entrained in a flue gas stream discharged from a dryer. The following information is supplied for the design: gas volumetric flow, V_{sec} = 1.8 m^3/s; particle density, ρ_P = 1100 kg/m^3; gas density, ρ = 1.2 kg/m^3; gas kinematic viscosity, ν = 2.25 × 10^{-5} m^2/s. The particles are approximately round with a shape factor of 0.77.

Solution

1. The relative dimensions (in terms of the barrel diameter) of one manufacturer's cyclone unit is given in Figure 10.12.
2. All dimensions of a cyclone of any design are selected depending on the width of inlet duct b or on the diameter of cyclone D_c. The problem is to

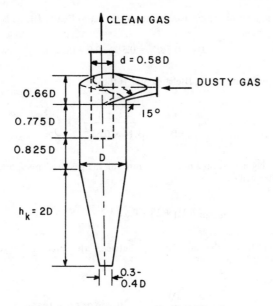

Figure 10.12 Schematic of cyclone for Example 10.1.

properly select a dimension from which secondary dimensions are pro-portionally evaluated. The cyclone diameter, settling velocity, gas velocity and parameters of the suspension to be separated are interrelated. Therefore, we select a preliminary diameter for approximate calculations and then refine our estimates to a more exact design. Examining the relative dimen-sions of a standard cyclone in Figure 10.1, $b = \alpha D_c$ and $h_{in} = \beta D_c$. For the chosen cyclone, $\alpha = 0.21$ and $\beta = 0.66$. The continuity equation for the inlet nozzle is

$$bh_{in} = \frac{V_{sec}}{W_{in}} \tag{10.30}$$

where W_{in} is the inlet gas velocity, which, for a primary cyclone operation, is typically 18–22 m/s. Expressing b and h_{in} in terms of the barrel diameter, D_c, Equation 10.30 is rearranged to solve for the cyclone diameter:

$$D_c = \sqrt{\frac{V_{sec}}{\alpha \beta W_{in}}} = \sqrt{\frac{1.8}{0.21 \times 0.66 \times 18}} = 0.85 \text{ m}$$

For design purposes, assume a value of 0.9 m for D_c.

3. The diameter of the discharge pipe (according to Figure 10.12) is

$$D_d = 0.58 D_c = 0.58 \times 0.9 = 0.52 \text{ m}$$

4. The gas velocity in the discharge pipe is thus

$$W_d = \frac{4 V_{sec}}{\pi D_d^2} = \frac{4 \times 1.8}{3.14 \times 0.52^2} = 8.5 \text{ m/s}$$

5. Specifying a wall thickness $\delta = 5$ mm for the gas discharge pipe, its outside diameter will be

$$D_{d.out} = D_d + 2\delta = 0.52 + 2 \times 0.005 = 0.53 \text{ m}$$

6. The width of the circular gap between the pipe and cyclone shell (Figure 10.12) is

$$\ell = \frac{D_c}{2} - \frac{D_{d.out}}{2} = 0.45 - 0.265 = 0.185 \text{ m}$$

7. The height of the circular gap from a spiral surface to the lower edge of discharge pipe in Figure 10.12 is

$$H = 0.775 D_c = 0.775 \times 0.9 = 0.7 \text{ m}$$

8. The calculated dimensions of the cyclone can be checked by comparing the particle settling time,

$$\tau_0 = \frac{R_c - R_{d.out}}{W_0} = \frac{\ell}{W_0} \qquad (10.31)$$

to the residence time of gas in the cyclone,

$$\tau = \frac{2\pi R_{avg} n}{W_g} \qquad (10.32)$$

where R_c and $R_{d.out}$ = the radii of the cyclone and discharge pipe, respectively
n = number of gas rotations around the discharge pipe (we may assume $n = 1.5$)

9. The peripheral velocity of gas is

$$W_g = \frac{V_{sec}}{H\ell} = \frac{1.8}{0.7 \times 0.185} = 13.9 \text{ m/s}$$

For cyclones, this value must be in the range of 12–14 m/s.
10. The average radius of the gas rotation is

$$R_{avg} = \frac{D_{d.out}}{2} + \frac{\ell}{2} = \frac{0.53}{2} + \frac{0.185}{2} = 0.357 \text{ m}$$

11. The centrifugal acceleration (at the average radius) is

$$a = \frac{W_g^2}{R_{avg}} = \frac{13.9^2}{0.357} = 542 \text{ m/s}^2$$

12. The separation criterion is

$$K_s = \frac{a}{g} = \frac{542}{9.81} = 55.2 \qquad (10.33)$$

i.e., in this case, the centrifugal field in the cyclone is 55.2 times more intensive than the gravitational.
13. The Archimedes number is

$$Ar = \frac{gd^3}{\nu^2} \times \frac{\rho_f}{\rho} = \frac{9.81 \times (10 \times 10^{-5})^3}{(2.25 \times 10^{-5})^2} \times \frac{1.1 \times 10^3}{1.2} = 17.8$$

where $\rho_f = \rho_P - \rho \simeq \rho_P$. The settling number is

$$S_1 = Ar \times K_s = 17.8 \times 55.2 = 980$$

14. Because $3.6 < S_1 < 82{,}500$, the flow regime through the cyclone is transitional. Therefore, the theoretical velocity of the particles is

$$W = 0.22d \sqrt[3]{\frac{(\alpha\rho_f)^2}{\mu\rho}} = 0.22d \sqrt[3]{\frac{(\alpha\rho_p^2)}{\nu\rho^2}}$$

$$= 0.22 \times 10 \times 10^{-5} \sqrt[3]{\frac{(5.42 \times 10^2 \times 1.1 \times 10^3)^2}{2.25 \times 10^{-5} \times 1.2^2}} = 1.07 \text{ m/s}$$

15. The particles have a shape factor of $\psi = 0.77$, and the inlet gas stream contains a low volume of solid particles. Based on the operating conditions specified, the settling velocity is

$$W_s = R\psi W = 0.77 \times 1.07 = 0.825 \text{ m/s}$$

Because the concentration of the suspension is low, we may assume $R = 1$.
16. The settling time is, therefore,

$$\tau_0 = \frac{\ell}{W_s} = \frac{0.185}{0.825} = 0.224 \text{ s}$$

17. The residence time for the gas is

$$\tau = \frac{2\pi R_{avg}n}{W_g} = \frac{2 \times 3.14 \times 0.357 \times 1.5}{13.9} = 0.24 \text{ s}$$

As $\tau_0 < \tau$, the diameter of the cyclone selected is acceptable and we may now specify the other dimensions as based on the recommended proportions in Figure 10.12.
18. As a final calculation for the design, we evaluate the hydraulic resistance of the cyclone:

$$\Delta P = \frac{1}{2} C_D \rho W_{in}^2 = 0.5 \times 7 \times 1.2 \times 18^2 = 1360 \text{ N/m}^2$$

where C_D for a typical cyclone is equal to 7.

Let us now direct our attention to a battery of cyclones. Normally, the allowable pressure drop is specified along with the conditional gas velocity for each cyclone element. For multiclones, typical pressure drop requirements and conditional inlet velocities are 350-600 N/m²(ΔP_{bc}) and 3-4 m/s (w_{cond}), respectively. A typical design problem often concerns determining the number of cyclone units required for a specified separation.

The continuity equation states that

$$V_{sec} = z \frac{\pi d_{el}^2}{4} w_{cond}, \text{ m}^3/\text{s} \qquad (10.34)$$

where z is the number of cyclone elements having a diameter d_{el}.

The hydraulic resistance of a cyclone battery at the conditional velocity is

$$\Delta P_{bc} = \frac{1}{2} \Sigma C_D \rho w_{cond}^2 \qquad (10.35)$$

From Equations 10.34 and 10.35 we obtain the number of elements with diameter d_{el}:

$$z = 0.9 \; \frac{V_{sec}}{d_{el}^2 \sqrt{\dfrac{\Delta P_{bc}}{\rho \Sigma C_D}}} \tag{10.36}$$

From information on ΔP_{bc}, ΣC_D and d_{el}, z can be computed. Equation 10.34 then may be used to compute the conditional velocity w_{cond}, and resistance ΔP_{bc} can be estimated from Equation 10.35. If values computed from Equations 10.34 and 10.35 are within the recommended operating limits, the number of cyclone elements chosen represents an acceptable design.

Based on the specified number of cyclones, the principal dimensions of the retaining vessel (see Figure 10.9) can be sized through proper layout of the individual cyclone elements. The simplest vessel to design for is rectangular, and a rough guideline for spacing between individual cyclones is 30–50 mm.

Assuming a gas velocity in the discharge pipe to be $w_{gd} = 14–18$ m/s, from the working expression

$$V_{sec} = \frac{\pi d_d^2}{4} z w_{gd} \tag{10.37}$$

we can determine the diameter of the discharge pipe and, from a specified pipe wall thickness, δ, compute the outside diameter of the discharge pipe $d_{d.out}$.

If z_1 cyclone elements are positioned perpendicular to the gas flow in the retaining vessel, the width of the chamber is

$$b = d_{el} z_1 + \delta(z_1 + 1) \tag{10.38}$$

The free passage in the entrance chamber is $b - z_1 d_{d.out}$.

The continuity equation then becomes

$$V_{sec} = w_{in} h_{in} (b - z_1 d_{d.out}) \tag{10.39}$$

From the specified gas entrance velocity, the chamber height can be computed.

Further examples of specific designs and design methodology are given by Cheremisinoff and Cheremisinoff [1974,1975], Rietema and Verver [1961], Taggart [1950], Trowbridge [1955], the Dallavalle Proceedings [1952], Kelsal [1953], Dahlstrom [1949], Krijgman [1951], Williams and Mesaros [1954], Scott and Lummus [1956], Bergman et al. [1956].

Hydroclones

Hydroclones are employed for the separation of solid particles from medium- to low-viscosity liquids. Like their cyclone counterparts, hydroclones are simple in design, and the degree of separation can be altered by either varying loading conditions or changing geometric proportions.

Unlike other types of separating equipment, they are better suited for classifying than for clarifying because high shearing stresses in a hydroclone promote the suspension of particles which oppose flocculation. However, by properly specifying dimensions and operating conditions, they can be used as thickeners in such a manner that the underflow contains mostly solid particles, while the clear overflow constitutes the largest portion of the liquid.

The fluid vortices and flow patterns characteristic of gas-cyclone operations are equally descriptive of liquid-hydroclones. However, the density differences between particles and liquids are significantly smaller than for gas-solid systems. For example, the density of water is approximately 800 times greater than that of air. This means that high fluid-spinning velocities cannot be employed in hydroclones as excessive pressure drop becomes a limitation. Obviously, the efficiency of hydroclones is low in comparison to gas cyclones.

The design features of a hydroclone are illustrated in Figure 10.13. It consists of an upper short cylindrical section (1) and an elongated conical bottom (2). The suspension is introduced into the cylindrical section (1) through the nozzle (3) tangentially, whence the fluid acquires an intensive rotary motion. The larger particles, under the action of centrifugal force, move toward the walls of the apparatus and concentrate on the outer layers of the rotating flow. Then they move spirally downward along the walls to the nozzle (4), through which the thickened slurry is evacuated. The largest portion of liquid containing small particles (clear liquid) moves in the internal spiral flow upward along the axis of the hydroclone. The cleared liquid is discharged through the nozzle (5) and fixed at the partition (6) and the nozzle (7). The actual flow pattern is more complicated than described because of radial and closed circulating flows. Because of peripheral flow velocities, the liquid column formed at the hydroclone axis has a pressure that is below atmospheric. The liquid bulk flow limits the upward flow of

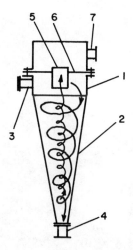

Figure 10.13 Design features of a hydroclone: (1) cylindrical section; (2) conical bottom; (3) feeding nozzle; (4) discharge nozzle; (5) nozzle; (6) partition; (7) nozzle for liquid discharge.

small particles from the internal side and has a significant influence on the separating effect. Hydroclones are used successfully for classification, clarification and thickening of suspensions containing particles from 5–150 μm in size.

The smaller the hydroclone diameter, the greater the centrifugal forces developed and, consequently, the smaller the size particles that can be separated. The following are typical hydroclone diameters used for various general applications: for classification and degritting process streams, D = 300–350 mm; for thickening of suspensions, D = 100 mm; for clarification (where it is necessary to apply powerful centrifugal fields), D = 10–15 mm. In the last case, multiclones are employed. Figure 10.14 shows an example of a hydroclone being used in a degritting operation. Sand is accumulated in a grit chamber for intermittent blowdown. Such an operation could be used off of a cooling tower installation.

Good separation of suspensions is achieved, especially in thickening and clarification, when hydroclones have an elongated shape with the slope of the cone equal to approximately 10–15°. At such a cone shape, the path of solid particles is increased as well as the residence time, which thus increases the separating efficiency. Design methodology for hydroclones and gas cyclones is not well developed. Current design principles are limited to determining capacity, approximate sizes of particles settled and

Figure 10.14   A hydroclone used for degritting process water (courtesy Krebs Engineers, Menlo Park, CA).

horsepower requirements. The flowrate of a suspension with density ρ through an inlet nozzle of diameter d_N, at pressure drop ΔP, may be calculated from the following formula:

$$V_{sec} = \tilde{\mu}\, \frac{\pi d_N^2}{4} \sqrt{2\,\frac{\Delta P}{\rho}} \tag{10.40}$$

where $\tilde{\mu}$ is a flowrate coefficient.

Introducing the hydroclone diameter, D, and the lower nozzle, $d_{\varrho N}$, to the RHS, we obtain

$$V_{sec} = C'(D d_{\varrho N}) \sqrt{\frac{\Delta P}{\rho}} \tag{10.41}$$

where factor C' is

$$C' = \tilde{\mu} \, \frac{\pi \sqrt{2}}{4} \, \frac{d_N^2}{Dd_{\varrho N}} \tag{10.42}$$

Coefficient C' is constant for geometrically similar cyclones. Defining the coefficient

$$K = \frac{C'}{\sqrt{\rho}} \tag{10.43}$$

and including K in Equation 10.41, we obtain the design formula for calculating the suspension flowrate in the following form:

$$V_{sec} = KDd_{\varrho N} \sqrt{\Delta P} \, , \, m^3/s \tag{10.44}$$

If values in Equation 10.44 are in SI units, then from experimental data with $D = 125–600$ mm and cone angle $38°$, the coefficient K is equal to 2.8×10^{-4}. The maximum size of particles in the cleared liquid is

$$d = \frac{1.33 \times 10^{-2} d_N^2}{\phi_x} \sqrt{\frac{\mu}{V_{sec} h \rho_f}} \, , \, m \tag{10.45}$$

where ϕ_x is the factor reflecting the change of liquid velocity. For hydroclones with a cone angle $\alpha = 38°$,

$$\phi_x = 0.103 \, \frac{D}{d_{\varrho N}} \tag{10.46}$$

where h is the height of the central flow, which is assumed to be equal to $\frac{1}{3}$ the cone height.

The power required for a hydroclone operation is the horsepower of a pump providing the desired capacity, V_{sec}, with an appropriate pressure.

The separating effect is influenced mostly by the ratio $d_N/d_{\varrho N}$ of nozzles, which may be assumed to be in the range of 0.37 to 0.40.

The diameter of the inlet nozzle, d_N, is assumed to be $(0.14–0.3)D$, and the diameter of discharge nozzle $d_d = (0.2–0.167)D$. The cone angle for hydroclones used as classifiers is $20°$; for thickeners it is $10–15°$.

The same operating arrangements used with gas cyclones are applicable to hydroclones, i.e., parallel and series operations. An example of a battery of multiclones is shown in Figure 10.15. The rubber block (1) is enclosed in the metal body (2). On the head of the cyclone body are 24 multiclones (3). The suspension is introduced through a central pipe (4), which is connected with a block (1) through three radial channels (5). The suspension enters into all cyclones at the same time from the feed pipe, and all products of separation are discharged to a common chamber.

Further discussions of hydroclone operations are given by Krijgman [1951], Williams and Mesaros [1954], Scott and Lummus [1956] and Bergman et al. [1956].

SEDIMENTATION CENTRIFUGES

General Principles

Another approach to the separation of solids from suspensions in a centrifugal field is through the use of sedimentation centrifuges. As in cyclones and hydroclones, particles of the heavier phase "fall" through

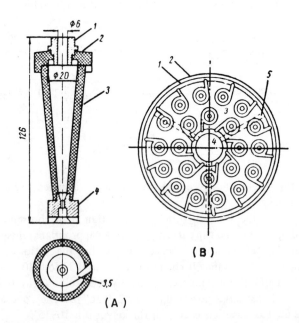

Figure 10.15   A multiclone arrangement. (A) Front view of single hydroclone: (1) bushing; (2) cover; (3) body; (4) endpiece. (B) Top view of a battery multihydroclone: (1) rubber block; (2) metal body; (3) multiclones; (4) central feeding pipe; (5) radial channels.

the lighter phase away from the center of rotation. In centrifuges, liquid (or gas) and solids are acted on by two forces: gravity acting downward and centrifugal force acting horizontally. In commercial units, however, the centrifugal force component is normally so large that the gravitational component may be neglected.

The magnitude of the centrifugal force component is defined by the ratio $R_c/G = \omega^2 r/g$, which is referred to as the "relative centrifugal force" (RCF) or *centrifugal number*, N_c, where $R_c = m\omega^2/2$ or $R_c = m\dot{v}^2/2$ is the centrifugal force, and $F_g = mg$ is the gravitational force. The RCF typically varies from 200 times gravity for large-basket centrifuges to 360,000 for high-speed tubular-gas centrifuges and ultracentrifuges.

For liquid-solid separations, centrifugal force may be applied in sedimentation-type centrifuges, centrifugal filters or in a combination of both.

Sedimentation-type centrifuges also are used for size or density classification of solids, the separation of immiscible liquids of different densities and for concentrating gases of different molecular weights. The principles of centrifugation are illustrated in Figure 10.16. In Figure 10.16(A) a stationary cylindrical bowl contains a suspension of solid particles in which the particle density is greater than that of the liquid. Because the bowl is stationary, the free liquid surface is horizontal and the particles settle due to the influence of gravity. In Figure 10.16(B) the bowl is rotating about its vertical axis. Liquid and solid particles are acted on by gravity and centrifugal forces, resulting in the liquid assuming a position with an almost vertical inner surface (free interface).

If the suspension consists of several components, each with different densities, they will stratify with the lightest component nearest the axis of rotation and the heaviest adjacent to the solid bowl wall (Figure 10.16(C)).

In Figure 10.16(D), the bowl wall is perforated and lined with a permeable membrane, such as filter cloth or wire screen, which will support and retain the solid particles but allow the liquid to pass through due to the action of centrifugal force.

The main components of a centrifuge are (1) a rotor or bowl in which the centrifugal force is applied to a heterogeneous system to be separated; (2) a means for feeding this system into the rotor; (3) a drive shaft; (4) axial and thrust bearings; (5) a drive mechanism to rotate the shaft and bowl; (6) a casing or "covers" to contain the separated components; and (7) a frame for support and alignment.

There are three main types of centrifuges, which may be classified according to the centrifugal number, and the range of throughputs or the solids concentration in suspension that can be handled. The first of these is the *tubular-bowl centrifuge*. This type has a centrifugal number in the range of 13,000 but is designed for low capacities (50–500 gph) and can handle only small concentrations of solids.

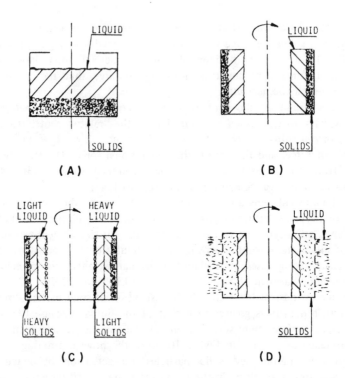

Figure 10.16 (A) Stationary cylindrical bowl (liquid plus heavy solid particles); (B) rotating cylindrical bowl (liquid plus heavy solid particles); (C) rotating cylindrical bowl (two liquids plus light and heavy solid particles; (D) rotating perforated bowl.

The second is the *solid-bowl centrifuge*, with maximum bowl diameters ranging from 4–54 in. The larger-diameter machines can handle up to 50 ton/hr of solids with a centrifugal number up to 3000. Similar centrifuges are manufactured with a perforated wall on the bowl. These machines operate exactly like filters described in Chapter 12, with the filtrate draining through the cake and bowl wall into a surrounding collector.

The third type is the *disk-bowl centrifuge*, which is larger than the tubular-bowl centrifuge and rotates at slower speeds with a centrifugal number up to 14,000. These machines can handle as much as 30,000 gph of feed containing moderate quantities of solid particles.

The migration of particles in sedimentation centrifuges is radially toward or away from the axis of rotation, depending on whether the density of the dispersed particles is greater, or less, than that of the continuous phase.

There must be a measurable difference between the density of the continuous and dispersed phases to provide effective separation.

In commercial machines the discharge of the liquid, or separated liquid phases is performed almost always in a continuous fashion. The heavy solid phase deposited against the bowl wall is discharged and recovered intermittently, manually, or by action of an unloader knife or skimmer; continuously by action of a differential screw conveyor; or intermittently or continuously with a portion of the continuous phase through openings in the wall of the bowl. These variations are summarized in Table 10.2. Manual solids removal units can operate continuously up to one hour and generally only require a few seconds for a fully automated intermittent operation. In systems in which solids have a lesser density than the continuous phase, particulates can be removed continuously from the surface of the liquid via a skimming tube or exit as an overflow from the bowl with a portion of the continuous phase.

Tubular-Bowl Centrifuges

Tubular-bowl centrifuges are used extensively for the purification of oils by separating suspended solids and free moisture from them; for removal of oversize particles from dye pastes, pigmented lacquers and enamels; for "polishing" citrus and other aromatic oils; and other small-scale separating applications.

Table 10.2 Classification of Sedimentation Centrifuges

Flow Arrangement	Centrifuge Types
Liquid-Batch	Analytical and clinical
Solid-Batch	Ultra and preparatory miscellaneous batch
Liquid-Continuous	Tubular bowl
Solids-Batch	Multipass clarifier Disk (solid wall) Basket type (solid wall)
Liquid-Continuous	Disk-valve discharge
Solids-Intermittent	Disk-automatic opening
Liquid-Continuous	Disk-peripheral nozzles
Solids-Continuous	Disk-light solid-phase skimmer, continuous decanter

The tubular-bowl clarifiers and separators are comprised of small-diameter cylinders (about 100 mm) which allows operation at very high velocities. Commercial machines typically work at 15,000–19,000 rpm, which corresponds to N_c = 13,000–18,000. For special applications (e.g., treatment of vaccines, etc.), the diameter of the tubular-bowl centrifuge is only several centimeters, with values of N_c as high as 50,000. Figure 10.17 shows the operating principle for this type of centrifuge. The tubular centrifuge rotor is suspended from its drive assembly on a spindle that has a built-in degree of flexibility. It essentially hangs freely with a sleeve bushing in a dampening assembly at the bottom. In some designs a similar dampening assembly is included at the upper end of the rotor. This permits the rotor to determine its own mass axis after it exceeds its critical speed. The details of a specific design are illustrated by the cutaway view in Figure 10.18.

The feed liquid is introduced to the rotor at the bottom through a stationary feed nozzle. The inlet feed is under sufficient pressure to create a standing jet, which ensures a clean entrance into the rotor. Often, an acceleration device is provided at the bottom of the rotor to bring the feed stream to the rotational speed of the bowl. The feed moves upward through the bowl as an annulus and discharges at the top. To effect this, the radius of the discharge at the top must be larger than the opening at the bottom through which the feed enters. Solids move upward with the velocity of the annulus and simultaneously receive a radial velocity that is a function of their equivalent spherical diameter d^2, their relative density $\Delta\rho$, and the applied centrifugal force $\omega^2 r$. If the trajectory of a particle intersects the cylindrical bowl wall, it is removed from the liquid; if it does not, the particle flows out with the effluent overflow.

Multichamber (Multipass) Centrifuges

This machine combines the process principles of a tubular clarifier with mechanical drive and the bowl contour of a disk centrifuge. The suspension flows through a series of nested cylinders of progressively increasing diameter. The direction of the flow from the smallest to the largest cylinders is in parallel to the axis of rotation, as in the tubular bowl. The basic design is illustrated in Figure 10.19.

Such a rotor usually contains six annuli, so that the effective length of suspension travel is approximately six times the interior height of the bowl. The multipass bowl can be considered as a multistage classifier because the centrifugal force acting in the machine is greater in each subsequent annulus. Consequently, larger, heavier particles are deposited in the first annulus (the zone of least centrifugal force), while smaller, lighter particles are deposited in the last annulus (the zone of greatest centrifugal force). The

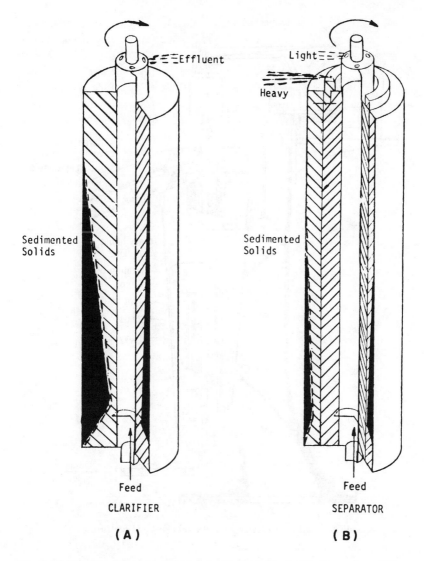

Figure 10.17   Operation of tubular-bowl centrifuges for clarification and separation.

radial distance particles must migrate to reach the cylinder wall is thus minimized. Multipass rotors typically have a total holding volume of up to 65 liters for the largest size, of which about 50% is available for the retention of collected solids before the clarification process is impaired. As with the tubular type, the collected solids must be removed manually. These

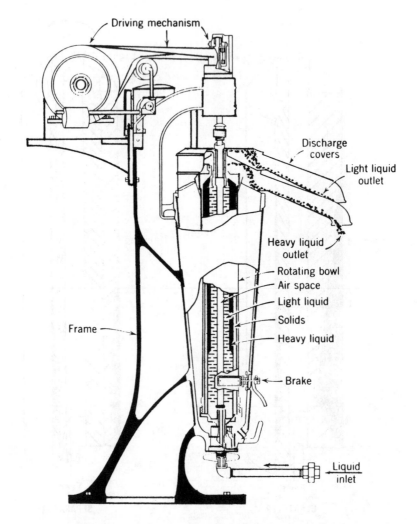

Figure 10.18 Cutaway view of a tubular-bowl centrifuge.

machines are applied to the clarification of fruit and vegetable juices, wine and beer.

Solid-Bowl Centrifuges

The continuous solid-bowl centrifuge basically consists of a solid-wall rotor, which may be tubular or conical in shape, or a combination of the

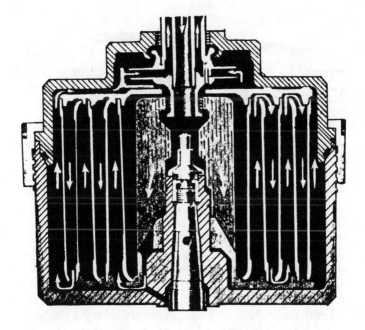

Figure 10.19   A multipass centrifuge.

two. The rotor may rotate about a horizontal or a vertical axis because the centrifugal force is many times that of gravitational force (for many units N_c is more than 3000).

A typical example of this kind of equipment is a continuous horizontal centrifuge, as shown in Figure 10.20. It consists of a cylindrical rotor with a truncated cone-shaped end and an internal screw conveyor rotating together. The screw conveyor often rotates at a rate of 1 or 2 rpm below the rotor's rate of rotation. The suspension enters the bowl axially through the feed tube to a feed accelerated zone, then passes through a feed port in the conveyor hub into the pond. The suspension is subjected to centrifugal force and thrown against the bowl wall where the solids are separated. The clarified suspension moves toward the broad part of the bowl to be discharged through a port.

The solid particles being scraped by the screw conveyor are carried in the opposite direction (to the small end of the bowl) across discharge ports through which they are ejected continuously by centrifugal force.

As in any sedimentation centrifuge, the separation takes place in two stages: settling (Figure 10.20, in the right part of the bowl), and thickening or pressing out of the sediment (left-hand side of the bowl). Because the

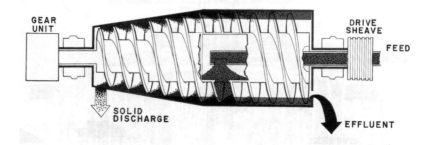

Figure 10.20   A continuous solid-bowl centrifuge.

radius of the solid discharge port is usually less than the radius of the liquid overflow at the broader end of the bowl, part of the settled solids is submerged in the pond. The remainder, closer to the center, is inside the free liquid interface, where they can drain before being discharged. The total length of the "settling" and "pressing out" zones depends on the dimensions of the rotor. Their relative length can be varied by changing the pond level through suitable adjustment of the liquid discharge radius. When the pond depth is lowered, the length of the pressing out zone increases with some sacrifice in the clarification effectiveness.

The critical point in the transport of solids to the bowl wall is their transition across the free liquid interface, where the buoyancy effect of the continuous phase is lost. At this point, soft amorphous solids tend to flow back into the pond instead of discharging. This tendency can be overcome by raising the pond level so that its radius is equal to, or less than, that of the solids discharge port. In reality, there are no dry settled solids. The solids form a dam, which prevents the liquid from overflowing. The transfer of solids becomes possible because of the difference between the rotational speed of the screw conveyor and that of the bowl shell. The flights of the screw move through the settled solids and cause the solids to advance. To achieve this motion, it is necessary to have a high circumferential coefficient of friction on the solid particles with respect to the bowl shell and a low coefficient axially with respect to the bowl shell and across the conveyor flights. These criteria may be achieved by constructing the shell with conical grooves or ribs and by polishing the conveyor flights. The conveyor or differential speed is normally in the range of 0.8% to 5% of the bowl's rotational speed.

The required differential is achieved by a two-stage planetary gear box. The gear box housing carrying two ring gears is fixed to, and rotates with, the bowl shell. The first stage pinion is located on a shaft that projects outward from the housing. This arrangement provides a signal that is proportional to the torque imposed by the conveyor. If the shaft is held rotational (for example, by a torque overload release device or a shear pin), the relative conveyor speed is equivalent to the bowl rotative speed divided by the gear box ratio. Variable differential speeds can be obtained by driving the pinion shaft with an auxiliary power supply or by allowing it to slip forward against a controlled breaking action. Both arrangements are employed when processing soft solids or when maximum retention times are needed on the pressing out zone. The solids handling capacity of this type centrifuge is established by the diameter of the bowl, the conveyor's pitch and its differential speed.

Feed ports should be located as far from the effluent discharge as possible to maximize the effective clarifying length. Note that the feed must be introduced into the pond to minimize disturbance and resuspension of the previously sedimented solids. As a general rule, the preferred feed location is near the intercept of the conical and cylindrical portions of the bowl shell. The angle of the sedimentation section with respect to the axis of rotation is typically in the range of 3 to 15°. A shallow angle provides a longer sedimentation area with a sacrifice in the effective length for clarification. In some designs, a portion of the conveyor flights in the sedimentation area is shrouded (as with a cone) to prevent intermixing of the sedimented solids with the free supernatant liquid in the pond through which they normally would pass.

In other designs, the clarified liquid is discharged from the front end via a centrifugal pump or an adjustable skimmer that sometimes is used to control the pond level in the bowl. Some displacement of the adhering virgin liquor can be accomplished by washing the solids retained on the settled layer, particularly if the solids have a high degree of permeability. Washing efficiency ranges up to 90% displacement of virgin liquor on coarse solids. Foust et al. [1980] note one configuration in which the settled layer has two angles; comparatively steep in the wetted portion (10–15°) and shallow in the dry portion (3–5°). A wash is applied at the intersection of these angles, which, in effect, forms a constantly replenished zone of pure liquid through which the solids are conveyed.

The longer section of a dry shallow layer provides more time for drainage of the washed solids. This system is especially effective for washing and dewatering such spherical particles as polystyrene. In either washing system, the wash liquid that is not carried out with the solids fraction returns to the pond and eventually discharges along with the effluent virgin liquor.

Separation Rates in Tubular- and Solid-Bowl Centrifuges

To evaluate the radial velocity of a particle moving toward a centrifuge wall, the expression developed earlier for particle settling in a gravitational field is applied with the acceleration term replaced by centrifugal acceleration, a:

$$a = \frac{u_r^2}{r} \tag{10.47}$$

where u_r is the peripheral velocity at a distance, r, from the axis of rotation. Expressing u_r in terms of the number of rotations, n,

$$u_r = 2\pi rn \tag{10.48}$$

The centrifugal acceleration is

$$a = 4\pi^2 rn^2 \tag{10.49}$$

Depending on the particle diameter and properties of the liquid, the radial motion of particles will be laminar, turbulent or transitional.

The motion of large particles at Re > 500 is turbulent. Therefore, their settling velocity in a gravitational field may be expressed as

$$u = 1.74 \sqrt{\frac{d(\gamma_p - \gamma)g}{\gamma}} \tag{10.50}$$

Replacing g by centrifugal acceleration, a,

$$u_r = 1.74 \sqrt{\frac{d(\gamma_p - \gamma)a}{\gamma}} \tag{10.51}$$

where γ_p = specific weight of the solids
γ = specific weight of the liquid

Substituting for "a" (from Equation 10.49) into this last expression, we obtain the particle velocity in the radial direction of the wall:

$$u_r = 10.94 \sqrt{\frac{d(\gamma_p - \gamma)r}{\gamma}} \tag{10.52}$$

Dividing Equation 10.52 by Equation 10.50, we determine the number of times the particle velocity in a centrifuge is greater than that in free particle settling, i.e., the separation number, K_s:

$$\frac{u_r}{u_g} = 2\pi n \sqrt{\frac{r}{g}} \tag{10.53}$$

For example, at n = 1200 rpm = 20 liter/s^{-1} and r = 0.5 m, the settling velocity in the centrifuge is almost 28 times greater than that in free settling. Note that Equation 10.52 used in the derivation is only applicable at Re > 500.

For small particles, Re < 2, migration toward the wall is laminar. The proper settling velocity expression for the gravitational field is

$$u = \frac{d^2(\gamma_p - \gamma)}{18\mu} \tag{10.54}$$

Substituting in ρg for γ, we obtain

$$u_r = \frac{d^2(\rho_p - \rho)a}{18\mu} \tag{10.55}$$

Replacing "a" by Equation 10.49 and noting that

$$u_r = \frac{2\pi^2 d^2(\gamma_p - \gamma)n^2 r}{9\mu g} \tag{10.56}$$

Dividing this expression by Equation 10.54 provides the separation number:

$$\frac{u_r}{u_g} = 4\pi^2 n^2 \frac{r}{g} \tag{10.57}$$

For the same case of n = 1200 rpm and r = 0.5, we obtain $u_r/u_g = 800$, whereas for the turbulent regime the ratio was only 28. This example demonstrates that the centrifugal process is more effective in the separation of small particles than of large ones. Note that after the radial velocity u_r is determined, it is necessary to check whether the laminar condition, Re < 2, is fulfilled.

For the transition regime, 2 < Re < 500, the sedimentation velocity in the gravity field is

$$u = \frac{0.153d^{1.14}(\gamma_p - \gamma)^{0.71}}{(\gamma/g)^{0.29}\mu^{0.43}} \tag{10.58}$$

In a similar development, the expression for particle radial velocity toward the wall is

$$u_r = \frac{0.153d^{1.14}(\rho_p - \rho)^{0.71}a^{0.71}}{\rho^{0.29}\mu^{0.43}} \tag{10.59}$$

or

$$u_r = \frac{1.36d^{1.14}(\gamma_p - \gamma)^{0.71}(n^2 r)^{0.71}}{\gamma^{0.29}(\mu g)^{0.43}} \tag{10.60}$$

Finally, we obtain the ratio of settling velocities in a centrifuge to that in the gravitational field:

$$\frac{u_r}{u} = \left(4\pi^2 n^2 \frac{r}{g}\right)^{0.71} \tag{10.61}$$

This ratio represents an average between similar ratios for the laminar and turbulent regimes.

For the most general case, $u_r = f(D, \rho_p, \rho, \mu, \omega, r)$, so we may ignore whether the particle displacement is laminar, turbulent or transitional. In this analysis, we may use the dimensionless centrifugal Archimedes number:

$$Ar_c = C_D Re^2 = \frac{4}{3} \frac{d^3\rho(\rho_p - \rho)\omega^2 r}{\mu^2} \tag{10.62}$$

If Ar_c is known, the plot given in Figure 10.21 may be used to evaluate the Reynolds number and, subsequently, u_r. The following example from Loncin [1961] illustrates the use of Figure 10.21.

Example 10.2

Oil droplets ($d_p = 10^{-4}$m, $\rho_p = 900$ kg/m^2) suspended in water ($\rho = 1000$ kg/m^3, $\mu = 10^{-3}$ dp) are to be separated in a sedimentation centrifuge designed to operate at 5000 rpm ($\omega = 2\pi \times 5000/60$). If the distance of a single droplet from the axis of rotation is 0.1 m, determine the droplet's radial settling velocity.

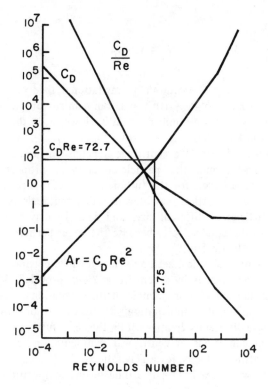

Figure 10.21   Plot of Archimedes number vs Reynolds number.

Solution

$$Ar_c = C_D \times Re^2 = \frac{4 \times 10^{-12} \times 1000 \times 100 \left(\dfrac{2\pi \times 5000}{60} \right)^2 \times 0.1}{3 \times 10^{-6}} = 3650$$

Note that we assume the absolute value of $\rho_p - \rho$. The negative value of this difference indicates that the droplet displacement is centripetal. The value of the Reynolds number corresponding to $Ar_c = 3650$, from Figure 10.21, is 45. The radial settling velocity is evaluated from Re:

$$45 = \frac{10^{-4} \times 1000 u_r}{10^{-3}}$$

or

$$u_r = 0.45 \text{ m/s}$$

If the separation were to occur in a gravitational field only, the droplet velocity would be $u = 5.45 \times 10^{-4}$ m/s. For laminar droplet motion, this corresponds to a separation number of 2800. However, in this case the flow is transitional and $K_s = 825$.

Particle settling velocity in a centrifuge depends on the particle location. For the laminar regime, the particle velocity is proportional to the centrifuge radius; however, for the turbulent regime, it is related to $\sqrt{r}$. In an actual operation, a particle may blow over different regimes depending on its location. Particles of different sizes and densities may be located at the same point with the same velocity because the larger particle diameter tends to compensate for lower density.

Two particles with identical settling velocities in a gravitational field will settle in the same manner in a centrifugal field, provided the regimes of motion remain unchanged. For example, if the gravitational sedimentation is laminar and centrifugal sedimentation is transitional or turbulent, then particles will have different velocities within the centrifuge.

Under certain conditions of centrifugal sedimentation (characterized by r and ω), the settling velocity for a gravitational field can be applied. This is exactly correct when gravitational settling occurs in the turbulent regime, where the velocity ratio is given by $\sqrt{K_s}$. However, when a particle passes from one regime to the other, or when it is in the transitional regime, one either must calculate the velocities or extrapolate from experimental results. Extrapolation is valid only in cases in which there are no changes in the flow regime. Loncin [1961] gives the example in which a particle ($d = 5 \times 10^{-4}$ m) located at a distance of 0.2 m from the axis of a centrifuge operating at 4000 rpm has a velocity of 1 m/s through water. In this case,

$$Re = \frac{1 \times 1000 \times 0.0005}{0.001} = 500$$

and, hence, the regime is turbulent. It is possible to extrapolate to conditions of greater radii and higher rotations. If the centrifuge operates at 5000 rpm and the distance is 0.25 m from axis of rotation, then the settling velocity is equal to 1 m/s times $\sqrt{K_s}$ (turbulent regime); that is,

$$u_r = 1 \times \frac{5000 \times \sqrt{0.25}}{4000 \times \sqrt{0.20}} = 1.4 \text{ m/s}$$

If the centrifuge operates at 4000 rpm and the particle's distance from the axis of rotation is 0.25 m, the settling velocity is only 0.01 m/s, which corresponds to

$$Re = \frac{1000 \times 10^{-4} \times 0.01}{0.001} = 1$$

In this latter case the flow is laminar and extrapolation can be only to much smaller rotation velocities and radii.

If the particle now settles in the centrifuge operating at 3000 rpm and the distance of the particle from the axis of rotation is 0.20 m, the settling velocity is

$$u_r = 0.01 \frac{3000^2 \times 0.20}{4000^2 \times 0.25} = 0.0045 \text{ m/s}$$

The reference velocity in this case is multiplied by the ratio K_s.

Note that the above calculations are based on nonhindered sedimentation and, therefore, should be modified for a hindered "fall."

Estimating Capacities of Tubular- and Solid-Bowl Centrifuges

When a rotating centrifuge is filled with suspension, the internal surface of liquid acquires a cylindrical geometry of radius R_1, as shown in Figure 10.22. The free surface is normal at any point to the resultant force acting on a liquid particle.

If the liquid is lighter than the solid particles, the liquid moves toward the axis of rotation while the solids flow toward the bowl walls. The flow of the continuous liquid phase is effectively axial. A simplified model of centrifuge operation is that of a cylinder of fluid rotating about its axis. The flow forms a layer bound outwardly by a cylinder, R_2, and inwardly by a free cylindrical surface, R_1 (Figure 10.22). This surface is, at any point, normal to a resulting force (centrifugal and gravity) acting on the solid particle in the liquid. The gravity force is, in general, negligible compared to the centrifugal force, and the surface of liquid is perpendicular to the direction of centrifugal force.

Consider a solid particle located at distance R from the axis of rotation. The particle moves centrifugally with a settling velocity, u_s, while liquid particles move in the opposite direction centripetally with a velocity u_f:

$$u_f = \frac{\frac{dV}{d\tau}}{2\pi R \ell} \tag{10.63}$$

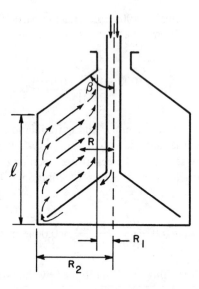

Figure 10.22   Schematic of a centrifuge operation.

where V = volume (m^3)
τ = time (s)
ℓ = height of a bowl (m)

The resulting velocity will be centrifugal, and the solid particles will be separated, provided

$$u_s > \frac{\dfrac{dV}{d\tau}}{2\pi R\ell} \qquad (10.64)$$

Thus, the capacity of a centrifuge will be

$$\frac{dV}{d\tau} = 2\pi R\ell u_s \qquad (10.65)$$

If the particle's density is lower than that of the liquid, the path of the liquid is centrifugal, as shown in Figure 10.23. Settling will occur when u_s (centripetal) is higher than the radial velocity, u_f (centrifugal).

Settling capacity for a given size of particles is a function of ℓ, R and u_s, which itself is proportional to R. In general, for the sedimentation of heavy

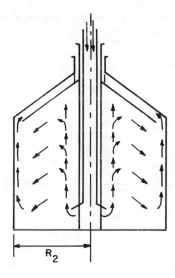

Figure 10.23   Operation when particle density is less than that of the liquid.

particles in a suspension it is sufficient that the radial component of u_f be less than u_s at a radius greater than R_2.

Because of turbulence effects, it is generally good practice to limit the settling capacity so that u_s again exceeds u_f near R_1. The same situation occurs when the particles are lighter than the continuous liquid.

The relation between u_s and R depends on the regime. In the laminar regime, u_s is proportional to R, whereas in the turbulent regime, it is related to $\sqrt{R}$. Most industrial sedimentation centrifuges operate in the transition regime.

Example 10.3

Determine the settling velocity of a particle ($d = 4 \times 10^{-5}$ m and density $\rho_p = 900$ kg/m^3) through water in a sedimentation centrifuge operating at a velocity $\omega = 420$ rad/s (4000 rpm). The particle velocity is a function of its distance from the axis of rotation. Data are provided in Table 10.3 and Figure 10.24.

Solution

The data in Table 10.3 indicate that the particle motion is laminar for values of R up to 0.34 m. For R values exceeding 0.04, u_s must be estimated

Table 10.3   Data of Settling Velocity vs Distance, R, for Example 10.3

R,m	u_s (m/s)	Re
0.01	0.0155	0.62
0.02	0.031	1.24
0.03	0.0465	1.90
0.04	0.060	2.4
0.05	0.070	2.8
0.10	0.1125	4.5
0.20	0.185	7.4
0.30	0.245	9.8
1.00	0.437	17.5

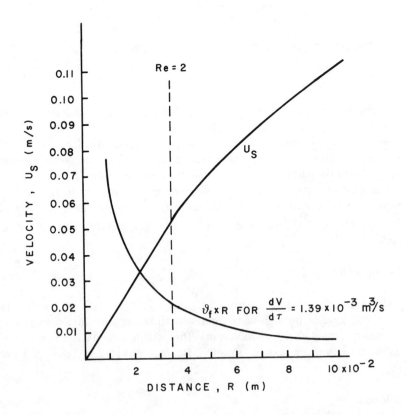

Figure 10.24   Plots of u_s and $u_f R$ vs R for Example 10.3.

from the Reynolds-Archimedes numbers relationship. As part of the analysis, let us consider similar spherical particles of diameter 20×10^{-5} m being centrifuged at the same conditions at a rotating velocity of $\omega = 523$ rad/min (5000 rpm). The settling velocities for such a particle as a function of distance R are given in Table 10.4. The settling velocity for these particles is plotted in Figure 10.25. As shown from the plot, those particles up to a distance of 0.44 m settle in the transition regime. At higher values of R, the regime is turbulent.

To evaluate the capacity or, more correctly, the behavior of a fixed capacity $dV/d\tau$, we write, from Equation 10.63,

$$u_f R = \frac{\dfrac{dV}{d\tau}}{2\pi \ell}$$

The curve $u_f = f(R)$ at a given $dV/d\tau$ and ℓ is an equilateral hyperbola (Figure 10.24). If, for example, $dV/d\tau = 1.39 \times 10^{-3}$ m^3/s (5000 ℓ/hr) and ℓ (height of the bowl) = 0.3 m, we have

$$u_f R = \frac{1.39 \times 10^{-3}}{2\pi \times 0.3} = 0.736 \times 10^{-3} \text{ m}^2/\text{s}$$

Tracing the curve $u_f = f(R)$ in Figure 10.24, we see that for all radii larger than 0.022 m, u_s is higher than the radial component of u_f and, consequently, the particles of diameter 4×10^{-5} m (40 μm) and density $\rho_p = 900$ kg/m^3 are centrifuged at 4000 rpm in water and are displaced centripetally.

Table 10.4 Settling Velocities for a 200-μm Particle in a Centrifuge

R (m)	u_s (m/s)	Re
0.01	0.1625	32.5
0.02	0.300	60.0
0.05	0.575	115.0
0.10	0.950	190
0.25	1.875	375
0.50	2.820	563
0.75	3.450	690
1.00	3.990	798

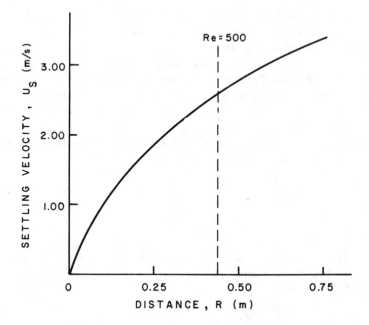

Figure 10.25   Settling velocity as a function of R for 200-μm particles.

For particles of 2×10^{-4} m (200 μm) at 5000 rpm, the total sedimentation occurs at the same conditions for $R < 0.01$ m. The capacity of a sedimentation centrifuge can be determined from Equation 10.65, where R_1 is the radius corresponding to the internal surface of the liquid, and u_s is the settling velocity at R_1.

For certain types of centrifuges with very long ℓ and small radius R_2 (Figure 10.26), the liquid motion cannot be assumed radial. Velocity u_f is mostly axial, and the condition of settling is that the maximum time of sedimentation $(R_2 - R_1)/u_s$ be less than the time of passage ℓ/u_f. The velocity of passage is

$$u_f = \frac{\ell u_s}{R_2 - R_1} \qquad (10.66)$$

The capacity is equal to u_f multiplied by the passage section

$$\pi(R_2^2 - R_1^2)$$

Figure 10.26   A long-barrel centrifuge.

and, consequently,

$$2\pi \frac{(R_2 + R_1)}{2} (R_2 - R_1)$$

Thus, the capacity is

$$\frac{dV}{d\tau} = \frac{2\pi \dfrac{(R_2 + R_1)}{2} (R_2 - R_1)\ell u_s}{R_2 - R_1} \tag{10.67}$$

and

$$\frac{dV}{d\tau} = 2\pi R \ell u_s \tag{10.68}$$

where

$$R = \frac{R_2 + R_1}{2}$$

Capacity also may be expressed as

$$\frac{dV}{d\tau} = 2\pi R_1 u_s \tag{10.69}$$

Analysis of the Centrifugal Process

The process of centrifugal sedimentation may be considered as a two-stage operation that first involves the concentration of particles onto the walls of the bowl and, secondly, involves compressing the formed sediment. The first of these stages proceeds according to the laws of hydrodynamics, whereas the second is based on the laws of soil mechanics (porous media).

At low concentrations of solids in a suspension (no more than 4% by volume), sedimentation is observed in the rotor without the formation of a distinct interface between clear liquid and a laminated suspension. At higher concentrations, a distinct interface is formed due to the hindered sedimentation of solid particles.

The separation process in settlers is significantly different than in sedimentation centrifuges. In settlers, the gravitational field is homogeneous, whereas in a centrifuge the intensity of the centrifugal field increases as the motion of the particle proceeds toward the periphery of the rotor. Thus, at the rotor's rotation the particle is acted on by an increasing centrifugal force at a constant rpm. In settlers, particles move through a constant cross-sectional area of a flat liquid layer. In a sedimentation centrifuge, however, they move through increasing cross-sectional areas of circular layers. Therefore, the relationships obtained in the settlers cannot be propagated on the processes taking place in sedimentation centrifuges.

The separating power of sedimentation centrifuges is characterized by the capacity factor, Σ_c, which is equal to the product of the area of a cylindrical surface of sedimentation, F, in the rotor and the centrifugal number, N_c (where $N_c = K_s$):

$$\Sigma_c = FN_c \tag{10.70A}$$

or

$$N_c = \frac{\Sigma_c}{F} \tag{10.70B}$$

It follows from this last expression that Σ_c is an equivalent area function. The capacity factor has the dimensions of area and represents the cross section of a gravity settling tank having a separation performance equal to that of the sedimentation centrifuge handling the same particles. The capacity factor is a characteristic of the centrifuge itself and does not characterize the settling behavior of the solid particles or liquid droplets in the suspension. However, the capacity factor can be used to compare the relative performance of different centrifuges.

If two centrifuges are to process the same quantities, then, from continuity,

$$\frac{Q_1}{\Sigma_{c1}} = \frac{Q_2}{\Sigma_{c2}} \tag{10.71}$$

We shall evaluate the capacity factor for a cylindrical centrifuge rotor containing a layer of liquid. Figure 10.27 illustrates the system under consideration. The thickness, h, of a liquid layer is considerably less than the rotor diameter, D. The centrifugal number, N_c, therefore may be related to an average diameter, $D - h$.

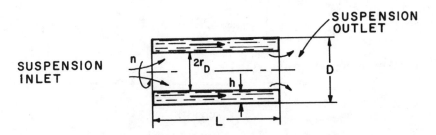

Figure 10.27   Scheme for a sedimentation centrifuge.

Centrifugal force is defined as

$$C = \frac{mu^2}{r} = \frac{Gu^2}{gr} \qquad (10.72)$$

where   m = the mass of a rotating body (kg)
          G = the weight of a rotating body (N)
          u = the peripheral velocity (m/s)
          r = the radius of rotation (m)
          C has units of N

The peripheral velocity is

$$u = \omega r = \frac{2\pi n}{60} r \qquad (10.73)$$

where ω = angular velocity (radians)
          n = number of rotations per minute

Substituting for u into Equation 10.72 gives

$$C = \frac{G}{rg}\left(\frac{2\pi n}{60} r\right)^2 \qquad (10.74)$$

or

$$C \simeq \frac{Grn^2}{900} \qquad (10.75)$$

This expression indicates that the centrifugal force is much more sensitive to the rotational speed of the rotor than it is to the rotor diameter. Recall that the centrifugal number is

$$N_c = \frac{u^2}{gr}$$

For G = 1 N, we note that

$$N_c = \frac{rn^2}{900}$$

And substituting into an average radius for r,

$$N_c = \frac{(D - h)n^2}{1800}$$

The area of the cylindrical surface of sedimentation in the rotor is

$$F = \pi(D - h)L$$

Hence,

$$\Sigma_c = FN_c = \pi L \frac{(D - h)^2 n^2}{1800} \tag{10.76}$$

Assuming that the liquid is displaced not over the total annular space occupied by a layer but only over the thin internal zone of the annular space, i.e., the surface region of flow, then the following approximate expressions are obtained:

$$N_c = \frac{r_0 n^2}{900} \tag{10.77}$$

and

$$F = 2\pi r_0 L \tag{10.78}$$

whence we obtain

$$\Sigma_c = FN_c = 2\pi L \frac{r_0^2 n^2}{900} \tag{10.79}$$

Equation 10.70 is applicable if the sedimentation process occurs in the laminar regime. However, as the forces acting on a particle in a centrifuge are quite large, sedimentation usually proceeds over the transition and turbulent regimes. Based on our earlier analysis of the settling velocity in the centrifugal field, the expressions for the capacity factor in the other flow regimes are as follows:

For the transition regime,

$$\Sigma_c = FN_c^{0.715} \tag{10.80}$$

For the turbulent regime,

$$\Sigma_c = FN_c^{0.50} \tag{10.81}$$

Hence, the required area of a settler, with the equivalent capacity of a centrifuge, is slightly less sensitive to the centrifugal number for the transition and turbulent regimes of sedimentation.

The actual capacity of a sedimentation centrifuge is less than that estimated on the basis of the settling velocity of solids in a centrifugal field. The reasons for this are as follows:

1. Retardation of the liquid rotational velocity from the rotation of a rotor results in a decrease in the centrifugal force acting on the particle.
2. There is nonuniformity of liquid flow along the rotor walls and entrainment of settled particles from its walls.
3. Eddies form that cause particle resuspension.

To evaluate these effects an *efficiency coefficient* is introduced:

$$\zeta_c = \frac{Q_a}{Q_T}$$

(10.82)

where Q_a and Q_T are the actual and calculated capacities of a centrifuge in m^3/s, respectively.

Values for ζ_c must be determined experimentally and Equation 10.82 used to obtain the actual capacity of the centrifuge. The capacity of a sedimentation centrifuge can be determined from the settling velocity of particles in a gravitational field, u_g, as follows:

$$Q_T = u_g \Sigma_c$$

(10.83)

where Σ_c may be determined from one of the above equations. From Equation 10.82, the actual capacity can be computed.

When the suspension layer thickness is large, one must account for the variation of the centrifugal field. Rearranging Equation 10.3 derived for cyclones and applying it to a solid-liquid suspension, an expression is obtained for the settling velocity of spherical particles located at distance r in a centrifugal field rotating at rate ω:

$$u_s = \frac{r\omega^2(\rho_p - \rho)d^2}{18\mu}$$

(10.84)

The radial distance traveled by the particle may be obtained by multiplying the expression by the differential time, $d\tau$:

$$u_s d\tau = dr = \frac{r\omega^2(\rho_p - \rho)d^2}{18\mu} d\tau$$

(10.85)

On integration, the following is obtained:

$$ln \frac{r_2}{r_1} = \frac{\omega^2(\rho_p - \rho)d^2}{18\mu} \tau = \frac{\omega^2(\rho_p - \rho)d^2}{18\mu} \frac{V}{Q} \qquad (10.86)$$

where V = volume of material in the centrifuge
Q = volumetric feedrate to the centrifuge
V/Q = residence time of a particle in the centrifuge

Diameter, d, is that of a particle settling over the distance r_1 to r_2 during the residence time of the centrifuge. If the liquid layer in the centrifuge is very small in comparison to the radius, then the centrifugal field may be assumed constant and Equation 10.85 may be expressed directly in terms of the residence time, V/Q:

$$u_s \tau = x = \frac{r\omega^2(\rho_p - \rho)d^2}{18\mu} \frac{V}{Q} \qquad (10.87)$$

where x is the radial distance traveled by a particle of diameter d in the available residence time.

If the thickness of the liquid layer in the bowl is divided into two equal parts $(r_2 - r_1)/2$ and initially we assign the same number of particles to each portion, then the particles of diameter d' present in the suspension will be separated while the other half escape. Diameter d' is referred to as the "critical diameter." Most of those particles larger than d' will settle out from the suspension, whereas particles smaller than d' will remain. Solving Equation 10.87 for d' and substituting $(r_2 - r_1)/2$ for x, the critical particle diameter is

$$d' = \sqrt{\frac{9\mu Q}{(\rho_p - \rho)\omega V}} \times \frac{r_2 - r_1}{r} \qquad (10.88)$$

where $r_2 - r_1$ = thickness of the suspension
d' = critical particle diameter

As noted earlier, if the liquid layer is thick, the variation of R_c over r is great. This may be accounted for by defining an effective value of $(r_2 - r_1)/r$:

$$\left(\frac{r_2 - r_1}{r}\right)_{eff} = 2 ln \frac{r_2}{r_1} \qquad (10.89)$$

This term is introduced into Equation 10.88 when the thickness of the liquid layer is not negligible compared to either r_1 or r_2.

By manipulating the expression for the settling velocity in a gravitational field,

$$u_g = \frac{(\rho_p - \rho)gd^2}{18\mu} \tag{10.90}$$

and solving Equation 10.88 for Q and the gravitational constant, we obtain

$$Q = \frac{(\rho_p - \rho)gd'^2}{9\mu} \times \frac{V\omega^2 r}{g(r_2 - r_1)} = 2u_g\Sigma_c \tag{10.91}$$

in which

$$u_g = \frac{(\rho_p - \rho)gd'^2}{18\mu} \tag{10.92}$$

is the particle terminal settling velocity in a gravitational field and

$$\Sigma_c = \frac{V\omega^2 r}{g(r_2 - r_1)} \tag{10.93}$$

is the capacity factor of the centrifuge itself and not the system being separated.

For a tubular-bowl centrifuge, application of Equation 10.89 gives

$$\Sigma_c = \frac{\pi\omega^2 L}{g} \times \frac{(r_2^2 - r_1^2)}{\ln{(r_2^2/r_1^2)}} \tag{10.94}$$

where   L = bowl length

Table 10.5 gives typical values for several types of centrifuges based both on calculations from geometry considerations and on laboratory and plant data.

The above analysis is somewhat simplified because the actual flow patterns in a centrifuge are complex. Finkelshtein [1962] showed that rapid fluid motion is essentially confined to a surface layer of liquid that is only about 1–2 mm thick. The remainder of the liquid exists in a stagnant "pond," with the exception of any agitation introduced by the use of a conveyor.

Table 10.5 Comparative Performances of Different Centrifuges (Data of Ambler [1951])

	Σ_c Values (ft^2)		
	Calculated from Geometry	From Experimental Data Clarifying Ideal Systems	Extrapolation on Commercial Systems (supercentrifuge tests)
Laboratory supercentrifuge (tubular bowl 1^3/$_4$ in. i.d. × 7^1/$_4$ in. long) operating at			
10,000 rpm	582	582	582[a]
16,000 rpm	1,485	1,485	1,290
23,000 rpm	3,070	3,070	Not used
50,000 rpm	14,520	14,520	
No. 16 supercentrifuge (tubular bowl 4^1/$_8$ in. i.d. × 29 in. long) operating at			
15,000 rpm	27,150	27,150	27,150
No. 2 disk centrifuge, 1^7/$_8$ in. r_1 × 5^3/$_4$ in. r_2 on disks			
52 disks, 35° half angle, 6000 rpm	178,800	98,000	89,400–178,800
50 disks, 45° half angle	134,000	72,600	67,900–134,000
Super-D-Cantor (solid-bowl centrifugal)			
PN-14 (conical bowl), 3250 rpm (D = 14 in.−8 in., L = 23 in.)	4,750	2,950	2,950[a]
PY-14 (cylindrical bowl), 3250 rpm (D = 14 in., L = 23 in.)	8,940	5,980	5,980[a]

[a]For relatively low throughput rates.

Outlet Dam Settings for Liquid-Liquid Separations

When centrifugal operations are applied to liquid-liquid separations, the position of the outlet dam becomes much more of an important consideration than for solid-liquid separations. In the latter case, the outlet dam controls both the volumetric holdup and the critical particle diameter in the centrifuge. However, in liquid-liquid separations, the dam position has a dramatic influence on whether a separation can be made at all. A comparison of the two operations is made in Figure 10.28. The radii defined in this figure are: r_1 = radius to top of light-liquid layer; r_2 = radius to liquid-liquid interface; r_3 = radius to outside edge of dam; and r_4 = radius to surface of heavy-liquid downstream from dam.

Note that the location of the interface is established by an equilibrium state existing between forces derived from the hydraulic heads of the two liquid layers. By means of a force balance,

$$\int_{r_i}^{r_f} dP = \int_{r_i}^{r_f} \frac{d\Upsilon}{dF} = \int_{r_i}^{r_f} \frac{adm}{g_c F} = \int_{r_i}^{r_f} \frac{\rho(\omega^2 r)(2\pi r \ell dr)}{(2\pi r \ell)g_c} \qquad (10.95)$$

where r_i, r_f are the initial and final radii, respectively.

After some simplification and integration, the following is obtained:

$$P = \frac{\rho \omega^2}{2g_c}(r_f^2 - r_i^2) \qquad (10.96)$$

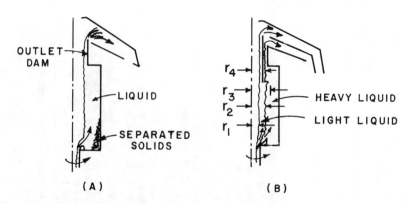

Figure 10.28   Overflow-dam arrangements in tubular-bowl centrifuges: (A) case for clarifying a liquid of entrained solids; (B) arrangement for separating two liquids.

Noting that the pressure on each side of the liquid-liquid interface is the same at r_2 (r_2 is the radius to the two-fluid interface),

$$\frac{\rho_h \omega^2}{2g_c}(r_2^2 - r_4^2) = \frac{\rho_1 \omega^2}{2g_c}(r_2^2 - r_1^2) \tag{10.97A}$$

or

$$\frac{r_2^2 - r_4^2}{r_2^2 - r_1^2} = \frac{\rho_1}{\rho_h} \tag{10.97B}$$

where r_1 = the radius to the top of the light liquid layer,
           r_4 = the radius to the surface of the heavy-liquid down-
                  stream from the dam
    ρ_1 and ρ_h = the densities of the light and heavy fluids, respectively

To achieve separation of the two fluids, the liquid-liquid interface must be located at a radius, r, in which

$$r_4 < r < r_3$$

As is often the case, one of the liquids is generally more difficult to clarify. This is somewhat compensated for by increasing the volume of the difficult phase relative to that of the easily clarified phase. In practice, this is accomplished by adjusting the heights of the two overflow dams in Figure 10.28(B).

Disk-Bowl Centrifuges

Disk-bowl centrifuges are used widely for separating emulsions, clarifying fine suspensions and separating immiscible liquid mixtures. More sophisticated designs can separate immiscible liquid mixtures of different specific gravities while simultaneously removing solids. Figure 10.29 illustrates the physical separation of two liquid components within a stack of disks. The light liquid phase builds up in the inner section, and the heavy phase concentrates in the outer section. The dividing line between the two is referred to as the "separating zone." For the most efficient separation this is located along the line of the rising channels, which are a series of holes in each disk, arranged so that the holes provide vertical channels through the entire disk set. These channels also provide access for the liquid mixture into the spaces between the disks. Centrifugal force causes the two liquids to separate, and the solids move outward to the sediment-holding space.

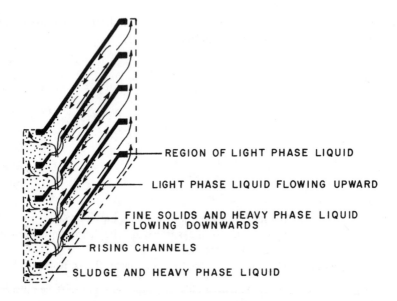

REGION OF LIGHT PHASE LIQUID

LIGHT PHASE LIQUID FLOWING UPWARD

FINE SOLIDS AND HEAVY PHASE LIQUID FLOWING DOWNWARDS

RISING CHANNELS

SLUDGE AND HEAVY PHASE LIQUID

Figure 10.29   The principle of separation using a stack of disks contained in a centrifuge.

The position of the separating zone is controlled by adjusting the back pressure of the discharged liquids or by means of exchangeable ring dams.

Figure 10.30 illustrates the main features of a disk-bowl centrifuge, which includes a bowl (2) with a bottom (13); a central tube (18), the lower part of which has a fixture (16) for disks; a stack of truncated cone disks (17), frequently flanged at the inside and outer diameters to add strength and rigidity; collectors (3 and 4) for the products of separation; and a feed tank (5) with a tube (6). The bowl is mounted to the tube (14) with a guide in the form of a horizontal pin. This arrangement allows the bowl to rotate along with the shaft.

The suspension is supplied from the feed tank (5) through the fixed tube (6), to the central tube (18), which rotates together with the bowl and allows the liquid to descend to the bottom. In the lower part of the bowl, the suspension is subjected to centrifugal force and, thus, directed toward the periphery of the bowl. The distance between adjacent disks is controlled by spacers that usually are radial bars welded to the upper surface of each disk. The suspension may enter the stack at its outside diameter or through a series of vertical channels cut through the disks, as described earlier.

The suspension is lifted up through vertical channels formed by the holes in the disks and distributed simultaneously under the action of centrifugal

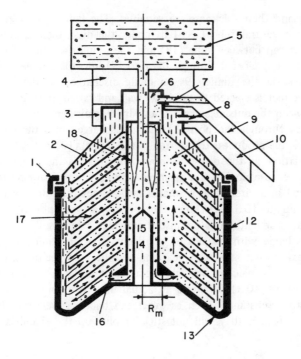

Figure 10.30   Schematic of a disk-bowl centrifuge: (1) ring; (2) bowl; (3 and 4) collectors for products of separation; (5) feed tank; (6) tube; (7 and 8) discharge nozzles; (9 and 10) funnels for collectors; (11) through channels; (12) bowl; (13) bottom; (14) thick-walled tube; (15) hole for guide; (16) disk fixator; (17) disks; (18) central tube.

force into the spacings between the disks (spacings are usually in the range of 0.5 to 3 mm).

Due to a larger diameter, the disk bowl operates at a lower N_c than the tubular design. Its effectiveness depends on the shorter path of particle settling. The maximum distance a particle must travel is the thickness of the spacer divided by the cosine of the angle between the disk wall and the axis of rotation. Spacing between disks must be wide enough to accommodate the liquid flow without promoting turbulence and large enough to allow sedimented solids to slide outward to the grit-holding space without interfering with the flow of liquid in the opposite direction.

The disk angle of inclination (usually in the range of 35 to 50°) generally is small to permit the solid particles to slide along the disks and be directed to the solids-holding volume located outside of the stack. Dispersed particles transfer from one layer to the other; therefore, the concentration in the layers and their thickness are variables. The light component from the spacing near central tube (18) falls under the disk; then it flows through the annular gap between tube (18) and the cylindrical end of the dividing disk, where it is ejected through the port (7) into the circular collector (4) and farther via the funnel (9) on being discharged to the receiver. The heavier product is ejected to the bowl wall and raised upward. It enters the space between the outside surface of the dividing disk and the cone cover (2); then passes through the port (8) and is discharged into the collector (3). From there, the product is transferred to the funnel (10).

One variation of the disk-type bowl centrifuge is the nozzle centrifuge, so named because nozzles are arranged on the periphery or on the bottom of the bowl in a circle that is smaller in diameter than the bowl peripheral diameter. Figure 10.31(A) shows the conceptual operation of such a unit. The design is advantageous because it provides a high solids concentration in the discharge with nozzles of relatively large diameters. As centrifugal force is less in that area than near the periphery of the bowl, the concentrated solids are ejected through the nozzles under a comparatively low pressure. Figure 10.31(B) shows an actual unit being employed as a yeast concentrator. Substances such as yeast and bacteria are very slippery and slide easily; hence, they will not stick or plug up the channels leading to the nozzles.

Analysis of Disk Centrifuge Operations

Now attention will be directed toward analyzing the controlling mechanisms for particle separation from a heterogeneous mixture in the spaces between the disks. Figure 10.32 is a vector diagram of a particle moving in this region. The mixture enters the vertical channel at section $A_1 A_2$ and is displaced farther down along the generatrixes of disks A_1, B_1 and A_2, B_2. The mixture then is discharged from the gap downward at section $B_1 B_2$.

The design equation for the capacity of this type centrifuge will be developed based on the following assumptions:

1. The flow in each disk space is laminar, i.e., there is no mixing between the layers.
2. Because a disk centrifuge usually separates emulsions or fine suspensions, the sedimentation process may be assumed to take place in the laminar regime and, therefore, the settling law is determined from the following equation:

$$w_s = \frac{d^2}{18\mu} \rho a$$

3. All spaces between disks are assumed to be filled with liquid moving in one direction.

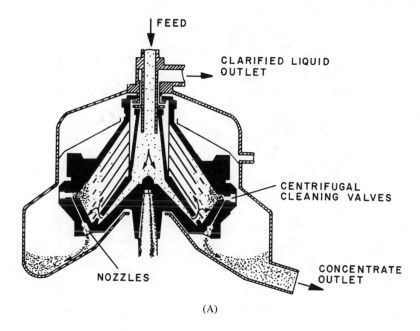

FEED

CLARIFIED LIQUID
OUTLET

CENTRIFUGAL
CLEANING VALVES

NOZZLES

CONCENTRATE
OUTLET

(A)

(B)

Figure 10.31   (A) Cutaway view of a disk-type bowl centrifuge fitted with nozzles on
its underside for concentrate discharge; (B) the same unit applicable as a yeast concentrator (courtesy Centrico Inc., Northvale, NJ).

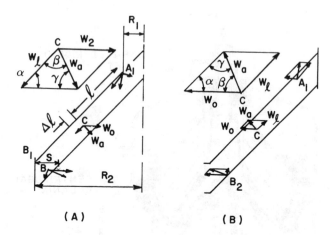

Figure 10.32   Triangle of vector velocities of a particle in the space between adjacent disks: (A) particle has density lighter than liquid; (B) particle has density heavier than liquid.

The radius of the annular gap between disks increases in the downward direction. Therefore, the flow area increases, resulting in a decrease in fluid velocity from w_{ϱ_1} at the inlet to w_{ϱ_2} at the outlet..

The velocity of the flow at a distance R from the axis of rotation is

$$w_\varrho = \frac{V_{sec}}{2\pi R \delta z} \tag{10.98}$$

where V_{sec} = centrifuge capacity
    z = number of disks
    δ = space between two adjacent disks

w_ϱ is a drift velocity, whereas the settling velocity is a relative velocity for the dispersed phase. Both velocities are unidirectional, but their magnitudes vary. This means that the absolute velocity of a particle, w_a, as a geometric sum is a variable by direction and magnitude. For a given set of conditions, $a = \omega^2 R$, we denote the constant complexes in the two velocity equations as follows:

$$b_1 = \frac{d^2(\rho_p - \rho)}{18\mu} \omega^2 \tag{10.99A}$$

$$b_2 = \frac{V_{sec}}{2\pi\delta z}$$ (10.99B)

Thus,

$$w_0 = b_1\omega^2$$ (10.100)

$$w_\varrho = \frac{b_2}{R}$$ (10.101)

and the ratio of these velocities is

$$\frac{w_0}{w_\varrho} = \frac{b_1}{b_2} R^2$$ (10.102)

Applying the sine theorem to the velocity vectors shown in Figure 10.32,

$$\frac{w_0}{w_\varrho} = \frac{\sin\beta}{\sin\gamma}$$ (10.103)

Consequently,

$$\frac{\sin\beta}{\sin\gamma} = \frac{b_1}{b_2} R^2$$ (10.104)

At a constant b_1/b_2 ratio, but increasing radius, the ratio of the sines increases. At constant angle α, β may be increased, but only after decreasing angle γ (Figure 10.32). The resulting velocity vector rotates about a pivot point C.

Figure 10.32(A) shows three positions of the velocity vector summation for the separation of a particle having a density less than that of the liquid. In separating particles with densities heavier than the liquid, the process is similar but with particle motion in the opposite direction, as shown in Figure 10.32(B).

For a unit with z disks and a bowl operating at an angular velocity ω, the capacity expression is developed as follows: the projection of an elementary sedimentation surface in the direction perpendicular to the settling velocity vector shown in Figure 10.32(A) is

$$dF_0 = 2\pi Rd\varrho z \sin\alpha$$ (10.105)

The elemental capacity is

$$dV_{sec} = w_0 dF_0 = 2\pi Rd\ell zw_0 \sin\alpha \qquad (10.106)$$

Substituting for w_0 (Equation 10.100), and expressing $d\ell = dR/\cos\alpha$, we obtain

$$dV_{sec} = 2\pi zb_1 R^2 dR \tan\alpha \qquad (10.107)$$

Integrating over the limits of 0 to V_{sec} and from R_2 to R_1, and substituting for b_1, the capacity expression is obtained:

$$V_{sec} = \frac{2\pi\omega^2 \tan\alpha(\rho_p - \rho)(R_2^3 - R_1^3)zd}{3 \times 18\mu} \qquad (10.108)$$

Denoting all the values characterizing the heterogeneous system and constants by one coefficient,

$$K' = \frac{2\pi d^2(\rho_p - \rho)}{54\mu} \qquad (10.109)$$

Thus, the final expression for centrifuge capacity is

$$V_{sec} = K'\omega^2(R_2^3 - R_1^3)z \tan\alpha \qquad (10.110)$$

As this expression is based on several idealized conditions, an efficiency coefficient must be included to correct for the actual capacity:

$$V_{sec} = \theta K'\omega^2(R_2^3 - R_1^3)z \tan\alpha \qquad (10.111)$$

Coefficient θ depends on the size and quality of the disks, the concentration of the suspension and products of separation, the suspension's physical properties, the rpm of the rotor, etc.

The concentration of separated products depends on the distance of the discharge ports from the axis of rotation. The internal space of a bowl may be considered as two communicating vessels filled with two liquids of different densities. The dam is the dividing disk. This becomes more obvious by rotating Figure 10.30 by 90° to the right and replacing the centrifugal field by gravity. Denoting the radii of the free surfaces of the light and heavy products by R_e and R_h, respectively, and distinguishing the respective densities of each product by the same subscripts, we may write the following equality of pressures on the interface:

$$\frac{\omega^2 \rho_\ell (R_b^2 - R_\ell^2)}{2} = \frac{\omega^2 \rho_h (R_b^2 - R_h^2)}{2} \tag{10.112}$$

Hence,

$$\frac{\rho_h}{\rho_\ell} = \frac{R_b^2 - R_\ell^2}{R_b^2 - R_h^2} \tag{10.113}$$

where R_b = radius of the bowl

At $R_\ell < R_h$,

$$R_b^2 - R_\ell^2 > R_b^2 - R_h^2$$

Consequently,

$$\rho_h > \rho_\ell$$

As follows from this analysis, the greater the difference between R_ℓ and R_h, the greater the difference between concentrations; the smaller the differences in radii, the smaller the density difference. Thus, if $R = R_h$, then $\rho_h = \rho_\ell$, i.e., the separation of components is impossible in this case.

To change the radii of the product free surfaces, which is a means of controlling concentrations, adjustable nozzles (7) and (8) (Figure 10.30) are provided at different radial positions.

In comparison to gravity sedimentation equipment, the disk-bowl centrifuge has the following advantages:

1.   It can be operated at speeds up to 10,000 rpm, producing a centrifugal force roughly 14,000 times greater than gravity. Handling capacities can exceed 30,000 gph. Hence, the settling velocity in a disk centrifuge is thousands of times higher than that in a gravitational field. For example, a milk separator with rotor radii R_2 = 64 and R_1 = 150 mm at n = 6000 rotations per minute provides a settling velocity 4000 times greater than in a sedimentation tank operating under identical loading conditions.
2.   The surface area for sedimentation in a disk-bowl centrifuge is large in comparison to the volume requirements of the unit due to the large number of disks often used. This large settling surface allows for a higher capacity over simple settlers.

Power Requirements for Centrifugation

The overall power requirements for a centrifuge are based on five contributions:

1. power needed to start the rotor;
2. power to set the material in the bowl in motion;
3. power required to overcome friction;
4. power required to transfer the sediment to discharge ports or nozzles; and
5. power required to discharge products from the bowl.

Power Required for Startup of a Rotor

First we will examine the power needed to start up the centrifuge rotor without any material. If the centrifuge bowl is started from rest to some angular velocity, w, over starting time, τ_s, the energy spent is

$$E = \frac{1}{2} Jw^2 \qquad (10.114)$$

where J is the inertia moment, i.e., the sum of mR^2 for the specific bowl under consideration.

If the cylindrical wall of the bowl is considered (that is, neglecting the other parts of the bowl, as shown in Figure 10.33, then as a first approximation the moment of inertia is

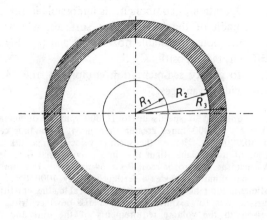

Figure 10.33   Cylindrical part of a bowl.

$$J = \int_{R_2}^{R_3} R^2 dm = \int_{R_2}^{R_3} R^2 2\pi R \ell \rho_b dR = \frac{\pi \ell \rho_b}{2} (R_3^4 - R_2^4) \qquad (10.115)$$

Substituting this expression into Equation 10.114 for J, we obtain

$$E = \frac{\pi \ell \rho_b}{4} (R_3^4 - R_2^4)\omega^2 \qquad (10.116)$$

Hence, the required horsepower for startup is

$$\mathcal{W}_{s.b.} = 10^{-3} \frac{\pi \ell \rho_b (R_3^4 - R_2^4)\omega^2}{\tau_s} \qquad (10.117)$$

Example 10.4

A steel bowl (ρ_b = 7800 kg/m^3) has the following characteristics:

- R_3 = 0.32 m
- R_2 = 0.30 m
- ℓ = 0.30 m

ω = 523 rad/s (or 5000 rpm). Determine the required horsepower to set the centrifuge rotor in motion.

Solution

The energy spent is

$$E = \frac{\pi \times 0.3 \times 7800}{4} (0.32^4 - 0.30^4) \times 523^2$$

If the required startup time is 600 s, then the horsepower is

$$\mathcal{W} = \frac{1.2 \times 10^6}{600} = 2000 \text{ W}$$

Starting Up the Material in the Bowl

We will direct our attention now to the moment of inertia of the material in the bowl. If the density of the material is ρ_m, then the mass of an elemental ring at distance R from the axis of rotation is $2\pi R dR H \rho_m$ (Figure 10.33), and the inertia moment of the material is

$$J_m = 2\pi H \rho_m \int_{R_m}^{R_b} R^3 dR = \frac{1}{2} \pi (R_b^4 - R_m^4) H \rho_m \qquad (10.118)$$

Note that

$$(R_b^4 - R_m^4) = (R_b^2 + R_m^2)(R_b^2 - R_m^2) \text{ and } \pi(R_b^2 - R_m^2)H\rho_m = M_m$$

where M_m = mass of the material

Hence, the expression may be rewritten as

$$J_m = \frac{1}{2} M_m (R_b^2 + R_m^2) \qquad (10.119)$$

Substituting into Equation 10.114 for J_m and dividing through by τ_s, we obtain

$$\mathscr{W}_s = \frac{1}{4} \times 10^{-3} \omega^2 \frac{M_m(R_b^2 + R_m^2)}{\tau_s} , kW \qquad (10.120)$$

Power Requirements Imposed by Friction

Bearings, belts, gears and seals transform small amounts of power into heat. The greatest loss of power is derived from the friction of the gas in contact with the surface of the rotating elements of the centrifuge, referred to as "windage." This value is determined as follows:

$$\mathscr{W}_f = \dot{F}_f w \qquad (10.121)$$

where the friction force is defined as

$$\dot{F}_f = P\lambda_f \qquad (10.122)$$

where P = force pressing the body to the surface
λ_f = friction coefficient
w = the linear velocity of the body relative to the surface considered

In rotational motion, the linear velocity is

$$w = \omega R = \frac{1}{2} \omega D$$

Consequently, the power required to overcome friction is

$$\mathcal{N}_f = \frac{1}{2} \times 10^{-3} P \lambda_f \omega D \qquad (10.123)$$

Let us now consider individual contributions.

Friction in Bearings. Force P is the sum of the weight of the rotating parts of the centrifuge (mostly the bowl), $M_b g$, and the weight of material, $M_m g$. The horsepower expended in overcoming the friction in bearings is

$$\mathcal{N}_{f.b.} = \frac{1}{2} \times 10^{-3} (M_b + M_m) g \omega D_b \lambda_b, \, kW \qquad (10.124)$$

where D_b = diameter of the shaft neck
λ_b = friction coefficient in the shaft journal

For ball bearings, λ_b is typically 0.02 to 0.03.

Friction in the Glands. In a horizontal, hermetically closed centrifuge, in which the shaft is clamped in seals, an additional friction force develops:

$$\mathcal{N}_{f.g.} = \frac{1}{2} \times 10^{-3} (M_b + M_m) g \omega D_s \lambda_s \qquad (10.125)$$

where D_s and λ_s are shaft diameter and friction, respectively.
The friction coefficient in the gland assembly typically ranges from 0.4 to 0.6.

Friction of Material Against the Conical Bowl of a Centrifuge. This is illustrated by the vector diagram in Figure 10.34. M_0 denotes the amount of sediment, in kg, transported along the generatrix, L, of the bowl in a unit of time, s. The material is loaded at approximately the middle of the bowl and, under the action of centrifugal force, is distributed uniformly over the entire surface of the bowl from point D along the generatrix to points A and B.
The sediment at point B is discharged immediately through the ports, whereas sediment close to point A must travel along the total length of generatrix L. Consequently, if an average path of travel for the sediment, ½ L, is considered, the work against the friction force is ½ $\dot{F}_b$ L.
To evaluate the friction force, the centrifugal force, R_c, is resolved about point D (in the middle of the bowl with an average radius $R_{avg} = (R_b + R_m)/2$)

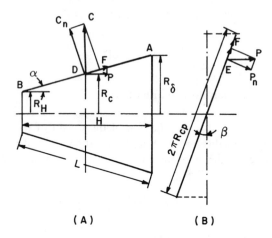

Figure 10.34   Sketch for determining the friction force imposed on the material against the bowl.

into two components: R_{c_n} normal to the bowl surface, and R_{c_t}, tangent to the bowl surface:

$$R_{c_n} = R_c \cos\alpha = M_0 \omega^2 R_{avg} \cos\alpha \qquad (10.126)$$

$$R_{c_t} = R_c \sin\alpha = M_0 \omega^2 R_{avg} \sin\alpha \qquad (10.127)$$

Multiplying the normal component, R_{c_n}, by a friction coefficient, λ_b, the friction force is obtained:

$$\dot{F}_b = M_0 \omega^2 R_{avg} \lambda_b \cos\alpha \qquad (10.128)$$

As M_0 has units of kg/s, the required horsepower is

$$\mathscr{H}_{f.B.} = \frac{1}{2} M_0 \omega^2 R_{avg} L \cos\alpha \lambda_b \qquad (10.129A)$$

and as $L = H/\cos\alpha$,

$$\mathscr{H}_{f.B.} = \frac{1}{2} \times 10^{-3} M_0 \omega^2 R_{avg} H \lambda_b, \text{ kW} \qquad (10.129B)$$

Friction Against the Screw Conveyor. According to Figure 10.34(A), the sediment is transported by a screw conveyor from right to left. Let β be the angle between the fillet of the screw conveyor and the vertical axis (Figure 10.34(B)) and z be the number of fillets. The length of one fillet of the screw conveyor is $2\pi R_{avg}/\cos\beta$, and the total length of the fillet is $2\pi R_{avg}z/\cos\beta$, where R_{avg} is an average radius. The force per unit mass of material acting on the surface of the screw along the axis of the bowl is

$$P = (R_{c_t} + F_b)\cos\alpha = M_0\omega^2 R_{avg}(\sin\alpha\,\cos\alpha + \cos^2\alpha\lambda_b) \qquad (10.130)$$

Resolving the force acting on the surface of the screw (Figure 10.34(B)) into components tangent P_t and normal P_n, and multiplying the last component by the friction coefficient λ_s between the screw and sediment, we obtain the friction force acting against the screw:

$$\dot{F}_s = M_0\omega^2 R_{avg}(\sin\alpha\,\cos\alpha + \lambda_b\cos^2\alpha)\lambda_s\cos\beta \qquad (10.131)$$

Multiplying this force by half the length of the screw, an expression for the power consumed for friction against the screw conveyor is

$$\mathcal{W}_{f.s.} = \pi\omega^2 R_{avg}^2\, zM_0(\sin\alpha\,\cos\alpha + \lambda_b\cos^2\alpha)\lambda_s \qquad (10.132)$$

or, finally,

$$\mathcal{W}_{f.s.} = \frac{1}{2}\times 10^{-3}\pi\omega^2 R_{avg}^2 M_0 z(\sin2\alpha + 2\lambda_b\cos^2\alpha)\lambda_s \qquad (10.133)$$

where $\mathcal{W}_{f.s.}$ is in units of kW.

Friction Against Air (Windage). To determine the power lost in windage, dimensional analysis is applied. The power lost due to windage is

$$\mathcal{W}_{f.w.} = Aw^x\rho^y D_{avg}^z\mu^\alpha H \qquad (10.134)$$

where D_{avg} = average diameter of the bowl
ρ = density of the medium
w = linear velocity of the bowl
H = height of the bowl

Writing this equation in terms of its dimensions,

$$kg\times m^2\times s^{-3} = (m\times s^{-1})^x\times(kg\times m^{-3})^y\times(kg\times m^{-1}\times s^{-1})^\alpha\times m^z\times m$$

whence,

$$-x - \alpha = -3 \; ; \quad x = 3 - \alpha$$

$$y + \alpha = 1 \; ; \quad y = 1 - \alpha$$

$$x - 3y + z - \alpha + 1 = 2 \; ; \quad z = 1 - \alpha$$

Equation 10.134 will take the following form:

$$\mathscr{W}_{f.w.} = A w^{3-\alpha} \rho^{1-\alpha} D_{avg}^{1-\alpha} \mu^{\alpha} H \tag{10.135}$$

Recognizing the Reynolds number, we obtain

$$\mathscr{W}_{f.w.} = A \left(\frac{w D_{avg} \rho}{\mu} \right)^{-\alpha} w^3 \rho D_{avg} H \tag{10.136}$$

as $A Re^{-\alpha} = C_D$,

$$\mathscr{W}_{f.w.} = C_D \rho D_{avg} H w^3 \tag{10.137}$$

As the density of the air is unlikely to vary significantly, the product of ρ and the drag coefficient may be represented by one coefficient. Expressing the linear rotational velocity in terms of angular velocity, ω, and average diameter, D_{avg}, we obtain the formula for the horsepower needed to overcome the resistance of the medium to the rotation of a bowl:

$$\mathscr{W}_{f.w.} = \xi \omega^3 D_{avg}^4 H \; ; \quad kW \tag{10.138}$$

Coefficient ξ contains the value $1/2$ as $w = 1/2 \; \omega D_{avg}$, along with a conversion from watts to kilowatts. From experiments, $\xi = 1.15 \times 10^{-6}$.

The power lost in windage can exceed that required to accelerate the feed, especially in a high-pressure operation. For instance, a rotor 30 cm in diameter rotating in air at 10,000 rpm at atmospheric pressure requires about 2 kW to overcome the windage. At 150 psi, the power requirement increases to 20 kW.

Power Requirements for Sediment Transportation to the Discharge Ports Using a Screw Conveyor

In the sediment displacement from point A to point B (Figure 10.34(A)), the kinetic energy of the sediment decreases. At point A there is a greater radius and, hence, the potential of the centrifugal field is greater than at point B, where the radius is smaller. Therefore, a certain amount of power is expended. In addition, energy is lost in overcoming friction forces associated with sediment against the bowl wall and screw conveyor. The power for the transportation of sediment M_0 (kg/s) along half the length of the generatrix ($\frac{1}{2}$ L) is:

$$\mathscr{N}_0 = \frac{1}{2} M_0 \omega^2 R_{avg}^2 L \; ; \quad W \qquad (10.139)$$

Power Requirements for the Discharge of Products from the Bowl

The products of separation—sediment and centrifuged effluent—entrain a certain amount of energy on leaving the bowl. The power requirement for the discharge of products is

$$\mathscr{N}_d = \frac{1}{2} \times 10^{-3} \omega^2 (M_0 R_0^2 M_e R_e^2) \; , \quad kW \qquad (10.140)$$

where M_0 = mass flow of sediment (kg/s)
               M_e = mass flow of effluent (kg/s)
       R_0 and R_e = radii of rotation of sediment before and after leaving the bowl, respectively

We now may evaluate the *total power requirements for a sedimentation centrifuge*. As most operations are continuous, startup power requirements often are ignored. The total power required to operate a centrifuge is the sum of the individual power contributions evaluated above:

$$\mathscr{N} = \mathscr{N}_d + \mathscr{N}_0 + \mathscr{N}_{f.b.} + \mathscr{N}_{f.s.} + \mathscr{N}_{f.g.} + \mathscr{N}_{f.B.} + \mathscr{N}_{f.w.} \qquad (10.141)$$

Additional minor energy requirements include that for initial mixing of the suspension for overcoming hydraulic resistances, for ventilation, and so on. For practical calculations, therefore, the computed total power requirements should be increased slightly.

In contrast, for a high-speed or high-inertia centrifuge, the minimum required motor size often is determined by the time needed to bring the centrifuge to full speed, rather than by the power consumed under normal operating conditions. The energy required to start the motor is in considerable excess of the value computed by Equation 10.114. Only about 20% of the energy conducted to the motor is transferred to the centrifuge rotor before the operating speed is reached. Even with special measures, such as a star-delta or depressed torque motors or clutch drives, the frequency of startups must be limited.

Analysis of Stresses in Centrifuge Walls

Although the subject of strength of materials is separate from that of unit operations, it is appropriate to evaluate the stresses in the wall of a centrifuge bowl as they may impose limitations on both the rpm and size of the unit.

The centrifugal rotating force acting on the walls of a revolving bowl may be viewed as a fictitious internal pressure. Consider an elemental section of the bowl along with the contained liquid of area dF. The system is illustrated in Figure 10.35. The centrifuge radius is R and the thickness of the wall is δ. Then the mass of the wall section is

$$dm = \frac{\gamma}{g} \times \delta \times dF \qquad (10.142)$$

where δ = specific gravity of the wall material

Similarly, the mass of liquid acting on this wall section is

$$dm_f = \frac{\gamma_f}{g} \delta_f dF_f \qquad (10.143)$$

where dF_f = surface across the middle of the liquid volume slice.

From geometric considerations, an approximate formula for this surface is

$$dF_f = dF \left(\frac{r}{R}\right) \qquad (10.144)$$

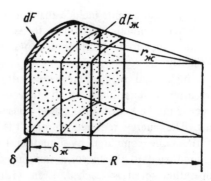

Figure 10.35   Elemental slice of rotating bowl wall and liquid.

and, hence, the liquid mass is

$$dm_f = \frac{\gamma_f}{g} \delta_f \left(\frac{r}{R}\right) dF \tag{10.145}$$

where γ_f = specific weight of the liquid
δ_f = thickness of the liquid layer

The centrifugal force acting on the wall element is

$$dC = \frac{\gamma}{g} \delta \frac{u^2}{R} dF \tag{10.146}$$

where   u = peripheral velocity of the bowl

Similarly, we derive the equation for the centrifugal force due to the liquid mass. Because the peripheral bowl velocity is equal to that of the liquid, we may write

$$\omega = \frac{u}{R} = \frac{u_f}{r} \tag{10.147}$$

Consequently, the centrifugal force due to the liquid is

$$dC_\varrho = \frac{\gamma_f}{g} \delta_f \frac{u^2}{R} \left(\frac{r}{R}\right)^2 dF \tag{10.148}$$

In the final analysis, the wall material must withstand the sum of all the forces, that is

$$dC_\varrho = \frac{u^2}{R} \frac{\gamma}{g} \left[\delta + \delta_f \frac{\gamma_f}{\gamma} \left(\frac{r}{R}\right)^2 \right] dF \tag{10.149}$$

The pressuring acting inside the bowl will be equal to the ratio of the force and surface, dC_ϱ/dF:

$$P = \frac{u^2}{R} \frac{\gamma}{g} \left[\delta + \delta_f \frac{\gamma_f}{\gamma} \left(\frac{r}{R}\right)^2 \right] \tag{10.150}$$

Now consider only a half section of the bowl (Figure 10.36). The force acting on the bowl will be equal to the product of pressure, P, and the area of the section:

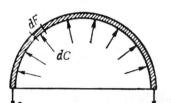

Figure 10.36   Half section of bowl.

$$T = (h \times 2R)p \tag{10.151}$$

where   h = height of the bowl

Here we consider the area of the section, but not the internal surface of half a ring, because the components of forces, dC, directed parallel to the plane of the section are compensating. Force T is balanced by the stresses in the cylinder shell. The product of this stress and the sectional area is equal to force T.

The section of the solid bowl on which the stress σ is acting is $h\sigma$; consequently, the following force acts on both halves of the bowl:

$$T = 2h\sigma\delta \tag{10.152}$$

As the internal force in the wall balances the centrifugal force tending to break up the wall, we equate Equations 10.151 and 10.152 to obtain

$$\sigma = \frac{RP}{\delta} \tag{10.153}$$

Taking into account Equation 10.150, we obtain the final expression for evaluating the stresses in the wall of the bowl:

$$\sigma = \frac{u^2\gamma}{g}\left[1 + \frac{\delta_f}{\delta}\frac{\gamma_f}{\gamma}\left(\frac{r}{R}\right)^2\right] \tag{10.154}$$

This equation shows that stress is proportional to $(nR)^2$. Consequently, for a given bowl material the product nR must be within defined limits. Recall that the settling velocity in centrifuges is proportional to the product n^2R. Hence, although it is most efficient to design for the highest rpm, from the point of view of the strengths of materials a maximum rpm limits the design basis. For high-speed centrifuges, this consideration also limits usage to small rotor radii.

Example 10.5

Particles with diameters greater than 3 μm and specific gravity γ_p = 1.10 g/cm^3 are to be separated from an aqueous suspension by centrifugation. The centrifuge is to be half-filled with the suspension. The radius of the bowl is 0.4 m. Determine the rpm required for the separation of these particles in 10 min. What is the stress in the bowl wall if its thickness is 4 mm?

Solution

We may assume the particle settling to be laminar and therefore can determine the sedimentation velocity in a relatively straightforward manner:

$$\frac{dr}{d\tau} = \frac{2\pi^2(\gamma_p - \gamma)}{9\mu g} d^2 n^2 r = \frac{2 \times 3.14^2(1100 - 100)(3 \times 10^{-6})^2 n^2 r}{9 \times 10^{-3}}$$

or

$$\frac{dr}{d\tau} = 1.98 \times 10^{-6} n^2 r$$

Integrating this equation over the limits from r to R, an expression for the settling time is obtained:

$$\tau = \frac{ln\left(\frac{R}{r}\right)}{1.98 \times 10^{-6} n^2}$$

If the centrifuge is half-filled with liquid, then

$$\pi R^2 h = 2(\pi R^2 - \pi r^2)h$$

Hence,

$$\frac{R}{r} = \sqrt{2}$$

Time, τ, is equal to 10 min, or 600 s. We then obtain

$$n = \sqrt{\frac{\ln \sqrt{2}}{1.98 \times 10^{-6} \times 600}} = 17 \text{ s}^{-1}$$

or

$$n = 17 \times 60 = 1020 \text{ rpm}$$

We check to see whether the particle motion is laminar. The maximum settling velocity is

$$w_s = 1.98 \times 10^{-6} \times 17^2 \times 0.4 = 2.3 \times 10^{-4} \text{ m/s}$$

and the Reynolds number is

$$\text{Re} = \frac{(2.3 \times 10^{-4})(3 \times 10^{-6}) \times 10^3}{15^3} = 7 \times 10^{-4} < 2$$

Thus, the assumed regime is laminar.

To determine the stresses in the bowl wall, the peripheral velocity is computed first:

$$u = 2\pi Rn = 2 \times 3.14 \times 0.4 \times 17 = 42.5 \text{ m/s}$$

The thickness of the liquid layer is obtained from the ratio $R/r = \sqrt{2}$:

$$\delta = (R - r) = 0.4\left(1 - \frac{1}{\sqrt{2}}\right) = 0.116 \text{ m}$$

The radius corresponding to half this thickness is

$$r = 0.4 - \frac{0.116}{2} = 0.34 \text{ m}$$

Hence, using Equation 10.154, the stress in the bowl wall is

$$\sigma = \frac{42.5^2 \times 7000}{1 \times 9.81}\left[1 + \frac{0.116}{0.004} \times \frac{1000}{7000}\left(\frac{0.34}{0.40}\right)^2\right] = 5.3 \times 10^6 \text{ kg/m}^2$$

or

$$\sigma = 530 \text{ kg/cm}^2$$

Continuing discussions, recall that settling velocity varies either as $\mathcal{v}_c = \omega^2 R$ or $\sqrt{N_c} = \sqrt{\omega^2 R}$, depending on the flow regime. Therefore, to increase the settling velocity it is more effective to increase the rpm rather than the bowl radius.

On the other hand, the stress in the bowl metal varies as $\omega^2 R^2$ and, obviously, an increase in the speed results in an increase in stress. Thus, an increase in the sedimentation velocity, ν_c, due to an increase in rpm produces less stress in the metal than does an increase in the radius.

For a laminar system we have

$$\nu_c = k\omega^2 R \qquad (10.155)$$
$$\sigma = k'\omega^2 R^2 \qquad (10.156)$$

And, as follows,

$$\sigma = k''R\nu_c \qquad (10.157)$$

Thus, the stress in the bowl metal varies simultaneously with ν_c and R.

For different centrifuges having the same ν_c, the stress in the metal is directly proportional to the radius. The same may be said for the turbulent regime:

$$\sigma = k'''R\nu_c^2 \qquad (10.158)$$

Additional references on centrifuge design and operation are given at the end of this chapter (see for example Flood [1955], Dufour [1956], Block [1931], Flowers [1950], Ambler [1952], Finkelshtein [1962]). For discussions on strengths of materials see the work of Azbel and Cheremisinoff [1982].

SELECTING SEPARATION EQUIPMENT

Before selecting the type of equipment needed for a given separation problem, the sedimentation velocity of the discrete phase must be determined either experimentally or by the calculations outlined in this chapter and previous ones. Based on this parameter and the volumetric flows to be handled, the residence time and capacity of the unit may be approximated.

To estimate sedimentation velocity, information on the physical properties of the system (particle or droplet sizes, densities, viscosity) and the equipment characteristics (e.g., rpm, characteristic size, etc.) must be known. It is always advisable to confirm estimates with laboratory data.

In all cases, after calculations and bench-scale experiments have been made, a check should be made on the actual sedimentation rate, hopefully in a geometrically similar machine. Such a confirmation is critical because it provides indicators of the negative influence of turbulence and the positive influence of internals (such as disks in a centrifuge).

For centrifuges that are to be applied to processing liquids containing solids and that must serve as clarifiers, the volume fraction of sediment in the liquid to be cleared must be determined. If this fraction approaches 1%, it generally is necessary to provide for pretreatment by a decanter or via hydroclones, or by centrifugation with continuous withdrawal of the sediment. In this case, equipment selection depends on the relative cost of processing liquid and sediment, the sedimentation velocity necessary for obtaining an acceptable capacity and in the required form of the sediment, its properties and the liquid capacity to be treated.

In practice, a clarifier must perform the separation within the same time as a separator. Often, it is necessary to obtain from the light liquid phase, the heavy phase and the sediment, which generally is of little or no process value. If, for example, one must process a mixture of oil and water containing an amount of sediment having no value, a conical machine often is selected when the amount of sediment is large. If only small amounts of sediment are to be processed, it may be possible to achieve a preliminary treatment in a hydroclone or in a large-diameter centrifuge. A final clarification then can be performed in a more powerful centrifuge or in a disk centrifuge. The selection of the specific machine depends on the amount of liquid to be processed and the desired settling velocity, w_s.

If the sediment is valuable and the liquid is a waste product, the equipment selected should produce a concentrated slurry, which may be processed further by filtration or direct drying. For preliminary sedimentation it is also possible to employ hydroclones. In some cases an incomplete clarification is desirable, for example, in the treatment of wastewater in large-diameter centrifuges with lower port discharges or automatic discharge. Products such as this can be processed by a preliminary separating stage to remove the large-sized solids. Examples of precleaning devices are gravity settlers, slow-speed centrifuges of large diameters, hydroclones or, in some cases, screening devices. In general, the efficiency of a centrifuge almost always depends on the concentration of the small particles that are settled, and all calculations made on the basis of gravitational sedimentation are valid for the final phase of the separation. An essential part of the operation of separators is the discharge level. In centrifuges designed for separating two liquids, the pureness of each fluid depends on this level.

For a very stable emulsion, one with very fine globules, centrifugal separation is often impossible. In this case, it is necessary to find a method of coalescence to partially break the emulsion or to make it less stable through moderate mixing, steam injection, neutralization of electrical charges, or by inactivation with a demulsifier. Methods of breaking emulsions are very different and often empirical. For example, the injection of steam and agitation can have very different and sometimes opposite effects, depending on the actual conditions. Emulsions first can be broken through pH changes or by adding certain surface-active agents. If the demulsifying agent is a product of a nonionic substance (e.g., an oxide of ethylene) it is possible to achieve separation by an increase in the emulsion temperature.

For the processing liquids that change in composition, centrifugation should be performed in hermetically sealed machines prior to pressurization.

Gas cyclones are widely used for the separation of particles in the size range of 25 μm to 1 mm. For particles larger than 1 mm in size, simple gravity settlers are employed.

For dusts with particles ranging from 1 to 25 μm, cyclones in series, or multiclones are used. For high removal efficiencies of small particles, filters and/or electrostatic precipitators are preferred.

Hydroclones greatly increase the sedimentation velocity of particles up to 50 μm in size. They may be used also for classification in cases in which the velocities of gravitational decantation are almost the same.

PRACTICE PROBLEMS

10.1 An airstream contains particles with a 50 wt% size of 110 μm and a density of 59 lb_m/ft^3. The mixture enters a cyclone at a linear velocity of 70 fps and at a temperature of 77°F. The cyclone diameter is 36 in., and the outlet diameter is 9 in. The barrel diameter corresponding to the centerline of the inlet is 24 in. Both the gas and solid exit pipes are 7 in. in diameter. Determine the terminal velocity of the particles at the radius of the outlet connection.

10.2 Particles suspended in an airstream have a 50 wt% particle size of 70 μm with a standard deviation of sizes of 0.158. The solids enter the above described cyclone at a solids loading of 0.01 gr/acfm and an inlet gas velocity of 65 fps. Write a computer program to calculate the fractional efficiency of the cyclone as a function of particle size. Present the results graphically.

10.3 Four identical cyclones are fed by a common plenum. Each cyclone has the following dimensions:

- Gas outlet diameter, 4.0 ft
- Diplet diameter, 1.5 ft
- Barrel diameter, 8 ft
- Cone height, 13 ft

- Barrel height, 9.7 ft
- Inlet width, 3.7 ft
- Inlet height, 3.7 ft

Total gas rate to the cyclone is 2.6×10^5 acfm and total solids loading is 6.5 lb_m/ft^2-s. The physical properties of the gas and solids are: gas viscosity = 0.020 cp, particle density = 80 lb/ft^3, gas density = 0.07 lb/ft^3. The particle size distribution for the solids is given in Figure 10.37. Develop a computer program that provides the following information: the amount of solids escaping and collected (in lb/hr) as a function of cumulative wt% less than stated particle size. What is the overall removal efficiency of this system?

10.4 Estimate the pressure drop across one of the cyclones described in the previous problem.

10.5 (a) For one of the cyclones described in Problem 10.3, calculate the average residence time for the 50 wt% particle size.
(b) The average residence time for the gas in one of the cyclones is 0.1 min. Determine the minimum size particle that can be collected.

10.6 A hydroclone is to be used to separate out sand and grit particles from process cooling water that will be recycled to the plant. The unit has the same relative proportions of the cyclone unit shown in Figure 10.12, with a body diameter, D_c, of 32 in. The average temperature of the water is 85°F, and the specific gravity of the solids is approximately 2.2. The volumetric flow to the hydroclone is 300 gpm with a 7% by weight solids concentration. The average size of the particles is 300 μm.
(a) Determine the overall efficiency of the hydroclone.
(b) Determine the minimum size pump (hp) required.
(c) If process requirements demand that the return water only contain 1% by weight solids, will additional units be necessary? If so, size the additional units.

10.7 For problem 10.6(c), size a simple settler to perform the same degree of separation.

10.8 A clarifying centrifuge is operating under the following set of conditions:

- Bowl diameter, 750 mm
- Depth of bowl, 600 m
- Speed, 1500 rpm
- Thickness of liquid layer, 95 mm
- Specific gravity of liquid, 1.5

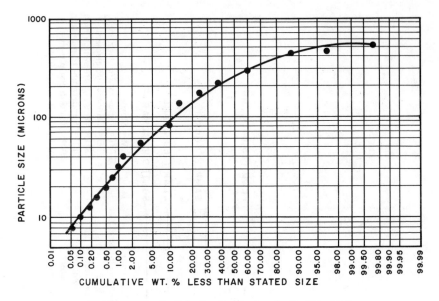

Figure 10.37   Particle size distribution for Problem 10.3.

- Specific gravity of solids, 1.8
- Cut size of particles, 40 μm
- Viscosity of liquid, 15 cp

Determine the capacity of the centrifuge in gpm.

10.9 (a) For the previous problem, determine the required horsepower to operate the unit.

(b) If the wall thickness of the bowl is 8 mm, determine the stresses that develop at the bowl wall.

10.10 The process flowsheet for separating titanium dioxide particles from a water slurry is shown in Figure 10.38. The centrifuge is a continuous solid-bowl type with a bowl diameter of 400 mm, a length/width ratio of 3.0 and a rotational speed of 2000 rpm. The feed contains 18% by weight particles and is fed at a rate of 25,000 ℓ/hr at a temperature of 95°F. The average particle size is 65 μm.

(a) Determine the amount of solids recovered per hour.

(b) Determine the solids concentration in the clarified liquid.

(c) Determine the horsepower requirements on the centrifuge.

(d) Size a gravity settler to remove an additional 15% of the solids from the clarified liquid.

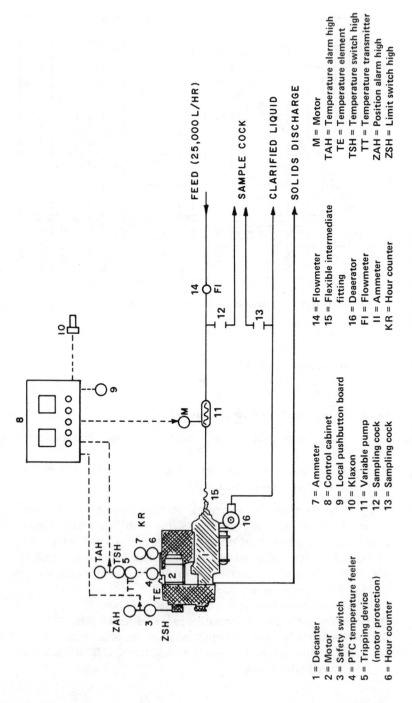

Figure 10.38   Process flowsheet for Problem 10.10.

1 = Decanter
2 = Motor
3 = Safety switch
4 = PTC temperature feeler
5 = Tripping device
      (motor protection)
6 = Hour counter

7 = Ammeter
8 = Control cabinet
9 = Local pushbutton board
10 = Klaxon
11 = Variable pump
12 = Sampling cock
13 = Sampling cock

14 = Flowmeter
15 = Flexible intermediate
       fitting
16 = Deaerator
FI = Flowmeter
II = Ammeter
KR = Hour counter

M = Motor
TAH = Temperature alarm high
TE = Temperature element
TSH = Temperature switch high
TT = Temperature transmitter
ZAH = Position alarm high
ZSH = Limit switch high

FEED (25,000 L/HR)

SAMPLE COCK

CLARIFIED LIQUID

SOLIDS DISCHARGE

NOTATION

A	coefficient
Ar	Archimedes number
a	acceleration, m/s^2
$\dot{a}$	cyclone constant, see Equation 10.28
b	width, m
C, C_ϱ	centrifugal force, N
C'	flow coefficient defined by Equation 10.42
C_D	drag coefficient
c	particle concentration, kg/m^3
D	equipment diameter, m
d	particle diameter, m or μm
d_{el}	diameter of single cyclone in a multiclone, m
d_N	nozzle diameter, m
E	energy, J
F	area, m^2
$\dot{F}_b$	friction force, N
$\dot{F}_f$	friction force, N
$\dot{F}_s$	friction against screw, N
Fr	Froude number
G	gravity force or weight, N
H	height, m
h	height, m
J	inertia moment, m^4
K	coefficient defined by Equation 10.43
K'	constant defined by Equation 10.109
K_0	resistance coefficient
K_s	separation number, a/g
$k, \ldots, k'''$	coefficients
L	rotor length, m
L	reference dimension, m
ℓ	dimension, m
M_0, M_e	mass flow of sediment and effluent, respectively, kg/s
M_M	mass of material, kg
m	mass, kg
$\mathcal{n}$	power, hp or W
N_c	centrifugal number
n	number of rotations
P	force, N
p	pressure, Pa
$Q_{1,2\ldots}$	total material processed per unit time or system capacity, m^3/s

R	radius of particle rotation, m
R_{avg}	average diameter, m
R_D	drag force, N
$R_{1,2\ldots}$	radius, m
Re	Reynolds number
RCF	relative centrifugal force, N
r	radius, m
S_1	settling number, ArK_s
Stk	Stokes number
u	peripheral velocity, m/s
V	volume, m^3
V_c	cyclone working volume, m^3
V_{sec}	volumetric flow, m^3/s
w	velocity, m/s
w_g	fluid rotational velocity, m/s
w_s	particle settling velocity, m/s
x	distance, m
z	number of elements
Z	total resistance, N/m^2

Greek Symbols

α	angle, °, or proportionality constant
β	angle, °, or proportionality constant
Υ	force, N
γ	specific gravity or angle, °
δ	spacing or wall thickness, mm
ζ_c	efficiency coefficient, Equation 10.82
η	collection efficiency
θ	efficiency coefficient
λ_f	friction coefficient
λ	friction factor
μ	viscosity, p
$\tilde{\mu}$	flowrate, coefficient in Equation 10.40
ν	kinematic viscosity, m^2/s
$\dot{\nu}$	centrifugal velocity, m/s
ν_c	sedimentation velocity in a centrifuge, m/s
ν_{sec}	gas flowrate, m^3/s
ρ	density, kg/m^3
ρ_{eff}	effective density, kg/m^3
Σ_c	capacity factor, defined by Equation 10.70
σ	stress, Pa

T	force, N
τ	time, s
ϕ_X	velocity correction factor, see Equation 10.46
ξ	coefficient in Equation 10.138
ψ	shape factor or friction factor
ψ_X	factor reflecting change in velocity, see Equation 10.45
ψ'	friction coefficient
ω	angular velocity, rad/s

REFERENCES

Ambler, C. M. (1952) *Chem. Eng. Prog.* 48:150.

Azbel, D., and N. P. Cheremisinoff (1982) *Chemical and Process Equipment Design: Vessel Design & Selection* (Ann Arbor, MI: Ann Arbor Science Publishers).

Bergman, W. E., C. J. Eagle, S. J. Marwil and W. E. Porter (1956) *Oil Gas. J.* 54:114.

Block, B. (1931) "Clarification et Separation des Liquides par Force Centrifuge" (Paris: Dunod).

Cheremisinoff, P. N., and N. P. Cheremisinoff (1974) "Cyclones: Fundamentals, Applications and Design," *Plant Eng.* 28(14 and 15).

Cheremisinoff, P. N., and R. A. Young (1975) *Pollution Engineering Practice Handbook* (Ann Arbor, MI: Ann Arbor Science Publishers).

Dahlstrom, D. A. (1949) *Trans. Am. Inst. Mining Met. Eng.* 184:33.

Dallavalle, M. (1952) *Proc. U.S. Tech. Conf. on Air Pollution* (Washington, DC).

Dufour, M. (1956) "Centrifugation," Brochure No. 2082, Centre de Perfectionnement technique" (Paris: Presses Documentaires).

Finkelshtein, G. A. (1962) *"Shnekovye Osaditelnye Ztentrifugy,"* Goskhimisdat, Moscow.

Flood, J. E. (1955) *Chem. Eng.* 6:2117.

Flowers, A. E., and S. H. Hull (1950) "Centrifuges," in *Chemical Engineer's Handbook,* 4th ed. J. Perry, Ed. (New York: McGraw-Hill Book Co.).

Foust, A. S., et al. (1980) *Principles of Unit Operations,* 2nd ed. (New York: John Wiley & Sons, Inc.).

Kelsall, D. F. (1952) *Trans. Inst. Chem. Eng.* 30:87.

Kelsall, D. F. (1953) *Chem. Eng. Sci.* 2:254.

Krijgman, C. (1951) *Chem. Eng. Tech.* 23:540.

Loncin, M. (1961) *Operations Unitaires du Genie Chimique* (Paris: Dunod).

Rietema, K., and C. G. Verver, Eds. (1961) *Cyclones in Industry* (New York: Elsevier North-Holland, Inc.).

Scott, P. P. Jr., and J. L. Lummus (1956) *Oil Gas J.* 54:188.

Sheppard, C. B., and C. E. Lapple (1939) *Ind. Eng. Chem.* 31:972.

Sheppard, C. B., and C. E. Lapple (1940) *Ind. Eng. Chem.* 32:1246.

Taggart, A. F. (1950) *Handbook of Mineral Dressing* Vol. I, 4th ed. (New York: John Wiley & Sons Inc.).

Trowbridge, M. E. (1955) *Ind. Chem.* 31:195.

Williams, R. W., and J. Mesaros (1954) *Pet. Eng.* 29:B44.

CHAPTER 11

MIXING THEORY AND PRACTICE

CONTENTS

675

INTRODUCTION

Mixing is a unit operation of chemical engineering that is practiced widely to meet a variety of process requirements. The specific mixing system design, operating arrangement and power requirements depend largely on the desired form of the intermediate or final products. Mixing is applied to achieve specified results in the following situations:

1. creating a suspension of solid particles;
2. blending miscible liquids;
3. dispersing gases through liquids;
4. blending or dispersing immiscible liquids in each other; and
5. promoting heat transfer between a fluid (liquid) and the coil or jacket of a heat exchanging device.

The operating characteristics and design configuration of a mixing system are established on the basis of the required energy expenditure to create or approximate a homogeneous fluid system. For example, in producing an emulsion one must supply sufficient energy to "break up" the dispersed phase. In doing so, high shear stresses, which depend on velocity gradients, are developed in the mixing medium. In the zones in which the velocity gradient approaches a maximum, an intensive breaking up of the dispersed phase occurs.

Mixing reduces concentration and temperature gradients in the processed system, thus exerting a favorable effect on the overall rates of mass and heat transfers. This applies in particular to dissolving applications, electrolysis, crystallization, absorption, extraction, heating or cooling, and heterogeneous chemical reactions, which proceed for the greater part in a liquid medium.

Increased turbulence of the fluid system caused by mixing leads to a decrease in the fluid's boundary layer thickness. This is derived from a continuous renewal of the surface contact area, resulting in a pronounced rate of increase in heat and mass transfer mechanisms. Regardless which medium is mixed with the liquid, i.e., gas, liquid or solid particles, two basic methods are employed. These are mechanical mixers, which utilize different types of impellers, and pneumatic mixers, which utilize air or an inert gas to effect mixing. In addition to these designs, mixing also is achieved in normal fluid handling operations, such as in pumps and jet flows.

Two major characteristics of all mixing devices that provide a basis for comparative evaluations are: (1) the efficiency of a mixing device, and (2) the intensity of mixing. The *efficiency* of a mixing device characterizes the quality of the process to be treated and may be expressed differently depending on the mixing purpose. For example, in producing suspensions mixing efficiency is characterized by the uniform distribution of the solid phase in the volume of equipment. For the intensification of thermal and diffusion processes, it is characterized by the ratio of mixed and unmixed heat and mass transfer coefficients, respectively. Mixing efficiency depends not only on the equipment design, but also on the amount of energy introduced in the liquid being agitated.

The *intensity* of mixing is determined by the time required to achieve a desired technological result or by the mixer rpm at fixed process conditions (for mechanical mixers). From an economical standpoint, it is beneficial to achieve the required mixing effect in the shortest possible time. In evaluating the energy required for a mixing operation, one must account for the total energy consumption during the time needed to achieve a specified mixing result.

Mechanical mixers, which for the most part comprise rotating devices, are employed for liquids almost exclusively. Because of the widespread practice, most of the discussions in this chapter will revolve around fundamentals pertaining to such applications. Because these mechanical devices involve the flow of fluids around immersed bodies, such as impellers, mixing constitutes a unit operation in the *external problems of hydrodynamics*. The governing laws of hydrodynamics presented in previous chapters therefore are equally applicable to mixing.

From an earlier presentation, the slow motion of a body immersed in a viscous fluid produces a laminar boundary layer, which is influenced by the presence of a solid boundary. The shape and thickness of the layer depend on the geometry and size of the body. Boundary-layer separation occurs whenever a change in fluid velocity (either in magnitude or direction) is so large that the fluid cannot adhere to the solid surface.

The maximum speed of an agitated fluid occurs at the periphery of the mixer impeller because it is proportional to the mixer diameter. As follows from Bernoulli's equation, a zone of reduced pressure develops in this region. Resulting flow patterns, including radial flows due to centrifugal forces, result in intensive mixing of materials contained in the reacting or process vessel.

Flow around bodies responsible for performing mixing may be described through application of the Navier-Stokes equations and continuity. However, their exact solution to mixing problems is complicated and, from an engineering standpoint, often applicable to a few specific cases only. Hence, the analyses presented here are based largely on similarity theory, which although less rigorous, affords a greater degree of flexibility in developing design-oriented formulations.

MECHANICAL MIXING DEVICES

Mechanical mixing devices comprise three basic parts: an impeller, a shaft and a speed-reducing gearbox. The impeller constitutes the working element of the device, mounted on a vertical, horizontal or inclined shaft. The drive may be connected directly to an electric motor (for high-speed mixers) or through a gearbox.

The multitude of impeller configurations can be grouped into five distinct categories, of which only the first four are of commercial importance. Discussions to follow concern only the three most widely used types, namely propeller, turbine and paddle mixers.

Propeller Mixers

The two basic propeller mixer configurations are fixed to a rotating vertical, horizontal or inclined shaft. The first is similar to an aircraft propeller, while the second resembles a marine propeller. Depending on the height of liquid layer, one shaft may carry one to three propellers.

Due to their more streamlined shape, a propeller mixer's power requirements are less than the other types of mixers at the same Reynolds number. Their transition in the self-modeling region is observed at relatively low values of Reynolds number, $Re_M \simeq 10^4$. They are capable of high-speed operation without the use of a gearbox and, hence, provide a more cost-effective operation because there are no mechanical losses in transmission. Propeller mixers produce an axial flow, which has a great pumping effect and provides short mixing times.

Disadvantages compared to paddle and turbine mixers are its high cost, the sensitivity of operation to the vessel geometry and its location within the tank. As a general rule, propeller mixers are installed with convex bottom vessels. They should not be used in square tanks or in vessels with flat or concave bottoms.

A rotating propeller traces out a helix in the fluid, from which a full revolution moves the liquid longitudinally to a fixed distance, depending on its pitch, i.e., the ratio of this distance to the propeller diameter. Pitch may be computed from the following formula:

$$s = 2\pi r \tan\psi \tag{11.1}$$

where r is the propeller blade radius and, therefore, also the radius of the cylinder created in the liquid as derived from the movement of the impeller, and ψ is the angle of tilt of the blade. Pumping and mixing efficiencies increase with pitch, as achieved by the axial flow of the liquid from the impeller. This flow results from the delivery head of the propeller and the helical turbulent flow of the entire contents of the vessel, which is caused by the radial velocity gradients in the liquid strata at different distances from the impeller. At high rotational speeds the entire fluid mass swirls despite the axial flow, and a central vortex begins to form around the shaft.

Draft tubes are employed to improve the mixing of large quantities of liquids by directing the motion of the liquid. Figure 11.1 shows such an arrangement, favorable for large ratios of liquid depth to mixer diameter. In such applications a high pumping capacity of the mixer is utilized, especially where mixtures of low viscosity are concerned. The draft tube directs the flow to the regions of the vessel that otherwise would not be agitated by the liquid stream. In the absence of draft tubes and at high rotational velocities of the propeller, baffles generally are located at various points in the vessel. Baffles minimize vortex formation and divide it into a number of local eddies, increasing the total turbulence of the tank.

Depending on the application, multiple impellers may be mounted on a single revolving shaft and more than one shaft may be employed in a given tank. In some applications it is desirable to have two adjacent impellers rotating in opposite directions, forming a beater. Sometimes the impellers actually touch the walls of the tank, giving a positive scraping action, which is desirable when thick layers of material tend to stick to the wall.

Propeller mixers are used for mixing liquids with viscosities up to 2000 cp (see Quillen [1954] for examples). They are suitable for the formation of low-viscosity emulsions, for dissolving applications and for liquid-phase chemical reactions. For suspensions, the upper limit of particle size is 0.1–0.5 mm, with a maximum dry residue of 10%. Propeller mixers are unsuitable for

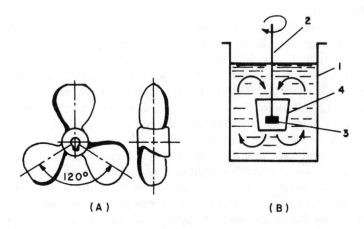

(A)   (B)

Figure 11.1   (A) A typical propeller-type mixer; (B) propeller mixer with a draft tube: (1) tank; (2) shaft; (3) propeller; (4) draft tube.

suspending rapid settling substances and for the absorption of gases. Propellers are designed on the basis of data obtained from properly executed modeling experiments. Sterbacek and Tausk [1965] have characterized the primary dimensions of commercial mixers as follows:

- d = (0.5 – 0.2)D
- s = (1.0 – 3.0)d
- h = (0.5 – 1.0)d
- A = (0.1 – 0.5)D
- H = (0.8 – 1.2)D for one propeller
- H = (1.0 – 5.0)D for more than one propeller
- depth of submersion of mixer = (2 – 4)d
- distance between two propellers on the shaft = d – 5d
- rotational speed = 440 – 2400 rpm
- peripheral velocity = 300 – 900 m/min

where   d = mixer diameter
         s = propeller pitch
         h = height of mixer above the vessel bottom
         D = diameter of the vessel
         H = depth of mixed liquid

Turbine Mixers

The AIChE "Standard Test Procedure for Impeller-Type Mixing Equipment" [1960] defines a turbine mixer as "an impeller with essentially constant blade angle with respect to a vertical plane, over its entire length or over finite sections, having blades either vertical or set at an angle less than 90° with the vertical." Blades may be curved or flat, as shown by the various configurations in Figure 11.2. A brief description of each type impeller is given in Table 11.1.

Turbine mixer operation is analogous to that of a centrifugal pump working in a vessel against negligible back pressure. The mixing action is accomplished by the turbine blades, which entrain and discharge the liquid. The predominantly radial flow from the impeller impinges onto the vessel walls, where it splits into two streams. These streams cause mixing by their energy.

Figure 11.3 illustrates the operation of a unit using an axial flow turbine. The design shown is mounted through the top of the vessel for a closed-tank operation. Details of the stuffing box and gear box are shown in Figure 11.4. The unit shown can cover the range of 1 to 250 hp. Both double and triple reduction gears are available with many designs. Mechanical seal agitators in closed-vessel operations provide sealing for pressures up to 5000 psi. Most units are supplied with a removable-cartridge single or double mechanical seal assembly for seal change simplicity. Seal housings usually are supplied with a cooling jacket. Access change gears are usually spline or taper hub mounted for rapid speed change in the field. High-speed shaft bearings normally are oil lubricated, whereas low-speed shaft bearings are grease lubricated. A

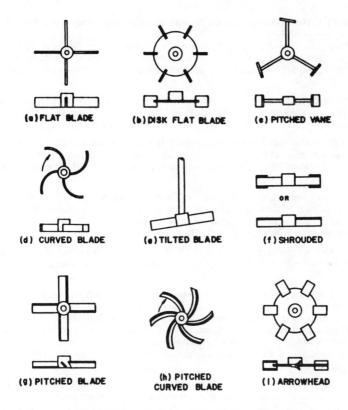

Figure 11.2   Various turbine impeller designs.

typical application is shown in Figure 11.5, in which flotation cell mixers are employed at a corn processing plant.

Open-tank agitators usually are supplied with mounting rails for bolting to mounting beams spanning the vessel top. Mounting rails often are designed to be removed and the drive adapts easily to a variety of special mounting requirements. Figure 11.6 gives an example of an open-tank turbine agitator drive.

When turbine mixers are operated at sufficiently high rotational speeds, both radial and tangential flows become pronounced, along with vortex formation. This flow situation warrants the installation of baffles to ensure a more uniform flow distribution throughout the mixing vessel. Typical flow patterns observed by Newitt et al. [1951] around wall-mounted baffles are illustrated in Figure 11.7.

Table 11.1 Description of Turbine Impeller Configurations Shown in Figure 11.2

Flat blade. Also termed a "straight-blade turbine," this impeller discharges radially, deriving suction from both top and bottom. Customary operation is in a peripheral speed range of 600 to 900 fpm. Blade widths generally are one-fifth to one-eighth of the diameter.

Disk flat blade. This turbine is widely used industrially and has been employed in many laboratory investigations. Although it has essentially the same performance characteristic as the flat-blade turbine, the difference in power consumption is significant.

Pitched vane. Also called a "radial propeller," it is an adaptation of the disk type, with the area reduced by pitching the blades to the vertical plane. Its advantage is its ability to obtain a high ratio of d/D and a high speed (for drive economy) without high power consumption.

Curved blade. Also called the "backswept," or "retreating blade," turbine, the blades curve away from the direction of rotation. This modification of the flat-blade style currently is thought to reduce mechanical shear effect at the impeller periphery. Industrial usage in suspensions of friable solids is widespread.

Tilted blade. In comparison to the curved-blade style, it has the effect of increasing the depth of the flow pattern and generally improving performance without an increase in power.

Shrouded. Addition of a plate, full or partial, to the top or bottom planes of a radial flow turbine is made to control the suction and discharge pattern. The upper unit usually has annular rings on the top and bottom. The lower design is fully shrouded on top to restrict suction to the lower side. The same impeller used in gas dispersion (the latter style) is referred to as a "vaned disk." A full shroud on the lower surface of an impeller, which is located near the liquid surface, will increase the vortex considerably, e.g., for gas reentrainment.

Pitched blade. This impeller has a constant blade angle over its entire blade length. Its flow characteristic is primarily axial, but a radial component exists and can predominate if the impeller is located close to the tank bottom. The blade slope can be anywhere from 0° to 90°, but 45° is the commercial standard. This impeller also is known as a "fan type."

Pitched curve blade. Sloping the blades of a curved-blade style is possible and has been practiced occasionally. No performance or power data are available, and the high cost of construction of this impeller limits it to only a few special applications.

Arrowhead. This is a mixed-flow (axial and radial) impeller. It is a laboratory design with no commercial application.

To concentrate the entraining action near the tank bottom, guide rings sometimes are mounted under the turbine. This is especially important with suspensions when the solids settle directly in the center of the vessel beneath the mixer. A stator ring is employed to eliminate the central vortex. Such an arrangement provides near-perfect radial flow from the turbine.

The optimum blade configuration depends on the properties of the materials to be mixed and the intended product state. For example, with mobile liquid mixtures, straight flat blades are suitable; if it is desirable to increase

Figure 11.3   Mixer shown on a phantom tank with an axial-flow turbine (courtesy Lightnin Mixing Equipment Co., Rochester, NY).

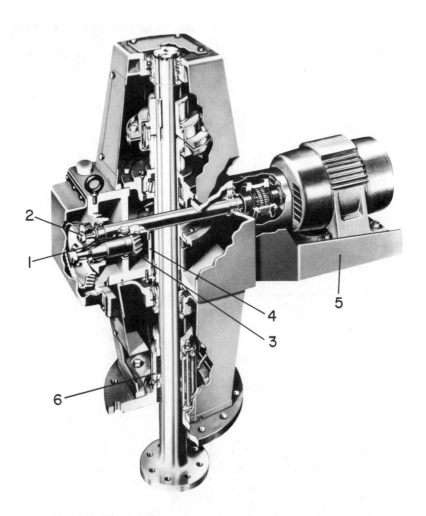

Figure 11.4   Cutaway view of a double-reduction gearing with stuffing box for closed-tank applications: (1) spiral bevel gears; (2) change gears; (3) splash lubricant; (4) oil dam; (5) motor mount; (6) integral stuffing box (courtesy of Lightnin Mixing Equipment Co., Rochester, NY).

the pumping effect, inclined blades are recommended. For viscous liquids, inclining the blades in the opposite direction of rotation is advantageous. Profiled and curved blades are recommended because their starting moment is smaller and they facilitate the transmission of energy from the impeller to the liquid.

Figure 11.5   Two 30-hp flotation cell mixers in a closed-vessel application at a corn processing plant (courtesy Lightnin Mixing Equipment Co., Rochester, NY).

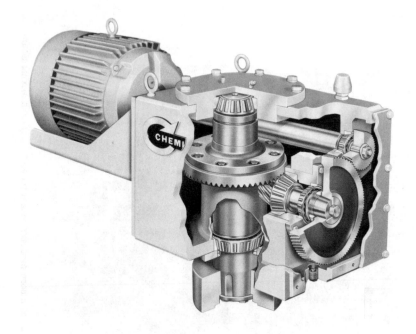

Figure 11.6   Cutaway view of a turbine agitator drive for open-tank applications (courtesy Chemineer-Kenics, Agitator Div., Dayton, OH).

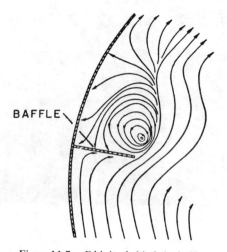

BAFFLE

Figure 11.7   Eddying behind the baffle.

Lyons [1954] has summarized general data on basic dimensions and the range of applications for turbine mixers. Some of this information is summarized in Table 11.2. The rotational speed of these mixers is 120–200 rpm, with peripheral velocities of 200–500 m/min.

Paddle Mixers

Paddle mixers are devices consisting of two or more blades mounted on a vertical or inclined shaft. The basic paddle impeller configurations are shown in Figure 11.8, and a description of each is given in Table 11.3. The main advantages of paddle mixers are their simplicity and low cost. A disadvantage is their small pumping capacity (a slow axial flow), which does not provide a thorough mixing of the tank volume. Perfect mixing is attained only in a relatively thin stratum of liquid in the immediate vicinity of the blades. The turbulence spreads outward very slowly and imperfectly into the entire contents of the tanks; hence, circulation of the liquid is slow. Therefore, paddle mixers are used for liquids with viscosities only up to about 1000 cp. Because of a concentration gradient that often is created in the liquid when these type mixers are used, they are unsuitable for continuous operation. This can be remedied by tilting the paddle blades 30–45° to the axis of the shaft,

Table 11.2 Basic Dimensions of Turbine Mixers
(Data of Lyons [1954])

Application	Basic Dimensional Ratios		Type of Mixing Element and its Location
	D/d	H/d	
Liquid-Liquid Mixing	3–6	Unlimited	
Dispersion	3–3.5	1–2	Simple or multiple impeller on, or below, the center line of the liquid charge
Reaction in a Solution	2.5–3.5	1–3	
Dissolution	1.6–3.2	1–2	
Suspension	2–3.5	1–2	According to particle size above or near the bottom
Absorption of Gases	2.5–4	4–1	Multiple, lowest impeller above the bottom
Mixing of Plastic Substances	1.5–2.5	1–2	Simple or multiple impeller
Crystallization or Precipitation	2–3.2	2–1	Simple impeller on or below center line of liquid charge

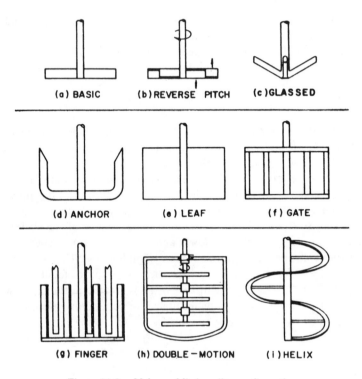

Figure 11.8   Major paddle impeller configurations.

resulting in an increase in axial flow and, consequently, a decrease in concentration gradients. Such a mixer can maintain particles suspended, provided settling velocities are not high. Mixers with tilted blades are used for processing slow chemical reactions, which are not limited by diffusion.

To increase the turbulence of the medium in tanks with a large height to diameter ratio, a configuration is employed that consists of several paddles mounted one above the other on a single shaft. The separation between individual paddles lies in the range of 0.3 to 0.8 d (where d is the diameter of the paddle) and is selected according to the viscosity of the mixture. For mixing liquids with viscosities up to 1000 cp, as well as for heated tanks in cases in which sedimentation can occur, anchor or gate paddle mixers are employed (Figure 11.8(D) and (F)). Paddle diameters are almost as large as the inside diameter of the tank in such applications, so that the outer and bottom edges of the paddle scrape (or clean) the walls and bottom.

Leaf-shaped (broad blade) paddle mixers provide a predominant tangential flow of liquid, but there is also turbulence at the upper and lower edges of

Table 11.3 Description of Paddle Impeller Configurations Shown in Figure 11.8

Basic paddle. The simplest design is a single horizontal flat beam. The ratio of impeller diameter to tank diameter usually is 0.5 to 0.9. The peripheral speed range is generally 250–450 fpm. Paddles used in the United States generally have had a width/vessel diameter (b/D) ratio of one-sixth to one-twelfth; European practice is generally one-fourth to one-sixth.

Reverse pitch. To improve the top-to-bottom turnover characteristic of a simple paddle, the reverse pitch design in Figure 11.8(B) is used. The 45° blade angle is reversed at a diameter of approximately three-fourths of the impeller diameter, and rotation generally is set to produce upflow in the outer section.

Glassed steel. This impeller is a three-blade style common in glass-coated vessel applications. Blade form is either a pitch of about 30° from the vertical, or curved. Usually, d/D is 0.55 to 0.65. For low-viscosity fluids, its operation is definitely in the turbine category, but the difficulty of achieving a fully baffled condition in a glass-lined vessel causes it to perform like a paddle in many cases.

Anchor. Contouring a simple paddle to the shape of a tank bottom gives the anchor or "horseshoe" style. Extent of the blade may be limited to the lower vessel tangent line or the blades may continue upward along the straight side. Clearance between blade and vessel shell may be from $\frac{1}{8}$ to 3 in., depending on tank diameter and heat transfer requirements.

Leaf. This has an extreme b/D ratio for a paddle. This geometry is used frequently in European practice.

Gate. A multiple-arm paddle with connecting vertical members, this design often is adopted for structural reasons in large tanks.

Finger. Also known as a "paste mixer," this combination of vertical blades meshing with stationary baffles has been in use for many years, but no data on power or performance have been published. The application is restricted to small batch sizes (less than 1000 gal) because of structural design difficulties.

Double motion. This design combines a gate and anchor and a multiple-pitched paddle. Rotation of the two assemblies is countercurrent. As a special drive with concentric counterrotating output shafts is required, use is restricted to a few applications requiring intensive mixing of very viscous or non-Newtonian fluids.

Helix. This configuration least resembles the basic paddle. It does operate in the laminar range, but at normally high d/D ratios, and is an important member of the paddle group. One traditional use of a helix or screw is in a vertical-draft tube.

---

the blade. Leaf-type blades are used for mixing low-viscosity liquids, intensifying heat transfer processes, promoting chemical reactions in a reactor vessel and for dissolution. For dissolving applications, leaf-blades generally are perforated. During the mixer rotation, jets are formed at the exits from the holes and promote the dissolution of solid materials.

The rotational velocity of paddle mixers is in the range of 15 to 45 rpm. Under these conditions, pumping action is small and there is no danger of vortex formation. Therefore, paddle mixers are most often used without baffles.

However, for broad-blade paddles, which operate at speeds up to 120 rpm, baffles are incorporated into the design to minimize vortex formation.

Based on the studies by Sterbacek and Tausk [1965] with liquids having viscosities up to 1000 cp, the recommended geometric proportions for a single paddle mixer are as shown in Figure 11.9: $d/D = 0.66 - 0.9$; $h/D = 0.1 - 0.2$; $A'/d = 0.0 - 0.3$; and $H/D = 0.8 - 1.3$.

Emulsifiers, Blenders and Planetary Mixers

There are a number of other mixing configurations that have widespread applications to both liquid-solid and solid-solid blending operations. Although design principles for these mixers are not treated in this chapter, their description merits attention.

Mixer emulsifiers sometimes are used as an alternative to slow-speed impeller mixing or high-pressure homogenization for a wide range of processing requirements. Typical applications include the preparation of adhesives (e.g., asphalts, carbon dispersions, clay dispersions, dyestuffs, paints and inks, lacquers), cosmetics (e.g., creams, emulsions, hand lotions, perfumes, shampoos, deodorants), foods (e.g., chocolate coatings, mustard, soft drinks, sugar emulsions), pharmaceuticals (e.g., antibiotics, ointments, reducing animal tissues), plastics (e.g., cold cutting resins, polyester dispersions, resin solutions) and various miscellaneous mixtures such as floor polishes, gum dispersions, lubricants, petroleum emulsions, etc. These types of mixers normally are used in dished or conical bottom vessels.

An example of a mixer-emulsifier is shown in Figure 11.10. The mixing process can be thought of as performed in three stages. In the first stage, the

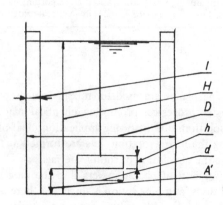

Figure 11.9   Fundamental geometric variables of a paddle mixer.

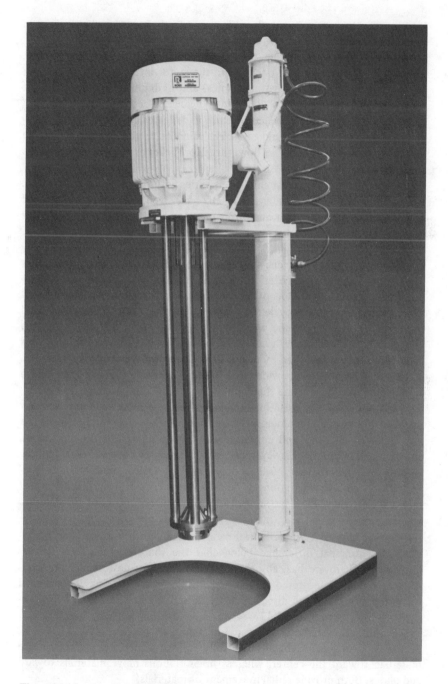

Figure 11.10   A mixer-emulsifier mounted on an adjustable lift (courtesy Charles Ross & Son Co., Hauppauge, NY).

high-speed rotor operating at close clearance to the stator draws material in from the bottom of the mixing vessel and subjects it to intense mixing.

In stage 2, the rotor accelerates the product toward the blade periphery. There it is expelled through openings in the stator into the body of the mix while undergoing an intensive mechanical and hydraulic shearing action. At the same time, new material is drawn into the center of the rotor.

In the third stage, the expelled mixture is deflected by the tank wall, completing the circulation. To increase circulation or to create a vortex for the incorporation of light solids, a downthrust propeller may be mounted on the rotor shaft.

The operating sequence for emulsifiers varies with the intended application. An operating guide for these types of mixers is given in Table 11.4. Standard batch units are available through 100 hp. For most applications the rotor head should be mounted two to three head diameters above the vessel floor. Obviously, with these designs a slender-shaped vessel generally is preferred.

Maximum mixing efficiency generally is achieved on mixtures under 10,000 cp. The principal criterion for achieving the best efficiency is to maintain a maximum circulation of all materials through the rotor/stator at all times during the process cycle. As viscosity increases, flow through the head decreases, thus lessening the work on a given volume of material while it is being circulated within the vessel. Inline emulsifiers are less susceptible to this problem as they may be fed by means of positive displacement pumps.

Note that because of the wide range of applications to which a single design is subjected, the density and rheology of the mixture are variables. The pumping capabilities of a unit on materials that have the same apparent viscosity may not be the same in actual practice. For example, many polymers have a low apparent viscosity and appear ideally suited to a particular unit design and application criteria. However, on testing, it may be observed that the flow is poor and that the end result is only marginal. Often, modified tank configurations, auxiliary agitation, different rotor speeds or head diameters will improve efficiency. Pilot testing almost always is recommended for accurate scaleup.

Ribbon blenders resemble helical paddle mixers both in design and operation, the principal difference being that the shaft orientation is strictly horizontal. There are basically three standard agitator designs, namely, continuous ribbon, interrupted ribbon and paddle type. They may be arranged for either center or end discharge from the mixing vessel. The principal designs are shown in Figure 11.11.

The continuous ribbon, arranged for center discharge (Figure 11.11(A)), produces homogeneous blends relatively quickly. The outer ribbons move materials toward the center, whereas inner ribbons move materials toward end plates. Both provide radial movement of materials.

Table 11.4 Operating Guidelines for Emulsifiers
(courtesy Charles Ross & Son Company, Hauppauge, NY)

Application	Procedure	Benefits
Dissolution	1. Charge all liquids into vessel.	1. Short turnaround time permits the use of smaller equipment to accomplish the required end result.
	2. Add any minor ingredients that require long dispersion time.	2. Complete wetting and subsequent dissolution of all solids.
	3. Add solids to be dissolved. If the solids are of low density, down-thrust propellers may be advantageous to create a vortex and assist in initial wetting.	3. Low heat buildup results from the short cycle time.
	4. Run unit until all powders are completely wetted out and dissolved in the liquid components.	4. Significant energy savings are experienced.
Suspensions	1. Charge all liquid components of the formulation into the tank.	1. Uniform particle size reduction is achieved.
	2. Start mixer and charge solid materials until a uniform particle size has been achieved.	2. A one-step operation suspends solids and adds the dispersant.
	3. Add the dispersant as rapidly as possible.	3. Interchangeable stator heads may be used with solids of varying size.
		4. There is a short cycle time.
Emulsification	1. Charge continuous phase and heat.	1. There is a short cycle time.
	2. Add surfactant.	2. Intimate and immediate contact of dispersed phases occurs.
	3. Externally heat and slowly add the dispersed phase through an open pipeline, terminating below the rotor/stator assembly.	3. Fine droplet size results.
	4. Mix until emulsion is completed.	4. There are potential surfactant savings through efficient use of high shear agitation.
Disintegration	1. Charge all liquids into vessel.	1. Involves a single-step process with potential materials handling and time savings.
	2. Add to liquids any solids that are liquefied easily and begin mixing.	2. Drastically reduced cycle time results.
	3. Add slowly the solids that are to be reduced in size.	3. Substantial savings are realized in electrical costs, resulting from shorter cycle times.
	4. Continue operation until solids are reduced to a desired uniform particle size.	4. Floorspace savings are realized, and less capital equipment is required.

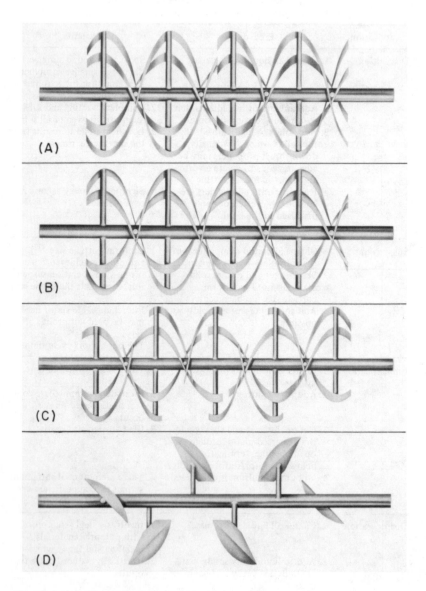

Figure 11.11   Principal types of ribbon blender agitator designs: (A) continuous ribbon, arranged for center discharge; (B) continuous ribbon, arranged for end discharge; (C) interrupted ribbon; (D) paddle-type (courtesy Charles Ross & Son Co., Hauppauge, NY).

A continuous ribbon, arranged for end discharge, is shown in Figure 11.11(B). Outer ribbons move materials towards the discharge end. The inner ribbons move materials in the opposite direction. Again, both provide radial transfer of materials.

The interrupted ribbon (C) provides the same basic action as a continuous ribbon agitator. It does require less power, however, and generally is applied to materials having higher bulk densities. The last design (paddle type (C)) often is used in such applications as for making instant beverage premixes.

Ribbon blenders are essentially self-contained mixers. An example of a unit with a variable drive is given in Figure 11.12. These mixers are employed in a

Figure 11.12   A 1-ft^3-capacity ribbon blender with vari-drive (courtesy Charles Ross & Son Co., Hauppauge, NY).

variety of solid-liquid, solid-solid and some liquid-liquid blending operations. Examples include foods, plastics, pigments, pharmaceuticals, specialty chemicals and confectionary.

Planetary mixers are used for a variety of liquid and solid mixing applications, from simple mixtures to sophisticated reactions, involving high temperature, vacuum or internal pressure. This type of mixer is employed in batch operations. During the mix cycle, two rectangular-shaped stirrer blades revolve around the tank on a central axis. Each blade revolves on its own axis simultaneously, at approximately the speed of the central rotation. With each revolution on its own axis, each stirrer blade advances forward along the tank wall. This movement provides homogeneity of the material being mixed and does not depend on the flow characteristics of the mix. Instead, the stirrers cover every point within the mix tank.

Double planetary mixers have no packing glands or bearings in the product zone. Hence, cleaning between batches is minimized. Figure 11.13 shows a planetary mixer unit. The machine is equipped with a hydraulic lift that permits the stirrer blades to be lowered and raised in and out of the mixing tank. The unit shown is capable of handling low-viscosity fluids to very high-viscosity pastes and doughlike materials. Typical capacities of these units range from 10 to more than 300 gal.

FLOW PATTERNS IN AGITATED TANKS

The forces applied by an impeller to the material contained in a vessel produce characteristic flow patterns that depend on the impeller geometry, properties of the fluid, and the relative sizes and proportions of the tank, baffles and impeller. There are three principal types of flow patterns: tangential, radial and axial.

Tangential flow is observed when the liquid flows parallel to the path described by the mixer (Figure 11.14). When the flow pattern in a mixed tank is primarily tangential, the fluid discharge from the impeller to the surroundings and its entrainment into the impeller are small. Also, fluid transfer in the vertical direction is at a minimum. The mixing effect is lowest when the rotational velocity of the liquid approaches that of the mixer.

With *radial flow*, the liquid discharges from the impeller at right angles to the mixer axis and moves along the radius. As soon as the centrifugal force, which depends on the impeller's diameter and speed, overcomes the resistance of the medium, it initiates radial flow from the impeller into the bulk fluid. Figure 11.15 shows the flow pattern of a mixer producing radial flow in two sections of the tank. In the bottom part of the tank, the impeller entrains the liquid in an upward direction and displaces it at right angles to the axis of the

Figure 11.13   A planetary mixer with a vacuum hood (courtesy Charles Ross & Son Co., Hauppauge, NY).

impeller. In the upper part of the tank the impeller entrains the liquid downward, also displacing it perpendicular to the impeller axis.

Axial flow, in which the liquid enters the impeller and discharges along a parallel path to the axis, is shown in Figure 11.16. The radial and longitudinal components are primarily responsible for the derived mixing action. The tangential component is important when the shaft has a vertical orientation and is positioned near the center of the tank.

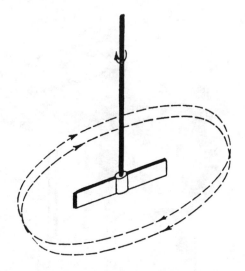

Figure 11.14   Tangential flow generated by a paddle mixer.

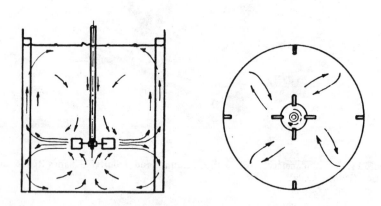

Figure 11.15   Radial flow pattern.

Based on the pitch of the impeller with regard to the direction of rotation, there are two possible axial flow patterns:

1.   that in which the impeller pumps the liquid from the bottom to the surface; and
2.   that in which the impeller pumps liquid from the surface to the bottom.

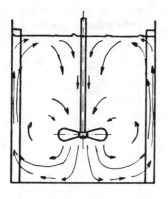 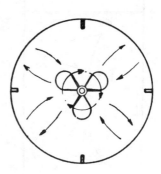

Figure 11.16 Axial flow pattern.

A combination of the three principal types of flow normally is encountered in mixing tanks. The tangential flow following a circular path around the shaft forms a vortex at the surface of the liquid. The vortex formation results from the influence of gravity forces, quantitatively determined by means of the Froude number, which increases at higher speeds, promoting vortex formation.

Figure 11.17 presents a three-dimensional flow pattern affording a clear image of the liquid flow in the tank obtained by projecting the path of a liquid particle in two planes. Part (A) shows the path that the particle takes at a given impeller speed. The particle undergoes four horizontal revolutions per single vertical revolution. As the impeller speed is increased, the particle performs a greater number of horizontal revolutions per one vertical revolution (Figure 11.17(B)), showing that the greater centrifugal force at higher speeds increases the radius of curvature of the paths of the particle, which thus is shifted farther from the center of the tank. Further increases of the rotational speed and, subsequently, the centrifugal force, result in higher ratios between the horizontal and vertical revolutions of the particle and, thus, larger radii of curvature for particle paths.

A particle subjected to gravitation also is acted on by tangential acceleration, as determined by the centrifugal force developed by the impeller, and influenced by the density and viscosity of the mixture. The direction of resultant acceleration is perpendicular to the surface of the liquid at any given point. If the tangential acceleration is negligible, the level of the fluid will be horizontal. Otherwise, the action of the centrifugal field becomes pronounced and the liquid level exhibits a depression that is referred to as the vortex.

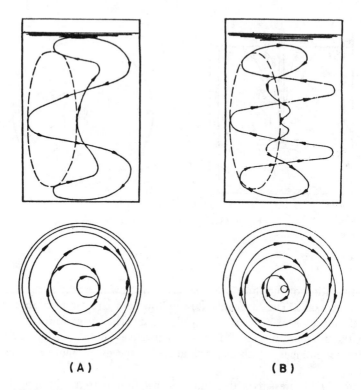

Figure 11.17   Three-dimensional flow pattern for tank mixing, as suggested by Peck [1955].

Vortex formation is a condition that arises from centrifugal acceleration acting on gravitational acceleration. The circular motion of the entire contents of the tank predominates over the flow of the liquid from the impeller. Flow orientation thus is important not only in cases of noticeable vortex formation, but also in mixers with tangential flow. In the latter case, at certain mixer dimensions and impeller speeds the circulating liquid in the tank may attain the same velocity as the impeller. Mixing becomes ineffective under such conditions. For instance, Sterbacek and Tausk [1965] note that the critical conditions for blade mixers are attained when the impeller surface is about one-fifth of the tank cross section.

Nagata et al. [1960] have shown that vortex formation leads to a considerable drop in mixing efficiency and should be suppressed as much as possible in practical applications to increase the homogenizing effects of mixers. The preferable method of vortex suppression is to install vertical baffles at the walls of the mixing tank. These impede rotational flow without interfering with the radial or longitudinal flow. Figure 11.18 illustrates such a system.

The distribution of velocity components (radial, u_r; tangential, u_t; and axial, u_z) under conditions of mixing with baffles in comparison with the conditions of vortex formation has been analyzed by Nagata et al. [1960] and Sachs and Rushton [1954]. The graphic results are shown in Figure 11.19. The dashed lines in Figure 11.19 indicate nonbaffled conditions. Comparison of the nonbaffled and fully baffled velocity curves (solid line) leads to the following set of conclusions on vortex suppression when dealing with perfectly miscible liquids:

1. Baffles are responsible for restricting the tangential velocity component, u_t, and augment the vertical component, u_z, while simultaneously increasing the radial velocity, u_r. The net result is that the liquid discharges from the impeller in a wider flow radius.
2. The streamlined distribution becomes more regular when baffles are used. Hence, the energy transmitted from the impeller to the liquid is utilized more uniformly.
3. The circulation increases, and the difference between the circulation rate under fully baffled conditions and at the vortex formation rises to two to four times the original value. This means that the power input increases considerably in the range of two to ten times the input without baffles.

Bissell et al. [1974] note for turbine mixers that the width of a baffle should not exceed more than one-twelfth the tank diameter and, for propeller mixers, no more than one-eighteenth the tank diameter. With side-entering, inclined or off-center propellers, as shown in Figure 11.20, baffles are not

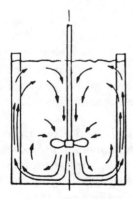
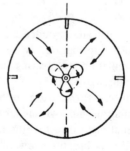

Figure 11.18   Flow patterns in a baffled tank with a centrally mounted propeller.

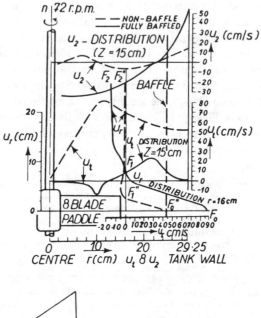

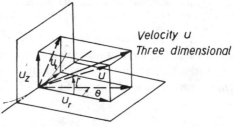

Figure 11.19   Variation of velocity distribution in a mixing tank on insertion of full side wall baffles.

required. Instead, shrouded impellers and diffuser rings may be used to suppress vortex formation. These devices contribute to flow resistance and reduce circulation by creating intense shear and abnormal turbulence at the impeller discharge.

The specific flow pattern in the tank depends on the type of impeller if vortex formation and swirling are eliminated. With *propeller mixers*, the liquid is driven straight down to the bottom of the tank, where the stream spreads radially in all directions toward the wall, flows upward along the wall, and returns to the suction of the propeller from the top (Figure 11.18).

Paddle mixers provide good radial flow in the immediate plane of the impeller blades but are poor in developing vertical currents.

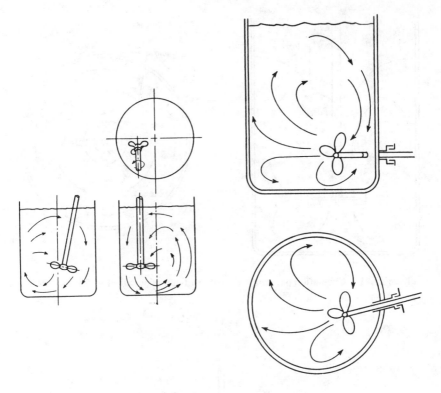

Figure 11.20   Off-center impeller operations.

Turbine mixers drive the liquid radially against the wall, where the stream divides into two portions. One portion flows downward to the bottom and back to the center of the impeller from below; the other flows upward toward the surface and back to the impeller from above. Figure 11.21 shows that there are two separate circulation currents generated. Turbines are especially effective in developing radial currents, but they also induce vertical flows, especially under baffled conditions.

In vertical cylindrical vessels, the ideal liquid depth for good mixing should be somewhat greater than the tank diameter. If greater depths are required, two or more impellers may be installed on the same shaft, with each impeller serving as a separate mixer. Two circulation currents are generated for each mixer, as shown in Figure 11.22. The bottom impeller should be mounted about one impeller diameter above the bottom of the tank.

As noted earlier, when the direction and velocity of the flow to the impeller suction are to be controlled, draft tubes are employed. These devices are

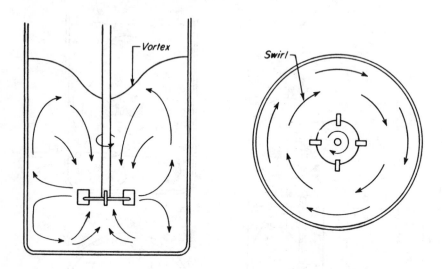

Figure 11.21   Circulation patterns in an agitated tank, as described by Lyons [1948].

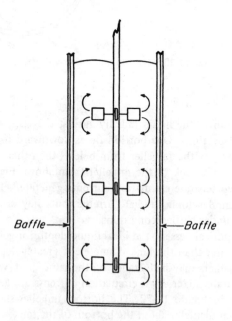

Figure 11.22   Multiple turbines in a tall tank.

designed to set in motion regions in the tank that otherwise would not be agitated. Typical flow profiles are shown in Figure 11.23. Draft tubes are mostly used with propellers and really amount to mixing by circulation of the contents through a thoroughly agitated region in the vicinity of the impeller. The turbulence induced in the tank by the flow of liquid from the draft tube only contributes to the blending action.

THE MIXING OF HETEROGENEOUS FLUID SYSTEMS

The above discussions concerned the simplest types of mixing problems, those of perfectly miscible fluids. However, more complicated cases often are encountered in which the resulting mixture is a heterogeneous system. Often, the substances initially exist in two different phases and then gradually form a single phase (e.g., dissolving a solid in a liquid).

The formation of heterogeneous fluid systems may be classified into one of the following areas:

1. liquid dispersed in gas—mist;
2. solids dispersed in gas—smoke;
3. gas dispersed in liquid;
4. immiscible liquids—one component dispersed in the other; or
5. solids dispersed in liquid.

1. **Liquid Dispersed in Gas**. The dispersion of a liquid into a gas requires a certain amount of work to overcome surface tension forces, in which the

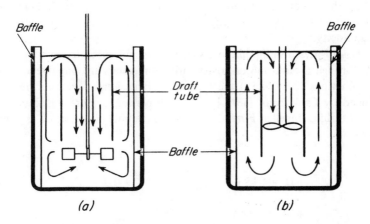

Figure 11.23   Draft tubes in baffled tanks: (A) turbine; (B) propeller.

liquid must take on a form occupying the smallest possible surface area. By droplet generation, the overall surface of the original amount of liquid is increased, and the required work to be expended for this purpose is

$$dW_e = \sigma dA \qquad (11.2)$$

where W_e = the work
σ = the surface tension
A = the overall surface area of the liquid

Because the specific gravity of the droplets is greater than that of the gas, settling occurs. For small droplets in the Stokes region, the free-fall velocity is

$$u_s = \frac{d_0^2(\gamma_1 - \gamma_2)}{18\ \mu} \qquad (11.3)$$

where d_0 = the diameter of the droplet
γ_1 = the density of the liquid
γ_2 = the density of the gas
μ = the viscosity of the gas

The settling of droplets can be counteracted by a gas flow in the opposite direction.

2. **Solids dispersed in gas.** Solids particles usually are dispersed in a gas by passing the gas at a sufficient velocity through a layer of particles in such a manner that the upward velocity of the gas exceeds that of the particle settling velocity.

3. **Gas dispersed in liquid.** To disperse a gas and circulate a gas-liquid mixture through a tank, motor-driven turbine impellers are employed. For low void fractions ($\psi' < 0.15$) the following dimensional equations are recommended by Calderbank [1967] for gas dispersion in pure liquids by a six-bladed turbine impeller. First, the average gas bubble diameter, $\overline{D}'_s$, is given by (mm)

$$\overline{D}'_s = 4.15 \frac{(\sigma g_c)^{0.6}}{(Pg_c/V)^{0.4}\rho_L^{0.2}} \psi'^{1/2} + 0.9 \qquad (11.4)$$

where Pg_c/V is the power input per unit volume of ungassed liquid. The interfacial area a' (mm^2) is

$$a' = 1.44 \frac{(Pg_c/V)^{0.4}\rho_L^{0.2}}{(\sigma g_c)^{0.6}} \left(\frac{\bar{u}_s}{u_t}\right)^{1/2} \tag{11.5}$$

And combining these expressions and taking into account that

$$\bar{D}_s \equiv \frac{6\psi'}{a'} \tag{11.6}$$

the following dimensional equation for the void fraction is obtained:

$$\psi' = \left(\frac{\bar{u}_s\psi'}{u_t}\right)^{1/2} + 0.0216 \frac{(Pg_c/V)^{0.4}\rho_L^{0.2}}{(\sigma g_c)^{0.6}} \left(\frac{\bar{u}_s}{u_t}\right)^{1/2} \tag{11.7}$$

where $\bar{u}_s$ and u_t are the superficial velocity of the fluid and particle terminal velocity, respectively.

4. **Immiscible liquids.** In batch mixing of two liquids, the dispersed one (i.e., that distributed as droplets in the continuous liquid) is first divided by the action of the impeller into small cylinders, which then are extended into thin strips. These give off globules subject to further disintegration into fine droplets. Because both liquids usually differ in density, the centrifugal acceleration in passing through the impeller creates velocity gradients both in the dispersed and continuous phases. These gradients are equalized by shear stresses and consequent momentum transfer between the phases. The greatest differences in the velocities are observed during passage through the mixer and, in particular, after the liquid, which discharges radially from the impeller, impinges onto the vessel wall. The deformation of droplets by shear is counteracted by interfacial tension, which influences the shape of the droplet and is also responsible for its breaking up. The most intense droplet splitting in the mixing tank takes place in the regions of the largest velocity gradients between both phases, i.e., at the impeller and at the wall of the tank.

In turbulent flow, droplets not only undergo disintegration into smaller particles but often collide and agglomerate. Thus, after a certain time at constant mixer speed, an equilibrium state is established between the numbers of combining and splitting droplets.

For the case of a droplet breaking up, Hinze [1955] recommends expressing the Weber number by the ratio between the shear and interfacial tension:

$$We = \frac{\tau_s d_0}{\sigma} \tag{11.8}$$

where τ_s = the shear stress

The shear stress between the droplets and continuous liquid is proportional to the intensity of mixing, which thus influences the Weber number. For direct representation of the mixing effect, Rodger et al. [1956] modified the Weber number to the following:

$$We = \frac{d^3 n^2 \rho_c}{\sigma} \qquad (11.9)$$

where d = the diameter of the impeller
n = the rotational speed
ρ_c = the density of the continuous phase

An increase in the Weber number thus indicates a higher shear stress as compared to interfacial tension forces and, consequently, greater deformation occurs. On achieving the critical value of the Weber number, droplets begin to break up.

In a continuous operation the tank is first charged with the continuous phase and then the phase to be dispersed is introduced. For a higher-density continuous phase, the phase to be dispersed is added at the bottom. On initiating mixer operation, rising droplets enter the region subjected to mixing, where they split into smaller particles. The influence of mixer speed on droplet diameter is determined from the following equation (see Vermeulen et al. [1955]:

$$d_0 = K' n^{-1.2} \qquad (11.10)$$

where d_0 = the diameter of the droplets
K' = a constant dependent on the properties of both liquids
n = the rotational speed

The splitting of droplets is most intensive in the region of the impeller itself and at the wall of the tank on which the radially discharged liquid impinges. Therefore, the distance between the periphery of the impeller and the wall of the tank is a very important factor.

For conditions termed acollisive (in which the amount of dispersed phase is small compared with that of the continuous phase), a plot of the dimensionless drop diameter versus Weber number can be used (Figure 11.24). Introducing the distance between the mixer and tank radius, Misek [1960] obtained the following relationship:

$$\frac{d_0 n^2 d^2 \rho_c}{\sigma \exp(0.087D)} = 16.3 \left(\frac{H}{D}\right)^{0.46} \qquad (11.11)$$

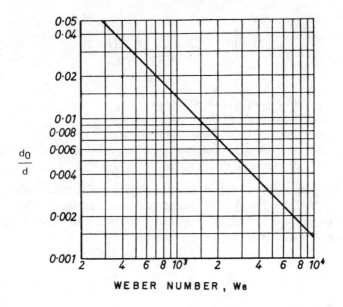

Figure 11.24   Plot of the ratio d_0/d versus Weber number, as recommended by Misek [1960].

And for conditions other than acollisive, the following expression is applicable:

$$\frac{d_a}{d_0} = \exp(zX) \tag{11.12}$$

where d_a = the droplet diameter under actual conditions, i.e., when collisions occur
   X = the fractional liquid holdup
   z' = the coalescence coefficient defined by

$$z' = 1.59 \times 10^{-2} \left[\frac{D}{\nu} \left(\frac{\sigma}{\rho_c d_0} \right)^{0.5} \right]^{0.5} \tag{11.13}$$

The viscosity of the emulsion is much higher than that of the initial continuous liquid and is determined from the following expression of Vermeulen et al. [1955]:

$$\eta' = \eta_c \left[1 + 2.5\psi'' \left(\frac{\eta_d + \frac{2}{5}\eta_c}{\eta_d + \eta_c} \right) \right] \tag{11.14}$$

where η' = the viscosity of the emulsion
η_d = the viscosity of the dispersed phase
η_c = the viscosity of the continuous phase
ψ'' = the ratio of the total volume of the dispersed liquid to the overall volume of the emulsion

4. **Solids suspended in a liquid.** For suspensions of fine solid particulates in a mixing vessel, the theory of turbulent diffusion is applicable, provided the size of the particles is small as compared to the turbulent mixing length, L_p. Turbulent diffusion is defined by the equation

$$\mathscr{D}_t \frac{dC}{dz} - u_p C = \text{const.} = q \tag{11.15}$$

where $\mathscr{D}_t$ = the coefficient of turbulent diffusion
C = the concentration of solid particles
u_p = the relative particle velocity, i.e., the settling velocity minus the velocity of the liquid
q = the rate of flow of particles
z = the axial coordinate

For a closed system, q is zero and the expression may be rewritten

$$\mathscr{D}_t \frac{dC}{dz} - u_p C = 0 \tag{11.16}$$

On integration, we obtain

$$\ln \frac{C}{C_0} = \frac{u_p z}{\mathscr{D}_t} \tag{11.17}$$

where C_0 is the concentration of particles in the upper part of the tank, i.e., where $z = 0$.

The average concentration of the particles in the tank is obtained from the following:

$$\bar{C} = \frac{1}{z}\int_0^z Cdz = \frac{\mathscr{D}_t C_0}{u_p z}\left[\exp(u_p z/\mathscr{D}_t) - 1\right] \qquad (11.18)$$

where z is the effective depth of the liquid in the tank, and $\bar{C}$ is the average concentration of particles, equaling their total number divided by the volume of liquid in the tank.

The expression $u_p z/\mathscr{D}_t$ is dimensionless and may be denoted by symbol ϕ, whence:

$$\frac{\bar{C}}{C_0} = \frac{1}{\phi}(e^\phi - 1) \qquad (11.19)$$

When the impeller is motionless, all solid particles are at rest on the bottom of the tank. The height of the bed of these particles may be denoted by z_1 and their concentration in this bed is C_1. The relationship of the total effective depth of the liquid in the tank, the height of the bed of solid particles, their average concentration, and their concentration in the bed is defined by the following expression:

$$z\bar{C} = z_1 C_1 \qquad (11.20)$$

When the impeller is set into motion, the solids begin to disperse in the liquid. Their distribution in the liquid column follows a logarithmic relationship according to Equation 11.17. The concentration of the solid particles increases with depth until it attains the value C_1 at the surface of the fluid.

As the impeller speed increases, more solid particles are suspended and the height of the fluid decreases. At some certain speed, when the concentration at the bottom is still C_1, the concentration at some arbitrary height is equal to $C < C_1$.

Eventually, at a high enough impeller speed, the concentration at the bottom is lower than C_1. This corresponds to the case in which all solid particles are suspended. It follows that

$$\frac{\phi e^\phi}{e^\phi - 1} \leqslant \frac{z}{z_1} \qquad (11.21)$$

For most cases encountered in practice, $\phi > 1$, so that the term

$$\frac{e^\phi}{e^\phi - 1} = 1$$

and Equation 11.21 simplifies to

$$\phi \leqslant \frac{z}{z_1} \quad \text{or} \quad \mathscr{D}_t \geqslant u_p z_1 \tag{11.22}$$

At this condition, all particles are maintained in suspension, provided the coefficient of turbulent diffusion is equal to the product of the velocity of the particles and the height of the bed that would be formed by sedimentation of all particles onto the vessel bottom.

Zwietering [1957] endeavored to derive a universal relation to describe the optimum mixer arrangement for suspending solid particles in a liquid for constant ratios of tank diameter, and the distance of the impeller from the bottom. This equation has the form

$$\frac{nd^{0.85}}{\nu^{0.1}d_p^{0.2}\left(g\frac{\Delta\rho}{\rho_f}\right)^{0.45}w^{0.13}} = k\left(\frac{D}{d}\right)^a \tag{11.23}$$

where n = rotational velocity
    d = impeller diameter
    d_p = particle size
    w = content of solids (wt%)
    ν = kinematic viscosity
    $\Delta\rho$ = difference in densities between particles and liquid
    D = tank diameter
    k = coefficient
    a = exponent

Splitting up the left-hand side of Equation 11.23 into more conventional groups, we obtain

$$(nd^2/\nu)^{0.1}(\rho_f n^2 d/g\Delta\rho)^{0.45}(d/d_p)^{0.2}$$

Here the first group is a Reynolds number and the second a modified Froude number. For each type of impeller, a plot can be prepared with D/d on the abscissa and the following dimensionless form on the ordinate:

$$S' = \frac{nd^{0.85}}{\nu^{0.1}d_p^{0.2}\left(g\,\dfrac{\Delta\rho}{\rho_f}\right)^{0.45}w^{0.13}}$$

The ratio D/C is considered to be a parameter. Figure 11.25 presents the curves obtained by Zwietering [1957] for various impellers:

(A) for complete suspension with d/w = 2 and Re = 5.9;
(B) for complete suspension with d/w = 4 and Re = 2.5;
(C) for complete suspension using a turbine with 6 flat blades and Re = 6.2;
(D) for complete suspension using a vaned disk, Re = 4.6; and
(E) for complete suspension using a propeller mixer, Re = 0.5.

The correlation for complete suspension using a propeller presets a break that corresponds to a change in the flow pattern. Experiments have shown that for d < 0.45D, the solid particles move radially outward along the bottom; and when d > 0.45D, the solids move inward and centrally up into the propeller. Depending on the shape of propeller blades, a sudden decrease in efficiency can occur at low values of D/d. A similar effect is observed with axial turbine mixers; however, the homogeneity of the suspension is not influenced noticeably by the shape of the bottom of the tank.

BASIC DIMENSIONLESS GROUPS FOR MIXED LIQUID FLOW

The basic dimensionless relationship for determining the flow in any mixing system is the Euler number:

$$Eu = f(Re, Fr, \Gamma_1, \Gamma_2 \dots) \qquad (11.24)$$

where Γ_1 and Γ_2 are the simplexes of geometrical similarity.

We will define the contents of each dimensionless group and further determine all other variables that may affect the mixing process. In mixing with rotating mechanical impellers, fluid velocity is expressed in terms of the linear speed of the tip of the impeller, that is

$$\omega = \pi nd$$

where   n = the number of revolutions per second
        d = the diameter of the impeller

The characteristic length is usually taken to be the impeller diameter. Thus, the Reynolds number group is

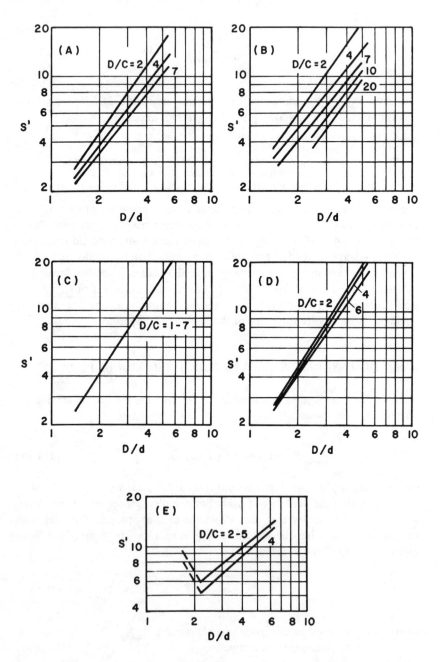

Figure 11.25 Correlations for complete particle suspension developed by Zwietering [1957]: W = vertical dimension of paddle impeller; Ne = $P/(\rho_f n^3 d^5)$ is the Newton number; ρ = power input to the impeller.

$$Re_M = \frac{\pi n d^2}{\nu} \qquad (11.25)$$

The modified Froude number is

$$Fr_M = \frac{n^2 d}{g} \qquad (11.26)$$

where constant π has been omitted.

The Euler number contains information on the required power input for mixing. The comparable expression based on pressure drop has the disadvantage that it is very difficult, if not impossible, to define the frictional losses across an impeller. Power input is, however, an easily measurable quantity for mixers. Therefore, the pressure drop is replaced in the Euler number by the power input. Recall that the conventional definition of the Euler number is

$$Eu = \frac{\Delta p}{\rho u^2} \qquad (11.27A)$$

where $u = \pi n d$

By leaving out constant π, we obtain

$$Eu_M = \frac{\Delta p}{\rho n^2 d^2} \qquad (11.27B)$$

If dimensional uniformity is to be achieved, the following dimensional adjustments must be made:

$$p = \frac{F'}{L^2} = \frac{F'L}{\theta} \cdot \frac{\theta}{L} \cdot \frac{1}{L^2} \qquad (11.28)$$

This means that the expression for power input, $\mathcal{P}_0$, must be divided by the speed, which is proportional to nd, and by the square of the basic dimension, d (the diameter of impeller):

$$Eu_M = \frac{\mathcal{P}_0}{\rho n^2 d^2 n d d^2} = \frac{\mathcal{P}_0}{\rho n^3 d^5} \qquad (11.29)$$

or, in terms of the specific weight,

$$Eu_M = \frac{\mathcal{N}_0 g}{\gamma n^3 d^5} = K_n \tag{11.30}$$

The Euler number expressed in the form of Equation 11.30 is called the *power number*, K_n.

HYDRODYNAMIC HEAD AND PUMPING CAPACITY

A turbine or propeller mixer is, in essence, a pump impeller operating without a casing and with undirected inlet and outlet flows. The governing relations studied for centrifugal pumps may be applied to mixers within limitation. The student may wish to refer to the works by Stepanoff [1957] and Wislicenus [1947] for in-depth treatments. The important quantities characterizing the action of a mixer which significantly influences turbulence and circulation are the volume of the liquid discharging from it per unit time, i.e., the *pumping capacity*, Q, and the overall *hydrodynamic head*, H. The latter form is calculated by dividing the energy expended for the motion of the impeller by the product of the liquid volume discharged per unit time from the impeller and its specific gravity (see Rushton and Oldshue [1953] for details).

The calculation of pumping capacity for mixers with radial flow is analogous to that of centrifugal pumps. Based on the work of Van de Vusse [1955] for the radial velocity, u_r, the following relationship is developed from the vector diagram of a turbine mixer (Figure 11.26):

$$u_r = c \sin\alpha = \omega \sin\beta \sqrt{1 - q'^2} \tag{11.31}$$

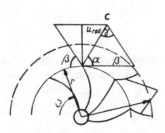

Figure 11.26   Vector diagram of a turbine mixer.

where   c = the absolute velocity of the liquid discharging from the blades

α = the angle included by the direction of the velocity c and the tangent to the circle described by the rotor of the mixer of radius r

β = the angle between the blade and the tangent to the circle described by the mixer

h = the height of the mixer blade

q' = the ratio of ω_ϱ/ω (ω_ϱ = angular velocity of the liquid, ω = angular velocity of the impeller), which indicates whether the flow prevailing in the tank is tangential or radial

At $q' = 1$, the flow is completely tangential; at $q' = 0$, it is completely radial.

The radial pumping capacity, Q, is defined as the product of the area of the peripheral ring, $2\pi rh$, and the radial velocity:

$$Q = 4\pi^2 nr^2 h\sin\beta\sqrt{1 - q'^2} = \pi^2 nd^2 h\sin\beta\sqrt{1 - q'^2} \qquad (11.32)$$

where   d = diameter of the rotor of the mixer

For paddle or turbine mixers with straight blades, $\beta = 90°$ and $\sin\beta = 1$, whence the expression simplifies to

$$Q = \pi^2 nd^2 h\sqrt{1 - q'^2} \qquad (11.33)$$

As, for tangential flow, $q' = 1$, the radial pumping capacity is $Q = 0$. Under conditions of radial flow, $q' = 0$, Equation 11.33 undergoes further simplification:

$$Q = \pi^2 nd^2 h \qquad (11.34)$$

This can be achieved by appropriately arranging baffles in the tank or using certain types of turbine mixers where the rotation of the liquid parallel to that of the impeller is prevented.

For turbine mixers with inclined blades, the resultant pumping capacity is the sum of radial pumping capacity, Q_{rad}, and axial pumping capacity, Q_{ax} (Figure 11.27).

The relative velocity, u_{rel}, between the blade, u_B, and the liquid, u_ϱ, in vector form is

$$\vec{u}_{rel} = \vec{u}_\varrho - \vec{u}_B \qquad (11.35)$$

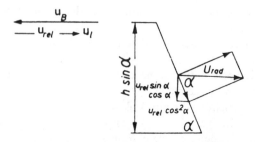

Figure 11.27 Vector diagram of a mixer with inclined blades.

Vector $\vec{u}_{rel}$ can be resolved into two components: one is perpendicular to the blade and the other parallel to it. Following Sterbacek and Tausk [1965], we may assume that the component perpendicular to the blade is cancelled. Hence, only the component of relative velocity in the direction of the rotation of the blade, $u_{rel} \cos\alpha$ (α is the angle of the incline of the blade), remains. The axial component of the relative velocity therefore is equal to $u_{rel} \cos\alpha \sin\alpha$, and the tangential component is $u_{rel} \cos^2\alpha$. The absolute velocity of the liquid in the direction tangential to the axis of rotation is

$$u_B - u_{rel} \cos^2\alpha = u_{rel} \sin^2\alpha + u_\varrho \tag{11.36}$$

Note that the tangential velocity is converted into radial via centrifugal action.

As the total mass of the moving liquid already has the tangential velocity u_ϱ, the tangential velocity in relation to the surrounding liquid becomes equal to $u_{rel} \sin^2\alpha$. The radial velocity is equal to the tangential velocity. Consequently, the total quantity of the liquid between the impeller blades consists of the axial velocity, which is equal to $u_{rel} \sin\alpha \cos\alpha$, and the radial velocity, equal to $u_{rel} \sin^2\alpha$.

The axial pumping capacity then is

$$Q_{ax} = \int_0^{1/2d} \omega_{rel} \left(\frac{1}{2}d\right) \sin\alpha \cos\alpha \pi d d\left(\frac{1}{2}d\right) = \omega_{rel} \sin\alpha \cos\alpha \frac{1}{12} \pi d^3 \tag{11.37}$$

and the radial pumping capacity equals

$$Q_{rad} = \omega_{rel} \left(\frac{1}{2} d \right) \sin^2 \alpha \, \alpha \pi dh \sin \alpha = \frac{\pi}{2} \, \omega_{rel} \sin^3 \alpha d^2 h \qquad (11.38)$$

If these equations are employed for mixers with straight blades, i.e., for $\alpha = 90°$, then $Q_{ax} = 0$ while for Q_{rad} the same value is obtained as derived for purely radial flow (namely, $Q_{rad} = \pi^2 n d^2 h$). To achieve predominantly axial flow, propeller mixers with blades of helical shapes and constant pitch, s, are employed. When applying the derivation for the pumping capacity of mixers with pitched blades, it must be remembered that the angle is not constant for propellers. The variation of the angle α can be expressed by the following relation:

$$\tan \alpha = \frac{s}{2\pi r} \qquad (11.39)$$

where r changes from 0 at the shaft to $\frac{1}{2}$ d at the tip of the propeller blade:

$$\sin \alpha = \frac{s}{\sqrt{s^2 + 4\pi^2 r^2}} \qquad (11.40)$$

and

$$\cos \alpha = \frac{2\pi r}{\sqrt{s^2 + 4\pi^2 r^2}} \qquad (11.41)$$

The expression for axial pumping capacity then assumes the following form:

$$Q_{ax} = \int_0^{d/2} 2\pi\omega \sin\alpha \cos\alpha r^2 dr = 4\pi\omega^2 s \int_0^{d/2} \frac{r^3 dr}{s^2 + 4\pi^2 r^2}$$

$$= \frac{1}{8} \, \omega s d^2 \left\{ 1 - \frac{s^2}{\pi^2 d^2} \left(1 + \frac{\pi^2 d^2}{s} \right) \right\} \qquad (11.42)$$

To calculate the radial pumping capacity of a propeller mixer, the radial velocity must be determined from the relation

$$\frac{1}{2} \rho u_{rad}^2 = \int_0^{d/2} \rho\omega^2 \sin^2\alpha r dr = \frac{1}{2} \rho\omega^2 \frac{s^2}{4\pi^2} \ln \left(1 + \frac{\pi^2 d^2}{s^2} \right) \qquad (11.43)$$

The radial pumping capacity of propeller mixers then follows as

$$Q_{rad} = u_{rad}\pi dh \sin\alpha = \frac{\omega dsh}{2\sqrt{1 + \frac{\pi^2 d^2}{s^2}}} \sqrt{ln\left(1 + \frac{\pi^2 d^2}{s^2}\right)} \qquad (11.44)$$

When the energy losses in the impeller itself are negligible, the product of the specific weight, γ, pumping capacity, Q, and the hydrodynamic head, H, gives the power input $\mathcal{W}_0$ required to establish liquid motion:

$$\mathcal{W}_0 = \gamma QH \qquad (11.45)$$

As follows from Equation 11.30, the general equation for the power input of a mixer has the form $\mathcal{W}_0 = (K_N/g)(\gamma n^3 d^5)$, which contains three easily measured variables. The combination of Equations 11.30 and 11.45 provides the interrelations of the individual variables.

To evaluate the conditions of mixing for a particular design, one of the variables in Equation 11.30 is kept constant. Data reported by Sterbacek and Tausk [1965] for this purpose are given in Table 11.5. These data comprise a survey of the relation of measured values ($\mathcal{W}_0$, n, d) to the hydrodynamic head, H, and pumping capacity, Q, which define the turbulence and circulation in the mixing tank.

Table 11.5 shows that by increasing the speed at a constant power input at the expense of impeller diameter, a higher hydrodynamic head, H, is obtained according to the relation $H \sim n^{4/5}$. However, the pumping capacity, Q, decreases ($Q \sim n^{-4/5}$), and the ratio Q/H diminishes according to $Q/H \sim n^{-8/5}$. This means that turbulence increases at the expense of circulation.

Increasing impeller diameter at the expense of speed (at constant power

Table 11.5 Typical Relations Between Mixer Variables and Flow
(From Sterbacek and Tausk [1965])

At Constant	$\mathcal{W}_0$	d	n	Q	H	Q/H
$\mathcal{W}_0$	–	$n^{-3/5}$	–	$\mathcal{W}_0 n^{-4/5}$	$n^{4/5}$	$n^{-8/5}$
	–	–	$d^{-5/3}$	$d^{4/3}$	$d^{-4/3}$	$d^{8/3}$
d	n^3	–	$\mathcal{W}_0^{1/3}$	n	n^2	n^{-1}
n	d^5	$\mathcal{W}_0^{1/5}$	–	d^3	d^2	d

input) produces a drop in the hydrodynamic head, H, according to the relation $H \sim d^{-4/3}$, while the pumping capacity increases, according to $Q \sim d^{4/3}$, and ratio Q/H rises ($Q/H \sim d^{8/3}$). Hence, circulation increases at the expense of turbulence.

By means of properly planned batch experiments, the time needed for a specified mixing effect can be determined for a certain ratio of Q/H. If this time is shortened by increasing Q/H, it is more advantageous to design a mixing system with a higher pumping capacity, Q, at the expense of turbulence. If, however, it is necessary to extend the mixing time by increasing the value of Q/H, a higher intensity of turbulence will be required. Table 11.5 shows that this may be attained by decreasing the mixer diameter and increasing its speed.

In cases such as the homogenization of mixtures, the highest possible number of passages through turbulent zones is more important than an overall high rate of turbulence. On the other hand, there are situations in which the attainment of a very intense turbulence is of principal importance. An example of this is gas absorption.

For centrally located impellers, the maximum volumetric flowrate is encountered at about two-thirds the distance between the center and the periphery of the tank (Sachs [1954]). Nagata et al. [1959] found that at low Reynolds numbers the discharge near the periphery of the impeller blades exhibits broad width and low velocity. With an increase in Reynolds number, the width of the flow contracts with a simultaneous increase in velocity. This corresponds to a sharper velocity profile. The coefficient of discharge is defined by the following relation:

$$K_Q = \frac{Q}{nd^3} \tag{11.46}$$

where Q is the pumping capacity of the impeller at the periphery of the blades, and n and d are defined as before.

Table 11.6 compares the discharge coefficients for three types of impellers at various Reynolds number values. From the data in this table, it is evident that the most intense discharge from the impeller occurs in the transition range.

POWER INPUT IN AGITATED TANKS

The basic functional relation between the dimensionless groups (Equation 11.24) for mixing processes may be expressed in the following form:

Table 11.6 Data on Discharge Coefficient as a Function of Reynolds Number
(From Nagata et al. [1959])

Type of Impeller	Re	K_Q
Impeller with Eight Straight Blades	6.7	0.28
	14	0.37
	39	0.80
	89	0.86
	830	0.57
	10^5	0.34
Impeller with Eight Swept-Back Blades	23	0.45
	43	0.65
	78	0.67
	700	0.63
	10^5	0.43
Eight-Blade Turbine Impeller	17	0.42
	38	0.58
	84	0.62
	630	0.60
	10^5	0.34

$$K_N = f(Re_M, Fr_M, \Gamma_1, \Gamma_2 \ldots) \tag{11.47A}$$

or

$$K_N = A Re_M^m Fr_M^n \Gamma_1^p \Gamma^q \ldots \tag{11.47B}$$

The gravity force influences vortex formation and waves on the free surface of the mixing liquid. If the tank has baffles or the shaft of the impeller is located eccentrically relative to the tank axis, it is possible to neglect the influence of gravity forces. In this case, the modified Froude number may be dropped from Equation 11.47B:

$$K_N = \psi(Re_M, \Gamma_1, \Gamma_2 \ldots) \tag{11.48A}$$

or

$$K_N = A' Re_M^{m'} \Gamma_1^{p'} \Gamma_2^{q'} \ldots \tag{11.48B}$$

Equations 11.47 and 11.48 are used for calculating the power consumed by a mixer. The values of coefficients A and A', as well as powers m, n, p, q and

m', n', p', q' must be determined experimentally. Rushton et al. [1950] have shown that they depend on the mixer type, its design and the mixing regime.

For convenience, experimental data on power input may be plotted as a function of the power number, K_N, versus the modified Reynolds number, Re_M, with the geometric simplexes Γ_1, Γ_2 . . . and the modified Froude number, Fr_M, as parameters. Stepanoff [1957] provides such a plot, a portion of which is presented in Figure 11.28. The numbers for each curve correspond to the respective turbine impeller configurations shown in Figure 11.29.

The geometric characteristics of the mixers and tanks for which the curves in Figure 11.28 are traced $[K_N = \psi(Re_M)]$ are summarized in Table 11.7. The character of the flow pattern is of great importance in mixing. It depends on the velocity of the flow, the fluid viscosity and the geometric arrangement of the space through which the liquid passes. Practical mixing applications are performed in either the laminar or turbulent regime.

The *laminar regime* ($Re_M < 30$) corresponds to nonintensive mixing in which liquid moves in streamlines along the edges of the mixer blades and is entrained by the blades and rotated along with them.

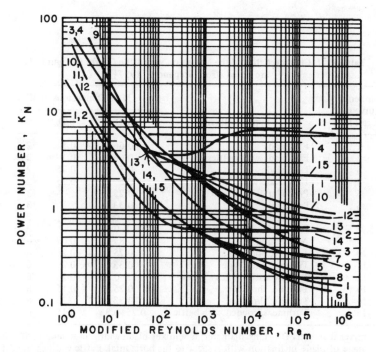

Figure 11.28   Power number, K_N, versus modified Reynolds number, Re_M, for mechanical agitators.

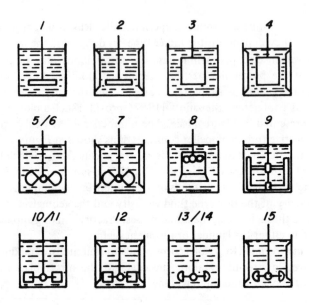

Figure 11.29   Turbine impeller designs.

Table 11.7 Characteristics of the Mixers on which the Curves in Figure 11.28
are Based (from Stepanoff [1957])

Number of Curve in Figure 11.28	Type of Mixer	Main Dimensions of Mixers[a]			
		d/D	b/D	n'	α
1	Flat blade	0.66	0.1	2	90°
2	Flat blade with baffles	0.66	0.1	2	90°
3	Leaf blade	0.5	0.75	2	90°
4	Leaf blade with baffles	0.5	0.75	2	90°
5	Propeller	0.25	–	3	40°
6	Propeller	0.33	–	3	40°
7	Propeller with baffles	0.25–0.33	–	3	40°
8	Propeller in draft tube	0.2–0.33	–	3	40°
9	Anchor and gate	0.87	0.07	–	90°
10	Turbine, open type	0.25	0.2	6	90°
11	Turbine, open type	0.33	0.2	6	90°
12	Turbine with baffles	0.25–0.33	0.2	6	90°
13	Turbine, shrouded	0.25	0.15	6	90°
14	Turbine, shrouded	0.33	0.15	6	90°
15	Turbine, shrouded with baffles	0.25–0.33	0.15	6	90°

[a]d = mixer diameter; D = tank diameter; b = mixer blade width; n' = number of blades;
α = angle of blade inclination with respect to the horizontal. Baffle widths were 0.1 D;
number of baffles = 4; liquid column in the tank corresponded to H = D.

Only those layers that are close to the mixer blade actually are mixed in the laminar regime. With an increase in mixer speed, the resistance of the medium to the rotation of the mixer increases as it is induced by turbulization of the boundary layer and the formation of a turbulent wake in the zone behind the moving blades.

At $Re_M > 10^2$, the turbulent mixing regime develops. Here, mixing is characterized by a less clearly defined relationship between power number K_N and Re_M. In the range of fully developed turbulence, $Re_M > 10^5$, the power number is practically independent of Re_M. In this region (the self-modeling regime), power consumption is determined only by inertia forces. Further increases in speed do result in more intensive mixing; however, the subsequent increase in power consumption is not compensated for by the resulting mixing effects.

Note that the critical Reynolds number values, which define the regime boundaries, are only approximations. Values can vary significantly depending on the specific designs and sizes of both mixer and tank. For heterogeneous systems, the value of continuous-phase density is used in computing the Reynolds and power numbers if the densities of the phases to be mixed differ by no more than 30%. For greater density differences, it is necessary to use an average mixture density, as determined by an additivity rule:

$$\rho_{avg} = \rho_d \psi' + \rho_c (1 - \psi') \tag{11.49}$$

where ρ_d, ρ_c = the densities of dispersed and continuous phases, respectively
ψ' = the void fraction of the dispersed phase

The equation to be used for estimating the viscosity of a mixture, μ_m, depends on the purpose and condition of the mixing process. If in the mixing of a liquid-liquid system the viscosity of the dispersed phase, μ_d, is higher than that of the continuous phase, μ_c, and the fraction of dispersed phase in the volume $\psi' \geqslant 0.3$, then the Reynolds number may be computed based on a mixture viscosity as calculated from the following relations of Laity and Treybal [1957]:

$$\mu_m = \frac{\mu_c}{1 - \psi'} [1 + 6\psi' \mu_d / \mu_c + \mu_d] \tag{11.50}$$

At $\mu_d < \mu_c$ and $\psi' \geqslant 0.3$:

$$\mu_m = \frac{\mu_c}{1 - \psi'} [1 - 1.5\psi'\mu_d/\mu_c + \mu_d] \qquad (11.51)$$

In the mixing of miscible liquids, if $\psi' \geqslant 0.4$ and the viscosity of the liquids differs by a factor of more than two, an effective viscosity defined by the geometric mean may be used (see Rushton and Oldshue [1953]):

$$\mu_m = \mu_c^{1-\psi'}\mu_d \qquad (11.52)$$

Suspending solid particles in a liquid increases the viscosity of the resulting mixture. The viscosity of suspensions is defined approximately by the Einstein relation:

$$\mu_m = \mu_c(1 + 2.5\psi') \qquad (11.53)$$

where ψ' is the ratio of the volume of the suspended particles to the total volume of the suspension. This equation holds for suspended particles that are spherical in shape and for low concentrations (up to 4%).
  For ψ' greater than 10%,

$$\mu_m = \mu_c(1 + 4.5\psi') \qquad (11.54)$$

In all other cases, the viscosity of the continuous phase, μ_c, may be used in the Reynolds number.
  When the liquid level is not equal to the tank diameter, the power number obtained from the plots in Figure 11.28 should be multiplied by a correction factor:

$$K'' = \left(\frac{H}{D}\right)^{0.5} \qquad (11.55)$$

If mixing takes place in tanks with rough walls, as well as in the presence of internals (e.g., coils, thermometer sockets, etc.), the power consumed for mixing is considerably increased only if the tank is not baffled. A simple heating coil in a tank can, for example, increase the consumed power 2 to 3 times. A socket for a thermometer or a liquid-level monitoring probe can increase power consumption by 1.1 to 1.2 times. The power consumption in tanks with rough walls is increased 10–20%.

Selection of an electric motor for a mixer drive is based on the value of the power consumption divided by the efficiency of the gearbox. Provision should be made to account for the increase in shaft torque during startup. The starting power usually is two times higher than the working power and, therefore, motors with phase rings are recommended. The power consumed by mixers may be presented in the following general form:

$$\mathcal{R} = K(Re_M)^a(Fr_M)^{a'}\left(\frac{D}{d}\right)^c\left(\frac{H}{d}\right)^e\left(\frac{t}{d}\right)^f\left(\frac{b}{d}\right)^g\left(\frac{h}{d}\right)^i\left(\frac{\ell}{d}\right)^j\left(\frac{s}{d}\right)^P\left(\frac{m_1}{m_0}\right)^q\left(\frac{n'}{m_0}\right)^r$$

(11.56)

where $Re_M = nd^2\rho/\mu$ = modified Reynolds number

$Fr_M = n^2d/g$ = modified Froude number

$\mathcal{R}$ = power input (kW)

K = proportionality constant

n = rotational speed (rpm)

D = diameter of vessel (m)

H = hydrodynamic head (m)

t = width of a baffle measured along the tank radius (m)

d = mixer diameter (m)

h = height of mixer blade (m)

s = propeller pitch (m)

ℓ = length of turbine blade (m)

b = space from lower edge of mixer to the tank bottom (m)

m_0 = dimensionless parameter, depending on the condition of experiments on which design formulas are based

m_1 = number of baffles

n' = number of blades of the turbine mixer

g = acceleration due to gravity (m/s^2)

μ = viscosity of medium (kg$_f$-s/m^2)

$\rho = \gamma/g$ = density of the medium (kg$_f$-s^2/m^4)

γ = specific gravity of the medium (kg$_f$/m^3)

Equation 11.56 is based on a significant amount of experimental data (Stepanoff [1957], Van de Vusse [1955], Nagata et al. [1959], Rushton et al. [1950]). It is the developed form of Equation 11.48B, with a series of invariants from geometric similarity.

Millon [1953] attempted to generalize the multitude of data and correlations for power input to provide a complete graphic representation for all conditions. Based on Millon's analysis, the results of all investigations are approximated by

$$\mathcal{n} = SQ'E'\Pi \left(\frac{A}{d}\right)^x \qquad (11.57)$$

where Π = the product of all parameters $(A/d)^x$, where A/d is any of the
         dimensionless parameters in Equation 11.56
     x = exponent (c, e, f . . .) in Equation 11.56 corresponding to the
         specific parameter A/d
     $S = \mu n^2 d^3$ = the power coefficient,$(kg_f\text{-}m/s)$
     $Q' = K(Re)$ = dimensionless number
     $E' = (Fr_M)^{a''}$ = dimensionless number

The influence of individual parameters entering in Equation 11.56 is different and depends on the mixer design, rotational speed, etc. For example, in the absence of baffles, exponent q = 0, and for a paddle mixer, p = 0, etc. The Froude number has no influence on power input when there is no vortex. It is sufficient to mount four baffles, t = (0.1 – 0.12)D, on the wall of the tank along its generatrix to avoid a vortex. The same effect may be achieved by eccentric location of the mixer or by employing an inclined propeller axis. In all these cases, $a'' = 0$ and $E' = 1$.

For the general case, Stepanoff [1957] recommends the following expression for the exponent:

$$a'' = \frac{K - \log Re}{w} \quad \text{and} \quad E' = (Fr_M)^{(K-\log Re)/w} \qquad (11.58)$$

Based on the results of different investigators, Millon obtained formulas estimating power input, $\mathcal{n}$, for many different types of commercial mixers. Nomograms based on these design formulas are given in Figures 11.30–11.34. Figure 11.30 allows determination of the power coefficient, S. Connecting the points corresponding to values n (rpm) and d (mixer diameter), we obtain on the auxiliary line through point A.

Tracing the line through point A and the point corresponding to the liquid viscosity (in $kg_f\text{-}s/m^2$), we find at the intersection with axis S the desired value of the power coefficient.

Figure 11.31 serves for the calculation of the Reynolds number, Re_M. Connecting the corresponding points of axes n and d, we determine point A

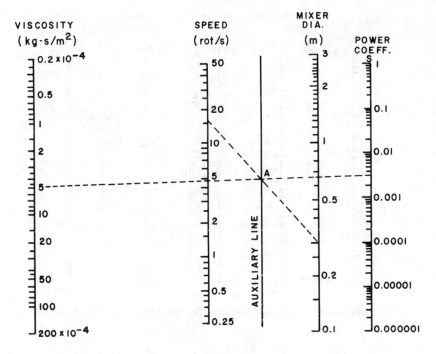

VISCOSITY
$(kg \cdot s/m^2)$

SPEED
(rot/s)

MIXER
DIA.
(m)

POWER
COEFF.

AUXILIARY LINE

Figure 11.30   Nomogram for determining the power coefficient.

on the first auxiliary line. Point A is then connected with the point on the axis of relative density ρ_0 (where $\rho_0 \simeq \gamma/100$ g), thus locating point B on the second auxiliary line. Then point B is connected with the point on the viscosity axis and the Reynolds number is determined.

Figure 11.32 is used for calculating Q. For each type of mixer, one of the points $(1, 2, 3 \dots 31)$ on the second line corresponds to the right and one of the points on the middle line A, B, C, ..., P, R. For example, let points 1 and N correspond to the given type of the mixer under consideration. The line is traced through the point on the Re_M-axis and point N to the intersection with the auxiliary axis at point A. Point A then should be connected with point 1, and the intersection of the line A-1 with the Q-axis gives the value of Q. The points that should be used in the calculation for a specified type of mixer are given in Tables 11.8 through 11.9 and in discussions on specific mixers in following sections.

Figure 11.33 serves to calculate E'. Because the determination of E' is different for each case, the technique of using this nomogram is illustrated for specific types of mixers in the example problems that follow.

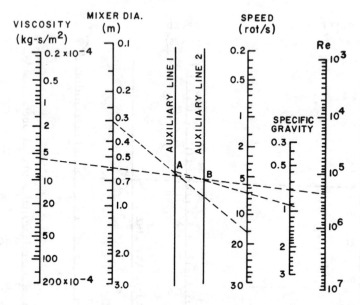

Figure 11.31   Nomogram for calculating the Reynolds number, Re_M.

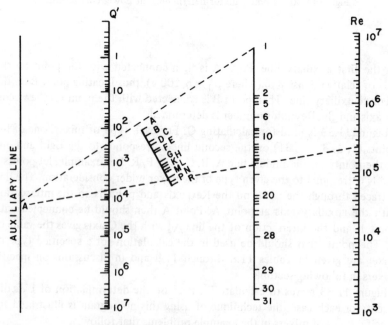

Figure 11.32   Nomogram for calculating Q'.

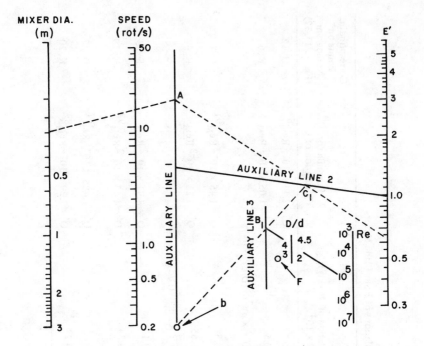

Figure 11.33   Nomogram for calculating the E'-number.

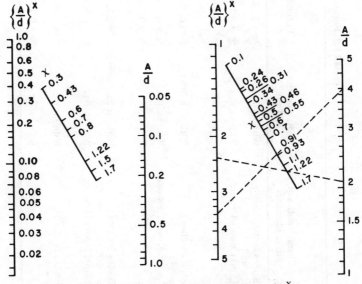

Figure 11.34   Nomogram for calculating $(A/d)^X$.

Table 11.8 Formulas for Calculations of Four-Blade Mixers

Characteristics of Mixer Elements	Characteristics of Tank	Formulas for Calculation of $\mathcal{N}$ (kW); $Re_M > 4 \times 10^4$	Formula for Calculation of Q'	Corresponding Points on Nomogram (Figure 11.32)
Blades Inclined $\alpha = 45°$; Liquid Moves Upward	Without baffles	$\mathcal{N} = 0.736 \, SQ' \, (D/d)^{0.26}$	$Q' = 2.11 \times 10^{-2} \, Re_M^{0.88}$	H; 12
	With baffles	$\mathcal{N} = 0.736 \, SQ' \, (D/d)^{0.5}$ $\mathcal{N} = 0.736 \times 105 \times 10^{-2} \times \rho n^3 d^5 (D/d)^{0.5}$	$Q' = 1.05 \times 10^{-2} \, Re_M$	R; 6
Blades Inclined $\alpha = 45°$; Liquid Moves Downward	Without baffles	$\mathcal{N} = 0.736 \, SQ' \, (D/d)^{0.31}$	$Q' = 1.45 \times 10^{-2} \, Re_M^{0.91}$	L; 7
	With baffles	$\mathcal{N} = 0.736 \, SQ' \, (D/d)^{0.55}$ $\mathcal{N} = 0.736 \times 7.2 \times 10^{-3} \times \rho n^3 d^5 (D/d)^{0.8}$	$Q' = 7.21 \times 10^{-3} \, Re_M$	R; 4
Blades Inclined $\alpha = 60°$; Liquid Moves Upward	Without baffles	$\mathcal{N} = 0.736 \, SQ' \, (D/d)^{0.1}$	$Q' = 8.91 \times 10^{-2} \, Re_M^{0.8}$	E; 23
	With baffles	$\mathcal{N} = 0.736 \, SQ' \, (D/d)^{0.34}$	$Q' = 4.43 \times 10^{-2} \, Re_M^{0.92}$	M; 19
Blades Inclined $\alpha = 90°$	Without baffles	$\mathcal{N} = 0.736 \, SQ'$	$Q' = 1.485 \times 10^{-1} \, Re_M^{0.77}$	D; 27
	With baffles	$\mathcal{N} = 0.736 \, SQ' \, (D/d)^{0.24}$	$Q' = 7.39 \times 10^{-2} \, Re_M^{0.89}$	K; 20

Table 11.9 Formulas for Calculations of Leaf-Paddle Mixers

Type Tank	Refer to Figure No.	Re_M	Formulas for Calculation of $\mathcal{N}$ (kW)	Formula for Calculation of Q'	Corresponding Points on Nomogram (Figure 11.32)
Tank Without a Coil	11.25 (A)	$<6.5 \times 10^4$	$\mathcal{N} = 0.736\, Q' \dfrac{h}{d}\left(\dfrac{H}{d}\right)^{0.46}$	$Q' = 2.013 \times 10^{-1}\, Re_M^{0.7}$	A; 28
		$>6.5 \times 10^4$	$\mathcal{N} = 0.736\, Q' \dfrac{h}{d}\left(\dfrac{H}{d}\right)^{0.6}$ $\mathcal{N} = 0.736 \times 6.52 \times 10^{-3}\, \rho n^3 d^4 h \left(\dfrac{H}{d}\right)^{0.6}$	$Q' = 6.52 \times 10^{-3}\, Re_M$	R; 3
Tank With a Coil	11.25 (B)		$\mathcal{N} = 0.736\, SQ' \dfrac{h}{d}\left(\dfrac{H}{d}\right)^{0.6}$ $\mathcal{N} = 0.736 \times 2.02 \times 10^{-2}\, \rho n^3 d^4 h \left(\dfrac{H}{d}\right)^{0.6}$	$Q' = 2.02 \times 10^{-2}\, Re_M$	R; 11
	11.25 (C)		$\mathcal{N} = 0.736\, SQ' \dfrac{h}{d}\left(\dfrac{H}{d}\right)^{0.6}$ $\mathcal{N} = 0.736 \times 2.17 \times 10^{-2}\, \rho n^3 d^4 h \left(\dfrac{H}{d}\right)^{0.6}$	$Q' = 2.17 \times 10^{-2}\, Re_M$	R; 13
		$>6.5 \times 10^4$			
	11.25 (D)		$\mathcal{N} = 0.736\, SQ' \dfrac{h}{d}\left(\dfrac{H}{d}\right)^{0.6}$ $\mathcal{N} = 0.736 \times 2.32 \times 10^{-2}\, \rho n^3 d^4 h \left(\dfrac{H}{d}\right)^{0.6}$	$Q' = 2.32 \times 10^{-2}\, Re_M$	R; 14
	11.25 (E)		$\mathcal{N} = 0.736\, SQ' \dfrac{h}{d}\left(\dfrac{H}{d}\right)^{0.6}$ $\mathcal{N} = 0.736 \times 2.71 \times 10^{-2}\, \rho n^3 d^4 h \left(\dfrac{H}{d}\right)^{0.6}$	$Q' = 2.71 \times 10^{-2}\, Re_M$	R; 16

Figure 11.34 serves to calculate the values of $(A/d)^x$. The nomogram consists of two parts: the values of A/d in the range from 1 to 5 are given on the right-hand side and the values of A/d in the range from 0.05 to 1 are on the left-hand side of the nomogram.

If, for a given case, the parameters of the system are beyond the stated limits, the Newton-Karman equation is recommended:

$$p = C' \frac{\gamma}{g} F\nu^2 \tag{11.59}$$

where parameter C' is defined from

$$C' = (0.2 - 0.3) \sqrt[7]{\nu} \tag{11.60}$$

where ν = kinematic viscosity (m^2/s)

The following sections outline design methodology and illustrate the use of the design nomograms for specific mixer types.

DESIGN OF PADDLE MIXERS

An example of a two-blade mixer is given in Figure 11.35. At $10^4 < Re_M < 10^7$, the power input for this type mixer is determined from the following formula:

$$\left. \begin{array}{l} \mathcal{n} = 0.736SQ' \left(\frac{D}{d}\right)^{1.1} \left(\frac{h}{d}\right)^{0.3} \left(\frac{H}{d}\right)^{0.6} \\ Q' = 2.954 \times 10^{-2}(Re_M)^{0.86} \end{array} \right\} \tag{11.61}$$

To evaluate Q' from the nomogram given in Figure 11.32, points G and 17 should be used.

If there are several two-blade mixers located along the height of the shaft, then

$$\mathcal{n}_k = 0.7 \times k_m \mathcal{n} \tag{11.62}$$

where k_m = the number of mixers

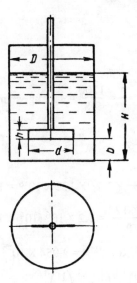

Figure 11.35 Scheme of a paddle mixer.

For mixers with inclined blades, $\alpha \neq 90°$, the power input is

$$\mathcal{N}_\alpha = \mathcal{N} \sin\alpha \qquad (11.63)$$

where $\mathcal{N}$ is determined from Equation 11.61.

Reported literature values for the power input of mixers with inclined blades, $\alpha = 45°$, are in the range of $0.55\,\mathcal{N}$ to $0.65\,\mathcal{N}$.

Example 11.1

Determine the power input for a mixer with the following characteristics:

- tank diameter, D = 2000 mm
- liquid height, H = 2000 mm
- mixer diameter, d = 1200 mm
- blade height, h = 150 mm
- rpm = 50, or 0.83 rot/s
- angular velocity = 5 rad/s
- specific gravity = 1200 kg_f/m^3
- viscosity = 122 $\times$ 10^{-6} $kg_f\text{-}s/m^2$

Solution

Liquid density is

$$\rho = \frac{\gamma}{g} = \frac{1200}{9.8} = 122 \text{ kg}_f\text{-s}^2/\text{m}^4$$

From Equation 11.57,

$$S = n^2 d^3 \mu = 0.83^2 \times 1.2^3 \times 122 \times 10^{-6} = 145 \times 10^{-6}$$

$$\text{Re}_M = \frac{0.83 \times 1.2^2 \times 122}{122 \times 10^{-6}} = 1.2 \times 10^6 \, (10^4 < \text{Re}_M < 10^7)$$

$$Q' = 2.954 \times 10^{-2} (\text{Re}_M)^{0.86} = 2.954 \times 10^{-2} (1.2 \times 10^6)^{0.86} = 4994$$

$$\mathcal{N} = 0.736 SQ' \left(\frac{D}{d}\right)^{1.1} \left(\frac{h}{d}\right)^{0.3} \left(\frac{H}{d}\right)^{0.6}$$

$$= 0.736 \times 145 \times 10^{-6} \times 4994 \left(\frac{2}{1.2}\right)^{1.1} \times \left(\frac{0.15}{1.2}\right)^{0.3} \times \left(\frac{2}{1.2}\right)^{0.6}$$

$$\mathcal{N} = 0.67 \text{ kW}$$

Let us now consider a *four-blade mixer with vertical inclined blades* (Figure 11.36). The formulas summarized in Table 11.8 are recommended for the design of this type mixer. These formulas may be applied when the following conditions are fulfilled:

- $\text{Re}_M > 4 \times 10^4$
- $h = D/12$
- $b = D/6$
- $2.3 < D/d < 5.2$
- $H \simeq D$
- $b = b_1$
- $t = D/12$

The influences of b and H are not great if the blades are completely covered with fluid.

Leaf-paddle mixers are shown in Figure 11.37. For design of this type of mixer, the formulas given in Table 11.9 are recommended. These formulas may be applied at the conditions of

- $D/d = 2$
- $h/d \simeq 0.9$
- $b/d = 0.36$
- $1 < H/d < 3$

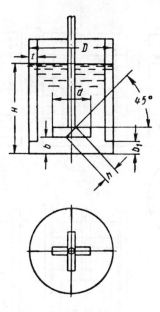

Figure 11.36   A four-blade mixer with inclined blades.

The influence of b on power input is not great, but at b/d = 0.36 it is at a minimum.

DESIGN OF ANCHOR MIXERS

A *simple anchor mixer* is shown in Figure 11.38. The power input is determined from the following formula:

$$\mathcal{N} = 0.736 SQ' \frac{h}{d} \qquad (11.64)$$

where $Q' = 1.052 \times 10^{-1} Re_M^{0.77} (10^2 < Re_M < 3 \times 10^5)$

To determine Q' from Figure 11.32, points D and 24 should be used.
Finger anchor mixers are illustrated in Figure 11.39. The power input is determined from the following formula:

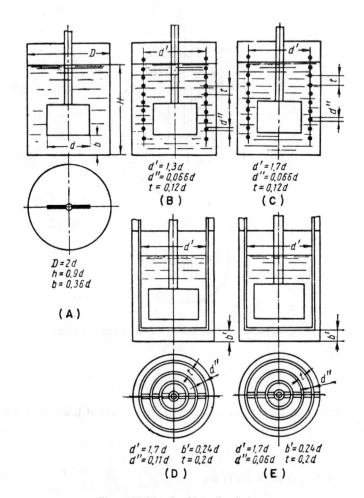

Figure 11.37   Leaf-impeller designs.

$$\mathcal{V} = 0.736 SQ' \frac{h}{d} \tag{11.65}$$

where $Q' = 1.046 \times 10^{-1} Re_M^{0.763} \; (10^2 < Re_M < 3 \times 10^5)$

For *finger-anchor mixers with baffles* (geometry shown in Figure 11.39(A)) with $\ell' = 0.06D$ and $b' = 0.06D$, use $Q' = 1.42 \times 10^{-1} Re_M^{0.736} \; (10^2 < Re < 6.5 \times 10^3)$ with Equation 11.65. To use Figure 11.32, apply points C and 26. If $Re_M > 6.5 \times 10^3$, then $Q = 1.75 \times 10^{-2} Re_M$ and

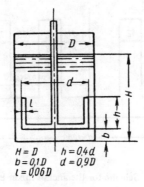

Figure 11.38   A simple anchor mixer.

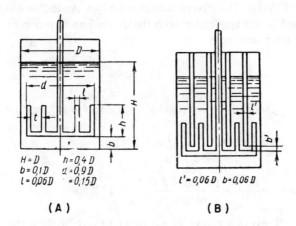

Figure 11.39   Finger-anchor mixers (A) with and (B) without stationary baffles.

$$\mathcal{N} = 0.736 \times 1.75 \times 10^{-2}\rho n^3 d^4 h \tag{11.66}$$

With Figure 11.32, points R and 10 should be used.

Example 11.2

A liquid with specific gravity $\gamma = 1000$ kg$_f$/m^3 and viscosity $\mu = 1020 \times 10^{-6}$ kg$_f$-s/m^2 is to be mixed in a tank (D = 1200 m). The liquid height in

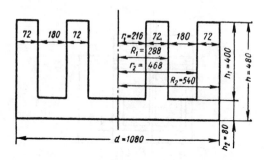

Figure 11.40   Scheme for the anchor mixer in Example 11.2.

the tank is H = D; the mixer diameter is d = 0.9D = 1080 mm; the clearance is b = 0.1D = 120 mm. The mixer operates at 60 rpm. An anchor mixer-impeller is to be used in this application with the dimensions shown in Figure 11.40. Determine the power input.

Solution

$$S = n^2 d^3 \mu = 1^2 \times 1.08^3 \times 1020 \times 10^{-6} = 1285 \times 10^{-6}$$

$$\rho = \frac{1000}{9.81} = 102 \text{ kg}_f\text{-s}^2/\text{m}^4$$

$$\text{Re}_M = \frac{nd^2\rho}{\mu} = \frac{1 \times 1.08^2 \times 102}{1020 \times 10^{-6}} = 1.17 \times 10^5$$

$$Q' = 0.142(\text{Re}_M)^{0.763} = 0.142(1.17 \times 10^5)^{0.763} = 1043$$

The power input is thus

$$\mathcal{N} = 0.736 \times S \times Q' \times \frac{h}{d} = 0.736 \times 1043 \times 1285 \times 10^{-6} \times \frac{0.48}{1.08} = 0.44 \text{ kW}$$

For *round anchor mixers*, consider an arch of a circular ring with radii R and r, which rotates with an angular velocity, ω (Figure 11.41). The area of a surface element is

$$dF = \rho d\psi d\rho \tag{11.67}$$

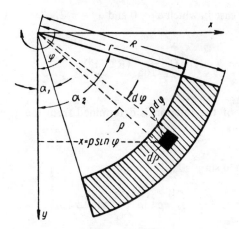

Figure 11.41   Scheme of the general case of a round anchor mixer.

The unit force acting on this element is

$$dP = C' \frac{\gamma}{g} \, dF\vartheta^2 = C' \frac{\omega^2\gamma}{g} \rho^3 \sin^2\psi d\psi d\rho \qquad (11.68)$$

where $\vartheta = \omega x = \omega\rho \sin\psi$, the velocity of the element

Integrating this expression, we obtain

$$P = C' \frac{\omega^2\gamma}{g} \int_r^R \rho^3 d\rho \int_{\alpha_1}^{\alpha_2} \sin^2\psi d\psi \qquad (11.69)$$

Taking into account that

$$\int_{\alpha_1}^{\alpha_2} \sin^2\psi d\psi = \left[\frac{\psi}{2} - \frac{\sin 2\psi}{4} \right]_{\alpha_1}^{\alpha_2} = \frac{\alpha_2 - \alpha_1}{2} - \frac{\sin 2\alpha_2 - \sin 2\alpha_1}{4}$$

we obtain

$$P = C' \frac{\gamma\omega^2}{4g} (R^4 - r^4) \left[\frac{\alpha_2 - \alpha_1}{2} - \frac{\sin 2\alpha_2 - \sin 2\alpha_1}{4} \right] \qquad (11.70)$$

For the specific case in which $\alpha_1 = 0$ and $\alpha_2 = \pi/2$,

$$P = C' \frac{\pi \gamma \omega^2}{16g} (R^4 - r^4) \qquad (11.71)$$

The coordinate of the fulcrum is determined from the general equations at the conditions that

$$x = \rho \sin \psi \; ; \quad y = \rho \cos \psi$$

$$x_0 = C' \frac{\gamma \omega^2}{gP} \int_r^R \rho^4 d\rho \int_{\alpha_1}^{\alpha_2} \sin^3 \psi d\psi \left.\right\} \qquad (11.72A)$$

$$y_0 = C' \frac{\gamma \omega^2}{gP} \int_r^R \rho^4 d\rho \int_{\alpha_1}^{\alpha_2} \sin^2 \psi \cos \psi d\psi \qquad (11.72B)$$

Noting that

$$\int_{\alpha_1}^{\alpha_2} \sin^3 \psi d\psi = \left[\frac{\cos^3 \psi}{3} - \cos \psi \right]_{\alpha_1}^{\alpha_2} = \frac{\cos \alpha_2^3 - \cos \alpha_1^3}{3} - (\cos \alpha_2 - \cos \alpha_1);$$

$$\int_{\alpha_1}^{\alpha_2} \sin^2 \psi \cos \psi d\psi = \left. \frac{\sin^3 \psi}{3} \right|_{\alpha_1}^{\alpha_2} = \frac{\sin^3 \alpha_2 - \sin^3 \alpha_1}{3}$$

and substituting these values into the last set of equations, we obtain

$$x_0 = \frac{4}{5} \frac{R^5 - r^5}{R^4 - r^4} \frac{\dfrac{\cos^3 \alpha_2 - \cos^3 \alpha_1}{3} - (\cos \alpha_2 - \cos \alpha_1)}{\dfrac{\alpha_2 - \alpha_1}{2} - \dfrac{\sin 2\alpha_2 - \sin 2\alpha_1}{4}} \qquad (11.73)$$

$$y_0 = \frac{4}{15} \frac{R^5 - r^5}{R^4 - r^4} \frac{\sin^3 \alpha_2 - \sin^3 \alpha_1}{\dfrac{\alpha_2 - \alpha_1}{2} - \dfrac{\sin 2\alpha_2 - \sin 2\alpha_1}{4}} \qquad (11.74)$$

For the case in which $\alpha_1 = 0$ and $\alpha_2 = \pi/2$, Equations 11.73 and 11.74 may be simplified to

$$x_0 = \frac{32}{15\pi} \frac{R^5 - r^5}{R^4 - r^4} \tag{11.75}$$

and

$$y_0 = \frac{16}{15\pi} \frac{R^5 - r^5}{R^4 - r^4} \tag{11.76}$$

The angle ψ_0 characterizing the fulcrum is determined from the expression

$$\tan \psi_0 = \frac{x_0}{y_0}$$

and for the specific case that is determined by the conditions of equations 11.75 and 11.76,

$$\tan \psi_0 = 2 \; ; \quad \psi_0 = 63°30'$$

The power input is

$$\mathcal{n} = \frac{P\vartheta}{102} = \frac{P\omega x_0}{102} \tag{11.77}$$

Hence, substituting the values of P from Equation 11.70 and x_0 from Equation 11.73, we obtain

$$\mathcal{n} = C' \frac{\gamma\omega^3}{510g} (R^5 - r^5) \left[\frac{\cos^3\alpha_2 - \cos^3\alpha_1}{3} - (\cos\alpha_2 - \cos\alpha_1) \right] \tag{11.78}$$

And, for $\alpha_1 = 0$ and $\alpha_2 = \pi/2$,

$$\mathcal{n} = C' \frac{2\gamma\omega^3}{1530g} (R^5 - r^5) \tag{11.79}$$

If the blade has the shape of a full quadrant ($\alpha_1 = 0$; $\alpha_2 = \pi/2$ and $r = 0$), then

$$\mathcal{n} = C' \frac{2\gamma\omega^3}{1530g} R^5 \; ; \quad x_0 = \frac{32}{15\pi} R$$

$$P = C' \frac{\pi\gamma\omega^2}{16g} R^4 \; ; \quad y_0 = \frac{16}{15\pi} R$$

$$(11.80)$$

The coefficient C' may be taken from Equation 11.60.

Example 11.3

Determine the power input for a round anchor mixer (Figure 11.42) from the following data:

- $C' = 0.5$
- $R = 500$ mm
- rpm = 30
- $\alpha_1 = 0$
- $r = 500$ mm
- $\gamma = 980$ kgf/m^3
- $\alpha_2 = 60° = \pi/3$
- $\cos\alpha_2 = 0.5$
- $\omega \simeq 3.1$ s^{-1}

Solution

Using formulas 11.70, 11.73, 11.74 and 11.78, we obtain

$$P = 0.5 \times \frac{980 \times 3.1^2}{4 \times 9.81} (0.6^4 - 0.5^4) \left(\frac{\pi}{6} - \frac{\sqrt{3}}{8}\right) = 2.49 \text{ kgf}$$

$$\mathcal{n}_1 = 0.5 \frac{980 \times 3.1^3}{510 \times 9.81} (0.6^5 - 0.5^5) \left[\frac{0.5^3 - 1}{3} - (0.5 - 1)\right] = 0.028 \text{ kW}$$

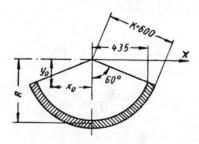

Figure 11.42   Scheme for the round anchor mixer in Example 11.3.

$$x_0 = \frac{4}{5} \frac{0.6^5 - 0.5^5}{0.6^4 - 0.5^4} \times \frac{0.33(0.5^3 - 1^3) - (0.5 - 1)}{0.5 \dfrac{\pi}{3} - \dfrac{\sqrt{3}}{8}} \simeq 0.37 \text{ m}$$

$$y_0 = \frac{4}{15} \frac{0.6^5 - 0.5^5}{0.6^4 - 0.5^4} \times \frac{0.86^3}{0.5\pi/3 - \sqrt{3}/8} \simeq 0.39 \text{ m}$$

The power input for the two blades is

$$\mathcal{N} = \frac{2\mathcal{N}_1}{\eta} = \frac{2 \times 0.028}{0.7} = 0.08 \text{ kW}$$

where η is the drive efficiency.

DESIGN OF MISCELLANEOUS MIXERS

The design formulas for *shrouded turbine mixers* are summarized in Table 11.10. These formulas may be used for designs having the same size ratio of elements as shown in Figure 11.43.

The design formulas for *open-blade turbine mixers* (Figure 11.44) are given in Table 11.11. Conditions for which these formulas are applicable (besides those given in Table 11.11) are as follows:

- $Re_M > 10^4$
- $2 < D/d < 7$
- $2 < H/d < 4.8$
- $0.7 < b/d < 1.6$
- $0.15 < t/d < 0.5$
- $\ell/h = 1.25$

If these conditions are not fulfilled, the power input is changed considerably. The parameter E' is determined from the formula

$$E' = (Fr)^{(1-\log Re)/40} \tag{11.81}$$

The dimensionless number E' also may be determined from the nomogram (Figure 11.33). The point of axis, $d(m)$, and axis, n (rot/s), should be connected and point A_1 determined on the first auxiliary straight line. Then the points of axis Re_M and D/d should be connected and determined through the intersection of point B_1 with the third auxiliary straight line. This last point is connected with the second point, b, and C_1 is determined on the second

Table 11.10 Design Formulas for Shrouded Turbine Mixers

Type of Mixer	Reynolds Number	Power Input (kW)	Q' Number	Points on the Nomogram (Figure 11.32)
Turbine with three blades; without baffles (Figure 11.43(A))	$5 \times 10^4 < Re_M < 4 \times 10^5$	$\mathcal{W} = 0.736\, SQ'$	$Q' = 1.65 \times 10^{-2} \times Re_M^{0.913}$	L and 9
Turbine with six blades; with 20 baffles (Figure 11.43(B))	$10^4 < Re_M < 3 \times 10^5$	$\mathcal{W} = 0.736\, SQ'$	$Q' = 2.58 \times 10^{-2}\, Re_M^{0.987}$	P and 15
Turbine with six blades; 20 baffles (Figure 11.43(C))	$Re_M > 10^4$	$\mathcal{W} = 0.736\, SQ'$ $\mathcal{W} = 0.736 \times 1.493 \times 10^{-2} \rho n^3 d^5$	$Q' = 1.493 \times 10^{-2}\, Re_M$	R and 8

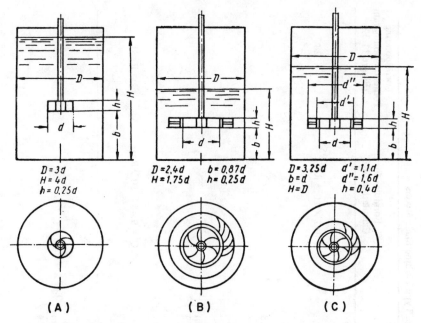

$D = 3d$
$H = 4d$
$h = 0.25d$

$D = 2.4d$ $b = 0.87d$
$H = 1.75d$ $h = 0.25d$

$D = 3.25d$ $d' = 1.1d$
$b = d$ $d'' = 1.6d$
$H = D$ $h = 0.4d$

(A)   (B)   (C)

Figure 11.43   Shrouded turbine mixer designs: (A) three-blade turbine; (B,C) six-blade mixers.

auxiliary line. Connecting points A_1 and C_1 obtained on the first and second auxiliary lines, we find the value E' on the intersection of this line with axis E'.

Example 11.4

Determine the power input for an open turbine mixer with flat baffles (Figure 11.45). The following information is supplied:

- $D = 1050$ mm
- $d = 350$ mm
- rpm $= 750$ (12.5 s^{-1})
- $\mu = 9180 \times 10^{-6} \text{ kg}_f\text{-s/m}^2$
- $b = 350$ mm
- $H = 1000$ mm
- $\gamma = 980 \text{ kg}_f/\text{m}^3$
- $\rho = 980/9.8 = 100 \text{ kg}_f\text{-s}^2/\text{m}^4$
- $Re_M = 16,200$

Solution

These general conditions permit the use of the formulas in Table 11.9.

Table 11.11 Design Formulas for Open-Type Turbine Blade Mixers

Characteristics of Tank and Mixer	Number of Blades	Formulas for Calculation of Power Input, $\mathcal{N}$ (kW)	Formulas for Calculation of Q'	Corresponding Points on Nomogram (Figure 11.32)
Flat Blades in a Tank without Baffles	<6	$\mathcal{N} = 0.736 \, SQ'E' \left(\frac{n'}{6}\right)^{0.8} \left(\frac{\varrho}{d}\right)^{1.5}$	$Q' = 2 \times 10^{-1} Re_M^{0.955}$	N; 28
	>6	$\mathcal{N} = 0.736 \, SQ'E' \left(\frac{n'}{6}\right)^{0.7} \left(\frac{\varrho}{d}\right)^{1.6}$		
Flat Blades in Tank with Baffles	<6	$\mathcal{N} = 0.736 \, SQ' \left(\frac{m_1}{4}\right)^{0.43} \left(\frac{n'}{6}\right)^{0.8} \times \left(\frac{\varrho}{d}\right)^{1.5} \left(\frac{t}{d}\right)^{0.3}$ $\mathcal{N} = 0.736 \times 9.48 \times 10^{-1} \rho n^3 d^5 \left(\frac{m_1}{4}\right)^{0.43} \times \left(\frac{n'}{6}\right)^{0.8} \left(\frac{\varrho}{d}\right)^{1.5} \left(\frac{t}{d}\right)^{0.3}$	$Q' = 9.48 \times 10^{-1} Re_M$	R; 31
	>6	$\mathcal{N} = 0.736 \, SQ' \left(\frac{m_1}{4}\right)^{0.43} \left(\frac{n'}{6}\right)^{0.7} \times \left(\frac{\varrho}{d}\right)^{1.5} \left(\frac{t}{d}\right)^{0.3}$ $\mathcal{N} = 0.736 \times 9.48 \times 10^{-1} \rho n^3 d^5 \left(\frac{m_1}{4}\right)^{0.43} \times \left(\frac{n'}{6}\right)^{0.7} \left(\frac{\varrho}{d}\right)^{1.5} \left(\frac{t}{d}\right)^{0.3}$		

Curved Blades in Tank with Baffles

<6

$$\mathcal{N} = 0.736 \, SQ' \left(\frac{m_1}{4}\right)^{0.43} \left(\frac{n'}{6}\right)^{0.8} \times \left(\frac{\varrho}{d}\right)^{1.5} \left(\frac{t}{d}\right)^{0.3}$$

$$\mathcal{N} = 0.736 \times 7.34 \times 10^{-1} \rho n^3 d^5 \left(\frac{m_1}{4}\right)^{0.43} \times \left(\frac{n'}{6}\right)^{0.8} \left(\frac{\varrho}{d}\right)^{1.5} \left(\frac{t}{d}\right)^{0.3}$$

>6

$$\mathcal{N} = 0.736 \, SQ' \left(\frac{m_1}{4}\right)^{0.43} \left(\frac{n'}{6}\right)^{0.7} \times \left(\frac{\varrho}{d}\right)^{1.5} \left(\frac{t}{d}\right)^{0.3}$$

$$\mathcal{N} = 0.736 \times 7.34 \times 10^{-1} \rho n^3 d^5 \left(\frac{m_1}{4}\right)^{0.43} \times \left(\frac{n'}{6}\right)^{0.7} \left(\frac{\varrho}{d}\right)^{1.5} \left(\frac{t}{d}\right)^{0.3}$$

$$Q' = 7.34 \times 10^{-1} \, Re_M \qquad R; \, 30$$

Arrow Blades, Tank with Baffles

Without limitation

$$\mathcal{N} = 0.736 \, SQ' \left(\frac{m_1}{4}\right)^{0.7} \left(\frac{t}{d}\right)^{0.3}$$

$$\mathcal{N} = 0.736 \times 8.41 \times 10^{-2} \rho n^3 d^5 \times \left(\frac{m_1}{4}\right)^{0.43} \left(\frac{t}{d}\right)^{0.3}$$

$$Q' = 8.41 \times 10^{-2} \, Re_M \qquad R; \, 22$$

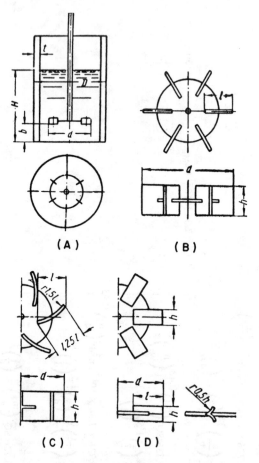

Figure 11.44 Turbine mixer designs: (A) disk mixer with radial flat blades; (B,C,D) disk mixers with blades.

$$\mathcal{N} = 0.736 \times 0.948 \times 100 \times 12.5^3 \times 0.35^5 \times \left(\frac{87.5}{350}\right)^{1.5} \times \left(\frac{105}{350}\right)^{0.3} \approx 62.3 \text{ kW}$$

This mixer is a solid disk (d = 0.4 D), having on its lower surface 16 radial flat blades. The length of each blade is ℓ = 0.35 d; its height is h = 0.1 d. For a tank with four baffles (t = 0.1 D) operating within the range $10^2 < \text{Re}_M < 10^5$, the power input is (from Table 11.10)

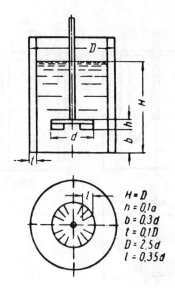

$H = D$
$h = 0.1a$
$b = 0.3d$
$t = 0.1D$
$D = 2.5d$
$l = 0.35d$

Figure 11.45   Turbine disk impeller with flat blades for Example 11.4.

$$\mathcal{N} = 0.736 \times SQ' = 0.736 \times 3.14 \times 10^{-2}\rho n^3 d^5$$

A *squirrel-cage mixer* consists of a hollow cylinder with a solid bottom or lid and rectangular blades welded to the circumference (Figure 11.46). For the dimensions shown in Figure 11.46 at $10^2 < Re_M < 3 \times 10^5$,

$$\mathcal{N} = 0.736 SQ' \frac{h}{d} \qquad (11.82)$$

where $Q' = 0.456\, Re_M^{0.723}$

In using Figure 11.32, points B and 29 should be applied. According to Kafarov [1949], the recommended ratio of the cage diameter to its height should be around 0.67; the ratio of the cage diameter to the tank diameter is 0.17–0.25, or more; and the ratio of the liquid height to the cage diameter may be as high as 10.

The following formulas are recommended for *propeller mixers*. For mixers operating in tanks without baffles, at $10^2 < Re_M < 10^4$,

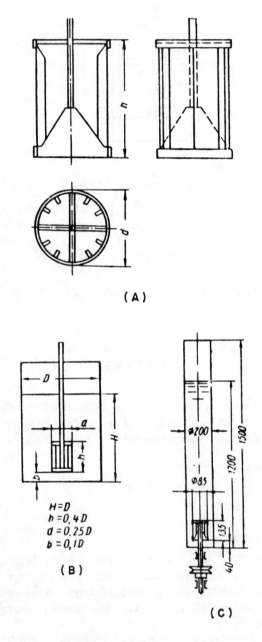

Figure 11.46   Squirrel-cage mixers.

$$\left.\begin{array}{l} Q' = 7.47 \times 10^{-3} Re_M^{0.81} \\ \mathcal{N} = 0.736 SQ' \left(\dfrac{D}{d}\right)^{0.93} \end{array}\right\} \qquad (11.83)$$

points F and 5 on the nomogram (Figure 11.32) at $Re_M > 10^4$.

$$\left.\begin{array}{l} Q' = 1.945 \times 10^{-3} Re_M^{0.96} \\ \mathcal{N} = 0.736 SQ'E' \left(\dfrac{D}{d}\right)^{0.91} \left(\dfrac{S}{d}\right)^{1.22} \end{array}\right\} \qquad (11.84)$$

points N and 1 on the nomogram (Figure 11.32).
   For mixers operating in tanks with baffles,

$$\left.\begin{array}{l} Q' = 6.12 \times 10^{-3} Re_M \\ \mathcal{N} = 0.736 SQ' \left(\dfrac{m_1}{4}\right)^{0.43} \left(\dfrac{t}{d}\right)^{0.3} \left(\dfrac{s}{d}\right)^{1.7} \\ \mathcal{N} = 0.736 \times 6.12 \times 10^{-3} \rho n^3 d^5 \left(\dfrac{m_1}{4}\right)^{0.43} \left(\dfrac{t}{d}\right)^{0.3} \left(\dfrac{s}{d}\right)^{1.7} \end{array}\right\} \qquad (11.85)$$

points R and 2 on Figure 11.32.
   These formulas may be used under the following conditions:

- $2 < D/d < 7$
- $2 < H/d < 4.8$
- $0.7 < b/d < 1.6$
- $0.15 < t/d < 0.5$
- $1 < s/d < 2$

The value E' is determined from the expression

$$E' = (FR_M)^{(\alpha - \log Re_M)/18}$$

where α depends on the ratio of D/d. Typical values are

$$D/d = 2.1, \ 2.7, \ 3.0, \ 3.3, \ 4.5$$
$$\alpha = 2.6, \ 2.3, \ 2.1, \ 1.7, \ 0$$

E' also may be determined from the nomogram of Figure 11.33 as follows. Connecting points on the axes n and d, we find point A_1 on the first auxiliary line. Then the points on the axes Re_M and D/d should be connected on a point on the third auxiliary line determined. The last point is connected with a constant point b, and the line of intersection with the second auxiliary line (point C_1) is found. Determined points A_1 and C_1 then are connected. The intersection of this line with axis E' gives the desired value.

Example 11.5

Determine the power input of a propeller mixer operating in a tank without baffles from the following data:

- D = 1200 mm
- d = 300 mm
- rpm = 900 (15 s^{-1})
- s = 600 mm
- ρ_0 = 0.9
- $Re_M > 10^4$
- μ = 510 × 10^{-6} kg-s/m^2

Solution

The power required is

$$\mathcal{n} = 0.736 SQ'E' \left(\frac{D}{d}\right)^{0.91} \left(\frac{s}{d}\right)^{1.22}$$

From the nomograms (Figures 11.30–11.34), we find

- Re_M = 2.35 × 10^5
- Q' = 2.8 × 10^2
- E' = 0.63
- $(D/d)^{0.91}$ = 3.53
- $(s/d)^{1.22}$ = 2.35
- S = 3.1 × 10^{-3}

Substituting these values into the formula, we obtain

$$\mathcal{n} = 0.736 × 3.1 × 10^{-3} × 2.8 × 10^2 × 0.63 × 2.35 × 3.53 = 3.34 \text{ kW}$$

Approximate power consumptions are given in Table 11.12 for different sizes of mixers operating in water.

When dealing with liquids other than water, appropriate correction coefficients (to be determined experimentally) should be applied to the values reported in Table 11.12.

Table 11.12 Approximate Power Consumptions for Different Sizes of Propellers

Diameter of Propeller (mm)	Rotations per Minute	Diameter and Height of a Tank (m)	Power Input, kW, at Filling– 0.66 Water is Mixed
200	270–900	0.9 × 0.85	0.147–0.368
300	210–420	1.2 × 1	0.29–0.736
400	180–360	1.6 × 1.25	0.59–1.600
500	160–320	2 × 1.5	1.17–3.68
750	115–230	3 × 2	2.35–7.36
1000	90–180	4 × 2.5	4.71–14.7

SELECTION OF ROTATIONAL SPEED

All the equations previously quoted hold for conditions of forced convection with turbulent flow in agitated systems. Mixer speed must be selected on the basis of process objective, type and the specific design of the mixing device.

Process of suspending particles. Uniform particle distribution in a liquid is achieved at a certain rpm, n_0, that allows the axial velocity component to become equal to, or somewhat higher than, the particle settling velocity. In this case, the upflow of liquid maintains the solid particles suspended. The number of rotations may be determined from the following equation:

$$Re_M = \frac{n_0 d^2 \rho}{\mu} = C_1 Ar_M \left(\frac{d_p}{d}\right)^{0.5} \left(\frac{D}{d}\right)^k \qquad (11.86)$$

where $Ar_M = (gd_p^3/\nu_f)(\Delta\rho/\rho)$ = Archimedes number
   $\Delta\rho$ = difference of phase densities
   ρ = density of continuous phase
   ν_f = kinematic viscosity of continuous phase
   d_p = particle diameter
   D/d = ratio of tank diameter to mixer diameter

The values of coefficient C_1 and exponent k are given below:

	D/d	C_1	k
Shrouded Turbine Mixer	1.5 –4.0	4.7	1.0
Propeller Mixer	1.5 –5.0	6.6	1.0
Paddle Mixer	1.33–1.5	14.8	0.0

Equation 11.86 is valid for the following values of variables:

$$Re_M = 5 \times 10^2 - 1.3 \times 10^5 \; ; \quad Ar_M = 2.4 \times 10^4 - 4.1 \times 10^{11}$$

$$\frac{d_p}{d} = 2.33 \times 10^{-4} - 1.2 \times 10^{-2}$$

Emulsification of liquids. In emulsification of mutually insoluble liquids, the mixer rpm may be calculated from the following formula:

$$Re_M = \frac{n_0 d^2 \rho}{\mu} = C_2 Ar_M^{0.315} \left(\frac{Re_M}{We_M}\right)^{0.185} \left(\frac{D}{d}\right)^{\ell} \tag{11.87}$$

where $We_M = n^2 d^3 \rho / \sigma$ = modified Weber number

The coefficient C_2 and exponent ℓ are as follows:

	D/d	C_2	ℓ
Shrouded Turbine Mixer	2–4	2.3	0.67
Propeller Mixer	2–4	2.95	0.67
Paddle Mixer	1.33–4	1.47	1.30

Equation 11.87 is valid for the following values of variables:
- $Re_M = 5 \times 10^2 - 2 \times 10^5$
- $Ar_M = 8.9 \times 10^3 - 3.4 \times 10^{10}$

$$\frac{Re_M}{We_M} = 6.15 - 1.18 \times 10^7$$

Homogenization of liquids. The number of mixer rotations when mixing in a one-phase system (to reduce concentration and/or temperature gradients) may be determined from the following relation:

$$n_0 \tau' = C_{\tau'} = \text{constant} \tag{11.88}$$

where τ' is the mixing time (the time for achieving the desired homogenization). The values of coefficient $C_{\tau'}$ for different mixing devices are given below:

	D/d	$C_{T'}$
Turbine Mixer (open type)	3	56
	4	99.5
Leaf-Paddle Mixer	2	20.5
	1.5	20.7
Paddle Mixer	3	96.5
Propeller Mixer (with draft tube)	3	66.2
	4	118
Propeller Mixer	3	96.5
	4	170
Anchor Mixer	1.15	30
Shrouded Turbine Mixer	3	46

RATE OF HOMOGENIZATION

Consider a mixer in a tank filled with a liquid of volume V and concentration Y (Figure 11.47). The tank is fed a pure solvent, Y = 0, and the mixed solution is removed at a rate at which the tank contents remain constant. The mixer maintains a constant concentration at any given moment, and this concentration will decrease with time. We will determine the law of concentration change with time by making a material balance over a time interval, $d\tau'$.

If the volumetric rate of the solvent is equal to q, then for the time $d\tau'$ the volume of solution to be evacuated is $qd\tau'$. The amount of solute is $Vqd\tau'$. The concentration in the tank changes over this time interval by dY, where dY < 0. Consequently, the decrease of the component concentration in the tank is −VdY, and the material balance is

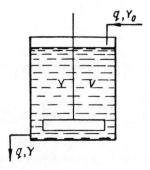

Figure 11.47 Sketch for determining the mixing rate.

$$-VdY = Yqd\tau' \tag{11.89}$$

After integration, the concentration, Y, in the tank for time τ' is

$$Y = Y_0 e^{-q\tau'/V} \tag{11.90}$$

where Y_0 = the concentration in the tank at $\tau' = 0$

Now consider liquid moving through a series of tanks (Figure 11.48) with capacities V_1, V_2, V_3 .... At a certain moment in some tank (e.g., n − 1 in Figure 11.48), the second liquid (or the mixture of both liquids) enters. Denote the concentration of a new liquid at the inlet of the first tank as $Y_0 = 1$. The concentration of the new liquid at the time interval, τ', in the tank, n, will be Y_n; the concentrations in the other tanks will be different. At time $\tau' = 0$, the concentration in any tank will be $Y_n = 0$. For an infinitely long time the concentration will reach $Y_n = 1$, and the total system will be filled by a new liquid. It is desirable to determine the concentration Y_n for any point in time and in any one of the tanks.

The material balance for tank n for time $d\tau'$ will be

$$V_n dY_n = q(Y_{n-1} - Y_n) \tag{11.91}$$

Integrating this expression from $\tau' = 0$ to τ', we obtain the concentration in tank n:

$$Y_n = \frac{q}{V_n} e^{-q\tau'/V_n} \int_0^{\tau'} e^{q\tau'/V_n} Y_{n-1} d\tau' \tag{11.92}$$

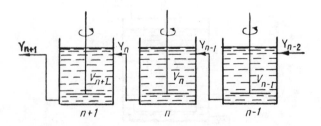

Figure 11.48   A series of mixing tanks.

or

$$Y_n = e^{-q\tau'/V_n} \int_1^{q\tau'/V_n} Y_{n-1} d(e^{q\tau'/V_n}) \qquad (11.93)$$

Thus, if the relationship between concentration change and time is known in the $(n - 1)$ tank, i.e., Y_{n-1}, then Equation 11.93 may be applied to determine the relationship between Y and time for tank n. If $V_1 = V_2 = V_3 \ldots = V_n$, the integral (Equation 11.93) is readily evaluated. Denoting $e^{q\tau'/V_n}$ as χ, we obtain

$$Y_n = \frac{1}{\chi} \int_1^\chi Y_{n-1} d\chi \qquad (11.94)$$

As $Y_0 = 1$ (new liquid entering in the first tank), Equation 11.94 takes the form

$$Y_n = \frac{1}{\chi} \int_1^\chi \frac{1}{\chi} \int_1^\chi \frac{1}{\chi} \ldots \int_1^\chi 1 d\chi^n \qquad (11.95)$$

As a result of successive integration, we obtain

$$Y_n = 1 - \frac{1}{\chi} \left[\frac{1}{(n-1)!} ln\chi^{n-1} + \frac{1}{(n-2)!} ln\chi^{n-2} + \ldots ln\chi + 1 \right] \qquad (11.96)$$

or

$$Y_n = 1 - e^{-q\tau'/V} \sum_{i=1}^{i=n} \frac{1}{(i-1)!} \left(\frac{q\tau'}{V} \right)^{i-1} \qquad (11.97)$$

where n is the number of tanks.

Using Equation 11.97, it is possible to compute concentration Y in any tank and at any time from the beginning of the charge of new liquid in the first tank.

OPTIMUM POWER INPUT

The equations presented for power input did not include the mixing time. However, to evaluate the power consumption of an impeller one must account for the total energy consumption during the time necessary for accomplishing a required mixing effect.

The minimum time necessary for complete mixing depends (in addition to the mixer shape) on its diameter, on the rotational speed and on the liquid properties, i.e., density and viscosity. The general expression for mixing time is

$$\psi(n, \rho, d, \mu, \Delta\rho, g, D, H, T) = 0 \qquad (11.98)$$

where T denotes time.

The number of π functions is 6 when the total number of variables is 9 and the FLΘ technical system of units is applied. As fundamental dimensional units, n, d and ρ are selected. The individual π terms then are obtained:

$$\pi_1 = \frac{nd^2\rho}{\mu} = \frac{nd^2}{\nu} = Re$$

$$\pi_2 = d^{x_2}n^{y_2}\rho^{z_2}\Delta\rho^{-1}$$

$$\pi = L^{x_2}\left[\frac{1}{\Theta}\right]^{y_2}\left[\frac{F\Theta^2}{L^4}\right]^{z_2}\left[\frac{F\Theta^2}{L^4}\right]^{-1} = F^0 L^0 \Theta^0$$

$$\Sigma L \equiv x_2 - 4y_2 + 4 = 0$$

$$\Sigma\Theta \equiv -y_2 + 2z_2 - 2 = 0$$

$$\Sigma F \equiv z_2 - 1 = 0$$

$$x_2 = 0$$

$$y_2 = 0$$

$$z_2 = 1$$

$$\pi_2 = \frac{\rho}{\Delta\rho} \qquad (11.99)$$

$$\pi_3 = \frac{n^2 d}{g} = Fr_1$$

$$\pi_4 = \frac{d}{D}$$

$$\pi_5 = \frac{H}{D}$$

$$\pi_6 = d^{x_6} n^{y_6} \rho^{z_6} T^{-1}$$

$$\pi_6 = L^{x_6} \left[\frac{1}{\Theta}\right]^{y_6} \left[\frac{F\Theta^2}{L^4}\right]^{z_6} \Theta^{-1} = F^0 L^0 \Theta^0$$

$$\Sigma L \equiv x_6 - 4z_6 = 0$$

$$\Sigma \Theta \equiv -y_6 + 2z_6 - 1 = 0$$

$$\Sigma F \equiv z_6 = 0$$

$$x_6 = 0$$

$$y_6 = -1$$

$$z_6 = 0$$

$$\pi_6 = \frac{1}{nT} \qquad\qquad (11.100)$$

For the terms π_4 and π_5, the diameter of the agitated vessel has been used. Equation 11.100 is inverted in the final relationship. This term is further enlarged by the volume, namely, by applying time, T, to a unit volume. The aquired term, nT/V, is not dimensionless, however; the dimension of volume L^3 must be compensated for by the same power on length dimension d, i.e., d^3.

The dimensionless mixing time, $\dot\vartheta$, then is obtained:

$$\dot\vartheta = \frac{Tnd^3}{V} = \frac{TQ}{V} \qquad\qquad (11.101)$$

This expression is adjusted with the use of the known relation for pumping effect of the impeller Q, which is proportional to nd^3.

On substitution for individual π terms, the basic relation (Equation 11.98) may be written in the following form:

$$\dot{\vartheta} = \frac{TQ}{V} = f\left(Re, \ Fr, \ \frac{\rho}{\Delta\rho}, \frac{d}{D}, \frac{H}{D}\right) \tag{11.102}$$

By adding up the second and third terms and by multiplying the resulting term by the quotient d/H, a modified Froude number is obtained. This Froude number expresses the relationship of dynamic pressure, $\rho n^2 d^2$, to static pressure, $\Delta\rho g H$; the final form of Equation 11.98 is, then,

$$\dot{\vartheta} = kRe^a\left(\frac{\rho n^2 d^2}{\Delta\rho g H}\right)^b \left(\frac{d}{D}\right)^\ell \left(\frac{H}{D}\right)^f \tag{11.103}$$

In steady flow without vortex formations, Fr is eliminated. With vortex formation the effect of Re is negligible. When maintaining geometric similarity the last two terms on the RHS of Equation 11.103 are eliminated, and a relation analogous to the equation derived for power input is obtained:

$$\dot{\vartheta} = kRe^a Fr_m^b \tag{11.104}$$

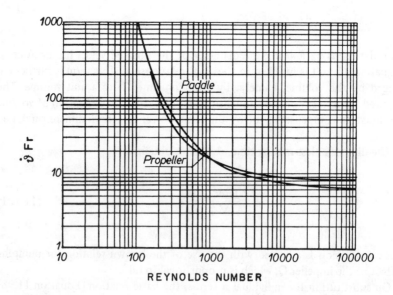

Figure 11.49   Plot of product of dimensionless time and Froude number vs Reynolds number (from Van de Vusse [1955]).

The values of exponent 'a' are not constant, as may be seen from Figure 11.49, where the dependence for propeller mixers and flat straight-blade mixers, including turbine impellers, is depicted. At further derivation, the value of 3 is taken for laminar flow and a = 0 for turbulent flow. Exponent 'b' can be obtained from the graph in Figure 11.50.

MIXER SELECTION

The optimum mixer for a specific application is one that performs a desired mixing function within a given time span, utilizing the least amount of power, or one that mixes the fastest for a specified amount of power. Moo-Young et al. [1972] have shown that significant differences exist between impeller

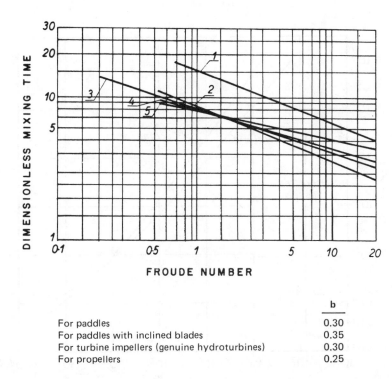

	b
For paddles	0.30
For paddles with inclined blades	0.35
For turbine impellers (genuine hydroturbines)	0.30
For propellers	0.25

Figure 11.50 Shows the influence of Froude number on dimensionless mixing time for various types of agitators (Van de Vusse [1955]). (1) turbine impeller, four-blade off-center; (2) turbine impeller with inclined blades, 45°; (3) shrouded turbine; (4) turbine impeller, centrally positioned; (5) propeller.

types when blending viscous liquids, but comparatively little difference is observed when handling low-viscosity liquids. Figure 11.51 shows the relationship between mixing time and power input for two different impellers (turbine and helical ribbon). The group $N_t = (t_T \mu / \rho D_t^2)$ is a *dimensionless mixing number*, where t_T is blending time and D_t is tank diameter. The second dimensionless group in Figure 11.50, N_L, is a *modified power number*, defined as $P g_c \rho^2 D_t / \mu^3$. N_t is formed from the product of the dimensionless groups $n t_T$, Re and $(S_1')^{-2}$. $S_1' = D_t/d$ is a shape factor and d is the diameter of the impeller. N_L is the product of N_t, Re^3 and S_1'. For highly viscous fluids at N_L values of $10^3 - 10^4$, the helix provides shorter mixing times at the same power input than the turbine. However, the data in Figure 11.51 suggest that for low-viscosity liquids, where N_L is $10^{12} - 10^{14}$, the turbine mixes faster than the helix. Baffles are required for good mixing when N_t is less than about 10^{-2}.

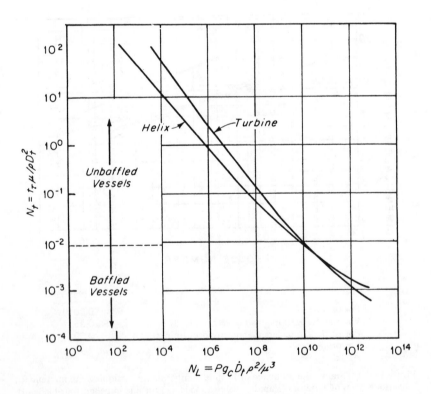

Figure 11.51 Mixing-time-power relationship for helical-ribbon and turbine impellers in agitated vessels (Moo-Young, Ticher and Dullien [1972]).

Approximate operating ranges for different types of impellers are given in Figures 11.52 and 11.53. Figure 11.52 gives typical ranges of operation for percent suspended solids content, heat and mass transfer. Note that mass transfer ranges are with respect to straight blade turbine agitators. Figure 11.53 gives typical ranges of rotational velocity, circumferential velocity and mixer volume. These charts can serve as a rough guide to

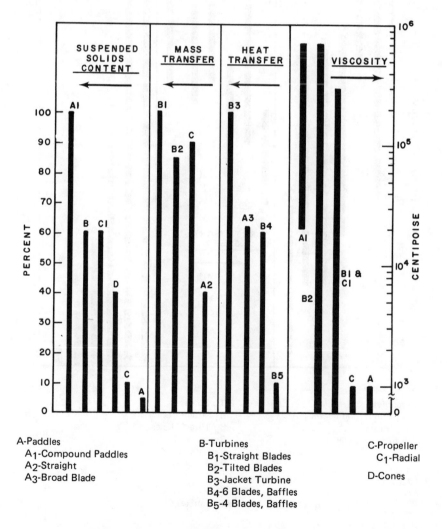

A-Paddles
  A_1-Compound Paddles
  A_2-Straight
  A_3-Broad Blade

B-Turbines
  B_1-Straight Blades
  B_2-Tilted Blades
  B_3-Jacket Turbine
  B_4-6 Blades, Baffles
  B_5-4 Blades, Baffles

C-Propeller
  C_1-Radial

D-Cones

Figure 11.52. Mechanical mixer selection chart.

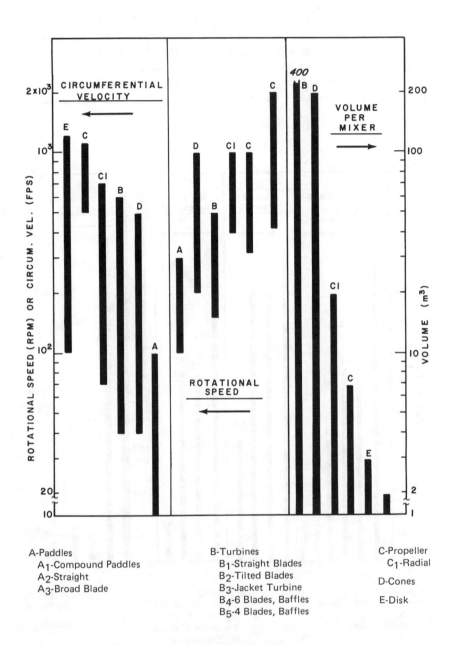

A-Paddles
  A₁-Compound Paddles
  A₂-Straight
  A₃-Broad Blade

B-Turbines
  B₁-Straight Blades
  B₂-Tilted Blades
  B₃-Jacket Turbine
  B₄-6 Blades, Baffles
  B₅-4 Blades, Baffles

C-Propeller
C₁-Radial

D-Cones

E-Disk

Figure 11.53   Mechanical mixer selection chart.

selecting a mixer configuration for a particular service. An extensive review of mixer designs and comparative operating ranges is given by Perry and Chilton [1973].

SCALEUP CRITERIA

Without Heat Transfer

Scaleup from a laboratory or a pilot plant mixer to a commercial size is a major consideration. The generalized correlations outlined above are applicable for scaleup purposes for many problems. However, mixing applications are often unique due to the complex nature of the hydrodynamic processes involved, which do not permit the solution of the general differential equations of flow.

According to the basic principles of the theory of similarity, the bases for hydrodynamic modeling of mixing processes are the dimensionless equations obtained by the method of similarity transformation of the Navier-Stokes equations. This enables calculations to be performed if the values of the constants and exponents of the dimensionless numbers are known. Unfortunately, these parameters must be evaluated for each individual problem by a model experiment, which thus becomes the principal source of data for scaleup calculations.

The principal consideration in scaleup is to define the primary mechanism controlling the mixing operation. That is, one must establish whether the process criterion is a solid suspension, mass transfer, heat transfer or chemical reaction. Each particular process operation has its own scaleup characteristics. Batch-operated pilot tests generally are the simplest approach to evaluating the process. For example, if the mixer design is to be evaluated on the basis of reaction kinetics, a batch test allows the progress of the reaction(s) to be studied over various times and at different concentrations of feed materials. By means of a carefully planned number of tests at different mixer speeds, the controlling mechanism of the overall process may be determined.

Recall that a condition for attaining hydrodynamic similarity in scaling up operations is identity of the instantaneous velocities, i.e., $\Sigma u_0 = \Sigma u_M$. If fulfilled, model systems can be scaled up on the basis of geometric similarity alone. However, it is not always advisable to insist on geometric similarity because the geometry of a large vessel may have to be changed to maintain the proper relationships among other parameters. The density of Reynolds numbers often is selected as the condition of hydrodynamic similarity, primarily because this dimensionless group contains easily measured and controlled variables. Rushton and Oldshue [1953], however, observed that this identity secures only the same overall pattern in both systems but not equality of instantaneous velocities, which may differ considerably even at equal Reynolds numbers.

Consequently, depending on the specific mixing application, other dimensionless relationships must be defined for scaling up. This is especially the case when it is necessary to extrapolate beyond the range of model experiments. The required relations between geometric enlargement and velocity criteria are based on relationships between dimensionless groups containing variables that influence the model system. These relationships generally follow an exponential form, such as the power equation.

Rushton and Oldshue [1953] did an extensive investigation of the velocity

Table 11.13 Principal Relations in Scaleup of

Variable	Dimension		Ratio to Remain Constant	Dimensionless Number, Forming Part of Criterial Equation
	In the FLΘ System	In Kinematic Expression		
i	ii	iii	iv	v
Power, $\mathcal{N}_0$	$\dfrac{FL}{\Theta}$	$\dfrac{L^5}{\Theta^3} = \dfrac{\mathcal{N}_0}{\rho}$	$\dfrac{a N a_\Theta^3}{a_\rho a_L^5} = C$	$C = \dfrac{\mathcal{N}_0 \Theta^3}{\rho L^5} = \dfrac{\mathcal{N}_0}{\rho n^3 d^5}$
Specific gravity, γ	$\dfrac{F}{L^3}$	$\dfrac{\gamma}{\rho} = \dfrac{L}{\Theta^2}$	$\dfrac{\gamma_D}{\gamma_M} = a_\gamma$	$Fr = \dfrac{u^2}{Lg} = \dfrac{n^2 d}{g}$
Viscosity, μ	$\dfrac{F\Theta}{L^2}$	$\dfrac{\mu}{\rho} = \dfrac{L^2}{\Theta}$	$\dfrac{\nu_D}{\nu_M} = a_\nu$	$Re = \dfrac{uL}{\nu} = \dfrac{nd^2\rho}{\eta}$
Interfacial Tension, σ	$\dfrac{F}{L}$	$\omega = \dfrac{\sigma}{\rho} = \dfrac{L^3}{\Theta^2}$	$\dfrac{\omega_D}{\omega_M} = a_\omega$	$We = \dfrac{u^2 L}{\omega} = \dfrac{n^2 d^3 \rho}{\sigma}$
Elasticity, e	$K_e = \dfrac{F}{L^2}$	$\dfrac{K_e}{\rho} = \dfrac{L^3}{\Theta^2}$	$\dfrac{e_D}{e_M} = a_e$	$Ca = \dfrac{u^2}{e} = \dfrac{n^2 d^2 \rho}{K_e}$
	–	–	$\dfrac{L_D}{L_M} = a_L$	a_L
Scaleup Ratio	–	–	$\dfrac{\Theta_D}{\Theta_M} = a_\Theta$	a_Θ

ratios between model and commercial-sized equipment. In this study, the basic parameter with which mixing variables were compared was the power number, K_n.

The derived relations as summarized by Rushton and Oldshue are given in Table 11.13. The information given under each column is as follows:

- **Column i**   denotes individual variables that may influence the mixing system.
- **Column ii**   denotes the dimensions of the technical system.
- **Column iii**   denotes the *kinematic expression* of the respective variable as given by its ratio to density, ρ.

Mixing Systems (Rushton and Oldshue [1953])

	Velocity Ratio in Scaleup				
		With Simultaneous Action of			
At Action of Respective Variable Only	Gravity and Viscous Forces	Gravity and Interfacial Forces	Viscous and Interfacial Forces	Viscous and Cohesive Forces	Gravity and Cohesive Forces
vi	vii	viii	ix	x	xi
$a_u = \left(\dfrac{a_\eta}{a_\rho a_L^2}\right)^{1/3}$	–	–	–	–	–
$a_u = a_L^{0.5}$	$a_u = a_L^{0.5}$	$a_u = a_L^{0.5}$	–	–	$a_u = a_L^{0.5}$
$a_u = \dfrac{a_\nu}{a_L}$	$a_u = \dfrac{a_\nu}{a_L}$	–	$a_u = \dfrac{a_\nu}{a_L}$	$a_u = \dfrac{a_\nu}{a_L}$	–
$a_u = \left(\dfrac{a_\omega}{a_L}\right)^{0.5}$	–	$a_u = \left(\dfrac{a_\omega}{a_L}\right)^{0.5}$	$a_u = \left(\dfrac{a_\omega}{a_L}\right)^{0.5}$	–	–
$a_u = a_e^{0.5}$	–	–	–	$a_u = a_e^{0.5}$	$a_u = a_e^{0.5}$
$a_u = const$	$a_L = a_\nu^{0.67}$	$a_L = a^{0.5}$	$a_L = \dfrac{a_\nu^2}{a_\omega}$	$a_L = \dfrac{a_\nu}{a_e^{0.5}}$	$a_L = a_e$
$a_\Theta = const$	$a_I = a_\nu^{0.67}$	$a_L = a_\omega^{0.5}$	$a = \dfrac{a_\nu^2}{a_\omega}$	$a_L = \dfrac{a_\nu}{a_e^{0.5}}$	$a_L = a_e$

- **Column iv**   denotes which dimensionless ratio should be kept constant when the variable under study acts on the system.
- **Column v**   denotes the *dimensionless groups* that are part of the criterial equation $K_n = kX^aY^b$ ... if the variable under study has an influence on the system to be scaled up.
- **Column vi**   denotes the required velocity ratios between the model and the full-sized unit $a_u = u_D/u_M$ for the case in which the operation is influenced only by the corresponding variable given in the same line of Column i.
- **Column vii–xi** denote similar ratios for the simultaneous influence of the two variables given in the head of the column. As an example, for the simultaneous action of viscous and interfacial forces, the criterial equation takes the form $K_n = kRe^aWe^b$. The velocity expression, given for the influence of either or both variables, must be identical for this example. By comparing both velocity expressions, the scaleup ratio, a_d, can be calculated directly as:

$$\frac{a_\nu}{a_L} = \left(\frac{a_W}{a_L}\right)^{0.5} \quad ; \quad a_L = a_d$$

whence $a_d = a_\nu^2/a_W$. The liquid used in the model must exhibit the same interfacial tension, but not the same viscosity, as in the full-sized unit.

Based on the example given, the simultaneous influence of viscous and interfacial forces in the system prohibits scaleup on the basis of using the same pair of liquids in the model and full-sized systems. In Chapter 2, an analogous case involving gravity and viscous forces in the formation of a vortex was considered in Practice Problem 2.5.

If the identical fluids are used in the model and full-sized units, scaleup can be performed only by either a distortion of the model geometry or by adjusting experimental conditions so that the influence of one of the variables is determined while the other is suppressed.

With Heat Transfer

Mixing systems involving heat transfer are described by the following equation:

$$\frac{h_\varrho L}{\lambda_v} = kRe^r Pr^s \tag{11.105}$$

Constant k and exponents r and s are dependent on the system geometry. When the heat transfer that occurs is by forced convection, the condition of thermal similarity is established by the Reynolds number. If the same liquid is employed in both the model and the commercial unit, the Prandtl numbers are identical. Defining J as a constant (the coefficient of heat transfer)

$$J = k\lambda_v \left(\frac{\rho}{\mu}\right)^r \left(\frac{C\mu}{\lambda_v}\right)^s \qquad (11.106)$$

Then Equation 11.105 may be rewritten as

$$h_\varrho = Jn^r d^{2r-1} \qquad (11.107)$$

where d is the impeller diameter in Re. For the same liquid used in both units, $J_D = J_M$ and

$$\frac{h_{\varrho D}}{h_{\varrho M}} = \frac{n_D^r d_D^{2r-1}}{n_M^r d_M^{2r-1}} \qquad (11.108)$$

where h_ϱ is the individual film heat transfer coefficient.

There are basically four scaleup conditions, any one of which may be specified to meet similarity. These are *dynamic similarity, constant liquid velocity, constant heat transfer coefficients* and *constant heat transfer per unit volume.*

1. *Dynamic similarity* involves maintaining constant Reynolds numbers in the model and full-sized system. That is,

$$\left(\frac{nd^2\rho}{\mu}\right)_D = \left(\frac{nd^2\rho}{\mu}\right)_M$$

If the same liquid is used in both systems, then $\nu_D = \nu_M$, and the ratio of the flow velocities becomes

$$\frac{u_D}{u_M} = \frac{(nd)_D}{(nd)_M} = \frac{d_M}{d_D}$$

Equation 11.108 then becomes

$$\frac{h_{\varrho D}}{h_{\varrho M}} = \frac{d_M}{d_D} = a_d^{-1}$$

Thus, the ratio of the heat transfer coefficients decreases proportionally to the linear scaleup ratio.

It can be shown that the power ratio for model and full-sized mixers using the same liquids is

$$\frac{\mathcal{P}_{OD}}{\mathcal{P}_{OM}} = \frac{(n^3 d^5)_D}{(n^3 d^5)_M} = \frac{d_M}{d_D} = a_d^{-1} \tag{11.109}$$

To the heat transfer coefficients, power is inversely proportional to the scale ratio and, hence, is lower for the full-sized unit than for the model.

Scaleup through equal Reynolds numbers is not desirable from consideration of the heat transfer coefficient. The heat transfer coefficients decrease as the surface area and, subsequently, the volume of equipment increases in scaleup. The only advantage is that power consumption is small, as shown by Equation 11.109.

2. By maintaining *constant flow velocities* in scaleup,

$$u_D = u_M , \quad (nd)_D = (nd)_M$$

or, in terms of the ratio of the heat transfer coefficients,

$$\frac{h_{\varrho D}}{h_{\varrho M}} = \left(\frac{d_D}{d_M}\right)^{r-1} = a_d^{r-1} \tag{11.110}$$

Note that the rotational speed ensures a constant peripheral velocity, whence the power ratio is

$$\frac{\mathcal{P}_{OD}}{\mathcal{P}_{OM}} = \left(\frac{d_D}{d_M}\right)^2 = a_d^2 \tag{11.111}$$

The heat transfer coefficients decrease considerably less than in the case of dynamic similarity. However, the power for the full-sized system increases as the square of the scaleup ratio.

3. By maintaining *constant heat transfer coefficients*, the dependence of speed on the scaleup ratio becomes

$$
\left. \begin{array}{l}
n_D^r d_D^{2r-1} = n_M^r d_M^{2r-1} \\[2mm]
\dfrac{n_D}{n_M} = \left(\dfrac{d_M}{d_D}\right)^{(2r-1)/r} = a_d^{1-2r}
\end{array} \right\}
\tag{11.112}
$$

Rushton and Oldshue [1953] found the value of exponent r to lie in the range of 0.5 to 0.67, depending on the mixer type. The ratio of the power input is

$$
\frac{\mathscr{P}_{OD}}{\mathscr{P}_{OM}} = \left(\frac{d_D}{d_M}\right)^{3/r-1} = a_d^{3/r-1}
\tag{11.113}
$$

This approach to scaleup provides good results for constant total heat flux (i.e., the heat transferred through a given area).

4. Maintaining the *rate of heat transfer per unit volume* constant is another approach to mixing scaleup with heat transfer. In this case, the ratio of the heat transfer coefficients is

$$
\frac{h_{\varrho D}}{h_{\varrho M}} = \frac{d_D}{d_M} = \frac{(nd)_D^r \; d_M^{1-r}}{(nd)_M^r \; d_D^{1-r}}
$$

$$
= \left(\frac{n_D}{n_M}\right)^r \left(\frac{d_M}{d_D}\right)^{1-2r}
\tag{11.114}
$$

The following relations are also easily derived:
Ratio of speeds in model and full-size:

$$
\frac{n_D}{n_M} = \left(\frac{d_D}{d_M}\right)^{2(1-r)/r} = a_d^{2(1-r)/r}
\tag{11.115}
$$

Ratio of Reynolds numbers:

$$
\frac{Re_D}{Re_M} = \left(\frac{d_D}{d_M}\right)^{2/r}
\tag{11.116}
$$

Ratio of powers:

$$\frac{\mathcal{N}_{OD}}{\mathcal{N}_{OM}} = \left(\frac{d_D}{d_M}\right)^{6/r-1} = a_d^{6/r-1} \tag{11.117}$$

From Equation 11.115 it is apparent that the speed of the model is lower than that of the full-sized unit and is proportional to the linear scaleup ratio. Equation 11.116 contradicts any hopes for hydrodynamic similarity. In fact, because of a reduction in Re, there is a strong possibility of attaining mixing in the transition, or even laminar regime in the model. If this occurs, the fundamental design equation (Equation 11.105) is no longer valid. Finally, Equation 11.117 shows for a typical value of r, that the increase of the power input in the full-scale system is the highest of the four scaleup possibilities.

The solution to Practice Problem 11.20 evaluates which of these four cases provides a reliable scaleup approach. Before proceeding to the chapter exercises, the student should study the following example problems, which summarize important principles examined throughout the chapter.

Example 11.6

Determine the power requirements of a propeller mixer operating at 900 rpm used for mixing a liquid with the following properties: $\rho = 900$ kg/m^3 and $\mu = 0.005$ N-s/m^2. The mixing tank is 1200 mm in diameter and is unbaffled. The propeller has a diameter of 300 mm, with a pitch of 600 mm.

Solution

The Reynolds number at n = 900/60 = 15 rps is

$$Re = \frac{nd^2\rho}{\mu} = \frac{15 \times 0.3^2 \times 900}{0.005} = 243,000$$

The expression for the power input is

$$K_N = 0.146 Re^{0.96} Fr^b \left(\frac{D}{d}\right)^{0.91} \left(\frac{s}{d}\right)^{1.22}$$

where

$$Fr = \frac{n^2 d}{g} = \frac{15^2 \times 0.3}{9.81} = 6.88$$

$$b = \frac{x - \log Re}{18} = \frac{0.7 - 5.385}{18} = -0.26$$

The following information is supplied:

$$D/d = 2, 2.5, 3, 3.5, 4$$

$$x = 2.8, 2.5, 2.1, 1.5, 0.7$$

From these data we obtain

$$K_N = 0.149 \times 243,000^{0.96} \times 6.88^{-0.26} \times \left(\frac{1.2}{0.3}\right)^{0.91} \times \left(\frac{0.6}{0.3}\right)^{1.22} = 107,200$$

The power input for mixing is, thus,

$$\mathscr{R} = K_N \mu n^2 d^3 = 107,200 \times 0.005 \times 15^2 \times 0.3^3 = 3260 \text{ W}$$

or

$$3.26 \text{ kW}$$

Example 11.7

Determine the horsepower required for a motor needed to operate a frame mixer consisting of two horizontal blades submerged in a 1000-mm-diameter tank, with a liquid level of 1500 mm. The liquid's viscosity and density are 5 N-s/m^2 and 1200 kg/m^3, respectively. The pressure in the tank is 25 $\times$ 10^{5}N/m^2. The diameter of the mixer's shaft is 40 mm, and the mixer's peripheral velocity is 2.4 m/s. The relative geometric proportions for this (as well as other mixer types) are given in Table 11.14.

Table 11.14 Recommended Geometric Proportions of Various Mixers

Type of Mixer	Size Ratios	Sphere of Application
Two Blades 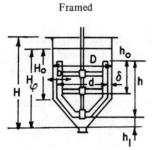	$\dfrac{d}{D} = 0.5–0.7$ $\dfrac{h}{d} = 0.1–0.3$ $\dfrac{h_1}{d} = 0.14–0.2$ $\dfrac{b}{D} = 0.08$	For mixing of liquids with viscosities up to 15 N-s/m^2 and dissolution of solid particles, $u = 1.5–3$ m/s
Four Blades	$\dfrac{d}{D} = 0.2–0.4$ $\dfrac{h}{d} = 0.2–0.4$ $\dfrac{h_1}{d} = 0.4–0.8$ $\dfrac{b}{D} = 0.08$	For mixing of liquids of medium viscosity and dissolution of solid particles, $u = 1.5–3$ m/s
Anchor	$\dfrac{H_\varphi}{H} = 0.75–0.85$ $\dfrac{h}{H} = 0.56$ $\dfrac{b}{d} = 0.07$ $\delta = 25–40$ mm	For mixing of liquids of viscosities up to 200 N-s/m^2 in tanks with heating elements, $u = 1–3$ m/s
Framed	$\dfrac{H_\varphi}{H} = 0.75–0.85$ $\dfrac{h}{H} = 0.6$ $\dfrac{h_0}{H_0} = 0.56$ $h_1 = 190–275$ mm $\delta = 25–40$ mm $\dfrac{b}{d} = 0.07$	See anchor mixers

Table 11.14, continued

Type of Mixer	Size Ratios	Sphere of Application
Turbine	$\dfrac{d}{D} = 0.3$	For intensive mixing of suspensions, dissolution and dispersion of liquids and gases.
	$\dfrac{H_\varphi}{H} = 0.75\text{--}0.85$	
	$\dfrac{h}{d} = 0.2\text{--}0.3$	May operate in the media with viscosities up to 25 N-s/m^2, u = 3–8 m/s
	$\dfrac{\ell}{d} = 0.25$	
	$\dfrac{h_1}{d} = 0.5\text{--}1.5$	
	$\dfrac{d_1}{d} = 0.65$	
	$\dfrac{b}{D} = 0.08$	
Shrouded Turbine	$\dfrac{D}{d} = 2.4$	
	$\dfrac{h}{d} = 0.25$	
	$\dfrac{h_1}{d} = 0.85$	
	$\dfrac{d_1}{d} = 1.1;\ \dfrac{d_2}{d} = 1.6$	
	$\dfrac{H_\varphi}{d} = 1.75$	
Propeller	$\dfrac{D}{d} = 2\text{--}4$	For mixing of liquids with viscosities up to 4 N-s/m^2, dissolution and suspending of solid particles
	$\dfrac{H_\varphi}{d} = 2\text{--}4$	
	$\dfrac{h_1}{d} = 0.7\text{--}1.6$	
	$\dfrac{b}{D} = 0.08$	
	$\dfrac{t}{d} = 1\text{--}2$	

Solution

Assume the clearance between the mixer blades and the tank wall to be $\delta = 25$ mm. The mixer diameter is then

$$d = D - 2\delta = 1000 - 2 \times 25 = 950 \text{ mm}$$

The mixer height (from Table 11.14) is

$$h = 0.6 \times H = 0.6 \times 1500 = 900 \text{ mm}$$

The number of rotations is

$$n = \frac{\omega}{\pi d} = \frac{2.4}{3.14 \times 0.95} = 0.8 \text{ s}^{-1}$$

The Reynolds number is

$$Re = \frac{nd^2\rho}{\mu} = \frac{0.8 \times 0.95^2 \times 1200}{5} = 173$$

For a frame mixer with two horizontal blades,

$$K_N = 12Re^{0.77}\left(\frac{h}{d}\right) = 12 \times 173^{0.77} \times \left(\frac{0.9}{0.95}\right) = 600$$

Hence, the power input is

$$\mathscr{P} = K_N\mu n^2 d^3 = 600 \times 5 \times 0.8^2 \times 0.95^3 = 1650 \text{ W}$$

The horsepower losses in the stuffing box are

$$\mathscr{P}_{s.b.} = 9.84(P + 0.98 \times 10^5)\lambda_f \ell n d^2$$
$$= 9.84(25 \times 10^5 + 0.98 \times 10^5) \times 0.2 \times 0.16 \times 0.8 \times 0.04^2 = 1050 \text{ W}$$

where P = excess pressure in the tank (N/m^2)
λ_f = friction coefficient of the packing
ℓ = length of stuffing box (m)
d = diameter of mixer shaft (m)
n = number of rev/s of the mixer

The length of the packing is

$$\ell = 4d = 4 \times 0.04 = 0.16 \text{ m}$$

Thus, the power required for the motor is

$$\mathcal{N}_{e.m.} = \frac{K_1 \mathcal{N} + \mathcal{N}_{s.b.}}{\eta} = \frac{1.125 \times 1650 + 1050}{0.85} = 3420 \text{ W}$$

or

$$3.5 \text{ kW}$$

where $K_1 = H_f/D = 0.75 \times H/D = 0.75 \times 1.5/1.0 = 1.125$

Example 11.8

A mixture of acids (relative density $\rho = 1.6$, viscosity $\mu = 20$ cp) is being prepared in a tank that is 1200 mm in diameter and 1500 mm high. The vessel is filled to 75% of capacity, and the propeller mixer being used operates at 210 rpm. The system is shown in Figure 11.54. Determine the motor horsepower.

Solution

The diameter of the mixer is

$$d = \frac{D}{3} = \frac{1.2}{3} = 0.4 \text{ m}$$

First determine the mixing regime:

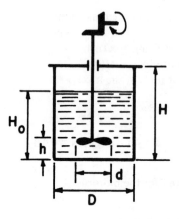

Figure 11.54   Mixing system for Example 11.8.

$$Re = \frac{\rho n d^2}{\mu} = \frac{1600 \times 210 \times 0.4^2 \times 9810}{9.81 \times 60 \times 20} = 44,800$$

Hence, mixing is turbulent.

Determine the power number:

$$K_N = 0.845 Re^{-0.05} = \frac{0.845}{44,800^{0.05}} = \frac{0.845}{1.71} = 0.49$$

This formula is applicable because

$$\frac{D}{d} = 3 \text{ and } \frac{H_0}{d} = \frac{0.75 \times 1.5}{0.4} = 2.8$$

Thus, the power input is

$$\mathcal{W} = K_N \rho n^3 d^5 = \frac{0.49 \times 1600 \times 210^3 \times 0.4^5}{9.81 \times 60^3}$$

$$= 35.2 \text{ kg}_f\text{-m/s} = \frac{35.2}{102} = 0.345 \text{ kW}$$

The horsepower of the motor (with an efficiency of 0.9 and a power reserve of 20%) is

$$\mathcal{N}_m = \frac{0.345 \times 1.2}{0.9} = 0.46 \approx 0.5 \text{ kW}$$

Example 11.9

A 12-kW motor drives a propeller mixer that agitates glycerine ($\gamma = 1200$ kg_f/m^3, $\mu = 16$ p) intensively in a tank 1630 mm in diameter. Determine the mixer's diameter if the unit operates at 500 rpm.

Solution

Because the mixing is intensive, we may use the turbulent regime formula:

$$d = \left\{ \frac{\mathcal{N}}{0.845\rho^{0.95}\mu^{0.05}n^{2.95}} \right\}^{1/4.9}$$

$$= \left\{ \frac{12 \times 102}{0.845\left(\dfrac{1200}{9.81}\right)^{0.95}\left(\dfrac{16}{98.1}\right)^{0.05}\left(\dfrac{500}{60}\right)^{2.95}} \right\}^{1/4.9} = 0.5 \text{ m}$$

The ratio of tank to mixer diameter is checked against Table 11.14 recommendations:

$$\frac{D}{d} = \frac{1.63}{0.5} = 3.26$$

Example 11.10

It was determined experimentally by mixing different liquids with propeller mixers that the power input, $\mathcal{N}$, is a function of liquid viscosity, μ, liquid density, ρ, rpm of the mixer, n, and mixer diameter, d. Determine the general dimensionless relationship among these variables.

Solution

The experimental relationship is

$$\mathcal{N} = \psi(\mu, \rho, n, d) \tag{a}$$

According to the π-theorem, for five variables ($\mathcal{v}$, μ, ρ, n, d) and three basic units of measure (kg$_f$, m, s), Equation a may be rewritten as:

$$\psi(\pi_1, \pi_2) = 0$$

or

$$\pi_1 = f(\pi_2)$$

where π_1 and π_2 are unknown dimensionless expressions composed of the values entering in Equation a.

For determining π_1 and π_2, we will write Equation a as follows:

$$\mathcal{v} = C\mu^x\rho^y n^z d^\nu \tag{b}$$

where   C = dimensionless coefficient
              x, y, z, ν = dimensionless exponents

Equation b now is expressed in terms of its dimensions:

$$\left[\frac{kg_f \times m}{s}\right] = \left[\frac{kg_f \times s}{m^2}\right]^x \times \left[\frac{kg_f \times s^2}{m^4}\right]^y \times \left[\frac{1}{s}\right]^z \times [m]^\nu$$

or

$$kg_f \times m \times s^{-1} = kg_f^{x+y} \times m^{-2x-4y+\nu} \times s^{x+2y-z}$$

Because the dimensions on both sides of the equation should be equal, we obtain three equations:

$$1 = x + y$$
$$1 = -2x - 4y + \nu$$
$$-1 = x + 2y - z$$

To solve this system of equations is impossible, i.e., there are three equations with four unknowns.

1. Thus, expressing y, z and ν through x,

$$y = 1 - x$$
$$z = 3 - x$$
$$\nu = 5 - 2x$$

Substituting the obtained expressions in Equation b,

$$\mathcal{N} = C\mu^x \rho^{1-x} n^{3-x} d^{5-2x} = C\mu^x \rho\rho^{-x} n^3 n^{-x} d^5 d^{-2x} = C\rho n^3 d^5 \left(\frac{\mu}{\rho n d^2}\right)^x$$

Hence,

$$\frac{\mathcal{N}}{\rho n^3 d^5} = C\left(\frac{\rho n d^2}{\mu}\right)^{-x} \tag{c}$$

Because

$$\frac{\mathcal{N}}{\rho n^3 d^5} = K_N$$

and

$$\frac{\rho n d^2}{\mu} = Re$$

the equation may be rewritten as

$$K_N = C Re^{-x} \tag{d}$$

2. Expressing x, z and ν through y,

$$x = 1 - y$$
$$z = 2 + y$$
$$\nu = 3 + 2y$$

Substituting the values obtained in Equation b,

$$\mathcal{V} = C\mu\mu^{-y}\rho^{y}n^{2}n^{y}d^{3}d^{2y} = C\mu n^{2}d^{3}\left(\frac{\rho n d^{2}}{\mu}\right)^{y}$$

$$\frac{\mathcal{V}}{\mu n^{2}d^{3}} = CRe^{y} \tag{e}$$

Transforming the left-hand side of Equation e,

$$\frac{\mathcal{V}}{\mu n^{2}d^{3}} \times \frac{\rho n d^{2}}{\rho n d^{2}} = \frac{\mathcal{V}}{\rho n^{3}d^{5}} \times \frac{\rho n d^{2}}{\mu} = K_{N}Re$$

Consequently, Equation e may be rewritten as

$$K_{N}Re = CRe^{y}$$
$$K_{N} = CRe^{y-1} \tag{f}$$

3. Expressing x, y and ν through z,

$$x = 3 - z$$
$$y = z - 2$$
$$\nu = 2z - 1$$

Substituting the values obtained in Equation b,

$$\mathcal{V} = C\mu^{3}\mu^{-z}\rho^{z}\rho^{-2}n^{z}d^{2z}d^{-1} = C\frac{\mu^{3}}{\rho^{2}d}\left(\frac{\rho n d^{2}}{\mu}\right)^{2}$$

$$\frac{\mathcal{V}\rho^{2}d}{\mu^{3}} = CRe^{z} \tag{g}$$

Transforming the left-hand side of Equation g,

$$\frac{\mathcal{V}\rho^{2}d}{\mu^{3}} = \frac{\mathcal{V}\rho^{2}d}{\mu^{3}} \times \frac{\rho^{3}n^{3}d^{6}}{\rho^{3}n^{3}d^{6}} = \frac{\mathcal{V}}{\rho n^{3}d^{5}} \times \frac{\rho^{3}n^{3}d^{6}}{\mu^{3}} = K_{N}Re^{3}$$

Substituting into Equation g,

$$K_N Re^3 = CRe^z$$

or

$$K_N = CRe^{z-3} \tag{h}$$

4. Expressing x, y and z through ν,

$$x = \frac{5 - \nu}{2}$$

$$y = \frac{\nu - 3}{2}$$

$$z = \frac{\nu + 1}{2}$$

Substituting the values obtained in Equation b,

$$n = C\mu^{5/2}\mu^{-\nu/2}\rho^{\nu/2}\rho^{-3/2}n^{\nu/2}n^{1/2}d^{\nu}$$

$$= C\frac{\mu^{5/2}n^{1/2}}{\rho^{3/2}}\left(\frac{\rho n d^2}{\mu}\right)^{\nu/2}$$

$$\frac{n^2\rho^3}{\mu^5 n} = C^2 Re^{\nu}$$

$$\frac{n^2\rho^3}{\mu^5 n} = \frac{n^2\rho^3}{\mu^5 n} \times \frac{\rho^2 n^5 d^{10}}{\rho^2 n^5 d^{10}} = \left(\frac{n}{\rho n^3 d^5}\right)\left(\frac{\rho n d^2}{\mu}\right)^5 = K_N^2 Re^5$$

Consequently,

$$K_N^2 Re^5 = C^2 Re^{\nu}$$

or

$$K_N = CRe^{(\nu-5)/2} \tag{i}$$

Comparing Equations c, f, h and i, we see that all the solutions bring us to the same formula:

$$K_N = Re^{-x}$$

This is the desired dimensionless equation. The numerical values of C and x may be determined experimentally.

Example 11.11

It is necessary to provide a uniform distribution of solid catalyst particles in a liquid (1.3 mm in size and specific gravity $\gamma = 245$ kg_f/m^3). The solid/liquid weight ratio (S/L) is $\frac{1}{4}$ in the reactor shown in Figure 11.55. The vessel is 1000 mm in diameter. The viscosity of the liquid is 150 cp and its specific weight is 1200 kg_f/m^3. Determine whether a three-blade propeller (Figure 11.55(B)) or a turbine mixer (Figure 11.55(C)) is best suited for this application.

Solution

The required mixer diameter is

$$d_m = (0.25 - 0.30)D$$

Assume

$$d_m = (0.3) \times 1 = 0.3 \text{ m}$$

To determine Re and the number of rotations, use the following equation from Paulushenko et al. [1957]:

$$Re = CGa^k S_\rho^e \Gamma_{dp}^\ell T_D^n \tag{a}$$

In this equation,

	C	k	e	ℓ	n
For a propeller mixer =	0.105	0.6	0.8	0.4	1.9
For a turbine mixer =	0.25	0.57	0.37	0.33	1.15

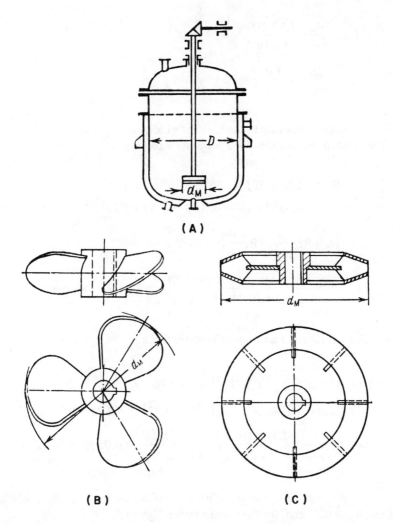

Figure 11.55   (A) Mixer-reactor for Example 11.11; (B) three-blade propeller; (C) turbine mixer.

$$Ga = \frac{d_m^3 \rho^2 g}{\mu^2} = \frac{d_m^3 \gamma^2 g}{\mu^2} \times 10^6 = \frac{3 \times 10^{-3} \times 1.2^2 \times 10^6 \times 9.81}{1.5^2 \times 10^4} \times 10^6$$

$$S_\rho = \frac{\rho_p}{\rho} = \frac{\gamma_p}{\gamma} = \frac{2.45 \times 10^3}{1.2 \times 10^3} = 2.04$$

$$\Gamma_{dp} = \frac{d_p}{d_m} = \frac{1.3 \times 10^{-3}}{3 \times 10^{-1}} = 4.33 \times 10^{-3}$$

$$\Gamma_D = \frac{D}{d_m} = \frac{1.0}{3 \times 10^{-1}} = 3.33$$

The determined values are in the limits of validity of Equation a.
The values of Re and rpm for a propeller mixer are

$$Re = 0.105 \times 10^{-1} \times 1.76^{0.6} \times 10^{4.2} \times 2.04^{0.8}$$

$$\times 4.33^{0.4} \times 10^{-1.2} \times 3.33^{1.9} = 4.61 \times 10^3$$

$$n_0 = Re \frac{\mu}{\rho d_m^2} = Re \frac{\mu_{cp}}{\gamma d_m^2} \times 10^{-3}$$

$$= \frac{4.61 \times 10^3 \times 1.5 \times 10^2 \times 10^{-3}}{1.2 \times 10^{-3} \times 3^2 \times 10^{-2}} = 6.4 \text{ rps} = 384 \text{ rpm}$$

The values of Re and rpm for a turbine mixer are

$$Re = 0.25 Ga^{0.57} S_\rho^{0.37} \Gamma_{dp}^{0.33} \Gamma_D^{1.15} = 2.5 \times 10^{-1} \times 1.76^{0.57} \times 10^{3.99}$$

$$\times 2.04^{0.37} \times 4.33^{0.33} \times 10^{-0.99} \times 3.33^{1.15} = 2.91 \times 10^3$$

$$n_0 = Re \frac{\mu_{cp}}{\gamma d_m^2} = \frac{2.91 \times 10^3 \times 1.5 \times 10^2 \times 10^{-3}}{1.2 \times 10^3 \times 3^2 \times 10^{-2}} = 4.05 \text{ rps} = 243 \text{ rpm}$$

For power input, from the graph of $K_N = f(Re)$ we find, for the propeller mixer, $K_N = 0.32$ and, for the turbine mixer, $K_N = 1.3$.
For the propeller mixer,

$$\mathcal{N} = K_N n^3 \rho d^5 = 3.2 \times 10^{-1} \times 6.4^3 \times 2.97 \times 10^{-1}$$
$$= 24.9 \text{ kg}_f\text{-m/s} \simeq 0.25 \text{ kW}$$

For the turbine mixer,

$$\mathcal{N} = K_N \times n^3 \rho d^5 = 1.3 \times 4.05^3 \times 2.97 \times 10^{-1} = 25.6 \text{ kg}_f\text{-m/s} \simeq 0.25 \text{ kW}$$

In this case, the power input is equal for both mixers. We therefore select a turbine mixer because it operates at a lower rpm.

Example 11.12

Carbon tetrachloride (30% by volume) must be distributed uniformly in water. For this process, a three-blade propeller mixer with a pitch of $s = 1$ is used. The tank diameter is $D = 1000$ mm and the mixer diameter is $d_m = 250$ mm. The available drives can provide 150 and 200 rpm. Determine the preferred rpm from the following information:

- $\rho_d = 1.63 \times 10^2 \text{kgf-s}^2/\text{m}^4$
- $\rho = 1.02 \times 10^2 \text{kgf-s}^2/\text{m}^4$
- $\mu_d = 9.8 \times 10^{-5} \text{kgf-s/m}^2$
- $\mu = 1.02 \times 10^{-4} \text{kgf-s/m}^2$
- $\sigma = 4.64 \times 10^{-3} \text{kgf/m}$

Solution

The following equation is taken from Paulushenko and Yanishevski [1958]:

$$Re = 68.9 \, Ga^{0.01} \left(\frac{Re^2}{We} \right)^{0.47} S_\mu^{0.03} S_{\Delta\rho}^{0.13} \qquad (a)$$

$$Ga = \frac{d_m^3 \rho^2 g}{\mu^2} = \frac{2.5^3 \times 10^{-3} \times 1.02^2 \times 10^4 \times 9.81}{1.02^2 \times 10^{-8}} = 1.53 \times 10^{11}$$

$$\frac{Re^2}{We} = \frac{d_m \rho \delta}{\mu^2} = \frac{2.5 \times 10^{-1} \times 1.02 \times 10^2 \times 4.64 \times 10^{-3}}{1.02 \times 10^{-8}} = 1.14 \times 10^7$$

$$S_\mu = \frac{\mu_d}{\mu} = \frac{9.81 \times 10^{-5}}{1.02 \times 10^{-4}} = 9.6 \times 10^{-1}$$

$$S_{\Delta\rho} = \frac{\rho_d - \rho}{\rho} = \frac{1.63 - 1.02}{1.02} = 5.98 \times 10^{-1}$$

The determined values are in the limit of the application of Equation a:

$$Re = 68.9 \, Ga^{0.01} \left(\frac{Re^2}{We} \right)^{0.47} S_\mu^{0.03} S_{\Delta\rho}^{0.13}$$

$$= 6.89 \times 10 \times 1.53^{0.01} \times 10^{0.11} \times 1.14^{0.47} \times 10^{3.29} \times 9.6^{0.03}$$

$$\times 10^{-0.03} \times 5.98^{0.13} \times 10^{-0.13} = 1.73 \times 10^5$$

$$n_0 = \text{Re} \frac{\mu}{\rho d_m^2} = \frac{1.73 \times 10^5 \times 1.02 \times 10^{-4}}{1.02 \times 10^2 \times 2.5^2 \times 10^{-2}} = 2.77 \text{ rps} = 166 \text{ rpm}$$

Therefore, we have to choose 200 rpm.

Further, we find the specific surface contact area, a', from the following equation (see Paulushenko and Yanishevski [1958]):

$$a' d_m = 2.12 \times 10^2 \text{Re} \left(\frac{\text{Re}^2}{\text{We}} \right)^{-0.56} S_{\Delta\rho}^{0.25} S_\mu^{-0.27} \Gamma_D^{-1.21} S_v^{0.32} \qquad \text{(b)}$$

$$\text{Re} = \frac{\rho n d_m^2}{\mu} = \frac{1.02 \times 10^2 \times 2 \times 10 \times 2.5^2 \times 10^{-2}}{1.02 \times 10^{-4} \times 6} = 2.08 \times 10^5$$

$$\frac{\text{Re}^2}{\text{We}} = 1.14 \times 10^7$$

$$S_\mu = 9.6 \times 10^{-1} \; ; \quad S_{\Delta\rho} = 5.98 \times 10^{-1}$$

$$\Gamma_D = \frac{D}{d_m} = \frac{1.0}{2.5 \times 10^{-1}} = 4$$

$$S_v = \frac{V_d}{V + V_d} = \frac{30}{100} = 3 \times 10^{-1}$$

$$a' d_m = 2.12 \times 10^2 \times 2.08 \times 10^5 \times 1.14^{-0.56} \times 10^{-3.92} \times 5.98^{0.25} \times 10^{-0.25}$$
$$\times 9.6^{-0.27} \times 4^{-1.21} \times 3^{0.32} \times 10^{-0.32} = 5.57 \times 10^2$$

$$a' = \frac{5.57 \times 10^2}{2.5 \times 10^{-1}} = 2.28 \times 10^3 \text{ m}^2/\text{m}^3$$

This corresponds to an average droplet size of

$$d_d = \frac{6 S_v}{a'} = \frac{6 \times 3 \times 10^{-1}}{2.28 \times 10^3} = 7.9 \times 10^{-4} \simeq 0.8 \text{ mm}$$

The reader should consult the references for more in-depth coverage and examples.

PRACTICE PROBLEMS

Solutions are provided in Appendix A for those problems with an asterisk (*) next to their numbers.

11.1–11.15
Determine the power required to operate each mixer for its corresponding set of operating conditions given in Table 11.15.

*11.16 A ¾-filled mixing tank has a diameter of 900 mm and a height of 1100 mm. The specific weight and viscosity of the liquid are 930 kg_f/m^3 and 180 p, respectively. Determine the power required for an electric motor needed to operate a three-blade propeller mixer at a speed of 180 rpm.

*11.17 A paddle mixer is used to dissolve mineral salt in water at a temperature of 64°C. The mixed solution is relatively dilute, i.e., assume the properties of water. Determine the mixer rpm if its diameter is 0.5 m and the motor requires 0.8 kW of energy.

*11.18 A paddle mixer with diameter $d_1 = D/3$ (where D is the tank diameter) is changed to $d_2 = D/4$. In both cases the mixing is laminar. Determine how the mixer rpm changes if both systems have the same power consumption.

*11.19 Determine the diameter of a propeller mixer for the intensive mixing of glycerine ($\gamma = 1200$ kg_f/m^3, $\mu = 16$ p). The tank is 1750 mm in diameter and the mixer rpm and power consumption are 500 rpm and 17 kW, respectively.

11.20 A full-sized mixer is used in a jacketed vessel heated by internal steam coils. The mixer operates under the following set of conditions: n = 250 rpm; mixer diameter d = 420 mm; Re = 6.7×10^5; power $\mathcal{W}_0$ = 5.1 kW; heat transfer coefficient h_ϱ = 800 kcal/hr-m²-°C. The liquid properties are 1100 kg/m^3 and 1 cp. The mixer exponent r = 0.65. Unfortunately, the unit exists in an older plant, and no scaleup information was ever developed. If a unit one-quarter the size of a large-scale mixer is to be built (i.e., d = 110 mm) to develop scaleup formulations, determine which of the four scaleup methods for mixing with heat transfer will provide reliable extrapolation of experimental results.

Table 11.15 Operating Conditions and Mixer Descriptions for Problems 11.1–11.15[a]

Problem	Type of Mixer	D	H	H_f	3	ρ	μ	d	$p \times 10^{-5}$	Special Conditions in Table 11.14
11.1.	Two blades	700	1100	900	2.8	1150	0.018	40	3	d/D = 0.6; h/d = 0.25; without horizontal blades
11.2.	Two blades	600	900	700	3.1	970	0.025	30	8	d/D = 0.7; h/d = 0.2; two horizontal blades
11.3.	Four blades	1200	1400	1200	2.5	800	0.038	45	5	d/D = 0.3; α = 45°; tank is without baffles; upward motion
11.4.	Four blades	1000	1500	1300	2.8	870	0.045	45	6	d/D = 0.35; α = 40°; unbaffled motion downward
11.5.	Four blades	1000	1400	1200	2.8	890	0.048	45	3	d/D = 0.35; α = 45°; baffled motion upward
11.6.	Anchor	1600	1800	1500	2.0	940	2.0	60	4	h/d = 0.93; δ = 40 mm; one blade
11.7.	Anchor	1800	2000	1600	1.5	1100	1.5	65	7	h/d = 0.9; δ = 25 mm; two horizontal blades

11.8.	Anchor	1200	1100	900	1.3	950	3.0	40	9	h/d = 0.67; δ = 40 mm; without horizontal blades
11.9.	Anchor	1000	1500	1200	1.4	870	4.0	40	16	h/d = 11; δ = 25 mm; two horizontal blades
11.10.	Anchor	1600	1900	1600	1.35	930	2.8	60	18	h/d = 0.95; δ = 40 mm; two horizontal blades
11.11.	Turbine	1200	1200	1000	7.8	1120	0.046	30	5	Flat blades, unbaffled; $n' = 6$
11.12.	Turbine	1400	1500	1300	6.8	1060	0.05	40	4	Flat blades; baffled; $n' = 8$; $m_1 = 6$
11.13.	Turbine	1000	1400	1100	6.0	950	0.047	30	7	Curved blades; baffled; $n' = 6$; $m_1 = 6$
11.14.	Propeller	1200	1500	1200	8.0	970	1.5	30	3	D/d = 3; unbaffled
11.15.	Propeller	1400	1600	1300	8.5	860	1.2	40	5	t/d = 1.5; d/D = 0.25; unbaffled

[a]Symbols: D = tank diameter, mm; H = tank height, mm; H_f = depth of mixed liquid, mm; ω = peripheral velocity, m/s; ρ = liquid density, kg/m^3; μ = liquid viscosity, $N\text{-}s/m^2$; d = mixer diameter, mm; P = excessive pressure in a tank, N/m^2; n' = number of blades; m_1 = number of baffles.

NOTATION

A	surface area, m^2, or coefficient
A'	height above vessel bottom, m, or coefficient
a	exponent or variable
a'	interfacial area, mm^2
a''	exponent defined by Equation 11.58
b	blade width, m
c	velocity of liquid discharge from blades, m/s
C	solids concentration, kg/m^3
$\overline{C}$	average concentration, kg/m^3
C'	parameter defined by Equation 11.60
C_1	coefficient in Equation 11.86
C_2	coefficient in Equation 11.87
C_τ'	mixing time coefficient
$\mathscr{D}_t$	turbulent diffusion coefficient, m^2/s
D	tank diameter, m
$\overline{D}_s'$	average bubble diameter, mm
d	mixer or paddle diameter, m
d_a	drop diameter after collision, mm
d_0	droplet diameter, mm
d_p	particle size, m
E	dimensionless number defined by Equation 11.81
E'	dimensionless number in Equation 11.57
Eu	Euler number
e	elasticity coefficient
F	area, m^2
F'	force, N
Fr	Froude number
Ga	Galileo number
g	gravitational acceleration, m/s^2
H	depth or head, m
h	height of mixer blade or depth, m
h_ϱ	film heat transfer coefficient, $kcal/m^2$-hr-°C
J	heat transfer coefficient, $kcal/m^2$-hr-°C
K'	properties constant in Equation 11.10
K	proportionality constant in Equation 11.56
k_m	number of mixers
K_n	power number
K_Q	discharge coefficient defined by Equation 11.46
K''	correction factor defined by Equation 11.55
k	coefficient in Equation 11.23

L	characteristic length, m
L_p	turbulent mixing length, m
LHS	left hand side
ℓ	turbine blade length, m
m_0	dimensionless parameter in Equation 11.56
m_1	number of baffles
m_2	number blades of a turbine mixer
$\mathscr{n}_0$	power input, kW
Ne	Newton number
N_t	mixing number defined in Figure 11.51
N_L	modified power number
n, n_0	rotational speed, rpm
n'	number of blades
P	pressure, Pa
Pr	Prandtl number
p	power input, W
Q	pumping capacity, m^3/hr
Q'	dimensionless group in Equation 11.57
q	mass or volumetric flowrate, kg/s or m^3/s
q'	ratio of angular velocity of liquid to that of the impeller
Re	Reynolds number
R,r	radius, m
RHS	right hand side
S	power coefficient, kg_f-m/s
s	pitch, turns/m
S'	dimensionless group, see Figure 11.25
S_1'	shape factor
T	time, s
t	baffle width, m
t_T	blending time, s
u	velocity, m/s
u_p	relative particle velocity, m/s
$u_{r,t,z}$	radial, tangential and axial velocity components, respectively, m/s
u_s	settling velocity or superficial velocity of fluid, m/s
u_t	terminal settling velocity, m/s
V	volume, m^3
W	vertical dimension of paddle mixer, m
W_e	work, J
We	Weber number
w	wt% solids
X	fractional liquid holdup
Y	concentration, kg/kg

x_0, y_0 parameters defined by Equations 11.75 and 11.76, respectively
z' coalescence coefficient defined by Equation 11.13
z axial coordinate

Greek Symbols

α angle, °
β angle, °
γ concentration or specific weight, kg/m^3
δ clearance, mm
λ_f stuffing box friction factor
λ_v heat of vaporization, J/kg
μ bulk fluid viscosity, p
η drive efficiency
η_c effective viscosity of continuous phase, p
η_d effective viscosity of disperse phase, p
η' effective viscosity of emulsion, p
Θ time, general, s
θ characteristic time, s
τ' mixing time, s
τ_s sheer stress, N/m^2
ν kinematic viscosity, m^2/s
Γ product of parameters in Equation 11.57 or simplex
ρ density, kg/m^3
σ surface tension, N/m
$\dot{\vartheta}$ fluid velocity, m/s
$\ddot{\vartheta}$ dimensionless mixing time, refer to Equation 11.101
ϕ parameter defined in Equation 11.19
χ dimensionless volume defined in Equation 11.94
ψ angle of tilt, °
ψ' void fraction
ψ'' ratio of dispersed phase to total emulsion volumes
ψ_0 angle, °

REFERENCES

American Institute of Chemical Engineering (1960) "Standard Testing Procedure: Impeller-Type Mixing Equipment," New York.

Bissell, E. S., H. C. Hesse, H. J. Everett and J. H. Rushton (1947) *Chem. Eng. Prog.* 43:649.

Calderbank, P. H. (1967) In: *Mixing: Theory and Practice*, Vol. 11, V. W. Uhl and J. B. Gray, Eds., (New York: Academic Press, Inc.) pp. 23, 28-29, 73.

Erdmenger, R., and S. Neidhardt (1952) *Chem. Eng. Technol.* 24:248-258.

Folsum, R. G., and C. K. Fergusson (1949) *Trans. ASME* 71:73.

Gray, J. B. (1963) *Chem. Eng. Prog.* 59(3):55.

Hinze, J. (1955) *AIChE J.* 1:289-295.

Kafarov, V. V. (1949) Processes Peremeshivanya v *Zhidkikh Sredakh, Goskhinizdat, Moscow.*

Laity, D. S., and R. E. Treybal (1957) *AIChE J.* 3:176-180.

Lyons, E. J. (1948) *Chem. Eng. Prog.* 44:341.

Lyons, J. B. (1954) *Chem. Eng. Prog.* 50:629-632.

Miller, S. A., and C. A. Mann (1944) *Trans. AIChE J.* 40:709.

Millon, R. (1953) *Chimie Ind.* 69:258-269.

Misek, T. (1960) *Hydrodynamic Behaviour of Agitated Liquid Extractors,* Ph.D. Thesis, Institute of Chemical Technology, Prague, Czechoslovakia.

Moo-Young, M., K. Tichar and F. A. L. Dullien (1972) *AIChE J.* 18:178.

Nagata, S., K. Yamamoto, K. Hashimoto and Y. Naruse (1959) *Mem. Fac. Eng., Kyoto Univ.* xxi(3):260-274.

Nagata, S., K. Yamamoto, K. Hashimoto and Y. Naruse (1960) *Mem. Fac. Eng., Kyoto Univ.* xxii(1):68-85.

Newitt, D. E., G. E. Shipp and C. R. Black (1951) *Trans. Inst. Chem. Eng.* 29:278.

Paulushenko, I. S., N. M. Kostin and S. F. Matveev (1957) *Zh. P. Khim.* 30:1160.

Paulushenko, I. S., and A. V. Yanishevski (1958) *Zh. P. Khim.* 32:1348.

Peck, W. C. (1955) *Ind. Chemist* 31(12):505-509.

Perry, R. H. and C. H. Chilton, Eds. (1973) *Chemical Engineer's Handbook,* 5th ed. (New York: McGraw Hill Book Co.).

Quillen, C. S. (1954) *Chem. Eng.* 61(6):178-224.

Rodger, W. A., V. G. Trice and J. H. Rushton (1956) *Chem. Eng. Prog.* 52:512-520.

Rushton, J. H. (1952) *Chem. Eng. Prog.* 48:33-38, 95-102.

Rushton, J. H., E. W. Costich and H. J. Everett (1950) *Chem. Eng. Prog.* 46:395-404, 467-476.

Rushton, J. H., J. B. Gallagher and J. Y. Oldshue (1956) *Chem. Eng. Prog.* 52:391-392.

Rushton, J. H., and J. Y. Oldshue (1953) *Chem. Eng. Prog.* 49:161-168, 267-275.

Sachs, J. P., and J. H. Rushton (1954) *Chem. Eng. Prog.* 50:507-603.

Smith, J. C. (1949) *Chem. Ind.* 64:339-404.

Stepanoff, A. J. (1957) *Centrifugal and Axial Flow Pumps,* 2nd ed. (New York: John Wiley & Sons, Inc.).

Sterbacek, Z., and P. Tausk (1965) *Mixing in the Chemical Industry* (Englewood Cliffs, NJ: Pergamon Press, Inc.).

Uhl, V. W., and J. B. Gray, Eds. (1966) *Mixing Theory and Practice* (New York: Academic Press, Inc.).

Van de Vusse, J. G. (1955) *Chem. Eng. Sci.* 4:178-200, 209-220.

Vermeulen, T., et al. (1955) *Chem. Eng. Prog.* 51:84-94.

Wislicenus, G. F. (1947) *Fluid Mechanics of Turbomachinery* (New York: McGraw-Hill Book Co.).

Zlokarnik, M. (1967) *Chem. Ing. Tech.* 39:539.

Zwietering, T. N. (1957) *Chem. Eng. Sci.* 8:244-253.

CHAPTER 12

MIXED PROBLEMS OF HYDRODYNAMICS: FILTRATION

CONTENTS

INTRODUCTION

The flow of fluids through a layer of packed elements, lumped and granular materials, or through fabric layers, is widely practiced in such unit operations as absorption, fluidization, adsorption, extraction, filtration and centrifugation, as well as in a wide variety of contact equipment. The hydrodynamic processes that take place when using this equipment may be divided into two general groups, both of which comprise the class of problems referred to as mixed hydrodynamics. The first involves fluid motion through a layer of fixed or constant thickness. That is, flow proceeds under constant hydraulic resistance (e.g., gas flowing through a catalyst layer, washing sediments after filtering).

The second class includes processes in which the hydrodynamics take place in the presence of an increasing layer thickness or height. That is, flow occurs against an increasing hydraulic resistance. Both classes are addressed in this chapter, with respect to the specific unit operations of filtration and filter centrifugation.

Emphasis in this chapter is placed on theory and design principles for liquid filtration applications; however, some of these principles can be extended to gas filtration, and appropriate discussions are included.

FLUID MOTION THROUGH LAYERS

Physical Concepts of Aero-Hydrodynamics

The motion of fluids (gas or liquid) through a layer of solid particles is quite different from that encountered in pipes and around single solid bodies. A layer formed from solid particles, packed in a random manner, contains winding channels of irregular configurations, which are abundant in abrupt narrowings and expansions, sharp turns and blind alleys. These varying flow passages are also responsible for the formation of local high-velocity streams, even if the overall flow is laminar. In analyzing such phenomena, the relative positions of the solid particles constituting the layer must be considered. Numerous factors influence the flow behavior; however, they may be characterized by three general groups.

Group I comprises those factors that characterize geometric and physical properties of single particles, namely, particle shape (sphere, ring, cylinder, etc.), determining size (equivalent diameter, specific surface area), surface properties (roughness), density, and compressibility of particles.

Group II comprises factors that characterize the layer as a geometric system of discrete particles. The specific parameters of this group are (1) the ratio of layer height to particle size (H/d); (2) the ratio of particle size to the layer diameter or apparatus size; (3) the relative particle location as character-ized by its void fraction, ψ'; and (4) particle specific surface area, a', and the equivalent diameter of a particle, d_{eq}.

Group III comprises those factors characterizing the physical properties of the fluid, as well as hydrodynamic characteristics of the flow. In particular, they are the flow regime (as defined by the Reynolds number), the average and maximum flow velocity through a layer, and others to be described below.

Accurate estimation of all these factors is impossible, as there is no method of assuring a perfectly uniform particle size distribution or arrangement throughout the layer.

Flow Through Layers of Constant Thickness

Fluid motion through and around regular packing materials (for example, in absorption columns) may be visualized simply as flow through parallel channels equipped with diaphragms. If the packing is installed in a retaining vessel or some piece of equipment in a random fashion, it is not possible to evaluate separately the head losses attributed to friction, ΔP_f, and local resistances, $\Delta P_{\varrho.r.}$. It is appropriate to characterize the hydraulic resistance of packings and granular materials by a coefficient, $f' = \Delta P_f + \Delta P_{\varrho.r.}$. This coefficient is a function of the Reynolds number:

$$f' = \psi(Re) \tag{12.1}$$

where the determining size in Re is an equivalent diameter of the channel, d_{eq}, and the fluid velocity is the actual velocity in the voids of the packing.

In this manner, the head losses associated with the fluid motion through the layer (say, for example, gas flowing through a packed tower) may be estimated from either the Darcy-Weisbach or Fanning equations. In this approach, the friction coefficient for smooth pipes, $\lambda/4$, and $\dot{F}$ are inter-changed by overall coefficients of hydraulic resistances for packings or granular materials:

$$\Delta P = \frac{4f'H\gamma w^2}{d_{eq} \times 2g} \tag{12.2}$$

where f' = dimensionless coefficient of hydraulic resistance
           H = layer height
           d_{eq} = equivalent diameter of the channels formed by the packing elements or granules
           w = actual velocity in the voids

The coefficient, f', for the turbulent regime is

$$f' = \frac{3.8}{Re^{0.2}}$$ (12.3)

and, for the laminar regime,

$$f' = \frac{100}{Re}$$ (12.4)

where

$$Re = \frac{d_{eq} \times w \times \rho}{\mu}$$ (12.5)

Some authors prefer to modify the Reynolds number by expressing the determining size in terms of the diameter of a packing element or a granule, and by defining velocity as the superficial velocity (i.e., the flow velocity based on the entire cross section of equipment). One expression that may be used to evaluate the hydraulic resistance in scrubbers has the following form:

$$Eu = CRe_g^n Re_f^m \Gamma_1$$ (12.6)

where subscripts g and f refer to gas and fluid, respectively. From experiments, an increase in void fraction of only 10% (e.g., from 0.5 to 0.55) for the same size of granules, decreases the hydraulic resistance by 50%.

Flow Through Layers of Variable Thickness

The filtration of suspensions and dusty gas mixtures is characterized by a growing layer thickness of solid particles. Filtration is a unit operation aimed at the separation of suspended solid particles from a liquid or gas stream. This process is performed by forcing the contaminated fluid through the voids of a porous mass called the *filtering medium*, thus permitting the flow of fluid (referred to as the filtrate in the case of liquid), but the retention of solid contaminants.

For liquids, there are basically two mechanisms that control filtration, namely, "cake filtration" and "depth filtration." In the former, solid particulates generate a cake on the surface of the filter medium. In depth filtration, solid particulates become entrapped within the medium. The filter medium for the latter case is either cartridges or granular media (such as sand or anthracite coal for liquid filtration). Depth filtration (also called "bed filtration") may be either gravity flow or a pressure operation.

Filtration thickening is either surface filtration or a modification of cake filtration. Cake accumulation on the medium surface is minimized by slurry cross flow, mechanical action, or electrical energy in the form of electrophoresis. Another approach involves a cyclic operation, which includes automatic intermittent cake formation, discharge and reslurrying. Centrifugation and ultrafiltration are additional methods of filtration. The latter involves a cross-flow filtration that is essentially classification based on molecular sieving.

Straining and screening operations are closely related to filtration. Often in these cases the objective is to separate out only a portion of the solids on the basis of size. Figure 12.1 gives one example of a screen liquid–solid separator. Note that the liquid flows through the screen while solids are removed from the screen surface.

Operations in which any successive layers of solids are accumulated, even if the cake thickness is negligible, are termed cake filtration processes. If no successive layers are formed, then it is either a straining operation (as in Figure 12.1) or a filter-thickening operation. Both are also referred to as *surface filtration*.

For liquids, filtration is practiced widely in treating water for the process industries, for processing various chemical end products and to transform waste into less objectionable materials before final disposal. Naturally, to meet these varied needs different types of filters are required. To account for the parameters governing filtration in designs and to select filtration equipment best suited to a particular application, two important variables must be considered: (1) the material that forms the filtering medium; and (2) the method used for forcing the liquid through this medium, which is largely determined by the resistances to the flow offered by the filter medium and the sediment formed on it. For relatively small resistance, only gravity force is required. Such a device is simply called a *gravity filter*.

If gravity is insufficient, the pressure of the atmosphere is allowed to act on one side of the filtering medium, while it is withdrawn from the other side; such a device is called a *vacuum filter*. The application of this type of filter is limited to 15 psi pressure; if greater force is needed, a positive pressure in excess of the atmosphere is applied to the suspension by means of a pump. This may be compressed air introduced in a *montejus*; the suspension may be forced directly through a pump acting against the filtering medium (as in a *filter press*); or centrifugal force may be used to drive the liquid through the filtering medium, as in *centrifuges*. In general, filters may be classified according to the nature of the driving force initiating filtration.

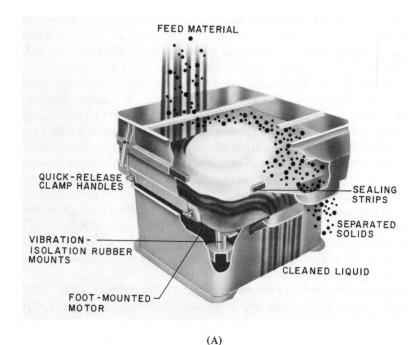

(A)

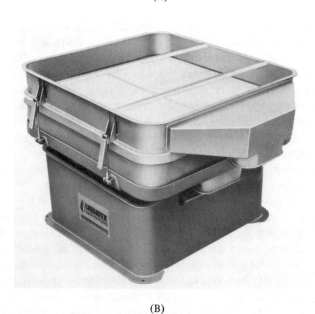

(B)

Figure 12.1 (A) Cutaway view of a screen, liquid-solid separator in operation: (B) full view of separator (courtesy Rotex Inc., Cincinnati, OH).

PRINCIPLES OF LIQUID FILTRATION

Operating Principles

Filtration operations handle sludges of varying characteristics ranging from granular, incompressible, free-filtering materials to slimes and colloidal suspensions that are compressible and tend to foul the filtering medium. The important characteristics of a sludge are its structure (granular and open or colloidal and dense) and its degree of compressibility. With an incompressible sludge, the resistance of the cake is essentially independent of the pressure. If the sludge is compressible, resistance increases rapidly with increasing pressure. In the process of suspension separation, a continuous accumulation of solids takes place, forming a wet filter *cake* and a *filtrate*. This type of filter operation provides a particle retention that exceeds that obtained based on the hole size of the filtering medium. Smaller particles tend to pass through the medium's pores. In contrast to this type of *filtering medium*, filtration in the *depth of the pores* of a filtering medium is responsible for capturing particles much smaller than the pore sizes in the actual passages. *Depth filtration* exhibits increased retentivity as particles travel from the upstream face of the filter plate to the downstream side. Fine particles are removed in the deeper portions of the filter, filling the pores.

The most common form of filtering in the chemical and process industries is *cake filtration* on the filtering medium. This involves handling the permeability of a bed of porous material, the schematic of which is shown in Figure 12.2. Filtration is based on a complicated interaction between the

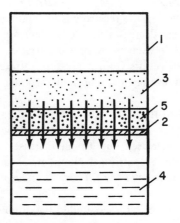

Figure 12.2 Operating scheme of a filtration process: (1) filter; (2) filtering medium; (3) suspension; (4) filtrate; (5) cake.

sludge and filtering medium. The solid particles entrained in the liquid flow through the filtering medium encountering different conditions. Particles larger than the pores on the filter media surface are captured. Many particles whose sizes are less than that of a pore at its narrowest cross section tend to pass through the filter altogether, along with the filtrate. However, a portion of these particles is captured inside the filter due to adsorption onto the walls of the pores or by impingement in the course of its twisted flow path. Such an arrested particle decreases the effective flow cross section, while increasing the probability of retaining the particles that follow. Some particles simply become lodged in the pore, making it impenetrable for other particles. With high solids concentration sludges, even particles that are relatively small in comparison to the pore size may not enter into the media and will tend to remain on the filter surface forming "bridges" over individual openings in the filter plate.

Filtrate flows through the filter medium and cake because of an applied pressure, the magnitude of which is proportional to the filtration resistance. This resistance results from the frictional drag on the liquid as it passes through the filter and cake.

Hydrostatic pressure varies from a maximum at the point where liquid enters the cake, to zero, where liquid is expelled from the medium; consequently, at any point in the cake the two are complementary. In other words, the sum of hydrostatic and compressive pressures on the solids always equals the total hydrostatic pressure at the face of the cake. As such, the compressive pressure acting on the solids varies from zero at the face of the cake to a maximum at the filter medium.

When the space above the sludge is connected to a source of compressed gas (for example, air) or the space under the filter plate is connected to a vacuum source, *filtering is accomplished under a constant pressure differential* because the pressure in the receivers is maintained constant. In this case, the rate of the process decreases due to an increase in the cake thickness and, consequently, its resistance. A similar filtration process, although rare in industrial practices, results from a pressure difference due to the hydrostatic pressure of a sludge layer of constant thickness located over the filter plate.

If the sludge is fed to the filter with a reciprocating pump having constant capacity, *filtering takes place under constant flowrate*. In this case, the pressure differential increases due to an increase in the cake resistance. If the sludge is fed by a centrifugal pump, its capacity decreases with an increase in cake resistance *and the filtering is under variable pressure differences and flowrates*.

In practice, filtration is achieved under the following pressures: Under vacuum,

$$5 \times 10^4 - 9 \times 10^4 \text{ N/m}^2 \ (0.5 \times 10^4 - 0.9 \times 10^4 \text{ kg}_f/\text{m}^2)$$

Under pressure of air, no more than

$$30 \times 10^4 \text{ N/m}^2 \ (3 \times 10^4 \text{ kg}_f/\text{m}^2)$$

Under feed with pumps, up to

$$50 \times 10^4 \text{ N/m}^2 \ (5 \times 10^4 \text{ kg}_f/\text{m}^2)$$

and more. Under hydrostatic pressure of the sludge layer, up to

$$5 \times 10^4 \text{ N/m}^2 \ (0.5 \times 10^4 \text{ kg}_f/\text{m}^2)$$

The most favorable filtration process with cake formation is the process in which there is no clogging of the filtering medium. Such a process is observed at sufficiently high concentrations of solid particles in sludge. This concentration may be assumed conditionally to be in excess of 1% by volume.

To prevent pore clogging in the filtering medium when handling relatively low solids concentrations (e.g., 0.1–1% by volume), general practice is to increase the solids concentration in thickeners before the sludge is fed to the filter. To improve filtration characteristics of sludges that are difficult to filter, *filter aids* and/or *flocculants* often are used to increase particle size. A filter aid is a finely divided solid material, consisting of hard, strong particles that are, in mass, incompressible. The most common filter aids (as an admix to the sludge) are diatomaceous earth, expanded perlite, solkafloc, fly ash or carbon. Filter aids may be applied in one of two ways. The first method involves the use of a *precoat* filter aid, which can be applied as a thin layer over the filter before the sludge is pumped to the apparatus. This prevents fine sludge particles from becoming so entangled in the filter plate that the resistance of the filtering medium itself becomes high. Furthermore, it facilitates the removal of the cake at the end of the filtration cycle.

The second application method involves the incorporation of a certain amount of the material with the sludge before introducing it to the filter. The addition of filter aids increases the porosity of the sludge, decreases its compressibility and reduces the resistance of the cake.

An acceptable filter aid should have a much lower filtration resistance than the material with which it is being mixed (see Tiller and Crump [1977] for examples). It should reduce the filtration resistance by two-thirds to three-fourths with the addition of no more than 25% by weight of filter aid as a fraction of total solids.

The addition of only a small amount of filter aid (e.g., 5% of the sludge solids) can cause a significant increase in filtration resistance. When the amount of filter aid is so small that the particles do not interact, they form a coherent structure and resistance may be affected adversely.

An increase in the particle size tends to improve the filtration process, which may be achieved by *flocculation*. Flocculants are filtered out more readily into a cake, which is permeable and does not clog the filtering medium. Flocculation is an application of the principles of *colloid chemistry*. Only when conditions are proper will particle flocs form. Mild agitation assists in bringing particles into contact for the action to take place. Many dispersions are often unstable and chemically ready to flocculate. Others require that the dispersing agent on their surfaces be neutralized or precipitated. There are others in which the particulates need to be flocculated with a precipitate (e.g., in lime-alum clarification of wastewater). Once chemical conditions are satisfied, flocculation action proceeds.

The operation of chemical pretreatment consists of flocculation or coagulation, which involves the manipulation of charges on the particulates. Small particulates, which are highly charged, form stable suspensions that neither readily settle nor filter.

The addition of trivalent cationic salts (e.g., $AlCl_3$, $FeCl_3$) or a change in the solution pH are common methods employed for changing the surface potential. Most natural substances are negatively charged. The addition of coagulants, in the form of salts, acids or bases, tends to neutralize surface charges, thus permitting attractive van der Waals-London-type forces to promote particle agglomeration.

High-molecular-weight linear polymers (which may be cationic, anionic or nonionic) combine with charged particulates to form large sedimenting flocs. Bentonite and kaolin clays are used principally as extender-aids in combination with flocculants. Because of active sites present on the extenders, they attach to nucleating sites. Their overall effect is to increase the density of the flocs and settling rates.

Filtration frequently is accompanied by hindered or free gravitational sedimentation of solid particles. The directions of action of gravity force and filtrate motion may be cocurrent, countercurrent or cross current, depending on the orientation of the filtering plate, as well as the sludge location over or below the filtering plate. The different orientations of gravity force and filtrate motion, with their corresponding distribution of cake, sludge, filtrate and clear liquid, are illustrated in Figure 12.3. Particle sedimentation complicates the filtration process and influences the controlling mechanisms. Furthermore, these influences vary depending on the relative directions of gravity force and filtrate motion. If the sludge is above the filter plate (Figure 12.3(A)), particle settling leads to more rapid cake formation with a clear filtrate, which can be evacuated from the filter by decanting. If the sludge is under the filter plate (Figure 12.3(B)), particle settling will prevent cake formation, and it is necessary to mix the sludge to maintain its homogeneity.

When the cake structure is composed of particles that are readily deformed or become rearranged under pressure, the structures generally are termed *compressible* cakes. Those not readily deformed are called *semicompressible*, and those that deform only slightly are considered *noncompressible*. Porosity (the ratio of pore volume to volume of cake) does not decrease with increasing pressure differential. The porosity of a compressible cake decreases under pressure, and its hydraulic resistance to the flow of the liquid phase increases with pressure differential across the filter media.

Cakes containing particles of inorganic substances, with sizes in excess of 100 μm, may be considered practically incompressible. Examples are sand particles and crystals of carbonates of calcium and sodium. The cakes containing particles of metal hydroxides, such as ferric hydroxide, cupric hydroxide and aluminum hydroxide, as well as sediments consisting of easily degradable aggregates, which are formed from primary fine crystals, are compressible.

Dickey [1961] notes that after completion of cake formation, treatment of the cake is dependent on the specific filtration objectives. For example, the cake itself may have no value, whereas the filtrate may. Depending on

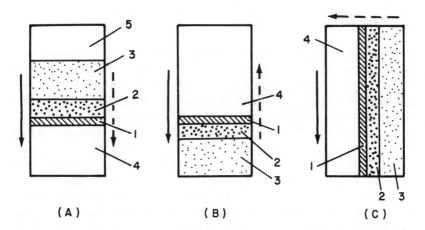

Figure 12.3   The direction of action of gravity force and filtrate motion in the filters: (A) cocurrent; (B) countercurrent; (C) cross current. The full arrows indicate the direction of action of gravity force. The dotted arrows indicate the direction of filtrate motion: (1) filtering plate; (2) cake; (3) sludge; (4) filtrate; (5) clear liquid.

the disposal method and properties of the particulates, the cake may be discarded in a dry form or as a slurry. In both cases the cake usually is subjected to washing following its formation, either directly or after a period of drying if the volume of wash liquor is limited, or to avoid filtrate dilution or excessive wash concentrations where valuable. In some cases a second washing is required, followed by a drying period in which (1) all possible filtrate must be removed from the cake; (2) wet discharge follows for disposal; (3) repulping and a second filtration occur; or (4) dry cake disposal is preferable. Similar treatment options are employed in cases in which the cake is valuable and all contaminating liquors must be removed, or in which both cake and filtrate are valuable. In the latter, cake-forming filtration is employed without washing, to dewater cakes in which a valueless, noncontaminating liquor forms the moisture content of the cake.

Crozier and Brownell [1952] describe the two basic methods of cake washing in detail, namely, *displacement* and *dilution*. The displacement method involves pouring the washing liquid onto the cake surface in the form of a liquid layer or by using atomizing devices. The liquid passes through the pores of the cake under the action of pressure difference, displacing the liquid phase from the pores and mixing with it. The method is employed for easily washed cakes that have no blind pores.

In the dilution method, the cake is removed from the filter and mixed with liquid. Suspension is then separated on the filter. This method is used for cakes that are difficult to wash. Additional discussions are given by Dickey and Bryden [1946], Poole and Doyle [1966], Chalmers et al. [1955] and Perry and Chilton [1973].

Filtration Analysis

In developing design formulations for a filtration process, a conceptual analysis is applied in two parts. The first half considers the mechanism of flow within the cake, while the second examines the external conditions imposed on the cake and pumping system, which brings the results of the analysis of internal flow in accordance with the externally imposed conditions throughout. The analysis presented below follows that of Tiller and Crump [1977], Grace [1953] and Tiller et al. [1972].

The characteristics of the pump relate the pressure applied to the cake to the flowrate at the exit face of the cake. It is the cake resistance that determines the pressure drop. During filtration, liquid flows through the porous filter cake in the direction of a decreasing hydraulic pressure gradient. The porosity, ϵ, is at a minimum at the point of contact between the cake and filter plate, i.e., where x = 0, and at a maximum at the cake surface, x = L, where the sludge enters (Figure 12.4).

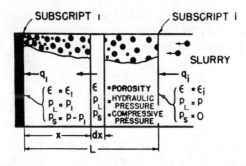

Figure 12.4   Important parameters in cake formation.

The drag imposed on each particle is transmitted to adjacent particles. Therefore, the net solid compressive pressure increases as the filter plate is approached, resulting in a decrease in porosity. Referring to Figure 12.5(A), it is assumed that particles are in contact at one point only on their surface, and that liquid surrounds each particle completely. Hence, the liquid pressure acts uniformly in a direction along a plane perpendicular to the direction of flow. As the liquid flows past each particle, the integral of the normal component of force leads to the form drag, and the integration of the tangential components results in the frictional drag. If the particles are nonspherical, we still may assume single point contacts between adjacent particles, as shown in Figure 12.5(B).

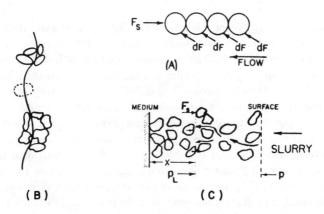

Figure 12.5   Frictional drag on particles in compressible cakes.

Now consider flow through a cake (Figure 12.5(C)) with the membrane located at a distance, x, from the filter plate. Neglecting all forces in the cake other than those created by drag and hydraulic pressure, a balance from x to L gives

$$F_s + AP_L = AP \tag{12.7}$$

where the applied pressure, P, will be a function of time but not of the distance, x. The term F_s is the cumulative drag on particles, increasing in the direction from x = L to x = 0. As single-point contact is assumed, the hydraulic pressure, P_L, is effective over the entire cross section (A) of the cake, for example, against the fictitious membrane shown in Figure 12.5(B). Dividing Equation 12.7 by A and denoting the compressive drag pressure by $P_s = F/A$, we obtain

$$P_L + P_s = P \tag{12.8}$$

The term P_s is a fictitious pressure because the cross-sectional area, A, is not equal to either the surface area of the particles or the actual contact areas. In actual cakes, there is a small area of contact, A_c, and the solid pressure may be defined as F_s/A_c.

Taking differentials with respect to x in the interior of the cake, we obtain

$$dP_s + dP_L = 0 \tag{12.9}$$

This means that drag pressure increases and hydraulic pressure decreases as fluid moves from the cake's outer surface toward the filter plate.

From Darcy's law, the hydraulic pressure gradient is linear through the cake if the porosity, ϵ, and specific resistance, α, are constant. The cake then may be considered "incompressible." This is shown by the straight line on the plot of flowrate per unit filter area versus pressure drop in Figure 12.6. The variations of porosity and specific resistance are accompanied by varying degrees of compressibility, also shown in Figure 12.6. Walas [1946] also illustrates this concept by a plot of fractional pressure drop versus fractional distance through the cake, as shown in Figure 12.7. Tiller and Green [1973] provide some data obtained from different investigators, showing porosity to decrease as α increases rapidly with pressure. The resistance to flow increases in proportion to pressure, and the flow per unit area levels off as shown in Figure 12.8 for highly compressible materials.

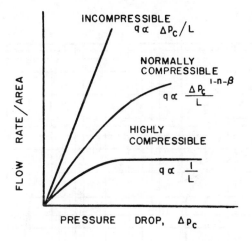

Figure 12.6   Plot of flowrate/area vs pressure drop across the cake.

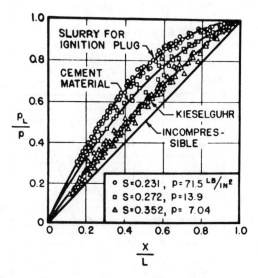

Figure 12.7   Data of Walas [1946] showing hydraulic pressure variation with fractional distance through cake.

Design calculations are outlined in the immediate discussions. Some of these calculations are specific to the filter operating mode. Further details are given by Tiller and Crump [1977], Carman [1937], Sharito et al. [1969], Tiller and Cooper [1962], Tiller [1953], Tiller et al. [1972], and Grace [1953].

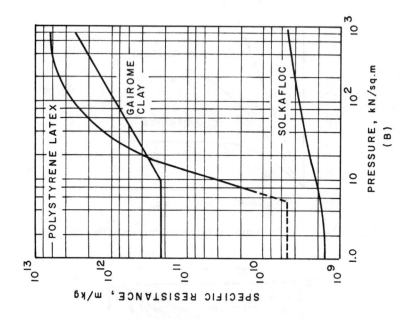

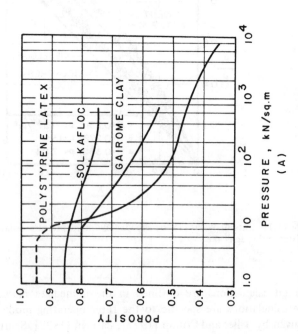

Figure 12.8   Data presented by Tiller and Green [1973]: (A) porosity as a function of drag pressure for different materials; (B) specific flow resistance as a function of drag pressure.

Rate of Filtration

Filtration is primarily a problem of fluid flow, that is, the flow of filtrate through a porous filter cake and filter plate. The rate of the filtration process is directly proportional to the driving force and inversely proportional to the resistance.

Because pore sizes in the cake and filter plate are small and the liquid velocity through the pores is low, the flow of filtrate may be considered laminar and Poiseuille's law is applicable. Filtration rate is directly proportional to the difference in pressure and inversely proportional to the fluid viscosity, as well as to the hydraulic resistance of the cake and filter plate. Because the pressure and hydraulic resistances of the cake and filter plate change with time, the variable rate of filtration may be expressed as

$$u = \frac{dV}{Ad\tau} \qquad (12.10)$$

where V = volume of filtrate (m^3)
A = filtration area (m^2)
τ = time of filtration (s)

Assuming laminar flow through the filter channels, the basic differential equation of filtration is

$$u = \frac{dV}{Ad\tau} = \frac{\Delta P}{\mu(R_c + R_f)} \qquad (12.11)$$

where ΔP = pressure difference (N/m^2)
μ = viscosity of filtrate $(N \cdot s/m^2)$
R_c = filter cake resistance (liter/min)
R_f = initial filter resistance (resistance of filter plate and filter channels) (liter/min)
u = filtration rate (m/s) (filtrate flow through cake and filter plate)
$dV/d\tau$ = filtration rate (m^3/s) filtrate flow

The value of R_f may be assumed constant during filtering. This assumption is based on any increase due to solids penetrating inside the pores. Filter cake resistance, R_c, is the resistance to filtrate flow per square meter of filtration area. R_c increases with cake thickness during filtration. At any instant, R_c depends on the mass of solids deposited on the filter plate as a result of the passage of V m^3 of filtrate. To determine the relationship between V and τ (residence time), Equation 12.11 must be integrated. This means that R_c must be expressed in terms of V.

Considering proportionality of volumes of cake and filtrate, the ratio of cake volume to filtrate volume is denoted by x_0. Then the cake volume is x_0V. The cake volume also may be expressed by the product h_cA, where h_c is the cake height in m. Consequently,

$$x_0V = h_cA \qquad (12.12)$$

Hence, the thickness of the cake, uniformly distributed over the filter plate, is

$$h_c = x_0 \frac{V}{A} \qquad (12.13)$$

And the filter cake resistance may be expressed as

$$R_c = r_0x_0 \frac{V}{A} \qquad (12.14)$$

where r_0 is the specific volumetric cake resistance, m^{-2}.

As follows from Equation 12.14, r_0 *characterizes the resistance to the liquid by a cake of 1-m thickness.*

Substituting for R_c from Equation 12.14 into Equation 12.11, we obtain

$$\frac{dV}{Ad\tau} = u = \frac{\Delta P}{\mu\left(r_0x_0 \dfrac{V}{A} + R_f\right)} \qquad (12.15)$$

Parameter x_0 often is expressed in terms of the ratio of the mass of solid particles settled on the filter plate to the filtrate volume, x_w, and, instead of r_0, a specific mass cake resistance, r_w, is used. That is, r_w is the resistance to the flow presented by a uniformly distributed cake in the amount of 1 kg/m^2. Replacing units of volume by mass, the term r_0x_0 in Equation 12.15 changes to r_wx_w. Neglecting the filter plate resistance ($R_f = 0$) and taking into account Equation 12.13, we obtain from Equation 12.15

$$r_0 = \frac{\Delta P}{\mu h_c u} \qquad (12.16)$$

At $\mu = 1$ N-s/m^2, $h_c = 1$ m and $u = 1$ m/s, the value $r_0 = \Delta P$. Thus, the specific cake resistance equals the pressure difference required by the liquid phase (with a viscosity of 1 N-s/m^2) to be filtered at a rate $u = 1$ m/s for a cake 1 m thick. This hypothetical pressure difference is, however, beyond a practical range.

For highly compressible cakes, the value r_0 reaches 10^{12} m^{-2} or more. Assuming $V = 0$ (at the start of filtration) where there is no cake over the filter plate, Equation 12.15 becomes

$$R_f = \frac{\Delta P}{\mu u} \tag{12.17}$$

At $\mu = 1$ N-s/m^2 and $u = 1$ m/s, $R_f = \Delta P$. This means that the filter plate resistance is equal to the pressure difference necessary for the liquid phase (with viscosity of 1 N-s/m^2) to pass through the filter plate at a rate of 1 m/s. For many filter plates, R_f is typically 10^{10} m^{-1}.

Filtration at Constant Pressure

For a constant pressure drop and temperature operation, Ruth [1946] and Reeves [1947] have shown that all the parameters in Equation 12.15, with the exception of V and τ, are constant. Integrating Equation 12.15 over the limits of 0 to V from 0 to τ, we obtain

$$\int_0^V \mu \left(r_0 x_0 \frac{V}{A} + R_f \right) dV = \int_0^\tau \Delta P A d\tau \tag{12.18A}$$

or

$$\mu r_0 x_0 \frac{V^2}{2A} + \mu R_f V = \Delta P A \tau \tag{12.18B}$$

Dividing both sides by $\mu r_0 x_0/2A$ gives

$$V^2 + 2\left(\frac{R_f A}{r_0 x_0} \right) V = \left(2 \frac{\Delta P A^2}{\mu r_0 x_0} \right) \tau \tag{12.19}$$

This shows the relationship between filtration time and filtrate volume. The equation is applicable to either incompressible or compressible cakes, because at constant ΔP the values r_0 and x_0 are constant.

It follows from Equation 12.15, for constant ΔP, that an increase in the filtrate volume produces a reduction in the rate of filtration. If we assume a definite filtering apparatus and set up a constant temperature and filtration pressure, then the values of R_f, r_0, μ and ΔP will be constant.

The terms in parentheses in Equation 12.19 contain constants that are denoted as filtration constants K and C:

$$K = \frac{2\Delta PA^2}{\mu r_0 x_0} \tag{12.20}$$

$$C = \frac{R_f A}{r_0 x_0} \tag{12.21}$$

Hence, the filtration process may be expressed by a simple relation:

$$V^2 + 2VC = K\tau \tag{12.22}$$

Once K and C are known (from experiments), the volume of filtrate obtained for a definite time interval (for a specified filter, at the same pressure and temperature) can be computed. If process parameters are changed, new constants K and C can be estimated from Equations 12.20 and 12.21.

Equation 12.22 may be simplified by denoting τ_0 as a constant, which depends on K and C:

$$\tau_0 = \frac{C^2}{K} \tag{12.23}$$

Substituting τ_0 into Equation 12.22, the equation of filtration at constant pressure is

$$(V + C)^2 = K(\tau + \tau_0) \tag{12.24}$$

Equation 12.24 is a parabolic relationship between filtrate volume and time. The expression is valid for any type of cake, i.e., compressible and incompressible.

From a plot of $(V + C)$ versus $(\tau + \tau_0)$, the filtration process may be represented by a parabola with its apex at the origin, as shown in Figure 12.9. Moving the axes to distances C and τ_0 provides the characteristic filtration curve for the system in terms of volume versus time. Because the parabola's

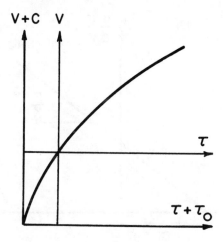

Figure 12.9   Typical filtration curve.

apex is not located at the origin of this new system, it is clear why the filtration rate at the beginning of the process will have a finite value, which corresponds to actual practice.

Constants C and τ_0 in Equation 12.24 have physical significance. They may be assumed equivalent to a fictitious layer of cake of equal resistance. The formation of this fictitious cake follows the same parabolic relationship. τ_0 denotes the time required for the formation of this fictitious cake and C is the volume of filtrate required. Differentiating Equation 12.22 gives

$$\frac{dV}{d\tau} = \frac{K}{2(V + C)} \qquad (12.25)$$

And rearranging in an inverse form,

$$\frac{d\tau}{dV} = \frac{2V}{K} + \frac{2C}{K} \qquad (12.26)$$

Figure 12.10 is a plot of this expression, showing a straight line. This expression is that of a straight line having slope 2/K and an intercept on the abscissa of C. The experimental determination of $d\tau/dV$ is simple due to the geometry of a parabola. Filtrate volumes V_1 and V_2 should be measured for time intervals τ_1 and τ_2. Then, according to Equation 12.22, the quotient $(\tau_2 - \tau_1)/(V_2 - V_1)$ is

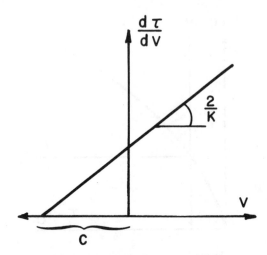

Figure 12.10   Plot of Equation 12.26.

$$\frac{\tau_2 - \tau_1}{V_2 - V_1} = \frac{1}{V_2 - V_1}\left[\frac{V_2^2 - V_1^2}{K} + \frac{2C(V_2 - V_1)}{K}\right]$$

$$= 2\frac{\left(\dfrac{V_1 + V_2}{2}\right)}{K} + \frac{2C}{K} \tag{12.27}$$

Examining the RHS, note that the quotient is equal to the inverse value of the rate at the moment of obtaining the filtrate volume, which is equal to the mean arithmetic value of volumes V_1 and V_2:

$$\frac{\tau_2 - \tau_1}{V_2 - V_1} = \left(\frac{d\tau}{dV}\right)_{(V_1 + V_2)/2} \tag{12.28}$$

Constants C and K are determined on the basis of several measurements of filtrate volumes for different time intervals. There are some doubts as to the actual constancy of C and K during constant pressure filtration. As follows from Equations 12.20 and 12.21, the values C and K depend on r_0 (specific volumetric cake resistance), which, in turn, depends on the pressure drop across the cake. This ΔP causes some changes in the cake, especially during the initial stages of filtration. When the cake is very thin, the main portion of the total pressure drop is exerted on the filter medium. As the cake becomes thicker, the pressure drop through the cake increases rapidly but then levels off to a constant value. Isobaric filtration shows insignificant deviation from Equation 12.22. For approximate calculations, it is possible to neglect the resistance of the filter plate, provided the cake is not too thin. Then the filter plate resistance, R_f, is equal to zero in Equation 12.21, $C = 0$ (Equation 12.21) and $\tau_0 = 0$ (Equation 12.23). Therefore, the simplified equation of filtration takes the form

$$V^2 = K\tau \tag{12.29}$$

However, in the case of thick cakes, Equation 12.29 gives results close to that of Equation 12.22. The following example illustrates the use of these equations.

Example 12.1

An aqueous slurry was filtered in a small laboratory filter press with a pressure drop of 0.5 atm and at a temperature of 20°C. After 10 minutes, 4.7 liters of filtrate were obtained; after 20 minutes, 7.0 liters were collected. From experimentation under other pressures, it was determined that the cake compression coefficient was s = 0.4. Determine the volume of filtrate expected after 30 minutes from a filter press having a filtering area 10 times greater than the laboratory press if the filtration is to be made under 1.5 atm pressure. The liquid temperature is 55°C. Also determine the rate of filtration at the end of the process.

Solution

Evaluate K and C from the experimental data at $\Delta P = 0.5$ atm through the use of Equation 12.22.

For $V = 4.7\ \ell$ and $\tau = 10$ min, we obtain

$$4.7^2 + 2 \times 4.7 \times C = 10\ K$$

For $V = 7.0\ \ell$ and $\tau = 20$ min, we obtain

$$7.0^2 + 2 \times 7 \times C = 20\ K$$

Hence,

$$C = 1.04$$

$$K = 3.18$$

These coefficients change as a function of the filtration area, temperature and pressure. According to Equation 12.21 and accounting for

$$r_0 = a \Delta P^s \qquad (12.30)$$

where a is a coefficient, C may be written in the following form:

$$C = \frac{R_f}{a x_0} \times \frac{A}{\Delta P^s}$$

The term $r/a x_0$ is constant and independent of the filtration conditions. Hence, we find the ratio of constant C_2 at pressure 1.5 atm to constant C_1 for the process under pressure of 0.5 atm:

$$\frac{C_2}{C_1} = \frac{A_2}{A_1} \left(\frac{\Delta P_1}{\Delta P_2} \right)^s$$

Then $A_2/A_1 = 10$ and s = 0.4:

$$C_2 = 1.04 \times 10 \left(\frac{0.5}{1.5} \right)^{0.4} = 8.85$$

Coefficient K is determined from Equations 12.20 and 12.30:

$$K = \frac{2}{a x_0} \times \frac{\Delta P^{1-s} A^2}{\mu}$$

$$\frac{K_2}{K_1} = \left(\frac{\Delta P_2}{\Delta P_1} \right)^{1-s} \left(\frac{A_2}{A_1} \right)^2 \frac{\mu_1}{\mu_2}$$

The viscosity of water at 20°C is $\mu_1 = 1$ cp; at 55°C, it is 0.5 cp. Therefore,

$$K_2 = 3.18 \left(\frac{1.5}{0.5} \right)^{1-0.4} \times 10^2 \times \frac{1}{0.5} = 1230$$

From Equation 12.23 and the values of K_2 and C_2, we determine r_0:

$$\tau_0 = \frac{(8.85)^2}{1230} = 0.0637$$

Consequently, Equation 12.24 becomes

$$(V + 8.85)^2 = 1230\,(\tau + 0.0637)$$

where volume V is in liters and time τ is in minutes. Hence, for time $\tau = 30$ min, we obtain

$$V = \sqrt{1230(30 + 0.064)} - 8.85 = 183 \text{ liters}$$

The final filtration rate (after 30 minutes) is obtained from Equation 12.25:

$$\frac{dV}{d\tau} = \frac{1230}{2(183 + 8.85)} = 3.2 \; \ell/\text{min}$$

The average rate of filtration is

$$\frac{V}{\tau} = \frac{183}{30} = 6.1 \; \ell/\text{min}$$

Constant-Rate Filtration

When sludge is fed to a filter by means of a positive displacement pump, the rate of filtration is nearly constant, i.e., $dV/d\tau$ = constant. During constant-rate filtration, the pressure increases with cake thickness. Therefore, the principal filtration variables are pressure and filtrate volume, or pressure and filtration time. Equation 12.15 is the principal design relation, which may be integrated for a constant-rate process. The derivative $dV/d\tau$ may be replaced simply by V/τ:

$$\Delta P = \mu r_0 x_0 \left(\frac{V^2}{A^2 \tau}\right) + \mu R_f \left(\frac{V}{A\tau}\right) \tag{12.31}$$

The ratios in parentheses express the constant volume rate per unit filter area. Hence, Equation 12.31 is the relation between time, τ, and pressure drop, ΔP.

For incompressible cakes, coefficient r_0 is constant and independent of pressure. For compressible cakes ($s \neq 0$) r_0 may be estimated from Equation 12.30. Substituting r_0 from Equation 12.30 into Equation 12.31, the relationship between time and pressure at constant-rate filtration is obtained:

$$\Delta P = \mu a x_0 \Delta P^s \left(\frac{V}{A\tau}\right)^2 \tau + \mu R_f \left(\frac{V}{A\tau}\right) \tag{12.32}$$

Filtration experiments in a small-scale device (e.g., at constant pressure or constant rate) permit determination of ax_0, as well as s and R_f, for a given sludge and filtering medium. Consequently, it is possible to predict the time required for the pressure drop to reach the desired level for a specified set of operating conditions. In the initial stages of filtration, the filter medium has no cake. Furthermore, ΔP is not zero but has a certain value corresponding to the filter medium resistance for a given rate. This initial condition is

$$\Delta P_0 = R_f \mu \left(\frac{V}{A\tau}\right) \tag{12.33}$$

For an incompressible cake ($s = 0$), Equation 12.32 takes the form

$$\mu a x_0 \left(\frac{V}{A\tau}\right)^2 \tau + R_f \mu \left(\frac{V}{A\tau}\right) = \Delta P \tag{12.34}$$

As follows for the filtration of incompressible sediment (at a constant rate), the pressure increases in direct proportion to time. However, Equation 12.32 shows that pressure increases faster than time.

In some cases the resistance of the filter medium may be neglected, especially for thick cakes. Assuming $R_f = 0$, Equation 12.32 simplifies to

$$\Delta P^{1-s} = \mu x_0 a \left(\frac{V}{A\tau}\right)^2 \tau \tag{12.35}$$

An increase in pressure influences not only coefficient r_0, but also the cake's porosity. Cake on the filter plate is compressed; consequently, the additional liquid is squeezed out of the cake. Thus, along with a constant feed of sludge, the flowrate will not be stable and will change somewhat in time.

The weight of dry solids in a cake is

$$W' = x_0'V \qquad (12.36)$$

where x_0' = weight of solids in the cake per unit filtrate volume

The concentration of solids in the feed sludge is expressed by weight fraction c. It also is possible to determine experimentally the weight ratio of wet cake to its dry content, m. Hence, a unit weight of sludge contains mc of wet cake. Denoting γ as the specific weight of filtrate, then $(1 - mc)/\gamma$ is the filtrate volume per unit weight of feed sludge. But this unit contains c amount of solids, and the ratio of solids in the cake to the filtrate volume is

$$x_0' = \frac{c\gamma}{1 - mc} \qquad (12.37)$$

Thus, from sludge concentration c and the weight of a wet cake per kg of dry cake solids, x_0' can be determined. If the suspension is dilute, then c is small and the product mc is small. This means that x_0' will be almost equal to c.

According to Equations 12.36 and 12.37, the weight ratio of wet to dry cake then will change. Equation 12.37 shows also that because x_0' depends on the product mc, at relatively moderate suspension concentrations this effect will not be great and thus may be neglected. However, for the filtration of concentrated sludges, the mentioned phenomena will play some role, i.e., at constant feed the filtrate changes with time. Silverblatt et al. [1974] provide further discussions.

Variable-Rate, Variable-Pressure Filtration

Figure 12.11 shows typical pressure profiles across a filter. According to this plot, the compression force in the cake section is

$$P = P_1 - P_{st}$$

where P_1 = pressure on the sludge over the entire cake thickness
P_{st} = static pressure over the same section of cake

Pressure P corresponds to the local specific cake resistance $(r_w)_x$. At the sludge-cake interface, $P_{st} = P_1$ and $P = 0$; and for the interface between the cake and filter plate, $P_{st} = P_{st}'$ and $P = P_1 - P_{st}'$. P_{st}' corresponds to the resistance of filter plate, ΔP_f, and is expressed by the equation

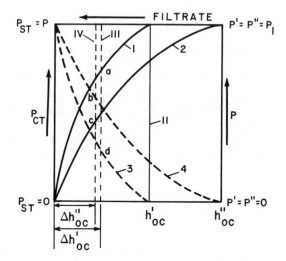

Figure 12.11   Distribution of static pressure, P_{st}, in liquid and P along the cake thickness and filter plate: (I, II) boundaries between the cake and sludge at τ'' and τ'; (III, IV) boundaries between cake layers or cake and filter plate at τ' and τ''; (V) boundary line between the cake and filter plate or free surface of the filter plate. (1, 3) curves $P_{st} = f(h_{oc})$ and $P = f(h_{oc})$ at τ'; (2, 4) curves $P_{st} = f(h_{oc})$ and $P = f(h_{oc})$ at τ''.

$$\Delta P_f = \mu R_f W \tag{12.38}$$

where   W = rate of filtration (m^3/m^2-s)

It is easy to see that ΔP_f is constant during the filtration process. The value P is also the driving force of the process. Therefore, starting from the governing filtration equations, the general expression for an infinitesimal increment of solid particles weight in a cake of unit of area is $x_w dq$; where q is the filtrate volume obtained from 1 m^2 filtering area, m^3/m^2. The corresponding increment, dP, may be expressed as

$$\frac{dq}{x_w dq} = \mu (r_w)_x W \tag{12.39}$$

The value x_w changes insignificantly with P. In practice, an average value can be assumed. Note that W is constant for any section of the cake. Hence, Equation 12.39 may be integrated over the cake thickness for the limits from $P = 0$ to $P = P_1 - P'_{st}$ and from $q = 0$ to $q = q$:

$$q = \frac{1}{\mu x_w W} \int_0^{P_1 - P'_{st}} \frac{dP}{(r_w)_x} \qquad (12.40)$$

Parameters q and W are variables when the filtration conditions are changed. Coefficient $(r_w)_x$ is a function of pressure:

$$(r_w)_x = f(P) \qquad (12.41)$$

The exact relationship can be derived from experiments in a device called a compression-permeability cell (Figure 12.12). Once this relationship is defined, the integral of the RHS of Equation 12.40 may be evaluated analytically. Or, if the relationship is in the form of a curve, the evaluation may be made graphically. The interrelation between W and P_1 is established by the pump characteristics, which define q = f(W) in Equation 12.40. Filtration time may then be determined from:

$$\frac{dq}{d\tau} = W \qquad (12.42)$$

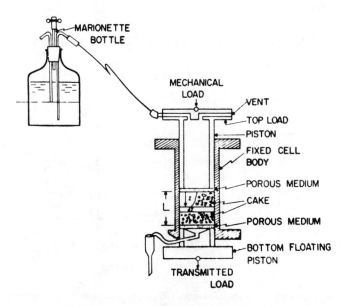

Figure 12.12   A compression-permeability cell.

Hence,

$$\tau = \int_0^q \frac{dq}{W} \qquad (12.43)$$

The following example illustrates this approach.

Example 12.2

The following data were obtained from separation tests for a particular suspension using a filter: $\mu = 0.00148$ N·s/m^2, $x_w = 3.05$ kg/m^3 and $R_f = 6.56 \times 10^{10}$ m^{-1}. The relationship $(1/r_w)_x = f(P)$ was obtained from a compression-permeability cell and the results are given by curve 1 in Figure 12.13. The pump characteristics in terms of capacity per unit filter area are given by curve 1 in Figure 12.14. Determine the relationship between filtrate volume and the time of filtration.

Solution

1. According to Equation 12.38, the static pressure at the interface between the cake and filter plate is numerically equal to the plate resistance provided that the pressure exerted on the free surface of the filter plate is equal to zero.

$$P'_{st} = \Delta P_f = 0.00148 \times 6.56 \times 10^{10} \, W = 97.1 \times 10^{10} \, W$$

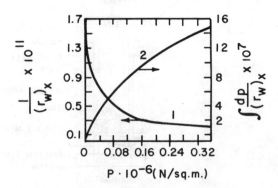

Figure 12.13   Local specific resistance $1/(r_w)_x$ vs P (curve 1) and integral of the right-hand side of Equation 12.40 (curve 2).

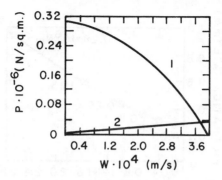

Figure 12.14 Pump characteristics (curve 1) and liquid static pressure on the boundary between the cake and filter plate, according to Equation 12.38 (straight line 2).

Line 2 in Figure 12.14 is obtained from this equation. It shows the relationship of liquid static pressure on the boundary between the cake and filter plate versus filtration rate. As seen from Figure 12.14, at any filtration rate, P_1 (ordinate AB of curve 1) is spent for overcoming the resistance of filter plate P'_{st} (ordinate AC of line 2) and cake resistance (line BC between curve 1 and line 2).

The intersection of curve 1 and line 2 corresponds to the start of filtration when there is no cake formed and the total pressure developed by the pump is spent for overcoming the resistance of the filter plate. It was assumed in this case that the filter plate resistance is independent of pressure. If it depends on pressure, the design should consider some average value, $P'_{st} = f(W)$, which may be determined from the experiment.

2. For the limits from $P = 0$ to the increasing values $P = P_1 - P'_{st}$ (refer to Figure 12.13), curve 2 is obtained by means of graphical integration to show the dependence of the integral (Equation 12.40) from forces compressing the cake.

3. To determine q versus W at given μ and x_w, use Equation 12.40. From Figure 12.14 we find the values $W_1 P_1$ and $P_1 - P'_{st}$; then for each value of $P - P'_{st} = P$ we determine from curve 2 the value of the integral of Equation 12.40. Substituting the corresponding values of W and the integral into Equation 12.40, q is computed. The results of these calculations are given in Figure 12.15.

To determine $\tau = f(q)$ from Equation 12.43 for different values of q by the method of graphical integration, use Figure 12.15. This determines the value of the integral in Equation 12.43 and, consequently, the value of τ. The data obtained in this manner are plotted in Figure 12.16.

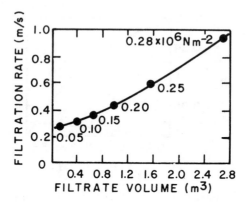

Figure 12.15   Rate of filtration versus filtrate volume according to Equation 12.40.

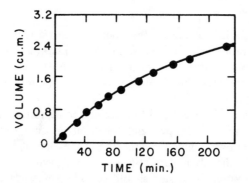

Figure 12.16   Filtrate volume vs filtration time.

Further examination of this and similar problems is given by Tiller [1958] and Zirkin and Zhuzhikov [1967]. See also the work of Rhodes [1934] and Meshengisser [1960].

Filtration at Constant Pressure and Constant Rate

This method of filtration is realized if a pure liquid is filtered through a cake of constant thickness at constant pressure difference. Cake washing by displacement when the washing liquid is located over the cake may be considered as filtering of a washing liquid through a constant cake thickness at constant pressure and flowrate.

The rate of washing is related to the rate of filtration during the last stages and may be expressed by Equation 12.15, where ΔP is the pressure at the final moment, and V is the filtrate volume obtained during filtration regardless of the filtration method (constant-pressure or constant-rate operation). Usually, in the final period, filtration is performed at constant pressure. Then the rate of this process may be presented by Equation 12.25. From filtration constants C and K, at constant pressure for a given system the filtration rate for the last period is determined. If the washing liquid passes through the filter in the same paths as the sludge and filtrate, then the difference between the washing rate and filtration rate for this last period will be due mostly to a difference in the viscosities of the wash liquor and filtrate. Therefore, Equation 12.25 is applicable using the viscosity of the washing liquid, μ_w. Denoting the rate of filtration in the last period as $(dV/d\tau)_f$, the washing rate is

$$\left(\frac{dV}{d\tau}\right)_w = \left(\frac{dV}{d\tau}\right)_f \frac{\mu}{\mu_w} \tag{12.44}$$

Additional discussions are given by Rhodes [1934], Meshengisser [1960], Brownell and Katz [1947], and Moncrieff [1965].

LIQUID FILTRATION EQUIPMENT

Continuous filters are essentially many elemental surfaces on which a series of different operations is performed: solids separation and cake formation, cake washing, dewatering, drying, cake removal and filter media washing. The specific equipment consists of two general classes of components: stationary components, which are supporting devices, such as the suspension vessel and scraping mechanisms, and movable devices such as the filter medium itself.

In *cake filtration*, either continuous or batch filters are applicable. In *filter medium filtration*, however, where particulates are retained within the framework of the filter medium, only batch systems are acceptable. Batch filters may be operated in any filtration regime, whereas continuous filters operate most often at constant pressure.

In an attempt to organize the almost overwhelming number of different types of filtration equipment, two classification schemes have evolved for continuous operations.

1. Continuous filtration equipment may be classified according to its method of generating pressure difference:

	Pressure Differential (N/m^2)
Hydrostatic pressure of the suspension layer to be separated	Usually no more than 5
Action of compressors	5–9
Action of pumps	Up to 50 and higher

2. Classification also may be based on the relative directions of gravity force and filtrate motion:

Relative Directions	Angle $(°)$
Opposite	180
Cocurrent	0
Perpendicular	90

Because particle sedimentation under the influence of gravity force plays a major role in the filtration process, the second classification basis is adopted in this book. From the above, we classify three types of filtration equipment:

1. Class A, in which gravity force and filtrate motion are opposing;
2. Class B, in which the relative directions of gravity force and filtrate are cocurrent; and
3. Class C, in which the relative directions are normal to each other.

Accordingly, the operating principles and important features of specific filtration equipment are described below. The reader is referred to the following references for in-depth reviews of the operating performances and detailed design features of the most frequently employed devices described below (see Kufferath [1954], Daniells [1964], Flood et al. [1966], Purchas [1964], Thomas [1965], Trawinski [1970]).

Class A Designs (Rotary Drum Filters)

Rotary drum filters are either vacuum or pressure type. They are operated most frequently as vacuum-type filters. Although operated under pressure, they are rarely subjected to excessive pumping pressures. The principal advantage of these filters is the continuity of their operation. Unfortunately, total filtration cycles are limited to narrow time intervals. As such, it is necessary to maintain near constant slurry properties. Changing slurry properties can lead to wide variations in the required times for completing individual operations of the filtration process. For separating low stratified suspensions, rotary drum filters normally are specified at a submergence rate of 50%. Such slurries require only mild mixing to prevent particle settling. These filters are less useful in handling polydispersions containing particles with wide size ranges. Fouling by small particles is a frequent problem in these latter cases.

Drum vacuum filters with external filtering surfaces are characterized by the rate at which the drum is immersed in the suspension. These are perhaps the most widely employed Class A filters in industry, an example of which is shown in Figure 12.17. As shown, the design consists of a hollow drum (1) with a slotted face, the outer periphery of which contains a number of shallow tray-shaped compartments (2). The filter cloth is supported by a grid or a heavy screen, which lies over these compartments. The drum rotates on a shaft with one end connected to the drive (3) and the other to a hollow trunion adjoining to an automatic valve. The drum surface is partially immersed in the suspension contained in the vessel (6).

The cake formed on the outer surface of the drum is removed by a scraper (7) as the drum rotates (as shown by the arrow). Figure 12.18 shows a longitudinal view of the system. Each compartment (2) of the drum (1) is connected through a pipe (3) passing through the hollow trunion (4) of the shaft (5), with the automatic valve (6). A stirring device (7) is mounted under the drum to prevent particle settling. A diagrammatic cross section of the filter is shown in Figure 12.19. As the drum rotates in a clockwise manner,

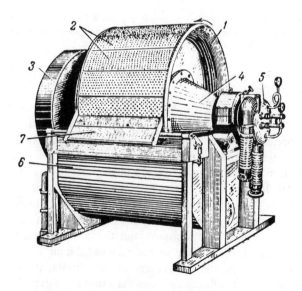

Figure 12.17   General view of a rotary drum vacuum filter with an external filtration surface: (1) hollow drum; (2) filtration compartments; (3) drive; (4) hollow trunion; (5) automatic valve; (6) tank for suspension; (7) knife for cake scraping.

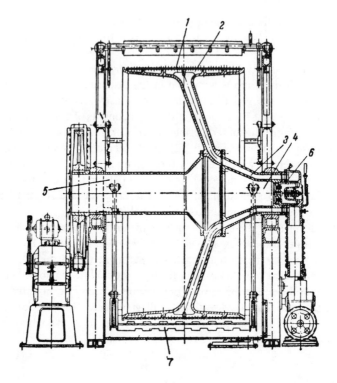

Figure 12.18   Longitudinal section of a rotary drum vacuum filter with an external filtration surface: (1) drum; (2) compartment; (3) connecting pipe; (4) hollow trunion; (5) shaft; (6) automatic valve; (7) stirring device.

each compartment is connected by the pipe (2) with different chambers of immobile parts of the automatic valve (4) and passes in series through the following operating zones: filtration, first dewatering, washing, second dewatering, cake removal and cloth regeneration. In the filtration zone, the compartment contacts the suspension in the tank (11), and is connected to a pipe (10) hooked up to a vacuum source. In this case, the filtrate is discharged through the pipe and space in the collector and the cake is formed on the compartment's surface. In the zone of "first dewatering," the cake comes in contact with the atmosphere and the compartment is connected to the same space (10). Because of the vacuum, the air is drawn through the cake and, for a maximum recovery of the filtrate, the compartment remains connected to a collection port on the automatic valve.

   In the washing zone the cake then is washed by a series of nozzles (8) or "wash headers." The compartment is connected through the port (6),

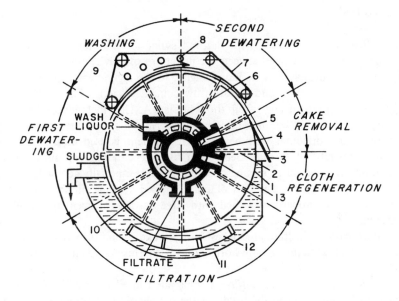

Figure 12.19   Diagrammatic cross section of a rotary drum vacuum filter: (1) drum; (2) connecting pipe; (3) scraper; (4) automatic valve; (5, 13) chambers of automatic valve connected to the source of compressed air; (6, 10) chambers of automatic valve connected with a source of vacuum; (7) endless belt; (8) wash header; (9) guiding roll; (11) tank for suspension; (12) stirring device.

which also is tied into a vacuum source. The wash liquor is removed in the other collector.

In the second dewatering zone, the cake is also in contact with the atmosphere, and the compartment is connected with the same port (6). Consequently, the washing liquid is displaced from the cake pores and delivered to the collector. To avoid cake cracking during washing and dewatering, an endless belt (7) is provided that moves over a set of guide rollers. In the discharge zone the compartment is connected with the port (5), which is in communication with a compressed air source. This reversal of pressure, or "blow," loosens the cake from the filter medium, whence it is removed by a scraper or doctor blade (3).

In the regeneration zone the cloth is blown through by compressed air, which enters the compartment through the pipe from the port (13). The automatic valve serves to activate the filtering, washing and cake discharge functions of the filter sections. It provides separate outlets for the filtrate and wash liquid and a connection by which the compressed air blowback can be applied.

Class B Designs

Top-feed filters are used with flat and cylindrical filtering media. In the former designs, the angle between the directions of gravity force and filtrate motion is 0° but may vary to larger angles. In the second case, the direction of gravity force action and filtrate flow coincide exactly. Filter designs in this class are quite different from those in Class A. They include sophisticated rotary drum filters, continuous-belt filters, Nutsch-batch filters and filter presses with horizontal chambers. This equipment is used most often for separating stratified slurries. Separation by filters of the first subgroup is based on intensive slurry mixing by agitators. It is especially advisable to use these filters for the separation of polydispersed systems. In this case, the cake formed is properly stratified with large particles adjacent to the filter medium.

Filters with flat filtering surfaces form a cake of uniform thickness and homogeneous structure at any horizontal plane. This permits highly effective washing.

The internal rotary drum filter is illustrated in Figure 12.20. The filter medium is contained on the inside of the periphery. This design is ideal for rapid-settling slurries, which do not require a high degree of washing.

Tankless filters of this design consist of multiple-compartment drum vacuum filters. One end is closed and contains an automatic valve with pipe connections to individual compartments. The other end is open for the feed entrance. For cake removal, the drum is supported on a tire and rigid rollers. The drum is driven by a motor and speed reducer connected to the riding roll shaft.

The feed slurry is discharged to the bottom of the inside of the drum from the distributor and is maintained as a pool by a baffle ring around the open end and the closed portion of the other end. As the drum revolves, the compartments successively pass through the slurry pool, where a vacuum is applied as each compartment becomes submerged. Slurry discharge is accomplished at the top center where the vacuum is cut off, and gravity (usually aided by blowback) allows the solids to drop off onto a trough. From there, a screw (or belt) conveyor removes the solids out of the drum. This filter is capable of handling heavy, quick-settling materials. Dickey [1961] notes that variations in feed consistency cause little or no difficulties.

Nutsch filters are one design type with a flat filtering plate. This configuration basically consists of a large false-bottom tank with a loose filter medium. Older designs employ sand or other loose, inert materials as the filtering medium and are still employed in water clarification operations. In vacuum filtration these false-bottom tanks are of the same general design as the vessels employed for gravity filtration. They are less widely used, however, being confined mostly to rather small units, particularly for acid

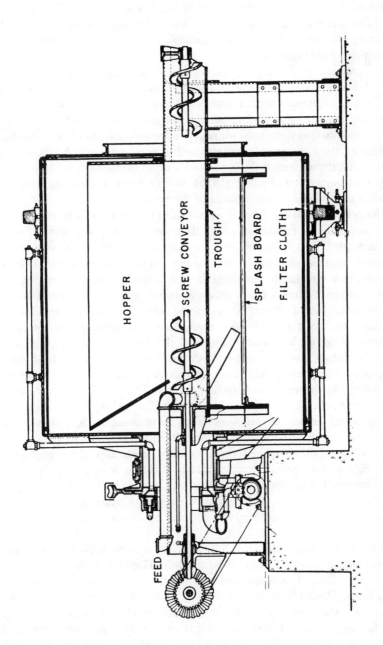

HOPPER

SCREW CONVEYOR

TROUGH

SPLASH BOARD

FILTER CLOTH

FEED

Figure 12.20   Sectional view of an internal rotary drum vacuum filter.

work. Greater strength and more careful construction is necessary to withstand the higher pressure differentials of vacuum over gravity. This naturally increases construction costs. However, when high filtering capacity or rapid handling is required and either is obtainable by vacuum, the advantages may more than offset higher costs.

Construction of the vacuum false-bottom tank is relatively simple, consisting of a single vessel divided into two chambers by a perforated section. The upper chamber operates under atmospheric pressure and retains the ulfiltered material. The perforated false bottom supports the filter medium. The lower chamber is designed for negative pressure and to hold the filtrate.

Nutsch filters can provide frequent and uniform washings. A type of continuous filter that essentially consists of a series of Nutsches is the rotating-tray horizontal filter. Berline [1955] and *Chemical Engineering Magazine* [1954a,b] give detailed descriptions of this system.

The *horizontal rotary filter* (Figure 12.21) is well adapted to the filtering of quick-draining crystalline solids. Due to its horizontal surface, solids are prevented from falling off or from being washed off by the wash water. As such, an unusually heavy layer of solids can be tolerated. The basic design consists of a circular horizontal table that rotates about a center axis. The table comprises a number of hollow pie-shaped segments with perforated or woven metal tops. Each section is covered with a suitable filter medium and is connected to a central valve mechanism that appropriately times the removal of filtrate and wash liquids and the dewatering of the cake during each revolution. Each segment receives the slurry in succession. Wash liquor is sprayed onto each section in two applications. Then the cake is dewatered by pulling air through it and, finally, the cake is scooped off of the surface by a discharge scroll.

Belt filters consist of series of Nutsches moving along a closed path. Nutsches are connected as an infinite chain so the longitudinal edge of each Nutsch has the shape of a baffle plate overlapping the edge of the neighboring unit. Each unit is displaced by driving and tensioning drums.

Nutsches are equipped with supporting perforated partitions covered with the filtering cloth. The washed cake is removed by turning each unit over. Sometimes a shaker mechanism is included to ensure more complete cake removal.

Belt filters have (instead of Nutsches) an endless supporting perforated rubber belt covered with the filtering cloth. The basic design is illustrated in Figure 12.22. The supporting and filtering partitions (1) are displaced by a driving drum (2) and maintained in a stretched condition by a tensioning drum (3), which rotates due to friction against the rubber belt. Belt edges (at the upper part of their path) slide over two parallel horizontal guide planks. An elongated chamber (4) is located between the guide planks. The chamber in the upper part has grids with flanges adjoining to the lower

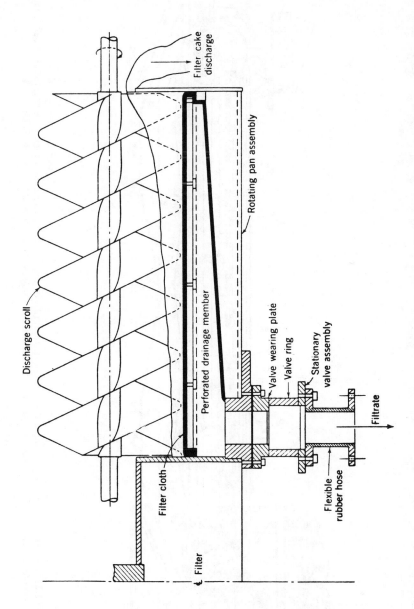

Figure 12.21   Cross-sectional view of a rotary horizontal vacuum filter showing filtrate-removal system, filter cloth and discharge scroll.

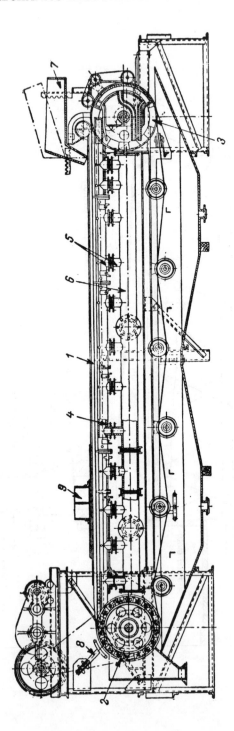

Figure 12.22   A belt filter: (1) supporting and filtering partition; (2) driving drum; (3) tensioning drum; (4) elongated chamber; (5) nozzles; (6) collector for filtrate; (7) trough for sludge feed; (8) device for washing cake off; (9) tank for washing liquid.

surface of the rubber belt. The region under the belt is connected by nozzles (5) to the filtrate collector (6), which is attached to a vacuum source. The chamber and collector are divided into sections from which filtrate and washing liquid may be discharged. The sludge is fed by the trough (7). The cake is removed from the drum (2) due to gravity force or blowing, or sometimes is washed off with liquid using a nozzle (8). The washing liquid is supplied from the tank (9), which can move along the filtering partition. It can be washed at its motion along the lower path.

The filtering partition, shown in Figure 12.23, consists of a riffled rubber belt (1) with slots (2) and grooves (3) and a filter cloth (4), which is fixed in grooves by cords (5). The slots (2) through which the filtrate passes are located over the grids of the elongated chamber. The edges of the rubber belt are bent upward by guides to form a gutter on the upper path of the belt.

The velocity of the filtering partition depends on the sludge property and filter length. The cake thickness may be 1-25 mm. The advantages of belt filters are their simplicity in design compared to filters with automatic valves and their ability to provide countercurrent cake washing and removal of thin layers of cake. Disadvantages include large area requirements, the existence of zones not used for filtration and poor washing conditions at belt edges.

Class C Designs

Filters in this group have a vertical flat or cylindrical filtering partition. In this case, filtrate may move inside the channels of the filtering elements along the surface of the filtering partition in the direction from upward to downward under gravity force action or rise along this partition upward under the action of pressure difference. In the separation of heterogeneous suspensions, nonuniform cake formation along the height can occur because larger particles tend to settle out. This results in a worsening of the cake

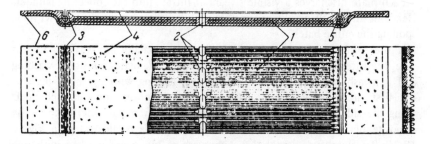

Figure 12.23   Filtering partition for belt filter: (1) rubber belt; (2) slots; (3) grooves; (4) filtering cloth; (5) cord; (6) edges of rubber belt.

washing due to different specific resistances over the partition height. The cake may creep down along the partition due to gravity, which is almost inevitable during the absence of a pressure gradient over both sides of the filtering partition. The vertical filtering partition makes these filters especially useful as thickeners because it is convenient to remove cake by inverse filtrate flow.

Of the *filter presses*, the most common is the plate-and-frame design, consisting of a metal frame made up of two end supports rigidly held together by two horizontal steel bars. On these bars are placed a varying number of flush plates with filtering surfaces. The number of plates depends on the capacity and desired cake thickness. The plates are clamped together to form a series of hollow chambers. Figure 12.24 shows the principal design features. The faces of the plates are grooved in either a pyramid or a ribbed fashion. The entire plate is covered with cloth, which forms the filtering surface. The filter cloth has holes that register with the connections on the plates and frames so that when the press is assembled these openings form a continuous channel over the entire length of the press and register with the corresponding connections on the fixed head. The channel opens only into the interior of the frames and has no openings on the plates. At the bottom of the plates, holes are cored so that they connect the faces of the plates to the outlet cocks. As the sludge to be filtered is pumped through the feed channel, it first fills all of the frames. As the feed pump continues to supply materials and builds up pressure, the filtrate passes through the cloth, runs down the face of the plate and passes out through the discharge cock. When the press is filled, it is opened and dumped. In such a press the cake cannot be washed and, therefore, is discharged containing a certain amount of filtrate with whatever valuable or undesirable material is entrained. Figure 12.25 shows a closeup view of the filter press plate and frame. The plate has a perforated screen on both sides of the plate. This perforated-type construction prevents tearing of the filter medium. The filter medium is placed against the screen on each side of the plate, fitting between the sludge frame and plate.

Each plate discharges a visible stream of filtrate into the collecting launder. Hence, if any cloth breaks or runs cloudy, that plate can be shut off without spoiling the entire batch.

If the solids are to be recovered, the cakes usually are washed. In this case, the filter has a separate wash feed line, and the plates consist of washing and nonwashing types arranged alternately, starting with the head plate as the first nonwashing plates. The wash liquor moves down the channels along the side of each washing plate and then across the filter cake to the opposite plate and drains toward the outlet. This is illustrated in Figure 12.26. To simplify assembly, the nonwashing plates are marked with one button and the washing plates with three buttons. The frames carry two buttons.

(A)

(B)

Figure 12.24 (A) A single-stage round filter press generally employed with a filter aid and capable of flows in excess of 100 gpm; (B) a common four-port, two-stage filter press with double inlet and double outlet (courtesy Star Tank & Filter Corp., Bronx, NY).

Figure 12.25   Details of fabricated filter press plate and frame (courtesy Star Tank and Filter Corp., Bronx, NY).

In open-delivery filters, the cocks on the one-button plates remain open and those on the three-button plates remain closed. In closed-delivery filters, a separate wash outlet conduit is provided. Figure 12.27 illustrates the basic design of a frame, a nonwashing plate and a washing plate. These plates and frames are used in open-delivery filters. The plates and frames used in closed-delivery filters are shown in Figure 12.28.

In terms of initial investment and floor area requirements, plate-and-frame filters are inexpensive in comparison to other filters. They can be operated at full capacity (all frames in use) or at reduced capacity by blanking off some of the frames by means of dummy plates. They can deliver reasonably well-washed and relatively dry cakes. However, the combination of labor

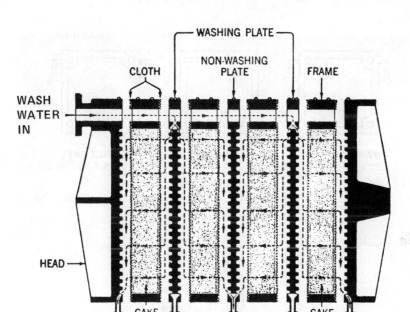

Figure 12.26   Wash water outlets.

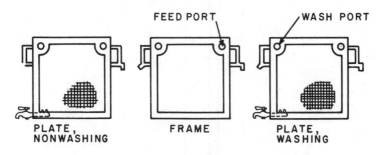

Figure 12.27   Plates and frame of an open-delivery, through-washing filter.

charges for removing the cakes and fixed charges for downtime may constitute a high percentage of the total cost per operating cycle.

Leaf filters are similar to plate-and-frame filters in that a cake is deposited
on each side of the leaf (Figure 12.29) and the filtrate flows to the outlet
in the channels provided by a coarse drainage screen in the leaf between the

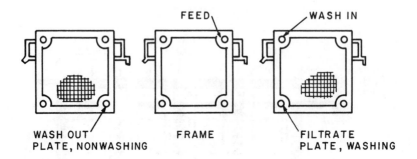

Figure 12.28   Plates and frame of a closed-delivery, through-washing filter.

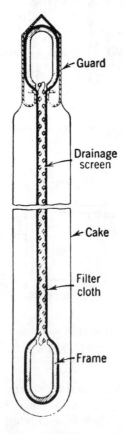

Figure 12.29   Sectional view of a filter leaf showing construction and approximate location of cake.

cakes. The leaves are immersed in the sludge when filtering, and in the wash liquid when washing. Therefore, the leaf assembly may be enclosed in a shell, as in pressure filtration, or simply immersed in sludge contained in an open tank, as in vacuum filtration.

In operating a pressure leaf filter, the sludge is fed under pressure from the bottom and distributed equally. The clear filtrate from each leaf is collected in a common manifold and carried away. In filters with an external filtrate manifold (Figure 12.30), the filtrate from each leaf is visible through a respective sight glass. This is not possible when the leaves are mounted on a hollow shaft, which serves as an internal filtrate-collecting manifold (such as the Vallez rotary leaf filter). The filter cakes are built on each side of the leaves, and filtration is continued until the required cake thickness is achieved. For washing, usually the excess sludge is drained, admitting compressed air (3–5 lb pressure) simultaneously, which serves mainly to prevent the cake from peeling off the leaves.

Disk filters consist of a number of concentric disks mounted on a horizontal rotary shaft and operate on the same principle as the rotary drum vacuum filters. The basic design is illustrated in Figure 12.31(A). The disks are formed by using V-shaped hollow sectors, which are assembled radially around the shaft. Each sector is covered with filter cloth and has an outlet

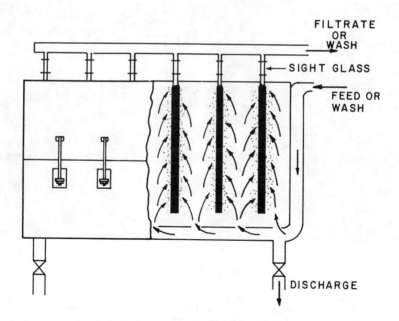

Figure 12.30   Sweetland pressure filter.

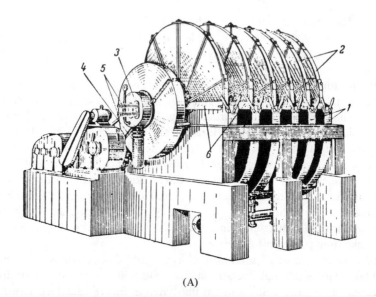

(A)

(B)

Figure 12.31   (A) A rotary disk vacuum filter: (1) section; (2) filtering disks; (3) automatic valve; (4) manifold for vacuum and filtrate discharge; (5) piping for compressed air; (6) doctor's knives for cake removal. (B) A high-pressure disk filter (courtesy Artisan Industries Inc., Waltham, MA).

nipple through which it is connected to a manifold extending along the length of the shaft and leading to a port on the filter valve. Each row of sectors is connected to a separate manifold. The sludge level in the tank should provide complete submergence to the lower-most sector of the disks.

Compared to drum vacuum filters, the greatest advantage of the disk filter is that for the same filtering area it occupies considerably less floor space. However, because of vertical filtering surfaces, cake washing is not as efficient as when a drum filter is used. The disk filter is ideal when the cake is not washed and floor space is at a premium. A high-pressure unit in operation is shown in Figure 12.31(B).

Cartridge filters are used extensively throughout the chemical and process industries in applications from laboratory-scale to commercial operations ranging to more than 5000 gpm. Table 12.1 lists typical filtering applications and operating ranges. Early designs, still used widely, include a series of thin metal disks that are 3–10 in. in diameter and set in a vertical stack with very narrow uniform spaces between them. The disks are supported on a vertical hollow shaft and fit into a closed cylindrical casing. Liquid is fed to the casing under pressure, whence it flows inward between the disks to openings in the central shaft and out through the top of the casing. Solids are captured between the disks and remain in the filter. As most of the solids are removed at the periphery of the disks, the unit is referred to as an *edge filter*. The accumulated solids are removed periodically from the cartridge.

More recent designs are simpler, experience lower pressure drop and have fewer maintenance problems. Figure 12.32 shows the operating principle behind one type of design. Unfiltered liquid enters the inlet (bottom) port. It flows upward, around and through the filter media, which is a stainless steel or fabric screen reinforced by a perforated, stainless steel backing. Filtered liquid discharges through the outlet (top port). Because of the outside-to-inside flow path, solids deposit onto the outside of the element so that screens are easy to clean. This external gasketing design prevents solids from bypassing the filter and contaminating the process downstream. There are no o-ring seals that can crack, channeling media that can fail or cartridges that can collapse or allow bypassing.

As with any filter, careful media selection is essential. Media that are too coarse, for example, will not provide the needed protection. However, specifying finer media than necessary can add substantially to both equipment and operating costs. Factors to be considered in media selection include solids content, type of contaminant, particle size and shape, amount of contaminant to be removed, viscosity, corrosiveness, abrasiveness, adhesive qualities, liquid temperature and required flowrate. Typical filter media are wire mesh (typically 10 mesh to 700 mesh), fabric (30 mesh to 1 μm), slotted screens (10 mesh to 25 μm) and perforated stainless steel screens

Table 12.1 Typical Applications of Cartridge Filters (courtesy Ronningen-Petter Div., Dover Corp., Portage, MI)

Industry and Liquid	Typical Filtration Range	Industry and Liquid	Typical Filtration Range
Chemical Industry		Gasoline	1-3 μ
Alum	60 mesh-60 μ	Hydrocarbon wax	25-30 μ
Brine	100-400 mesh	Isobutane	250 mesh
Ethyl alcohol	5-10 μ	MEA	200 mesh to 5-10 μ
Ferric chloride	30-250 mesh	Naphtha	25-30 μ
Herbicides/pesticides	100-700 mesh	Produced water for injection	1-3 to 15-20 μ
Hydrochloric acid	100 mesh to 5-10 μ	Residual oil	25-50 μ
Mineral oil	400 mesh	Sea water	5-10 μ
Nitric acid	40 mesh to 5-10 μ	Vacuum gas oil	25-75 μ
Phosphoric acid	100 mesh to 5-10 μ	Steam injection	5-10 μ
Sodium hydroxide	1-3 to 5-10 μ		
Sodium hypochlorite	1-3 to 5-10 μ	Pulp and Paper	
Sodium sulfate	5-10 μ	Calcium carbonate	30-100 mesh
Sulfuric acid	250 mesh to 1-3 μ	Clarified white water	30-100 mesh
Synthetic oils	25-30 μ	Dye	60-400 mesh
		Fresh water	30-200 mesh
Drugs and Cosmetics		Groundwood decker recycle	20- 60 mesh
Acetic acid	40 mesh-150 mesh	Hot melt adhesives	40-100 mesh
Aerosol	60 mesh-200 mesh	Latex	40-100 mesh
Bath oil	400-700 mesh	Millwater	60-100 mesh
Citric acid	60 mesh to 1-3 μ	Paper coating	30-250 mesh
Glycerine	5-10 μ	River water	20-400 mesh
Lipstick	60-150 mesh	Starch size	20-100 mesh
Shampoo	100-250 mesh	Titanium dioxide	100-200 mesh
Soap	10-250 mesh		

Suntan lotion	15-20 μ
Tallow	700 mesh to 25-30 μ
Toothpaste	100 mesh

Food and Beverage

Apple juice	5-10 μ
Beer	250 mesh-400 mesh
Brine	400 mesh to 15-20 μ
Chocolate	10-400 mesh
Corn syrup	80 mesh to 5-10 μ
Fructose syrup	5-10 to 25-30 μ
Fruit juices with pulp	10-100 mesh
Jelly	700 mesh
Lard	500 mesh to 5-10 μ
Lemon effluent	60-150 mesh
Liquors	700 mesh to 15-20 μ
Vegetable oil	150 mesh to 5-10 μ
Wash water	20-250 mesh

Petroleum Industry

Atmospheric reduced crude	25-75 μ
Completion fluids	200 mesh to 1-3 μ
DEA	250 mesh to 5-10 μ
Deasphalted oil	200 mesh
Decant oil	60 mesh
Diesel fuel	100 mesh
Gas oil	25-75 μ

All Industries

Adhesives	30-150 mesh
Boiler feed water	5-10 μ
Caustic soda	250 mesh
Chiller water	200 mesh
City water	500 mesh to 1-3 μ
Clay slip (ceramic and china)	20 mesh-700 mesh
Coal-based synfuel	60 mesh
Condensate	200 mesh to 5-10 μ
Coolant water	500 mesh
Cooling tower water	150-250 mesh
Deionized water	100-250 mesh
Ethylene glycol	100 mesh to 1-3 μ
Floor polish	250 mesh
Glycerine	5-10 μ
Inks	40-150 mesh
Liquid detergent	40 mesh
Machine oil	150 mesh
Pelletizer water	250 mesh
Phenolic resin binder	60 mesh
Photographic chemicals	25-30 μ
Pump seal water	200 mesh to 5-10 μ
Quench water	250 mesh
Resins	30-150 mesh
Scrubber water	40-100 mesh
Wax	20-200 mesh
Well water	60 mesh to 1-3 μ

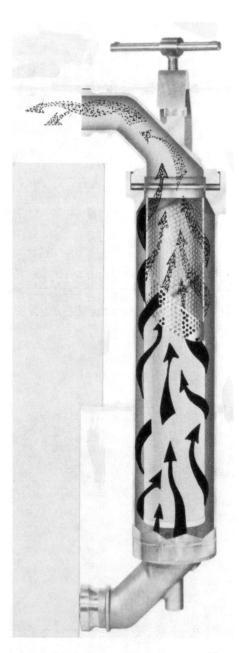

Figure 12.32   Operating principle behind a single cartridge-type filter (courtesy Ronningen-Petter Div., Dover Corp., Portage, MI).

(10 mesh to 30 mesh). Table 12.2 provides typical particle retention sizes for different media.

Figure 12.33 shows a full view of a single unit. Single filters may be piped directly into systems requiring batch or intermittent service. Using quick-coupling connectors, the media can be removed from the housing, inspected

Table 12.2 Typical Filter Retentions (courtesy Ronningen-Petter Div., Dover Corp., Portage, MI)

	Mesh or Mesh Equivalent	Nominal Particle Retention		Percentage of Open Area
		(in.)	(μ)	
Wire Mesh	10	0.065	1650	56
	20	0.035	890	46
	30	0.023	585	41
	40	0.015	380	36
	60	0.009	230	27
	80	0.007	180	32
	100	0.0055	140	30
	150	0.0046	115	37
	200	0.0033	84	33
	250	0.0024	60	36
	400	0.0018	45	36
	700	0.0012	30	25
Perforated	10	0.063	1575	15
	20	0.045	1125	18
	30	0.024	600	12
Slotted	10	0.063	1600	50
	15	0.045	1140	43
	20	0.035	890	36
	30	0.024	610	30
	40	0.015	380	20
	60	0.009	230	18
	80	0.007	180	25
	100	0.006	150	13
	120	0.005	125	11
	150	0.004	100	9
	200	0.003	75	7
	325	0.002	50	5
	–	0.001	25	3
Fabric	60	0.009	230	
	80	0.007	180	Percentage
	100	0.0055	140	of
	150	0.0046	115	open area
	250	0.0024	60	not
	500	0.0016	40	applicable
	–	0.0010–0.0012	25–30	to
	–	0.0006–0.0008	15–20	fabric
	–	0.0002–0.0004	5–10	media.
	–	0.00004–0.00012	1–3	

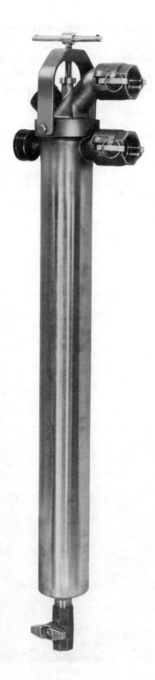

Figure 12.33   A standard single-unit cartridge filter (courtesy Ronningen-Petter Div.,
Dover Corp., Portage, MI).

or cleaned. Also, filtering elements are interchangeable. Hence, while one is being cleaned, another can be placed in service.

Multiple filters are also commercially available, consisting of two or more single filter units valved in parallel to common headers (Figure 12.34). The distinguishing feature of these filters is the ability to sequentially backwash each unit in place while the others remain onstream. Hence, these systems

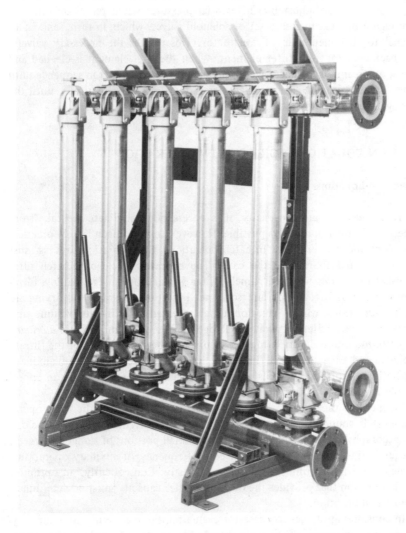

Figure 12.34   An internal-backwashing, multiplex-filtering unit (courtesy Ronningen-Petter Div., Dover Corp., Portage, MI).

are essentially continuous filters. These units can be automated fully to eliminate manual backwashing. Backwashing can be controlled by changes in differential pressure between the inlet and outlet headers. One possible arrangement consists of a controller and solenoid valves that supply air signals to pneumatic valve actuators on each individual filter unit.

As solids collect on the filter elements, flow resistance increases. This increases the pressure differential across the elements and, thus, between the inlet and outlet headers on the system. When the pressure drop reaches a preset level, an adjustable differential pressure switch relays information through a programmer to a set of solenoid valves, which, in turn, send an air signal to the pneumatic valve actuator. This rotates the necessary valve(s) to backwash the first filter element. When the first element is cleaned and back onstream, each successive filter element is backwashed in sequence until they are all cleaned. The programmer is then reset automatically until the rising differential pressure again initiates the backwashing cycle.

DESIGN EQUATIONS FOR OPTIMUM FILTRATION

Principal Equations

To increase filtration capacity, it is necessary to eliminate the cake from the filter plate as quickly as possible. However, this approach is only practical for continuous operations. In batch operations, the separation process must be interrupted to remove the cake. The operating cycle of a batch filter consists of a series of steps, namely, filter preparation, sludge loading, filtration, cake washing, drying and scraping. Filtration, washing and drying are *basic operations* whose duration time increases with filtrate volume and cake thickness. Filter preparation, sludge loading and removing are *auxiliary operations*, whose duration may be assumed to be independent of the filtrate volume and cake thickness. For any given filter, the duration of auxiliary operations is practically constant. Such an assumption can result in appreciable error in estimating the maximum filter capacity in some cases.

To increase the capacity of batch filters (with respect to the basic operations) it is sound practice to repeat the cycle as often as possible. This can be accomplished by feeding the filter with small portions of sludge. However, frequent filtration cycles mean an equal frequency of auxiliary operations, which is obviously time consuming and costly. Consequently, an optimum cycle duration that provides the maximum filter capacity must be determined for each application.

In considering the general case for evaluating the maximum filter capacity at constant pressure, the operating cycle can be viewed as having four steps in series: filtration, washing, cake drying and removal. The filtration step may be described by Equation 12.19, and that of washing by the following expression given by Komarovski [1950]:

$$\frac{G}{G_0} = 1 - \frac{0.25}{V_{w.f.}/V_0} \tag{12.45}$$

where   G = weight of dissolved substances in washing liquid (N)
G_0 = weight of dissolved substances in filtrate retained in cake before washing (N)
$V_{w.f.}$ = volume of washing liquid (m^3)
V_0 = filtrate volume in pores before washing (m^3)

Equation 12.45 is valid at $V_{w.f.}/V_0 \geqslant 0.5$.

Brownell and Katz [1947] give the following equation for drying:

$$\frac{\tau_{dew}}{C_{dew}} = \left[\frac{(1 - m_0)^2 + 1}{2} \right] \times \frac{m_e^{1-y}}{y - 1} \tag{12.46}$$

where τ_{dew} = dewatering time (s)
C_{dew} = dewatering factor (s)
m_0 = residual cake saturation with liquid equal to the ratio of stationary liquid in the end of dewatering to the pore volume (fractions of unity)
m_e = effective saturation with liquid equal to ratio of moving liquid volume to the total volume the same liquid and air (fractions of unity)
y = exponent, depending on the particle size in the cake

C_{dew} is the dewatering factor:

$$C_{dew} = \frac{\epsilon \mu r_0 h_c^2}{\Delta P} \tag{12.47}$$

The dewatering factor, C_{dew}, combines the variables influencing the motion of liquids in the cake pores. Brownell and Katz [1947] evaluated exponent y in Equation 12.46 experimentally, observing it to range from 2.0 (for particles 1.8 mm in size) to 3.0 (for particles 0.09 mm in size). An average value of 2.5 may be assumed.

A plot of τ_{dew}/C_{dew} versus m_s for different values of m_0 is shown in Figure 12.35. In this plot, τ_{dew}/C_{dew} is determined from Equation 12.46 at y = 2.5 and corresponding values of m_s (m_s = cake saturation with liquid,

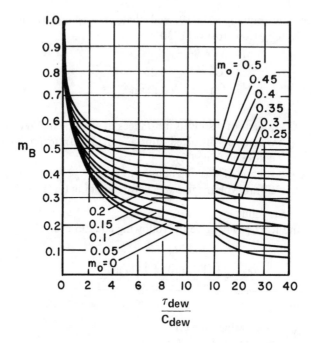

Figure 12.35   Diagram for determining cake saturation vs dewatering time.

which is equal to the ratio of total stationary and moving liquid to pore volume, fractions of unity) are obtained from

$$m_s = \frac{m_e - 2m_e m_0 + m_0}{1 - m_e m_0} \qquad (12.48)$$

The curves shown permit evaluation of cake saturation by liquid rate versus dewatering time for a given set of conditions. During the dewatering stage, the volume of air sharply increases from zero to a maximum corresponding to the situation of air blowing through a dry cake. The amount of air passing through the pores as a function of dewatering time is given in Figure 12.36.

In evaluating the optimum filtration time, we shall assume that the filter plate resistance is negligible. Furthermore, the duration of each stage will be evaluated as a function of filtrate volume on the assumption that all filtration conditions are constant. The time of filtration may be computed from Equation 12.19 for

$$R_f = 0$$

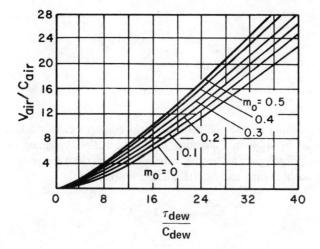

Figure 12.36 Volume of blowing air vs dewatering time (from Brownell and Katz [1947]).

Then

$$\tau = \frac{1}{2} Bq^2$$

where

$$B = \frac{\mu r_0 x_0}{\Delta P}$$

Substituting for the ratio $V_{w.f.}/V_0$ with $q_{w.f.}/q_0$ from Equation 12.45, we obtain

$$q_{w.f.} = \frac{0.25 q_0}{1 - \dfrac{G}{G_0}} \tag{12.49}$$

As $q_0 = \epsilon x_0 q$, then

$$q_{w.f.} = D'' q \tag{12.50}$$

where

$$D'' = \frac{0.25\epsilon x_0}{1 - \dfrac{G}{G_0}} \tag{12.51}$$

Because washing is essentially filtration at constant cake thickness, q may be replaced by $q_{w.f.}$ in Equations 12.20 and 12.29. Assuming the viscosity of the washing liquid to be the same as that of the filtrate, a relationship between washing time and filtrate amount is obtained:

$$\tau_w = B \times D''^2 \times q^2 \tag{12.52}$$

Drying time as a function of filtrate amount may be determined from Equations 12.46 and 12.49 with $h_c = x_0 q$ as follows:

$$\tau_{dew} = BC' \times q^2 \tag{12.53}$$

where

$$C' = \left[\frac{(1 - m_0)^2 + 1}{2} \right] \times \frac{m_e^{1-y} - 1}{y - 1} \, \epsilon x_0 \tag{12.54}$$

The total time of all three steps is

$$\tau_b = \tau + \tau_w + \tau_{dew} = Eq^2 \tag{12.55}$$

or

$$q = \sqrt{\frac{\tau_b}{E}} \tag{12.56}$$

where

$$E = B \left(\frac{1}{2} + D''^2 + C' \right) \tag{12.57}$$

The average filter capacity per unit time for a total filtration cycle may be characterized by a conditional average filtration rate obtained by dividing filtrate volume per cycle per unit filter area by the total cycle duration ($\tau_t = \tau_b + \tau_{aux}$):

$$u_{cond} = \frac{\sqrt{\tau_b/E}}{\tau_b + \tau_{aux}} \tag{12.58}$$

To determine the maximum value of u_{cond}, the first derivative of this expression is set equal to zero:

$$\frac{du_{cond}}{d\tau_b} = \frac{\sqrt{B}\,(\tau_{aux} - \tau_b)}{2\sqrt{\tau_b}\,(\tau_b + \tau_{aux})^2} \tag{12.59}$$

Setting the numerator equal to zero,

$$\tau_{aux} - \tau_b = 0$$

or

$$\tau_b = \tau_{aux} \tag{12.60}$$

At $\tau_b = \tau_{aux}$, the second derivative is negative. Thus, neglecting the resistance of the filter plate, the maximum capacity of a batch filter is achieved when the duration time of the basic operation is equal to that of the auxiliary ones. If, however, the filter plate resistance is significant, optimum filter capacity can be achieved, but at $\tau_b > \tau_{aux}$. Zhuzhikov [1953] gives the following expression for filtration without washing and drying steps:

$$\tau_b = \tau_{aux} + 2\sqrt{\frac{\mu R_f^2}{2\Delta Pr_0 x_0}}\,\tau_{aux} \tag{12.61}$$

This expression can be represented graphically in a dimensionless form to simplify calculations. Dividing both sides by τ_{aux},

$$\dot{S} = 1 + 2F_{\ell 1}^{0.5} \tag{12.62}$$

where $\dot{S} = \tau_b/\tau_{aux}$ and

$$F_{\ell_1} = \frac{\mu R_f^2}{2\Delta P \tau_{aux} r_0 x_0} \equiv \text{Filtration Number} \qquad (12.63)$$

F_{ℓ_1} is dimensionless and varies from zero at $R_f = 0$ to a large value when there is an increase in the viscosity of the sludge and R_f or a decrease in pressure drop, auxiliary time, specific cake resistance and the ratio of cake volume to filtrate volume. It may be assumed in practice that $F_{\ell_1} = 0 - 10$. If washing and drying times are constant and independent of filtration time, they may be added directly to the auxiliary time. In this case, the calculation of maximum filter capacity is similar to the case without washing and drying steps; as such, Equations 12.60, 12.61 and 12.62 are applicable with τ_{aux} assumed to be the sum of τ_b, τ_w and τ_{dew}.

From the times of the basic filtration steps, the filtrate volume per cycle can be computed along with the corresponding maximum capacity for different cases. Thus, if τ_b is determined from Equation 12.60, q may be estimated from Equation 12.56. τ_b is obtained from the following expression,

$$q^2 + 2 \frac{R_f}{r_0 x_0} q = 2 \frac{\Delta P}{\mu r_0 x_0} \tau \qquad (12.64)$$

after substituting for τ by τ_b.

From q, the cake thickness corresponding to a maximum filter capacity can be computed from Equation 12.13, noting that

$$V/A = q$$

For a cycle consisting of filtration and washing stages, and accounting for the filter plate resistance, the cake thickness corresponding to a maximum filter capacity is given by Zhuzhikov [1953]:

$$h_c = \sqrt{\frac{2\Delta P x_0 \tau_{aux}}{\mu r_0 (1 + 2x_0 L)}} \qquad (12.65)$$

where

$$L = \frac{\log(c_a/c_0)}{K_w'} \times \frac{\mu_{w.f.}}{\mu} \qquad (12.66)$$

where c_a is the instantaneous concentration in washing liquid and c_0 is the concentration of solute in filtrate and K'_w is a constant, which depends on the properties of cake, filtrate and washing liquid.

Example 12.3

A batch Nutsch filter operates under constant pressure and maximum capacity with three stages: filtration, washing and drying. The Nutsch filter with a 1-m^2 filtration area operates under the following set of conditions:

- Viscosity of filtrate and washing liquid μ (N-s/m^2) = 10^{-3}
- Air viscosity, μ_{air} (N-s/m^2) = 1.83×10^{-5}
- Surface tension of washing liquid σ (N/m) = 0.04
- Cake specific resistance r_0 (m^{-2}) = 2×10^{12}
- Ratio of cake volume to filtrate volume, x_0 = 0.1
- Pressure drop ΔP (N/m^2) = 40,000
- Porosity, ϵ = 0.45
- Weight ratio G/G_0 dissolved substance extracted by washing liquid to the substance in the cake before washing = 0.98
- Residual liquid saturation of cake m_0 (approximate, considering the specific cake resistance) = 0.5
- Effective cake saturation with liquid at the end of drying, m_e = 0.1
- Exponent y in dewatering equations = 2.5
- Auxiliary time of operations τ_{aux} (min) = 30
- Maximum permissible cake thickness h_c (m) = 0.05

The resistance of the filter plate is negligible.

Determine the filtration, washing and drying times, the cake thickness, the volumes of filtrate, washing liquid and drying air.

Solution

1. The constant, E, in Equation 12.56 is determined as follows: from Equation 12.19 we obtain

$$B = \frac{\mu r_0 x_0}{\Delta P} = \frac{10^{-3} \times 2 \times 10^{12} \times 0.1}{40,000} = 5 \times 10^3$$

From Equation 12.51,

$$D'' = \frac{0.25 \epsilon x_0}{1 - \dfrac{G}{G_0}} = \frac{0.25 \times 0.45 \times 0.1}{1 - 0.98} = 0.56$$

From Equation 12.54,

$$C' = \left[\frac{(1 - m_0)^2 + 1}{2}\right] \times \frac{m_e^{1-y} - 1}{y - 1} \epsilon x_0$$

$$= \left[\frac{(1 - 0.5)^2 + 1}{2}\right] \times \frac{0.1^{1-2.5} - 1}{2.5 - 1} \times 0.45 \times 0.1 = 0.575$$

Then, from Equation 12.57,

$$E = B\left(\frac{1}{2} \times D''^2 + C'\right) = 5 \times 10^3 (0.5 + (0.56)^2 + 0.575) = 6945$$

2. The filtrate volume is obtained from Equation 12.56 at $\tau_b = \tau_{aux}$, which corresponds to the maximum filter capacity:

$$q = \sqrt{\frac{\tau_b}{E}} = \sqrt{\frac{30 \times 60}{6945}} = 0.51 \text{ m}$$

3. Cake thickness:

$$h_c = x_0 q = 0.1 \times 0.51 = 0.051 \text{ m}$$

which corresponds to the given maximum cake thickness.

4. Filtration time is determined from Equation 12.19:

$$\tau = \frac{1}{2} Bq^2 = 0.5 \times 5 \times 10^3 (0.51)^2 = 650 \text{ s}$$

5. The volume of washing liquid is obtained from Equation 12.50:

$$q_{w.f.} = D''q = 0.56 \times 0.51 = 0.286 \text{ m}$$

6. The washing time is obtained from Equation 12.52:

$$\tau_w = B \times D''^2 \times q^2 = 5 \times 10^3 \times (0.56)^2 \times (0.51)^2 = 410 \text{ s}$$

7. Drying time is obtained from Equation 12.53:

$$\tau_{dew} = BC' \times q^2 = 5 \times 10^3 \times 0.575 \times (0.51)^2 = 745 \text{ s}$$

Note that the total time of the filtration, washing and dewatering steps is approximately equal to the auxiliary time.

8. The amount of drying air is determined as follows. From Equation 12.47 we obtain

$$C_{dew} = \frac{\epsilon\mu r_0 h_c^2}{\Delta P} = \frac{0.45 \times 10^{-3} \times 2 \times 10^{12} \times (0.051)^2}{40,000} = 58.5 \text{ s}$$

The ratio $\tau_{dew}/C_{dew} = 745/58.5 = 12.7$. From Figure 12.36 (at $m_0 = 0.5$ and $\tau_{dew}/C_{dew} = 12.7$), it follows that $V_a/C_a = 8.0$. The blowing factor (from Brownell and Katz [1947]) is

$$C_a = \epsilon h_c \frac{\mu}{\mu_a} = 0.45 \times 0.051 \times \frac{10^{-3}}{1.83 \times 10^{-5}} = 1.25 \text{ m}^3/\text{m}^2$$

Hence, the volume of blowing air is

$$V_a = 8 \times 1.25 = 10 \text{ m}^3/\text{m}^2$$

Because the value $r_0 > 10^{11}$ m^{-2}, there is no need to introduce a correction factor for turbulence.

9. Checking the assumed value of residual saturation, m_0, using the *capillary number*,

$$K_p = \frac{\Delta P}{r_0 h_c \sigma} = \frac{40,000}{2 \times 10^{12} \times 0.51 \times 0.04} = 9.8 \times 10^{-6}$$

Hence,

$$m_0 = aK_p^{-0.264} = 0.025(9.8 \times 10^{-6})^{-0.264} = 0.55$$

a = 0.025; constant considering end effects, which corresponds to the assumed value.

It is important to evaluate the decrease in average filter capacity per unit time for the total operating cycle if the total time of filtration, washing and drying changes from zero to a value several times larger than the maximum filter capacity. We trace the curve obtained by a plot of τ_b versus u_{cond}

(from Equation 12.58) shown in Figure 12.37. This is a three-step process—filtration, washing and drying—under conditions of maximum capacity. The curve shows a shallow slope towards the right from a maximum. This means that even for a large increase in the basic steps, for $\tau_b = \tau_{aux}$ (Equation 12.60), no significant decrease in filter capacity is observed. Thus, the increase in τ_b from 1800 to 9000 s results in a decrease in filter capacity from 1.42×10^{-4} to 1.05×10^{-4} m^3/m^2-s, or a 1.35 times reduction.

Optimum Capacity of Batch Filters at Constant Rate

Following Orlicek et al. [1954] and Egorov [1950], consider a filter whose operating cycle consists of a filtration step at constant rate and auxiliary steps for filter preparation, which include sludge feeding and cake removal. Assuming that the cake and filter plate are incompressible, two common cases are examined below.

Case 1

Consider a constant filtration process whose filtration time changes along with filtrate volume and cake thickness. Then

$$\tau_b = \frac{q}{u} \tag{12.67}$$

where u denotes a constant rate of filtration.

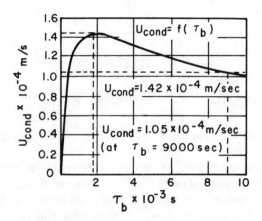

Figure 12.37   Graph of the Nutsch design.

The average filtration rate is

$$u_{cond} = \frac{q}{\tau_b + \tau_{aux}} \qquad (12.68)$$

Substituting τ_b into Equation 12.68, we obtain

$$u_{cond} = \frac{1}{\dfrac{1}{u} + \dfrac{\tau_{aux}}{q}} \qquad (12.69)$$

For $q \rightarrow 0$ or $\tau_b \rightarrow 0$, $u_{cond} \rightarrow 0$, and at $q \rightarrow \infty$ or $\tau_b \rightarrow \infty$, $u_{cond} \rightarrow u$. That is, to increase the filter capacity, filtration time should be increased. However, such an increase is limited by the maximum allowable pressure drop which at constant operating conditions, establishes a maximum cake thickness.

Example 12.4

Determine the capacity of a batch filter operating at a constant rate. The rate of filtration is $q = 0.1 \times 10^{-3}$ m^3/m^2-s and the auxiliary time is 900 s.

Solution

Filter capacity is characterized by the average conditional filtration rate, q_{cond}. Values of u_{cond} calculated from Equation 12.69 are shown plotted in Figure 12.38. As shown, an increase in the amount of filtrate causes a sharp

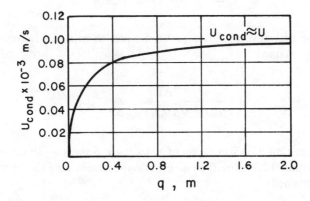

Figure 12.38   Graph for the solution of Example 12.4.

increase in filter capacity at first, with a limiting value of u achieved eventually.

Case 2

Consider a process in which filtration rate changes from one cycle to another; however, a constant rate is maintained during each cycle. Furthermore, the filtration is terminated when the pressure difference reaches a maximum allowable value. The amount of filtrate and cake thickness for different cycles will be different, as in Case 1, because the pressure difference depends not only on cake thickness but also on the filtration rate. For the given system, in the limits of one cycle, Equation 12.67 is still applicable. From Equations 12.67 and 12.68, we obtain

$$u_{cond} = \frac{qu}{q + \tau_{aux}u} \qquad (12.70)$$

Solving Equation 12.15 for q, we obtain

$$q = \frac{\Delta P - \mu R_f u}{\mu r_0 x_0 u} \qquad (12.71)$$

ΔP is constant, which corresponds to the end of the filtration step. Substituting from Equation 12.71 into 12.70,

$$u_{cond} = \frac{\Delta P u - \mu R_f u^2}{\Delta P - \mu R_f u + \mu r_0 x_0 \tau_{aux} q^2} \qquad (12.72)$$

Setting the derivative of this expression equal to zero, the maximum filter capacity is obtained:

$$u = \frac{\Delta P}{\mu R_f + \sqrt{\mu r_0 x_0 \tau_{aux} \Delta P}} \qquad (12.73)$$

Solving Equation 12.15 in terms of τ, letting $\tau = \tau_b$, and substituting for u from Equation 12.73, the time of filtration corresponding to the maximum u_{cond} is obtained:

$$\tau_b = \tau_{aux} + \sqrt{\frac{\mu R_f^2}{\Delta P r_0 x_0} \tau_{aux}} \qquad (12.74)$$

Comparison of Equations 12.61 and 12.74 reveals that they differ by a factor of $\sqrt{2}$. Note that ΔP in Equation 12.74 is the maximum allowable pressure difference at the end of filtration and is numerically equal to the constant pressure difference in Equation 12.61. If the filter plate resistance is negligible, Equations 12.61 and 12.74 are the same.

Example 12.5

Determine a constant rate of filtration and the time of operation corresponding to the maximum capacity of a batch filter for the following conditions:

- maximum permissible pressure difference $\Delta P = 9 \times 10^4$ N/m^2
- sludge viscosity $\mu = 10^{-3}$ N $\times$ s/m^2
- filter plate resistance $R_f = 56 \times 10^{10}$ m^{-1}
- specific cake resistance $r_0 = 3 \times 10^{13}$ m^{-2}
- $x_0 = 0.333$
- auxiliary time $\tau_{aux} = 600$ s
- maximum permissible cake thickness $h_c = 0.025$ m

Solution

1. The constant filtration rate from Equation 12.73 is

$$u = \frac{9 \times 10^4}{10^{-3} \times 56 \times 10^{10} + \sqrt{10^{-3} \times 3 \times 10^{13} \times 0.333 \times 600 \times 9 \times 10^4}}$$

$$= 0.0695 \times 10^{-3} \text{ m/s}$$

2. The filtration time is computed from Equation 12.74:

$$\tau_b = 600 + \sqrt{\frac{10^{-3}(56 \times 10^{10})^2}{9 \times 10^4 \times 3 \times 10^{13} \times 0.333} \times 600} = 1056 \text{ s}$$

3. The amount of filtrate is obtained from Equation 12.67:

$$q = \tau_b \times u = 0.0695 \times 10^{-3} \times 1056 \approx 0.073 \text{ m}$$

4. The cake thickness is

$$h_c = q \times x_0 = 0.073 \times 0.333 \simeq 0.025 \text{ m}$$

This corresponds to the maximum allowable value.

If the value of h_c obtained from calculation is more than the allowable value of h_c, it is necessary either to decrease the rate of filtration or the time of operation. This will result in a decrease in the average filter capacity.

Maximum Capacity of Batch Filters at Variable Pressure Difference and Process Rates

Two processes are considered: (1) at constant rate and increasing pressure difference, and then at constant pressure difference and decreasing rate; and (2) at continuously decreasing rate and increasing pressure difference.

The first process is realized when sludge is fed by a piston pump, first at a constant rate, then at constant pressure differential. The second process is realized when sludge is fed by a centrifugal pump.

From Kruglikov [1965], the filtration time at constant rate, τ_1, is

$$\tau_1 = \frac{\mu r_0 x_0 q_1^2 + \mu R_f q_1}{\Delta P} \tag{12.75}$$

Specifying a filtration rate for the first stage of the process, a definite filtrate volume q_1 is obtained for achieving the maximum permissible pressure difference. Then the filtrate volume obtained for the second part of the process (at constant pressure difference) is $q_2 = q = q_1$, where q is the total filtrate volume.

The filtration time at constant pressure difference is

$$\tau_2 = \frac{\mu r_0 x_0 (q - q_1)^2}{2 \Delta P} + \frac{\mu (R_f + r_0 x_0 q_1 (q - q_1))}{\Delta P} \tag{12.76}$$

To determine the washing time, assume $\mu_w/\mu = M$ and experimentally evaluate $q_w/q = N$. Then

$$\tau_w = M \times N \frac{\mu r_0 x_0 q^2 + \mu R_f q}{\Delta P} \tag{12.77}$$

For this case, $\tau_b = \tau_1 + \tau_2 + \tau_w$ and, on substitution of τ_1, τ_2 and τ_w into Equation 12.68 and setting the derivative equal to zero, we obtain the filtrate volume per cycle for the maximum filter capacity:

$$q = \sqrt{\frac{\tau_{aux} + Bq_1^2}{B + 2BMN}} \qquad (12.78A)$$

where

$$B = \frac{\mu r_0 x_0}{2\Delta P}$$

When $q_1 = 0$ and $N = 0$, the expression simplifies to

$$q = \sqrt{\frac{2\Delta P \tau_{aux}}{\mu r_0 x_0}} \qquad (12.78B)$$

Note that the conditional capacity is computed from the ratio q/τ_{total}. Additional discussions are given by Purchas [1957], Zhuzhikov [1950], Fan-Young [1953], Strelzov [1955], Komarovski [1958], and Bolen [1961].

Optimum Filtration Cycle

The operation of batch filters at maximum capacity usually does not coincide with the most economically favorable filtration cycle. This derives from the fact that for attaining the optimum capacity it is often necessary to perform auxiliary operations of sludge feeding and cake unloading, imposing labor and energy expenses. Consider, for example, a filtration process at constant pressure differential, in which the cycle consists of only one step—filtration. The resistance of the filter plate is taken into account. To analyze this system, Equation 12.19 is the starting basis. Introducing the following notations,

$$A' = \frac{R_f}{r_0 x_0} \qquad (12.79)$$

$$B' = \frac{2\Delta P}{\mu r_0 x_0} \qquad (12.80)$$

Solving Equation 12.19 with respect to q,

$$q = -A' \sqrt{A'^2 + B'\tau} \tag{12.81}$$

In this case, $\tau = \tau_b$ and, therefore,

$$u_{cond} = \frac{-A' + \sqrt{A'^2 + B'\tau_b}}{\tau_b + \tau_{aux}} \tag{12.82}$$

Consider one of the typical filtration processes that takes place under the following conditions:

- pressure difference $\Delta P = 9 \times 10^4 \, N/m^2$
- auxiliary time $\tau_{aux} = 600 \, s$
- sludge viscosity $\mu = 10^{-3} \, N\text{-}s/m^2$
- specific volume cake resistance $r_0 = 3 \times 10^{13} \, m^{-2}$
- $x_0 = 0.333$
- filter plate resistance is variable and subsequently equal to 0, 10^{10}, 20×10^{10}, 30×10^{10}, 40×10^{10}, 50×10^{10} and $60 \times 10^{10} \, m^{-1}$

Parameter B' is independent of R_f and is determined from Equation 12.80:

$$B' = \frac{2 \times 9 \times 10^4}{10^{-3} \times 3 \times 10^{13} \times 0.333} = 0.18 \times 10^{-4} \, m^2/s$$

A' depends on R_f and is determined from Equation 12.79. For different R_f values, A' is equal to 0, 1×10^{-2}, 2×10^{-2}, 3×10^{-2}, 4×10^{-2}, 5×10^{-2} and $6 \times 10^{-2} \, m$, respectively.

Using the determined values A' and B', Equation 12.82 may be applied to obtain values of q_{cond} and τ_b at different R_f. A family of curves can be obtained from a plot of τ_b versus q_{cond}. The student should prepare such a plot and evaluate the maximum filter capacities for each R_f.

All curves giving the relation q versus τ_b at R_f as parameter have a blunt apex at the maximum and a relatively small slope to the right from the maximum. Therefore, we may assume the time of filtration is several times more than the time at the maximum filter capacity. In this case, the filter capacity decreases by a relatively small value compared with the maximum capacity, but the auxiliary operations must be performed less frequently.

The filtration time may be assumed to be four to six times more than the filtration time corresponding to the maximum filter capacity. A general method for determining the filtration time corresponding to the optimum economical regime consists of finding the minimum of the function that determines the relationship of the cost of a unit filtrate volume or weight of cake versus the sum of the expenses for basic and auxiliary operations, including amortization and maintenance. Akselrod and Leichkis [1971] proposed the following equation for analyzing filtration with pore plugging caused by the introduction of a filter aid:

$$\tau_b = \sqrt{\frac{2[K_{c_2}p'S + K_{c_1}(\tau_{cov} - \tau_{aux})]}{K'K_{c_1}q_0}} \qquad (12.83)$$

where S = weight ratio of liquids to solids in cake
τ_{cov} = covering time for filter aid (s)
p' = amount of filter aid (N/m^2)
K_{c_1} = operating expenses $(\$/s)$
K_{c_2} = cost of filter aid $(\$/N)$
K' = constant, determined as

$$\frac{K'\tau}{2} = \frac{\tau}{q} - \frac{1}{u_0} \qquad (12.84)$$

where $q = m^3/m^2$, filtrate volume
τ = filtration time, s
u_0 = initial filtration rate $(m^3/m^2\text{-s})$

Capacity of Continuous Filters

For any filter, the cake must be removed in layers that are as thin as possible to increase capacity. With continuous filters, it is necessary to increase the displacement of the filtering medium in a closed cycle. Cake thickness decreases with increasing rpm of rotary drum or disk filters, or with the velocity of a belt-filtering device. However, an increase in the filter medium velocity is limited by the ability to scrape thin cakes. For rotary drum and disk filters partially submerged in a suspension, the increase in rpm is also limited by the possible washing off of the fresh cake from the filtering cloth.

A continuous filter may be viewed conceptually as a series of elemental batch filters, where the cake thickness increases from zero to some maximum value. As such, the basic equations described for batch filters also are applicable to continuous filtration. This is more evident when we consider that the principles of filtration remain valid, whether the elemental filter is stationary or moving relative to the suspension. Autenrieth [1953] proposed the following equation for the capacity of a continuous filter:

$$\frac{G_s}{\tau A_{sub}} = \frac{f_{exp}(S - \ell) + \sqrt{f_{exp}^2(\ell - S)^2 + 4f_c(\ell - S)}}{2(\ell - S)} \qquad (12.85)$$

where G_s = weight of solids in the cake (N)

A_{sub} = filter medium surface submerged in suspension (m^2)

τ = time of submergence of surface element in suspension (hr)

f_{exp} = experimental value characterizing the resistance of the filter medium $(N/m^2\text{-hr})$

f_c = experimental value characterizing cake permeability $(N/m^4\text{-hr}^2)$

S = weight ratio of liquid to solids in the cake

ℓ = weight ratio of liquid to solids in suspension

Equation 12.85 can be used to construct a series of curves of $G_s/\tau A_{sub}$ versus ℓ at constant S and different values of f_{exp} and f_c. Such curves will show that the capacity increase of the filter is a strong function of the suspension concentration. At lower suspension concentrations, the influence of filter medium resistance is larger than at higher concentrations. Therefore, it is generally advisable to separate low concentration suspensions with filtering media having relatively large pores. The obtained filtrate then may be thickened and recirculated.

By integration of the differential equation for an elemental filter surface, along with the data from a material balance, the equation for continuous filtration was developed. Holland and Woodham [1956] have shown that this equation is similar, in principle, to that of a batch filter operation. They found that filter capacity increases by $\sqrt{K}$ times with an increase in K times the submerged surface of the filter medium, rpm and pressure difference:

$$\left(\frac{V}{\tau A_{drum}}\right)^2 + 2\frac{R_f}{r_0 x_0} \times \frac{V}{\tau A_{drum}} \times N_{drum} = 2\frac{\Delta P K_{sub}}{\mu r_0 x_0} N_{drum} \qquad (12.86)$$

where A_{drum} = total filtration area of the drum

K_{sub} = ratio of drum filtration area submerged into suspension to the total drum filtration area

and

$$N_{drum} = \frac{K_{sub}}{\tau} \qquad (12.87)$$

Substituting for N_{drum} in Equation 12.86 and multiplying both sides by $(\tau A_{sub})^2$, an expression analogous to the batch process is obtained:

$$V^2 + 2 \frac{R_f A_{drum} K_{sub}}{r_0 x_0} V = 2 \frac{\Delta P A_{drum}^2 K_{sub}^2}{\mu r_0 x_0} \tau \tag{12.88}$$

Equation 12.86 provides the basis for evaluating the capacity of a rotary-drum filter as a function of its rpm and includes the filter medium resistance. This can be performed by constructing a curve of $V/\tau A_{drum}$ versus N_{drum}.

Additional discussions and example problems are given by Gottner [1954], Skelland [1962], Bruk and Gainzeva [1970] and Kelsey [1965].

Example 12.6

A filter press with a hydraulic drive has 50 frames, with internal dimensions of 1000 × 1000 mm. The unit operates under the following conditions:

- Viscosity of filtrate, μ (N-s/m^2) = 2 × 10^{-3}
- Viscosity of washing liquid $\mu_{w.\varrho.}$ (N-s/m^2) = 1 × 10^{-3}
- Specific cake resistance, r_0 (m^{-2}) = 5 × 10^{13}
- Ratio volume of cake to filtrate volume, x_0 = 0.0333
- Permissible pressure difference, ΔP (N/m^2) = 4 × 10^5
- Cake porosity, ϵ = 0.55
- Ratio of weights G/G_0 of dissolved substance extracted by washing liquid to that before washing = 0.98
- Time of auxiliary operations, τ_{aux} (min) = 30

The cake is incompressible, and the resistance of the filter plate is negligible. The suspension passes through the frames and cake in one direction and the cake is not dewatered in the frames. Determine:

1. the rate of filtration and washing under the conditions of maximum capacity;
2. the times of filtration and washing;
3. the capacity of the filter press;
4. the rate of washing liquid; and
5. the cake thickness.

We shall first develop the principal design equations applicable to this problem.

Derivation of Design Equations

1. The relation between rate of filtration and washing is determined as follows. We denote $\mu_{w.c.}/\mu$ as K_b. The ratio of cake thicknesses at washing to that at the end of the filtration stage is 2. Considering that the liquid velocity through the cake is inversely proportional to liquid viscosity and cake thickness, we find that for a maximum permissible pressure difference during the washing stage the following relation holds:

$$u_w = \frac{u}{2K_b} \tag{i}$$

2. The cake thickness at the end of the filtration stage is

$$h_c = x_0 \frac{V}{A} \tag{ii}$$

where A is the effective area of filtration.

3. The cake thickness during the washing stage is

$$h_{c.w.} = 2x_0 \frac{V}{A} \tag{iii}$$

4. The volume of filtrate obtained may be determined from the following equation of filtration with a constant rate:

$$\Delta P = \mu r_0 x_0 \frac{V^2}{\tau A^2} + \mu R_f \frac{V}{A\tau}$$

if we assume $R_f = 0$ and change the ratio $V/\tau A$ by U and solve the equation with respect to V:

$$V = \frac{A}{A''u} \tag{iv}$$

where $A'' = \mu r_0 x_0 / \Delta P$.

5. The time of filtration may be determined if we assume in the equation that

$$\Delta P = \mu r_0 x_0 u^2 \tau + \mu R_f u$$

where $R_f = 0$ and solve this equation with respect to τ:

$$\tau = \frac{1}{A''u^2} \tag{v}$$

6. The volume of washing liquid is found as follows. At $V_c = x_0 V$, the volume of filtrate in the cake pores before washing is $V_0 = \epsilon x_0 V$. Hence,

$$q_{w.\ell.} = B''q$$

$$B'' = 0.25\epsilon x_0 \Big/ \left(1 - \frac{G}{G_0}\right)$$

$$V_{w.\ell.} = B''V \tag{vi}$$

7. Washing time may be determined from the following equation:

$$V = \frac{\Delta PA}{\mu(r_0 h_c + R_f)}$$

which gives the relation between the volume and time at filtration of pure liquid through the cake obtained before. Substituting $V_{w.\ell.}$ for V, 0.55 for ϵ, and replacing h_c by $h_{c.w.} = 2x_0 V/A$ and τ by τ_w,

$$\tau = \frac{4\mu_{w.\ell.} r_0 x_0 V V_{w.\ell.}}{\Delta PA^2}$$

Considering the definition of A'', using Equation vi and taking into account that

$$\mu_{w.\ell.} = K_b \mu$$

we obtain

$$\tau_n = 4 K_b A'' B'' \frac{V^2}{A^2}$$

Substituting in the last equation the value V from Equation iv, we finally obtain

$$\tau = \frac{4 K_b B''}{A'' u^2} \tag{vii}$$

8. From continuity of basic stages,

$$\tau_t = \tau + \tau_w = \frac{1 + 4K_bB''}{A''} = \frac{1}{u^2} \qquad \text{(viii)}$$

9. The conditional average filtration rate or average capacity of the filter per unit time for the total cycle is

$$u_{cond} = \frac{q}{\tau_b + \tau_{aux}}$$

substituting the value τ_t from Equation viii and substituting the value q by V in Equation iv,

$$u_{cond} = \frac{Au}{1 + 4K_bB'' + A''\tau_{aux}u^2} \qquad \text{(ix)}$$

Note that u_{cond} is now expressed in m^3/s because the calculations are based on the total filter area.

10. The rate of filtration (during the filtration stage) corresponding to the maximum average filter capacity per unit time for the cycle is obtained by differentiation of Equation ix and equalizing the derivative to zero:

$$u = \sqrt{\frac{1 + 4K_bB''}{A''\tau_{aux}}} \qquad \text{(x)}$$

11. The filtration time of the stage of filtration corresponding to the maximum of the average capacity of the filter is obtained from Equations v and x:

$$\tau = \frac{\tau_{aux}}{1 + 4K_bB''} \qquad \text{(xi)}$$

12. The filtrate volume corresponding to the maximum average filter capacity from Equations x and xi is

$$V = Au\tau = A\sqrt{\frac{\tau_{aux}}{A''(1 + 4K_bB'')}} \qquad \text{(xii)}$$

13. The time of washing corresponding to the maximum average capacity is obtained from Equations vii and x:

$$\tau_w = \frac{4K_b B'' \tau_{aux}}{1 + 4K_b B''} \qquad \text{(xiii)}$$

14. The volume of washing liquid corresponding to the maximum average filter capacity is determined from Equations vi and xii:

$$V_{w.\ell.} = B''V = B''A \sqrt{\frac{\tau_{aux}}{A''(1 + 4K_b B'')}} \qquad \text{(xiv)}$$

Solution

1. Determine constants K_b, A'' and B'':

$$K_b = \frac{1 \times 10^{-3}}{2 \times 10^{-3}} = 0.5$$

From the equation,

$$A'' = \frac{\mu r_0 x_0}{\Delta P} = \frac{2 \times 10^{-3} \times 5 \times 10^{13} \times 0.0333}{4 \times 10^5} = 8.32 \times 10^3$$

From the equation,

$$B'' = \frac{0.25 \epsilon x_0}{1 - \dfrac{G}{G_0}} = \frac{0.25 \times 0.55 \times 0.0333}{1 - 0.98} = 0.229$$

2. The rate of filtration from Equation x is

$$u = \sqrt{\frac{1 + 4 \times 0.5 \times 0.229}{8.32 \times 10^3 \times 1800}} = 0.000312 \text{ m/s}$$

3. The rate of washing from Equation i is

$$u_w = \frac{0.000312}{2 \times 0.5} = 0.000312 \text{ m/s}$$

4. The time of the filtration stage from Equation xi is

$$\tau = \frac{1800}{1 + 4 \times 0.5 \times 0.229} = 1235 \text{ s} \simeq 20.5 \text{ min}$$

5. The time of the washing stage from Equation xiii is

$$\tau_w = \frac{4 \times 0.5 \times 0.229 \times 1800}{1 + 4 \times 0.5 \times 0.229} = 565 \text{ s} \simeq 9.5 \text{ min}$$

6. The time of the basic stages is

$$\tau_b = 20.5 + 9.5 = 30 \text{ min}$$

7. The time of the filtration cycle is

$$\tau_{cyc} = \tau_b + \tau_{aux} = 30 + 30 = 60 \text{ min}$$

8. Filtration area (50 frames; size 1 X 1 m) is

$$A = 2 \times 1 \times 1 \times 50 = 100 \text{ m}^2$$

9. The capacity of the filter press is determined as follows. From Equation xii, the filtrate volume obtained per filtration cycle ($\tau_{cyc} = 1$ hr) is

$$V = \frac{100 \times 0.000312 \times 1235}{1} = 38.5 \text{ m}^3/\text{hr}$$

and from the relation $V_c = x_0 V$, we determine the cake volume:

$$V_c = \frac{0.0333 \times 38.5}{1} = 1.28 \text{ m}^3/\text{hr}$$

10. The volume of washing liquid per cycle from Equation xiv is

$$V_{w.\ell.} = \frac{0.229 \times 38.5}{1} = 8.82 \text{ m}^3/\text{hr}$$

11. The thickness of the cake from Equation ii is

$$h_c = 0.0333 \frac{38.5}{100} = 12.8 \times 10^{-3} \text{ m} = 12.8 \text{ mm}$$

12. Selection of frame thickness minus the total thickness of two cakes in the frame at the end of the filtration stage and during washing is equal to $2 \times 12.8 = 25.6$ mm. Thus, the frames may be 25 mm thick.

It is interesting to evaluate the decrease in average filter press capacity per cycle if the rate of filtration is changed from zero to a value several times higher than the maximum rate of filtration. To make this evaluation, a plot of u versus u_{cond} (using Equation ix) should be prepared:

$$u_{cond} = \frac{u}{1.458 \times 10^{-2} + 0.15 \times 10^6 u^2} \qquad \text{(xv)}$$

The student should prepare this plot, from which a maximum will be observed. This means that the deviation of the filtration rate from the value calculated from Equation x leads to a significant decrease in capacity. Thus, for example, an increase in the rate of filtration from 0.312×10^{-3} to 1×10^{-3} m/s, i.e., 3.2 times the filter capacity, will cause a decrease from 10.7×10^{-3} to 6.1×10^{-3} m^3/s (i.e., almost 1.8 times).

Example 12.7

In a sugar production operation, the saturated liquor is thickened in a cartridge thickener under the following set of conditions:
For suspension properties,

1. Solid particles concentration before thickening, $C_{b.th.}$ (N per 1 m^3 of liquid phase) = 482.
2. Solid particles concentration after thickening, $C_{a.th.}$ (N per 1 m^3 of liquid phase) = 2246.
3. Viscosity of liquid phase, μ (N-s/m^2) = 0.483×10^{-3}.
4. Specific gravity of liquid phase, γ_f (N/m^3) = 1.023×10^4.
5. Specific gravity of solid particles, γ_p (N/m^3) = 2.170×10^4.

For cake properties,

1. Solid particles content, C_c (N per N cake) = 0.44.
2. Specific volumetric resistance, r_0 (m^{-2}) = 101.3 × 10^{12}.

Dimensions of the cartridge thickener are as follows:

1. Length of cartridge (m) is 2.065.
2. Diameter of upper base (m) is 0.215.
3. Diameter of lower base (m) is 0.185.

Other pertinent data are as follows:

1. Capacity of the thickener in suspension, G_s (N/hr) is 51.46 × 10^4.
2. Pressure difference at the stage of cake removal, ΔP_R (N/m^2) is 10 × 10^4.
3. Time of cake removal, τ_R (s) is 6.
4. Ratio of filtration time to the total cycle of thickener $\alpha = \tau/\tau_{cyc} = 11/12 = 0.9167$.
5. Resistance of filter medium, R_f (m^{-1}) is 0.196 × 10^{12}.

The pressure difference in the filtration stage (accounting for the back pressure of filtrate column in the vertical section of discharge piping) is $\Delta P = 3.24 \times 10^4$ N/m^2. Determine (1) the capacity of the thickener in filtrate and thickened suspension; (2) the time of the thickening cycle corresponding to the minimum filtering area; (3) the minimum filtering area; and (4) the cake thickness.

Derivation of Design Equations

The cake removal from the cartridge surface by inverse filtrate flow requires a certain minimum time during which part of the filtrate obtained will be spent for dissolution of thickened suspension. If we assume the minimum time to be constant and independent of the time of the cycle, then at the given pressure difference the filtrate volume entering per cycle into the thickened suspension will be constant. At variable filtration time and a given pressure difference, the amount of filtrate entering per cycle from the thickened suspension into the cartridges will be variable.

At constant cake removal time and a constant ratio of filtration time to the thickener cycle time, there is a period in which the capacity of the thickener is at a maximum. The time of this "unworking" stage, $\tau_{u.w.}$ (when the cartridge is not connected to a vacuum and there is no filtration) is equal to, or greater than, the stage of cake removal, τ_R (when the cartridge is connected with the source of pressurized air). That is, it is necessary to account for $\tau_R/\tau_{u.w.} = \beta < 1$. Under the condition that $\beta < 1$ during the time interval $\tau_{u.w.} - \tau_R$, the cartridge is not connected either with a vacuum or with pressurized air. The design expression for this problem must be based on a finite filter medium resistance because at $R_f = 0$, the amount of filtrate for the stage of cake removal would be infinite.

The derivation is based on determining the maximum value of the ratio of filtrate volume obtained per cycle to the time of the cycle expressed as a function of variable β.

1. The times $\tau_{u.w.}$, τ and τ_{cyc} as functions of τ_R are evaluated from

$$\tau_{u.w.} = \frac{\tau_R}{\beta} \tag{xvi}$$

$$\tau = \tau_{cyc} - \tau_{u.w.} = \frac{\tau_{u.w.}}{1 - \alpha} - \tau_{u.w.}$$

$$= \tau_{u.w.} \frac{\alpha}{1 - \alpha} = \frac{\tau_R \alpha}{(1 - \alpha)\beta} \tag{xvii}$$

$$\tau_{cyc} = \tau + \tau_{u.w.} = \frac{\tau_R \alpha}{(1 - \alpha)\beta} + \frac{\tau_R}{\beta} = \frac{\tau_R}{(1 - \alpha)\beta} \tag{xviii}$$

2. The filtrate volume entering during the filtration stage from the thickened suspension is determined from the equation

$$V^2 + 2 \frac{R_f A}{r_0 x_0} V = 2 \frac{\Delta P A^2}{\mu r_0 x_0} \tau$$

Substituting for τ from Equation xvii,

$$V = A \left[\sqrt{A''^2 + \frac{B''}{\beta}} - A'' \right] \tag{xix}$$

where

$$A'' = \frac{R_f}{r_0 x_0} \tag{xx}$$

and

$$B'' = \frac{2\Delta P \tau_R \alpha}{\mu r_0 x_0 (1 - \alpha)} \tag{xxi}$$

3. We find the filtrate volume entering during cake removal on the basis of the following considerations. Because during this stage we have essentially

the filtration of a pure liquid through the filtering medium, i.e., no cake, Equation xxi is applicable under the condition that $h_c = 0$. Substituting for V, ΔP and τ by V_R, ΔP_R and τ_R, respectively, we obtain

$$V_R = AC = \text{const} \tag{xxii}$$

where

$$\tilde{C} = \frac{\Delta P_R \tau_R}{\mu R_f} \tag{xxiii}$$

4. The filtrate volume obtained per cycle is equal to the difference in filtrate volumes passing through the filtering medium during filtering and cake removal:

$$V_{cyc} = V - V_R = A\left[\sqrt{A''^2 + \frac{B''}{\beta}} - A'' - \tilde{C} \right] \tag{xxiv}$$

5. The average thickener capacity in filtrate per unit time for the total cycle is obtained by dividing Equation xxiv by Equation xviii:

$$u_{cond} = \frac{A(1 - \alpha)}{\tau_R}\left[\sqrt{A''^2 + \frac{B''}{\beta}} - A'' - \tilde{C} \right]\beta \tag{xxv}$$

6. The time of the thickener cycle corresponding to the minimum filtration area is determined as follows. The minimum filtration area corresponds to the maximum average capacity of 1 m^2 of filtering surface. Dividing both sides of Equation xxv by A and differentiating both sides of the equation by β and equating the derivative to zero, we find the ratio β at which the thickener will work at maximum capacity:

$$\beta = \frac{B''E}{2D}\left[\sqrt{1 + \frac{D}{E^2}} - 1 \right] \tag{xxvi}$$

where

$$D = 2A''^2C + A''^2\tilde{C}^2 \tag{xxvii}$$

and

$$E = 2A''\widetilde{C} + \widetilde{C}^2 \qquad \text{(xxviii)}$$

Substituting β into Equations xvi, xvii and xviii, we find the time of the unworking period, the time of filtration and the time of the cycle.

7. The minimum area of filtration is determined as follows. Noting the given capacity of the thickener in filtrate (m^3/s) by G_F and multiplying this value by the time of the cycle, τ_{cyc}, we express the volume of filtrate per cycle:

$$V_{cyc} = G_F \tau_{cyc} \qquad \text{(xxix)}$$

Comparing Equation xxiv with Equation xxix we obtain

$$G_F \tau_{cyc} = A\left[\sqrt{A''^2 + \frac{B''}{\beta}} - A'' - \widetilde{C}\right]$$

Hence,

$$A = \frac{G_F \tau_{cyc}}{\sqrt{A''^2 + \frac{B''}{\beta}} - A'' - \widetilde{C}} \qquad \text{(xxx)}$$

where τ_{cyc} is determined from Equations xviii and xxvi.

Solution

1. The capacity of the thickener in the thickened suspension is computed as follows: The weight of particles in suspension before thickening is

$$\frac{C_{b.th.}}{C_{b.th.} + 1 \times \gamma_f} = \frac{482}{482 + 1 \times 1.023 \times 10^4} = 0.045 \text{ N per N suspension}$$

The weight of solid particles entering the thickener is

$$0.045 \, G_s = 0.045 \times 51.46 \times 10^4 = 2.316 \times 10^4 \text{ N/hr}$$

The weight of solid particles in suspension after thickening is

$$\frac{C_{a.th.}}{C_{a.th.} + 1 \times \gamma_f} = \frac{2246}{2246 + 1 \times 1.023 \times 10^4} = 0.180 \text{ N per N suspension}$$

The weight of the thickened suspension after the thickener is

$$\frac{2.316 \times 10^4}{0.180} = 12.87 \times 10^4/\text{hr}$$

2. Thickener capacity in filtrate is

$$G_f = \frac{51.46 \times 10^4 - 12.87 \times 10^4}{1.023 \times 10^4} = 37.72 \text{ m}^3/\text{hr} = 0.01048 \text{ m}^3/\text{s}$$

3. The ratio of cake volume to filtrate volume is found on the basis of a material balance related to dividing the thickened suspension into wet cake and pure filtrate and made for a definite weight of solids.

The average suspension concentration may be assumed equal to the mean arithmetic value of the concentration before and after suspension thickening:

$$\frac{C_{b.th.} + C_{a.th.}}{2} = \frac{482 + 2246}{2} = 1364 \text{ N/m}^3 \text{ liquid}$$

The weight of solid particles in the cake related to 1 m³ of liquid in the pores is

$$\frac{C_c \gamma_f}{1 - C_c} = \frac{0.44 \times 1.023 \times 10^4}{1 - 0.44} = 8040 \text{ N}$$

The volume of the liquid phase in the cake containing 1364 N of solid particles is

$$\frac{1 \times 1364}{8040} = 0.1697 \text{ m}^3$$

Filtrate volume obtained at cake formation containing 1364 N of solid particles is

$$1 - 0.1697 = 0.8303 \text{ m}^3$$

The volume of cake containing 1364 N solid particles is

$$\frac{1364}{2.170 \times 10^4} + 0.1697 = 0.2326 \text{ m}^3$$

The ratio of cake volume to filtrate volume is

$$x_0 = \frac{0.2326}{0.8303} \simeq 0.280$$

4. Values A'', B'', $\widetilde{C}$, D and E are determined as follows. From Equation xx,

$$A'' = \frac{0.196 \times 10^{12}}{101.3 \times 10^{12} \times 0.280} = 6.91 \times 10^{-3}$$

From Equation xxi,

$$B'' = \frac{2 \times 3.24 \times 10^4 \times 6 \times 0.9167}{0.483 \times 10^{-3} \times 101.3 \times 10^{12} \times 0.280(1 - 0.9167)} = 0.312 \times 10^{-3}$$

From Equation xxiii,

$$\widetilde{C} = \frac{10 \times 10^4 \times 6}{0.483 \times 10^{-3} \times 0.196 \times 10^{12}} = 6.34 \times 10^{-3}$$

From Equation xxvii,

$$D = 2(6.91 \times 10^{-3})^2 \times 6.34 \times 10^{-3} + (6.91 \times 10^{-3})^2 \times (6.34 \times 10^{-3})^2$$
$$= 0.6074 \times 10^{-8}$$

From Equation xxviii,

$$E = 2 \times 6.91 \times 10^{-3} \times 6.34 \times 10^{-3} + (6.34 \times 10^{-3})^2 = 12{,}780 \times 10^{-8}$$

5. Continue the unworking cycle, filtration stage and working cycle of the thickener corresponding to the minimum area of filtration: From Equation xxvi,

$$\beta = \frac{0.312 \times 10^{-3} \times 12,780 \times 10^{-8}}{2 \times 0.6074 \times 10^{-8}} \left[\sqrt{1 + \frac{0.6074 \times 10^{-8}}{(12,780 \times 10^{-8})^2}} - 1 \right] = 0.57$$

From Equation xvi,

$$\tau_{u.w.} = \frac{6}{0.57} = 10.5 \text{ s}$$

From Equation xvii,

$$\tau = \frac{6 \times 0.9167}{(1 - 0.9167) \times 0.57} = 115.8 \text{ s}$$

From Equation xviii,

$$\tau_{cyc} = \frac{6}{(1 - 0.9167) \times 0.57} = 126.3 \text{ s}$$

6. The minimum filtration area from Equation xxx is

$$A = \frac{0.01048 \times 126.3}{\sqrt{(6.91 \times 10^{-3})^2 + \frac{0.312 \times 10^{-3}}{0.57}} - 6.91 \times 10^{-3} - 6.34 \times 10^{-3}}$$

$$= 119 \text{ m}^2$$

7. The filtrate volume obtained in filtration stage (for 1 m^2 of filtering surface) from Equation xix is

$$\frac{V}{A} = \sqrt{(6.91 \times 10^{-3})^2 + \frac{0.312 \times 10^{-3}}{0.57}} - 6.91 \times 10^{-3} = 0.0175 \text{ m}^3/\text{m}^2$$

8. Cake thickness is

$$h_c = x_0 \frac{V}{A} = 0.280 \times 0.0175 = 0.0049 \text{ m} = 4.9 \text{ mm}$$

9. The ratio of filtrate volume entering from the cartridge into the thickening suspension and the volume of cartridge are computed from the following considerations. The filtrate volume passing through 1 m^2 of cartridge surface into the suspension is determined from Equations xxii and xxiii:

$$\frac{V_{sp}}{A} = \tilde{C} = 6.34 \times 10^{-3} \text{ m}^3/\text{m}^2$$

The ratio of the cartridge volume to its surface (from geometry) is equal to $52 \times 10^{-3} \text{ m}^3/\text{m}^2$.

Thus, there is enough filtrate in the cartridge to remove the cake from the filtering surface.

10. The graphic relationship between β and the area of filtration can be obtained as follows: For a series of values $0.1 \leqslant \beta \leqslant 1.0$, using Equation xviii we determine τ_{cyc}. From Equation xxx, we obtain A. A plot of β versus A then may be prepared. The resulting curve shows that at an increase in β to 1, when the cake removal time becomes equal to the unworking time, the filtration area of the thickener increases to 126 m^2, i.e., only 9.6% in comparison to the minimum filtration area.

CENTRIFUGAL FILTRATION

Filtering centrifuges are distinguished from standard centrifugation because a filtering medium is incorporated into the design. Slurry is fed to a rotating basket or bowl having a slotted or perforated wall covered with a filtering medium, such as canvas or metal-reinforced cloth. Centrifugal action produces a pressure that forces the liquor through the filtering medium, thus leaving the solids deposited on the filter medium surface as a cake. When the feed stream is stopped and the cake spun for a short time, residual liquid retained by the solids drains off. This results in processed solids that are considerably drier than those obtained from a filter press or vacuum filter.

Principal types of filtering centrifuges are suspended batch machines, automatic short-cycle batch machines and continuous-conveyor centrifuges. In suspended centrifuges, the filter medium is usually canvas or a similar fabric, or woven metal cloth. Automatic machines employ fine metal screens. In contrast, the filter medium in conveyor centrifuges is usually the slotted wall of the bowl itself.

Figure 12.39 shows still another widely used design. The system combines the features of a centrifuge and a screen. Feed enters the unit at the top and is immediately brought up to speed and distributed outward to the screen surface by a set of vanes. Water or other liquid is forced by the sudden ·centrifugal action through the screen openings into an effluent housing. As solids are accumulated, they are moved gently down the screen by the slightly faster rotating helix. With the increase in screen diameter higher centrifugal forces are encountered and solids are dispersed over a gradually increasing area, thus forming a thin, compact cake from which the remaining liquid is extracted. The relatively dry solids are blown out the bottom of the rotor by a set of vanes into a conical collection hopper.

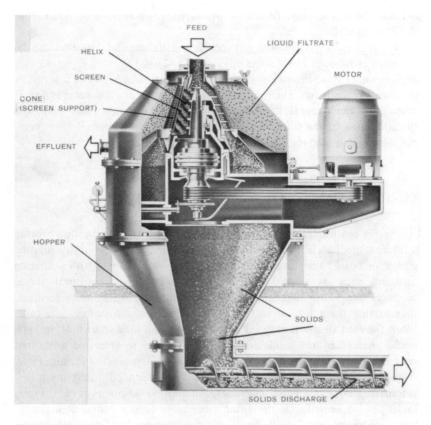

Figure 12.39   Cutaway view of one type of filter centrifuge (courtesy Merco Centrifuges Div., Dorr-Oliver Inc., Stamford, CT).

The theory of constant-pressure filtration may be applied approximately to filtration in a centrifuge. We apply the following assumptions:

1. Effects of gravity and changes in the liquid kinetic energy are negligible.
2. The pressure drop developed from centrifugal action is equivalent to the drag of the liquid flowing through the cake.
3. Particle voids in the cake are filled completely with liquid.
4. The resistance of the filter medium is constant.
5. Liquid flow is laminar.
6. The cake is incompressible.

Based on these assumptions, a simplified analysis is applied to describing the cake after it has been deposited and during the flow of clear filtrate through it, can be developed. The linear velocity of the liquid through the cake is

$$w = \frac{dV/d\tau}{A} = q'/A \qquad (12.89)$$

where V = the volume of filtrate
q' = the volumetric flowrate

From Chapter 10, the pressure drop from centrifugal action is

$$-\Delta P = \frac{\rho\omega^2(r_2^2 - r_1^2)}{2g_c} \qquad (12.90)$$

where ω = the angular velocity (rad/s)
r_2 and r_1 = the radii of the inner surfaces of the liquid and face of the cake, respectively

Combining these two expressions yields

$$q = \frac{\rho\omega^2(r_2^2 - r_1^2)}{2\mu(\alpha' m_c/\overline{A}_L\overline{A}_a + R_m/A_2)} \qquad (12.91)$$

where A_2 = area of filter medium (inside area of centrifuge bowl), (m^2)
$\overline{A}_a$ = arithmetic mean cake area (m^2)
$\overline{A}_L$ = logarithmic mean cake area (m^2)
α' = specific cake resistance (m/kg)
R_m = filter medium resistance (m^{-1})
m_c = mass of solid in filter cake (kg)

The average cake areas are defined as follows:

$$\overline{A}_a = (r_i + r_2)\pi b$$

$$\overline{A}_L = \frac{2\pi b (r_2 - r_i)}{ln(r_2/r_i)} \tag{12.92}$$

where   b = height of the bowl

Equation 12.92 is only applicable to a cake of definite mass, i.e., the expression is not integrated over an entire filtration starting from an empty centrifuge, and is accurate within the limitations of the derivation assumptions.

GAS FILTRATION

Gas filtration is accomplished by several different mechanisms, some of which are not important in liquid filtration. Filter fabrics have the ability to capture particles smaller than the smallest opening in the cloth. For example, even when the spaces between certain fibers of cloth measure 100 μm or more, particles as small as 1 μm can be captured. The principal mechanisms responsible for this phenomenon are interception, impingement, diffusion, gravity settling and electrostatic attraction, in addition to simple particle sieving. These forces retain particles on the cloth fibers, forming a filter cake or mat, which then leads to finer sieving.

Sieving refers simply to the fact that large particles will not pass through small holes. Particle velocity determines how far the solid will penetrate the cloth before it comes to rest. As most dust particles are shaped irregularly, it is easy for many of the larger particles to build up and form a matrix of increasingly smaller holes. These smaller holes then enable the capture of smaller particles. In fine dusts and fumes, however, there may be no particles larger than a few microns, and a precoat of coarse dust on the filter medium is applied, as in the case of liquid filtration.

When an obstruction, such as a cloth fiber, is placed in the path of a fluid stream, the streamlines will curve around the obstruction. Particles suspended in, and traveling with the gas stream will, depending on their inertia, either follow a single streamline or leave it. Fabric filters generally are designed to operate in the laminar regime and, hence, small inertialess particles will remain on single streamlines. As these streamlines pass close to the fibers of the cloth and within a distance equal to the radii of the particles, these particles will contact and adhere to the fibers due to van der Waals forces, i.e., *particle inception*. Van der Waals forces exist between molecules of nonpolar compounds and are accounted for by quantum mechanics. As two nonbonded atoms are brought together, the attraction between them increases and reaches a maximum when the distance between the nuclei is equal to the sum of the van der Waals radii. If forced closer together, the attraction will be replaced by repulsion and the particle will want to back off to a comfortable distance. Particles smaller than 1 μm can be considered inertialess without serious error.

In laminar flow, streamlines are not affected by velocity, but rather the size of an obstruction or, in this case, the fiber strand. The streamlines will pass closer to smaller-diameter fibers than to larger ones. Large particles are collected easily due to *impingement or inertial impact* because their streamlines need not pass as close to the fiber strand as smaller ones for the van der Waals forces to take effect. As particle size increases, however, so does the mass, and the particles tend not to follow their streamlines. Rather, they leave them when a bending of the streamlines occurs. At this point, the particle's high inertia carries it out of the gas stream in the original direction of motion. This condition is favored by high particle mass and high filtering velocity.

At low velocities, random particle movement (or Brownian motion) is a factor in bringing about the impact of fine particles in a gas stream in accordance with Stokes' law. The particle shape and weight, along with the gas viscosity and velocity, determine the particle's settling velocity on the filter medium. Hence, *diffusion* plays an important role in the capturing of small particulates.

Electrostatic attraction and repulsion are mechanisms that, while understood qualitatively, are as yet mysteries quantitatively where fabric filtration is concerned. These forces are factors in particle agglomeration and they often determine the ease or difficulty of media cleaning. But to what degree electrostatic forces affect filtration efficiency is undefined. Polarity, charge intensity and dissipation rate of both the filter media and the dust particles are all important factors. Electrostatic charging of fabrics has been tried successfully in laboratory experiments but as yet has not found extensive use in industrial gas cleaning.

Fabric filters, or *baghouses*, as they more commonly are known, have been in use since the turn of the century in the mining industry. Today's applications extend throughout the CPI, with primary emphasis in industrial air pollution control. Dry dust filters are available in sizes ranging from a few square feet up to several hundred thousand square feet of cloth. Gas flows that can be handled by individual units range from under 100 cfm to more than 1,000,000 cfm. The fabric filter's design is similar to that of a large vacuum cleaner. It consists of bags of various shapes constructed from a porous fabric. Filter bags are available in two major configurations, namely, flat (envelope) bags and round (tubular) bags. The tubular-type configuration for a large unit is shown in Figure 12.40. These bags are coated with Teflon®* "B" fluorocarbon resin for high-temperature resistance.

Figure 12.41(A) illustrates the operation of a baghouse. The dust-laden gas enters the module through an inlet diffuser that breaks up the stream and evenly disperses the dust. The heavier dust particles settle to the hopper and the fine particles rise through the tube sheet into the bags. Particles as small as 0.5 μm are collected on the inside of the bags, while the cleaned gas passes through the fabric.

Dust is removed from the bags by periodic shaking. The frequency of cleaning depends on the type of dust, the concentration and the pressure drop that must be overcome. The dust shaken from the bags falls into the hopper below and is removed by a rotary airlock, screw conveyor or other device. Figure 12.41(B) shows specific design features.

Series modules can be joined to provide any desired capacity. When two or more modules are joined together, a single module can be shut down for bag cleaning and then returned to service. The simple closing of an inlet or outlet damper diverts the dirty gas stream to other modules. Thus, the gas is filtered continuously.

Multimodule installations typically employ a large single fan or small individual fans mounted on each module. Small fan arrangements are more flexible and eliminate the need for outlet ductwork and the foundation, which are required for a large fan. Individual fans often simplify maintenance and permit fan, motor, drive or other components to be changed readily without interrupting normal service. Any one module can be shut down and isolated from the rest of the system while still maintaining full operation and efficiency levels.

The particles to be removed play an important role in the selection of a fabric and filter efficiency. Specifically, the particulate density, concentration, velocity and size are important. Each of these properties is interrelated with the pressure drop of the system, which is one of the most significant points affecting efficiency. Principal variables directly related to pressure drop are (1) gas velocity, (2) cake resistance coefficient, (3) weight of cake per unit area, and (4) air-to-cloth ratio.

The cake resistance coefficient is dependent on the particle size and shape, range of the particle sizes and humidity. Weight of the cake per unit area is related to the concentration of particulates.

Cheremisinoff and Young [1975] note that the design of fabric filters must satisfy two criteria: high efficiency and low pressure drop. Attempts have been made to correlate the filtration efficiency to the operating conditions

---

*Registered trademark of E. I. du Pont de Nemours and Company, Inc. Wilmington, Delaware.

Figure 12.40   The glass fabric filter bags used by Nebraska Public Power Co. to help control particulate emissions (courtesy Du Pont Co., Wilmington, DE).

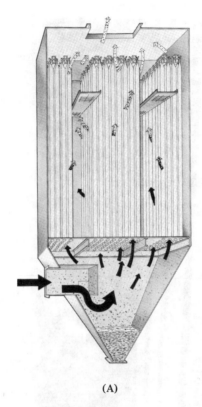

(A)

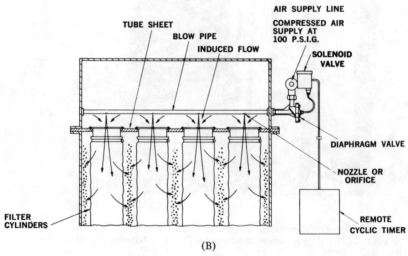

(B)

Figure 12.41   (A) Operation of a baghouse; (B) principal design

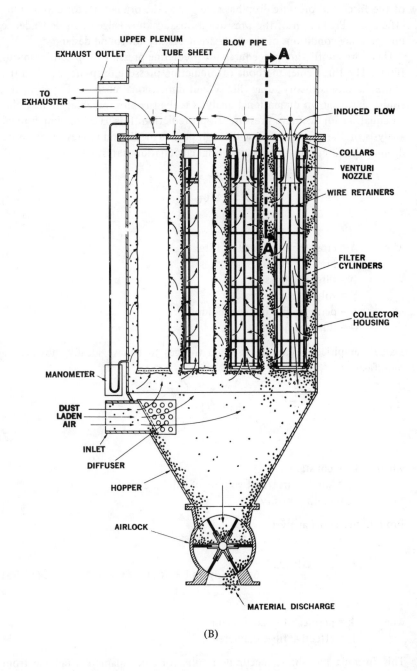

(B)

features of the baghouse (courtesy MikroPul Corp., Summit, NJ).

of the filter, but pressure drop has emerged as the primary factor determining efficiency. Prediction of the pressure drop and knowledge of its dependence on operating conditions of the filter are necessary for sound design.

There are conflicting procedures for predicting pressure drop across fibrous filters. The three chief methods of predicting pressure drop are based on the hydraulic radius theory, drag theory and dimensional analysis. Experimental studies have shown dimensional analysis to be the most reliable, and it yields an equation for accurate pressure drop prediction. It is based on dimensional analysis of Darcy's law of flow through porous media and relates the pressure drop to the filter porosity. Darcy's original equation states:

$$\frac{\Delta PA}{hq'} = kw \tag{12.93}$$

where A = cross-sectional area of filter
P = pressure drop
h = filter thickness
q' = volumetric flowrate
k = permeability of the medium
w = fluid (gas) velocity

Darcy's empirical equations assumed the gas to behave ideally; however, a modification of this is

$$\frac{PA}{hq'} = \frac{k_3 w s_0^2 (1 - \epsilon)^2}{\epsilon^3} \tag{12.94}$$

where k_3 = constant
s_0 = surface area/unit volume of solid material
ϵ = cloth porosity

From dimensional analysis,

$$k = \frac{\Delta PA d_e^2}{hq'w} = 64(1 - \epsilon)^{1.5}(1 + 56(1 - \epsilon)^3) \tag{12.95}$$

where k = permeability coefficient
d_e = effective fiber diameter

This formula has shown accurate results for fiber diameters ranging from 1.6 to 80 μm and filter porosities ranging from 0.700 to 0.994.

Still another formula obtained from dimensional analysis is:

$$\frac{\Delta P A \bar{d}^2}{hq'w} = k''(1 - \epsilon)^{1.5} \tag{12.96}$$

where $\bar{d}^2$ = mean square fiber diameter
k'' = resistance coefficient

This expression predicts pressure drop for filters with porosities ranging from 0.88 to 0.96 and fiber diameters ranging from about 0.1 to 3 μm.
A more useful formula is

$$\Delta P = w\xi(k_0 + k_1 W_A) \tag{12.97}$$

where ΔP = pressure drop
w = gas velocity
ξ = air-cloth ratio
k_0 = weave resistance coefficient
k_1 = cake resistance coefficient (dependent on shape, concentration of particle, humidity)
W_A = weight of cake per square foot of surface

Use of these formulas (especially the last formula) makes possible accurate prediction of the pressure drop of a particular unit. Cheremisinoff and Young [1975,1977] provide detailed design procedures for sizing fabric filters and filter cloth selection.

PRACTICE PROBLEMS

Solutions are provided in Appendix A for those problems with an asterisk (*) next to their numbers.

*12.1   A suspension of aluminum hydroxide in water is to be filtered under constant pressure in a batch Nutsch filter having a filtering area of 1 m². Each filter cycle is estimated to separate out 0.5 m³ of suspension. The temperature of the suspension is 20°C. The following expression for the cake resistance was developed from batch studies over the pressure drop range of 4 × 10⁴ to 8 × 10⁴ N/m².

$$r_0 = r_0'(\Delta P)^{s'}$$

where r_0 = specific cake resistance (m^{-2})
r_0', s' = constants determined from experiments, 0.5×10^{10} and 0.95, respectively

The volume ratio of cake to filtrate (x_0) is 0.01. If the resistance of the filter plate is negligible, determine the required filtering time for the same pressure range.

*12.2 A suspension of solid particles in water is to be separated in a Nutsch-filter where $A = 1$ m^2, $\mu = 10^{-3}$ N-s/m^2 and the temperature of the suspension is 20°C. The allowable pressure drop for the system is 20×10^4 N/m^2. The filter plate resistance, R_f, is 5×10^{10} m^{-1} and is independent of the pressure drop. The volume ratio of cake to filter, x_0, is 0.025, and the filter capacity is 0.2×10^{-3} m^3/s of filtrate. The relationship between cake resistance and pressure drop is the same as described in Problem 12.1, with $r_0' = 0.126 \times 10^{12}$ and $s' = 0.5$. Determine the following information:
(a) filtration time,
(b) filtrate volume, and
(c) cake thickness.

12.3 For the previous problem, prepare a plot of pressure drop versus filtration time.

*12.4 A suspension is filtered with a filter cloth media. The resistance of the cloth is negligible. The following information is supplied:

- $R_{c.in}' = 0.050$ m
- $R_{c.out}' = 0.100$ m
- $r_0 = 6.0 \times 10^{10}$ N-s/m^4
- $x_0 = 0.2$
- $\Delta P_t = 20 \times 10^4$ N/m^2

Determine the difference in filtration times between a cylindrical and flat filter medium.

*12.5 An aqueous solution of sodium bicarbonate (t = 20°C, $\mu_w = 10^{-3}$ N-s/m^2) is being filtered under vacuum in a Nutsch filter. The filtration area is 1 m^2. The operation is performed in two stages: the first filtration stage is at constant rate and the second is at constant pressure drop. The allowable pressure drop is 80,000 N/m^2 and the permissible cake thickness is 0.2 m. The filter capacity at constant rate is 0.5×10^{-3} m^3/s of filtrate. The cake is incompressible, with the following characteristics:

- $x_0 = 0.1$
- $r_0 = 2 \times 10^{12}$ m^{-2}
- $R_f = 10^{10}$ m^{-1}

Determine the total filtration time of both steps and the volumes of filtrate, cake and separated slurry. The operation is performed without any washing or dewatering steps.

*12.6 The system described in Problem 12.5 has an auxiliary time of $\tau_{aux} = 1800$ s. Determine the total filtration time at constant pressure drop, cake thickness and filtrate volume.

12.7 A filter press with a hydraulic drive has 35 frames with internal dimensions of 1900 × 1200 mm. The unit operates under the following set of conditions:

- Filtrate viscosity, μ (N-s/m^2) = 1.8 × 10^{-3}
- Washing liquid viscosity, $\mu_{w\varrho}$ (N-s/m^2) = 1.0 × 10^{-3}
- Specific cake resistance, r_0 (m^{-2}) = 9 × 10^{14}
- Volume ratio of cake to filtrate, x_0 = 0.059
- Allowable pressure drop, ΔP (N/m^2) = 7 × 10^6
- Cake porosity, ϵ = 0.49
- Weight ratio of dissolved solids extracted to material before washing, G/G_0 = 0.85
- Auxiliary operating time, τ_{aux} (min) = 20

The cake is incompressible and the average plate resistance, R_f, is 2×10^9 m^{-1}. The suspension passes through all the frames and in one direction. The cake is not dewatered. Determine the following:

(a) the rate of filtration and washing under conditions of maximum filter press capacity;
(b) the required times of filtration and washing;
(c) the capacity of the filter press;
(d) the rate of washing liquor; and
(e) the cake thickness.

12.8 A saturated liquor is thickened in a cartridge filter at maximum capacity and constant pressure drop. The properties of the suspension and cake are as follows:

Suspension	Cake
• Solids concentration before thickening = 700 N/m^3	• Solids concentration = 0.6 N/N cake
• Solids concentration after thickening = 3510 N/m^3	• Specific volumetric resistance = 90 × 10^{10} m^{-2}
• Liquid viscosity = 0.6 × 10^{-3} N-s/m^2	
• Liquid specific weight = 1.04 × 10^4 N/m^3	
• Solids specific weight = 2.7 × 10^4 N/m^3	

The cartridge filter has the following dimensions: length, 3.4 m; upper base diameter, 0.3 m; lower base diameter, 0.21 m. The unit is operated under the following set of conditions:

- Capacity in suspension = 60×10^5 N/hr
- Pressure loss at the stage of cake removal = 15×10^5 N/m^2
- Time of cake removal = 8 s
- Ratio of filtration time to total thickness cycle = 0.99
- Filter medium resistance = 0.4×10^{12} m^{-1}

Determine the following information:

(a) capacity of the thickener in the filtrate and the thickened suspension;

(b) thickening cycle time corresponding to the minimum filtering area;

(c) minimum filter area required; and

(d) cake thickness.

12.9 Estimate the blower requirements for a rotary vacuum filter used for filtering solids from a liquid suspension under the following set of conditions:

Properties of Slurry:

- Weight fraction of suspended solids, $x' = 0.3$
- Filtrate density, $\rho = 90$ lb/ft^3
- Density of solids in slurry, $\rho_p = 101$ lb/ft^3
- Filtrate viscosity, $\mu_f = 7$ cp
- Viscosity of air, $\mu_a = 0.018$ cp
- Surface tension of filtrate, $\sigma = 0.006$ lb$_f$/ft

Properties of Cake:

- Cake porosity, $\epsilon = 0.55$
- Particle sphericity, $\varphi = 0.80$
- Mean particle size, $d_p = 210 \ \mu$m

The drum has a diameter of 4.5 ft and is 2.7 ft wide. The washing ratio is 0.12. The unit operates at 2 rpm and under 5 in. Hg vacuum pressure. The cake thickness, h_c, is 3 in.

*12.10 The operating cycle of a belt vacuum filter consists of filtration, washing and dewatering stages. The solids suspension in water is separated under the following set of conditions:

- Liquid-phase viscosity, μ (N-s/m^2) = 0.9×10^{-3}
- Specific volume cake resistance, r_0 (m^{-2}) = 4×10^{12}
- Volume ratio of cake to filtrate, $x_0 = 0.25$
- Pressure difference, ΔP (N/m^2) = 5×10^4
- Minimum allowable cake thickness, h_c (m) = 0.008
- Washing time, τ_w (s) = 15
- Dewatering time, τ_d (s) = 25
- Belt effective length, ℓ (m) = 6.0
- Belt effective width, b (m) = 0.5

The resistance of the filter medium is negligible. Determine the filter's capacity in filtrate and the velocity of the belt. (Hint: The design of a belt filter is basically the same as a rotary drum filter.)

12.11 A homogeneous slurry is to be filtered at a constant rate of 10 cfm in a leaf filter. Constant-pressure condition tests indicate that the cake has a compressibility coefficient of 0.4 and that at a pressure of 75 psi the volume of filtrate (ft^3) at any given time (min) is described by the expression $V^2 - 5000$ t. The time of the constant rate cycle is 150 mm:
(a) Compute the horsepower of the motor required.
(b) Determine the theoretical work done by the pump per cycle.

*12.12 A batch cartridge filter contains 248 cartridges and is 1.8 m long. The outside cartridge radius (equal to the inside radius of the cake, $r'_{c.in}$) is 0.03 m. The unit is to be operated at a constant pressure drop of 2.94×10^5 N/m^2. The permissible cake thickness on the cartridge (considering the distance between cartridges, i.e., $R'_{c.out} - r'_{c.in}$) is 0.0225 m. The specific resistance of the cake (as related to the filtrate), r_0, is 617×10^9 $N\text{-}s/m^4$. The required time for auxiliary operations, τ_{aux}, is 1800 s. Determine the average filter capacity. The resistance of the filter medium may be neglected.

12.13 Determine the filtration constants a and b from the following experimental data:

Time, τ (s):	70	225	455	770
Filtrate volume (ℓ):	10	20	30	40

Additional information includes the following: filtration surface, $A = 0.05$ m^2; pressure drop $= 4.91 \times 10^4$ N/m^2; temperature, $t = 20°C$.

12.14 For the conditions given in Problem 12.13, determine the specific cake resistance and filter medium resistances if $x_0 = 0.01$ m^3/m^3 and the slurry viscosity is 1 cp.

12.15 Determine the filtration constants for a water suspension at $t = 20°C$ and a pressure drop of 6 kg_f/cm^2. The following parameters are known:

- $x_0 = 0.07$ m^3/m^3
- $r_0 = 2.86 \times 10^{15}$ m^{-2}
- $R_f = 4.3 \times 10^{10}$ m^{-1}

12.16 A suspension containing 50% solids is to be filtered in a filter press to a cake containing 40% water (weight basis). The filter capacity in the cake is $G_c = 700$ kg/hr. Determine the amount of suspension, filtrate and the ratio, x_0, if the density of the solid phase is $\rho_1 = 1600$ kg/m^3 and the density of the liquid phase is $\rho_2 = 1000$ kg/m^3.

12.17 For the conditions given in Problem 12.15, determine the filtration area. The filtration constants are a = 1.19×10^6 s/m^2 and b = 51 s/m.

12.18 For the conditions given in Problem 12.16, determine the washing time if 2 kg of water are used per 1 kg cake. The filtrate viscosity is μ_f = 0.7 cp and the viscosity of the washing liquor is 1 cp.

12.19 Determine the required surface area of a Nutsch filter operating under the following conditions:

- Amount of suspension is Q_s = 2500 kg
- Solid phase concentration = 10%
- Cake wetness = 60%
- Filtrate density is ρ_2 = 1040 kg/m^3
- Cake density is ρ_c = 1100 kg/m^3
- Specific cake resistance is r_0 = 1.324×10^{14} m^{-2}
- Filter plate resistance is R_f = 5.69×10^{11} m^{-1}
- Amount of washing water = 1.75 m^3/m^3 of wet cake
- Filtrate viscosity, μ_f = 1.1 cp
- Washing water viscosity is μ_w = 1.0 cp
- Pressure drop is ΔP = 1.96×10^5 N/m^2

The total time for cake removal and filter preparation is 20 min.

12.20 Determine the amount of wet cake obtained from filtering 10 m^3 of suspension with a relative specific weight of 1.12. The suspension is 20% by weight solids and the cake wetness is 25%.

12.21 Approximately 15 m^3 of filtrate was removed from a suspension containing 20% by weight solids. The cake wetness is 30%. Calculate the amount of dry cake obtained.

12.22 The following data were obtained at different pressures for filtration in a press having a filtering area of 900 cm^2:

Pressure (kg/cm^2)	Time from Starting (min)	Filtrate Volume Obtained (ℓ)
1.75	13.3	1.8
1.75	49.1	3.57
3.5	12.2	1.9
3.5	42.6	3.6
5.3	11.7	2.0
5.3	35.5	3.5
7.0	11.3	2.1
7.0	32.8	3.6

Determine the constants in the filtration equation, K (m^2/hr) and C (m^3/m^2). From the given data, prepare a plot of K versus pressure.

12.23 Determine the required filtration times for the suspension described in Problem 12.22 using a commercial filter press at the following pressures: 1.75, 3.5, 5.3 and 7.0 atm. The unit has 40 frames. The cake thickness is 25 mm and its wetness is 40%. The suspension contains 10% by weight of solids with a relative specific weight of 3. The inside dimensions of the frames are 0.85 X 0.85 m. Also, compute the washing time for 1 m^3 of water.

12.24 A filter press has 26 frames (62 X 62 cm; frame thickness, 25 mm). The filtration time is 2 hr. Washing is performed with water using 10% from the filtrate volume. Pressure during filtration and washing is constant and the same. Compute the required time for washing. The cake is homogeneous and incompressible and its volume is 5% of the filtrate volume. The calculation can be made with the following equation:

$$V^2 + 2VC = K$$

assuming $C = 0$.

12.25 The filtration time of 20 m^3 of a suspension in a filter press is 2.5 hours. Determine the approximate washing time of the cake for 2 m^3 of water, assuming that the washing rate is 4 times less than that of filtration at the end of process. The resistance of the filter cloth may be neglected. The viscosity of filtrate is equal to that of the washing liquor.

12.26 By how much will the washing time change in Problem 12.25 if filtrate viscosity is 1.5 cp and the viscosity of the washing water is 1 cp?

12.27 Determine the required washing time under the following conditions:

- Washing intensity = 6 ℓ/m^2-min
- Cake thickness = 30 mm
- Initial solids concentration in filtrate of washing water = 120 g/ℓ
- Final concentration = 2 g/ℓ

The washing rate constant, K, is 350 cm^3/ℓ.

12.28 Determine the washing rate constant, K, for the following conditions:

- Washing intensity = 10 ℓ/m^2-min
- Cake thickness = 25 mm
- Initial solids concentration in the filtrate of washing water = 40 g/ℓ
- Final solids concentration = 0.5 g/ℓ

The washing time is 1 hr and 40 min.

12.29 The following data were obtained in a 0.1-m^2 laboratory-scale filter press:

Amount of Filtrate (ℓ)	At P = 1 atm for time (s)	At P = 2 atm for time (s)
4.5	181	98
13.5	1443	788

Cake wetness was 32% in both cases, with temperature, t = 20°C. Determine the increase of cake-specific resistance on changing pressure from 1 atm to 2 atm.

12.30 For the conditions described in Problem 12.22, prepare a graphic relationship between specific cake resistance and pressure. The initial solids concentration in suspension is 10%, with temperature, t = 23°C and an average cake wetness of 40%.

*12.31 Calculate the required filtration time for 10 ℓ of liquid through 1 m^2 of filter area if tests have shown that 1 ℓ of filtrate is obtained after 2.25 min and 3 ℓ after 14.5 min from the start of the process.

*12.32 For the conditions described in Problem 12.31, determine the washing time if the amount of wash water is 2.4 ℓ/m^2 and the washing is cocurrent with the filtrate.

*12.33 During the filtration test of a water suspension containing 13.9% calcium carbonate in a filter press (A = 0.1 m^2; 50 mm cake thickness; temperature ~20°C), the following data were obtained:

Pressure (atm)	Filtrate Obtained (ℓ)	Time from Start of Test (s)
0.35	2.92	146
	7.80	888
1.05	2.45	50
	9.80	660

Determine the filtration constants, K (m^2/hr) and C (m^3/m^2), at the two pressures.

*12.34 For the conditions described in Problem 12.33, compute the specific cake resistance of calcium carbonate. The wetness of the cake at 0.35 atm is 37%, and at 1.05 atm, 32% of wet cake by weight.

*12.35 Determine the filter cloth resistance for the filtration of calcium carbonate under conditions given in Problems 12.33 and 12.34.

*12.36 A suspension is to be processed in a filter press to obtain 6 m³ of filtrate in 3 hr. A filtration test in the lab under the same pressure and cake thickness showed that the filtration constants for 1 m² of filter area are as follows: K = 20.7 × 10⁻⁴ m²/hr and C = 1.45 × 10⁻³ m³/m². Determine the size of the filter.

*12.37 A rotary drum vacuum filter is fed with 8.5 m³/hr of a suspension containing 17.6% solids by weight. The final wetness of the cake must be 34%. The vacuum is 600 mm Hg. During a filtration test on a model at a vacuum of 510 mm Hg, it was found that the desired wetness is achieved in 32 s when operating in the filtering zone. The filtration constants related to 1 m² of area are K = 11.2 ℓ²/m⁴-s and C = 6 ℓ/m². The specific weight of the suspension is 1120 kg/m³ and that of the filtrate is 1000 kg/m³. Determine the filter area and its number of rotations.

*12.38 How long will it take to wash a cake of NaCl in a filter press to achieve 5 g/ℓ NaCl in washing water? Washing is performed with clean water at an intensity of 0.33 m³/m²-hr. The cake thickness is 35 mm and the washing constant, K, is 520 cm³/ℓ. At the beginning of the washing, the NaCl concentration in the wash water is 143 g/ℓ.

*12.39 For the conditions of Problem 12.38, calculate the NaCl concentration in the wash water after 50 min of washing.

*12.40 The following data were obtained in a filtration test using a CaCO₃ suspension on a filter of area A = 500 cm² at constant pressure:

Time (s):	6.8	19.0	36.4	53.4	76.0	102	131.2	163
Filtrate volume (ℓ):	0.5	1.0	1.5	2	2.5	3	3.5	4

Determine the filtration constants a and b, where 'a' is the constant characterizing cake resistance related to 1 m² of filter area in s/m², and 'b' characterizes the resistance of the filter medium related to 1 m² of filter area in s/m.

*12.41 Using the filtration constants determined from Problem 12.40, determine the filtration time of a 5 m³ CaCO₃ suspension containing 5% solids on the filter with 10 m² area. The cake wetness is 40% and solids density is 2200 kg/m³. The density of the liquid is 1000 kg/m³.

*12.42 Using the data of Problems 12.40 and 12.41, determine the cake specific resistance and resistance of the filter medium if the pressure difference at filtering is 5 × 10⁴ N/m² at 20°C.

*12.43 For the system described in Problem 12.41, determine the final filtration rate after 2 hr of operation.

*12.44 A water suspension is filtered in a media of 50 m^2 area at 20°C. Approximately 7 m^3 of filtrate are obtained. The filtration constants are a = 1.44 $\times$ 10^6 s/m^2 and b = 9 $\times$ 10^3 s/m. Determine the washing time at 40°C if the water washing rate is 10 ℓ/m^2.

*12.45 A suspension of 15% solids is filtered over an area of 50 m^2. Cake wetness (by weight) is 40%. Determine the cake thickness if the filtration and washing time is 10 hr. The wash water's volume is one-fifth that of the filtrate volume, and the filtration constants are a = 1.08 $\times$ 10^5 s/m^2 and b = 2.16 $\times$ 10^3 s/m. The densities of the solid and liquid phases are 3000 kg/m^3 and 1000 kg/m^3, respectively.

12.46 Design a Nutsch filter for processing a suspension containing 15% solids. The cake wetness is 45% and the filter capacity for filtrate is 5 m^3/hr. The pressure drop is 500 mm Hg. Cake specific resistance, r_0, is 9 $\times$ 10^{11} m^{-2} and filter plate resistance, R_f, is 2 $\times$ 10^9 m^{-1}. Solid- and liquid-phase densities are 2000 kg/m^3 and 1000 kg/m^3, respectively. The cake is washed with water at 20°C using 1 kg of water per kg of cake.

NOTATION

A'	parameter defined by Equation 12.79
A''	parameter defined in Examples 12.6 and 12.7
A	effective area, m^2
A_c	contact area, m^2
A_{drum}	total filtration area of drum, m^2
A_{sub}	filter medium surface submerged, m^2
a	coefficient
a'	specific surface area, m^2
B	parameter defined in Equation 12.52
B'	parameter defined by Equation 12.80
B''	parameter defined in Examples 12.6 and 12.7
b	height of centrifuge bowl, m
C	filtration constant defined by Equation 12.21
C'	parameter defined by Equation 12.54
$\tilde{C}$	filtration parameter defined in Example 12.7
C_a	blowing factor
C_{dew}	dewatering factor defined by Equation 12.47

c	weight fraction of solids in feed
c_a, c_0	solute concentration in washing liquor and filtrate, respectively, kg/m^3
D''	parameter defined by Equation 12.51
D	parameter in Example 12.7
d	particle size, m or mm
d_{eq}	equivalent diameter of pores, mm
d_e	effective fiber diameter, m
E	parameter defined in Equation 12.57 and Example 12.7
Eu	Euler number
$\dot{F}$	total friction losses, m or N
F_s	cumulative drag on particles, N
F_{ϱ_1}	filtration number, refer to Equation 12.63
f_c	experimental value of cake permeability, N/m^4-hr
f_{exp}	resistance of filter medium based on experiments, N/m^2-hr
f'	coefficient of friction losses
G	weight of dissolved solids, N
H	height, m
h	thickness, m
h_c	cake height, m
K	filtration constant defined by Equation 12.20
K'	filtration constant defined by Equation 12.84, m^{-1}
K_b	solute to wash liquor viscosity ratio
K_{c_1}	operating cost factor, see Equation 12.83, $/s
K_{c_2}	filter aid cost factor, $/N
K_p	capillary number
K'_w	properties constant in Equation 12.66
k_0	weave resistance coefficient, refer to Equation 12.96
k_1	cake resistance coefficient, refer to Equation 12.96
k_2	filter operating expenses per unit time, $/s
k_3	constant in Equation 12.94
K_{sub}	ratio of submerged filtration area to total drum area
k'_w	cake property constant, Equation 12.66
k''	resistance coefficient, Equation 12.96
L	parameter in Equation 12.65
ℓ	weight ratio of liquid to solids in suspension
M	viscosity ratio, see Equation 12.77
m	dry mass concentration of solids, kg/m^3, or weight ratio
m_c	mass of solids in filter cake, kg
m_s	cake saturation factor defined by Equation 12.48
N	flow capacity ratio, see Equation 12.77
N_{drum}	fractional submergence per unit time, s^{-1}

P	pressure, Pa
p'	amount of filter aid, N/m^2
q	filter volume obtained per unit area filtering medium, m^3/m^2
q'	volumetric flow, m^3/s
R_c, R_f	cake and filter resistance, respectively, m^{-1}
Re	Reynolds number
RHS	right hand side
r_0	volumetric cake resistance, m^{-2}
r_w	specific cake resistance, kg/m^2
$r_{1,2}...r_i$	radius, m
s	cake compression coefficient
S	weight ratio of liquid to solids in cake
$\dot{S}$	ratio of total filtration to auxiliary operating times
s_0	surface area per unit volume of solid material, m^2/m^3
t	temperature, °C
u	filtration rate, m/s, or m^3/s where noted
V	volume, m^3
W	filtration rate, m^3/m^2-s
W'	weight on dry basis, N
W_A	weight of cake per unit area, kg/m^2
w	velocity, m/s
x_0'	weight of solids in cake per unit filtrate volume, kg/m^3
x_0	ratio cake to filtrate volumes
x_w	mass ratio of solids to filtrate
y	exponent

Greek Symbols

α	specific resistance, m^{-1} or ratio of filtration to total operating cycle times
α'	specific cake resistance, m/kg
β	ratio of actual filtration to unworking times
Γ_1	dimensionless mass rate
γ	specific weight, N
ϵ	porosity
λ	friction factor
μ	viscosity, p
ξ	air to cloth ratio
ρ	density, kg/m^3
σ	surface tension, N/m
τ	time, s
τ_{cov}	filter aid covering time, s

τ_{dew} dewatering time, s
τ_t total filtration cycle duration, s
ψ particle shape factor
ψ' void fraction

REFERENCES

Akselrod, L. S., and I. M. Leichkis (1971) *Khim. Prom.* 5:392.

Autenrieth, H. (1953) *Chem. Ing. Tech.* 25(12):731.

Berline, R. (1955) *Genie Chim.* 73(5):130.

Bolen, M. (1961) *Tonind.* 85(6):135.

Brownell, L. E., and D. L. Katz (1947) *Chem. Eng. Prog.* 43:537,601,703.

Bruk, O. L., and R. A. Gainzeva (1970) *Khim. Prom.* 9:61.

Carman, P. C. (1937) *Trans. Inst. Chem. Eng.* 15:150.

Chalmers, J. M., L. R. Elledge and H. F. Porter (1955) *Chem. Eng.* 62(6):191.

"Better Washing on a New Turntable Filter" *Chem. Eng.* (1954a) 61(6):128.

"Uses of a Novel Rotating-Tray Filter" *Chem. Eng.* (1954b) 61(6):128.

Cheremisinoff, P. N., and R. A. Young (1975) *Pollution Engineering Practice Handbook* (Ann Arbor, MI: Ann Arbor Science Publishers, Inc.).

Cheremisinoff, P. N., and R. A. Young (1977) *Air Pollution Control & Design Handbook*, Part 1 (New York: Marcel Dekker Inc.).

Crozier, H. E., and L. E. Brownell (1952) *Ind. Eng. Chem.* 44:631.

Daniells, K. J. (1964) *Filtration* 1(4):196.

Dickey, G. D. (1961) *Filtration* (New York: Van Nostrand Reinhold Co.).

Dickey, G. D., and C. L. Bryden (1946) *Theory and Practice of Filtration* (New York: Van Nostrand Reinhold Co.).

Egorov, N. N. (1950) *Neft. Khoz.* 9:26.

Fan-Young, A. F. (1953) *Khim. Prom.* 10:31.

Flood, J. E., H. F. Porter and F. W. Rennie (1966) *Chem. Eng.* 73(13):163.

Gottner, G. H. (1954) *Erdol und Kohle* 7(5):286.

Grace, H. P. (1953) *Chem. Eng. Prog.*, 49:303-318; 367-376.

Holland, C. D., and J. F. Woodham (1956) *Petrol. Ref.* 35(2):149.

Kelsey, G. B. (1965) *Trans. Inst. Chem. Eng.* 43(8):248.

Komarovski, A. A. (1950) *Trudy Novoch. Politekh. Inst.* 20:3.

Komarovski, A. A., and V. V. Strelzov (1958) *Khim. Prom.* 3:173.

Kruglikov, P. M. (1965) *Khim. Neft. Mash.* 7:18.

Kufferath, A. (1954) *Filtration and Filter*, Berlin, Germany.

Meshengisser, M. Y. (1960) *Khim. Mash.* 3:15.

Moncrieff, A. G. (1965) *Filtration Washing Theory, Filtration and Separation* 2(2):88.

Orlicek, A. F., A. Schmidt and H. Schmidt (1954) *Chem. Zeit.* 78(8):266.

Perry, R. H., and J. F. Chilton, Eds. (1973) *Chemical Engineer's Handbook*, 5th ed. (New York: McGraw-Hill Book Co.).

Poole, J. B., and D. Doyle (1966) *Solid-Liquid Separation* (London: Her Majesty's Stationery Office).

Purchas, D. B. (1957) *Chem. Prod.* 20(4):149.

Purchas, D. B. (1964) *Filtration* 1(6):316.

Reeves, E. J. (1947) *Ind. Eng. Chem.* 39:203.

Rhodes, F. H. (1934) *Ind. Eng. Chem.* 26(12):1331.

Ruth, B. F. (1935) *Ind. Eng. Chem.* 27(708):806.

Ruth, B. F. (1946) *Ind. Eng. Chem.* 38:564.

Sharito, M., M. Sambuichi, H. Kato and A. Aragaki (1969) *AIChE J.* 15:405-409.

Silverblatt, C. E., H. Risbud and F. M. Tiller (1974) *Chem. Eng.* 81:127-136.

Skelland, A. H. (1962) *Chem. Proc. Eng.* 43(2):78.

Strelzov, V. V. (1955) *Khim. Prom.* 5:35.

Thomas, C. M. (1965) *Trans. Inst. Chem. Eng.* 43(8):233.

Tiller, F. M. (1958) *AIChE J.* 2:171-174.

Tiller, F. M. (1973) *Chem. Eng. Prog.* 49:467-479.

Tiller, F. M., and H. R. Cooper (1962) *AIChE J.* 8:445-449.

Tiller, F. M., and J. R. Crump (1977) *Chem. Eng. Prog.* 73(10):65.

Tiller, F. M., and T. C. Green (1973) *AIChE J.* 19:1266-1269.

Tiller, F. M., S. Haynes and W. M. Lu (1972) *AIChE J.* 8:569-572.

Trawinski, H. (1970) *Chem. Ing. Tech.* 42(23):1453.

Walas, S. M. (1946) *Trans. Am. Inst. Chem. Eng.* 42(5-6):783.

Zhuzhikov, V. A. (1950) *Khim. Prom.* 1:12.

Zhuzhikov, V. A. (1953) "Processy i apparaty Khimicheskoy Technologii," *Goskhimidat, Moscow.*

Zirkin, I. I., and V. A. Zhuzhikov (1967) *Khim. Prop.* 7:63.

CHAPTER 13

FLUIDIZATION AND FLUID–PARTICLE TRANSPORT

CONTENTS

INTRODUCTION

In this chapter we shall discuss one of the comparatively less developed areas of chemical engineering, namely, "fluidization." An overview of the present state of knowledge is presented, but as the entire scope of the subject cannot be covered in a single chapter, certain areas are excluded. Among the aspects covered are the onset of fluidization, its importance, advantages and disadvantages as compared to packed beds, pressure drop and effect of bed height, among others. Such recent developments as centrifugal fluidization and three-phase fluidization are discussed in the latter part of the chapter, and the last section deals with entrainment.

Growth in fluidized-solids techniques has occurred almost exponentially. At the same time, theoretical studies aimed at better comprehension of the internal fluid mechanics have proliferated. Such studies could lead to proper definition of reaction kinetics, heat transfer, particle-dispersion rates and associated phenomena of importance in the design of industrial plants. One now can describe quantitatively most of these internal mechanics because of the work carried out in the commercial development of applications in processes involving, for example, vinyl chloride, phthalic anhydride, melamine, amporitvile, polyethylene, polypropylene, and fluidized catalytic cracking of petroleum. There is hardly a coal gasification, combustion or liquefaction process that does not involve a fluidized bed. Incineration, nuclear fuel preparation, drying, coating and similar physical applications are other well-established operations. The application of fluidization falls into one of two general classes: (1) chemical reactions and catalysis, or (2) physical and mechanical processes.

Fluidization is a method of contacting granular solids with fluids, and understanding it is critical to the current efforts to produce clean, inexpensive energy from coal resources. A myriad of large- and small-scale fluidized-bed processes are in use and/or under development today, and current and future applications that affect both energy and the environment certainly will include fluidized-bed processes. There have been enormous advances toward understanding the basic mechanics of fluidization and the physical and chemical processes occurring in them during the past 40 years. However, a large number of related areas remain to be developed to their fullest extent. Although subjects such as the equation of motion, elutriation, attrition, stability, heat transfer, mass transfer, chemical reaction, etc., are as important as ever, newer research interests such as the mechanics in the distributor plate and freeboard regions of fluidized beds, the effects of electric and magnetic fields, the turbulent fluidization of large particles, fast fluidization, transport and spouting are all of growing importance.

ONSET OF FLUIDIZATION

At a certain rate of fluid flow, a packed bed of granular solids will expand to such a point that the granules may move within the bed. This condition is known as the onset of fluidization, or the minimum fluidization point. A packed bed expands only when the pressure drop due to the upward flow of fluid through a granular unrestricted bed equals the weight of the packing. As the bed expands, it retains its top horizontal surface with the fluid passing through the bed much as it did when the bed was stationary. Now, however, the porosity is much greater, and the individual particles move under the influence of the passing fluid. The column of solids has a characteristic hydrostatic head. The fluidized bed has similar properties to those of liquids; therefore, it is considered to be a "rendered fluid." Hence, the operation achieving this is termed "fluidization."

The experience gained so far indicates that a wholesale substitution of fixed beds by fluidized beds is by no means in sight. Each side has its advantages and disadvantages. The principal advantages of fluidized systems over fixed-bed processes include the following:

1. Local temperatures and particle distributions are much more uniform than in fixed beds.
2. As particle sizes are an order of magnitude smaller for many processes (e.g., coal) in fluidized beds than in fixed beds, there is less resistance to diffusion through the particle.
3. Solids in the fluidized state are relatively easy to handle, and this favors continuous flow and recirculation systems.
4. Fluidized beds are mechanically simple and suited to large-scale operations.
5. Fluidized beds tend to be more compact than alternative gas/solid contacting techniques.
6. Owing to the motion of the particles past internal or external heat and mass transfer surfaces, respective transport coefficients in fluidized beds are higher than in the fixed-bed operations under comparable flow conditions.
7. Owing to the high particle-gas heat and mass transfer rates, fluidized particles lend themselves more readily to the recovery of heat from waste solids than do the generally larger solid particles in fixed beds.

Principal disadvantages include the following:

1. The average flow of solids and gases in the single-bed fluidized reactor is cocurrent. This has an unfavorable effect on the driving force. To approach countercurrent flow, a multicompartment reactor is required. This is much more expensive than a fixed bed.
2. From low height/diameter ratios there may result appreciable longitudinal mixing of gas and solids in the reactor. This can lead to low conversion rates.
3. Not all solids fluidize well. This disadvantage can be overcome by further preparation or special techniques, such as spouting for large particles or stirring and pulsing for fine particles.
4. Scaleup from laboratory- to commercial-sized reactors is not simple because different regimes of fluidization tend to occur at different scales of operation, i.e., small-diameter beds tend to slug, moderate-sized beds are freely bubbling and large-diameter beds exhibit large-scale circulation cells of solids and preferred bubble paths. These regimes must be predicted and allowed for.
5. The gas velocity must be closely coordinated with the properties of particles so that adequate fluidization results. Thus, the fluidized reactor is in this respect restricted, and the fixed bed offers a greater degree of freedom and adjustment of space velocity.
6. Fluidization is efficient only if no liquids or waxes form during the reaction. This is a severe restriction and a great disadvantage as compared with the fixed bed.
7. In equipment for fluidized operations, erosion may be serious. Special and generally expensive designs may be required to eliminate or minimize wear in gasifiers and transfer lines.
8. Owing to solids carryover, installation for fines recovery may be required.

DENSE-PHASE FLUIDIZATION

The various operating ranges of a fluidized bed, as the gas superficial velocity up through it increases, are illustrated in Figure 13.1. For (A) and (B) the bed is stable, and the pressure drop and Reynolds number are related. At (B) the pressure drop essentially balances the solids weight. From (B) to (C) the bed of solids is unstable, and particles adjust their position to present as little resistance to flow as possible. At (C) the loosest possible arrangement is obtained in which the particles are in contact. Beyond this point the particles begin to move freely but collide frequently, so that the motion is similar to that of particles in hindered settling. Figure 13.1(C) is referred to as the *point of fluidization* or minimum fluidization. Just beyond this point the particles are all in motion. Increases in Re result from very small increases in ΔP as the bed continues to expand and the particles move in more rapid, independent motion. Ultimately, the particles will stream with the fluid and the bed will cease to exist. This occurs in (E). Figure 13.1(A) shows a column that is traversed countergravity-wise by a gas with a superficial velocity, u_d. A pressure drop equal to ΔP_d will result, and the magnitude of the drop will be determined by the fluid rate and the

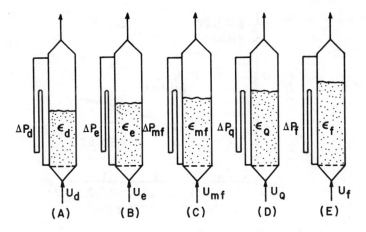

Figure 13.1   The various operating stages of a dense-phase fluidized bed.

characteristics of the bed. As the gas velocity is increased, the pressure drop will rise. There will be a condition for which

$$\Delta P_e = L_e(1 - \epsilon_e)(\rho_s - \rho) \tag{13.1}$$

where L_e = bed height
  ρ_s, ρ = solids and fluid densities, respectively

The voidage terms for each bed in Figure 13.1 are as follows:

- ϵ_d = voidage of packed bed
- ϵ_e = voidage in expansion
- ϵ_{mf} = voidage at minimum fluidization
- ϵ_Q = voidage at quiescent condition
- ϵ_f = voidage of fluidized state

The order of magnitude over which the gas velocity spans in transforming the bed from incipient fluidization to the dilute phase is shown in Figure 13.2. Note that for d_p = 0.002 in., for instance, the velocity range is nearly one hundredfold. As the particle diameter increases, the range narrows considerably. However, even for d_p = 0.012 in. it is still about fortyfold. On the average, the ratio of the terminal velocity, u_t, to the minimum fluidization velocity is about 70.

Two main types of fluidization have been observed experimentally. In cases in which the fluid and solid densities are not too different with small particles and the gas velocity is low, the bed fluidizes evenly, with each

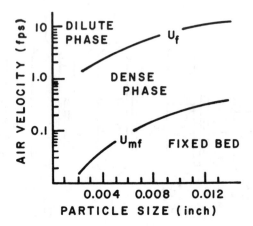

Figure 13.2   Typical fluid velocities and velocity ranges encountered in the fluidization spectrum.

particle moving individually through a relatively uniform mean free path. The dense phase has many of the characteristics of a liquid. This is called *particulate fluidization*.

Where the fluid and solid densities are greatly different or the particles are large, the velocity of flow must be relatively high. In this case fluidization is uneven, and the fluid passes through the bed mainly in the form of large bubbles. These bubbles burst at the surface, spraying solid particles above the bed. Here, the bed has many of the characteristics of a liquid, with the fluid phase acting as a gas bubbling through it. This is called *aggregative fluidization*. In aggregative fluidization the gas rises through the bed primarily in the form of bubbles. The value of the Froude number, $u_f^2/d_p g$ (in which u_f is the superficial gas velocity, d_p is the particle size and g is the gravitational constant), is indicative of this type of fluidization. For Fr >1.0, aggregative fluidization is said to prevail. In addition to these classifications, fluidized beds show considerable bypassing and slugging (i.e., intermittent and unstable flow of large gas bubbles through the bed).

Fluidization may be described as incipient buoyancy because the particles are still so close as to have essentially no mobility, whereas the usual desire in fluidization is to create bed homogeneity. Such homogeneity can be achieved only by violent mixing. This is brought about by increasing the fluid velocity to the point of blowing "bubbles" or voids into the bed, which mix the bed as they rise. The increased fluid velocity at which bubbles form first is referred to as the incipient (or minimum) bubbling velocity.

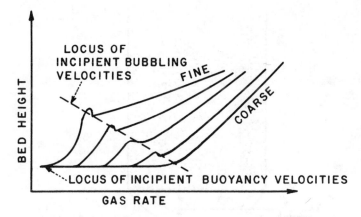

Figure 13.3   Particle size effects on the minimum fluidization and bubbling velocities.

Fine powders can exist over a wide range of bulk densities and, therefore, exhibit substantial differences between incipient buoyancy and incipient bubbling. For coarser granular solids, no distinction can be made between incipient buoyancy and incipient bubbling, as illustrated qualitatively in Figure 13.3. From a practical point, the incipient bubbling velocity is the more significant one in reactor design.

The terms "particulate" and "aggregative" were coined in the 1940s to differentiate between bubbling beds (aggregative) and nonbubbling beds (particulate). In general, liquid-fluidized beds are nonbubbling, whereas gas-fluidized beds bubble. It is presently recognized that bubbling is related to fluid and particle properties in a manner permitting the prediction of a system's maximum attainable bubble size, which, if negligible, leads to the observation of so-called particulate fluidization. Rather than employ the terms aggregative and particulate, it is more correct to refer to the maximum stable bubble size for a particular system.

BUBBLE PHENOMENA IN FLUIDIZATION

Most gas-fluidized commercial reactors operate under conditions of relatively profuse bubbling, as illustrated in Figure 13.4. Bubbles form at the ports where the fluidizing gas enters the bed (distributor plate). They form because the velocity at the interface of the bed, just above the entrance hole, represents a gas input rate in excess of what can pass through the interstices

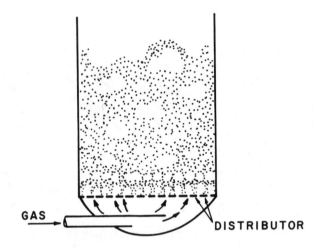

Figure 13.4   A freely bubbling gas-fluidized bed.

at a frictional resistance less than the bed weight. Hence, the layers of solids above the holes are pushed aside until they represent a void through whose porous surface the gas can enter at the incipient fluidization velocity.

If the void grows larger, the interface velocity becomes insufficient to hold back the walls of the void. Hence, the walls cave in from the sides, cutting off the void and presenting a new surface to the incoming gas. This sequence is illustrated in Figure 13.5. The size of the initial bubble resulting from a detached void is typically on the order of about half the penetration depth. Bubbles or gas voids rise in a fluidized bed by being displaced with an inflow of solids from their perimeters.

Minimum Gas Voidage

The voidage required for the onset of fluidization is termed the *minimum gas voidage* and is determined readily by subjecting the bed to a rising gas stream and recording the bed height, L_{mf}, that coincides with incipient particle motion. Then from a knowledge of the buoyant weight of the bed, w, the bed cross section, F, and solids and gas densities, we obtain

$$\epsilon_{mf} = 1 - \frac{w}{L_{mf}F(\rho_s - \rho_g)} \tag{13.2}$$

Values of ϵ_{mf} for a variety of materials are given in Figure 13.6.

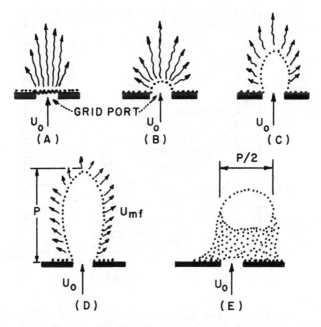

Figure 13.5   Bubble formation from bed penetrating gas jets.

A good indication of fluidization behavior of granular material may be obtained from the pressure drop-flow relationships of the expanded bed. Figure 13.7 displays ideal fluidization characteristics of a material. Branch ϵ_e' to ϵ_e on the plot pertains to the fixed-bed pressure drop of the bed at a voidage ϵ_e. At ϵ_e expansion occurs and continues along the horizontal line. Particle motion sets in when the voidage has slightly surpassed a value ϵ_{mf} (i.e., the minimum fluidization). Fixed-bed conditions are reestablished as the flowrate is lowered again. This occurs at the pressure drop that is exactly equal to the value that may be calculated from the buoyant weight of the bed, designated as $\Delta\rho_w$.

Channeling

Very few gas-solid particles follow the behavior presented in Figure 13.7. The pressure drop-flow relationship in Figure 13.8 pertains to a system that exhibits channeling. This is an abnormality characterized by the establishment of flowpaths in the bed of solids through which large amounts of gas will pass up the column. Branch $\epsilon_e' - \epsilon$ defines the relationship for the fixed bed. The point of initial bed expansion usually will occur at a somewhat

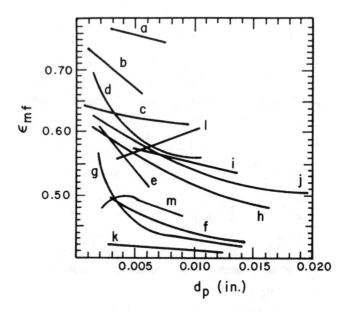

Figure 13.6   Typical values of ϵ_{mf} in relation to d_p: (a) soft brick; (b) absorption carbon; (c) broken Raschig rings; (d) coal and glass powder; (e) carborundum; (f) sand; (g) round sand, ψ = 0.86; (h) sharp sand, ψ = 0.67; (i) Fischer-Tropsch catalyst, ψ = 0.58; (j) anthracite coal, ψ = 0.63; (k) mixed round sand, ψ = 0.86; (l) coke; (m) carborundum. (ψ is the particle shape factor.)

higher pressure drop than that calculated from the weight gradient of the bed. As particles "unlock" from each other, the pressure drop decreases rather suddenly. After the minimum pressure drop at point C has been passed, there is always a pressure drop recovery, but in channeling beds the theoretical value is never quite reached. The amount by which the pressure drop is lower should be indicative of the channeling tendencies of the solids.

There are two very common cases of channeling. In *through channeling* the flowpaths extend through the entire bed, and *intermediate channeling* involves only a portion of the bed. The intermediate channeling bed will yield a higher fluidizing pressure drop with increasing gas rates. Because with channeling the resulting bed density is not homogeneous, there will be local space velocities inside the bed that greatly differ from the overall planned space velocity. This will lead to erratic temperature profiles, as well as inefficiency and larger equipment designs than would be required for a homogeneously operating system.

Shape and density of solids are factors that affect channeling, as well as the fluidization chamber diameter. In general, an increase in channeling occurs

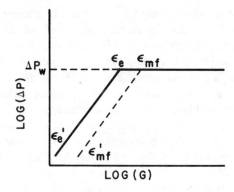

Figure 13.7   Pressure drop-flow diagram for ideally fluidizing solids.

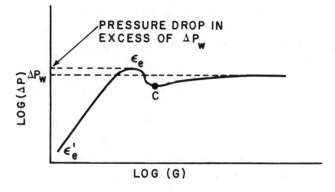

Figure 13.8   Typical pressure drop-flow diagram for moderately channeling solids.

with decreasing particle size. Furthermore, the design of the gas-inlet device has a profound effect on channeling. With porous plates the gas distribution into a bed tends initially toward uniformity. Channeling tendencies are always smaller with multiorifice distributors. In this arrangement the gas is introduced through a relatively small number of geometrically spaced holes.

Slugging is an abnormality frequently encountered in fluidization. It is considerably affected by the choice and design of the equipment. Characteristics of particles also are involved, although to a lesser extent than in channeling. Slugging is described as the condition in which bubbles of gas coalesce to a size approaching the order of magnitude of diameter of the confining vessel. Particles form layers, or slugs, between the large gas pockets and tend to move upward in a piston-like manner. On reaching a certain height, they then disintegrate and fall downward in a raining fashion. Local space velocities may differ widely from the overall velocity. This has an erratic effect on yields and temperature distributions. In addition, slugging may accelerate the rate of mechanical attrition of particles.

The pressure drop-flow diagram of solid particles reflects the slugging behavior. Referring to Figure 13.9, the particles approach ideal behavior up to and considerably beyond the point of the onset of fluidization. Beyond a certain flow range, however, the pressure drop will increase above the value calculated from the weight of the bed (point S). This condition is defined as the onset of slugging.

The evidence is that the pressure drop excess over the theoretical value is due to friction between the solid slugs and the wall of the fluidized bed. This implies that the height to diameter ratio of the bed is important.

The particle size effect on slugging is shown in Figure 13.10. Smooth operation increases very markedly as particle size decreases. There are, unfortunately, numerous correlations in the literature for predicting the onset of slugging, many of which are specific to the type of solids and exact system in which the test is made. One correlation for fluidizing particles is

$$u_s = 6.8 \frac{u_{mf}^{0.6}}{L_e^{0.8}} \tag{13.3}$$

where u_s = gas velocity for incipient slugging (fps)
u_{mf} = gas velocity at minimum fluidization (fps)
L_e = bed height (ft)

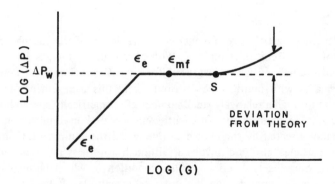

Figure 13.9   Pressure drop-flow diagram for slugging solids.

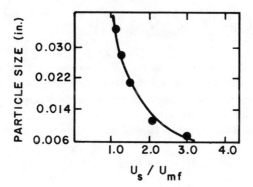

Figure 13.10   Relation between particle size and reduced slugging velocities for coal particles: D_t = 1.64 in.; settled bed depth = 10 in.

Equation 13.3 is based on tests using 72–100 mesh coal with a maximum bed depth of 6 in.

Geldart and Abrahamsen [1978] proposed the following formulations for the minimum fluidization and bubbling velocities based on an examination of several types of solids in various-sized columns:

$$u_{mf} = 0.0008g(\rho_p - \rho)d_{p50}^2/\mu \tag{13.4}$$

$$u_{mb} = 33(\rho/\mu)^{0.1}d_{p50} \tag{13.5}$$

where d_{p50} is the 50 wt% particle size, in m, and all other units are in SI.

PRESSURE DROP

The pressure drop across a fluidized bed is equal to the particle layer weight, G_p, divided by the cross-sectional area, F. The volume of the layer is FH, where H is the height of the layer. If the layer porosity is ϵ, then the volume occupied by the particles in the layer will be FH(1 − ϵ), and taking into account the gas lifting force, the particle's weight is

$$G_p = FH(1 - \epsilon)(\rho_p - \rho)g \tag{13.6}$$

Thus, the pressure drop across a fluidized bed is

$$\Delta P = \frac{G_p}{F} = H(\rho_p - \rho)g(1 - \epsilon) \ , \quad N/m^2 \tag{13.7}$$

The height of the layer and its porosity increase with an increase in the superficial gas velocity. The value $(1 - \epsilon)$ is decreased, but the product $H(1 - \epsilon)$ remains constant because the pressure drop across the bed is independent of the superficial gas velocity.

If the height of a fixed layer is H_0 and its porosity is ϵ_0, then

$$H(1 - \epsilon) = H_0(1 - \epsilon_0)$$

and

$$\epsilon = 1 - \frac{H_0}{H}(1 - \epsilon_0) = 1 - \frac{1 - \epsilon_0}{K} \ , \quad \text{where } K = \frac{H}{H_0} \tag{13.8}$$

K is called the expansion coefficient of a fluidized layer. The fluidization velocity may be found by equating the pressure drop of a fixed bed to that of the fluidized bed:

$$\frac{3}{4} \lambda \frac{H(1 - \epsilon_0)\rho(u_0)^{1/2}}{d_p \psi \epsilon_0^3} = H(\rho_p - \rho)g(1 - \epsilon_0) \tag{13.9}$$

After transformation we obtain

$$\lambda(Re_0')^2 = \frac{4}{3} \psi E_0^3 \frac{gd_\rho^3(\rho_p - \rho)}{\mu^2} = \frac{4}{3} \psi E_0^3 Ar \tag{13.10}$$

where λ = dimensionless coefficient (friction factor)
            Re_0' = modified Reynolds number corresponding to minimum fluidization

where in general, the Reynolds number is defined as

$$Re = \frac{u_0 D_{eq}\rho}{\mu} = \frac{4u_0\rho}{a'\mu} = \frac{4G}{a'\mu}$$

where a' = specific surface area, m^2/m^3
            G = gas mass rate, kg/m^2-s
   $D_{eq} = \frac{4\epsilon_0}{a'}$ = equivalent diameter, m

Dimensionless groups important in the fluidization process are the Galileo number defined as

$$Ga = \frac{Re^2}{Fr} = \frac{g\ell^3\rho^2}{\mu^2} = \frac{g\ell^3}{\nu^2} \tag{13.11}$$

and the Archimedes number:

$$Ar = Ga\frac{\rho_s - \rho}{\rho} = \frac{g\ell^3\rho(\rho_s - \rho)}{\mu^2} = \frac{g\ell^3}{\nu^2}\frac{\rho_s - \rho}{\rho} \tag{13.12}$$

At $Re < 50$;

$$\lambda = \frac{220}{Re}$$

At $Re = 50 - 7200$;

$$\lambda = \frac{11.6}{Re^{0.25}}$$

At $Re > 7200$;

$$\lambda = 1.26$$

Introducing a dimensionless shape factor coefficient ψ', the following conditions are evaluated: At $\psi' Ar < 18,500$,

$$Re'_0 = \frac{0.00404(1 - \epsilon_0)}{\psi}(\psi'Ar) \tag{13.13}$$

At $\psi'Ar = 18,500 - 1.1 \times 10^8$

$$Re'_0 = \frac{0.275(1 - \epsilon_0)}{\psi}(\psi'Ar\rho \times 0.57) \tag{13.14}$$

At $\psi'Ar > 1.1 \times 10^8$,

$$Re'_0 = \frac{1.03(1 - \epsilon_0)}{\psi}(\psi'Ar\rho \times 0.5) \tag{13.15}$$

where

$$\psi' = \frac{\psi^3 \epsilon_0^3}{(1 - \epsilon_0)}$$

(13.16)

For any regime of motion and particle shape, the resistance force may be expressed as

$$R_f = \xi F \rho \frac{u_0^2}{2}$$

(13.17)

where ξ = resistance coefficient
   F = surface area normal to the particle motion (m^2)
   ρ = density of fluid (kg/m^3)
   u_0 = velocity (m/s)

For spherical particles,

$$R_f = \xi \frac{F d_p^2}{4} \cdot \frac{\rho u_0^2}{2}$$

(13.18)

where d_p = particle diameter
   $\xi \dfrac{\pi}{8} = \phi$ = resistance coefficient

Hence,

$$R_f = \phi u_0^2 d_p^2 \rho$$

(13.19A)

$$\phi = \frac{R_f}{u_0^2 d_p^2 \rho}$$

(13.19B)

This leads to the following set of expressions for the resistance coefficient:

Flow Regime	Reynolds Number	Resistance Coefficient	
Laminar (Stokes equation)	Re < 2	$\phi = \dfrac{3\pi}{Re}$	(13.20)
Intermediate (Allen equation)	Re = 2 – 500	$\phi = \dfrac{7.27}{Re^{0.6}}$	(13.21)
Turbulent (Newton equation)	Re > 500	$\phi = 0.173$	(13.22)

The resistance coefficient, ϕ, for nonspherical particles depends on Re and ψ, where

$$\psi = \left(\frac{f_{sph}}{f_p}\right)$$

(13.23)

where f_{sph} = sphere surface area
        f_p = particle surface area

The porosity of the fluidized bed may be determined using the following approximate equation:

$$\epsilon = \left\{\frac{18Re'_0 + 0.36Re'_0}{Ar}\right\}^2$$

(13.24)

Note that $P' - R_f = ma$, where a is particle acceleration and P' is the force exerted on the solids.

When $u_0 = 0$, $R_f = 0$ and $R_f = P'$, $a = 0$, where initially $a = \frac{P'}{m}$. $P' > R_f$ corresponds to the condition of sedimentation. This means that

$$u_0 = \frac{1}{d_p}\left(\frac{\sqrt{P'}}{\phi\rho}\right)$$

(13.25)

Or on rearranging terms:

$$\frac{P'\rho}{\mu^2} = \phi\frac{u_0^2 d_p^2\rho^2}{\mu^2} = \phi Re^2$$

(13.26)

When a particle of diameter d_p is falling under the influence of gravity, force P' is equal to the weight of a particle in gas.

$$P' = \frac{\pi d_p^3}{6}g(\rho_p - \rho)$$

(13.27)

Thus,

$$\phi Re^2 = \frac{\pi d_p^3 g(\rho_p - \rho)\rho}{6\mu^2} = \frac{\pi}{6}Ar$$

(13.28)

At $Re < 2$ or $Ar < 36$,

$$Re = 0.056 \, Ar \qquad (13.29)$$

At $Re = 2 - 500$ or $Ar = 36 - 83 \times 10^3$,

$$Re = 0.152 \, Ar^{0.715} \qquad (13.30)$$

At $Re > 500$ or $Ar > 83 \times 10^3$,

$$Re = 1.74 \, Ar^{0.5} \qquad (13.31)$$

The last three formulas were obtained by substituting the values of ϕ obtained from the Stokes, Allen and Newton equations. The entrainment velocity is equal to the velocity of sedimentation. It may be assumed that the settling velocity of an irregularly shaped particle is 0.75 of that of a spherical particle having the same density.

The void fraction, or density of the fluidized state, may be estimated from the discussions presented earlier. Azbel [1981] proposed another method based on dimensional analysis of the system of gas bubbling through a liquid. From this analysis the height of the fluidized bed is

$$x_i = h(Fr^{1/2} + 1) \qquad (13.32)$$

where h = height of the packed bed
Fr = Froude number, u_0^2/gh

where u_0 is the superficial velocity of fluid and g is the acceleration due to gravity. The average void fraction of a fluidized bed can be estimated from:

$$\epsilon_{avg} = \frac{1}{1 + (1/Fr^{1/2})} \qquad (13.33)$$

The above equations are obtained by making an energy balance of a unit cross section of the differential layer during fluidization. The analysis, developed for gas bubbling through a liquid, is presented in Chapter 14.

CHEMICAL REACTIONS IN FLUIDIZED BEDS

The most important application of fluidization is in the field of chemical reaction engineering. Indeed, within the last four decades fluidization has evolved to become one of the most useful reactor concepts available in industry.

Reaction Applications

The potential for using fluidized beds as chemical reactors is best illustrated by considering the many applications at the commercial and semicommercial levels. A convenient classification for gas-fluidized reactors is to consider them as either gas-phase reactors (which may be subdivided into catalytic and noncatalytic), or gas/solid reactors for gas-phase reactions. Examples of catalytic fluidized-bed reactors are petroleum cracking and hydroforming. Examples of gas/solid fluidized-bed reactors are coal combustion, coal carbonization, coal gasification, incineration (oxidation) of waste solids, uranium processing, nuclear materials processing, calcination of ores, etc.

Reaction Zones

The idealized fluid-bed reactor, a fluidized bed that is not homogeneous, can be considered as having three distinct-reaction zones. At the bottom of the bed is the *grid region*, which contains vertical or horizontal gas jets and/or small formation-sized bubbles; the type of grid will determine the form of the gas voids. Above the grid region is the *bubbling zone*. In this zone bubbles grow by coalescence and may even approach a diameter similar to the vessel width, forming slugs (typical of laboratory and small pilot-plant reactors). As bubbles break the surface of the bed, particles are entrained in the upward-flowing gas stream. Some of the solids are elutriated, while others fall back to the bed. This is referred to as *freeboard region* and it affords an opportunity for lean-phase reactions. Entrainment in the freeboard is dealt with in detail in the latter part of this chapter.

Reactions will occur in the freeboard for most practical situations where moderate to high gas flowrates are employed. The temperature may be higher or lower depending on the heat of the reaction. For example, with the exothermic oxidation reactions occurring in the freeboard of a coal combustor, Gibbs and Hedlay [1978] report gas temperatures 20°C above the bed temperature.

Catalyzed Gas-Phase Reactions in Freeboard

The presence of catalyst particles in the freeboard offers additional opportunity for gas conversion. A simple model of the freeboard region is given by Yates and Rowe [1977]. It is assumed that the particles are well dispersed and traveling upward in the gas stream at a velocity of $u - u_t$ (where u_t is the average terminal velocity of the particles). The model does not account for particles that are entrained but subsequently fall back to the bed surface. This simplification is not unreasonable for moderate to high gas flowrates and for narrowly sized catalyst beds. For a first-order reaction and plug flow of gas, the freeboard model predicts

$$\frac{C_{AF}}{C_{Aout}} = (1 - \text{conversion in freeboard})$$

$$= \exp(-N_{RFB}) \tag{13.34}$$

where N_{RFB} is the number of reaction units in the freeboard,

$$N_{RFB} = \frac{(0.33f_e)k'L_{FB}f_e}{(u - u_t)} \tag{13.35}$$

and

$$k' = \cfrac{1}{\cfrac{d_p}{6k_f(1 - \epsilon_{mf})} + \cfrac{1}{k}}$$

with C_{AF}, C_{Aout} = gas concentration at top of freeboard and at the bed surface
f_e = fraction of wake solids ejected
k = first-order reaction rate constant
k_f = fluid particle mass transfer coefficient
L_{FB} = height of freeboard zone

SPOUTED BEDS

Spouting is a relatively new technique for contacting fluids with coarse granular solids. It is generally used for contacting fluids with solids of sizes and other physical properties that render them difficult to fluidize. Spouted beds are gaining in process industry use. First used for drying materials such as grains and peas, spouted beds have been extended to other uses, such as for the transportation of solids, humidification, for heat and mass transfer operations and catalytic reactions.

Apparatus construction is of more importance than spouting or fluidization of the solid material. Certain granular solids spout rather than fluidize when they come in contact with a fluid. In this respect, particle size plays a very important role. The motion of solids in a spouted bed is entirely different from that in a freely-bubbling fluidized bed. In the case of a spouted bed, there is an upward movement in the central spout in the form of a dilute fluidized phase and a downward movement in the outside annulus. Spouted beds and channeling fluidized beds are similar from the viewpoint of non-homogeneous fluid distribution over the bed cross section. The final qualitative equations necessary to predict solids movement in a spouted bed still have not been developed.

Description

From the description so far available, it appears that a spouted bed is, in a sense, a combination of a dilute fluidized phase and a coexistent moving fixed bed of solids. Figure 13.11 shows a schematic diagram of a spouted bed. The fluid enters the apparatus through an opening in the apex of a conical inlet. The fluid entrance is so abrupt that there is essentially no opportunity for an appreciable lateral distribution over the total apparatus cross section. Hence, a central channel is formed, and in it solids are entrained upward. The solids enter the channel mainly in the cone section.

The solids concentration in the central channel increases with height and, at the same time, the physical outline of the channel becomes increasingly less distinct. However, for all practical purposes the solids to gas ratio in the central channel is of the same order of magnitude as found in a typical dilute-phase fluidized system. At the upper end of the channel the solids spill over radially into an annulus that defines a column of descending solids. In this ring the solids move downward, without eventually varying their relative positions. The prevailing bed voidage and pattern of solids movement are similar to those found and already described in connection with aerated moving-solids beds.

Coexistence of the two phases produces a characteristic solids circulation pattern by which the material is entrained upward in the center and is caused to descend by gravity through the surrounding dense bed contained in the annular space. The fluid distribution over the bed cross section must, of necessity, be highly uniform. This is similar to conditions found in fluidized beds with strong channeling tendencies.

In the beds shown in Figure 13.11, a jet of fluid enters a cylindrical vessel containing the solids along its vertical axis of symmetry. Under the proper conditions the jet penetrates the bed of particles, creating a central spout zone and an annular region surrounding the spout. The gas passes upward through the spout and annulus, while the solids are conveyed up the spout and down the annulus. When the total flow is divided between the spout and annular regions, the bed is called a spout-fluid bed. The spouted and fluidized beds form the asymptotes of the spout-fluid bed.

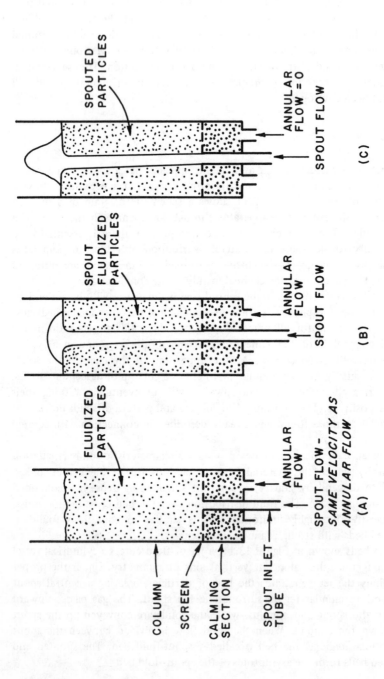

Figure 13.11   Schematic diagram of a spouted fluid bed: (a) fluidized bed—annular flow almost entirely at inlet; (b) spout-fluid bed—both annular and spout flow at inlet; (c) spouted bed—spout flow only at inlet.

Spouted beds commonly are thought of as being useful only for processes involving coarse particles (of the order of 1 mm and up), but smaller particles are spoutable using draft tubes (artificial spout tubes). Hattori and Takeda [1978] developed a side outlet spouted bed with a draft tube capable, for example, of eliminating spout gas bypassing (bypassing or inefficient contacting due to bubbles in fluidized beds cannot be totally eliminated) because all the gas leaving at the top of the spout is redirected down the annulus. Calculated conversions for a first-order solid catalyzed reaction in the side outlet system show that it is superior to any spouted system. Because of their unique characteristics, side outlet-spouted beds have the potential to cut into applications previously reserved for fluidized-bed equipment.

Various models have been developed to describe the flowfield and pressure distribution in spouted beds:

1. Mamuro and Hattori [1968] Model
2. Lefroy and Davidson [1969] Model
3. Yamaguchi et al. [1963] Model
4. Axisymmetric Annular Flow Model

Models 1 and 2 provide accurate descriptions of the distribution and, hence, our discussions are limited to them. For discussion of models 3 and 4, the reader should refer to the work of Azbel [1981].

Mamuro and Hattori Model

This is the most successful model to date for describing the flowfield and pressure distribution in spouted beds. It is based on the force balance analysis on a differential length of the annulus and the assumption that Darcy's law is applicable to the fluid flow in the annulus. A model of the balance of forces acting in the annulus and the boundary conditions there for a bed at its maximum spoutable height are given in Figure 13.12. The force balance based on Figure 13.12 is

$$-dP_b = (1 - \epsilon a')(\rho_p - \rho)gdz - (-dP_a) \qquad (13.36)$$

with $P_b = K'\rho U_r^2/2$ (called the Janssen assumption) $\qquad (13.37A)$

$$-dP_a = K'u_e dz \quad \text{(Darcy's law)} \qquad (13.37B)$$

u_r and u_e are related using the material balance on the element and, solving the resulting differential equation for u_r, we obtain

$$u_r = C_1(z + C_2)^2 \qquad (13.38)$$

where subscript r denotes radial location.

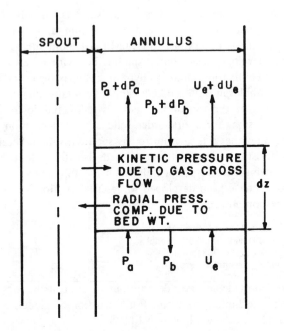

Figure 13.12   Forces acting on the annulus region of solids in a spouted bed.

and

$$u_e = \frac{\pi D_s}{3a'} C_1 [(z + C_2)^3 - C_2^3] \tag{13.39}$$

assuming $u_e(0) = 0$. Note that a' denotes the cross-sectional area of the annulus gas region.

The coefficients C_1 and C_2 are evaluated from the boundary conditions $u_r(H_m) = 0$ and $u_e(H_m) = u_{mf}$, giving the annular velocity as

$$\frac{u_e}{u_{mf}} = 1 - \left(1 - \frac{z}{H_m}\right)^3 \tag{13.40}$$

where H_m is the maximum spoutable bed depth and u_e is the gas velocity in the annulus region of the bed. For beds shallower than H_m, the Mamuro and Hattori [1968] model suggests Equation 13.40 in an arbitrary modified form:

$$\frac{u_e}{u_{eH}} = 1 - \left(1 - \frac{z}{H}\right)^3 \tag{13.41}$$

Equation 13.41 and Darcy's law (Equation 13.37B) give the pressure distribution in the annulus:

$$\frac{P_a}{\Delta P_{ms}} = 1 - \frac{1}{3}\left[6\left(\frac{z}{H}\right)^2 - 4\left(\frac{z}{H}\right)^3 + \left(\frac{z}{H}\right)^4\right] \tag{13.42}$$

and

$$\left.\frac{\Delta P_{ms}}{\Delta P_{mf}}\right|_{H=H_m} = 0.75 \tag{13.43}$$

The boundary conditions were modified by Gabavcic et al. [1976] to apply to beds shorter than H_m. They obtained the following result:

$$\frac{u_e}{u_{eH}} = \frac{1 - \left(1 - \dfrac{z}{H_m}\right)^3}{1 - \left(1 - \dfrac{H}{H_m}\right)^3} \quad ; \quad \frac{u_e}{u_{mf}} = 1 - \left(1 - \frac{z}{H_m}\right)^3 \tag{13.44}$$

Equation 13.44 provides better agreement with experiments than does Equation 13.41, as shown in Figures 13.13 and 13.14. They concluded that $u_e = u_{mf}$ when $H = H_m$. Equation 13.44 was derived under the assumptions that (1) the radial velocity (spout to annulus) is zero only at the top of a bed of height H_m, and (2) the velocity at the top of the spout in the minimum spouting conditions is always equal to the minimum fluidizing velocity for any bed height. (Generally it is assumed that the velocity in the annulus does not change as the inlet fluid flowrate is increased above the minimum required for spouting.) According to Epstein et al. [1978],

$$\left.\frac{-dP_a}{dz}\right|_{\substack{H_m = \text{constant} \\ z = \text{constant}}} \quad \text{or} \quad u_e\big|_{\substack{H_m = \text{constant} \\ z = \text{constant}}} = \text{constant} \pm f(H) \tag{13.45}$$

This means that the annular velocity in a spouted bed at a given distance above the spout inlet is unaffected by changing the bed height (Figure 13.15).

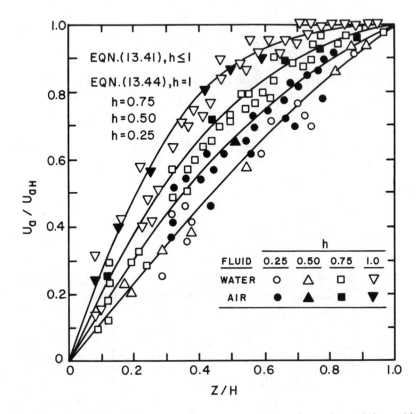

Figure 13.13 Annular velocity distribution—comparison of experimental data with Equations 13.41 and 13.44.

Draft tube systems are designed on the basis of the assumptions obtained from Equation 13.45.

Gabavcic et al.'s [1976] pressure equations are as follows:

$$\frac{P_a}{\Delta P_{mf}} = 1 - \left[\frac{\left(\frac{z}{H}\right) - \left(\frac{0.25}{h}\right)\left[1 - \left(1 - \frac{hz}{H}\right)^4\right]}{1 - \left(\frac{0.25}{h}\right)\left[1 - (1-h)^4\right]} \right] \tag{13.46A}$$

$$\frac{\Delta P_{ms}}{\Delta P_{mf}} = \frac{0.75 - (1-h) + 0.25(1-h)^4}{h} \quad ; \quad h \leqslant 1 \tag{13.46B}$$

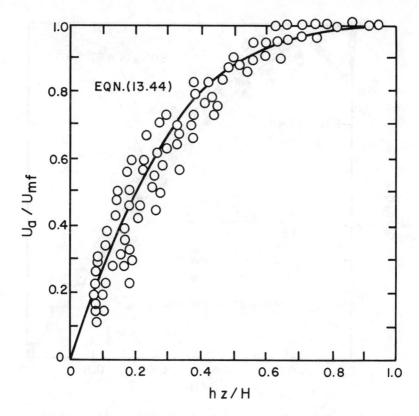

Figure 13.14   Annular velocity distribution—comparison of experimental results with Equation 13.44.

$$\frac{\Delta P_{ms}}{\Delta P_{mf}} = 0.75 \; ; \quad h = 1 \tag{13.46C}$$

$$\frac{\Delta P_{ms}}{\Delta P_{mf}} = 1 - \frac{0.25}{h} \; ; \quad h \geqslant 1 \tag{13.46D}$$

These equations agree well with experimental data, as shown in Figures 13.15 and 13.16. As expected, the poorest fit is near the spout inlet.

Both the Mamuro and Hattori theory and the modification of it by Gabavcic et al. assume unidirectional Darcy flow in the annulus. For the coarse particles generally used in spouted beds, Darcy's law is inapplicable to

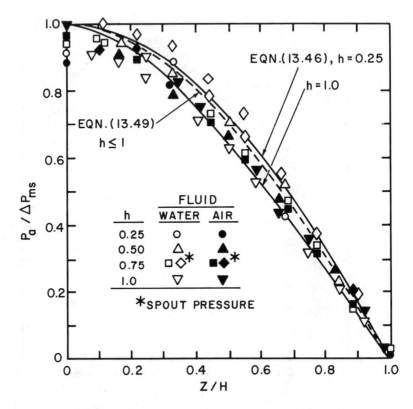

Figure 13.15   Fluid pressure in the annulus as a function of bed level—comparison of experimental data with Equations 13.46A and 13.46D.

a major fraction of the annulus and the nonlinear form of the pressure gradient-velocity relation is clearly applicable.

Lefroy and Davidson Model

Lefroy and Davidson [1969] determined the annular velocity distribution from measurements of the pressure distribution in the spout. According to them,

$$\frac{P_{sa} - P_0}{\Delta P_{ms}} = \cos\left(\frac{\pi}{2}\frac{z}{h}\right) \tag{13.47}$$

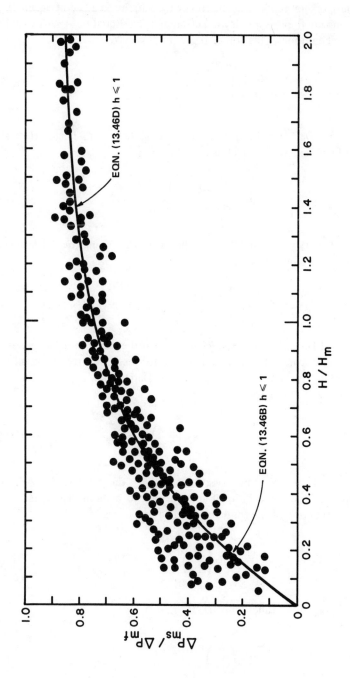

Figure 13.16   Variation of $\Delta P_{ms}/\Delta P_{mf}$ with H/H_m—comparison of experimental data with Equations 13.46B and 13.46D.

and the pressure gradient at the top of the annulus in a bed of height H_m was the fluidization pressure drop. Differentiating Equation 13.46A and then using the condition at the top of the annulus gives

$$\frac{\Delta P_{ms}}{\Delta P_{mf}}\bigg|_{H\,=\,H_m} = \frac{2}{\pi} = 0.637 \qquad (13.48)$$

Lefroy and Davidson [1969] replaced ΔP_{ms} is beds of height $H < H_m$ in terms of a parameter B, which is defined as

$$B = \frac{\Delta P_{ms}/H}{H_m(\Delta P_{ms})_{H\,=\,H_m}} \qquad (13.49)$$

$B = 1$ when $H = H_m$ and $B < 1$ when $H < H_m$. Using the definition of B, Equation 13.46 becomes

$$\frac{P_{sa} - P_0}{\Delta P_{mf}} = \frac{2}{\pi} B \cos\left(\frac{\pi}{2}\frac{z}{H}\right) \qquad (13.50)$$

Lefroy and Davidson [1969] recognized that the flow is not one-dimensional in the annulus near the spout inlet but assumed the pressure was uniform horizontally, for convenience. Assuming Darcy's law is applicable in the annulus, they obtained

$$\frac{u_e}{u_{mf}} = B \sin\left(\frac{\pi}{2}\frac{z}{H}\right) \qquad (13.51)$$

According to McNab and Bridgwater [1977] the ratio $(\Delta P_{ms}/\Delta P_{mf})$ is a function of D_i/D_t (the ratio of inlet orifice to column diameters) and also may be a function of the fluid and particle properties:

$$B = \left(\frac{H}{H_m}\right)^{0.38} \qquad (13.52)$$

If the exponent is a function of D_i/D_t, then

$$B = \left(\frac{H}{H_m}\right)^{0.1(D_i/D_t)^{-0.6}} \qquad (13.53)$$

The expressions for B obtained from the Gabavcic et al. [1976] and Lefroy and Davidson [1969] models are, respectively,

$$B = \frac{1.5\,h - h^2 + 0.25\,h^3}{0.75} \qquad (13.54)$$

and

$$B = \sin\left(\frac{\pi}{2}h\right) \qquad (13.55)$$

Equation 13.52 predicts higher values of B than either Equations 13.54 or 13.55. There is enough scatter in the data to conclude that B and, therefore, the pressure condition at the spout-annular interface are not precisely known. Still the proper correlating variables for B and ΔP_{ms} have not been identified.

Designing Spouted Beds

In designing a spouted bed, one must

1. estimate the maximum spoutable bed depth, which is important because it represents the length scaling parameter;
2. estimate the minimum fluidizing velocity for spouting; and
3. determine the mean spout diameter.

A number of equations have been developed for estimating these parameters. Three have been selected as giving the most reliable results. The first, for estimating maximum spoutable bed depth, was developed by Malek and Lu [1965]:

$$H_m/D_c = 0.105(D_c/d_p)^{0.75}(D_c/D_i)^{0.4}(\psi^2/\rho_s^{1.2}) \qquad (13.56A)$$

The particle shape factor is given as follows:

$$\psi = 0.205(A_\phi/\dot{V}_0^{2/3}) \qquad (13.56B)$$

where A_ϕ is the surface area of a single particle (cm^2) and $\dot{V}_0$ is the volume of a single particle (cm^3). Typical values for ψ range from 1 (for spheres, sand, millet, etc.) to 1.65 for gravel.

The second equation, for estimating the minimum fluidization velocity for spouting, is that of Mathur and Gishler [1955]:

$$u_{ms} = (d_p/D_c)(D_i/D_c)^{1/3}(2gH(\rho_s - \rho)/\rho)^{1/2} \tag{13.57A}$$

For air as the fluidizing gas at room temperature ($\rho = 0.0766$ lb/ft^3), Equation 13.57A becomes:

$$u_{ms} = 2g(d_p)(D_i^{1/3})(D_c^{-4/3})(H^{1/2})(\rho_s - 0.0766)^{1/2} \tag{13.57B}$$

The above formula is valid only at room temperature. However, it can be used at higher temperatures by multiplying the expression by the appropriate correction factor given in Table 13.1.

The mean spout diameter can be computed from the following equation of Mikhailik [1966]:

$$D_s = 14.5(0.115 \log D_c - 0.192)(G/\rho_s)^{1/2} \tag{13.58}$$

where G is the gas mass rate in units of kg/hr·m^2. The above formulas are invalid for the following ranges: 5 cm $< D_c <$ 50 cm; $d_p >$ 0.1 cm. Otherwise, they are accurate to within ±12% of the measured values.

Littman et. al. [1977,1979] also give an equation relating H_m to the mean spout diameter, D_s, for spherical particles:

$$\frac{H_m D_s}{D_c^2 - D_s^2} = 0.345 \left(\frac{D_s}{D_c}\right)^{-0.384} \tag{13.59}$$

Based on experimental values of D_s, Equation 13.59 is shown to provide predictions of H_m to within ±8.5%. H_m is also a function of fluid inlet orifice diameter, column diameter and the properties of the fluid-particle system:

Table 13.1 u_{ms} Factors at Different Temperatures

Air Temperature (°C)	Factor
50	1.06
100	1.14
200	1.28

$$\frac{H_m D_i}{D_c^2} = 0.218 + \frac{0.0005}{A} \; ; \quad A > 0.02 \qquad (13.60)$$

where $A = [\rho/(\rho_p - \rho)] \, [u_{mf} u_t / g D_i]$

where u_t is the particle terminal setting velocity.

For nonspherical particles, Morgan and Littman [1978] presented an equation similar to Equation 13.60:

$$\frac{H_m D_i}{D_c^2} = 0.218 + \frac{0.005}{A_\phi} + \frac{2.5 \times 10^{-5}}{(A_\phi)^2} \; ; \quad A_\phi > 0.014 \qquad (13.61A)$$

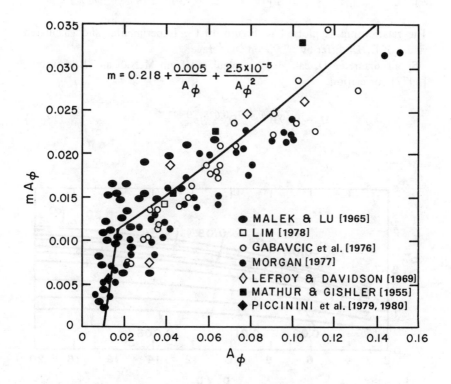

Figure 13.17   Plot of mA_ϕ vs A_ϕ.

or

$$\frac{H_m D_i}{D_c^2} = 175(A_\phi - 0.01) \; ; \quad 0.01 \leqslant A_\phi \leqslant 0.014 \qquad (13.61B)$$

Figure 13.17 correlates literature data for spherical and nonspherical particles in the form of a plot of $\left[\dfrac{H_m D_i}{D_c^2}\right] A_\phi$ versus A_ϕ. Figure 13.17 clearly shows that in the low range of A_ϕ $(0.01 < A_\phi < 0.02)$ a discontinuity in the correlation exists. There are no specific reasons for this but it clearly shows that predictions in this range do not agree with experimental values.

A relationship between the mean spout diameter and the inlet orifice diameter is:

$$\frac{D_s}{D_i} = \left[\frac{2.1e^{-(0.018/A_\phi)} + 1.0}{3.1}\right]\left[0.862 + 0.219\left(\frac{D_c}{D_i} - 0.0053\left(\frac{D_c}{D_i}\right)^2\right)\right] \quad (13.62)$$

The relationship is plotted in Figure 13.18. Experimental and calculated values of D_s/D_i differ by ±7.9% on an average.

Spout diameter D_s can also be calculated from McNab and Bridgwater's [1977] correlation:

$$D_s = 1.993G^{0.489}D_c^{0.678}/\rho_{blk}^{0.411} \qquad (13.63)$$

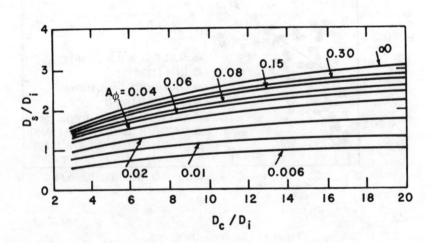

Figure 13.18   Plot of D_s/D_i as a function of D_c/D_i and A_ϕ.

where D_s and D_c are in units of m, G is in $kg/s\text{-}m^2$ and the solids bulk density, ρ_{blk}, is in kg/m^3. Equation 13.63 can be used to calculate D_s if G is taken as the gas mass velocity at minimum fluidization conditions. The correlation gives very good predictions in gas-spouted systems but leads to erroneous conclusions with respect to the effect of D_c/D_i and D_s/D_i; e.g., if there are two similar columns with the same superficial gas velocity but with the fluid inlet orifice diameters in the ratio of 2:1, McNab and Bridgwater's [1977] correlation shows that D_s would be the same for both columns but D_s/D_i would be in the ratio of 1:2. An experimentally unconfirmed correlation which may prove useful in sizing different orifice diameters includes a velocity correction dependency:

$$\frac{D_s}{D_i} = \left[\frac{U}{U_{ms}}\right]^{1/2} \frac{D_s'}{D_i'} \qquad (13.64)$$

where the primes (') denote alternate conditions. Many of the coefficients in the correlations given in this subsection are dimension specific. We therefore summarize the new notation introduced before leaving this topic.

- A_ϕ = surface area of a single particle, cm^2
- D_c = column i.d., cm (in Equation 13.58, mm; in Equation 13.57, ft)
- D_i = fluid inlet orifice diameter, cm (in Equation 13.57, ft)
- D_s = mean spout diameter, mm
- d_p = average particle diameter, cm (in Equation 13.57, ft)
- G = fluid mass flowrate, $kg/m^2\text{-}hr$
- g = acceleration of gravity, 32.2 ft/s^2
- H = depth of the bed, ft
- H_m = maximum spoutable bed depth, cm
- u_{ms} = minimum superficial fluid velocity for spouting, ft/s
- V_0 = volume of a single particle, cm^3
- ρ = fluid density, Mg/m^3 (in Equation 13.57, lb/ft^3)
- ρ_s = particle density, Mg/m^3 (in Equation 13.57, lb/ft^3)

Note that 1 Mg (megagram) equals a metric ton. Mg/m^3 is the SI unit for density, and is numerically equal to g/cm^3. Unfortunately, a large amount of basic fluid mechanical information is required for the design of spouted beds and spouted-bed reactors.

CENTRIFUGAL FLUIDIZATION

Description

The centrifugal fluidized bed (CFB) is a promising new fluidization concept under active study in the United States and abroad. The system has a number of operating advantages over conventional fluidized beds, which make it attractive for a wide range of applications. A rotating bed is mechanically more complex than a conventional one; however, recent work on solids feed and removal, design of seals and heat transfer suggests that the mechanical problems can be solved.

Sufficient work has been done to show that the system fluidizes qualitatively like a conventional bed with a well-defined transition from a packed to a fluidized state. Theoretical equations derived for pressure drop and minimum fluidization in the CFB are in good agreement with available data. Preliminary experimental results also have been reported on the use of CFB for coal combustion. These experiments show that it is possible to achieve a sustained stable coal combustion reaction in a rotating fluidized bed.

A CFB is cylindrical in shape and rotates about its axis of symmetry, as shown in Figure 13.19. As a consequence of the circular motion, the bed material is forced into the annular region at the circumference of the container and fluid flows radially inward through the porous surface of the

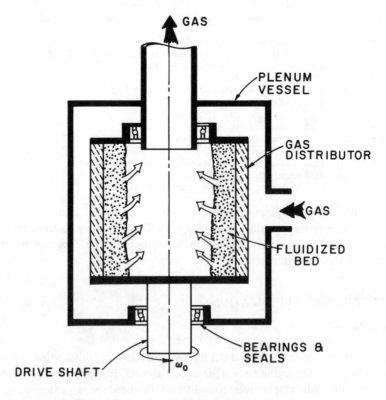

Figure 13.19   Sketch of a centrifugal fluidized bed.

cylindrical distributor, fluidizing the bed material against the centrifugal forces generated by the rotation. Radial accelerations many times in excess of gravity can be generated with modest speeds of rotation, permitting much larger gas flowrates at minimum fluidization than are possible with the conventional fluidized bed operating against the vertical force of gravity.

The advantages of a CFB over a conventional fluidized bed are as follows:

1. The CFB has much higher gas flowrates per unit volume than can be obtained with conventional systems. This results in a smaller distributor surface area and a smaller mass of bed material, reducing the need for a large number of bed modules to achieve large system capacity.
2. In applications in which solids need to be fed to the bed continuously, the smaller distributor surface area, cylindrical geometry and rotation of the CFB should lead to fewer solids feedpoints per unit of capacity than are needed in a conventional bed.
3. The CFB is generally more flexible to operate with a faster response time than can be obtained in a conventional fluidized bed. The gas flowrate to the CFB can be varied over an extremely wide range by adjusting the speed of rotation of the bed and the fluidizing velocity.
4. The rotational operation of a CFB leads to a vortex motion in the freeboard which tends to inhibit particle loss by elutriation.

Because of the relatively compact nature of the CFB and the operating flexibility provided by the rotational motion, the CFB has been proposed for a variety of applications including coal combustion, flue gas desulfurization, gas combustion, coal liquefaction and food drying.

Experiments were performed at Brookhaven National Laboratory (BNL) with a bed 0.25 m in diameter and 0.25 m high. The vortex motion was generated by rotating the distributor at velocities to 2000 rpm, producing radial accelerations in the range from 140 to 3450 g's. Results on minimum fluidization and bed pressure drop were reported. High-speed photographs indicate the CFB behaves qualitatively like a conventional fluid bed, expanding in thickness as the bed is fluidized and as bubbles of gas start to form within the bed. Investigators at Brookhaven also performed analyses on the use of centrifugal fluidized beds for coal liquefaction.

Work is underway on CFB combustion at two universities in Great Britain. Metcalfe and Howard [1977] performed fluidization and combustion experiments with anthracite coal, in which the coal was fed to the bed from a nozzle in the freeboard region, with a bed 200 mm diameter and 67 mm high. In a separate study, Demircan et al. [1978] performed propane, oil and coal combustion experiments in a CFB. This unit had a 200 mm i.d. and a variable axial length of up to 200 mm. The bed was cooled by twelve 25.4-mm-diameter water-cooled tubes immersed in the bed, each extending vertically along the length of the combustor. Stable combustion was achieved with all three fuels, with a combustion efficiency of over 90% with coal. Investigators at both British universities have constructed equipment for experiments on CFB coal combustion at elevated pressures.

These recent experiments demonstrate that stable coal combustion in a

CFB is feasible, where the coal is fed continuously to the bed surface from nozzles in the freeboard region. Much more work is needed to understand the mechanics of the various feed and removal concepts and to relate them to bed bubbling.

Heat Transfer

Studies at Sheffield University demonstrated the ability to extract heat from a CFB with vertical tubes, which rotate with the bed (see Gibbs et al. [1978] for details). However, as is the case with much of the CFB work that has been done to date, the information obtained in the study is only qualitative. Much information on tube to bed heat transfer already exists for 1-g beds, but because of influences of rotation on bubble behavior, heat transfer correlations for standard fluidized beds may not apply to the CFB. Basic work is needed to determine the influence of rotation on heat transfer to surfaces immersed in the CFB. The heat transfer coefficient at the surface of a particle in a CFB is much larger than occurs in a conventional fluidized bed. Additional work is needed to gather qualitative information on this phenomenon.

Chemical Kinetics

Centrifugal fluidized beds generally tend to operate with higher gas velocities and thinner beds than comparable 1-g systems. The effect of the relatively short gas residence time on reaction kinetics needs to be determined. For example, in the coal combustion area, studies of conventional fluidized-bed combustors have shown that if the gas residence time is too low, poor sulfur dioxide capture results. Research is needed to determine the ability of the CFB to achieve adequate desulfurization.

The CFB is a relatively new approach to fluidization, and has been studied only superficially. Until a better understanding of the behavior of centrifugal fluidized beds is achieved, it is difficult to predict for which applications this concept will prove useful in a commercial sense.

THREE-PHASE FLUIDIZATION

Classification

The term "three-phase" fluidization requires some explanation, as it can be used to describe a variety of rather different operations. The three phases are gas, liquid and particulate solids, although recently investigations have been performed using two immiscible liquids and particulate solids. As in the case of a fixed-bed operation, both cocurrent and countercurrent gas-liquid flow are permissible and, for each of these, both bubble flow, in which the liquid is the continuous phase and the gas dispersed, and trickle flow, in which the gas forms a continuous phase and the liquid is more or less dispersed, takes place.

A well established device for countercurrent trickle flow, in which low-density solid spheres are fluidized by an upward current of gas and irrigated by a downward flow of liquid, is variously known as the "turbulent bed," "mobile bed" and "fluidized packing" contactor, or the "turbulent contact absorber" when it is specifically used for gas absorption and/or dust removal. A more recent variation is the three-phase spouted bed contactor.

Cocurrent three-phase fluidization is commonly referred to as "gas-liquid fluidization." Cocurrent gas-continuous (i.e., upward trickle) flow has been observed by Mukherjee et al. [1974] but is otherwise almost totally unexplored.

Bubble flow, whether cocurrent or countercurrent, is conveniently subdivided into two modes: mainly liquid-supported solids, in which the liquid exceeds the minimum liquid-fluidization velocity, and bubble-supported solids, in which the liquid is below its minimum fluidization velocity or even stationary and serves mainly to transmit to the solids the momentum and potential energy of the gas bubbles, thus suspending the solids.

Countercurrent bubble flow with liquid-supported solids, which can be affected by downward liquid fluidization of particles having a density lower than that of the liquid, has been referred to as "inverse three-phase fluidization." The mass transfer potential of such a countercurrent operation is worthy of study, especially for cases in which dispersion of the gas rather than the liquid is called for and the required gas-liquid ratio and throughputs can be effected without flooding. In contrast, the corresponding cocurrent mode has received more attention than all other cases and constitutes the majority of the literature on three-phase fluidization.

Pressure Drop

For gas-liquid fluidization, the total axial pressure gradient at any bed level is simply the bed weight per unit volume at that level:

$$-\frac{dP}{dz} = (\epsilon_s \rho_s + \epsilon_\ell \rho_\ell + \epsilon_g \rho_g)g \qquad (13.65)$$

where the individual phase holdups are interrelated as

$$\epsilon_s + \epsilon_\ell + \epsilon_g = 1 \qquad (13.66)$$

Subscripts s, ℓ, g refer to solid, liquid and gas, respectively.

The total pressure drop across a bed of height H is then given by

$$-\Delta P = g \int_0^H (\epsilon_s \rho_s + \epsilon_\ell \rho_\ell + \epsilon_g \rho_g) dz \qquad (13.67)$$

in which $\epsilon_g \rho_g$ usually can be neglected relative to the other terms to which it is added. When the liquid is the continuous phase, the dynamic pressure gradient as measured by differential manometry is the total pressure gradient corrected for the hydrostatic head of liquid:

$$-\frac{dP}{dz} = \left[-\frac{dP}{dz}\right] - \rho_\ell g \qquad (13.68)$$

Substituting Equations 13.65 and 13.66 into Equation 13.67

$$-\frac{dP}{dz} = [\epsilon_s(\rho_s - \rho_\ell) - \epsilon_g(\rho_\ell - \rho_g)] g \qquad (13.69)$$

when $\epsilon_g = 0$, $\epsilon_s = 1 - \epsilon_1$ and Equation 13.68 reduces to the familiar relationship for liquid fluidization:

$$-\frac{dP}{dz} = \epsilon_s(\rho_s - \rho_\ell)g = (1 - \epsilon_\ell)(\rho_s - \rho_\ell)g \qquad (13.70)$$

The frictional pressure gradient is the total pressure gradient corrected for the hydrostatic head of two-phase fluid:

$$-\frac{dP_f}{dz} = \left[-\frac{dP}{dz}\right] - \rho_f g \qquad (13.71)$$

where ρ_f is the composite fluid density given by

$$\rho_f = \frac{\epsilon_\ell \rho_\ell + \epsilon_g \rho_g}{\epsilon_\ell + \epsilon_g} = \frac{\epsilon_\ell \rho_\ell + \epsilon_g \rho_g}{1 - \epsilon_s} \qquad (13.72)$$

Substituting for ρ_f into Equation 13.70

$$-\frac{dP_f}{dz} = \left[-\frac{dP}{dz}\right] - \frac{\epsilon_\varrho\rho_\varrho + \epsilon_g\rho_g}{1-\epsilon_s}g \tag{13.73}$$

The integral form of this expression, neglecting $\epsilon_g\rho_g$ relative to $\epsilon_\varrho\rho_\varrho$, can be used for expressing pressure drop. Substituting Equations 13.65 and 13.71 into Equation 13.72,

$$-\frac{dP_f}{dz} = \epsilon_s(\rho_s - \rho_f)g = (\rho_s - \rho_f)g \tag{13.74}$$

which has the same form as Equation 13.69, to which it reduces identically in the absence of gas.

Equations 13.65, 13.68 and 13.73 assume that the buoyed weight of the solid particles is supported by the upward fluid drag on these particles, that the flow of gas causes negligible additional losses by friction and that wall friction also may be neglected. In addition, radial pressure gradients are commonly assumed to be negligible relative to axial gradients, so that the pressure drop experienced by each fluid phase is essentially the same.

The frictional pressure gradient at minimum fluidization is given by Equation 13.74, with $\epsilon = \epsilon_{mf}$. The upward liquid superficial velocity required to initiate fluidization in the presence of an upward gas flow is lower than in its absence.

Solids Holdups and Bed Expansion

The solids holdup, ϵ_s, is defined as the volume fraction of the fluidized bed occupied by particulate solids and therefore is given at any bed level by

$$\epsilon_s = \frac{dm/dz}{\rho_s A_b} \tag{13.75A}$$

and for the bed as a whole by

$$\epsilon_s = \frac{M}{\rho_s A_b H} \tag{13.75B}$$

assuming a univalued solids density, ρ_s, and a column of constant cross-sectional area, A_b. Where the upper bed surface is distinctly defined, as occurs when relatively coarse and/or dense solids are used, then H can be measured by direct visual observation through a transparent column. For finer and/or lighter solids (e.g., glass beads smaller than 1 mm and fluidized by air and water), the upper bed surface becomes increasingly diffuse at higher fluid flowrates, primarily due to particle entrainment, but also due to stratification of solids by size where a significant particle size variation exists. A reproducible value of H then can be obtained as the intersection of two straight lines, one of positive slope representing the pressure drop profile in the homogeneous portion of the three-phase bed, and the other of negative slope representing the pressure drop profile in the solids-free two-phase region above the bed.

A procedure for measuring ϵ_s locally, thus avoiding the use of Equation 13.75B and the necessity for estimating H, is presented by Begovich and Watson [1978]. The method depends on a local measuring of ϵ_ϱ by an electroconductivity technique and of local pressure gradient via a pressure profile. Equations 13.65 and 13.66 are then solved simultaneously for ϵ_g and ϵ_s. For a known bed solids mass, M, a check on the local values of ϵ_s can be obtained by integrating Equation 13.75A as follows:

$$\int_0^\infty \epsilon_s dz = \frac{1}{\rho_s A_b} \int_0^M dm = \frac{M}{\rho_s A_b} \tag{13.75C}$$

Under certain circumstances, the introduction of gas to a liquid-fluidized bed or the increase of gas velocity to a gas-liquid fluidized bed results in a contraction of the bed. This counterintuitive result has intrigued many investigators and undoubtedly has prompted more studies on bed expansion or solids holdup than on any other aspect of three-phase fluidization. The explanation lies in the fact that some of the liquid that otherwise would give support to the solid particles is diverted to the solids-deficient wakes behind the gas bubbles. (Two models of such wakes have been proposed by Darton and Harrison [1974,1975,1976].) Where the wake flux is large relative to the remaining liquid flux, as in the case of small and/or light particles in viscous liquids, the resulting contraction effect usually overrides the expansion caused by the presence of the gas bubbles. Where the bubble wake flux is relatively small, as for large and/or heavy particles in nonviscous liquids, the expansion effect tends to predominate. For glass beads in water as the continuous phase, the transition particle size is about 3 mm when the dispersed fluid is a gas and somewhat smaller when the dispersed fluid is either kerosene or toluene. Other three-phase systems show different transition sizes.

The behavior for wettable solids can be rationalized most consistently by the "generalized wake model" of Bhatia et al. [1972] and Bhatia and Epstein [1976], which can be reduced to:

$$1 - \epsilon_s = \left[\frac{u_\ell - u_g k''(1 - x)}{v_1(1 - \epsilon_g - k''\epsilon_g)}\right]^{1/n} [1 - \epsilon_g(1 + k'' - k''x)] + \epsilon_g(1 + k'' - k''x)$$

(13.76)

where $k'' = \epsilon_\ell/\epsilon_g$

$x = \epsilon_{s\ell}/\epsilon_{sf}$

u_ℓ and u_g are the liquid and gas superficial velocities, respectively.

Many investigators, after measuring ϵ_s and ϵ_g, have calculated k'' from Equation 13.76 on the assumption that $x = 0$, for which the expression reduces to an equation derived by Darton and Harrison [1975]:

$$\epsilon_i = \left[\frac{u_\ell - k''u_g}{v_i}\right]^{1/n} (1 - \epsilon_g - k''\epsilon_g)^{1-1/n} + k''\epsilon_g$$

(13.77)

To avoid this arbitrary assumption, El-Temtamy and Epstein [1978] computed k'' by treating each bubble wake as the sphere-completing volume of a spherical-cap bubble in a viscous medium with due allowance being made for the hydrodynamic interaction between bubbles. The partition coefficient, x, was then calculated from Equation 13.76. By applying this procedure to a large amount of experimental data on wettable solids, the following empirical, but rational, correlation for x was generated:

$$x = 1 - 0.877 \frac{v_1}{u_{g\ell}} , \quad 0 \leqslant u_{g\ell} \leqslant 1.14$$

(13.78A)

and when $x = 0$,

$$v_1/u_{g\ell} > 1.14$$

(13.78B)

where v_1 = Richardson-Zaki intercept for liquid fluidization = $(u_\ell/\epsilon^n)u_g$

$u_{g\ell}$ = velocity of gas relative to liquid = $u_g/\epsilon_\ell - u_\ell/\epsilon_\ell$

Equations 13.76 and 13.78 have been used to predict successfully the initial expansion or contraction characteristics not only of relatively large and/or heavy particles (e.g., water-fluidized glass beads 1 mm or larger), for which $x \cong 0$, but also of finer and/or lighter particles for which $0 < x < 1$. Equation 13.76 can be used to predict solids holdup for three-phase fluidization. A simpler approach to the prediction of solids holdup is through the use of empirical equations for ϵ_s or $\epsilon(= 1 - \epsilon_s)$ as a function of fluid fluxes and particle and fluid properties. A more fruitful approach is the use of Equation 13.76, with further refinements on the methods for predicting k'' and x, as well as the gas holdup, ϵ_g.

Gas and Liquid Holdups

If the solids holdup is known, e.g., from Equation 13.75B, then a pressure gradient measurement allows simultaneous solution of Equations 13.65 and 13.66 for ϵ_g and ϵ_ϱ. A point bed density measurement can be used instead of the pressure gradient. One simple, but elegant method for arriving at ϵ_g employed by El-Temtamy [1974] involves measuring the pressure gradient for the three-phase, solid-liquid bed at the same level H (and, therefore, of ϵ_s) using differential manometry. The three-phase pressure gradient, given by Equation 13.69, is subtracted from the two-phase gradient, given by Equation 13.70, and the result is $\epsilon_g(\rho_\varrho - \rho_g)$, from which ϵ_g can be determined. The liquid holdup is, of course, $1 - \epsilon_g - \epsilon_s$.

A more direct method of measuring ϵ_g in the test section is by simultaneously shutting two quick-closing valves and measuring the fraction of the isolated volume occupied by the gas. A precaution must be taken not to include the two-phase gas-liquid volume above the three-phase bed in the isolated section. Other methods of measuring ϵ_g that do not require knowledge of ϵ_s, are the use of electroresistivity (or impedance) probes, which yield local measurements of ϵ_g. Alternatively, measurements of ϵ_ϱ may be effected by the electroconductivity technique. The electroconductivity and electroresistivity methods offer the greatest promise.

The most interesting generalization on gas holdup in three-phase liquid-supported beds is that ϵ_g for relatively small particles is lower than ϵ_g for the corresponding (same u_ϱ and u_g) solids-free system, while the reverse is the case for relatively large particles. Characteristics of three-phase and two-phase (gas-liquid) systems are alternated as the liquid flux is increased, i.e., as the bed is expanded, it becomes more dilute in solids. These phenomena are directly related to bubble characteristics.

Bubble Characteristics

The most striking effects of introducing fluidized solids to an upward-flowing two-phase gas-liquid system are the enhanced bubble growth, which occurs in beds of fine particles, and the increased bubble splitting, which takes place in beds of coarse particles, especially at low bed expansions. The bubble growth effect has been attributed to the fact that when the solid particles are much smaller than the bubbles, the latter experience the liquid-fluidized bed as a pseudohomogeneous medium of density and viscosity greater than the density and viscosity of the pure liquid. The bubble disintegration phenomenon has provoked more controversy. Lee et al. [1974] state that when the solid particles are similar in size to that of the bubbles, the latter will break up if the particles have sufficient inertia to penetrate the roof of a bubble. The numerical criterion for bubble breaking is that the Weber number, $We = \rho_s V_b^2 d_p / \sigma$, exceeds about 3. However, different theories have been presented by different authors.

The similarity between the value of the particle diameter for transition from a bed-contracting to a bed-expanding system on introducing gas, to that for transition from a bubble-coalescing to a bubble-disintegrating system (about 3 mm for air-water-sand) is tantalizing. A mechanistic linkage between the two phenomena still remains to be established.

Heat and Mass Transfer

The difference between bubble-coalescing and bubble-disintegrating systems is given by the experimental data of Ostergaard and Suchozebrski [1971] on volumetric liquid-phase mass transfer coefficients, $k_1 a'$, for gas liquid mass transfer. These data, especially for the larger particles, are in agreement with those of many subsequent studies, most recently with those of Dhanuka and Stepanek [1978,1980]. According to Nishikawa et al. [1977] $k_1 a'$ for cocurrent gas-liquid spouting of solids is considerably larger than for cocurrent gas-liquid fluidization of solids at comparable conditions of operation.

The volumetric mass transfer coefficient for the dispersed fluid phase has been evaluated along with that of the continuous phase only in the recent liquid-liquid-solid study of Roszak and Gawronski [1979]. Solid-liquid mass transfer in three-phase fluidized beds, unlike gas-liquid transfer, has barely been investigated.

For a wall-to-bed heat transfer at low bed expansions, the heat transfer coefficient for a liquid-fluidized bed increases significantly on the introduction of gas bubbles, presumably due to the stirring effect of the latter. As the gas rate increases, the heat transfer coefficient eventually reaches a maximum value.

There are no reported studies of gas-liquid or particle-liquid heat transfer in three-phase fluidized beds, although the mass-heat transfer analogy can be used to generate credible values of volumetric heat transfer coefficients where the corresponding mass transfer coefficients have been determined, as in the gas-liquid case.

ENTRAINMENT

General Description

After a fluidized bed has been set up, if the fluid velocity is increased further, the bed expands and the solids tend to mix readily. This condition leads to a *turbulent fluidized bed*. If the fluid velocity is increased still further, the bed expands considerably and a condition of large amounts of solids dilution is reached. The solids then are entrained with the fluid. Thus, a condition of an entrained bed is achieved. In general, entrainment refers to the removal of solids from the bed by fluidizing gas in both single component and multicomponent systems.

The various factors, such as the rate of entrainment of solids from a bed, size distribution of the entrained solids, size distribution of solids in the bed, etc., are important for design.

The understanding of entrainment today is similar to that of dense bed fluidization a few decades ago. Commercial designs have evolved on the basis of data for specific systems at specific operating conditions. The data available to date are adequate for reasonably good engineering designs; however, at present several variables associated with entrainment design equations have not been studied to the fullest extent.

Generally, entrainment studies are conducted at low gas velocities with large average particle sizes and no fines present. The onset of entrainment indicates an undesired boundary condition to the scope of the experiment. In industry, gas velocities are usually high to keep the fluid bed diameter at a minimum. In most aspects of fluidization, commercial experience with entrainment is far ahead of theoretical understanding.

Entrainment generally is expressed as pounds of dust per cubic foot of gas leaving the system. Entrainment is dependent on the following variables:

- superficial gas velocity to the 4th to 6th power
- particle density to the −4th power
- particle size to the −2nd power
- gas density to the 1st power
- a factor of 10 in raising disengaging height from 5 ft to 25 ft, above which height negligible change may occur.

One difficulty encountered with entrainment data is the poor reproducibility. A range of ±25% is very common in a series of replica experiments in the same apparatus with identical charges of particles. Scatter is slightly less at high gas velocities and may be greater at low entrainment rates or low disengaging heights. Commercial entrainment data do not provide a good basis for general correlations, although they are of limited value in validating correlations specific to the particular equipment. Most of the entrainment studies have not included disengaging height, and it is believed that various literature reported data that were calculated for conditions above the transport disengaging height (TDH), at which point entrainment becomes constant with increasing height. Column diameter greatly affects entrainment. Gas velocity and disengaging height also have significant effects on entrainment changes with bed diameter. At bed diameters greater than a few feet, any wall effect is probably small, and entrainment above TDH probably does not change much with diameter. Entrainment experiments generally are conducted in large equipment. Geometric factors are of secondary importance compared to variation of effects of operating and physical properties. The particle properties important to entrainment are average size, size distribution, shape factor and density.

Particles of narrow size ranges, up to 1800 μm average diameter, have been studied by various investigators. Entrainment of very large particles at high velocities is becoming of interest in fluid bed combustion. Entrainment of single-sized particles is important from the viewpoint of generating various correlations and to theoretical understanding.

It has been observed that the entrainment rate of a narrow size fraction high above a bed is proportional to the weight percent of that fraction in the bed. It also has been found that the entrainment rate of any size fraction decreases with height until TDH is reached. The fine component entrainment rate decreases with height in proportion to its concentration in the bed, even though there are essentially no coarse particles in the dilute phase. Single-particle terminal velocity, u_t, often assumes a prominent role in the correlations. For uniform suspensions, particle slip velocity becomes progressively less than the terminal velocity as the void fraction decreases. It is also possible that particles can cluster in dilute suspensions and achieve velocities higher than a single-particle terminal velocity. According to definition, however, entrainment occurs at voidages less than 1. Further, with the possibility of cluster formation, it is clear that particle slip velocity could be expected to depart from u_t. With particles of mixed sizes, the situation becomes more hazy because different-sized particles will move at different velocities and exert drag effects on each other.

Similarity Analysis

A total analytical treatment of the phenomenon of entrainment is very difficult. According to Azbel [1981] it is convenient to collect and interpret data making use of dimensionless groups obtained from the method of similarity analysis. Assuming that the gas phase is turbulent and considering the gas to be incompressible, and neglecting gravitational force, the equivalent equation of motion is

$$\frac{\partial \bar{u}_i}{\partial t} + \bar{u}_k \frac{\partial \bar{u}_i}{\partial x_k} = -\frac{1}{\rho_g} \frac{\partial \bar{P}_g}{\partial x_i} - \frac{\overline{\partial u_i u_k'}}{\partial x_k} \tag{13.79}$$

where $\bar{u}_i$ is the time-averaged gas velocity vector, and u_i is the fluctuating gas velocity vector (where the gas velocity vector is $u_i = \bar{u}_i + u_i'$). The term $\bar{P}_g$ is the time-averaged gas pressure. The continuity equation for the gas is

$$\frac{\partial \bar{u}_i}{\partial x_i} = 0 \qquad (13.80)$$

The dimensionless fractions are

$$\bar{U}_i = \frac{\bar{u}_i}{u_0} \ , \quad U_i' = \frac{u_i}{u_0}$$

$$\bar{P}_g = \frac{\bar{P}_g}{\Delta \beta}$$

$$T = \frac{t}{\tau} \ , \quad X = \frac{x}{h_s} \ , \quad R_{0i} = \frac{R_i}{h_s}$$

where u_0 is the gas superficial velocity, $\Delta \beta$ is a typical flow system pressure difference (e.g., between the walls of a chamber), τ is a characteristic time, and h_s is the distance above the free solid level. Using these in Equations 13.79 and 13.80, we get

$$\left. \begin{array}{c} \dfrac{u_0}{\tau}\left(\dfrac{\partial \bar{U}_i}{\partial T}\right) + \dfrac{u_0^2}{h_s}\left(U_{gk}\dfrac{\partial U_i}{\partial x_k}\right) = -\dfrac{\Delta\beta}{\rho_g h_s}\left(\dfrac{\partial \bar{P}_g}{\partial x_i}\right) - \dfrac{u_0^2}{h_s}\left(\dfrac{\overline{\partial U_i' U_R'}}{\partial x_R}\right) \\[4mm] \dfrac{u_0}{h_s}\left(\dfrac{\partial \bar{U}_i}{\partial x_i}\right) = 0 \end{array} \right\} \qquad (13.81)$$

Azbel [1981] presented various nondimensional numbers based on the above equations, such as the Euler number, Froude number, Homochronity number, etc.:

$Eu = \dfrac{\Delta \beta}{\rho_s u_0^2}$ Euler number ratio of pressure to inertia forces

$Fr = \dfrac{u_0^2}{gh_s}$ Froude number ratio of inertia to gravitational forces

$Ho = \dfrac{u_0 \tau}{h_s}$ Homochronity number ratio of spatial to temporal inertia forces

In most applications the Homochronity number is approximately unity. Thus,

$$\tau = \frac{h_s}{u_0}$$

Using these dimensionless numbers, one can determine the total solids entrainment per unit cross sectional area and time.

The Freeboard Region

This is defined as the space between the surface of the bed and the top of the reactor. Primarily, its purpose is to provide disengagement of solids and gas. Sometimes, the freeboard has a larger cross section than the bed zone. This helps in lowering the superficial gas velocity and, thus, fewer particles are elutriated.

When gas bubbles erupt at the surface of a fluidized bed, particles are ejected upwards. How ejection actually occurs is still a matter of dispute. Ejection may be occurring due to the combination of the following:

1. Some of the solids in the wakes of fast-rising bubbles are thrown upward as bubbles burst at the surface (Leva and Wen [1971], George and Grace [1978], Lin [1978]).
2. Solids at the surface become a part of the erupting dome of a bubble and are thrown upward as the dome breaks (Do et al. [1972]).

The first mechanism dominates. Solids that are ejected either return to the bed against the upward gas stream or are carried out of the bed. Larger particles that return to the bed will acquire trajectories depending on their size, initial projection velocity, the background superficial velocity, etc. Actually, erupting gas bubbles can and do splash solids far into the freeboard region above the average surface of the bed. If the gas exit is situated immediately above the bed surface, a large amount of solids become entrained in the gas. The amount of entrainment is considerably smaller when the gas exit is higher and, finally, a level is reached above which entrainment becomes almost constant. The amount of entrainment can range from negligible to large amounts, especially if the solids contain a large quantity of fines or if the gas velocity is very high. This height of exit above the top of the bed where entrainment becomes almost constant is known as the TDH, shown in Figure 13.20. Very little information is available on TDH. For given solids and vessels the entrainment is very sensitive to gas velocity, varying approximately as $u^{2.5}$ to u^4. However, TDH is not sensitive to gas velocity; it increases by about 75% for a doubling in the gas velocity. For a given gas velocity the TDH increases with vessel size, as shown in Figure 13.21.

For a freeboard height less than the TDH, the size distribution of solid in the freeboard changes with position, and entrainment decreases as the freeboard height approaches the TDH. When the gas stream exits above the TDH,

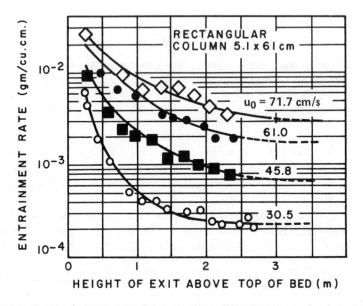

Figure 13.20   Entrainment of solids from different heights above upper surface of dense bubbling fluid bed (Zenz and Othmer [1977]).

both the entrainment rate and the size distribution become constant and are established by the saturation-carrying capacity of the stream of gas under pneumatic transport conditions. This term also can be applied to the entrainment above the TDH for monosize solids. In addition, another phenomenon may occur either below or above TDH, i.e., elutriation, which is the separation or removal of the fines from a mixture. The various terms involved are explained in Figure 13.22.

TDH depends on the properties of fines and gas in addition to the hydrodynamics. Let u_{i0} be the initial velocity of the fine particles with diameter d_{pi}. When there is eruption of bubbles at the surface of the bubbling bed, TDH can be estimated using the following equation (Leva and Wen [1971]):

$$TDH = \frac{\rho_s d_{pi}^2}{18\mu} \{u_{i0} - (u_{ti} - u) \, ln \, [1 + u_{i0}/(u_{ti} - u)]\} \qquad (13.82)$$

An estimation of the entrainment rate in the freeboard is given by Chen and Wen [1979]. For particles of diameter d_{pi}, the entrainment rate is described by the following equation:

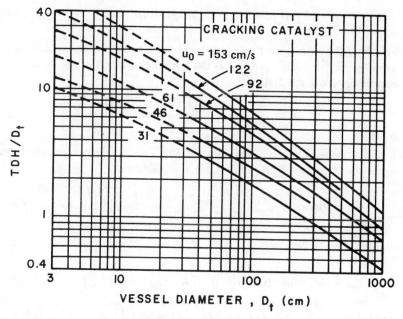

Figure 13.21 Empirical correlation for estimating the transport disengaging height (TDH) (from Zenz and Weil [1958]).

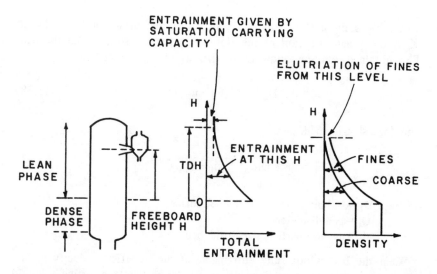

Figure 13.22 Terms employed to account for the removal of solids from a bed.

$$F_i = F_{i\infty} + (F_{i0} - F_{i\infty}) \exp(-a_f h) \qquad (13.83)$$

where F_i is the entrainment rate at a point h cm above the bed surface, a_f is a constant in freeboard, and $F_{i\infty}$ is the rate of elutriation of the fines with diameter d_{pi} above the TDH:

$$F_{i\infty} = E_{i\infty} X_i \qquad (13.84)$$

where $E_{i\infty}$ is the elutriation rate constant and is obtained from the following correlations (Chen and Wen [1979]), for a bubbling bed:

$$E_{i\infty} = 20G \exp\{-32.6(u_t/u_0)^{0.5}[u_{mf}/(u_0 - u_{mf})]^{0.25}\} \qquad (13.85)$$

and, for a slugging bed:

$$E_{i\infty} = 13G \exp\{-10.4(u_t/u_0)^{0.5}[u_{mf}/(u_0 - u_{mf})]^{0.25}\} \qquad (13.86)$$

Constant a_f is independent of the bed composition (Large et al. [1977]) and is obtained from the experimental data on F_i versus h. F_{i0} is the entrainment rate of the particles at the bed surface and is given by

$$F_{i0} = E'_0 X_i \qquad (13.87)$$

where E'_0 is the entrainment rate constant at the surface of the bed, and X_i is the weight fraction of the closely cut particle size, d_{pi}, in the bed. E'_0 is obtained from the following correlations, for a bubbling bed:

$$E'_0 = 0.3 D_B d_p^{-1.17}(u_0 - u_{mf})^{0.66} \qquad (13.88)$$

and, for a slugging bed:

$$E'_0 = 36.8 D_t d_p^{-2.47}(u_0 - u_{mf})^{1.69} \qquad (13.89)$$

where D_B = bubble diameter at the bed surface
 D_t = bed diameter
 d_p = average particle size of the bed

When the bed is not fully entrainable at the superficial velocity in the vessel, i.e., when the superficial velocity is lower than the setting velocity of the coarse particles, coarse particles stay in the bed and in the dilute phase up to the TDH. Thus, they have plenty of time to undergo reactions. The fines, on the other hand, spend very little time in the bed and so do not have much time to react. Their main contribution is probably in maintaining small bubble sizes in the grid region. Thus, in units in which the solids take part in the chemical reaction, e.g., in coal combustion and gasification, fines are difficult to react.

Entrainment at and above the TDH

Single Size Particles

It is known that entrainment from a bed of closely sized solids is not significant until a superficial velocity, u_0, considerably in excess of the terminal velocity, u_t, is reached. Under entrainment conditions, the bed can be considered as a saturation feed device such that the freeboard above the TDH is a pneumatic conveying tube for the transportation of solids. Several results are shown in Figure 13.23. In pneumatic conveying there is a maximum particle concentration that can be held in suspension by the flowing gas without collapse of solids into a dense slugging mass. This limiting

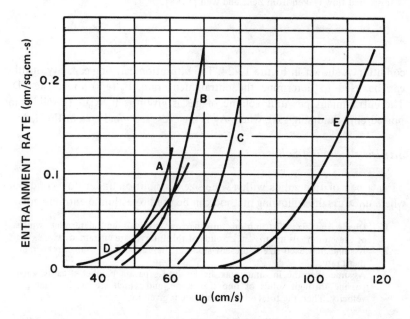

Figure 13.23   Entrainment versus superficial air velocity, u_g, in a bed with a single size solids (by Lewis et al. [1962]): vessel diameter, 5.08 cm; static bed height, 10.2 cm; freeboard, 112 cm. Glass spheres: (A) 0.051 mm; (B) 0.074 mm; (C) 0.094 mm. Cracking catalyst: (D) 0.070 mm; (E) 0.162 mm.

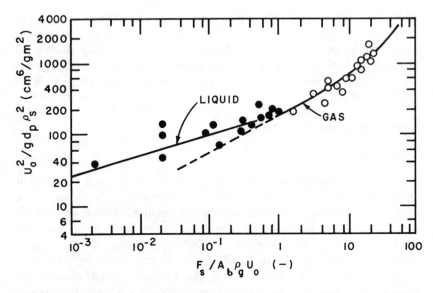

Figure 13.24   Empirical correlation of the saturation carrying capacity with superficial gas velocity, u_g, for uniformly sized particles in either horizontal or vertical cocurrent dispersed flow (taken from Zenz and Weil [1958].

condition is shown in Figure 13.24. The saturation carrying capacity of tubes can be used to determine the entrainment rate, F_s (g/s) above the TDH. The entrainment per unit volume of gas is given as $F_s/A_b u_0$ (g solid transported/cm^3 fluidizing gas), where A_b is the cross sectional area of the bed.

Size Distribution of Solids

For a bed of fine solids with a wide size distribution under flow conditions where $u_0 > u_t$, the following method can be used to estimate entrainment.

1.   Divide the size distribution into narrow intervals and determine which of these intervals has $u_t < u_0$. Solids of this interval are entrained.
2.   Assume that each size is present alone in the bed and determine its entrainment rate.
3.   Assume that the incoming gas divides into separate parallel streams, each passing through solids of one size alone, and which are of the same gas velocity. Then the total entrainment rate is given by

$$F_t = \begin{pmatrix} \Sigma \\ \text{all fraction} \\ \text{of solids where} \\ u_t < u_0 \end{pmatrix} \begin{pmatrix} \text{entrainment rate} \\ \text{for a particular} \\ \text{size} \end{pmatrix} \begin{pmatrix} \text{fraction of that} \\ \text{size in the bed} \end{pmatrix} \text{[g/s]}$$

A more straightforward estimation is as follows: Take the 50% point by weight of all solids where $u_t < u_0$. Estimate the entrainment rate for this size and then multiply this rate by the weight fraction of solids that have $u_t < u_0$. A comparison between experimental and predicted entrainment rates using this method for different pressures is shown in Figure 13.25.

Entrainment below the TDH

Bubbles of gas rise through the dense phase, erupt at the surface and project solids into the freeboard region. It has been determined that the entrainment rate decreases exponentially with the freeboard height. It has been noted by Zenz and Weil [1958] that the intermittent bursting action of bubbles causes sharp velocity fluctuations just above the surface of the dense bed. These fluctuations dissipate with height, and the gas velocity smoothes to the average velocity at the TDH. It is proposed that the effective velocity, u_j, corresponding to the jet velocity be used instead of u_0 to calculate the entrainment for any height of freeboard. Figure 13.26 shows that u_j is a maximum at the surface of the dense bed and decreases to u_0 at the TDH.

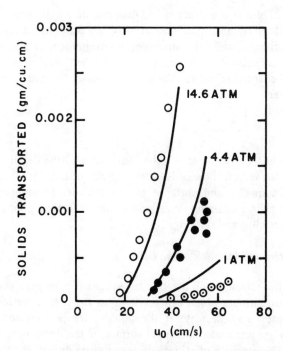

Figure 13.25 Comparison between calculated and experimental entrainment at various pressures (from Zenz and Weil [1958]). Solid lines are predicted.

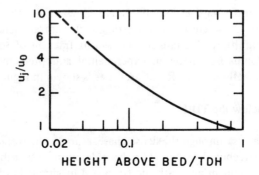

Figure 13.26   Velocity dissipation above a fluidized bed correlated from experimental data (by Zenz and Weil [1958]). Section of a two-dimensional bed: 5.08 cm × 61 cm.

Effect of Physical Parameters on Entrainment

Bed Diameter

Figure 13.27 shows the effect of bed diameter on entrainment at a given u_0 and H. The sharp rise in entrainment at small bed diameters is associated with slugging conditions, while the minimum in entrainment at intermediate bed diameters is associated with channeling. For bed diameters greater than 8–10 cm, the entrainment rate becomes constant and independent of the vessel diameter.

Bed Depth

For fine solids, entrainment is rather insensitive to the depth of the dense bed, except for very shallow beds, where the size of bubbles reaching the surface becomes smaller and entrance effects intrude. Entrainment increases with deeper beds containing dense solids, which fluidize poorly, giving rise to severe channeling and slugging.

Baffles, Stirrers and Internals in the Dense Bed

The effective diameter of the bed is reduced by the presence of baffles and wire obstructions, which can result in an increase in entrainment. Entrainment rates may decrease if the bubble size is decreased. When the obstructions are just under the upper surface of the dense bed, entrainment is reduced. The presence of stirrers decreases entrainment, their effect being more pronounced at high gas velocities.

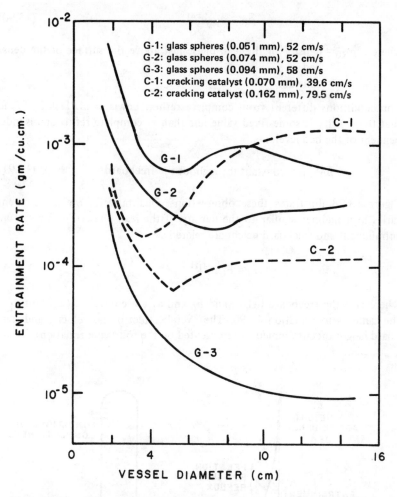

Figure 13.27   Effect of vessel diameter on entrainment rate (Lewis et al. [1962]).

Density of the Dispersed Phase

For a given freeboard, the density of the dispersed phase decreases with height. With the rise in freeboard, density increases. If the freeboard is above the TDH, or high enough so that entrainment is negligible, then the density at any level becomes a maximum, $\bar{\rho}_R$. These conditions are known as complete reflux. At complete reflux conditions, the density at any level ℓ above the dense phase is given as

$$\bar{\rho}_R = \bar{\rho}_{R0}e^{-a_1\ell} \qquad (13.90)$$

where $\bar{\rho}_{R0}$ = density of the lean phase just above the surface of the dense
bed

a_1 = constant

For conditions different from complete reflux, Lewis et al. [1962] found
that the density is some fixed value less than at complete reflux and is independent of the bed level, i.e.,

$$\bar{\rho}_R - \bar{\rho} = \text{constant throughout the freeboard} \qquad (13.91)$$

Figure 13.28 illustrates these observations. Qualitatively, the entrainment
varies in a manner similar to the density of the lean phase. For a given u_0,
entrainment and freeboard height are related as

$$F_t = F_0e^{-a_1H} \qquad (13.92A)$$

where H is the freeboard height and F_0 and a_1 are constants. Constant a_1 is
the same as in Equation 13.90. The overall effect of gas velocity and freeboard height on entrainment is represented by the following equation:

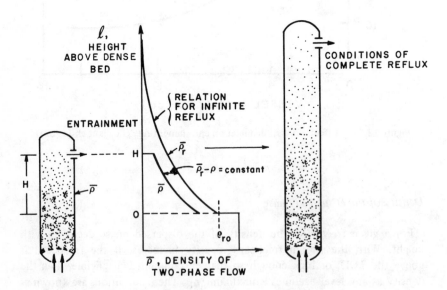

Figure 13.28   Density of lean phase as a function of level in the bed and the freeboard
height.

Table 13.2 Values of B' (g/cm^3) in Equation 13.92B

Column Diameter (cm)	Cracking Catalyst, 0.070 mm	Glass Spheres, 0.075 mm
5.1	–	0.036
7.1	0.031	0.053
14.6	0.020	0.083

$$\frac{F_t}{A_b u_0} = B' \exp\left\{-\left[\left(\frac{b}{u_0}\right)^2 + aH\right]\right\} (g/cm^3) \qquad (13.92B)$$

where $b = 8.86 \times 10^4 \, \rho_s^{1/2} d_p$ (cm/s) and values for parameter B' are given in Table 13.2.

Entrainment Model

Lewis et al. [1962] have presented an entrainment model from a dense fluidized bed, which describes the appearance of the entrainment process. Bursting bubbles of gas eject agglomerates of particles into the space above the bed and, as the air flowrate is increased, this action becomes more evident with agglomerates ejected successively higher into the freeboard. These agglomerates are broken up many times to form a dispersed phase, as well as streams of particles in a random motion. Measurable entrainment occurs only if $u_0 \gg u_{mf}$. At these conditions practically all the gas passes through the bed as large bubbles with velocities considerably in excess of u_t. A rather simple model showing various aspects of entrainment for a dense bubbling bed is illustrated in Figure 13.29. There are three distinct phases in the freeboard above the dense fluidized bed.

- Phase 1: Gas stream containing completely dispersed solids. Solids velocity is u_1.
- Phase 2: Projected agglomerates moving upward with velocity u_2.
- Phase 3: Descending agglomerates and a thick dispersion moving downward with velocity u_3.

At any level in the bed the rate of dissipation of agglomerates to form the dispersed solid of phase 1 is proportional to the concentration of agglomerate solids at that level. Upward-moving agglomerates occasionally reverse direction and move downward. The frequency of this change from phase 2 to phase 3 is proportional to the solids concentration in phase 2. Let F_1, F_2, F_3 (g/s) be the mass flowrate of each phase and C_{s1}, C_{s2}, C_{s3} (g/cm^3) be the weight of each phase per unit volume of freeboard. The net upward flow of solids F is

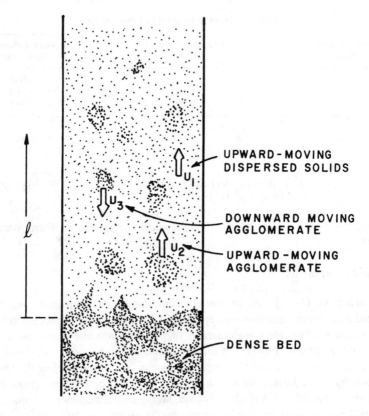

Figure 13.29   Model to account for elutriation and entrainment from fluidized beds.

$$F = F_1 + F_2 - F_3 \tag{13.93}$$

where

$$F_1 = A_b C_{s1} u_1 \, , \quad F_2 = A_b C_{s2} u_2 \, , \quad F_3 = A_b C_{s3} u_3 \tag{13.94}$$

the average solids concentration is

$$\bar{\rho} = C_{s1} + C_{s2} + C_{s3} \tag{13.95}$$

Mass balances for various phases are as follows:

- Phase 1: Increase of solids in phase 1
  = transfer of solids from phases 2 and 3 into 1
- Phase 2: Decrease of solids in phase 2
  = transfer of solids from phase 2 to 1 and 3
- Phase 3: Increase of solids in phase 3
  = transfer of solids from phase 2 to 3 and from phase 3 to 1.

Symbolically, the mass balances are

$$
\left.
\begin{aligned}
u_1 \frac{dC_{S1}}{d\ell} &= \hat{K}(C_{S2} + C_{S3}) \\[2mm]
-u_2 \frac{dC_{S2}}{d\ell} &= (\hat{K} + K^*)C_{S2} \\[2mm]
u_3 \frac{dC_{S3}}{d\ell} &= \hat{K}C_{S3} - K^*C_{S2}
\end{aligned}
\right\}
\tag{13.96}
$$

where $K(s^{-1})$ is the mass transfer coefficient for the transfer of mass from phases 2 and 3 to 1, and $K^*(s^{-1})$ is the mass transfer coefficient for the transfer of mass from phase 2 to phase 3. At the freeboard height, H, the gas stream leaves the vessel and so there is no downward flow. Therefore,

$$F_3 = 0$$

and

$$C_{S3} = 0 \quad \text{at} \quad \ell = H \tag{13.97}$$

We also have

$$C_{S1} = 0$$

$$C_{S2} = \frac{F_0}{A_b u_0} \quad \text{at} \quad \ell = 0 \tag{13.98}$$

where F_0 is the flowrate of solids projected from the bed surface. Solving Equation 13.96 using Equations 13.97 and 13.98, we obtain

$$\frac{1 - \dfrac{F}{F_0}}{1 - \dfrac{F_s}{F_0}} = 1 - e^{-cH} \tag{13.99}$$

where

$$\frac{F_s}{F_0} = \frac{F_K}{1 + F_K} \tag{13.100}$$

$$F_K = \frac{\left(1 + \dfrac{u_2}{u_3}\right)\hat{K}}{K^*} \tag{13.101}$$

$$a_1 = \frac{K^*}{u_2}\left[1 + \left(1 + \frac{u_2}{u_3}\right)\frac{\hat{K}}{K^*}\right] \tag{13.102}$$

F_s is the mass flowrate that corresponds to the saturation capacity of the flowing gas stream.

For normal entrainment, a large quantity of solids projected from the bed returns to the bed. Therefore,

$$\frac{F_s}{F} \ll 1 \tag{13.103}$$

Hence, from Equation 13.100 F_K is approximately zero and Equation 13.99 reduces to

$$\left. \begin{array}{c} F = F_0 C^{-a_1 H} \\[2ex] a_1 = \dfrac{K^*}{u_2} \end{array} \right\} \tag{13.104}$$

If we take u_2 proportional to u_0, then $a_1 \propto u_0^{-1}$. From the above, it can be shown that

$$F_0 = A_b u_0 B' \exp\left[-\left(\frac{b}{u_0}\right)^2\right] \tag{13.105}$$

It has been found that for a new process, even with a perfect correlation of laboratory-determined parameters, entrainment cannot be predicted for large size equipment to a high degree of confidence. Better correlations are needed for scale-up purposes.

Specific examples where better correlations are needed are:

1. Entrainment predictions, where operation without cyclones is desired. Most beds have cyclones, giving a design that is insensitive to entrainment rate. However, fluid bed combustion processes often operate without cyclones, and such designs are sensitive to carryover.
2. Confirmation of effects at very high gas density and viscosity is needed. New processes such as pressurized fluid bed combustion and coal gasification operate at gas properties greatly different from the existing entrainment data bases and correlations.
3. Prediction of dilute phase holdup, which surely must be linked intimately with entrainment is needed. With fast reactions, such as riser cat cracking, a very substantial fraction of catalyst is present in the dilute phase.

For further discussions on fluidization, the reader is referred to the compilation of references at the end of this chapter.

PRACTICE PROBLEMS

13.1   Monosized glass beads are fed to a fluidized bed at a rate of 3.2 kg/min. The bed contains 80 kg of solids. Determine the mean residence time of the solids if carryover is negligible.

13.2   A mixture of solids (60 kg/hr of solids A and 85 kg/hr of B) is continuously fed to a fluidized bed containing 150 kg of solids. The bed is operated at such a high superficial velocity that approximately 10%/hr of solids A is lost as overhead. The remaining solids are removed by means of an overflow pipe. Determine the mean residence times of each solid and of the total mixture.

13.3   Monosized coal particles (300 μm) are fed at a uniform rate to a fluidized bed operating at a constant gas flowrate. Approximately 30% of the solids feed is lost as overhead. To minimize the losses, the column height is doubled. Assuming the bed density to remain unchanged, determine the following:
(a) the solids mean residence time;
(b) the elutriation constant; and
(c) the fraction of feed blown out as carryover.
The gas is nitrogen at ambient conditions.

13.4   Coal with a particle size distribution as shown in Figure 10.36 is charged to a 3-ft-diameter column. The initial bed height prior to fluidization is 7 ft. The coal is fluidized with hot air at 1200°F, in which the gas is introduced through a porous plate distributor. The superficial velocity of the gas through the column is 0.9 fps. Determine
(a) the relative elutriation velocity as a function of solids diameter;
(b) the elutriation rate as a function of particle size;
(c) the average rate of entrainment; and

(d) the expanded bed height and the density of the fluidized bed of solids.

(Hint: As an approximation, assume coal particles do not disintegrate when combusted).

13.5 Batch fluidized bed tests are used for evaluating solids attrition and dilute phase holdup. A quantity of 200 kg of solids is fluidized in column at a specified gas velocity and the overhead fines captured by means of a cyclone-baghouse series arrangement and weighed at various time intervals. From the data recorded below, determine the solids attrition constant (i.e., kg fines generated/kg of solids in the bed/s) and the elutriation constant for the system.

Time (s)	Solids Collected (g)	Time (s)	Solids Collected (g)
0–100	950	400–450	105
100–170	730	450–500	60
170–250	610	500–550	20
250–320	400	550–600	20
320–400	315	600–650	40

13.6 The open-cycle fluidized system shown in Figure 13.30 is being used to dry polyvinyl chloride (PVC) pellets. The pellets are uniform in size with a diameter of 280 μm. The fluidized bed is approximately 8 ft in diameter, and a 10-ft level (dumped solids) is maintained at all times. The feed solids are metered into the fluid bed at a rate of 10,000 lb/hr (dry weight basis) and contain 15% moisture by weight. The average temperature of the feed solids is 110°F. The superficial gas velocity through the bed is 0.3 fps, and the gas temperature is 5°F below the softening point of PVC.

(a) Determine the amount of water evaporated per unit time and, hence, the drying efficiency of the unit.

(b) What is the final water concentration in the product solids if the fines captured by the cyclone are returned to the bed?

(c) What gas superficial velocity would be necessary to achieve 90% drying efficiency?

13.7 For the previous problem, determine the amount of solids carryover. Establish the required dimensions of the cyclone separator to achieve 90% solids recovery.

13.8 A fluidized-bed coal burner feeds high-sulfur coal to a hot bed of limestone and coal ash. The flue gas produced contains a high concentration of SO_2 and is removed by reaction with the limestone in the bed according to the following reaction:

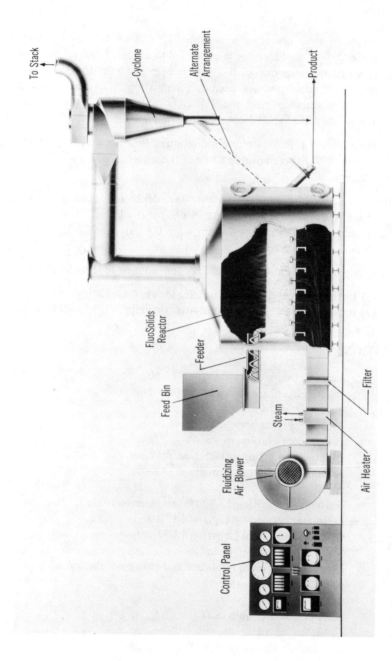

Figure 13.30 The Dorr-Oliver FluoSolids plastics drying system without recycle (courtesy Dorr-Oliver Inc., Stamford, CT).

$$CaCO_3 + SO_2 + \frac{1}{2}O_2 \rightarrow CaSO_4 + CO_2$$

At steady-state operation approximately 75% of the SO_2 is removed, while 20% of the limestone is converted to sulfate. The fraction of SO_2 captured by limestone in this design does not meet the 95% capture level established by air pollution regulations. Determine the size of bed required to meet the 95% SO_2 capture level if the same coal and gas feedrates, and limestone utilization are maintained.

13.9 For Problem 13.8, determine the required limestone feedrate that will provide 99% SO_2 removal in a bed of the same size and unchanged coal feedrate.

13.10 Estimate the maximum spoutable bed depth for a bed of spherical glass beads fluidized by air at ambient conditions. The following information is given: $D_i = 4.0$ cm, $d_p = 0.8$ cm, $\rho_s = 2.6$ Mg/m^3, $D_c = 29$ cm.

13.11 Estimate the minimum fluid velocity for spouting for Problem 13.10.

13.12 (a) Estimate the mean spout diameter for Problem 13.10.
      (b) Repeat the exercise for a fluid-mass flowrate of $G = 2500$ kg/hr-m^2.

NOTATION

A	parameter defined for Equation 13.60
A_b	cross sectional area of bed or column, m^2
A_ϕ	surface area of single particle, cm^2
Ar	Archimedes number
a	particle acceleration, m/s^2
a'	effective contact or surface area, m^2
a_f	constant in freeboard, Equation 13.83
a_1	coefficient
B'	constant in Equation 13.92 (values given in Table 13.2), g/cm^3
B	parameter defined by Equation 13.49
b	velocity constant in Equation 13.92, m/s
C	solids concentration, g/cm^3
C_{AF}, C_{out}	gas concentration in freeboard and at bed surface, kg/m^3
C_1, C_2	coefficients in Equation 13.38
D_B	bubble diameter, m or cm
D_c	column diameter, mm or ft
D_{eq}	equivalent diameter, m
D_i	inlet orifice diameter, cm or ft
D_s	mean spout diameter, m

D_t	column diameter, m
d_p	particle size, μm
d_{pi}	ith particle size, μm
d_{p50}	50 wt% particle size, μm
E_0'	entrainment rate constant at bed surface
E_0	transformation parameter defined in Equation 13.10
$E_{i\infty}$	elutriation rate constant
Eu	Euler number
$F_{0,1,2,3}$	solids mass flowrates (i.e., entrainment), g/s
F_i	entrainment rate of ith particle size, kg/s
F_{i0}	entrainment rate at bed surface, kg/s
F_K	entrainment coefficient, see Equation 13.101
F_s	entrainment rate above TDH (g/s)
F_t	total entrainment, kg/s
Fr	Froude number
f_e	fraction of wake solids ejected
f_p	particle surface area, m^2
f_{sph}	sphere surface area, m^2
G	gas mass rate, kg/s-m^2
G_p	particle layer weight, kg
Ga	Galileo number
g	gravitational acceleration, m/s^2
H	height, m
H_o	homochronity number
h	height of packed bed, m
h_s	height above free solid level, m
$\hat{K}, K^*$	mass transfer coefficients, s^{-1} or kmol/m^2-s
K'	coefficient in Equation 13.37B
K	expansion coefficient, refer to Equation 13.8
k_f	fluid particle mass transfer coefficient, kg/m^2-s
K	parameter in Equation 13.76
k	first-order reaction rate constant, s^{-1}
k'	freeboard parameter in Equation 13.35
k''	ratio of liquid to gas voidage, see Equation 13.76
k_1	mass transfer coefficient, kg-mol/m^2-s
L_e	expanded bed height, m
L_{FB}	height of freeboard zone, m
L_{mf}	bed height at minimum fluidization, m
ℓ	unit bed height, m
M	solids mass, kg
N_{RFB}	number of reaction units in freeboard
n	exponent
P	pressure, N/m^2

P'	force, N
$\bar{P}_g$	time-averaged gas pressure, Pa
R_f	resistance force, N
Re_0'	modified Reynolds number corresponding to minimum fluidization
TDH	transport disengaging height, m
T	dimensionless time
t	time, s
$\bar{U}_i$	average dimensionless velocity
u	velocity, m/s
u_j	effective jet velocity, m/s
u_{mb}	minimum bubbling velocity, m/s
u_{mf}	minimum fluidization velocity, fps or m/s
u_0	gas superficial velocity, m/s
u_s	gas velocity at incipient slugging, fps or m/s
u_t	terminal settling velocity, m/s
$u_{g\varrho}$	velocity of gas relative to liquid, m/s
v_1	Richardson-Zaki intercept for liquid fluidization, m/s
V_b	bubble rise velocity, m/s
V_0	volume of single particle, cm^3
V_r	average terminal settling velocity, m/s
We	Weber number
w	weight, N
x	partition coefficient defined in Equations 13.76 and 13.78A.
X_i	weight fraction of ith particle size

Greek Symbols

β	flow system pressure, N/m^2
ϵ	void fraction
$\epsilon_s, \epsilon_\varrho, \epsilon_g$	solids, liquid and gas holdup, respectively
λ	friction factor in Equation 13.10
ν	kinematic viscosity, m^2/s
ξ	resistance coefficient
ρ_{blk}	bulk density of fluidized state, kg/m^3
ρ_s, ρ	solid and fluid phase densities, kg/m^3, respectively
$\bar{\rho}_R$	density at reflux, kg/m^3
τ	characteristic time, s
ϕ	resistance coefficient, see Equation 13.19B
ψ	shape factor
ψ'	parameter defined by Equation 13.16

REFERENCES

Azbel, D. S. (1981) *Two-Phase Flows in Chemical Engineering* (Cambridge, England: Cambridge University Press).

Baker, C. A. J., S. D. Kim and M. A. Bergougna (1977) *Powder Technol.* 18:201.

Begovich, J. M., and J. S. Watson (1978) "Fluidization," in *Proc. 2nd Eng. Found. Conf.*, J. F. Davidson and D. L. Keairns, Eds. (Cambridge, England: Cambridge University Press), p. 190.

Begovich, J. M., and J. S. Watson (1978) *AIChE J.* 24:351.

Bhatia, V. K., and N. Epstein (1976) In: *Fluidization and its Applications*, H. Angelino et al., Eds. (Toulouse, France: Cepadues Editions).

Bhatia, V. K., K. A. Evans and N. Epstein (1972) *Ind. Eng. Chem. Proc. Des. Dev.* 11:151.

Blum, D. B., and J. J. Toman (1977) *AIChE Symp. Series* 73(161):115.

Chen, L. H., and C. Y. Wen (1979) paper presented at METC.

Dakshinamurtz, P., K. Veerabhadrarao and A. L. Venkatarao (1978) *Ind. Eng. Chem. Proc. Des. Dev.*

Darton, R. C., and D. Harrison (1974) *Instr. Chem. Eng. Symp. Ser.* 38, paper B1, London.

Darton, R. C., and D. Harrison (1975) *Chem. Eng. Sci.* 30:581.

Darton, R. C., and D. Harrison (1976) *Fluidization Technology*, D. L. Keairns, Ed., V. I, p. 399.

Davidson, J. F., and D. Harrison (1963) *Fluidized Particles* (Cambridge, England: Cambridge University Press).

Davidson, J. F., and D. L. Keairns, Eds. (1978) *Fluidization* (Cambridge, England: Cambridge University Press).

Demircan, N., et al. (1978) "Rotating Fluidized Bed Combustor," *Proc. 1978 Int. Fluidization Conf.* Cambridge, England.

Dhanuka, V. R., and J. B. Stepanek (1978) In: *Fluidization, Proc. 2nd Eng. Found. Conf.* J. F. Davidson and D. L. Keairns, Eds. (Cambridge, England: Cambridge University Press), p. 179.

Dhanuka, V. R., and J. B. Stepanek (1980) "Gas-Liquid Mass Transfer in a Three-Phase Fluidized Bed," paper presented at the 3rd Eng. Found. Conf. on Fluidization, Henniker, NH.

Do, H. T., J. R. Grace and R. Chift (1972) *Powder Technol.* 6:195.

Efremov, A. J., and I. A. Vakhrustov (1970) *Int. Chem. Eng.* 10:37.

El-Temtamy, S. A. (1974) Ph.D. Thesis, Cairo University.

El-Temtamy, S. A., and N. Epstein (1978) *Int. J. Multiphase Flow* 4:19.

Epstein, N., C. J. Lim and K. B. Mathur (1978) *Can. J. Chem. Eng.* 56:436.

Foust, et al. (1959) *Principles of Unit Operations* (New York: John Wiley & Sons, Inc.).

Gabavcic, Z. B., D. V. Vukovic, F. K. Adanski and H. Littman (1976) *Can. J. Chem. Eng.* 54:33.

Geldart, D., and A. R. Abrahamsen (1978) "Homogeneous Fluidization of Fine Powders Using Various Gases and Pressures," *Powder Technol.*, 19:133-136.

George, S. E., and J. R. Grace (1978) *AIChE Symp. Series 74* 176:67-74.

Gibbs, B. M., and A. B. Hedley (1978) In: *Proc. Eng. Found. Conf.*, J. Davidson and D. L. Keairns, Eds. (Cambridge, London: Cambridge University Press).

Hattori, H., and K. J. Takeda (1978) *J. Chem. Eng.*, Japan 11:125.

Kim, S. D., C. A. J. Baker and M. A. Bergougna (1975) *Can. J. Chem. Eng.* 53:134.

Kunii, D. and O. Levenspiel (1969) *Fluidization Engineering* (New York: John Wiley & Sons, Inc.).

Large, J. F., Y. Martinie and M. A. Bergougnou (1977) *J. Powder Bulk Solids Tech.* 1:15.

Lee, J. C., A. J. Sherrard and P. S. Buckley (1974) *Fluidization and its Applications*, H. Angelino et al., Eds. (Toulouse, France: Cepadues-Editions).

Lefroy, G. A., and J. F. Davidson (1969) *Trans. Instr. Chem. Eng.* 47:T120.

Leva, M. (1959) *Fluidization* (New York: McGraw-Hill Book Company).

Leva, M., and C. Y. Wen (1971) *Fluidization*, J. F. Davidson and D. C. Harrison, Eds. (New York: Academic Press, Inc.).

Levy, E., et al. (1979) "Operation of a Centrifugal Fluidized Bed with Continuous Feed and Removal of Bed Material," in *Proc. 14th Intersoc. Energy Conversion Eng. Conf.*, Boston, MA.

Lewis, W. K., E. R. Gilliland and P. M. Lang (1962) *Chem. Eng. Prog. Symp. Series 58* 38:65.

Lin, L. (1978) "The Elutriation of Char from a Large Fluidized Bed," M. S. Thesis, West Virginia University, Morgantown, WV.

Littman, H., Ed. (1979) "Fluidization and Fluid-Particle Systems, Research Needs and Priorities," Conference, RPI, NY.

Littman, H., M. H. Morgan III, D. V. Vukovic, F. K. Zdanski and Z. B. Gabavcic (1977) *Can. J. Chem. Eng.* 55:497.

Malek, M. A., and C. Y. Lu Benjamin (1965) *Ind. Eng. Chem. Proc. Des. Dev.* 4(1):123-128.

Mamuro, T., and H. J. Hattori (1968) *J. Chem. Eng., Japan* 1:1.

Mathur, K. B., and N. Epstein (1974) *Spouted Beds* (New York: Academic Press, Inc.).

Mathur, K. B., and P. E. Gishler (1955) *AIChE J.* 1:157.

McNab, G. S., and J. Bridgwater (1974) "Spouted Beds-Estimation of Spouting Pressure Drop and the Particle Size for the Deepest Bed," paper presented at Nuremberg Congress on Particle Technology.

Metcalfe, C., and J. Howard (1977) *App. Energy* 3:65.

Michelson, M. L., and K. Ostergaard (1970) *Chem. Eng. J.* 1:37.

Mikhailik, V. D. (1966) "Research on Heat and Mass Transfer in Technological Processes," *Nauk. Tekh. USSR, Minsk* 37.

Mukherjee, R. N., P. Bhattacharya and D. K. Taraphdar (1974) "Fluidization and its Applications," H. Angelino et al., Eds. (Toulouse, France: Cepadues-Editions).

Nishikawa, M., K. Kosaka and K. Hashimoto (1977) *AIChE J.* 2:1389.

Ostergaard, K. (1969) "Studies of Gas-Liquid Fluidization," (Copenhagen: Danish Technical Press).

Ostergaard, K. (1971) *Fluidization*, J. F. Davidson and D. Harrison, Eds. (New York: Academic Press, Inc.).

Ostergaard, K., and W. Suchozebrski (1971) *4th Europ. Symp. Chem. React. Eng.* (Elmsford, NY: Pergamon Press, Inc.).

Piccinini, N. (1980) In: *Proc. Int. Fluid Conf.*, J. R. Grace and J. M. Matsen, Eds. (New York: Plenum Publishing Corporation).

Piccinini, N., J. R. Grace and K. B. Mathur (1979) *Chem. Eng. Sci.* 34:1257.

Rigby, A. R., K. P. Van Blockland, W. H. Park and C. E. Capes (1970) *Chem. Eng. Sci.* 25:1729.

Roszak, J., and R. Gawronski (1979) *Chem. Eng. J.* 17:101.

"Rotating Fluidized Bed Reactor for Space Nuclear Propulsion" (1971/172) Brookhaven National Laboratory, Reports BNL 50321 and BNL 50362.

Soung, W. Y. (1978) *Ind. Eng. Chem. Proc. Des. Dev.* 17:33.

Wolk, R. H., and E. S. Johanson (1970) U.S. Patent 3,540,995, assigned to U.S. Secretary of the Interior and Hydrocarbon Research Inc.

Yamaguchi, I., S. Kabuta and S. Nagata (1963) *Chem. Eng.*, Japan, 27(8):576.

Yates, J. G., and P. N. Rowe (1977) *Trans. Inst. Chem. Eng.* 55:137.

Zanker, A. (1977) *Chem. Eng.* 84(25):207.

Zenz, F. A., and D. F. Othmer (1977) *Chem. Eng.* 84(27):81.

Zenz, F. A., and N. A. Weil (1958) *AIChE J.* 4:472.

CHAPTER 14

HYDRODYNAMICS OF GAS-LIQUID FLOWS

CONTENTS

INTRODUCTION

The subject of two-phase, gas-liquid flows is diverse and spans an immense number of different unit operations. Examples include absorption (packed towers), distillation columns, heat exchangers, thermosiphon reboilers, atomization and a variety of heat and mass transfer operations in the nuclear field. Despite long established operating and design methodology for two-phase systems, it remains a subject of controversy steeped in empiricism.

As a single chapter on the varying unit operations dealing with these flows would be an obvious injustice, we will depart from the overall book's format of discussing design formulations and emphasize here the more fundamental concepts. We will make a further break from the traditional approach in most textbooks by presenting a theoretical analysis, employing the method of calculus of variation. This type of an approach provides both a more fundamental understanding of the mechanics of all two-phase flows and a basis for generalizing problem analysis.

A two-phase flow is one in which we have dynamic, and sometimes chemical reactions between two phases or components in a flow system. Sometimes the two phases consist of the same chemical substances, as in distillation equipment. In all cases the operation involves the flow of vapor (gas) and liquid mixtures. These flows may occur in tubes and various heat exchange equipment, during liquid atomization by spray nozzles, and in other equipment involving direct contact between gases and liquids or in the breakup/entrainment of a liquid by a gas stream. Emphasis here is placed on one class of problems, namely, those associated with a liquid layer through which gas is blown, that is, a dynamic two-phase layer. Bubbling through a liquid layer is used in various industrial processes such as bubble columns, the Bessemer process in metallurgy, scrubbing in steam boilers and others.

REGIMES OF FLOW

Perhaps the most obvious example of a two-phase flow is that of gas and liquid flowing through a tube. The typical flow patterns encountered in cocurrent flow through a horizontal tube are illustrated in Figure 14.1. At very low gas and liquid flowrates, the liquid flows along the bottom of the tube and the gas along the top. This flow pattern is designated as stratified flow. At higher gas flowrates, waves are found on the gas-liquid interface, giving rise to "wavy flow."

At higher liquid flowrates than those seen during stratified flow, "bubble flow" is encountered if the gas flowrate is very low. Here, small bubbles of gas tend to flow in the upper portion of the tube. When the gas flowrate is increased, large bullet-shaped bubbles called plugs are formed. In plug flow, gas pockets move through the liquid along the upper portion of the tube. An increase in gas flow leads to "slug flow," in which frothy slugs of liquid move across the upper region of the tube.

At still higher gas velocities "annular flow" appears, in which the liquid flows around the outside of the tube while the gas flows in a central core. Further increases in gas velocities lead to entrainment of some of the liquid as droplets are carried along in the central gas core.

At very high mass flowrates, dispersed flow is encountered. At low gas

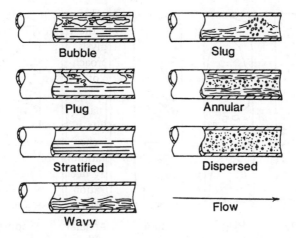

Figure 14.1   Flow pattern configurations in horizontal flows.

qualities this is seen as a froth of tiny bubbles essentially uniformly distributed in the liquid. At high qualities it is seen as a mist of fine droplets suspended in the vapor. A significant change in flow patterns occurs when a tube is inclined slightly to the horizontal. The stratified flow pattern disappears and is replaced by intermittent flow.

Wavy flow is still observed in inclined lines but the wavy flow region is restricted. As the inclination angle increases, the gas flowrate at which wavy flow begins also increases. At a sufficiently sharp angle of inclination, wavy flow disappears altogether; however, the bubble flow area expands as the tube's inclination is increased.

In vertical tubes and tubes at very sharp angles of inclination, slug flow disappears and the only flow patterns observed are bubbly, plug, churn and annular (Figure 14.2). Churn flow is a chaotic mixture of large pockets of gas and liquid, which seem to have a churning motion. Both churn and plug flow may be considered as intermittent.

Experimental flow pattern observations normally are presented in a flow pattern map. The coordinates of such maps are generally superficial mass or linear velocities for the vapor and liquid. A symbol, indicating the flow pattern, is placed at each location where an observation has been made. Figure 14.3 shows such an experimental flow pattern map for observations made with air and water in a horizontal tube. The shaded regions represent the location of the flow pattern transitions based on the observations recorded, and the solid lines represent the investigator's prediction of the location of the flow pattern transitions.

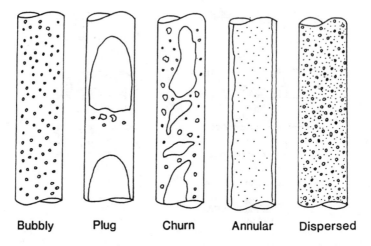

| Bubbly | Plug | Churn | Annular | Dispersed |

Figure 14.2   Flow pattern configurations in upwardly inclined and vertical flows.

HYDRODYNAMICS OF BUBBLING MIXTURES

To simplify the analysis of the hydrodynamics of bubbling mixtures and to demonstrate the range of flowrates for which viscous and inertia effects appear, two limiting regimes are considered: *rapid bubbling* and *slow bubbling*. The regime of rapid bubbling is applicable to shallow liquid pools (about 3–15 cm deep) and at relatively large superficial gas velocities (more than 0.5 m/s). This system translates into Froude number values (Fr = v_s^2/gh) greater than unity. Such liquid pools are encountered in the use of sieve and bubble plates in distillation and absorption operations. In contrast, the regime of slow bubbling is characterized by a deep pool, which is often termed a bubble column (with a depth more than 15 cm), and by a superficial gas velocity less than 0.2 m/s, which implies a Froude number much less than unity. This regime occurs extensively in gas-liquid reactors. Examples of reactors in which this takes place are oxidation, hydrogenation and fermentation vessels.

The flow in the dynamic two-phase mixture consists of numerous discrete elements (i.e., bubbles or droplets) enclosed in a continuous fluid (which may be liquid or gas, depending on the gas void fraction). The formulation of fundamental equations of hydrodynamics is only possible for single discrete elements of the two-phase flow. In this chapter, where a theoretical analysis of the total two-phase flow is considered, the approach of calculus of variations is adopted as the method for formulating flow characteristics.

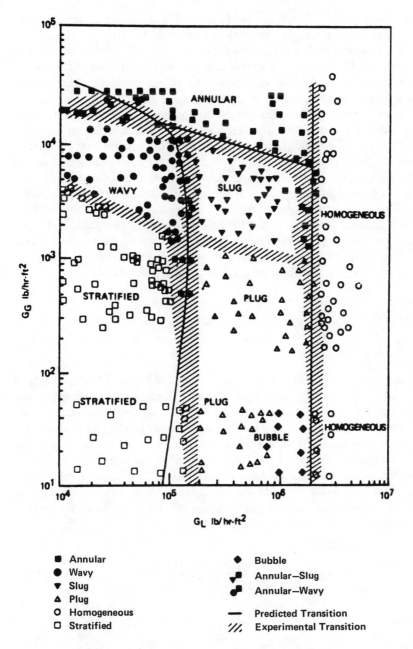

Figure 14.3 Experimental flow pattern map for the air-water system in horizontal flow. Tube diameter = 2 in. (from Choe et al. [1978]).

By taking account of the geometric characteristics of the system in which bubbling takes place, the physical properties of both phases, the gas and liquid flowrates and also the boundary conditions, the fundamental characteristics of the dynamic two-phase mixture can be established. The specific characteristics of interest vary depending on the problem. One property of general interest is certainly the relative gas and liquid holdups.

As an example of how the physical properties of each phase may affect the dynamics of the flow, consider a gas-liquid flow in which surface-active substances are present (i.e., a contaminant at the interface of the bubbles and the liquid). The presence of such a substance can result in a change in the capillary force, as well as in the appearance of additional surface forces, leading to a significant change in the hydrodynamic regime of the bubble process. When surface-active substances are present in the gas-liquid mixture, the bubble surfaces are covered by a monolayer of surface-active molecules, and the energy dissipation in this layer is known to be small compared to the dissipation of energy in the two-phase mixture (see Landau and Lifshitz [1959]). The form drag, F_0, of a bubble, considered as a sphere for simplicity, is

$$F_0 \simeq C_D \frac{\pi r_b^2 v_r^2}{2} \qquad (14.1)$$

where C_D is the drag coefficient. As a result of the flow of the liquid around the bubble, a saturated monolayer of the surface-active substance is accumulated at the rear of the bubble, and this area, s_0 (less than $4\pi r_b^2$), of the bubble remains covered by a nondeforming monolayer of the surface-active substance while the bubble rises. During the entire period of motion this monolayer on the rear of the bubble will cause the liquid velocity on the bubble surface to become zero, just as on the surface of a solid body. Separation of liquid from the bubble surface then occurs, accompanied by a new factor in the form drag (Levich [1962]). On this basis, the ratio of viscous drag, F_D, to form drag, F_0, is

$$\frac{F_D}{F_0} \simeq \frac{12\pi\mu_f v_r r_b}{C_D s_0 \rho_f v_r^2} = \frac{12\pi r_b^2}{C_D s_0 Re} \qquad (14.2)$$

where

$$Re = \frac{\rho_f v_r r_b}{\mu_f} \qquad (14.3)$$

The ratio F_D/F_0 will be less than unity when

$$s_0 > \frac{12\pi r_b^2}{C_D Re} = \frac{3 s_b}{C_D Re} \tag{14.4}$$

where s_b is the bubble surface area. The following inequality is then obtained:

$$\frac{s_0}{s_b} > \frac{3}{C_D Re} \tag{14.5}$$

This condition is satisfied where the bubble surface, s_0, covered by a mono-layer of surface-active substances is more than the fraction 6/Re of the total bubble surface area (i.e., when Re $\gg$ 6). In this case, the problem of the motion of the gas bubble with surface contaminants in a continuous liquid may be reduced to the problem of the motion of a solid sphere in an ideal, infinite liquid, as shown by Landau and Lifshitz [1959] and Kochin et al. [1955].

Rapid Bubbling Through an Ideal Liquid

The gas void fraction, ϕ, defined as the ratio of gas volume to total volume, is an important characteristic parameter of any two-phase flow. As will be shown, the void fraction is a function of the system properties for the case of rapid bubbling of gas through a liquid. In this limiting case the liquid is ideal and, hence, its viscosity is negligible.

For moderate and large values of the Reynolds number, which are typical of bubbling processes utilized in commercial distillation and absorption equipment, viscous forces are small compared to inertia forces. Thus, viscous forces have little influence on the hydrodynamics of the gas-liquid mixture and, therefore, may be ignored in a study of such flows. Various investigations (see Landau and Lifshitz [1959], Akselrod and Dillman [1956], Kutateladze and Styrikovich [1958], Kasatkin et al. [1961] and Aizenbud and Dillman [1961]) concerning the influence of viscosity on the behavior of the two-phase flow in a bubbling process confirm this conclusion. Various experimental studies on mass bubbling operations in shallow pools report three apparent groups of parameters that influence the gas-liquid flow:

1. geometric characteristics of the flow system;
2. physical properties of the gas and liquid; and
3. dynamic factors.

The height of the liquid in a flow system in which the liquid is static, the diameter of the equipment and the geometry of the gas distributing device (i.e., the orifices) belong in the first group. To the second group belong the viscosity, density and surface tension of the gas and liquid, and to the third group belong the gas and liquid flowrates.

Consider the region of flow removed from the vessel boundaries, i.e., there is no influence of wall effects on the two-phase flow. The system then may be considered as one dimensional. At a distance, x, from the gas inlet, consider a differential layer located perpendicular to the direction of motion of the gas stream having a thickness dx. The fraction of the volume of this differential layer occupied by the gas is

$$\phi = \frac{4}{3} \pi r_b^3 n \qquad (14.6)$$

where n is the number of bubbles per unit volume. An energy balance of a unit cross section of the differential layer during the bubbling process is given by

$$dE = dE_1 + dE_2 + dE_3 \qquad (14.7)$$

where dE is the total energy of the layer, dE_1 is the potential energy of the liquid, dE_2 is the kinetic energy of the layer, and dE_3 is the energy of surface tension of the bubbles in the layer. When buoyancy forces are balanced by resistance forces, it is assumed that the potential energy of the gas does not change during the motion of a bubble. The potential energy of the liquid per unit cross-sectional area is

$$dE_1 = (1 - \phi) \rho_f g x dx \qquad (14.8)$$

The kinetic energy of the layer comprises the kinetic energy of the bubbles of gas [equal to $(\rho_g v^2/2) \, 4/3 \, \pi r_b^3 n dx$, where v is the velocity of the gas in the differential layer, and ρ_g is the density of the gas] and the kinetic energy of the liquid carried along by the bubbles. To account for both components, it is simpler to assume that each gas bubble is a rigid sphere. The mass of a single bubble consists of that of the gas enclosed in it plus the mass of a volume of liquid equal to half the volume of the bubble (the so-called "additional mass") (Kochin et al. [1955], Loitsyanskii [1957]). The total kinetic energy in the layer is then given as

$$dE_2 = \left(\rho_g + \frac{\rho_f}{2} \right) \phi \frac{v^2}{2} dx \qquad (14.9)$$

From the mass conservation of the gas,

$$\phi v = v_s \tag{14.10}$$

where v_s is the gas velocity in the region of the flow that is free of liquid (i.e., the gas superficial velocity). Equation 14.9 becomes

$$dE_2 = \left(\rho_g + \frac{\rho_f}{2}\right)\frac{v_s^2}{2\phi} \, dx \tag{14.11}$$

Finally, the energy of surface tension can be calculated using the expression

$$dE_3 = 4\pi r_b^2 \, \sigma n dx = \frac{3\phi}{r_b} \, \sigma dx \tag{14.12}$$

where σ is the surface tension coefficient.

Substituting the expressions for dE_1, dE_2 and dE_3 into Equation 14.7 gives

$$dE = \left[(1 - \phi)\rho_f g x + \left(\rho_g + \frac{\rho_f}{2}\right)\frac{v_s^2}{2\phi} + \frac{3\sigma}{r_b} \, \phi\right] dx \tag{14.13}$$

The total energy of a two-phase mixture of height x_1 is then

$$E = \int_0^{x_1} \left[(1 - \phi)\rho_f g x + \left(\rho_g + \frac{\rho_f}{2}\right)\frac{v_s^2}{2\phi} + \frac{3\sigma}{r_b} \, \phi\right] dx \tag{14.14}$$

It is a fundamental axiom of physical theory that, for any system, the steady state of the system is that for which the available energy is at a minimum. Hence, the steady-state distribution of the gas void fraction, ϕ, occurs when this energy is at a minimum. Thus, to find ϕ it is necessary to determine the minimum of the integral in Equation 14.14. Smirnov [1941] has shown that this problem can be reduced to that of finding the minimum of the integral, E, under the condition of invariability of the amount of liquid in the system:

$$h = \int (1 - \phi)dx = \text{constant} \tag{14.15}$$

where h is the static liquid height. Equation 14.15 is a boundary condition on the variation of Equation 14.14. The method of the calculus of variations is now considered by starting with the following function of a function:

$$E = \int f[\phi(x), \phi'(x), x] \, dx \qquad (14.16)$$

where f is some known function of its argument, $\phi(x)$ is an unknown function, and $\phi'(x) \equiv d\phi/dx$. The calculus of variations is a method to obtain a function $\phi(x)$ such that the variation of E is zero:

$$\delta E = 0 \qquad (14.17)$$

which also means that the function, E, is at a minimum (or possibly a maximum) for that function, $\phi(x)$. This is done subject to the constraint that

$$\psi(\phi, x) = 0 \qquad (14.18)$$

where ψ is some known function. Using the method of variations on Equations 14.16 through 14.18, the Euler equation is obtained:

$$\frac{\partial f}{\partial \phi} - \frac{d}{dx}\left(\frac{\partial f}{\partial \phi'}\right) + \lambda \frac{\partial \psi}{\partial \phi} - \lambda \frac{d}{dx}\left(\frac{\partial \psi}{\partial \phi'}\right) = 0 \qquad (14.19)$$

where λ is the Lagrange multiplier and is determined by the constraint of Equation 14.18. Equation 14.19 then is used to obtain that function, $\phi(x)$, that gives a minimum of E.

When we apply this analysis to Equation 14.14 with the constraint of Equation 14.15, we obtain

$$f(\phi, \phi', x) = (1 - \phi)\rho_f g x + \left(\rho_g + \frac{\rho_f}{2}\right)\frac{v_s^2}{2\phi} + \frac{3\sigma}{r_b}\phi \qquad (14.20A)$$

$$\psi(\phi, x) = (1 - \phi) \qquad (14.20B)$$

and, for the Euler equation,

$$-\rho_f g x - \left(\rho_g + \frac{\rho_f}{2}\right)\frac{v_s^2}{2\phi^2} + \frac{3\sigma}{r_b} - \lambda = 0 \qquad (14.21A)$$

The variation due to the unknown upper limit, x_1, has been ignored in the above analysis. This upper limit is the point at which the two-phase nature of the flow breaks down (i.e., when $\phi = 1$). We next show that it is acceptable to ignore the variation due to x_1. Note that

$$E = \int_0^{x_1} f(\phi,\phi',x)dx \qquad (14.21B)$$

$\phi(x)$ gives an extremum of E when $\phi(0) = \phi_0$ and is set (the lower limit is fixed), and $\phi(x_1)$ is not set (no condition is made on the upper limit). Then the value of δE will be given not only by the variation of f under the integral, but also by the variation at the upper limit. The most general expression of the variation δE can be written as

$$\delta E = [f_{\phi'}\delta\phi]_0^{x_1} + \int_0^{x_1}\left[f_\phi - \frac{d}{dx}(f_{\phi'})\right]\delta\phi dx \qquad (14.21C)$$

In this case $f = f(\phi,x)$, that is the term $\phi' = d\phi/dx$, $f_{\phi'} = 0$, and the term integrated in the brackets is equal to zero, $[f_{\phi'}\delta\phi]_0^{x_1} = 0$. Therefore, the floating limit of x has no effect and Euler's equation is as shown in Equation 14.19. As the derivative ϕ' does not enter into Equation 14.19, Equation 14.21 is not differential, but algebraic. Hence,

$$\phi(x) = \left[\frac{v_s^2\left(\rho_g + \frac{\rho_f}{2}\right)}{2\left(\frac{3\sigma}{r_b} - \rho_f g x - \lambda\right)}\right]^{1/2} \qquad (14.22)$$

Further, as the derivative $\phi'(x)$ does not appear in Equation 14.21A we cannot impose any additional conditions on the values of $\phi(x)$ at the limits of the interval 0 to x_1. Therefore, this function may become discontinuous at the ends of the interval. However, from physical considerations it must be the case that the function $\phi(x)$ is continuous at all values of x between the limits and, therefore, can be determined from Equation 14.22. Thus, the value of x_1 can be determined from the equality

$$1 = \left[\frac{v_s^2\left(\rho_g + \frac{\rho_f}{2}\right)}{2\left(\frac{3\sigma}{r_b} - \rho_f g x_1 - \lambda\right)} \right]^{1/2} \tag{14.23}$$

The undetermined Lagrange multiplier must be eliminated. To do this, we use ϕ from Equation 14.22 in the constraint Equation 14.15:

$$h = \int_0^{x_1} (1 - \phi)dx = \int_0^{x_1} \left\{ 1 - \left[\frac{v_s^2\left(\rho_g + \frac{\rho_f}{2}\right)}{2\left(\frac{3\sigma}{r_b} - \rho_f g x - \lambda\right)} \right]^{1/2} \right\} dx \tag{14.24}$$

By integrating this expression and eliminating λ using Equation 14.23, we get

$$h = x_1 + \frac{\left(\rho_g + \frac{\rho_f}{2}\right)}{\rho_f} \frac{v_s^2}{g} - \frac{2}{\rho_f g} \left\{ \left[\frac{\left(\rho_g + \frac{\rho_f}{2}\right)v_s^2}{2} \right] \left[\rho_f g x_1 + \frac{\left(\rho_g + \frac{\rho_f}{2}\right)v_s^2}{2} \right] \right\}^{1/2} \tag{14.25A}$$

Defining,

$$A = 1 + \frac{2\rho_f g x_1}{\left(\rho_g + \frac{\rho_f}{2}\right)v_s^2} \tag{14.25B}$$

which means that

$$x_1 = \frac{1}{\rho_f g}\left(\rho_g + \frac{\rho_f}{2}\right)v_s^2(A - 1) \tag{14.25C}$$

By substituting this value of x_1 into Equation 14.25A, we obtain, after simplifying,

$$\frac{A}{2} - \frac{1}{2} + 1 - A^{1/2} = \frac{\rho_f gh}{\left(\rho_g + \dfrac{\rho_f}{2}\right)v_s^2} \tag{14.26A}$$

We now define

$$B = \frac{2\rho_f gh}{\left(\rho_g + \dfrac{\rho_f}{2}\right)v_s^2} \tag{14.26B}$$

and Equation 14.26A takes the form

$$A - 2A^{1/2} + 1 - B = 0 \tag{14.26C}$$

and solving with respect to $A^{1/2}$,

$$A^{1/2} = 1 \pm B^{1/2} \tag{14.26D}$$

Replacing A and B and, after some manipulation, we find

$$\left[\rho_f g x_1 + \frac{\left(\rho_g + \dfrac{\rho_f}{2}\right)v_s^2}{2}\right]^{1/2} = \left[\frac{\left(\rho_g + \dfrac{\rho_f}{2}\right)v_s^2}{2}\right]^{1/2} \pm [\rho_f gh]^{1/2} \tag{14.26E}$$

In the above equation the positive sign (+) is selected because when $v_s = 0$, the condition $x_1 = h$ must be fulfilled.

The equation for the height of the dynamic two-phase mixture in terms of the static liquid level is then

$$x_1 = 2\left[\frac{h\left(\rho_g + \dfrac{\rho_f}{2}\right)v_s^2}{2\rho_f g}\right]^{1/2} + h \tag{14.26F}$$

Assuming $\rho_g \ll \rho_f$,

$$x_1 = \left[\frac{v_s^2 h}{g}\right]^{1/2} + h = h(Fr^{1/2} + 1) \tag{14.27A}$$

It follows from Equation 14.27 that the height of the dynamic two-phase mixture varies linearly with the superficial velocity of the gas, v_s. The value of $\phi(x)$ will be determined next. From Equation 14.23, one can write the following expressions:

$$\lambda = \frac{3\sigma}{r_b} - \rho_f g\left(x_1 + \frac{v_s^2}{4g}\right) \tag{14.27B}$$

where we have again used the inequality $\rho_g \ll \rho_f$. By substituting the value of x_1 from Equation 14.27A into Equation 14.27B, one obtains

$$\lambda = \frac{3\sigma}{r_b} - \rho_f g\left(hFr^{1/2} + h + \frac{v_s^2}{4g}\right) \tag{14.27C}$$

Note that

$$\frac{hv_s^2}{4gh} = \frac{h}{4} Fr \tag{14.27D}$$

Hence,

$$\frac{3\sigma}{r_b} - \lambda = \rho_f gh\left(Fr^{1/2} + \frac{Fr}{4} + 1\right) \tag{14.27E}$$

and using this in Equation 14.22 the following expression is obtained:

$$\phi(x) = \left\{\frac{\rho_f v_s^2}{4\left[\rho_f gh\left(Fr^{1/2} + \frac{Fr}{4} + 1\right) - \rho_f gx\right]}\right\}^{1/2} \tag{14.28A}$$

or

$$\phi(x) = \left[\frac{F}{(1 + F^{1/2})^2 - \dfrac{x}{h}} \right]^{1/2} \tag{14.28B}$$

where

$$F = (Fr/4)$$

Thus, we have obtained the local gas void fraction, $\phi(x)$, solely in terms of the Froude number, Fr, and the location, x, for the case of an ideal liquid. Predictions of Equation 14.28B are compared against the air-water data of Vinokur and Dillman [1959] in Figure 14.4. As shown, reasonably good agreement between experiment and theory is obtained.

The average gas content (or void fraction) of the two-phase mixture may be determined as follows:

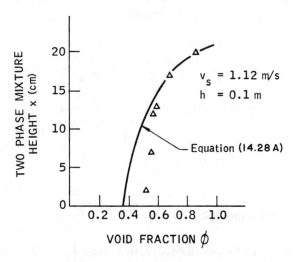

Figure 14.4   Plot of void fraction versus height, using data of Vinokur and Dillman [1959].

$$\bar{\phi} = \frac{\int_0^{x_1} \phi(x)dx}{\int_0^{x_1} dx} = \frac{\int_0^{x_1} \frac{F^{1/2}}{[(1 + F^{1/2})^2 - x/h]^{1/2}} dx}{x_1}$$

$$= -\frac{2h}{x_1} F^{1/2}\{[(1 + F^{1/2})^2 - x_1/h]^{1/2} - (1 + F^{1/2})\} \qquad (14.29)$$

After substitution of x_1 into Equation 14.29 and noting that $F = (Fr/4)$, we obtain

$$\bar{\phi} = \frac{1}{1 + 1/Fr^{1/2}} \qquad (14.30)$$

This shows that the average void fraction in rapid bubbling gas flow, through an ideal liquid, is a simple function of the Froude number alone. Equation 14.30 is compared against the air-water data of Kashnikov [1965] obtained for various orifice and sieve configurations in Figure 14.5. Again, the theoretical expression provides good results regardless of the flow geometry.

The average relative density of the gas-liquid mixture (the "specific gravity of the foam") is determined by the expression

$$\psi = \frac{h}{x_1} \qquad (14.31)$$

If we replace x_1 according to Equation 14.27A, we have

$$\psi = \frac{1}{1 + Fr^{1/2}} \qquad (14.32)$$

Rapid Bubbling Through a Real Liquid

Azbel [1981] has further developed an expression for the void fraction, ϕ, in rapid bubbling by using the same notion of energy minimizing for the case of a real liquid. In this case, no allowance is made for the energy dissipa-

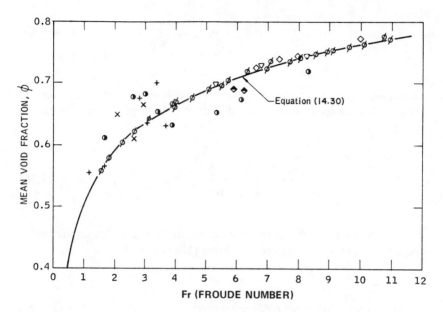

Figure 14.5   Mean void fraction versus Froude number for air-water flows studied by Kashnikov [1965]. Symbols denote different sieve plate geometries.

tion due both to the viscosity of the liquid and liquid turbulence. The derivation is based on the use of the energy balance for the differential layer, (Equation 14.7), except that dE_2 now represents the energy dissipation in the layer, while the potential energy, dE_1, of the liquid is given by Equation 14.8. The expression for the dissipative energy, dE_2, is developed through the use of the model of a uniform bubble distribution in the differential layer, with each bubble being located in the center of a "spherical compartment" formed by adjacent bubbles. Calculation of drag based on this model agrees fairly well with experimental data on liquid-liquid and gas-liquid systems. With this model the energy dissipation of the bubble is obtained by using the following expression:

$$dE_2 = 12\pi\mu_f r_b v^2 \left[\frac{1 - \phi^{5/3}}{(1 - \phi)^2} \right] \tau n dx \qquad (14.33)$$

where τ is the time scale of turbulent eddies, and the number of bubbles, n, per unit volume is

$$n = \frac{\phi}{\frac{4}{3}\pi r_b^3} \tag{14.34}$$

Combining these two equations results in the following expression for calculating the energy dissipation produced by motion of the bubbles in the layer:

$$dE_2 = \frac{9\mu_f v^2 \phi (1 - \phi^{5/3})}{r_b^2 (1 - \phi)^2}\, \tau dx \tag{14.35}$$

For the time scale of the eddies most likely to have an effect on the bubble motion, an expression developed by Azbel [1981] is used:

$$\tau = \frac{k r_b^2 (1 - \phi)^2}{9\nu_f (1 - \phi^{5/3})} \tag{14.36}$$

where k is an added mass coefficient and can be taken to be constant. Using Equation 14.36 and the continuity condition, $v_s = v\phi$ in Equation 14.35, the expression of dE_2 takes the form

$$dE_2 = k\rho_f \frac{v_s^2}{\phi}\, dx \tag{14.37}$$

The surface tension energy, dE_3, is given in Equation 14.12. Substituting for dE_1, dE_2 and dE_3 into Equation 14.7 we obtain

$$dE = \left[(1 - \phi)\rho_f gx + k\rho_f \frac{v_s^2}{\phi} + \frac{3\sigma}{r_b}\phi \right] dx \tag{14.38}$$

Consequently, the total energy of a two-phase layer of height, x_1, is

$$E = \int_0^{x_1} \left[(1 - \phi)\rho_f gx + k\rho_f \frac{v_s^2}{\phi} + \frac{3\sigma}{r_b}\phi \right] dx \tag{14.39A}$$

As in the previous section, the equilibrium distribution of the gas void fraction, ϕ, is established when this energy is a minimum. Therefore, to determine ϕ, we must find the minimum of the integral (Equation 14.39A).

Following the analysis of the previous section, the Euler equation (with the variation with respect to x_1 being neglected as before) in this case is

$$-\rho_f g x - k\rho_f \frac{v_s^2}{\phi^2} + \frac{3\sigma}{r_b} - \lambda = 0 \qquad (14.39B)$$

Hence,

$$\phi(x) = \left[\frac{k\rho_f v_s^2}{\frac{3\sigma}{r_b} - \lambda - \rho_f g x}\right]^{1/2} \qquad (14.40)$$

With $\phi(x_1) = 1$, where x_1 is the point where the flow is totally gaseous, the value of x_1 may be determined using the above equation:

$$1 = \left[\frac{k\rho_f v_s^2}{\frac{3\sigma}{r_b} - \lambda - \rho_f g x_1}\right]^{1/2} \qquad (14.41)$$

As before, we can eliminate the undetermined Lagrange multiplier by using the constraint Equation 14.15:

$$h = \int_0^{x_1} (1 - \phi)dx = \int_0^{x_1} \left\{1 - \left[\frac{k\rho_f v_s^2}{\frac{3\sigma}{r_b} - \rho_f g x - \lambda}\right]^{1/2}\right\}dx \qquad (14.42)$$

Integrating, while making allowance for Equation 14.41 and solving with respect to x_1, an equation suitable for calculating the height of a dynamic two-phase flow is obtained:

$$x_1 = h[2(kFr)^{1/2} + 1] \qquad (14.43)$$

Equation 14.43 shows that the height of the two-phase flow is a linear function of the superficial gas velocity, v_s.

From Equations 14.41 and 14.43 it follows that

$$\lambda = \frac{3\sigma}{r_b} - \rho_f g \left[h(4kFr)^{1/2} + h + \frac{kv_s^2}{g} \right] \tag{14.44}$$

and, because $(kv_s^2/gh)h = khFr$, we may write

$$\frac{3\sigma}{r_b} - \lambda = \rho_f gh \left[(4kFr)^{1/2} + 1 + kFr \right] \tag{14.45}$$

Equation 14.45 is used in expression 14.40, and the functional form of $\phi(x)$ becomes

$$\phi(x) = \left[\frac{kFr}{(kFr)^{1/2} + kFr + 1 - \dfrac{x}{h}} \right] \tag{14.46}$$

From this equation it follows that the gas content does not depend on the shape and the size of the gas bubbles because the bubble radius does not occur in the equation. It is noteworthy that the analysis of rapid bubbling for a real liquid gives a formula for the void fraction very similar to that for an ideal liquid (Equation 14.28B). The average gas content, $\bar{\phi}$, of the flow is given by

$$\bar{\phi} = \frac{\displaystyle\int_0^{x_1} \phi(x)dx}{\displaystyle\int_0^{x_1} dx} = \frac{2(kFr)^{1/2}}{1 + 2(kFr)^{1/2}} \tag{14.47}$$

Allowing for dissipative forces, we result in a simple formula for the mean void fraction in terms of the Froude number and a constant. Note the similarity between this equation and Equation 14.30 for an ideal liquid.

The relative density of the two-phase flow is

$$\psi = \frac{h}{x_1} \tag{14.48}$$

and, on substituting the expression for x_1 from Equation 14.43, we find

$$\psi = \frac{1}{1 + 2(kFr)^{1/2}} \tag{14.49}$$

For air-water systems, the constant k has been shown to be of the order of unity. Hence, we can write Equations 14.43, 14.47 and 14.49 as

$$x_1 = h(2Fr^{1/2} + 1) \tag{14.50}$$

for the height of the two-phase mixture.

$$\bar{\phi} = \frac{2Fr^{1/2}}{1 + 2Fr^{1/2}} \tag{14.51}$$

for the average void fraction, and

$$\psi = \frac{1}{1 + 2Fr^{1/2}} \tag{14.52}$$

for the relative density.

It follows from the last three equations that the main hydrodynamic parameters of the flow in rapid bubbling depend neither on the physical properties of the liquid and gas nor on the geometric characteristics of the gas-distributing devices. They are determined by the ratio of the liquid and gas flowrates (where v_s is a characteristic gas velocity) and are characterized by the ratio between the inertia and gravity forces. That is, they are functions only of the Froude number. Equations 14.51 and 14.52 are illustrated in Figures 14.6 and 14.7, respectively, in comparison to the air-water data of Kashnikov [1965] and Rodionov et al. [1964]. All the data are for flow systems whose characteristic dimension is greater than 20 cm in diameter.

TWO-PHASE FLOW PARAMETERS IN THE
SLOW BUBBLING REGIME

We will now examine the main hydrodynamic parameters of the two-phase flow in the *slow bubbling* regime. In this regime geometric specifications of the gas-distributing devices are known to have no practical

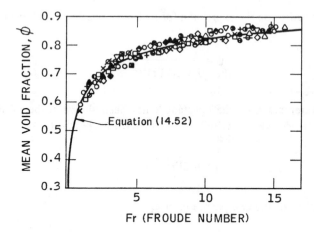

Figure 14.6   Mean void fraction vs Froude number for the air-water system. Symbols denote different sieve plate geometries.

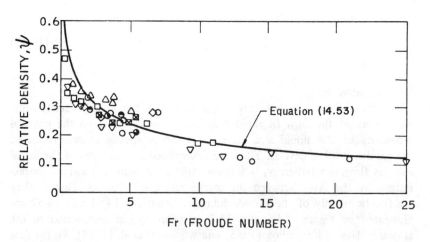

Figure 14.7   Relative density vs Froude number for the air-water system. Symbols denote different orifice sizes.

influence on the gas voidage. Key references in this analysis are works of Kasatkin [1961], Kurbatov [1953], Sterman [1956] and Sterman and Surnov [1955].

The gas void fraction, ϕ, by definition is a measure of the amount of gas per unit volume. From continuity, the void fraction is

$$\frac{\partial \phi}{\partial t} + \frac{\partial}{\partial x_i}(u_i \phi) \tag{14.53}$$

where u_i ($i = 1,2,3$) is the gas velocity vector at vector coordinate x_1, and the cartesian tensor convention of summation of repeated indices is implied.

Because of the turbulence of the two-phase flow, the gas content, ϕ, and velocity, u_i, change their values in space and time, and we can decompose them into two components: the mean component and the fluctuating component:

$$\phi = \bar{\phi} + \phi'$$
$$u_i = \bar{u}_i + u_i' \tag{14.54}$$

where $\bar{\phi}$ is the mean and ϕ' is the fluctuation component of the void fraction, and $\bar{u}_i$ is the mean and u_i' is the fluctuation component of the velocity vector. The turbulent exchange coefficient is defined by

$$\overline{u_i' \phi'} = -D_t \frac{\partial \bar{\phi}}{\partial x_i} \tag{14.55}$$

where D_t is the coefficient of turbulent diffusion. And, substituting the value of $\overline{u' \phi'}$ from the above equation into Equation 14.53 and time-averaging gives

$$\frac{\partial \bar{\phi}}{\partial t} + \frac{\partial}{\partial x_i}(\bar{u}_i \bar{\phi}) = \frac{\partial}{\partial x_i}\left(D_t \frac{\partial \bar{\phi}}{\partial x_i}\right) \tag{14.56}$$

This modified equation of continuity accounts for the gas mass transfer as a result of the interaction of the fluctuating velocity and void fraction.

As a first approximation, assume that the gas content, ϕ, can be averaged over the flow cross section so that it depends only on the height, x, in the two-phase flow. Then, for a steady-state flow, Equation 14.56 becomes

$$\frac{d}{dx}\left(\bar{v}\bar{\phi} - D_t \frac{d\phi}{dx}\right) = 0 \tag{14.57}$$

where $\bar{v}$ is now the mean gas velocity in the x direction. The expression in parentheses in this equation is obviously a constant, and in a liquid-free zone $\bar{v} = v_s$ (the superficial gas velocity) and $d\bar{\phi}/dx = 0$ because $\phi = 1$, so that

$$\bar{v}\bar{\phi} - D_t \frac{d\bar{\phi}}{dx} = v_s \qquad (14.58)$$

Taking as a boundary condition

$$\bar{\phi}(x_i) = 1 \qquad (14.59)$$

the solution of Equation 14.58, assuming as a first approximation that v and D_t are independent of the coordinate x, will be

$$\bar{\phi}(x) = \frac{v_s}{\bar{v}} + \left(1 - \frac{v_s}{\bar{v}}\right) \exp\left\{-\frac{\bar{v}}{D_t}(x_1 - x)\right\} \qquad (14.60)$$

It follows from Equation 14.60 that the gas void fraction has only a weak dependency on the position of the two-phase mixture when the liquid height, h, is sufficiently high ($\sim$1 m) that $(\bar{v}/D_t)(h - x) \gg 1$. Thus, this equation implies that if the undisturbed liquid level h($<x_1$) is large, then $\phi \simeq$ constant over most of the flow. Studies by Sterman [1956] on the gas content confirm this conclusion.

In the lower region (see Sterman [1956], Sterman and Surnov [1955] and Vinokur and Dillman [1959]) of the two-phase flow, in its initial section, the void fraction is defined by the cross-sectional area of the gas-distributing device for example, the orifices. The main region of the two-phase flow, which has a fairly constant void fraction, is upstream of this initial section, and the void fraction increases in the upper part of the main region; approaching unity at the "interface" of the two-phase-flow and the free gas region. The gas void fraction is higher in the upper part because the surface bubbles burst more slowly than the rate of rise of the newly generated bubbles. For flow systems in which h is large and the superficial gas velocity is small, the influence of the upper region on the space-average gas void fraction of the two-phase flow is not great. However, in flow systems in which h has a moderate value (5 cm $<$ h $<$ 120 cm), the mean void fraction, $\bar{\phi}$, depends on the value of h, and there may be a significant effect on the average. The location at which the void fraction moves increasingly rapidly to unity is basically determined by the superficial gas velocity. For $v_s < 0.2$ m/s, the transition is at a height of 1–2 cm, and for $v_s > 0.2$ m/s it may be extended to 30–35 cm.

The energy balance of a differential layer, dx (of unit cross section), for a regime of slow bubbling is given by Equation 14.7, where dE is the total energy of the layer, dE_1 is the potential energy of the liquid, dE_2 is the dissipative energy of the layer, and dE_3 is the energy of surface tension. The expression of dE_1 is given by Equation 14.8 and that of dE_2 has the form

$$dE_2 = C_D \frac{\rho_f v^2}{2} \pi r_b^2 n x dx \tag{14.61}$$

where C_D is the drag coefficient, v is the local gas velocity, and n is the number of bubbles per unit volume. From the law of conservation of mass,

$$\phi v = v_s \tag{14.62}$$

and the fraction of the volume of the differential layer occupied by the gas is

$$\phi = \frac{4}{3} \pi r_b^3 n \tag{14.63}$$

Taking the above two equations into consideration,

$$dE_2 = \left(\frac{3}{8}\right) \frac{C_D \rho_f v_s^2}{r_b \phi} x dx \tag{14.64}$$

The expression for the energy of surface tension is given by Azbel [1981] as

$$dE_3 = 4\pi r_b^2 \sigma n dx = \frac{3\sigma}{r_b} \phi dx \tag{14.65}$$

Substituting for the expression of dE_1, dE_2 and dE_3 into Equation 14.7, the following expression for the total energy of the mixture of height x_1 is obtained:

$$E = \int_0^{x_1} \left[(1 - \phi)\rho_f g x + \frac{3}{8} \frac{C_D \rho_f v_s^2}{r_b \phi} x + \frac{3\sigma}{r_b} \sigma \right] dx \tag{14.66}$$

The kinetic energy of the two-phase flow is ignored because its value is usually three orders of magnitude less than the energy of dissipation for slow bubbling.

As in the analysis of rapid bubbling, the steady-state distribution of the gas content, ϕ, sets in when the energy given by Equation 14.66 is at a minimum (see Lavrentev and Luysternak [1950]). Thus, to find the value of ϕ, it is necessary to determine the minimum of the integral under the condition that the height of the liquid in the equipment is constant (Equation 14.15). Ignoring variations due to x_1, Euler's equation in this case becomes

$$-\rho_f g x - \frac{3C_D \rho_f v_s^2 x}{8 r_b \phi^2} + \frac{3\sigma}{r_b} - \lambda = 0 \ , \tag{14.67}$$

where, as in the earlier cases, λ is the Lagrange multiplier. Rearranging terms,

$$\phi = \left[\frac{3C_D \rho_f v_s^2 x}{8 r_b \left(\dfrac{3\sigma}{r_b} - \rho_f g x - \lambda \right)} \right]^{1/2} \tag{14.68}$$

The height of the two-phase layer, x_1, can be determined from Equation 14.15 by substituting into it the value of ϕ from the above equation:

$$h = \int_0^{x_1} \left\{ 1 - \left[\frac{3C_D \rho_f v_s^2 x}{8 r_b \left(\dfrac{3\sigma}{r_b} - \rho_f g x - \lambda \right)} \right]^{1/2} \right\} dx \tag{14.69}$$

On integrating, we obtain

$$h = x + \frac{v_s}{g} \left(\frac{3C_D}{8 r_b \rho_f} \right)^{1/2} \left[\left(\frac{3\sigma}{r_b} - \lambda \right) x - \rho_f g x^2 \right]^{1/2}$$

$$- \frac{v_s}{2} \left(\frac{3\sigma}{r_b} - \lambda \right) \left(\frac{3C_D \rho_f}{8 r_b} \right)^{1/2} \frac{1}{(\rho_f g)^{3/2}} \sin^{-1} \left[\frac{2\rho_f g x - \left(\dfrac{3\sigma}{r_b} - \lambda \right)}{\left(\dfrac{3\sigma}{r_b} - \lambda \right)} \right] \Bigg|_0^{x_1}$$

$$\tag{14.70}$$

Noting the case of $\phi(x_1) = 1$, and after some manipulation of terms, the height of the two-phase mixture is

$$x_1 = \frac{h}{(1 + ab)\left\{1 - \frac{(ab)^{1/2}}{2}\left[\sin^{-1}\left(\frac{1 - ab}{1 + ab}\right) + \frac{\pi}{2}\right]\right\}} \tag{14.71}$$

and the average gas void fraction is

$$\bar{\phi} = \frac{x_1 - h}{x_1} = 1 - \frac{h}{x_1} = ab - \frac{(ab^{1/2})}{2}(1 + ab)\left[\sin^{-1}\left(\frac{1 - ab}{1 + ab}\right) + \frac{\pi}{2}\right]$$

$$\tag{14.72A}$$

where

$$a = \frac{3C_D}{8}, \quad \text{and } b = \frac{v_s^2}{gr_b} \tag{14.72B}$$

Equation 14.72A is plotted in Figure 14.8 along with some of the data of Kutateladze and Styrikovich [1958] and Aizenbud and Dillman [1961]. For superficial gas velocities less than 0.1 m/s, the drag coefficient can be estimated by the following formula given by

$$C_D = 0.82\left(\frac{g\mu_f^4}{\rho_f\sigma^3}\right)^{1/4} Re \tag{14.73}$$

Equations 14.72A and 14.73 can be used to obtain the height and average void fraction, respectively, in the slow bubbling regime. The radius of the bubbles is given by the expression

$$r_b = \frac{C_D v_b^2}{4g} \tag{14.74}$$

where v_b is the velocity of the bubble. The bubble velocity may be estimated from the velocity of buoyancy (Levich [1962], Azbel [1981]):

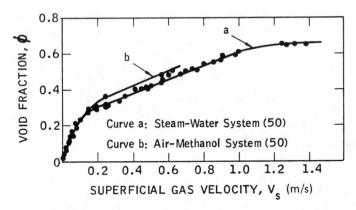

Figure 14.8   Void fraction vs superficial gas velocity using Equation 14.72A.

$$v_b = \frac{2}{3}\left(\frac{4\sigma^2 g}{3s\rho_f \mu_f}\right)^{1/5}$$ (14.75)

The factor s is a shape factor (for spherical bubbles s = 1).

TWO-PHASE FLOW PARAMETERS FOR SYSTEMS WITH DISSIPATION AND INERTIA

The influences of the dissipative and inertia forces have been examined separately above. We now consider the general problem of evaluating the average void fraction when dissipative and inertia forces are both present. For this case we shall use the equation for the relative motion of a gas bubble in the two-phase flow. whose form is given as follows:

$$g(\rho_f - \rho_g)V - C_D \frac{\rho_f v^2}{2}A_b - (\rho_g + k\rho_f)V\frac{dv}{dt} = 0$$ (14.76)

where ρ_f is the density of liquid, ρ_g is the density of gas, V is the volume of the bubble, C_D is the drag coefficient, A_b is the cross-sectional area of a bubble, and k is the "apparent additional mass" coefficient (for a spherical bubble k = 0.5). Multiplying Equation 14.76 by v, we obtain an expression for the change of energy of a single bubble during its motion, written in its integral form:

$$E_b = (\rho_f - \rho_g)gVx - C_D \frac{\rho_f A_b}{2} \int_0^x v^2 dx - \frac{(\rho_g + k\rho_f)Vv^2}{2} \tag{14.77}$$

where E_b is the energy of a single bubble.

Assuming that the differential layer, dx, contains n bubbles per unit volume (whose volume is a fraction $\phi = \frac{4}{3}\pi r_b^3 n$ of the total volume in the layer), we can express the energy of all the bubbles in a unit cross section of the layer dx as follows:

$$E_b n dx = \left[\left(\rho_g + \frac{\rho_f}{2}\right)\frac{v^2}{2}\phi + C_D \frac{\rho_f}{2}\phi\frac{A_b}{V}\int_0^x v^2 dx - g(\rho_f - \rho_g)\phi x\right]dx \tag{14.78}$$

Note that, as in Equation 14.11, $k = \frac{1}{2}$. If we combine the potential energy of the liquid and the energy of surface tension with the energy of the bubbles, the total energy of the two-phase layer will be

$$E = \int_0^{x_1}\left[(\rho_f - \rho_g)gx(1 - \phi) + \frac{1}{2}\left(\rho_g + \frac{\rho_f}{2}\right)v^2\phi\right.$$

$$\left. + C_D \frac{3}{4r_b}\phi\frac{\rho_f}{2}\int_0^x v^2 dx + \frac{3\sigma}{r_b}\phi\right]dx \tag{14.79}$$

We shall ignore the influence of the transition zones (lower and upper) described in the previous section on the average void fraction of the two-phase flow because these zones essentially balance each other out. With this condition of negligible change in the void fraction with height, the energy of the two-phase flow per unit liquid height, h, becomes

$$\frac{E}{h} = \rho_f gh(1 - \phi)\frac{\eta_1^2}{2} + \frac{\rho_f v_s^2 \eta_1}{4\phi} + \frac{3C_D\rho_f v_s^2 \eta_1^2 h}{16r_b\phi} + \frac{3\sigma\phi\eta_1}{r_b} \tag{14.80}$$

when $\eta_1 = x_1/h$ (h is the height of the static liquid and x_1 is the height of the two-phase flow), and v_s is the superficial gas velocity. Note that we have used the relation $v = v_s/\phi$ and the approximation $(\rho_f - \rho_g) \simeq \rho_f$.

Noting that $\bar{\phi} \simeq \phi \simeq$ constant by neglecting the transition regions, we find that

$$\eta_1 \simeq \frac{1}{1 - \phi}$$

and Equation 14.80 can be rewritten as

$$\frac{E}{h} = \frac{\rho_f gh}{2(1 - \phi)} + \frac{\rho_f v_s^2}{4\phi(1 - \phi)} + \frac{3C_D \rho_f v_s^2 h}{16 r_b \phi(1 - \phi)^2} + \frac{3\sigma\phi}{r_b(1 - \phi)} \qquad (14.81)$$

For the balance of our discussions, ϕ will denote the average void fraction. The equilibrium distribution of the void fraction, ϕ, is established when the specific energy E/h of the two-phase flow is at a minimum. Consequently, to determine the equilibrium ϕ we must vary Equation 14.81 with respect to ϕ to obtain

$$\frac{\rho_f gh}{2(1 - \phi)^2} - \frac{\rho_f v_s^2(1 - 2\phi)}{4\phi^2(1 - \phi)^2} - \frac{3C_D \rho_f v_s^2 h(1 - 3\phi)}{16 r_b \phi^2(1 - \phi)^3} + \frac{3\sigma}{r_b(1 - \phi)^2} = 0 \qquad (14.82)$$

After algebraic transformations, we can rewrite this equation as

$$\phi^3 - B\phi^2 - 3A\phi + A = 0 \qquad (14.83)$$

where A and B are the dimensionless quantities:

$$A = \frac{\frac{1}{2} v_s^2 \left(1 + \frac{3C_D h}{4 r_b}\right)}{\left(gh + \frac{6\sigma}{\rho_f r_b}\right)} \qquad (14.84)$$

and

$$B = 1 - \frac{v_s^2}{\left(gh + \frac{6\sigma}{\rho_f r_b}\right)} \qquad (14.85)$$

In many practical applications quantity B approaches unity and, therefore, Equation 14.85 simplifies to

$$A \simeq \frac{\phi^2(1 - \phi)}{1 - 3\phi} \tag{14.86}$$

Figure 14.9 shows a plot of $A^{1/2}$ versus ϕ, according to Equation 14.86. The curve can be used to determine the void fraction for a known A at $B \simeq 1$.

Foregoing some of the mathematics, Equation 14.83 can be expressed in the following form by eliminating parameters A and B:

$$\phi^3 - B\phi^2 - 3\alpha(1 - B)\phi + (1 - B)\alpha = 0 \tag{14.87}$$

or

$$B = \frac{\phi^3 - 3\phi\alpha + \alpha}{\phi^2 - 3\phi\alpha + \alpha} \tag{14.87A}$$

For this equation it follows that for values of ϕ, which are the roots of the equation,

$$\phi^2 - 3\alpha\phi + \alpha = 0 \tag{14.88}$$

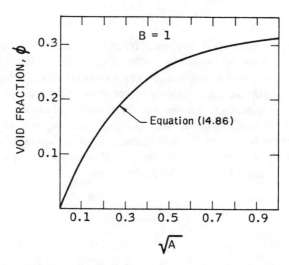

Figure 14.9   Plot of void fraction vs factor $(A)^{1/2}$ from Equation 14.86.

$B = \infty$ (because $0 < \phi < 1$), which means that the superficial velocity, v_s, is infinite when h is constant.

Equation 14.88 is applicable for flows with large superficial velocities. The roots of this equation become

$$\phi = \frac{3\alpha}{2}\left[1 \pm \left(1 - \frac{4}{9\alpha}\right)^{1/2}\right] \tag{14.89}$$

Because ϕ must be less than, or equal to, unity, we have only one value of ϕ from this equation, which is

$$\phi = \frac{3\alpha}{2}\left[1 - \left(1 - \frac{4}{9\alpha}\right)^{1/2}\right] \tag{14.90}$$

For the above equation, for $\alpha = 1$ (about the smallest possible value of α because $h/r_b \gg 1$), ϕ achieves a maximum value, equal to 0.382. This, then, is the value of the void fraction in the central zone, which cannot be surpassed either by decreasing the height, h, or by increasing the gas velocity, v_s, beyond a certain (high) value.

Further, for $\alpha \rightarrow \infty$, we find

$$\phi \simeq \frac{3\alpha}{2}\left(1 - 1 + \frac{2}{9\alpha}\right) = \frac{1}{3} \tag{14.91}$$

and this is the limit that the void fraction reaches when there are no restrictions on the upper limits of gas velocity and static liquid height.

It should be noted that, according to experimental data, the void fraction in this center zone actually achieves higher values than we have estimated because of the effect of the two transition zones, which we neglected in the analysis. Nevertheless, for high gas velocities, we can see that the analysis gives the result that the void fraction, ϕ, lies in the range 0.3 to 0.4.

Figure 14.10 provides a plot of quantity B versus ϕ for different values of parameter A, using Equation 14.83 in the form

$$B = \phi + \frac{A(1 - 3\phi)}{\phi^2} \tag{14.92}$$

As shown, the possible values of ϕ and B are limited from the above by line $B = 1$, from the left by $\phi = 0$ and from the right by $\phi = 1$. The lines

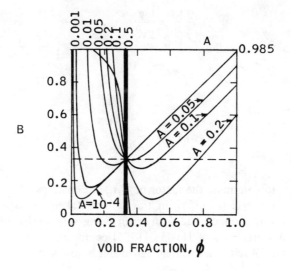

Figure 14.10   Plot of void fraction vs factors A,B.

$B = \phi$ and $\phi = {}^1/_3$ divide the field into four sections. For $A > 0$, which is physically necessary, the section limited by lines $B = 1$, $B = \phi$ and $\phi = {}^1/_3$, and the section limited by the lines $B = 0$, $B = \phi$ and $\phi = {}^1/_3$, are eliminated.

Equations 14.83 through 14.90 agree fairly well with experimental data in flow systems where the liquid depth, h, is large and the superficial gas velocities in sufficiently high regions of the liquid are small. If the extent of the transition zones created by larger superficial gas velocities (more than 0.5–0.6 m/s) is 35–40 cm in total, the prediction of ϕ by these equations will still hold good when the liquid depth is large (in excess of 1 m). The upper limit for using the equations (divergence from experimental data being no more than ±10%) is for $B \leqslant 0.985$.

To use this set of equations, a means of determining the bubble radius, r_b, and the drag coefficient, C_D, is needed. The bubble radius and drag coefficient have been shown (Azbel [1964,1981]) to be related in the following form:

$$r_b = \frac{C_D v_b^2}{4g} \qquad (14.93)$$

where v_b is the velocity of the bubble. Consequently,

$$\frac{3}{8}\frac{C_D}{r_b} = \frac{3g}{2v_b^2} \tag{14.94}$$

where

$$v_b = \frac{2}{3}\left[\frac{4\sigma^2 g}{3s\rho_f\mu_f}\right]^{1/5}$$

We still have to determine the factor $6\sigma/\rho_f r_b$ in Equations 14.84 and 14.85. To simplify the calculation, r_b in this term is taken to be 0.2 cm (the influence of this term on the final result being very small). Figure 14.11 shows the air-water data of Hughes [1955] compared to the predictions of Equation 14.83. As shown, excellent agreement is observed.

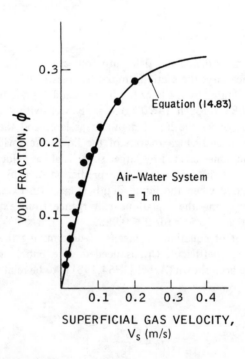

Figure 14.11   Void fraction vs superficial gas velocity for the air-water system.

EFFECTS OF STATIC LIQUID HEIGHT
AND EQUIPMENT DIAMETER ON
VOID FRACTION

In Equation 14.72A it is seen that the average gas void fraction can be defined as

$$\bar{\phi} = \frac{x_1 - h}{x_1} = 1 - \frac{h}{x_1} = 1 - \psi \tag{14.95}$$

where ψ is the average relative density of the two-phase flow, i.e., the ratio of the density of the gas-liquid mixture to that of the liquid. Thus, if ψ (or x_1 and h) is known, $\bar{\phi}$ can be obtained. However, when the diameter, D, of the equipment is less than 0.2 m and the static liquid level, h, is less than 1 m, we must introduce correction factors into the above equation to obtain

$$\bar{\phi} = 1 - k_c k_h \psi \tag{14.96}$$

where k_d is the correction factor for the equipment diameter, and k_h is the correction factor for the static liquid level.

Analyzing experimental data with the use of similarity theory yields

$$k_d = 1 - \exp\left\{-1.1\left[\frac{v_s}{v_b}\right]^{1/2}\left[\frac{(gd)^{1/2}}{v_s}\right]^{3/4}\right\} \tag{14.97}$$

and

$$k_h = 1 - \exp\left\{-0.405\left(\frac{v_s}{v_b}\right)^{0.7}\frac{(gh)^{1/2}}{v_s}\right\} \tag{14.98}$$

where v_b is the rising velocity of a single bubble, given by

$$v_b = 1.18\left[\frac{g\sigma(\rho_f - \rho_g)}{\rho_f^2}\right]^{1/4} \tag{14.99}$$

Additional references are included at the end of this chapter for more advanced studies.

PRACTICE PROBLEMS

14.1 Nitrogen and water at ambient conditions are fed to a 2.5-in.-diameter line at 15,000 cfm and 12,000 cfm, respectively. Determine the flow regime configuration in the tube.

14.2 Carbon dioxide is being bubbled at a rate of 12 lb/hr through a 10-in.-high pool of glycerine at 60°F. Determine the average size of the bubbles, their rise velocity and the average gas void fraction. The diameter of the vessel is 3.8 ft.

14.3 A plant wastewater stream is aerated in a large settling pond before being discharged to a nearby river. The plant water enters the pond at 120°F and the river water is typically 65°F. Determine how many spray nozzles and what surface area is needed to cool 15,000 gpm of water for an average wet bulb temperature of 60°F.

14.4 A small gas field operation will require disposal of 0.5×10^6 lb/hr of waste CO_2, of which 0.3% by weight is H_2S. One proposed method is to bubble the waste gas into a natural seal well offshore, from pipes on the bottom at a depth of 280 ft. Although there is little concern that the toxic H_2S will have a significant effect on aquatic life at these concentrations (the H_2S will oxidize to H_2SO_4, but very slowly), concern is raised over whether part of it will escape into the atmosphere at the water surface. The well is relatively stagnant and has an annual mean temperature of 65°F:

   (a) Determine the minimum size gas bubbles that can be kept from reaching the surface.
   (b) Determine the number of nozzles (or orifices on the discharge pipe on the well floor) needed.
   (c) Size the compressor needed for this operation.

14.5 Estimate the mass transfer coefficient for CO_2 bubbling through a 1-m-high pool of water at 90°F if the average bubble sizes are 1.5 in.

14.6 Benzene is to be absorbed from a benzene-air mixture containing 15 mol of benzene per 100 mol of air. For each 150 mol of the gas mixture there are 20 mol of an absorption oil. The operation is at 70°F and 1 atm pressure, and Raoult's law is applicable. Determine the percentage of benzene absorbed in each of the following cases and compare:

(a) The gas mixture and oil are brought into contact with each other and allowed to stand for three days.

(b) The gas mixture is bubbled through a pool of oil that is 2 m high by 3 m in diameter.

(c) The gas mixture is bubbled through an infinitely high pool of oil.

14.7 By means of similarity theory, develop an expression for the total diffusion flux on the surface of a spherical bubble rising through a stagnant pool of liquid.

NOTATION

A	dimensionless parameters defined by Equations 14.25B and 14.84
A_b	cross-sectional area of bubble, m^2
B	dimensionless parameters defined by Equations 14.26B and 14.85
a, b	parameters defined by Equation 14.72B
C_D	drag coefficient
D_t	turbulent diffusivity, m^2/hr
d	diameter of equipment, m
E	total energy of two-phase mixture, J
E_1	potential energy of layer, J
E_2	kinetic energy of layer, J
E_2	dissipation in layer, J
E_3	surface tension energy, J
E_b	energy of a bubble, J
F	one-fourth of Froude number, see Equation 14.28B
F_D	drag force, N
F_0	form drag, N
Fr	Froude number
g	gravitational acceleration, m^2/s
h	liquid column height, m
k	coefficient of apparent additional mass
k_d	correction factor for equipment diameter
k_h	correction factor for equipment height
n	number of bubbles per unit volume, m^{-3}
r_b	bubble radius, m
Re	Reynolds number
s	shape factor (distance of bubble center below surface)
s_0	contaminated area of bubble, m^2
s_b	bubble surface area, m^2
t	time, s
u_i	gas velocity vector, m/s

$\bar{u}_i$ mean velocity vector, m/s
u_i' fluctuating velocity vector, m/s
V bubble volume, m^3
v local gas velocity, m/s
$\bar{v}$ mean gas velocity in the x direction, m/s
v_b bubble velocity, m/s
v_r velocity of the bubble relative to the liquid, m/s
v_s superficial gas velocity, m/s
x direction of flow (vertical), m
x_1 height of two-phase mixture, m

Greek Symbols

n_1 ratio of expanded to stagnant fluid levels
λ Lagrange multiplier
μ_f liquid dynamic viscosity, p
ν_f liquid kinematic viscosity, m^2/s
ρ_f liquid density, kg/m^3
ρ_g gas density, kg/m^3
σ surface tension, N/m^2
τ eddy time scale
ϕ gas void fraction
$\bar{\phi}$ mean void fraction
ϕ' fluctuating void fraction
ψ average relative density of two-phase mixture

REFERENCES

Aizenbud, M. B., and V. V. Dillman (1961) *Khim. Prom.* 3:199.

Akselrod, L. S., and V. V. Dillman (1954) *Khim. Prom.* 1:8.

Akselrod, L. S., and V. V. Dillman (1954) *Zh. Prikl. Khim.* 27:15.

Akselrod, L. S., and V. V. Dillman (1956) *Zh. Prikl. Khim.* 29(12):1803.

Artomonov, D. S. (1961) *Candidate's Dissertation*, MIKhM, Moscow.

Azbel, D. S. (1962) *Khim. Prom.* 11:854.

Azbel, D. S. (1964) *Khim. Prom.* 1:43.

Azbel, D. S. (1981) *Two-Phase Flows in Chemical Engineering* (New York: Cambridge University Press).

Azbel, D. S., and A. N. Zeldin (1971) *Teoret. Osnovy Khim. Teknol.* 5:863.

Bell, R. L. (1972) *AIChE J.* 18(3):498.

Bhaga, D., and M. E. Weber (1972a) *Can. J. Chem. Eng.* 50(3):323.

Bhaga, D., and M. E. Weber (1972b) *Can. J. Chem. Eng.* 50(3):329.

Brown, R. W., A. Gomezplata and J. D. Price (1969) *Chem. Eng. Sci.* 24:1483.

Chekhov, O. S. (1960) *Candidate's Dissertation*, MIKhM, Moscow.

Chhabra, P. S., and S. P. Mahajan (1974) *Indian Chem. Eng.* 16(2):16.

Choe, W. G., L. Weinberg and J. Weisman (1978) "Observation and Correlation of Flow Pattern Transitions in Horizontal Cocurrent Gas-Liquid Flow," in *Two-Phase Transport and Reactor Safety*, T. N. Veziroglu and S. Kahac, Eds. (Washington, DC: Hemisphere Press).

Gomezplata, A., R. E. Munson and J. D. Price (1972) *Can J. Chem. Eng.* 50:669.

Hughes, R. R. (1955) *Chem. Eng. Prog.* 51(12):555.

Jackson, R. (1953) *Ind. Chemist* 336:16.

Jackson, R. (1953) *Ind. Chemist* 338:109.

Kasatkin, A. G., Yu I. Dyinerskii and D. M. Popov (1961) *Khim. Prom.* 7:482.

Kashnikov, A. M. (1965) *Candidate's Dissertation*, Moscow, D. I. Mendeleev Institute of Chemical Technology, Moscow.

Kochin, N. E., I. A. Kibel and N. N. Roze (1955) *Teoret. Gidrodinamika.*

Kodie, K., T. Hirahara and H. Kubota (1967) *Kogaku Koguku* 5(1):38.

Kurbatov, A. V. (1953) *Trudy MEI* 2:82.

Kutateladze, S. S., and M. A. Styrikovich (1958) "Hydraulics of Gas-Liquid Systems," Wright Field, Trans. T-TS-9814v.

Kuz'minykh, I. N., and G. A. Koval (1955) *Zh. Prinkl. Khim.* 28(1):21.

Landau, L. D., and E. M. Lifshitz (1959) *Fluid Mechanics* (Elmsford, NY: Pergamon Press, Inc.).

Laudie, H. A. (1969) *M. A. Sci. Thesis*, Ottawa, Canada.

Lavrent'ev, M. A., and L. A. Luysternak (1950) *A Course of Variation Calculus* (in Russian), GITTL, Moscow.

Levich, V. G. (1941) *K Teorii Poverkhnostnykh Yavlenii* (Theory of Surface Phenomena) (Moscow: Izd. Sov. Nauka).

Levich, V. G. (1962) *Physico-Chemical Hydrodynamics* (Englewood Cliffs, NJ: Prentice-Hall, Inc.).

Loitsyanskii, L. G. (1957) *Mekhanika Zhidkosti i Gaza, Gostekhizdat,* Moscow.

Marucci, G. (1965) *Ind. Eng. Chem. Fund.* 2:224.

Noskov, A. A., and V. N. Sokolov (1957) *Trudy L.T.I. im. Lensoveta* 39:110.

Peebles, F. N., and H. J. Garber (1953) *Chem. Eng. Prog.* 49(2):88.

Popov, D. M. (1961) *Moscow D. I.*, Mendeleev Institute of Chemical Technology, Moscow.

Pozin, M. E. (1957) *Sbornik: Voprosy massoperedachi* 148.

Pozin, M. E., I. P. Mukhlenov and E. Ya Tarat (1957) *Zh. Prinkl. Khim.* 30(1):45.

Pruden, B. B., W. Hayduk and H. Laudie (1974) *Can. J. Chem. Eng.* 52:64.

Pruden, B. B., and M. E. Weber (1970) *Can. J. Chem. Eng.* 48:162.

Rigby, G. R., and C. E. Capes (1970) *Can. J. Chem. Eng.* 48:343.

Rodionov, A. I., A. M. Kashnikov and V. M. Radikovsky (1964) *Khim. Prom.* 10:17.

Shepherd, E. B. (1956) *Ind. Chemist* 4:175.

Smirnov, V. I. (1941) *Kurs Vysshey Matematiki*, 4 Gostechisdat.

Solomakha, G. P. (1957) *Candidate's Dissertation*, MIKhM, Moscow.

Stepanek, J. (1970) *Chem. Eng. Sci.* 25:751.

Sterman, L. S. (1956) *Zh. Tekh. Phys.* 26(7):1512.

Sterman, L. S., and A. B. Surnov (1955) *Teploenergetika* 8:39.

Teletov, S. G. (1938) *Candidate's Dissertation*, Moscow University, USSR.

Teletov, S. G. (1945) *DAN SSSR* 4.

Teletov, S. G. (1948) *Doctoral Dissertation*, ENIN, AN SSSR, Moscow.

Teletov, S. G. (1958) *Vest. Mosk. Univ., N2 seria mekhaniki.*

Vinokur, Ya. G., and V. V. Dillman (1959) *Khim. Prom.* 7:619.

Zuber, N., and J. A. Finlay (1965) *J. Heat Transfer Trans. ASME* C87:453.

ANSWERS AND DETAILED SOLUTIONS TO SELECTED CHAPTER PROBLEMS

CHAPTER 2: SOLUTIONS

2.4 The boundary conditions for each tank are as follows:

For the Prototype Mixing Tank	For the Model Tank
$W = 0$ at $Z = 0$	$W = 0$ at $Z = 0$
for $0 < r < D_1/2$	for $0 < r < D_2/2$
at $r = D_1/2$	at $r = D_2/2$
for $0 < Z < H_1$	for $0 < Z < H_2$
$p = p_0$ at $S_1 (r,Z)$	$p = p_0$ at $S_2 (r,Z)$

where p_0 = atmospheric pressure

S_1, S_2 = vortex surfaces in the prototype and model vessels, respectively

As the tanks are in steady-state operation, no initial boundary conditions need be specified. The reference dimension chosen is the impeller diameter, I, and the reference velocity chosen is IN (rpm), where N is the rate of impeller rotation. The important dimensionless groups for this system are the Reynolds number and Froude number:

$$Re = I^2 N\rho/\mu$$

$$Fr = IN^2/g$$

Rewriting the above parameters in terms of appropriate dimensionless ratios:

Prototype Tank	Model Tank

$w^* = 0$ at $z^* = 0$

for $0 < r^* < \dfrac{D_1}{2I_1}$

$w^* = 0$ at $r^* = \dfrac{D_1}{2I_1}$

for $0 < z^* < \dfrac{H_1}{I_1}$

$w^* = 0$ at $z^* = 0$

for $0 < r^* < \dfrac{D_2}{2I_2}$

$w^* = 0$ at $r^* = \dfrac{D_2}{2I_2}$

for $0 < z^* < \dfrac{H_2}{I_2}$

(The relationships for the vortex surfaces are left to the student.)

To satisfy similarity conditions of flow patterns and, thus, mixing efficiency between the two systems, the following equalities must exist:

$$\frac{D_1}{I_1} = \frac{D_2}{I_2}$$

$$\frac{H_1}{I_1} = \frac{H_2}{I_2}$$

$$\frac{I_1^2 N_1 \rho_1}{\mu_1} = \frac{I_2^2 N_2 \rho_2}{\mu_2}$$

$$\frac{I_1 N_1^2}{g} = \frac{I_2 N_2^2}{g}$$

This last relationship is of special interest because it provides the scale factor for mixer rotational speed (and, hence, power input):

$$\frac{N_2}{N_1} = \sqrt{\frac{I_1}{I_2}}$$

CHAPTER 5: ANSWERS AND SOLUTIONS

5.10 $w = \left(\dfrac{\Delta P}{2KL}\right)^{1/n} \dfrac{n}{n+1} (R^{(n+1)/n} - r^{(n+1)/n})$

5.11 $w = \left(\dfrac{\Delta P}{KL}\right)^{1/n} \dfrac{n}{n+1} \left[\left(\dfrac{\delta}{2}\right)^{(n+1)/n} - h^{(n+1)/n}\right]$

5.30 37.3 hp

5.36 **Detailed Solution**

The critical velocity occurs at $Re_{cr} = 2300$. Consequently, we must use the following equation:

$$W_{cr} = \frac{2300\,\mu g}{d\gamma}$$

(a) For air,

$$W_{cr} = \frac{2300 \times 0.018 \times 9.81}{0.046 \times 9810 \times 1.2} = 0.75 \text{ m/s}$$

where   0.018 cp = viscosity of air at $t = 20°$
   p = 1 atm and
   1.2 = specific gravity of air obtained from the ideal gas law:

$$\gamma = \gamma_0 \frac{T_0 P}{T P_0} = \frac{M}{22.4} \times \frac{273 P}{T_0 P_0}, \quad \text{kg/m}^3$$

(b) For oil,

$$W_{cr} = \frac{2300 \times 35 \times 9.81}{0.046 \times 9810 \times 963} = 1.8 \text{ m/s}$$

5.37 **Detailed Solution**

As an approximation, the Bernoulli equation is applied to an incompressible fluid:

$$\frac{p_1}{\gamma} + \frac{w_1^2}{2g} = \frac{p_2}{\gamma} + \frac{w_2^2}{2g}$$

or

$$p_1 - p_2 = \frac{w_2^2 - w_1^2}{2g} \gamma$$

We now determine velocities of methane in sections I and II, assuming that its pressure is p = 1 atm:

$$w_1 = \frac{1700 \times 303}{3600 \times 273 \times 0.785 \times 0.2^2} = 16.7 \text{ m/s}$$

$$w_2 = w_1 \left(\frac{f_1}{f_2}\right)^2 = 16.7 \left(\frac{200}{100}\right)^2 = 66.8 \text{ m/s}$$

The specific weight of methane is

$$\gamma = \frac{MT_0}{22.4 \, T} = \frac{16 \times 273}{22.4 \times 303} = 0.645 \text{ kg/m}^3$$

The pressure difference is

$$p_1 - p_2 = \frac{(66.8^2 - 16.7^2) \times 0.645}{2 \times 9.81} = 138 \text{ kg/m}^2$$

Hence,

$$p_2 = p_1 - 138 \, , \quad \text{kg/m}^2$$

Therefore, the U-tube will show 138 − 40 = 98 mm H_2O.
The absolute pressure in section I is 10,330 + 40 = 10,370 kg/m^2.
In section II, 10,330 − 98 = 10,232 kg/m^2.

5.38 Detailed Solution

Use the following equation:

$$\tau = \frac{2F \sqrt{H}}{\alpha F_0 \sqrt{2g}}$$

where   F = the storage tank's cross-sectional area $(0.785 \, D^2)$
        F_0 = the orifice's cross-sectional area $(0.785 \, d^2)$
        H = initial liquid level (m)
        α = discharge coefficient $(\alpha \simeq 0.61)$

$$\tau = \frac{2 \times 0.785 \times 1^2 \times \sqrt{2}}{0.61 \times 0.785 \times 0.03^2 \sqrt{2 \times 9.81}} = 1180 \text{ s}$$

$$\simeq 20 \text{ min}$$

5.39 Detailed Solution

(a) Rate coefficient for the orifice meter, α, is a value in the following equation:

$$w_0 = \alpha \sqrt{2g \frac{\Delta P}{\gamma}}$$

where w_0 = velocity in the orifice meter's hole (m/s)
        ΔP = pressure difference before and after the orifice meter (kg/m^2)
        γ = specific gravity of the fluid (kg/m^3)

The Bernoulli equation for an ideal fluid is

$$\frac{P_1}{\gamma} + \frac{w_1^2}{2g} = \frac{P_2}{\gamma} + \frac{w_2^2}{2g}$$

or

$$\sqrt{w_2^2 - w_1^2} = \sqrt{2g \frac{\Delta P}{\gamma}} \tag{A}$$

Taking into account velocity coefficient ϕ,

$$\sqrt{w_2^2 - w_1^2} = \phi \sqrt{2g \frac{\Delta P}{\gamma}}$$

The continuity equation states

$$w_1 F_1 = w_2 F_2 = w_0 F_0$$

or

$$w_1 = w_0 \frac{F_0}{F_1} = w_0 \left(\frac{d_0}{d_1}\right)^2$$

$$w_2 = w_0 \frac{F_0}{F_2} = \frac{w_0}{\epsilon}$$

where $\epsilon = F_2/F_0$ = coefficient of flow constriction

Substituting w_1 and w_2 into Equation A, we obtain

$$\frac{w_0}{\epsilon} \sqrt{1 - \epsilon^2 \left(\frac{d_0}{d_1}\right)^4} = \phi \sqrt{2g \frac{\Delta P}{\gamma}}$$

or

$$w_0 = \frac{\phi \epsilon}{\sqrt{1 - \epsilon^2 \left(\frac{d_0}{d_1}\right)^4}} \sqrt{2g \frac{\Delta P}{\gamma}}$$

The group $\dfrac{\phi \epsilon}{\sqrt{1 - \epsilon^2 \left(\frac{d_0}{d_1}\right)^4}}$ is the rate coefficient α.

In our case,

$$\alpha = \frac{0.97 \times 0.62}{\sqrt{1 - 0.62^2 \left(\frac{86.5}{152}\right)^4}} = 0.61$$

(b) ΔP may be obtained from the following equation:

$$w_0 = \alpha \sqrt{2g \frac{\Delta P}{\gamma}}$$

If we substitute $w_1 \left(\dfrac{d_1}{d_0}\right)^2$ instead of w_0,

$$\Delta P = \frac{w_1^2}{\alpha^2}\left(\frac{d_1}{d_0}\right)^4 \frac{\gamma}{2g}$$

$$= \frac{1.3^2}{0.61^2} \times \left(\frac{152}{86.5}\right)^4 \times \frac{1000}{2 \times 9.81} = 2200 \text{ kg/m}^2$$

Neglecting the specific gravity of water as compared to mercury, the reading of the differential manometer will be

$$h = \frac{\Delta P}{\gamma_m - \gamma} = \frac{2200}{13,600 - 1000} = 0.176 \text{ m} = 176 \text{ mm Hg}$$

5.40 Detailed Solution

The Reynolds number is

$$Re = \frac{wd\gamma}{\mu g} = \frac{1 \times 0.038 \times 1000 \times 9810}{0.8 \times 9.81} = 47,500$$

where $\mu_{H_2O} = 0.8$ cp at 30°C

Hence, $e/d = 0.25/38 = 0.0066$.
From the graph given in Figure 5.49, $Re = 47,500$ and $e/d = 6.6 \times 10^{-3}$ we find $Eu/\Gamma = 0.017$ or $\Delta P/\rho w^2 = 0.017 \, L/d$.
The length of the coil is

$$L \simeq \pi DN = 3.14 \times 1 \times 10 = 31.4 \text{ m}$$

Then, for a straight pipe,

$$\Delta P = 0.017 \frac{Lw^2\gamma}{dg} = \frac{0.017 \times 31.4 \times 1^2 \times 1000}{0.038 \times 9.81}$$

$$= 1430 \text{ kg/m}^2$$

The correction factor, considering the coil resistance, is

$$x = 1 + 3.54\frac{d}{D} = 1 + 3.54\frac{0.038}{1} = 1.134$$

Consequently,

$$\Delta P_{coil} = \Delta P \times x = 1430 \times 1.134 = 1620 \text{ kg/m}^2$$

5.41 Detailed Solution

The velocity of the solution through the piping is

$$w = \frac{0.7}{60 \times 0.785 \times 0.0945^2} = 1.66 \text{ m/s}$$

The velocity pressure is

$$P_{vel} = \frac{\gamma w^2}{2g} = \frac{1100 \times 1.66^2}{2 \times 9.81} = 1.54 \text{ kg/m}^2$$

The pressure losses due to friction and local resistances can be obtained from the Darcy equation, provided conditions are turbulent:

$$\Delta P_{fr+\varrho r} = \frac{\lambda(L + L_{eq})w^2\gamma}{2gd}$$

$$Re = \frac{1.66 \times 0.0945 \times 1100 \times 9810}{1.1 \times 9.81} = 157{,}000$$

Hence, the flow is turbulent.
From Figure 5.25, and assuming a smooth pipe, we obtain the friction factor $\lambda = 0.021$. Now compute the equivalent length of piping corresponding to the local resistances. From Table 5.7, for a bend 90° and diameter from 76 to 152 mm, $Le/d = 40$; for a valve, $Le/d = 120$. Hence,

$$L_{eq} = \Sigma \frac{Le}{D} d = (2 \times 120 + 4 \times 40)0.0945 = 400 \times 0.0945 = 37.8 \text{ m}$$

$$= \frac{0.021(25 + 37.8)}{0.0945} \times 154 = 2150 \text{ kg/m}^2$$

The pressure required for lifting the solution is

$$\Delta P_{lift} = \gamma H = 1100 \times 16 = 17,600 \text{ kg/m}^2$$

Total pressure losses (including losses for liquid atomization) are

$$\Delta P = 154 + 2150 + 17,600 + 3500 = 23,400 \text{ kg/m}^2$$

Hence, the power input is

$$N = \frac{V_{sec} \times \Delta P}{102\eta} = \frac{0.7 \times 23,400}{60 \times 102 \times 0.6} = 4.45 \text{ kW or } \sim 6 \text{ hp}$$

5.42 Detailed Solution

At 50°C,

$$\Delta P_f = 715,000 \frac{187}{3430} = 39,000 \text{ kg/m}^2 = 3.9 \text{ atm}$$

This is the pressure needed for lifting the oil:
At 15°C,

$$\Delta P_{lift} = \gamma \times H = 20 \times 960 = 19,200 = 1.92 \text{ atm}$$

At 50°C,

$$\Delta P_{lift} = \gamma \times H = 20 \times 890 = 17,800 = 1.78 \text{ atm}$$

The pressure spent for generating sufficient velocity is

$$\Delta P_{vel} = \frac{w^2 \gamma}{2g} = \frac{960 \times 1.414^2}{2 \times 9.81} = 98 \text{ kg/m}^2$$

We may neglect this value as well as local pressure losses. Then the power input is as follows.

(a) At 15°C,

$$N = \frac{V_{sec} \times \Delta P}{102\eta} = \frac{40(715{,}000 + 19{,}200)}{3600 \times 102 \times 0.5} = 160 \text{ kW}$$

(b) At 50°C,

$$N = \frac{40(39{,}000 + 17{,}800)}{3600 \times 102 \times 0.5} = 12.3 \text{ kW}$$

Hence, to pump the colder oil an additional power consumption of

$$160 - 12.3 = 147.7 \text{ kW}$$

is required.

5.43 Detailed Solution

The oil velocity is

$$w = \frac{40}{0.785 \times 0.1^2 \times 3600} = 1.414 \text{ m/s}$$

The Reynolds number is

At 15°C,

$$Re = \frac{0.1 \times 1.414 \times 960 \times 9810}{3430 \times 9.81} = 39$$

At 50°C,

$$Re = \frac{0.1 \times 1.414 \times 890 \times 9810}{187 \times 9.81} = 670$$

Thus, in both cases we are in the laminar regime. Pressure losses due to friction are as follows:

(a) At 15°C,

$$\Delta P_f = \frac{32 L \mu w}{d^2} = \frac{32(430 + 20) \times 1.414 \times 3430}{187 \times 9.81}$$

$$= 715{,}000 \text{ kg/m}^2 = 71.5 \text{ atm}$$

The cost of additional energy for pumping is

$$147.7 \times 0.40 = 59 \text{ \$/hr}$$

Heat consumption for heating the oil from $15°$ to $50°$ (oil heat capacity $c_p = 0.4$ kcal/kg°C) is

$$Q = 40 \times 960 \times 0.40(50 - 15) = 538{,}000 \text{ kcal/hr}$$

The heat of condensation of saturated steam ($P = 1$ atm) is 540 kcal/kg. Hence, the amount of steam required to heat the oil from $15°$ to $50°$C is

$$D = \frac{Q}{r} = \frac{538{,}000}{540} \simeq 1000 \text{ kg/hr}$$

Hence, the cost of steam heating is

$$1000 \frac{\text{kg}}{\text{hr}} \times \frac{\text{lb}}{0.454 \text{ kg}} \times \frac{\text{ton}}{2000 \text{ lb}} \times \$20/\text{ton} \simeq \$22/\text{hr}$$

Therefore, it is considerably more economical to preheat the oil than to provide a larger pump, with a net savings of \$37/hr.

5.45 Detailed Solution

For isothermal flow of a real fluid, the increase in internal energy is equal to zero and density is constant. Hence, an energy balance between points 1 and 2 can be written as follows:

$$g(Z_1 - Z_2) + \frac{1}{2}(w_1^2 - w_2^2) + \frac{P_1 - P_0}{\delta} - F = 0$$

The pressure P_2 is equal to atmospheric (P_0). In this problem, $Z_1 - Z_2 = -H$, $w_1 = 0$ and $F = \Sigma\xi w_2^2/2$, where w_2 is the liquid velocity in the pipe. Therefore, the energy balance may be rearranged to give

$$\frac{P_1 - P_0}{\rho} = gH + (1 + \Sigma\xi)\frac{w_2^2}{2}$$

The coefficients of local resistances are:

- pipe entrance $\xi = 0.5$
- valve $\xi = 3.5$ (given)
- bend (90°) $\xi = 1.1$

Friction due to flow through the piping alone is (refer to Figure 5.51 for piping lengths)

$$\xi = \lambda\frac{\ell}{d} = \lambda\frac{49.5 + 8 + 0.5}{0.021}$$

Therefore,

$$\Sigma\xi = 0.5 + 3.5 + 2 \times 1.1 + \lambda\frac{58}{0.021} = 62.2 + \lambda\frac{58}{0.021}$$

and

$$\frac{P_1 - P_0}{P} = gH + \left(7.2 + \lambda\frac{58}{0.021}\right)\frac{w_2^2}{2} \qquad (\alpha)$$

(a) The minimum pressure providing liquid motion corresponds to the limiting velocity $w_2 = 0$. Hence,

$$P_{1,min} = P_0 + \rho gH = 1.013 \times 10^5 + 1100 \times 9.81 \times 8$$
$$= 1.878 \times 10^5 \text{ N/m}^2 = 1.91 \text{ atm}$$

(b) The critical value of velocity corresponding to the transition from laminar to the turbulent regime can be obtained from the following:

$$\text{Re}_{cr} = 2300 = \frac{w_{cr}d}{\nu} = \frac{w_{cr}d\rho}{\mu}$$

or

$$w_{cr} = \frac{\text{Re}_{cr}\mu}{d\rho} = \frac{2300 \times 2.5 \times 10^{-3}}{0.021 \times 1100} = 0.249 \text{ m/s}$$

where $\mu = 2.5 \text{ cp} = 2.5 \times 10^{-3} \text{N·s/m}^2$
$\rho = 1100 \text{ kg/m}^3$
$d = 0.021 \text{ m}$

The pressure in the montejus corresponding to the transition to turbulent flow in the piping is

$$P_{1,cr} = P_0 + \rho gH + \rho \left(7.2 + \lambda \frac{58}{0.021}\right) \frac{w_{cr}^2}{2}$$

$$= 1.878 \times 10^5 + 1100 \left(7.2 + \frac{64}{2300} \times \frac{58}{0.021}\right) \frac{0.249^2}{2}$$

$$= 1.9 \times 10^5 \text{ N/m}^2$$

For laminar flow, $\lambda = 64/\text{Re}$. We therefore assume $\lambda = 64/\text{Re}_{cr}$ for the critical point.

Hence, to achieve a flow in the laminar regime, the following condition must be fulfilled:

$$1.878 \times 10^5 \text{ N/m}^2 < P_1 < 1.9 \times 10^5 \text{ N/m}^2$$

(c) To determine the liquid velocity under the given conditions, we rearrange Equation α to the following form:

$$w_2 = \sqrt{\frac{\left[\dfrac{P_1 - P_c}{P} - gH\right] 2}{7.2 + \lambda \dfrac{58}{0.021}}}$$

Because $P_1 = 2.2 \text{ atm} = 2.229 \times 10^5 \text{ N/m}^2 > P_{1,cr}$, the motion is turbulent. The friction coefficient, λ, is determined from Figure 5.25.

Assuming $\lambda = 0.04$, then

$$w_2 = \sqrt{\frac{2[(2.229 - 1.013)10^5/1100 - 9.81 \times 8]}{7.2 + 0.04 \dfrac{58}{0.021}}} = 0.74 \text{ m/s}$$

and the Reynolds number is

$$Re = \frac{0.74 \times 2.1 \times 10^{-2} \times 1100}{2.5 \times 10^{-3}} = 6850$$

From Figure 5.25 at $Re = 6850$ and $e/d = 0.15/21$, we obtain $\lambda = 0.043$.

For $\lambda = 0.043$ we have $w_2 = 0.715$ m/s and $Re = 6600$. Checking Figure 5.25 again, $\lambda = 0.043$. Thus, the velocity is $w_2 = 0.715$ m/s, and the liquid rate is

$$G = \rho Sw = 1100 \times \frac{\pi}{4} \times 0.021^2 \times 0.715$$

$$= 0.272 \text{ kg/s}$$

5.47 Detailed Solution

The pressure losses are computed taking into consideration correction factors f and f_1, which correspond to nonisothermal liquid flow in nonstraight pipes:

$$\Delta P = \lambda \frac{\ell}{d} \frac{w^2}{2} \rho f f_1$$

The liquid velocity is

$$w = \frac{4G}{\rho \pi d^2} = \frac{4 \times 0.1}{998\pi(0.012)^2} = 0.882 \text{ m/s}$$

where $\rho = 998$ kg/m^3 at 20°C.

Hence, the Reynolds number is

$$Re = \frac{wd}{\nu} = \frac{0.882 \times 0.012}{10^{-6}} = 1.06 \times 10^4$$

where the kinematic viscosity of water at 20°C is $\nu = 1 \times 10^{-6}$ m²/s. Because the motion is turbulent, the friction coefficient depends on Re and e/d. We assume e = 0.025 mm. From Figure 5.25 for Re = 1.06 × 10⁴ and e/d = 0.025/12 = 0.00208, the value of the friction coefficient is λ = 0.0335.

The correction factor for nonisothermal flow is determined from the following equation:

$$f = \left(\frac{Pr_w}{Pr}\right)^{1/3} = \left(\frac{4.3}{7.06}\right)^{1/3} = 0.85$$

Note that the Prandtl numbers at the temperature of the wall and the bulk liquid are $Pr_w = 4.3$ and $Pr = 7.06$, respectively.

The correction factor, f_1, for flow over curved surfaces is

$$f_1 = 1 + 3.54 \frac{d}{2R} = 1 + 3.54 \frac{0.021}{0.5} = 1.148$$

The total length of tubing is

$$\ell = \pi Dn = \pi \times 0.5 \times 40 = 62.8 \text{ m}$$

Hence, pressure losses are

$$\Delta P = 0.0335 \frac{62.8}{0.012} \times \frac{0.882^2}{2} \times 998 \times 0.85 \times 1.148$$

$$= 6.65 \times 10^4 \text{ N/m}^2 = 0.667 \text{ atm}$$

5.48 Detailed Solution

We first compute the pressure losses in the tubular space. The volume rate is

$$G_v = \frac{G}{\rho} = \frac{40}{1100} = 3.64 \times 10^{-2} \text{ m}^3/\text{s}$$

The nozzle cross-sectional area is

$$S_1 = \frac{\pi}{4} \times 0.14^2 = 1.54 \times 10^{-2} \text{ m}^2$$

The liquid velocity in the nozzle is

$$w_1 = \frac{G}{S_1} = \frac{3.64 \times 10^{-2}}{1.54 \times 10^{-2}} = 2.36 \text{ m/s}$$

The head's cross-sectional area is

$$S_2 = \frac{\pi}{4} 0.6^2 = 0.283 \text{ m}^2$$

The resistance coefficient at the heat exchanger inlet (assuming a sudden expansion) is

$$\xi = \left(1 - \frac{S_1}{S_2}\right)^2 = \left(1 - \frac{0.0154}{0.283}\right)^2 = 0.89$$

Pressure losses at the inlet of heat exchanger are

$$\Delta P_1 = \xi \frac{w_1^2}{2} \rho = 0.89 \times \frac{2.36^2}{2} \times 1100 = 2.71 \times 10^3 \text{ N/m}^2$$

Head losses at the entrance of the heat exchanger pipes are

$$\Delta P_2 = \left(\xi_1 + \lambda \frac{\ell}{d} + \xi_2\right) \frac{w_2^2}{2} \rho$$

The resistance coefficient at the pipe entrance is assumed: $\xi = 0.5$.

For computing the friction coefficient, we determine the Reynolds number.

The cross-sectional area of the pipe is

$$S_3 = 253 \frac{\pi}{4} (2.1 \times 10^{-2})^2 = 8.73 \times 10^{-2} \text{ m}^2$$

The velocity of the solution in the pipes is

$$w_3 = \frac{G_v}{S_3} = \frac{3.64 \times 10^{-2}}{8.73 \times 10^{-2}} = 0.416 \text{ m/s}$$

Hence, the Reynolds number is

$$\text{Re} = \frac{w_3 d \rho}{\mu} = \frac{0.416 \times 0.021 \times 1100}{1.2 \times 10^{-3}} = 8020$$

The relative roughness is

$$\frac{e}{d} = \frac{0.1}{21} = 0.0048$$

For Re = 8020 and e/d = 0.0048, we find, from Figure 5.25, that λ = 0.0385. The resistance coefficient at the pipe's exit is

$$\xi_2 = \left(1 - \frac{S_3}{S_2}\right)^2 = \left(1 - \frac{8.73 \times 10^{-2}}{0.283}\right)^2 = 0.48$$

Substituting the above values into our expression for ΔP_r,

$$\Delta P_r = \left(0.5 + 0.0385 \frac{3.5}{0.021} + 0.48\right) \frac{0.416^2}{2} \times 1100$$

$$= 0.71 \times 10^3 \text{ N/m}^2$$

Pressure losses in the outlet nozzle are

$$\Delta P_2 = \left(0.5 + 0.0385 \frac{3.5}{0.021} + 0.48\right) \frac{0.416^2}{2} \times 1100$$

$$= 0.71 \times 10^3 \text{ N/m}^2$$

Pressure losses in the outlet nozzle are

$$\Delta P_3 = \xi \frac{w_1^2}{2} \rho = 0.485 \frac{2.36^2}{2} \times 1100 = 1.48 \times 10^3 \text{ N/m}^2$$

(At $S_1/S_2 = \dfrac{1.54 \times 10^{-2}}{0.283} = 0.055$ for a sudden constriction, $\xi = 0.485$.)

Hence, *the total pressure losses* are

$$\Delta P = \Delta P_1 + \Delta P_2 + \Delta P_3 = 2.71 \times 10^3 + 0.71 \times 10^3 + 1.48 \times 10^3$$

$$= 4.9 \times 10^3 \text{ N/m}^2$$

We must compute the pressure losses in the intertubular space. The distance between partitions is

$$L = \frac{\ell}{n+1} = \frac{3.5}{10+1} = 0.318 \text{ m}$$

where   n = number of partitions

The mean cross section of flow is

$$S_m = \frac{DL(t-d)}{t} = \frac{0.6 \times 0.318(0.032 - 0.025)}{0.032}$$

$$= 4.16 \times 10^{-2} \text{ m}^2$$

Thus, the average liquid velocity is

$$w_{avg} = \frac{G_v'}{S_m} = \frac{G'}{\rho S_m} = \frac{50}{992 \times 4.16 \times 10^{-2}} = 1.21 \text{ m/s}$$

The density of water at 40°C is 992 kg/m^3.
The equivalent diameter for a triangular pipe location is

$$d_{eq} = \frac{4\left(0.86t^2 - \frac{\pi d^2}{4}\right)}{\pi d} = \frac{4\left(0.86 \times 0.032 - \frac{\pi}{4} 0.025^2\right)}{\pi \times 0.025}$$

$$= 7.02 \times 10^{-2} \text{ m}$$

The Reynolds number is, thus,

$$Re = \frac{w d_{eq}}{\nu} = \frac{1.21 \times 7.02 \times 10^{-2}}{0.66 \times 10^{-6}} = 1.29 \times 10^5$$

(Note that the kinematic viscosity of water at $t = 40°C$ is $\nu = 0.66 \times 10^{-6}$ m^2/s.) From Figure 5.25, for $Re = 1.29 \times 10^5$ the friction coefficient is $\lambda' = 0.295$. The total losses in the intertubular space are

$$\Delta P' = \lambda' \frac{D(n+1)}{d_{eq}} \cdot \frac{w_{avg}^2}{2} \rho = \frac{0.6(10+1)}{0.0702} \times \frac{1.21^2}{2} \times 992$$

$$= 2.03 \times 10^4 \text{ N/m}^2$$

5.49 Detailed Solution

The liquid discharge from the orifice is

$$G_t = C \frac{\pi d^2}{4} \sqrt{2gH} = \frac{dV}{d\tau} \tag{A}$$

where $H = Z_1 - Z_2$
V = liquid volume to be discharged

$$dV = -S_1 dZ_1 = S_2 dZ_2$$

Level Z_1 is decreased, level Z_2 is increased. Because

$$H = Z_1 - Z_2$$

and

$$dH = dZ_1 - dZ_2$$

then

$$dV = \frac{S_1 S_2}{S_1 + S_2} dH \qquad (B)$$

After substituting B into A, we obtain

$$C \frac{\pi d^2}{4} \sqrt{2gH} = - \frac{S_1 S_2}{S_1 + S_2} \frac{dH}{d\tau}$$

After dividing the variables and integrating over the limits of 0 to τ for H_0 to H_{final}, we obtain

$$\tau = \frac{8 S_1 S_2}{C \pi d^2 (S_1 + S_2) \sqrt{2g}} (\sqrt{H_0} - \sqrt{H_{final}})$$

Hence, for Part (a), the time it takes for the level difference between the two tanks to fall from 2 m to 1 m will be

$$\tau = \frac{8 \times 8 \times 6 \times (\sqrt{2} - \sqrt{1})}{0.62 \times 0.02^2 (8 + 6) \sqrt{2 \times 9.81}} = 3260 \text{ s} = 0.906 \text{ hr}$$

where the rate coefficient $C = 0.62$.

For Part (b), the time it takes for the levels in the two tanks to become equal occurs at $H_{final} = 0$. Hence,

$$\tau' = \tau \frac{\sqrt{2}}{\sqrt{2} - \sqrt{1}} = 3260 \frac{\sqrt{2}}{\sqrt{2} - \sqrt{1}} = 11,200 \text{ s} = 3.11 \text{ hr}$$

CHAPTER 6: ANSWERS AND SOLUTIONS

6.1 The velocity of water in the suction line is

$$w_s = \frac{12}{60 \times 0.785 \times (0.35)^2} = 2.08 \text{ m/s}$$

and, in the discharge line,

$$w_d = \frac{12}{60 \times 0.785 \times (0.3)^2} = 2.83 \text{ m/s}$$

Pressure in the discharge pipe is

$$P_d = (3.8 + 1.03)10^4 = 48,300 \text{ kg/m}^2$$

Pressure in the suction pipe is

$$P_s = (0.76 - 0.21)13,600 = 7500 \text{ kg/m}^2$$

Hence, the pump head is

$$H = \frac{P_d - P_s}{\gamma} + H_0 + \frac{w_d^2 - w_s^2}{2g}$$

$$= \frac{48,300 - 7500}{1000} + 0.41 + \frac{(2.83)^2 - (2.08)^2}{2 \times 9.81}$$

$$= 40.8 + 0.41 + 0.19$$

$$H = 41.4 \text{ m } H_2O$$

6.6 From Figure 6.59, the suction height may be determined from the following expression:

$$H_s \leqslant P_a - h_t - \Sigma h , \quad \text{m} \tag{i}$$

where P_a = atmospheric pressure (refer to Table 6.9)
   h_t = pressure of saturated vapor of liquid to be pumped at the process temperature
   h = sum of losses over the suction height, including the energy consumption for creating the liquid velocity and

overcoming the inertia of the liquid column in the suction line, as well as overcoming friction and local resistances

The pressure exerted by the saturated vapor of the pumped liquid at different temperatures is given in Table 6.10. Finally, Table 6.11 provides information on the permissible suction height (in m) for pumping water with piston-type pumps. Solving Equation i with $h_t = 2.02$ m (from Table 6.10),

$$\frac{736 \times 13.6}{1000} - 2.02 - 6.5 = 1.48 \text{ m}$$

Hence, the theoretical suction head, H_s, cannot exceed 1.48 m. From Table 6.11, we note that for n = 150 rpm, the suction height is practically zero. *That is, the pump must be installed below the liquid level in the supply tank* ("under flooding").

6.7   The plunger displacement volume for one rotation is

$$(2F - f)s = (2 \times 0.785 \times 0.125^2 - 0.785 \times 0.035^2)0.272$$
$$= 0.00637 \text{ m}^3$$

(Note that the value 0.272 is the length of the plunger stroke, which is equivalent to the length of a double radius of the crankshaft). The theoretical pump delivery at 65 rpm is

$$0.00637 \times 65 = 0.413 \text{ m}^3/\text{min}$$

The actual delivery is

$$\frac{22.8}{60} = 0.38 \text{ m}^3/\text{min}$$

Hence, the delivery coefficient is

$$\eta_v = \frac{0.38}{0.413} = 0.92$$

6.9 The volumetric flow is

$$Q_{min} = \eta F s n$$

whence the number of rpm can be computed (i.e., $n = Q_{min}/\eta F s$):

$$Q_{min} = \frac{430}{1000} = 0.43 \text{ m}^3/\text{min}$$

$$F = 0.785 \times 0.16^2 = 0.0201 \text{ m}^2$$

Hence,

$$n = \frac{0.43}{0.85 \times 0.021 \times 0.2} = 126 \text{ rpm}$$

The head delivered by the pump is

$$H = \frac{P_2 - P_1}{\gamma} + H + h_n = \frac{3.2 \times 10,000}{930} + 19.5 + 10.3 = 64.2 \text{ m}$$

The horsepower consumed by the electric motor is obtained from Equation 6.2:

$$N_e = \frac{Q_{sec} \gamma H}{102 \eta} = \frac{0.43 \times 930 \times 64.2}{60 \times 102 \times 0.72} = 5.82 \text{ kW}$$

where 0.72 = total efficiency of the pump.

Hence, from Equation 6.6,

$$\eta_p \times \eta_{tr} \times \eta_m = 0.8 \times 0.95 \times 0.95 = 0.72$$

According to Table 6.1, an electric motor which accounts for overloading must be selected:

$$5.82 \times 1.17 = 6.8 \text{ kW}$$

6.11 The necessary head for the pump is 23.8 m for a capacity of 32 ℓ/s. A higher-speed pump will have to be used, where N = 13.1 kW. Details are left to the student.

6.12 The pump capacity is

$$Q = \eta \, \frac{2fbzn}{60} = \frac{2 \times 0.00096 \times 0.042 \times 12 \times 440}{60} = 0.00708 \text{ m}^3/\text{s}$$

where η = delivery coefficient
    f = tooth cross-sectional area (m²)
    b = tooth width (m)
    z = number of teeth in the gear
    n = rpm

The actual delivery in this case is

$$Q = 312 \text{ ℓ/min} = \frac{312 \times 0.001}{60} = 0.0052 \text{ m}^3/\text{s}$$

Hence, the delivery coefficient is

$$\eta = \frac{Q}{Q_{Th}} = \frac{0.0052}{0.00708} = 0.735$$

6.13 Writing Bernoulli's equation for Sections I and II (neglecting losses),

$$z_1 + \frac{P_1}{\gamma} + \frac{w_1^2}{2g} = z_2 + \frac{P_2}{\gamma} + \frac{w_2^2}{2g}$$

For a horizontal pump, $Z_1 = Z_2$. And

$$w_1 = \frac{f_2}{f_1} \, w_2 = \left(\frac{50}{23}\right)^2 \, 2.7 = 12.8 \text{ m/s}$$

From Bernoulli's equation, we obtain

$$P_1 = P_2 + \frac{(w_2^2 - w_1^2)\gamma}{2g} = 10,330 + \frac{(2.7^2 - 12.8^2)1000}{2 \times 9.81}$$

$$= 10,330 - 8000 = 2330 \text{ kg/m}^2$$

The theoretical vacuum is 0.8 atm.

6.14 The work performed by the pump is

$$7.8 \times 1020 \times 4 = 31,800 \text{ kg-m/s}$$

The pump horsepower is

$$9.6 \times 1000 (22 - 4) = 172,500 \text{ kg-m/hr}$$

Hence, the pump efficiency is

$$\eta = \frac{31,800}{172,500} \times 100 = 18.4\%$$

6.15 (a) Consider a level, s, above the air injection point in the riser leg. If the cross-sectional area of the leg is F, then the total pressure at that point is $(h_a + s)\rho g$. For an infinitesimal length of pipe, the total volume of the two-phase mixture is

$$dV_b = Fds \left(1 + \frac{m}{M} \rho v_a \frac{h_a}{h_a + s}\right)$$

Hence, the total volume is

$$V_b = \int_0^{V_b} dV_b = \int_0^s F \left(1 + \frac{m}{M} \rho v_a \frac{h_a}{h_a + s}\right) ds$$

or

$$V_b = F \left(s + \frac{m}{M} \rho_a v_a h_a \, ln \, \frac{h_a + s}{h_a}\right)$$

Note that this volume of fluid is contained in a length ℓ of pipe:

$$\ell = s + \frac{m}{M} \rho \nu_a h_a \, ln \, \frac{s + h_a}{h_a}$$

and the entire leg is $\ell = h_r + h_s$, where $s = h_s$. Hence,

$$h_r = \frac{m}{M} \rho \nu_a h_a \, ln \, \frac{h_s + h_a}{h_a}$$

or

$$\frac{m}{M} = \frac{h_r g}{P_a \nu_a \, ln \left(\frac{h_s + h_a}{h_a} \right)}$$

because $P_a = h_a \rho g$. This expression is identical to Equation 6.56.

(b) Examining the above equation, as $h_s \rightarrow 0$, $\frac{m}{M} \rightarrow \infty$. Hence, the pump will not work.

6.16 Assuming that the supply tank is at the same level as the final discharge point, we may use the following equation developed back in Chapter 5:

$$W = F = \lambda \frac{\ell}{d} \frac{w^2}{2} \frac{\Delta P}{\rho}$$

W represents the energy added by the pump to a unit mass of liquid delivered. The horsepower consumed by the pump is

$$N = \frac{WG}{\eta} = \frac{\Delta PG}{\rho \eta} \qquad \text{(i)}$$

To determine the pressure loss, ΔP, the friction coefficient, λ, first must be determined. As $\lambda = f(Re)$, we can express the Reynolds number in terms of the oil's viscosity:

$$Re = \frac{wd}{\nu} = \frac{4Gd}{\pi d^2 \mu} = \frac{4 \frac{12,800}{3600}}{\pi \times 0.054\mu} = \frac{836}{\mu}$$

Using the viscosity data reported in Table 6.13, the following Reynolds number values are computed for different temperatures:

Temperature, t (°C)	Re	λ	$\Delta P \times 10^{-5}$ (N/m^2)	Weight (N)
20	600	0.107	14	9780
30	840	0.076	10	7060
40	1120	0.057	7.7	5500
50	1530	0.042	5.7	4080
60	2000	0.032	4.4	3160
70	2630	0.045	6.2	4500
80	3360	0.042	5.9	4250

The above tabulated values show that for temperatures up to t = 60°C the flow is laminar and, hence, the friction factor may be computed from $\lambda = 64/Re$. For higher temperatures, $\lambda = 0.3164/Re^{0.25}$. To determine pressure losses, use the following formula:

$$\Delta P = \lambda \frac{\ell}{d} \frac{w^2}{2} \rho = \lambda \frac{\ell}{2d} \left(\frac{4G}{\pi d^2}\right)^2 \frac{1}{\rho}$$

$$= \frac{500}{2 \times 0.054} \left(\frac{4 \frac{12,800}{3600}}{\pi \times 0.054^2}\right)^2 \frac{\lambda}{\rho} = 1.15 \times 10^{10} \frac{\lambda}{\rho}$$

Sufficient information is now available to evaluate Equation i for the required horsepowers at different ΔP values:

$$N = \frac{\Delta P \frac{12,800}{3600}}{\rho \times 0.6} = 5.93 \frac{\Delta P}{\rho}$$

The student should now prepare a plot of horsepower, N, versus temperature, whence the desired temperature of the oil can be obtained. In this example, the answer is 52°C.

6.19 Assume the water surfaces in the feed tank and discharge point are $Z_1 = 0$ and $Z_2 = Z$, respectively. Then the difference in heights is

$$Z = \frac{P_1 - P_2}{\rho g} - \frac{w_2^2}{2g} - \frac{F}{g} = \frac{P_1 - P_2}{\rho g} - \left(1 + \lambda \frac{\ell}{d} + \Sigma \xi\right) \frac{w_2^2}{2g}$$

At the given value of capacity, the maximum suction height corresponds to a minimum value of pressure at the inlet. For providing suction, pressure, P_2, must be greater than that of the vapors of water at the working temperature. The physical parameters of water at the working temperature (60°C) are:

- pressure of saturated vapor, P_2', atm = 0.2031
- density, ρ, kg/m^3 = 983
- kinematic viscosity, ν, m^2/s = 0.479 × 10^{-5}

The water velocity is

$$w_2 = \frac{4G}{\pi d^2 \rho} = \frac{4 \times 2}{\pi \times 0.038^2 \times 983} = 1.79 \text{ m/s}$$

The Reynolds number is

$$Re = \frac{wd}{\nu} = \frac{1.79 \times 0.038}{0.479 \times 10^{-6}} = 1.42 \times 10^5$$

At $Re = 1.42 \times 10^5$ and e/d = 0.003, the friction coefficient according to Figure 5.25 is $\lambda = 0.027$. Substituting values into Equation i for Z, the maximum height of suction is determined:

$$Z = \frac{(1.033 - 0.2031) \times 9.81 \times 10^4}{983 \times 9.81} - \left(1 + 0.027 + \frac{15}{0.038} + 3.5\right)$$

$$\times \frac{1.79^2}{2 \times 9.81} = 6 \text{ m}$$

6.20 N = 1.3 kW

6.21 Z_{max} = 1.8 m

6.27 $t = 43°C$

6.29 $\eta_d = 0.89$

6.32 (a) $\eta = 0.59$
    (b) $Q = 71.2 \text{ m}^3/\text{hr}$
    (c) $H = 68 \text{ m}$
    (d) $N = 22.5 \text{ kW}$

6.36 3.75 atm gauge

CHAPTER 7: ANSWERS AND SOLUTIONS

7.1 The pressure difference between the discharge and suction points is

$$P_2 - P_1 = 74 - 60 = 14 \text{ kg/m}^2$$

Losses in the discharge and suction lines are

$$\Delta P_s + \Delta P_d = 19 + 35 = 54 \text{ kg/m}^2$$

The velocity pressure at the discharge is

$$\frac{W^2\gamma}{2g} = \frac{(11.2)^2 \, 1.2}{2(9.81)} = 7.7 \text{ kg/m}^2$$

Hence, the pressure created by the fan is

$$\Delta P = 14 + 54 + 7.7 \simeq 76 \text{ mm H}_2\text{O}$$

7.3 (a) The pressure developed may be determined from the following formula:

$$P = \left(P_{st.d.} + \frac{W_d^2 \gamma}{2g} \right) - \left(P_{st.s.} - \frac{W_s^2 \gamma}{2g} \right)$$

Because the suction and discharge ductwork are the same size, they have the same velocity head:

$$\Delta P = P_{st.d.} - P_{st.s} = 20.7 - (-15.8) = 36.5 \text{ kg/m}^2$$

(b) The fan capacity is

$$Q_{sec} = \frac{3700}{3600} = 1.03 \text{ m}^3/\text{s}$$

The theoretical horsepower is

$$N_{Th} = \frac{1.03 \times 36.5}{102} = 0.368 \text{ kW}$$

Fan efficiency is

$$\eta = \frac{N_{Th}}{N} = \frac{0.368}{0.77} = 0.48$$

(c) The new capacity at $n_2 = 1150$ rpm is

$$Q_2 = Q_1 \frac{n_2}{n_1} = 3700 \frac{1150}{960} = 4430 \text{ m}^3/\text{hr}$$

(d) The consumed horsepower at the new rpm is

$$N_2 = N_1 \left(\frac{n_2}{n_1}\right)^3 = 0.77\left(\frac{1150}{960}\right)^3 = 1.33 \text{ kW}$$

7.4 The solution to this problem requires determination of the operating point on the system performance diagram, i.e., the intersection of the fan and piping system characteristics' curve. The piping characteristics may be described by a parabolic expression:

$$\Delta P = aQ^2 + b \tag{i}$$

where (aQ^2) includes pressure losses $\Delta P_v + \Delta P_{fr} + \Delta P_{\varrho r}$. The overall pressure drop in Equation i is proportional to the square of capacity. Term (b) is not a function of Q but represents the difference between discharge and suction losses, ΔP_{ad}. Several values are computed and tabulated below:

Q (m³/hr)	aQ²	b	ΔP (kg/m²)
1350	38.1	13	51.1
$\dfrac{1350}{1.5} = 900$	$\dfrac{38.1}{1.5^2} = 16.9$	13	29.9
$\dfrac{1350}{2} = 675$	$\dfrac{38.1}{2^2} = 9.5$	13	22.5
$\dfrac{1350}{2.5} = 540$	$\dfrac{38.1}{2.5^2} = 6.1$	13	19.1
0	0	13	13

The student should prepare a plot of the fan characteristics from the test data given in Table 7.1 and on the same plot prepare the above-computed characteristics of the piping system, i.e., a plot of ΔP versus Q. The intersection of these two curves will show the actual capacity of the fan. (Answer: 1170 m³/hr air.)

7.5 (a) $P_1 = 10,000$ kg/m² and $P_2 = 11,000$ kg/m².

The work of compression for 1 m³ of gas (under suction conditions) is

$$L_{ad} = \frac{\kappa}{\kappa - 1} P_1 \left[\left(\frac{P_2}{P_1} \right)^{(\kappa-1)/\kappa} - 1 \right] = \frac{1.4}{0.4} \, 10^4 \, [1.1^{0.4/1.4} - 1]$$

$$= 970 \, \frac{\text{kg-m}}{\text{m}^3}$$

where $\kappa = 1.4$ for air

Work based on the hydraulic formula is

$$L_h = Q\Delta P = 1(11,000 - 10,000) = 1000 \text{ kg/m}^2$$

(b) P_1 = 10,000 kg/m^2 and P_2 = 50,000 kg/m^2.

From the thermodynamic formula,

$$L_{ad} = \frac{1.4}{0.4} \, 10^4 (5^{0.4/1.4} - 1) = 20{,}500 \text{ kg-m/m}^3$$

From the hydraulic formula,

$$L_h = 1 \times (50{,}000 - 10{,}000) = 40{,}000 \text{ kg-m/m}^3$$

Comparing the results obtained for cases (a) and (b), we see that in the first case the results obtained from the thermodynamic and hydraulic formulas differ only by 3%. This case (P_2/P_1 = 1.1) corresponds to the limiting compression ratio for fans, which therefore is designed by use of the hydraulic formula. In case (b) (P_2/P_1 = 5), which corresponds to the compression of air in a compressor, the results obtained from the thermodynamic and hydraulic formulas differ by 100%. Horsepower calculations for compressors always must be based on the thermodynamic formulas.

This is best illustrated on the operating diagram shown in Figure A.1, a theoretical diagram of a piston compressor. The area abce (representing the work spent for adiabatic compression at P_2 = 1.1) is approximately equal to the area abde, but area afge (for P_2 = 5 atm) is far away from the area afhe.

7.6   The adiabatic work of compression is

$$L_{ad} = \frac{1.29}{0.29} \, 49.8 \times 263 \left[\left(\frac{12}{2.5} \right)^{0.29/1.29} - 1 \right] = 24{,}500 \text{ kg-m/kg}$$

where κ = 1.29 [from R. H. Perry and C. H. Chilton, Eds. *Chemical Engineer's Handbook*, 5th ed. (New York: McGraw-Hill Book Co., 1973).]

$$R = \frac{848}{17} = 49.8 \text{ kg-m/kg}°K$$

An alternative approach to computing the adiabatic work is to use the following formula:

$$L_{ad} = 427(i_2 - i_1) \text{ kg-m/kg} \quad ,$$

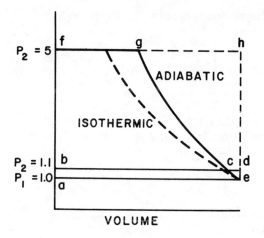

Figure A.1   Theoretical diagram of a piston compressor (solution to Problem 7.5).

and the T-S diagram, where i_2 and i_1 are initial and final gas enthalpies. On the T-S diagram in Figure 7.43, we locate the point −10°C and 2.5 atm, and note that $i_1 = 343.5$ kcal/kg. Tracing from this point along the vertical line (i.e., S = constant) to the intersection with the isobar ($P_2 = 12$ atm), point 2 is located, where $i_2 = 399$ kcal/kg. Hence, $L_{ad} = 427 (399 − 343.5) = 23,700$ kg-m/kg. This value is very close to that predicted by the chapter equation (the discrepancy is only 3%). The weight capacity of NH_3 is

$$G = 460 \times 0.76 = 350 \text{ kg/hr}$$

where 0.76 = specific weight of ammonia at normal conditions. Therefore,

$$\gamma_0 = \frac{M}{22.4} = \frac{17}{22.4} = 0.76 \text{ kg/Nm}^3$$

The compressor's horsepower is, thus,

$$N = \frac{GL_{ad}}{3600 \times 102\eta} = \frac{G(i_2 − i_1)}{860\eta} = \frac{350 \times 24,500}{3600 \times 102 \times 0.7} = 33.4 \text{ kW}$$

The temperature at the end of compression is

$$T_2 = T_1 \left(\frac{P_2}{P_1}\right)^{(\kappa-1)/\kappa} = 263 \left(\frac{12}{2.5}\right)^{0.29/1.29} = 374°K = 101°C$$

Note that the temperature can be determined directly from the T-S diagram by locating point 2. From Figure 7.43,

$$t_2 = 104°C$$

7.7 The volumetric efficiency of the compressor is

$$\eta_v = 1 - \epsilon_0 \left[\left(\frac{P_2}{P_1}\right)^{1/m} - 1\right] = 1 - 0.05(5.5^{1/1.25} - 1) = 0.854$$

The delivery coefficient is

$$\eta_d = 0.85\eta_v = 0.85 \times 0.854 = 0.725$$

The compressor's capacity is

$$Q = \eta_d \frac{Fsn}{60} = \frac{0.725 \times 0.18^2 \times 3.14 \times 0.02 \times 240}{4 \times 60} = 0.0147 \text{ m}^3/\text{s}$$

or

$$0.0147 \times 3600 = 53 \text{ m}^3/\text{hr}$$

Supposing that the compressor sucks atmospheric air at t = 20°C with specific weight 1.2 kg/m³. The compressor capacity is, therefore,

$$53 \times 1.2 = 63.6 \text{ kg/hr}$$

Consequently, the compressor does not provide the desired capacity (80 kg/hr). However, the desired capacity can be achieved if we increase the rpm from 240 to (80/63.6)240 = 302 rpm and provide a supercharger, which will increase the air from atmospheric pressure to a pressure of 80/63.6 = 1.26 atm and supply the compressor with air under this pressure.

7.11 The compressor's capacity will be equal to zero when the volumetric efficiency is zero:

$$\eta_v = 1 - \epsilon_0 \left[\left(\frac{P_2}{P_1} \right)^{1/m} - 1 \right] = 1 - 0.085(P_2^{1/1.31} - 1) = 0$$

Hence,

$$P_2^{0.763} = 12.8$$

or

$$P_2 \simeq 28 \text{ atm}$$

Consequently, the compressor's capacity will be equal to zero when the discharge pressure is 28 atm.

7.12 (a) For the single-stage compression, the temperature at the end of compression is

$$T_2 = T_1 \left(\frac{P_2}{P_1} \right)^{(\kappa-1)/\kappa} = 293 \times (9)^{0.4/1.4} = 293 \times 1.88 = 551°\text{K} = 278°\text{C}$$

The theoretical horsepower is

$$L_{ad} = \frac{\kappa}{1 - \kappa} RT_1 \left[\left(\frac{P_2}{P_1} \right)^{(\kappa-1)/\kappa} - 1 \right] = \frac{1.4}{0.4} \times 29.3 \times 293(1.88 - 1)$$

$$= 26,500 \frac{\text{kg-m}}{\text{kg}}$$

The compressor's volumetric efficiency is

$$\eta_v = 1 - \epsilon_0\left[\left(\frac{P_2}{P_1}\right)^{1/m} - 1\right] = 1 - 0.08(9^{1/1.4} - 1) \simeq 0.7$$

assuming the expansion from clearance is adiabatic.

(b) For the double-stage compression, the number of stages is

$$x^n = \frac{P_{fin}}{P_1}$$

$$x^2 = 9 \ ; \quad x = 3$$

The temperature at the end of compression for each stage is

$$T_2 = T_1\left(\frac{P_2}{P_1}\right)^{(\kappa-1)/\kappa} = 293 \times 3^{0.4/1.4} = 293 \times 1.37$$

$$= 402°K = 129°C$$

The total theoretical horsepower in two stages is

$$L_{ad} = nRT_1 \frac{\kappa}{\kappa - 1}\left[\left(\frac{P_{fin}}{P_1}\right)^{(\kappa-1)/\kappa n} - 1\right]$$

$$= 2 \times 29.3 \times 293 \times \frac{1.4}{0.4}(1.88^{1/2} - 1)$$

$$= 22,200 \text{ kg-m/kg}$$

where $1.88 = \left(\frac{P_{fin}}{P_1}\right)^{(\kappa-1)/\kappa}$

The volumetric efficiency is

$$\eta_v = 1 - 0.08(3^{1/1.4} - 1) = 0.905$$

Comparing the results for single- and two-stage compression, we obtain

	One-Stage Compression	Two-Stage Compression
Temperature at the end of compression $T_2, °C =$	278	129
Theoretical horsepower, L_{ad}, kg-m/kg =	26,500	22,200
Volumetric efficiency, η_v =	0.7	0.905

The comparison given shows the advantage of two-stage compression. The higher the ratio P_{fin}/P_1, the more pronounced the advantages of multistage compression.

7.13 (a) An allowable compression ratio for a single stage would be approximately 4. Hence, the required number of stages would be

$$\eta = \frac{\log P_{fin} - \log P_1}{\log x} = \frac{\log 55}{\log 4} = 2.9$$

Neglecting pressure losses between stages, we select a more precise compression ratio in each stage of a three-stage compressor:

$$x = \sqrt[3]{55} = 3.8$$

Thus, the approximate pressure distribution by stages is

	P_{in} (atm)	P_{fin} (atm)
Stage (I)	1	3.8
Stage (II)	3.8	14.45
Stage (III)	14.45	55

(b) The theoretical work consumed is

$$L_{ad} = nRT_1 \frac{\kappa}{\kappa - 1} \left[\left(\frac{P_{fin}}{P_1} \right)^{(\kappa-1)/\kappa n} - 1 \right]$$

$$= 3 \times 52.9 \times 303 \times \frac{1.31}{0.31} (55^{0.31/(1.31 \times 3)} - 1)$$

$$= 76,000 \text{ kg-m/kg}$$

where the following values were obtained from *Chemical Engineer's Handbook*, 5th ed., R. H. Perry and C. H. Chilton, Eds. (New York: McGraw-Hill Book Co., 1973).

- $\kappa = 1.31$
- $R = 52.9$ kg-m/kg-°K
- $\gamma_0 = 0.717$ kg/N-m^3

Compressor's horsepower:

$$N = \frac{GL_{ad}}{3600 \times 102 \times \eta} = \frac{2100 \times 0.717 \times 76,000}{3600 \times 102 \times 0.7} = 44.6 \text{ kW}$$

(c) To determine the amount of cooling water in the compressor's coolers, we find the temperature at the end of compression in stages II and III, assuming that the temperature of the methane is 303°K. The T-S diagram for the system is shown in Figure A.2. In the compressor's cylinder (stage I), the temperature at the end of

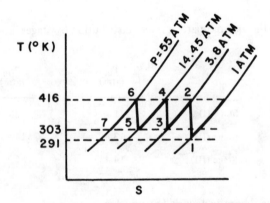

Figure A.2   T-S diagram for the solution to Problem 7.13.

compression will be somewhat lower (point 2) as the methane is
sucked in stage I, not at 30°C but at 18°C:

$$T_2 = T_1 \left(\frac{P_2}{P_1}\right)^{(\kappa-1)/\kappa} = 303 \times 3.8^{0.31/1.31} = 416°K = 143°C$$

Assuming that the heat capacities of methane at pressures 3, 8, 14.45
and 55 atm are approximately equal to 0.531 kcal/kg-°C, we find that
in three coolers (after stages I, II, and III) the water should absorb the
following amount of heat:

$$Q = 3 \times 210 \times 0.717 \times 0.531 (143 - 30) = 27,100 \frac{kcal}{hr}$$

This amount alternatively may be computed as the heat that is
equivalent to the work of compression:

$$Q = \frac{L_{ad} \, V \gamma_0}{427} = \frac{76,000 \times 210 \times 0.717}{427} = 268,000 \text{ kcal/hr}$$

If the water temperature is increased by 10°C, then the amount of
water required is

$$G_{H_2O} = \frac{27,100}{10} = 2710 \text{ kg/hr} = 2.71 \text{ m}^3/\text{hr}$$

7.14 At the pressure specified, it is possible to use either a piston compressor or a turbocompressor. According to Figure 7.44, the capacity
under suction conditions is $Q = 230 \times 60 = 13,800 \text{ m}^3/\text{hr}$, and the
discharge pressure is P = 5 atm gauge. Hence, it is necessary to select a
turbocompressor. The same result is obtained from the empirical
expression determining rationality of application of turbocompressors:

$$\frac{Q_{min}^2}{P_d} > 1000$$

where Q_{min} = capacity (m^3/min)
        P_d = discharge pressure (atm gauge)

In our case, $\dfrac{230^2}{5} = 10,600 \gg 1000$.

7.15 The theoretical work for 1 m^3 of gas to be sucked is

$$L = \frac{m}{m-1} P_1 \left[\left(\frac{P_2}{P_1} \right)^{(m-1)/m} - 1 \right]$$

(a) $\dfrac{P_2}{P_1} = \dfrac{1}{0.9} = 1.11$; $\dfrac{m}{m-1} = \dfrac{1.25}{0.25} = 5$; $\dfrac{m-1}{m} = 0.2$

$$L = 5 \times 0.9 \times 10^4 (1.11^{0.2} - 1) = 945 \text{ kg-m/m}^3$$

(b) $\dfrac{P_2}{P_1} = \dfrac{1}{0.3} = 3.33$

$$L = 5 \times 0.3 \times 10^4 (3.33^{0.2} - 1) = 4080 \text{ kg-m/m}^3$$

(c) $\dfrac{P_2}{P_1} = \dfrac{1}{0.1} = 10$

$$L = 5 \times 0.1 \times 10^4 (10^{0.2} - 1) = 2920 \text{ kg-m/m}^3$$

We see that the work spent goes through a maximum. The horsepower of the electric motor is designed for this maximum.

7.23 (a) The initial capacity of methane is

$$G_{v,0} = \frac{G}{\rho_0} \frac{T}{273} = \frac{1.5}{0.717} \times \frac{293}{273} = 2.25 \text{ m}^3/\text{s}$$

The theoretical horsepower at isothermic compression is

$$N_{iso} = P_0 G_{v,0} \, ln \frac{P_{fin}}{P_0} = 1.033 \times 9.81 \times 10^4 \times 2.25 \, ln \frac{3.5}{1.0}$$

$$= 2.86 \times 10^5 \text{W}$$

where $P_0 = 1$ atm $= 1.033 \times 9.81 \times 10^4$ N/m^2

The theoretical horsepower for adiabatic compression is

$$N_{ad} = P_0 G_{v,0} \frac{\kappa}{\kappa - 1} \left[\left(\frac{P_{fin}}{P_0} \right)^{(\kappa-1)/\kappa} - 1 \right]$$

$$= 1.033 \times 9.81 \times 10^4 \times 2.25 \times \frac{1.31}{1.31 - 1} [3.5^{(1.31-1)/1.31} - 1]$$

$$= 3.38 \times 10^5 \text{ W}$$

The indicator horsepower is

$$N_{ind} = N\eta_m = 430 \times 0.91 = 391 \text{ kW} = 3.91 \times 10^5 \text{ W}$$

The isothermal and adiabatic efficiencies are

$$\eta_{iso} = \frac{N_{iso}}{N_{ind}} = \frac{2.86 \times 10^5}{3.91 \times 10^5} = 0.732$$

$$\eta_{ad} = \frac{N_{ad}}{N_{ind}} = \frac{3.38 \times 10^5}{3.91 \times 10^5} = 0.865$$

(b) Determine the polytropic exponent by using the following equation:

$$\frac{T_{fin}}{T_0} = \left(\frac{P_{fin}}{P_0} \right)^{(n-1)/n}$$

After taking the logarithm from this equation we obtain

$$\frac{n - 1}{n} = \frac{\log (T_{fin}/T_0)}{\log (P_{fin}/P_0)} = \frac{\log \left(\dfrac{273 + 80}{273 + 20} \right)}{\log \dfrac{3.5}{1}} = 0.149$$

Hence,

$$n = \frac{1}{1 - 0.149} = 1.175$$

7.24 (a) The number of compression stages is estimated from

$$Z^m = \frac{P_{fin}}{P_0}$$

Assuming the compression ratio in each stage to be 3, we obtain

$$m = \frac{\log \dfrac{P_{fin}}{P_0}}{\log Z} = \frac{\log \dfrac{110}{1.033}}{\log 3} = 4.26$$

We assume four compression stages ($m = 4$) and, therefore, the compression ratio is

$$Z = \left(\frac{P_{fin}}{P_0}\right)^{1/4} = \left(\frac{110}{1.033}\right)^{1/4} = 3.22$$

The volumetric capacity at standard conditions depends on the temperature at the inlet of the corresponding compression stage:

$$G_{v,0} = \frac{G}{\rho_0} \times \frac{273 + t}{273} = \frac{0.5}{1.293} \times \frac{273 + t}{273} = 0.387 \frac{273 + t}{273} \quad , \quad m^3/s$$

The horsepower in the first stage is

$$N_1 = P_0 G_{v,0} \frac{n}{n - 1} (Z^{(n-1)/n} - 1)$$

$$= 1.033 \times 9.81 \times 10^4 \times 0.387 \times \frac{273 + 20}{273}$$

$$\times \frac{1.3}{1.3 - 1} (3.22^{(1.3-1)/1.3} - 1) = 5.65 \times 10^4 \, W$$

The horsepower for the other stages is determined as follows:

$$N_{2\text{-}4} = 3P_0 G'_{v,0} \frac{n}{n-1} (Z^{(n-1)/n} - 1)$$

$$= 3N_1 \frac{273+35}{273+20} = 17.82 \times 10^4 \text{ W}$$

where $\quad G'_{v,0} = G_{v,0} \dfrac{273+35}{273+20}$

The total horsepower is, therefore,

$$N = N_1 + N_{2\text{-}4} = (5.65 + 17.82) \times 10^4 = 2.35 \times 10^5 \text{ W}$$

(b) The horsepower for a single-stage compressor for isothermic compression is

$$N_{iso} = P_0 G_{v,0} \, ln \frac{P_{fin}}{P_0}$$

$$= 1.033 \times 9.81 \times 10^4 \times 0.387 \frac{273+20}{273} \, ln \frac{110}{1.033}$$

$$= 1.96 \times 10^5 \text{ W}$$

Thus, for multistage polytropic compression the horsepower is

$$\frac{2.35 - 1.96}{1.96} \times 100 = 19.9\%$$

more than that of the isothermic compression. The horsepower for a single-stage polytropic compression is

$$N_p = P_0 G_{v,0} \frac{n}{n-1} \left[\left(\frac{P_{fin}}{P_0}\right)^{(n-1)/n} - 1 \right]$$

$$= 1.033 \times 9.81 \times 10^4 \times 0.387 \times \frac{293}{273}$$

$$\times \frac{1.3}{1.3-1} \left[\left(\frac{110}{1.033}\right)^{(1.3-1)/1.3} - 1 \right] = 3.54 \times 10^5 \text{ W}$$

Thus, at the four-stage compression, the horsepower is decreased by

$$\frac{3.54 - 2.35}{3.54} \times 100 = 33.6\%$$

compared to that of a single-stage polytropic compression.

7.27 4.55%

7.28 For winter service, 17,090 kg. For summer service, 14,600 kg.

7.29 80 mm H_2O

7.30 4.2 kW

7.31 (a) $Q = 4170 \text{ m}^3/\text{hr}$
$\Delta P = 74.8 \text{ kg/m}^2$
$N = 1.78 \text{ kW}$
(b) $\eta = 0.48$

7.32 117°C

7.33 $L_{ad} = 12,000$ kg-m

7.34 7 kW

7.35 $\eta_v = 0.89$

7.36 $Q = 3.21 \text{ m}^3/\text{min}$
$N = 12.9$ kW

7.37 Q increases to 5.67 m^3/min
N increases to 14 kW

7.39 $P_d = 3.67$ atm for air
$P_d = 9.3$ atm for ethane gas

7.40 4 stages

7.44 (a) Single-stage piston compressor
(b) Rotary compressor

CHAPTER 9: ANSWERS AND SOLUTIONS

9.2 Stokes' formula is applicable for Re < 0.2. Therefore, the largest particle whose settling velocity can be computed from Stokes' law will have a diameter

$$d = \frac{0.2\mu g}{u_s \gamma}$$

Hence,

$$u_s = \frac{0.2\mu g}{d\gamma}$$

On the other hand,

$$u_s = \frac{d^2(\gamma_p - \gamma)}{18\mu}$$

Equating both expressions through u_s, we obtain

$$d = \sqrt[3]{\frac{0.2 \times 18 \times \mu^2 g}{(\gamma_p - \gamma)\gamma}}$$

The viscosity of water at 20°C is $\mu = 1$ cp $\simeq 10^{-4}$ kg-s/m^2. Then,

$$d = \sqrt[3]{\frac{0.2 \times 18 \times 10^{-8} \times 9.81}{(2650 - 1000)1000}} = 60 \times 10^{-6} m = 60\,\mu m$$

The settling velocity corresponding to this diameter is

$$u_s = \frac{0.2 \times 10^{-4} \times 9.81}{60 \times 10^{-6} \times 1000} = 3.27 \times 10^{-3} \text{ m/s}$$

9.3 First determine the Archimedes number:

$$Ar = \frac{d^3(\gamma_p - \gamma)\gamma}{\mu^2 g} = \frac{0.9^3 \times 10^{-9}(2650 - 1000)1000 \times 9810^2}{1^2 \times 9.81}$$

$$= 1.18 \times 10^4$$

where $\mu = 1$ cp for water

Using the computed value of Ar and using the appropriate curve for spherical particles in Figure 9.23, we find a value for the Reynolds number, Re = 140. From the Reynolds number, the settling velocity can be evaluated for 0.9-mm-diameter spherical particles:

$$u_s = \frac{Re\mu g}{d\gamma} = \frac{140 \times 1 \times 9.81}{0.0009 \times 1000 \times 9810} = 0.15 \text{ m/s}$$

9.4   The Lyachshenko number is

$$Ly = \frac{u_s^3 \gamma^2}{\mu(\gamma_p - \gamma)g^2} = \frac{0.5^3 \times 1000^2 \times 9180}{1.3(2710 - 1000)9.81^2} = 5.72 \times 10^3$$

where water viscosity at $t = 10°$ is $\mu = 1.3$ cp.

From the calculated value of Ly = 5.72×10^3, the value of the Reynolds number is obtained from Figure 9.23, Re = 1750. We then compute the maximum particle diameter entrained by the water:

$$d = \frac{Re\mu g}{u_s\gamma} = \frac{1750 \times 1.3 \times 9.81}{0.5 \times 1000 \times 9.81 \times 10^3} = 4.55 \times 10^{-3} \text{m} = 4.55 \text{ mm}$$

9.5 $u_s = 0.271$ m/s

9.6   The size of particles can be computed from the formula

$$d_{eq} = \sqrt[3]{\frac{Ar g \mu^2}{(\gamma_p - \gamma)\gamma}}$$

The value of Ar can be obtained from Figure 9.23, for the respective particle shapes. For the elongated coal particles,

$$Ly_1 = \frac{u_s^3 \gamma^2}{(\gamma_1 - \gamma)g^2} = \frac{10^{-3} \times 10^6 \times 9.81 \times 10^3}{1 \times 0.4 \times 10^3 \times 9.81^2} = 255$$

For plate-like shale particles,

$$Ly_2 = \frac{u_s^3 \gamma^2}{\mu(\gamma_2 - \gamma)g^2} = \frac{10^{-3} \times 10^6 \times 9.81 \times 10^3}{1 \times 1.2 \times 10^3 \times 9.81^2} = 85$$

The value of $Ly_1 = 255$ corresponds to $Ar_1 = 9 \times 10^4$ for elongated particles. The value of $Ly_2 = 85$ corresponds to $Ar_2 = 7 \times 10^4$ for plate-like particles. The equivalent diameter of the coal particles is

$$d_{eq} = \sqrt[3]{\frac{Ar_1 g \mu^2}{(\gamma_1 - \gamma)\gamma}} = \sqrt[3]{\frac{9 \times 10^4 \times 9.81 \times 1}{9.81^2 \times 10^6 \times 0.4 \times 10^3 \times 10^3}}$$

$$= 2.82 \times 10^{-3} \text{m} = 2.82 \text{ mm}$$

The equivalent diameter for shale particles is

$$d_{eq} = \sqrt[3]{\frac{Ar_2 \times g \times \mu^2}{(\gamma_2 - \gamma)\gamma}} = \sqrt[3]{\frac{7 \times 10^4 \times 9.81 \times 1}{9.81^2 \times 10^6 \times 1.2 \times 10^3 \times 10^3}}$$

$$= 1.81 \times 10^{-3} \text{m} = 1.81 \text{ mm}$$

9.7 We determine the Archimedes number:

$$Ar = Ga \frac{\rho_p - \rho}{\rho} = \frac{g\ell^3 \rho(\rho_p - \rho)}{\mu^2} = \frac{g\ell^3}{\nu^2} \cdot \frac{\rho_p - \rho}{\rho}$$

$$= \frac{(25 \times 10^{-6})^3 \times 9.81(2750 - 1200)}{(2.4 \times 10^{-3})^2} = 0.0497$$

In this case $Ar < 2$; consequently, the settling takes place in the laminar range ($Re < 2$) and, from the formula,

$$Re = 0.056 \, Ar = 0.056 \times 0.0497 = 0.00278$$

Therefore, the settling velocity of the particle is:

$$u_s = \frac{\mu Re}{d\rho} = \frac{2.4 \times 10^{-3} \times 0.00278}{25 \times 10^{-6} \times 1200} = 2.22 \times 10^{-4} \text{ m/s}$$

or

$$u_s = 2.22 \times 10^{-4} \times 3600 = 0.8 \text{ m/hr}$$

9.8   The volume fraction of solids in the suspension is

$$q = 1 - \epsilon = \frac{x\rho_s}{\rho_p} = \frac{0.3 \times 1440}{2750} = 0.157$$

$$\epsilon = 1 - 0.157 = 0.843$$

$$\Phi(\epsilon) = 10^{-1.82(1-\epsilon)} = 10^{-1.82(0.157)} = 10^{-0.286} \simeq 0.52$$

According to Problem 9.2, the free settling velocity is $u_s = 0.8$ m/hr. The velocity of hindered sedimentation is, therefore,

$$u_s = u_{fis.} \times \epsilon^2 \times \Phi(\epsilon) = 0.8 \times 0.843^2 \times 0.52 \simeq 0.3 \text{ m/hr}$$

9.9   The capacity of the thickener for the solid phase is

$$G_{s.ph} = 20{,}000 \times 0.20 = 4000 \text{ kg/hr}$$

The capacity of the thickener for the thickened sludge is

$$G_{th} = \frac{G_{s.ph}}{x_2} = \frac{4000}{0.5} = 8000 \text{ kg/hr}$$

Correspondingly, the capacity of the thickener for the clarified liquid is

$$G_\varrho = G_s - G_{th} = 20{,}000 - 8000 = 12{,}000 \text{ kg/hr}$$

The ratio of solids in suspension to sludge is

$$\beta = \frac{x_1}{x_2} = \frac{20}{50} = 0.4$$

The cross-sectional area of the thickener is, hence,

$$F = \frac{1.3 \times G_s}{\rho u_s} (1 - \beta) = \frac{1.3 \times 20{,}000(1 - 0.4)}{1050 \times 0.5} = 29.6 \text{ m}^2$$

Thus, the diameter of the thickener is

$$D = \sqrt{\frac{4F}{\pi}} = \sqrt{\frac{4 \times 29.6}{3.14}} = 6.15 \text{ m}$$

The design recommendation can assume a diameter of 7 m.

9.15 The critical diameter of particles (i.e., the maximum diameter of particles) that can settle in the given zone of settling is computed based on Stokes law:

$$d_{cr} = 2.62 \sqrt[3]{\frac{\mu^2}{g(\rho_1 - \rho_2)\rho_2}} = 2.62 \sqrt[3]{\frac{(1.8 \times 10^{-5})^2}{9.81(900 - 1.2)}}$$

$$= 2.62 \times 3.1 \times 10^{-5} = 8.1 \times 10^{-5} \text{m} = 81 \text{ } \mu\text{m}$$

For the transitional zone the critical diameter will be

$$d_{cr} = 69.1 \times 3.1 \times 10^{-5} = 2.1 \times 10^{-3} \text{m} = 2.1 \text{ mm}$$

9.16 As follows from Problem 9.15, the maximum settling diameter of droplets following Stokes' law is 81 μm. For a droplet diameter of 15 μm, the settling will follow Stokes' law. The settling velocity is

$$u_s = \frac{1}{18} d^2 \frac{\rho_1 - \rho_2}{\mu} g = \frac{(1.5 \times 10^{-5})^2}{18} \times \frac{900 - 1.2}{1.8 \times 10^{-5}} \times 9.81$$

$$= 6.15 \times 10^{-3} \text{ m/s}$$

9.19 First compute the gas capacity at the operating conditions through the ideal gas law:

$$V_{sec} = \frac{0.6(273 + 427)}{273} = 1.54 \text{ m}^3/\text{s}$$

The linear gas velocity through the unit (neglecting the thicknesses of the shelves) is

$$w_g = \frac{1.54}{2.8 \times 4.2} = 0.131 \text{ m/s}$$

The gas residence time is

$$\tau = \frac{L}{w_g} = \frac{4.1}{0.131} = 31.3 \text{ s}$$

The theoretical settling velocity of spherical particles (neglecting gas density) is (from Stokes' law)

$$w_s = \frac{1}{18} = \frac{(8 \times 10^{-6})^2 \times 4000 \times 9810}{0.034} = 0.0041 \text{ m/s}$$

We wish some overdesign in the system, so the actual settling velocity is assumed to be

$$0.5 \times 0.0041 = 0.002 \text{ m/s}$$

The distance between the shelves is, therefore,

$$h = w_s \times \tau = 0.002 \times 31.3 = 0.06 \text{ m} = 60 \text{ mm}$$

Computing the Reynolds number based on the theoretical settling velocity,

$$Re = \frac{w_s d \rho_p}{\mu} = \frac{8 \times 10^{-6} \times 0.0041 \times 1000}{0.034} = 0.00048$$

Because $Re = 0.00048 < 0.2$, the application of Stokes' formula is correct.

CHAPTER 11: SOLUTIONS

11.16 0.25 kW

11.17 240 rpm

11.18 The mixer rpm will be increased by a factor of 1.4.

11.19 0.54 m

CHAPTER 12: ANSWERS AND SOLUTIONS

12.1 Filtrate volume per unit area from a single batch operation is

$$q = \frac{V}{S} = \frac{V_s}{S(1 + x_0)}$$

where V = volume of filtrate (m^3)
V_s = volume of suspension (m^3)
S = filtration area (m^2)
x_0 = ratio of cake to filtrate volumes (m^2/m^3)

$$q = \frac{0.5}{1(1 + 0.01)} = 0.495 \text{ m}^3/\text{m}^2$$

At $R_f = 0$ and $S = 1 \text{ m}^2$, the following expression is applicable:

$$\tau = \frac{\mu r_0 x_0}{2\Delta P} \, q^2$$

and $r_0 = r_0'(p)^{s'}$. Hence,

$$\tau = \frac{\mu r_0 x_0}{2(\Delta P)^{1 - s'}} \, q^2$$

or

$$\tau = \frac{10^{-3} \times 0.5 \times 10^{10} \times 0.01}{2(\Delta P)^{1 - 0.95}} \, (0.495)^2 \simeq 6.12 \times 10^3 (\Delta P)^{-0.05}$$

Based on this expression, a doubling of the ΔP results in only a 3.3% reduction in τ (i.e., from 3.60×10^3 to 3.48×10^3 s). The problem demonstrates that a significant reduction in filtration time does not occur for heavily compressible cakes.

12.2 (a) Time required for a single cycle to achieve the maximum ΔP of 20×10^4 N/m^2 is $\tau \simeq 56$ min.
(b) 0.674 m^3 filtrate volume.
(c) $h_c = 0.017$-m cake thickness.

12.4 Assuming a length of 1 m for the filter media:

- Cylindrical surface = 0.0234 m^3
- Flat filter plate = 0.0157 m^3

12.5
- Total filtration time = 5953 s
- Total filtrate volume = 2.0 m^3
- Volume of cake = 0.2 m^3
- Volume of separated slurry = 2.2 m^3

12.6 Total filtration time $\tau = 3360$ s

12.10 6.9 m^3/hr and 7.2 m/min

12.12 12.1 m^3/hr

12.31 140 min

12.32 65 min

12.33 At P = 0.35 atm: $K = 278 \times 10^{-4}$ m^2/hr
$C = 4.7 \times 10^{-3}$ m^3/m^2
At P = 1.05 atm: $K = 560 \times 10^{-4}$ m^2/hr
$C = 3.78 \times 10^{-3}$ m^3/m^2

12.34 **Solution**

Specific cake resistance is computed from

$$r = \frac{2\Delta P(1 - mx)}{K\mu\gamma x} \quad , \quad \text{m/kg of dry cake} \qquad \text{(A)}$$

Filtration pressure $\Delta P = 3500$ kg/m^2

Filtrate specific gravity $\gamma = 1000$ kg/m^3

Filtrate viscosity at 20°C, $\mu = \dfrac{1}{9810 \times 3600} = 2.83 \times 10^{-8}$ kg-hr/m^2

Filtration constant (at $P_1 = 0.35$ atm) $K = 278 \times 10^{-4}$ m^2/hr

Weight fraction of solids in suspension $x_0 = 0.139$

Weight ratio of wet cake to dry, $m = \dfrac{1}{1 - 0.37} = 1.59$

Amount of filtrate per 1 kg of suspension, $1 - mx_0 = 1 - 1.59 \times 0.139 = 0.779$

The obtained values are substituted in Equation A:

$$r_0 = \frac{2\Delta P(1 - mx_0)}{K\mu\gamma x_0} = \frac{2 \times 3500 \times 0.779}{278 \times 10^{-4} \times 1000 \times 2.83 \times 10^{-8} \times 0.139}$$

$r_0 = 5 \times 10^{10}$ m/kg dry cake

At pressure $P_2 = 1.05$ atm.

$$m = \frac{1}{1 - 0.32} = 1.47$$

$$1 - mx_0 = 1 - 1.47 \times 0.139 = 0.795$$

The other values are the same as for $P_1 = 0.35$ atm. The new specific cake resistance may be calculated by substituting the numbers in Equation A or from the ratio

$$\frac{r_2}{r_1} = \frac{\Delta P_2 K_1 (1 - mx_0)_2}{\Delta P_1 K_2 (1 - mx_0)_1} = \frac{1.05 \times 278 \times 10^{-4} \times 0.795}{0.35 \times 560 \times 10^{-4} \times 0.779} = 1.52$$

Hence,

$$r_2 = 5 \times 10^{10} \times 1.52 = 7.6 \times 10^{10} \text{ m/kg dry cake}$$

Thus, the threefold increase in resistance results in the increase in specific cake pressure of 52%.

12.35 For $P_1 = 0.35$ atm, $R_f = 4.2 \times 10^{10}$ m/m^2
For $P_2 = 1.05$ atm, $R_f = 5 \times 10^{10}$ m/m^2

12.36 77.5 m^2

12.37 $F = 11.4$ m^2 and $n = 0.655$ rpm

12.38 41 min

12.39 2.53 g/ℓ

12.40 $a = 1.9 \times 10^{-4}$ s/m^2; $b = 570$ s/m

12.41 1.14 hr

12.42 $r_0 = 2.76 \times 10^{13}$ m^{-2}
$R_f = 3.6 \times 10^{10}$ m^{-1}

12.43 4.6×10^{-5} m^3/m^2-s

12.44 0.75 hr

12.45 0.096 m

APPENDIX B

GENERAL CONVERSION FACTORS AND PHYSICAL PROPERTIES DATA

Data and conversion factors given in
this Appendix were compiled by:

Durametallic Corp.
Kalamazoo, Michigan

CONTENTS

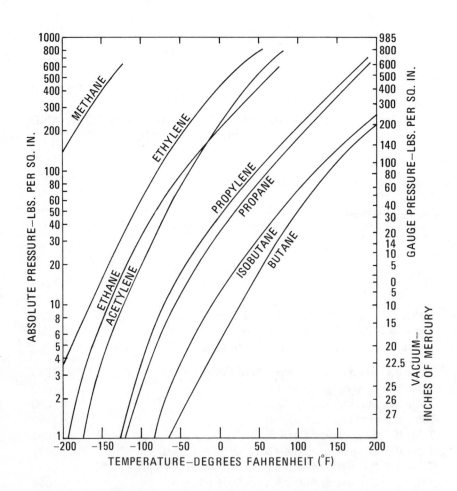

Figure B.1   Vapor pressures of light hydrocarbons.

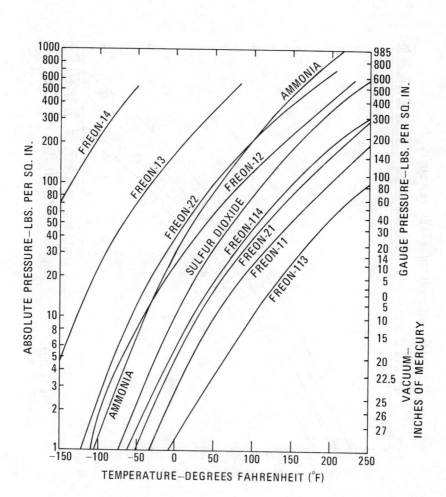

Figure B.2   Vapor pressures of refrigerants.

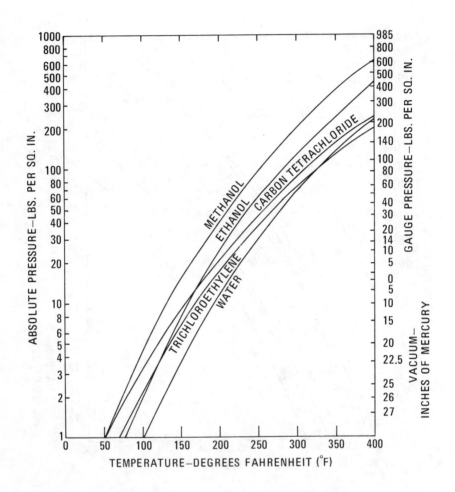

Figure B.3   Vapor pressures of common liquids.

Table B.1   Areas of Circumferences of Circles

$$Circumference = 3.14159265359 \times D$$
$$Area = 0.7853981634 \times D^2$$

Diameter	0 in. Circumference	0 in. Area	1 in. Circumference	1 in. Area	2 in. Circumference	2 in. Area	3 in. Circumference	3 in. Area
0			3.1416	0.7854	6.2832	3.1416	9.425	7.0686
1/64	0.04909	0.000192	3.1907	0.8101	6.3323	3.1909	9.474	7.1424
1/32	0.09817	0.000767	3.2398	0.8352	6.3814	3.2405	9.523	7.2166
3/64	0.1473	0.001726	3.2889	0.8607	6.4304	3.2906	9.572	7.2912
1/16	0.1963	0.003068	3.3379	0.8866	6.4795	3.3410	9.621	7.3662
5/64	0.2454	0.004794	3.3870	0.9129	6.5286	3.3918	9.670	7.4415
3/32	0.2945	0.006903	3.4361	0.9396	6.5777	3.4430	9.719	7.5173
7/64	0.3436	0.009396	3.4852	0.9666	6.6268	3.4946	9.768	7.5934
1/8	0.3927	0.01227	3.5343	0.9940	6.6759	3.5466	9.817	7.6699
9/64	0.4419	0.01553	3.5835	1.0218	6.7250	3.5989	9.867	7.7468
5/32	0.4909	0.01917	3.6325	1.0500	6.7741	3.6516	9.916	7.8241
11/64	0.5400	0.02320	3.6816	1.0786	6.8231	3.7048	9.965	7.9017
3/16	0.5890	0.02761	3.7306	1.1075	6.8722	3.7583	10.014	7.9798
13/64	0.6381	0.03240	3.7797	1.1369	6.9213	3.8121	10.063	8.0582
7/32	0.6872	0.03758	3.8288	1.1666	6.9704	3.8664	10.112	8.1370
15/64	0.7363	0.04314	3.8779	1.1967	7.0195	3.9211	10.161	8.2162
1/4	0.7854	0.04909	3.9270	1.2272	7.0686	3.9761	10.210	8.2958
17/64	0.8345	0.05541	3.9761	1.2577	7.1177	4.0315	10.259	8.3757
9/32	0.8836	0.06213	4.0252	1.2893	7.1668	4.0873	10.308	8.4561
19/64	0.9327	0.06922	4.0743	1.3209	7.2158	4.1435	10.357	8.5368

Table B.1, continued

Diameter	0 in. Circumference	0 in. Area	1 in. Circumference	1 in. Area	2 in. Circumference	2 in. Area	3 in. Circumference	3 in. Area
5/16	0.9817	0.07670	4.1233	1.3530	7.2649	4.2000	10.407	8.6179
21/64	1.0308	0.08456	4.1724	1.3854	7.3140	4.2570	10.456	8.6994
11/32	1.0799	0.09281	4.2215	1.4182	7.3631	4.3143	10.505	8.7813
23/64	1.1290	0.1014	4.2706	1.4513	7.4122	4.3720	10.554	8.8636
3/8	1.1781	0.1104	4.3197	1.4849	7.4613	4.4301	10.603	8.9462
25/64	1.2272	0.1198	4.3688	1.5188	7.5104	4.4886	10.652	9.0292
13/32	1.2763	0.1296	4.4179	1.5532	7.5595	4.5475	10.701	9.1126
27/64	1.3254	0.1398	4.4670	1.5879	7.6085	4.6067	10.750	9.1964
7/16	1.3744	0.1503	4.5160	1.6230	7.6576	4.6664	10.799	9.2806
29/64	1.4235	0.1613	4.5651	1.6584	7.7067	4.7264	10.848	9.3652
15/32	1.4726	0.1726	4.6142	1.6943	7.7558	4.7868	10.897	9.4501
31/64	1.5217	0.1843	4.6633	1.7305	7.8049	4.8476	10.946	9.5354
1/2	1.5708	0.1963	4.7124	1.7671	7.8540	4.9088	10.996	9.6212
33/64	1.6199	0.2088	4.7615	1.8042	7.9031	4.9703	11.045	9.7072
17/32	1.6690	0.2217	4.8106	1.8415	7.9522	5.0322	11.094	9.7937
35/64	1.7181	0.2349	4.8597	1.8793	8.0012	5.0946	11.143	9.8806
9/16	1.7671	0.2485	4.9087	1.9175	8.0503	5.1573	11.192	9.9678
37/64	1.8162	0.2625	4.9578	1.9560	8.0994	5.2203	11.241	10.0554
19/32	1.8653	0.2769	5.0069	1.9949	8.1485	5.2838	11.290	10.1435
39/64	1.9144	0.2916	5.0560	2.0342	8.1976	5.3477	11.339	10.2318

5/8	1.9635	0.3068	5.1051	2.0739	8.2467	5.4119	11.388	10.3206
41/64	2.0126	0.3223	5.1542	2.1140	8.2958	5.4765	11.437	10.4098
21/32	2.0617	0.3382	5.2033	2.1545	8.3449	5.5415	11.486	10.4994
43/64	2.1108	0.3545	5.2524	2.1953	8.3939	5.6069	11.536	10.5893
11/16	2.1598	0.3712	5.3014	2.2365	8.4430	5.6727	11.585	10.6796
45/64	2.2089	0.3883	5.3505	2.2782	8.4921	5.7388	11.634	10.7703
23/32	2.2580	0.4057	5.3996	2.3201	8.5412	5.8054	11.683	10.8614
47/64	2.3071	0.4236	5.4487	2.3623	8.5903	5.8723	11.732	10.9528
3/4	2.3562	0.4418	5.4978	2.4053	8.6394	5.9396	11.781	11.0447
49/64	2.4053	0.4604	5.5469	2.4484	8.6885	6.0073	11.830	11.1369
25/32	2.4544	0.4794	5.5960	2.4929	8.7376	6.0753	11.879	11.2295
51/64	2.5035	0.4987	5.6451	2.5359	8.7866	6.1438	11.928	11.3236
13/16	2.5525	0.5185	5.6941	2.5802	8.8357	6.2126	11.977	11.4159
53/64	2.6016	0.5386	5.7432	2.6248	8.8848	6.2819	12.026	11.5096
27/32	2.6507	0.5591	5.7923	2.6699	8.9339	6.3515	12.075	11.6038
55/64	2.6998	0.5800	5.8414	2.7153	8.9830	6.4215	12.125	11.6983
7/8	2.7489	0.6013	5.8905	2.7612	9.0321	6.4918	12.174	11.7933
57/64	2.7980	0.6230	5.9396	2.8074	9.0812	6.5626	12.223	11.8885
29/32	2.8471	0.6450	5.9887	2.8540	9.1303	6.6337	12.272	11.9842
59/64	2.8962	0.6675	6.0377	2.9010	9.1793	6.7052	12.321	12.0803
15/16	2.9452	0.6903	6.0868	2.9483	9.2284	6.7771	12.370	12.1768
61/64	2.9943	0.7135	6.1359	2.9961	9.2775	6.8494	12.419	12.2736
31/32	3.0434	0.7371	6.1850	3.0442	9.3266	6.9221	12.468	12.3708
63/64	3.0925	0.7610	6.2341	3.0927	9.3757	6.9952	12.517	12.4684

Table B.1, continued

Diameter	4 in.		5 in.		6 in.		7 in.	
	Circumference	Area	Circumference	Area	Circumference	Area	Circumference	Area
0	12.566	12.566	15.708	19.635	18.850	28.274	21.991	38.485
1/64	12.615	12.665	15.757	19.753	18.889	28.422	22.040	38.656
1/32	12.665	12.763	15.806	19.881	18.948	28.570	22.089	38.829
3/64	12.714	12.863	15.855	20.005	18.997	28.718	22.138	39.002
1/16	12.763	12.962	15.904	20.129	19.046	28.866	22.187	39.175
5/64	12.812	13.062	15.953	20.253	19.095	29.015	22.237	39.348
3/32	12.861	13.162	16.002	20.378	19.144	29.165	22.286	39.522
7/64	12.910	13.263	16.052	20.503	19.193	29.315	22.335	39.696
1/8	12.959	13.364	16.101	20.629	19.242	29.465	22.384	39.871
9/64	13.008	13.465	16.150	20.755	19.291	29.615	22.433	40.046
5/32	13.057	13.567	16.199	20.881	19.340	29.766	22.482	40.222
11/64	13.106	13.669	16.248	21.008	19.390	29.917	22.531	40.398
3/16	13.155	13.772	16.297	21.135	19.439	30.069	22.580	40.574
13/64	13.205	13.875	16.346	21.263	19.488	30.221	22.629	40.750
7/32	13.254	13.978	16.395	21.391	19.537	30.374	22.678	40.927
15/64	13.303	14.082	16.444	21.519	19.586	30.526	22.727	41.105
1/4	13.352	14.186	16.493	21.648	19.635	30.680	22.777	41.282
17/64	13.401	14.291	16.542	21.777	19.684	30.833	22.826	41.461
9/32	13.450	14.396	16.592	21.906	19.733	30.987	22.875	41.639
19/64	13.499	14.501	16.641	22.036	19.782	31.141	22.924	41.818

5/16	13.548	14.607	16.690	22.166	19.831	31.296	22.973	41.997
21/64	13.597	14.713	16.739	22.297	19.880	31.451	23.022	42.177
11/32	13.646	14.819	16.788	22.428	19.929	31.607	23.071	42.357
23/64	13.695	14.926	16.837	22.559	19.979	31.763	23.120	42.537
3/8	13.744	15.033	16.886	22.691	20.028	31.919	23.169	42.718
25/64	13.794	15.141	16.935	22.823	20.077	32.076	23.218	42.899
13/32	13.843	15.249	16.984	22.955	20.126	32.233	23.267	43.081
27/64	13.892	15.357	17.033	23.088	20.175	32.390	23.317	43.263
7/16	13.941	15.466	17.082	23.221	20.224	32.548	23.366	43.445
29/64	13.990	15.575	17.131	23.355	20.273	32.706	23.415	43.628
15/32	14.039	15.684	17.181	23.489	20.322	32.865	23.464	43.811
31/64	14.088	15.794	17.230	23.623	20.371	33.024	23.513	43.995
1/2	14.137	15.904	17.279	23.758	20.420	33.183	23.562	44.179
33/64	14.186	16.015	17.328	23.893	20.469	33.343	23.611	44.363
17/32	14.235	16.126	17.377	24.029	20.519	33.503	23.660	44.548
35/64	14.284	16.237	17.426	24.165	20.568	33.663	23.709	44.733
9/16	14.334	16.349	17.475	24.301	20.617	33.824	23.758	44.918
37/64	14.383	16.461	17.524	24.438	20.666	33.985	23.807	45.104
19/32	14.432	16.574	17.573	24.575	20.715	34.147	23.856	45.290
39/64	14.481	16.687	17.622	24.713	20.764	34.309	23.906	45.477
5/8	14.530	16.800	17.671	24.850	20.813	34.472	23.955	45.664
41/64	14.579	16.914	17.721	24.989	20.862	34.634	24.004	45.851
21/32	14.628	17.028	17.770	25.127	20.911	34.798	24.053	46.039
43/64	14.677	17.142	17.819	25.266	20.960	34.961	24.102	46.227
11/16	14.726	17.257	17.868	25.406	21.009	35.125	24.151	46.415

Table B.1, continued

Diameter	4 in. Circumference	4 in. Area	5 in. Circumference	5 in. Area	6 in. Circumference	6 in. Area	7 in. Circumference	7 in. Area
45/64	14.775	17.372	17.917	25.546	21.058	35.289	24.200	46.604
23/32	14.824	17.488	17.966	25.686	21.108	35.454	24.249	46.793
47/64	14.873	17.604	18.015	25.826	21.157	35.619	24.298	46.983
3/4	14.923	17.721	18.064	25.967	21.206	35.785	24.347	47.173
49/64	14.972	17.837	18.113	26.108	21.255	35.951	24.396	47.363
25/32	15.021	17.954	18.162	26.250	21.304	36.117	24.446	47.554
51/64	15.070	18.072	18.211	26.392	21.353	36.283	24.495	47.745
13/16	15.119	18.190	18.261	26.535	21.402	36.450	24.544	47.937
53/64	15.168	18.308	18.310	26.678	21.451	36.618	24.593	48.129
27/32	15.217	18.427	18.359	26.821	21.500	36.787	24.642	48.321
55/64	15.266	18.546	18.408	26.964	21.549	36.954	24.691	48.514
7/8	15.315	18.665	18.457	27.109	21.598	37.122	24.740	48.707
57/64	15.364	18.785	18.506	27.252	21.648	37.291	24.789	48.900
29/32	15.413	18.906	18.555	27.398	21.697	37.461	24.838	49.094
59/64	15.463	19.026	18.604	27.543	21.746	37.630	24.887	49.288
15/16	15.512	19.147	18.653	27.688	21.795	37.800	24.936	49.483
61/64	15.561	19.268	18.702	27.834	21.844	37.971	24.985	49.678
31/32	15.610	19.390	18.751	27.981	21.893	38.142	25.035	49.874
63/64	15.659	19.512	18.800	28.127	21.942	38.313	25.084	50.069

Table B.2 Conversion Factors

To convert from . . .	To . . .	Multiply by . . .
Atmospheres	Inches of mercury, at 0°C	29.92
Atmospheres	Pounds/square inch	14.70
Btu	Foot-pounds	778.3
Btu	Horsepower-hours	3.931×10^{-4}
Btu/hour	Foot-pounds/second	0.2162
Centimeters	Inches	0.3937
Cubic centimeters	Cubic feet	3.531×10^{-5}
Cubic centimeters	Cubic inches	0.06102
Cubic centimeters	Gallons (U.S. liquid)	2.642×10^{-4}
Cubic centimeters	Quarts (U.S. liquid)	1.057×10^{-3}
Cubic feet	Cubic centimeters	28320.0
Cubic feet	Cubic inches	1728.0
Cubic feet	Gallons (U.S. liquid)	7.48052
Cubic feet/minute	Cubic centimeters/second	472.0
Cubic feet/second	Gallons/minute	448.831
Cubic inches	Cubic centimeters	16.39
Degrees/second	Revolutions/minute	0.1667
Dynes	Kilograms	1.020×10^{-6}
Dynes	Pounds	2.248×10^{-6}
Feet of water	Inches of mercury	0.8826
Feet of water	Pounds/square inch	0.4335
Feet/minute	Centimeters/second	0.5080
Foot-pounds	Horsepower-hours	5.050×10^{-7}
Foot-pounds/minute	Foot-pounds/second	0.01667
Foot-pounds/second	Horsepower	1.818×10^{-3}
Gallons	Cubic centimeters	3785.0
Gallons	Cubic feet	0.1337
Gallons	Cubic inches	231.0
Gallons (U.S.)	Gallons (Imperial)	0.83267
Gallons/minute	Cubic feet/hour	8.0208
Horsepower	Foot-pounds/second	550.0
Inches	Centimeters	2.540
Inches of mercury	Kilograms/square centimeter	0.03453
Inches of mercury	Pounds/square inch	0.4912
Liters	Cubic inches	61.02
Liters	Gallons (U.S. liquid)	0.2642
Liters/minute	Gallons/second	4.403×10^{-3}
Millimeters	Inches	0.03937
Pounds/square inch	Atmospheres	0.06804
Pounds/square inch	Feet of water	2.307
Pounds/square inch	Inches of mercury	2.036
Pounds/square inch	Kilograms/square meter	703.1
Radians	Degrees	57.30
Radians	Minutes	3438.0
Radians	Seconds	2.063×10^{5}
Radians/second	Revolutions/minute	9.549
Radians/second	Revolutions/second	0.1592
Revolutions	Radians	6.283
Square centimeters	Square inches	0.1550
Square inches	Square centimeters	6.452
Square inches	Square millimeters	645.2
Square millimeters	Square inches	1.550×10^{-3}
Temperature (°F) −32	Temperature, °C	5/9

Table B.3 Specific Gravity Conversion

Relationship Between Baume Degrees and Specific Gravity

Liquids Lighter than Water

Formula—Specific Gravity $= \dfrac{140}{130 - °\text{Baume}}$

10	1.00000	30	0.87500	50	0.77778	70	0.70000
11	0.99291	31	0.86957	51	0.77348	71	0.69652
12	0.98592	32	0.86420	52	0.76923	72	0.69307
13	0.97902	33	0.85890	53	0.76503	73	0.68966
14	0.97222	34	0.85366	54	0.76087	74	0.68627
15	0.96552	35	0.84848	55	0.75676	75	0.68293
16	0.95890	36	0.84337	56	0.75269	76	0.67961
17	0.95238	37	0.83832	57	0.74866	77	0.67633
18	0.94595	38	0.83333	58	0.74468	78	0.67308
19	0.93960	39	0.82840	59	0.74074	79	0.66986
20	0.93333	40	0.82353	60	0.73684	80	0.66667
21	0.92715	41	0.81871	61	0.73298	81	0.66351
22	0.92105	42	0.81395	62	0.72917	82	0.66038
23	0.91503	43	0.80925	63	0.72539	83	0.65728
24	0.90909	44	0.80460	64	0.72165	84	0.65421
25	0.90323	45	0.80000	65	0.71795	85	0.65117
26	0.89744	46	0.79545	66	0.71428	86	0.64815
27	0.89172	47	0.79096	67	0.71066	87	0.64516
28	0.88608	48	0.78652	68	0.70707	88	0.64220
29	0.88050	49	0.78212	69	0.70352	89	0.63927

Liquids Heavier than Water

Formula—Specific Gravity $= \dfrac{145}{145 - °\text{Baume}}$

Baume Degrees	Specific Gravity 60°–60°F	Baume Degrees	Specific Gravity 60°–60°F	Baume Degrees	Specific Gravity 60°–60°F	Baume Degrees	Specific Gravity 60°–60°F
0	1.00000	10	1.07407	20	1.16000	30	1.26087
1	1.00694	11	1.08209	21	1.16935	31	1.27193
2	1.01399	12	1.09023	22	1.17886	32	1.28319
3	1.02113	13	1.09848	23	1.18852	33	1.29464
4	1.02837	14	1.10687	24	1.19835	34	1.30631
5	1.03571	15	1.11538	25	1.20833	35	1.31818
6	1.04317	16	1.12403	26	1.21849	36	1.33028
7	1.05072	17	1.13281	27	1.22881	37	1.34259
8	1.05839	18	1.14173	28	1.23932	38	1.35514
9	1.06618	19	1.15079	29	1.25000	39	1.36792

Table B.3, continued

Relationship between Baume Degrees and Specific Gravity

Liquids Heavier than Water

Formula—Specific Gravity = $\dfrac{145}{145 - °\text{Baume}}$

Baume Degrees	Specific Gravity 60°–60°F	Baume Degrees	Specific Gravity 60°–60°F	Baume Degrees	Specific Gravity 60°–60°F	Baume Degrees	Specific Gravity 60°–60°F
40	1.38095	50	1.52632	60	1.70588	70	1.93333
41	1.39423	51	1.54255	61	1.72619	71	1.95946
42	1.40777	52	1.55914	62	1.74699	72	1.98630
43	1.42157	53	1.57609	63	1.76829	73	2.01389
44	1.43564	54	1.59341	64	1.79012	74	2.04225
45	1.45000	55	1.61111	65	1.81250	75	2.07143
46	1.46465	56	1.62921	66	1.83544	76	2.10145
47	1.47959	57	1.64773	67	1.85897	77	2.13235
48	1.49485	58	1.66667	68	1.88312	78	2.16418
49	1.51042	59	1.68605	69	1.90789	79	2.19697

Relation of API Hydrometer Scale to Specific Gravity

Degrees API	Specific Gravity	Degrees API	Specific Gravity	Degrees API	Specific Gravity	Degrees API	Specific Gravity
10	1.0000	30	0.8762	50	0.7796	70	0.7022
11	0.9930	31	0.8708	51	0.7753	71	0.6987
12	0.9861	32	0.8654	52	0.7711	72	0.6953
13	0.9792	33	0.8602	53	0.7669	73	0.6919
14	0.9725	34	0.8550	54	0.7628	74	0.6886
15	0.9659	35	0.8498	55	0.7587	75	0.6852
16	0.9593	36	0.8448	56	0.7547	76	0.6819
17	0.9529	37	0.8398	57	0.7507	77	0.6787
18	0.9465	38	0.8348	58	0.7467	78	0.6754
19	0.9402	39	0.8299	59	0.7428	79	0.6722
20	0.9340	40	0.8251	60	0.7389	80	0.6690
21	0.9279	41	0.8203	61	0.7351	81	0.6659
22	0.9218	42	0.8156	62	0.7313	82	0.6628
23	0.9159	43	0.8109	63	0.7275	83	0.6597
24	0.9100	44	0.8063	64	0.7238	84	0.6566
25	0.9042	45	0.8017	65	0.7201	85	0.6536
26	0.8984	46	0.7972	66	0.7165	86	0.6506
27	0.8927	47	0.7927	67	0.7128	87	0.6476
28	0.8871	48	0.7883	68	0.7093	88	0.6446
29	0.8816	49	0.7839	69	0.7057	89	0.6417

Brinell	Brinell	Rockwell C	Rockwell A	Rockwell B	Shore Scleroscope
10-mm Ball, 3000-kg Diameter of Impression		Brale, 150-kg Load	Brale, 60-kg Load	$1/16$ in. Ball, 100-kg Load	
–	–	80	92.0	–	–
–	–	79	91.5	–	–
–	–	78	91.0	–	–
–	–	77	90.5	–	–
–	–	76	90.0	–	–
–	–	75	89.5	–	–
–	–	74	89.0	–	–
–	–	73	88.5	–	–
–	–	72	88.0	–	–
–	–	71	87.0	–	–
2.20	780	70	86.5	–	106
–	760	69	86.0	–	103
2.25	745	68	85.5	–	100
–	725	67	85.0	–	97
2.30	712	66	84.5	–	95
–	697	65	84.0	–	93
2.35	682	64	83.5	–	91
–	668	63	83.0	–	89
2.40	653	62	82.5	–	87
–	640	61	81.5	–	86
2.45	627	60	81.0	–	84
–	615	59	80.5	–	82
2.50	601	58	80.0	–	81
–	590	57	79.5	–	80
2.55	578	56	79.0	–	78
–	567	55	78.5	–	77
2.60	555	54	78.0	120	75
–	545	53	77.5	120	74
2.65	534	52	77.0	119	72
2.70	514	51	76.5	119	70
2.75	495	50	76.0	118	67
2.80	477	49	75.5	117	65
–	470	48	74.5	117	64
2.85	461	47	74.0	116	63
2.90	444	46	73.5	115	61
2.95	429	45	73.0	115	59
3.00	415	44	72.5	114	57
–	408	43	72.0	114	56
3.05	401	42	71.5	113	55
3.10	388	41	71.0	112	54
3.15	375	40	70.5	111	52
3.20	363	39	70.0	111	51
3.25	352	38	69.5	110	49
–	346	37	69.0	110	48
3.30	341	36	68.5	109	47

Hardness Conversion

Brinell	Brinell	Rockwell C	Rockwell A	Rockwell B	Shore Scleroscope
10-mm Ball, 3000-kg Diameter of Impression		Brale, 150-kg Load	Brale, 60-kg Load	$1/16$ in. Ball, 100-kg Load	
3.35	331	35	68.0	109	46
3.40	321	34	67.5	108	45
3.45	311	33	67.0	108	44
3.50	302	32	66.5	107	43
3.55	293	31	66.0	106	42
3.60	285	30	65.5	105	40
3.65	277	29	65.0	104	39
3.70	269	28	64.5	104	38
3.75	262	27	64.0	103	38
–	258	26	63.5	102	37
3.80	255	25	63.0	102	37
3.85	248	24	62.5	101	36
3.90	241	23	62.0	101	35
–	238	22	61.5	100	34
3.95	235	21	61.0	99	34
4.00	229	20	60.0	98	33
4.05	223	–	59.5	97	32
4.10	217	–	59.0	96	31
4.15	212	–	58.0	95	31
4.20	207	–	57.5	94	30
4.25	202	–	57.0	93	30
4.30	197	–	56.5	92	29
4.35	192	–	56.0	91	28
4.40	187	–	55.5	90	28
4.45	183	–	55.0	89	27
4.50	179	–	54.0	88	27
4.55	174	–	53.5	87	26
4.60	170	–	53.0	86	26
4.65	166	–	52.5	85	25
4.70	163	–	52.0	84	25
4.75	159	–	51.0	83	24
4.80	156	–	50.5	82	24
4.85	153	–	50.0	81	23
4.90	149	–	49.5	80	23
4.95	146	–	49.0	79	22
5.00	143	–	48.5	78	22
5.05	140	–	48.0	77	21
5.10	137	–	47.0	76	21
5.15	134	–	46.5	75	21
5.15	134	–	46.0	74	21
5.20	131	–	45.5	73	20
5.20	131	–	45.0	72	20
5.25	128	–	44.5	71	20
5.30	126	–	44.0	70	20

Table B.5   Temperature Conversion[a]

(Centigrade To Fahrenheit—Fahrenheit To Centigrade)

-459.4° to 0°			1° to 60°			61° to 290°			300° to 890°			900° to 3000°		
C	Centigrade or Fahrenheit	F	C	Centigrade or Fahrenheit	F	C	Centigrade or Fahrenheit	F	C	Centigrade or Fahrenheit	F	C	Centigrade or Fahrenheit	F
-273	-459.4		-17.2	1	33.8	16.1	61	141.8	149	300	572	482	900	1652
-268	-450		-16.7	2	35.6	16.7	62	143.6	154	310	590	488	910	1670
-262	-440		-16.1	3	37.4	17.2	63	145.4	160	320	608	493	920	1688
-257	-430		-15.6	4	39.2	17.8	64	147.2	166	330	626	499	930	1706
-251	-420		-15.0	5	41.0	18.3	65	149.0	171	340	644	504	940	1724
-246	-410		-14.4	6	42.8	18.9	66	150.8	177	350	662	510	950	1742
-240	-400		-13.9	7	44.6	19.4	67	152.6	182	360	680	516	960	1760
-234	-390		-13.3	8	46.4	20.0	68	154.4	188	370	698	521	970	1778
-229	-380		-12.8	9	48.2	20.6	69	156.2	193	380	716	527	980	1796
-223	-370		-12.2	10	50.0	21.1	70	158.0	199	390	734	532	990	1814
-218	-360		-11.7	11	51.8	21.7	71	159.8	204	400	752	538	1000	1832
-212	-350		-11.1	12	53.6	22.2	72	161.6	210	410	770	549	1020	1868
-207	-340		-10.6	13	55.4	22.8	73	163.4	215	420	788	560	1040	1904
-201	-330		-10.0	14	57.2	23.3	74	165.2	221	430	806	571	1060	1940
-196	-320		-9.4	15	59.0	23.9	75	167.0	227	440	824	582	1080	1976
-190	-310		-8.9	16	60.8	24.4	76	168.8	232	450	842	593	1100	2012
-184	-300		-8.3	17	62.6	25.0	77	170.6	238	460	860	604	1120	2048
-179	-290		-7.8	18	64.4	25.6	78	172.4	243	470	878	616	1140	2084
-173	-280		-7.2	19	66.2	26.1	79	174.2	249	480	896	627	1160	2120
-169	-273	-459.4	-6.7	20	68.0	26.7	80	176.0	254	490	914	638	1180	2156

C		F	C		F	C		F	C		F	C		F
−168	−270	−454	−6.1	21	69.8	27.2	81	177.8	260	500	932	649	1200	2192
−162	−260	−436	−5.6	22	71.6	27.8	82	179.6	266	510	950	660	1220	2228
−157	−250	−418	−5.0	23	73.4	28.3	83	181.4	271	520	968	671	1240	2264
−151	−240	−400	−4.4	24	75.2	28.9	84	183.2	277	530	986	682	1260	2300
−146	−230	−382	−3.9	25	77.0	29.4	85	185.0	282	540	1004	693	1280	2336
−140	−220	−364	−3.3	26	78.8	30.0	86	186.6	288	550	1022	704	1300	2372
−134	−210	−346	−2.8	27	80.6	30.6	87	188.4	293	560	1040	732	1350	2462
−129	−200	−328	−2.2	28	82.4	31.1	88	190.4	299	570	1058	760	1400	2552
−123	−190	−310	−1.7	29	84.2	31.7	89	192.2	304	580	1076	788	1450	2642
−118	−180	−292	−1.1	30	86.0	32.2	90	194.0	310	590	1094	816	1500	2732
−112	−170	−274	−0.6	31	87.8	32.8	91	195.8	316	600	1112	843	1550	2822
−107	−160	−256	0.0	32	89.6	33.3	92	197.6	321	610	1130	871	1600	2912
−101	−150	−238	0.6	33	91.4	33.9	93	199.4	327	620	1148	899	1650	3002
−96	−140	−220	1.1	34	93.2	34.4	94	201.2	332	630	1166	927	1700	3092
−90	−130	−202	1.7	35	95.0	35.0	95	203.0	338	640	1184	954	1750	3182
−84	−120	−184	2.2	36	96.8	35.6	96	204.8	343	650	1202	982	1800	3272
−79	−110	−166	2.8	37	98.6	36.1	97	206.6	349	660	1220	1010	1850	3362
−73	−100	−148	3.3	38	100.4	36.7	98	208.4	354	670	1238	1038	1900	3452
−68	−90	−130	3.9	39	102.2	37.2	99	210.2	360	680	1256	1066	1950	3542
−62	−80	−112	4.4	40	104.0	37.8	100	212.0	366	690	1274	1093	2000	3632
−57	−70	−94	5.0	41	105.8	43	110	230	371	700	1292	1121	2050	3722
−51	−60	−76	5.6	42	107.6	49	120	248	377	710	1310	1149	2100	3812
−46	−50	−58	6.1	43	109.4	54	130	266	382	720	1328	1177	2150	3902
−40	−40	−40	6.7	44	111.2	60	140	284	388	730	1346	1204	2200	3992
−34	−30	−22	7.2	45	113.0	66	150	302	393	740	1364	1232	2250	4082

Table B.5, continued

−459.4° to 0°			1° to 60°			61° to 290°			300° to 890°			900° to 3000°		
C	Centigrade or Fahrenheit	F	C	Centigrade or Fahrenheit	F	C	Centigrade or Fahrenheit	F	C	Centigrade or Fahrenheit	F	C	Centigrade or Fahrenheit	F
−29	−20	−4	7.8	46	114.8	71	160	320	399	750	1382	1260	2300	4172
−23	−10	14	8.3	47	116.6	77	170	338	404	760	1400	1288	2350	4262
−17.8	0	32	8.9	48	118.4	82	180	356	410	770	1418	1316	2400	4352
			9.4	49	120.2	88	190	374	416	780	1436	1343	2450	4442
			10.0	50	122.0	93	200	392	421	790	1454	1371	2500	4532
			10.6	51	123.8	99	210	410	427	800	1472	1399	2550	4622
			11.1	52	125.6	100	212	413.6	432	810	1490	1427	2600	4712
			11.7	53	127.4	104	220	428	438	820	1508	1454	2650	4802
			12.2	54	129.2	110	230	446	443	830	1526	1482	2700	4892
			12.8	55	131.0	116	240	464	449	840	1544	1510	2750	4982
			13.3	56	132.8	121	250	482	454	850	1562	1538	2800	5072
			13.9	57	134.6	127	260	500	460	860	1580	1566	2850	5162
			14.4	58	136.4	132	270	518	466	870	1598	1593	2900	5252
			15.0	59	138.2	138	280	536	471	880	1616	1621	2950	5342
			15.6	60	140.0	143	290	554	477	890	1634	1649	3000	5432

aLocate temperature in middle column. If in degrees Centigrade, read Fahrenheight equivalent in right-hand column; if in degrees Fahrenheit, read Centigrade equivalent in left-hand column.

$$F° = \frac{9}{5}C° + 32° \qquad C° = \frac{5}{9}(F° - 32°)$$

Table B.6 Length Conversion–Millimeters to Inches

Millimeters	0	1	2	3	4	5	6	7	8	9
0	0.00000	0.03937	0.07874	0.11811	0.15748	0.19685	0.23622	0.27559	0.31496	0.35433
10	0.39370	0.43307	0.47244	0.51181	0.55118	0.59055	0.62992	0.66929	0.70866	0.74803
20	0.78740	0.82677	0.86614	0.90551	0.94488	0.98425	1.02362	1.06299	1.10236	1.14173
30	1.18110	1.22047	1.25984	1.29921	1.33858	1.37795	1.41732	1.45669	1.49606	1.53543
40	1.57480	1.61417	1.65354	1.69291	1.73228	1.77165	1.81102	1.85039	1.88976	1.92913
50	1.96850	2.00787	2.04724	2.08661	2.12598	2.16535	2.20472	2.24409	2.28346	2.32283
60	2.36220	2.40157	2.44094	2.48031	2.51968	2.55905	2.59842	2.63779	2.67716	2.71653
70	2.75590	2.79527	2.83464	2.87401	2.91338	2.95275	2.99212	3.03149	3.07086	3.11023
80	3.14960	3.18897	3.22834	3.26771	3.30708	3.34645	3.38582	3.42519	3.46456	3.50393
90	3.54330	3.58267	3.62204	3.66141	3.70078	3.74015	3.77952	3.81889	3.85826	3.89763
100	3.93700	3.97637	4.01574	4.05511	4.09448	4.13385	4.17322	4.21259	4.25196	4.29133
110	4.33070	4.37007	4.40944	4.44881	4.48818	4.52755	4.56692	4.60629	4.64566	4.68503
120	4.72440	4.76377	4.80314	4.84251	4.88188	4.92125	4.96062	4.99999	5.03936	5.07873
130	5.11810	5.15747	5.19684	5.23621	5.27558	5.31495	5.35432	5.39369	5.43306	5.47243
140	5.51180	5.55117	5.59054	5.62991	5.66928	5.70865	5.74802	5.78739	5.82676	5.86613

Table B.6, continued

Millimeters	0	1	2	3	4	5	6	7	8	9
150	5.90550	5.94487	5.98424	6.02361	6.06298	6.10235	6.14172	6.18109	6.22046	6.25983
160	6.29920	6.33857	6.37794	6.41731	6.45668	6.49605	6.53542	6.57479	6.61416	6.65353
170	6.69290	6.73227	6.77164	6.81101	6.85038	6.88975	6.92912	6.96849	7.00786	7.04723
180	7.08660	7.12597	7.16534	7.20471	7.24408	7.28345	7.32282	7.36219	7.40156	7.44093
190	7.48030	7.51967	7.55904	7.59841	7.63778	7.67715	7.71652	7.75589	7.79526	7.83463
200	7.87400	7.91337	7.95274	7.99211	8.03148	8.07085	8.11022	8.14959	8.18896	8.22833
210	8.26770	8.30707	8.34644	8.38581	8.42518	8.46455	8.50392	8.54329	8.58266	8.62203
220	8.66140	8.70077	8.74014	8.77951	8.81888	8.85825	8.89762	8.93699	8.97636	9.01573
230	9.05510	9.09447	9.13384	9.17321	9.21258	9.25195	9.29132	9.33069	9.37006	9.40943
240	9.44880	9.48817	9.52754	9.56691	9.60628	9.64565	9.68502	9.72439	9.76376	9.80313
250	9.84250	9.88187	9.92124	9.96061	9.99998	10.03935	10.07872	10.11809	10.15746	10.19683
260	10.23620	10.27557	10.31494	10.35431	10.39368	10.43305	10.47242	10.51179	10.55116	10.59053
270	10.62990	10.66927	10.70864	10.74801	10.78738	10.82675	10.86612	10.90549	10.94486	10.98423
280	11.02360	11.06297	11.10234	11.14171	11.18108	11.22045	11.25982	11.29919	11.33856	11.37793
290	11.41730	11.45667	11.49604	11.53541	11.57478	11.61415	11.65352	11.69289	11.73226	11.77163
300	11.81100	11.85037	11.88974	11.92911	11.96848	12.00785	12.04722	12.08659	12.12596	12.16533

Table B.7 Pressure Conversion Table—Kilogram per Square Centimeter to Pounds per Square Inch

(1 kg/cm² = 14.233 P.S.I.)

kg

kg/cm²	0	1	2	3	4	5	6	7	8	9	10	11	12	13	14
0.0		14.22	28.45	42.67	56.89	71.11	85.34	99.56	113.8	128.0	142.2	156.5	170.7	184.9	199.1
0.1	1.4223	15.65	29.87	44.09	58.31	72.54	86.76	100.98	115.2	129.4	143.7	157.9	172.1	186.3	200.5
0.2	2.8446	17.07	31.29	45.51	59.74	73.96	88.18	102.41	116.6	130.9	145.1	159.3	173.5	187.7	202.0
0.3	4.2669	18.49	32.71	46.94	61.16	75.38	89.61	103.83	118.1	132.3	146.5	160.7	174.9	189.2	203.4
0.4	5.6892	19.91	34.14	48.36	62.58	76.80	91.03	105.25	119.5	133.7	147.9	162.1	176.4	190.6	204.8
0.5	7.1115	21.34	35.56	49.78	64.00	78.23	92.45	106.67	120.9	135.1	149.3	163.6	177.8	192.0	206.2
0.6	8.5338	22.76	36.98	51.20	65.43	79.65	93.87	108.10	122.3	136.5	150.8	165.0	179.2	193.4	207.7
0.7	9.9561	24.18	38.40	52.63	66.49	81.07	95.29	109.52	123.7	138.0	152.2	166.4	180.6	194.9	209.1
0.8	11.3784	25.60	39.82	54.05	68.27	82.49	96.72	110.94	125.2	139.4	153.6	167.8	182.0	196.3	210.5
0.9	12.8007	27.02	41.25	55.47	69.69	83.92	98.14	112.36	126.6	140.8	155.0	169.3	183.5	197.7	211.9

kg/cm²	15	16	17	18	19	20	21	22	23	24	25	26	27	28	29
0.0	213.4	227.6	241.8	256.0	270.2	284.5	298.7	312.9	327.1	341.4	355.6	369.8	384.0	398.2	412.5
0.1	214.8	229.0	243.2	257.4	271.7	285.9	300.1	314.3	328.6	342.8	357.0	371.2	385.4	399.7	413.9
0.2	216.2	230.4	244.6	258.9	273.1	287.3	301.5	315.8	330.0	344.2	358.4	372.6	386.9	401.1	415.3
0.3	217.6	231.8	246.1	260.3	274.5	288.7	303.0	317.2	331.4	345.6	359.8	374.1	388.3	402.5	416.7
0.4	219.0	233.3	247.5	261.7	275.9	290.2	304.4	318.6	332.8	347.0	361.3	375.5	389.7	403.9	418.2
0.5	220.5	234.7	248.9	263.1	277.4	291.6	305.8	320.0	334.2	248.5	362.7	376.9	391.1	405.4	419.6
0.6	221.9	236.1	250.3	264.6	278.8	293.0	307.2	321.4	335.7	349.9	364.1	378.3	392.6	407.8	421.0
0.7	223.3	237.5	251.7	266.0	280.2	294.4	308.6	322.9	337.1	351.3	365.5	379.8	394.0	408.2	422.4
0.8	224.7	239.0	253.2	267.4	281.6	295.8	310.1	324.3	338.5	352.7	367.0	381.2	395.4	409.6	423.9
0.9	226.2	240.4	254.6	268.8	283.0	297.3	311.5	325.7	339.9	354.2	368.4	382.6	396.8	411.0	425.3

Table B.7, continued

kg

kg/cm²	30	31	32	33	34	35	36	37	38	39	40	41	42	43	44
0.0	426.7	440.9	455.1	469.4	483.6	497.8	512.0	526.3	540.5	554.7	568.9	583.1	597.4	611.6	625.8
0.1	428.1	442.3	456.6	470.8	485.0	499.2	513.4	527.7	541.9	556.1	570.3	584.6	598.8	613.0	627.2
0.2	429.5	443.8	458.0	472.2	486.4	500.7	514.9	529.1	543.3	557.5	571.8	586.0	600.2	614.4	628.7
0.3	431.0	445.2	459.4	473.6	487.9	502.1	516.3	530.5	544.7	559.0	573.2	587.4	601.6	615.9	630.1
0.4	432.4	446.6	460.8	475.1	489.3	503.5	517.7	531.9	546.2	560.4	574.6	588.8	603.1	617.3	631.5
0.5	433.8	448.0	462.3	476.5	490.7	504.9	519.1	533.4	547.6	561.8	576.0	590.3	604.5	618.7	632.9
0.6	435.2	449.5	463.7	477.9	492.2	506.3	520.6	534.8	549.0	563.2	577.4	591.7	605.9	620.1	634.4
0.7	436.7	450.9	465.1	479.3	493.5	507.8	522.0	536.2	550.4	564.7	578.9	593.1	607.3	621.6	635.8
0.8	438.1	452.3	466.5	480.7	495.0	509.2	523.4	537.6	551.9	566.1	580.3	594.5	608.7	623.0	637.2
0.9	439.5	453.7	467.9	482.2	496.4	510.6	524.8	539.1	553.3	567.5	581.7	595.9	610.2	624.4	638.6

kg/cm²	45	46	47	48	49	50	51	52	53	54	55	56	57	58	59
0.0	640.0	654.3	668.5	682.7	696.9	711.1	725.4	739.6	753.8	768.0	782.3	796.5	810.7	824.9	839.2
0.1	641.5	655.7	669.9	684.1	698.4	712.6	726.8	741.0	755.2	769.5	783.7	797.9	812.1	826.4	840.6
0.2	642.9	657.1	671.3	685.6	699.8	714.0	728.2	742.4	756.7	770.9	785.1	799.3	813.6	827.8	842.0
0.3	644.3	658.5	672.8	687.0	701.2	715.4	729.6	743.9	758.1	772.3	786.5	800.8	815.0	829.2	843.4
0.4	645.7	660.0	674.2	688.4	702.6	716.8	731.1	745.3	759.5	773.7	788.0	802.2	816.4	830.6	844.9
0.5	647.2	661.4	675.6	689.8	704.0	718.3	732.5	746.7	760.9	775.2	789.4	803.6	817.8	832.1	846.3
0.6	648.6	662.8	677.0	691.2	705.5	719.7	733.9	748.1	762.4	776.6	790.8	805.0	819.2	833.5	847.7
0.7	650.0	664.2	678.4	692.7	706.9	721.1	735.3	749.6	763.8	778.0	792.2	806.4	820.7	834.9	849.1
0.8	651.4	665.6	679.9	694.1	708.3	722.5	736.8	751.0	765.2	779.4	793.6	807.9	822.1	836.3	850.5
0.9	652.8	667.0	681.3	695.5	709.7	724.0	738.2	752.4	766.6	780.8	795.1	809.3	823.5	837.7	852.0

	60	61	62	63	64	65	66	67	68	69	70	71	72	73	74
0.0	853.4	868	882	896	910	924	939	953	967	981	996	1010	1024	1038	1053
0.1	854.8	869	883	897	912	926	940	954	969	983	997	1011	1025	1040	1054
0.2	856.2	870	885	899	913	927	942	956	970	984	998	1013	1027	1041	1055
0.3	857.7	872	886	900	915	929	943	957	971	986	1000	1014	1028	1043	1057
0.4	859.1	873	888	902	916	930	944	959	973	987	1001	1016	1030	1044	1058
0.5	860.5	875	889	903	917	932	946	960	974	988	1003	1017	1031	1045	1060
0.6	861.9	876	890	905	919	933	947	961	976	990	1004	1018	1033	1047	1061
0.7	863.3	878	892	906	920	934	949	963	977	991	1006	1020	1034	1048	1062
0.8	864.8	879	893	907	922	936	950	964	979	993	1007	1021	1035	1050	1064
0.9	866.2	880	895	909	923	937	952	966	980	994	1008	1022	1037	1051	1065

	75	76	77	78	79	80	81	82	83	84	85	86	87	88	89
0.0	1067	1081	1095	1109	1124	1138	1152	1166	1181	1195	1209	1223	1237	1252	1266
0.2	1070	1084	1098	1112	1126	1141	1155	1169	1183	1198	1212	1226	1240	1254	1269
0.4	1072	1087	1101	1115	1129	1144	1158	1172	1186	1200	1215	1229	1243	1257	1272
0.6	1075	1089	1104	1118	1132	1146	1161	1175	1189	1203	1217	1232	1246	1260	1274
0.8	1078	1092	1107	1121	1135	1149	1163	1178	1192	1206	1220	1235	1249	1263	1277

	90	91	92	93	94	95	96	97	98	99
0.0	1280	1294	1309	1323	1337	1351	1365	1380	1394	1408
0.2	1283	1297	1311	1326	1340	1354	1368	1382	1397	1411
0.4	1286	1300	1314	1328	1343	1357	1371	1385	1400	1414
0.6	1289	1303	1317	1331	1345	1360	1374	1388	1402	1417
0.8	1291	1306	1320	1334	1348	1363	1377	1391	1405	1419

Table B.8 Volume Conversion

Cubic Inches to Cubic Centimeters

Cubic Inches	0	1	2	3	4	5	6	7	8	9
0	–	16.3871	32.7741	49.1612	65.548	81.935	98.322	114.709	131.097	147.484
10	163.871	180.258	196.645	213.032	229.419	245.806	262.193	278.580	294.967	311.354
20	327.741	344.128	360.515	376.902	393.290	409.677	426.064	442.451	458.838	475.225
30	491.612	508.00	524.39	540.77	557.16	573.55	589.93	606.32	622.71	639.10
40	655.48	671.87	688.36	704.64	721.03	737.42	753.80	770.19	786.58	802.97
50	819.35	835.74	852.13	868.51	884.90	901.29	917.68	934.06	950.45	966.84
60	983.22	999.61	1016.00	1032.39	1048.77	1065.16	1081.55	1097.93	1114.32	1130.71
70	1147.09	1163.48	1179.87	1196.26	1212.64	1229.03	1245.42	1261.80	1278.19	1294.58
80	1310.97	1327.35	1343.74	1360.13	1376.51	1392.90	1409.29	1425.67	1442.06	1458.45
90	1474.84	1491.22	1507.61	1524.00	1540.38	1556.77	1573.16	1589.55	1605.93	1622.32

Cubic Centimeters to Cubic Inches

Cubic Centimeter	0	1	2	3	4	5	6	7	8	9
0	–	0.061024	0.122047	0.183071	0.244095	0.305119	0.366142	0.427166	0.488190	0.54921
10	0.61024	0.67126	0.73228	0.79331	0.85433	0.91536	0.97638	1.03740	1.09843	1.15945
20	1.22047	1.28150	1.34252	1.40355	1.46457	1.52559	1.58662	1.64764	1.70866	1.76969
30	1.83071	1.89174	1.95276	2.01378	2.07481	2.13583	2.19685	2.25788	2.31890	2.37993
40	2.44095	2.50197	2.56300	2.62402	2.68504	2.74607	2.80709	2.86812	2.92914	2.99016
50	3.05119	3.11221	3.17323	3.23426	3.29528	3.35631	3.41733	3.47835	3.53938	3.60040
60	3.66142	3.72245	3.78347	3.84450	3.90552	3.96654	4.02757	4.08859	4.14961	4.21064
70	4.27166	4.33269	4.39371	4.45473	4.51576	4.57678	4.63780	4.69883	4.75985	4.82088
80	4.88190	4.94292	5.0039	5.0650	5.1260	5.1870	5.2480	5.3091	5.3701	5.4311
90	5.4921	5.5532	5.6142	5.6752	5.7362	5.7973	5.8583	5.9193	5.9803	6.0414

Gallons to Liters

Gallons	0	1	2	3	4	5	6	7	8	9
0	–	4.5460	9.0919	13.6379	18.1839	22.7298	27.2758	31.8217	36.3677	40.9137
10	45.4596	50.006	54.552	59.098	63.643	68.189	72.735	77.281	81.827	86.373
20	90.919	95.465	100.011	104.557	109.103	113.649	118.195	122.741	127.287	131.833
30	136.379	140.925	145.471	150.017	154.563	159.109	163.655	168.201	172.747	177.293
40	181.839	186.384	190.930	195.476	200.022	204.568	209.114	213.660	218.206	222.752
50	227.298	231.844	236.390	240.936	245.482	250.028	254.574	259.120	263.666	268.212
60	272.758	277.304	281.850	286.396	290.942	295.488	300.493	304.580	309.125	313.671
70	318.217	322.763	327.309	331.855	336.401	340.947	345.493	350.039	354.585	369.131
80	363.877	368.223	372.769	377.315	381.861	386.407	390.953	395.499	400.045	404.591
90	409.137	413.683	418.229	422.775	427.321	431.866	436.412	440.958	445.504	450.050

Liters to Gallons

Liters	0	1	2	3	4	5	6	7	8	9
0	–	0.219975	0.439951	0.65993	0.87990	1.09988	1.31985	1.53983	1.75980	1.97978
10	2.19975	2.41973	2.63970	2.85968	3.07965	3.29963	3.51960	3.73958	3.95956	4.17953
20	4.39951	4.61948	4.83946	5.0594	5.2794	5.4994	5.7194	5.9393	6.1593	6.3793
30	6.5993	6.8192	7.0392	7.2592	7.4792	7.6991	7.9191	8.1391	8.3591	8.5790
40	8.7990	9.0190	9.2390	9.4589	9.6789	9.8989	10.1189	10.3388	10.5588	10.7788
50	10.9988	11.2187	11.4387	11.6587	11.8787	12.0986	12.3186	12.5386	12.7586	12.9785
60	13.1985	13.4185	13.6385	13.8584	14.0784	14.2984	14.5184	14.7383	14.9583	15.1783
70	15.3983	15.6182	15.8382	16.0582	16.2782	16.4981	16.7181	16.9381	17.1581	17.3780
80	17.5980	17.8180	18.0380	18.2579	18.4779	18.6979	18.9179	19.1379	19.3578	19.5778
90	19.7978	20.0178	20.2377	20.4577	20.6777	20.8977	21.1176	21.3376	21.5576	21.7776

Table B.9 Unit Volume Conversion

U.S. Gallons per Minute to Liters per Minute

U.S. GPM	0	1	2	3	4	5	6	7	8	9
0	00	3.78	7.57	11.35	15.14	18.92	22.71	26.49	30.28	34.06
10	37.85	41.63	45.42	49.20	52.99	56.77	60.56	64.34	68.13	71.91
20	75.70	79.48	83.27	87.05	90.84	94.62	98.40	102.19	105.97	109.76
30	113.54	117.33	121.11	124.90	128.68	132.47	136.25	140.04	143.82	147.61
40	151.39	155.18	158.96	162.75	166.53	170.32	174.10	177.89	181.67	185.46
50	189.24	193.02	196.81	200.59	204.38	208.16	211.95	215.73	219.52	223.30
60	227.09	230.87	234.66	238.44	242.23	246.01	249.80	253.58	257.37	261.15
70	264.94	268.72	272.51	276.29	280.08	283.86	287.64	291.43	295.21	299.00
80	302.78	306.57	310.35	314.14	317.92	321.71	325.49	329.28	333.06	336.85
90	340.63	344.42	348.20	351.99	355.77	359.56	363.34	367.13	370.91	374.70
100	378.48	382.26	386.05	389.83	393.62	397.40	401.19	404.97	408.76	412.54
110	416.33	420.11	423.90	427.68	431.47	435.25	439.04	442.82	446.61	450.39
120	454.18	457.96	461.75	465.53	469.32	473.10	476.88	480.67	484.45	488.24
130	492.02	495.81	499.59	503.38	507.16	510.95	514.73	518.52	522.30	526.09
140	529.87	533.66	537.44	541.23	545.01	548.80	552.58	556.37	560.15	563.94
150	567.72	571.50	575.29	579.07	582.86	586.64	590.43	594.21	598.00	601.78

APPENDIX C

SI UNITS AND CONVERSION FACTORS

The tables contained in this Appendix were reprinted from API Publication 2564—*Manual of Petroleum Measurement Standards:* Chapter 15, "Guidelines for the Use of the International System of Units (SI) in the Petroleum and Allied Industries," 2nd edition (December 1980), pp. 7-31, courtesy of the American Petroleum Institute, 2101 L Street, Northwest, Washington, DC.

CONTENTS

NOTES ON USE OF TABLES

Metric units recommended for general use are given under the heading "API preferred metric unit." Other units that also may be needed are shown in the "other allowable" column. Preferred units do not preclude the use of other multiples or submultiples, as the choice of such unit-multiple is governed by the magnitude of the numerical value.

Notation used conforms to SI practice, i.e., groups of three digits to the left or right of the decimal marker are separated by spaces. No commas or other triad spacers are used. E-notation is used for convenience because it is a standard method of display in many calculators and because of the inability of computers to print out or transmit superscripts. An asterisk (*) denotes that

1105

all of the succeeding digits would be zeroes. If a conversion factor ends in zero but does not have an asterisk, then any subsequent digits would not necessarily be zeroes.

For example, then:

$$3.056\ 0\ E + 00 = 3.056\ 0 \times 10^0 = 3.056\ 0$$

$$3.056*E - 01 = 3.056\ 000 \times 10^{-1} = 0.305\ 600\ 0$$

$$9.290\ 304\ E + 02 = 9.290\ 304 \times 10^2 = 929.030\ 4$$

Table C.1   Nomenclature for Conversion Tables

Symbol	Name	Quantity	Definition		Type of Unit
			In Terms of Other Units	In Terms of Base Units	
A	Ampere	Electric current	—	—	Base
a	Annum (year)	Time	365 d	$3.153\,600 \times 10^7$ s	Allowable
bar	Bar	Pressure	10^5 Pa	10^5 kg/(m·s^2)	Allowable
Bq	Becquerel	Activity (of a radionuclide)	—	$1\ s^{-1}$	Derived
C	Coulomb	Quantity of electricity, electric charge	—	1 A·s	Derived
°C	Degree Celsius	Celsius temperature	—	—	Derived
cd	Candela	Luminous intensity	—	—	Base
cP	Centipoise	Dynamic viscosity	1 mPa·s	10^{-3} kg/(m·s)	Allowable
cSt	Centistokes	Kinematic viscosity	1 mm^2/s	10^{-6} m^2/s	Allowable
d	Day	Time	24 hr	8.640×10^4 s	Allowable
F	Farad	Capacitance	1 C/V	$1\ s^4\ A^2/(m^2 \cdot kg)$	Derived
g	Gram	Mass	—	10^{-3} kg	Allowable (submultiple of base unit)
Gy	Gray	Absorbed dose, specific energy imparted, kerma, absorbed dose index	1 J/kg	1 m^2/s^2	Derived
h	Hour	Time	60 min	3.6×10^3 s	Allowable
H	Henry	Inductance	1 Wb/A	$1\ m^2 \cdot kg/(s^2 \cdot A^2)$	Derived
ha	Hectare	Area	—	10^4 m^2	Allowable
Hz	Hertz	Frequency (of periodic phenomenon)	—	$1\ s^{-1}$	Derived
J	Joule	Work, energy, quantity of heat	1 N·m	$1\ m^2 \cdot kg/s^2$	Derived
K	Kelvin	Thermodynamic temperature	—	—	Base
kg	Kilogram	Mass	—	—	Base
kn	Knot	Velocity	1.852 km/h	5.144 444 m^2/s	Allowable
L	Liter	Volume	1 dm^3	10^{-3} m^3	Allowable

Table C.1, continued

			Definition		
Symbol	Name	Quantity	In Terms of Other Units	In Terms of Base Units	Type of Unit
lm	Lumen	Luminous flux	—	$1\ cd \cdot sr$	Derived
lx	Lux	Illumination	$1\ lm/m^2$	$1\ cd \cdot sr/m^2$	Derived
m	Meter	Length	—	—	Base
min	Minute	Time	—	$60\ s$	Allowable
mol	Mole	Amount of substance	—	—	Base
N	Newton	Force	—	$1\ m \cdot kg/s^2$	Derived
naut. mi	Nautical mile	Distance	$1.852\ km$	$1.852 \times 10^3\ m$	Allowable
Pa	Pascal	Pressure	$1\ N/m^2$	$1\ kg/(m \cdot s^2)$	Derived
r	Revolution	Angular displacement	$360°$	$2\pi\ rad$	Allowable
rad	Radian	Plane angle	—	—	Supplementary
S	Siemens	Conductance	$1\ A/V$	$1\ s^3 \cdot A^2/(m^2 \cdot kg)$	Derived
s	Second	Time	—	—	Base
sr	Steradian	Solid angle	—	—	Supplementary
Sv	Sievert	Dose equivalent	$1\ J/kg$	$1\ m^2/s^2$	Derived
t	Metric ton	Mass	$1\ Mg$	$10^3 kg$	Allowable
T	Tesla	Magnetic flux density	$1\ Wb/m^2$	$1\ kg/(s^2 \cdot A)$	Derived
V	Volt	Electric potential, potential difference, electromotive force	$1\ W/A$	$1\ m^2 \cdot kg/(s^3 \cdot A)$	Derived
W	Watt	Power, radiant flux	$1\ J/s$	$1\ m^2 \cdot kg/s^3$	Derived
Wb	Weber	Magnetic flux	$1\ V/s$	$1\ m^2 \cdot kg/(s^2 \cdot A)$	Derived
Ω	Ohm	Electric resistance	$1\ V/A$	$1\ m^2 \cdot kg/(s^3 \cdot A^2)$	Derived
°	Degree	Plane angle		$\pi/180\ rad$	Allowable
'	Minute	Plane angle	$(1/60)°$	$2.908\ 882 \times 10^{-4}$ rad	Allowable
"	Second	Plane angle	$(1/60)'$	$4.848\ 137 \times 10^{-6}$ rad	Allowable

Table C.2 Tables of Recommended SI Units and Conversion Factors

Quantity	SI Unit	Customary Unit	Metric Unit		Conversion Factor (multiply quantity expressed in customary units by factor to get metric equivalent)	
			API Preferred	Other Allowable		
		Space, Time				
Length	m	naut. mi	km	naut. mi	1.852*	E+00
					1	
		mi	km		1.609 344*	E+00
		mi (U.S. statute)	km		1.609 347	E+00
		Chain	m		2.011 684	E+01
		Rod	m		5.029 210	E+00
		Fathom	m		1.828 804	E+00
		m	m		1	E+00
		yard	m		9.144*	E−01
		ft	m		3.048*	E−01
		ft (U.S. survey)	m		3.048 006	E−01
		link	m		2.011 684	E−01
		in.	mm	cm	2.54*	E+01
					2.54*	E+00
		cm	mm	cm	1.0*	E+01
					1	
		mm	mm		1	
		mil	μm		2.54*	E+01
		micron (μ)	μm		1	

Table C.2, continued

Quantity	SI Unit	Customary Unit	Metric Unit		Conversion Factor (multiply quantity expressed in customary units by factor to get metric equivalent)
			API Preferred	Other Allowable	
Space, Time (continued)					
Surface Texture	m	μin.	μm		2.54* E-02
		nm	μm		1.0* E-03
Length/Length	m/m	ft/mi	m/km		1.893 939 E-01
Length/Volume	m/m^3	ft/U.S. gal	m/m^3		8.051 964 E+01
		ft/ft^3	m/m^3		1.076 391 E+01
		ft/bbl	m/m^3		1.917 134 E+00
Length/Temperature	m/K	see Temperature, Pressure, Vacuum			
Area	m^2	mi^2	km^2		2.589 988 E+00
		mi^2 (U.S. statute)	km^2		2.589 998 E+00
		ha	m^2		1.0* E+04
		acre	ha		4.046 873 E-01
				m^2	4.046 873 E+03
		sq chain	m^2		4.046 873 E+02
		sq rod	m^2		2.529 295 E+01
Area		yd^2	m^2		8.361 274 E-01
		ft^2	m^2		9.290 304* E-02
		ft^2 (U.S. survey)	m^2		9.290 341 E-02

Quantity	To convert from	to	to (alt.)	Multiply by
	in.²	mm²	cm²	6.451 6* E+02 / 6.451 6* E+00
	cm²	mm²	cm²	1.0* E+02 / 1
	mm²	mm²		1
Area/Volume (m²/m³)	ft²/in.³	m²/cm³		5.669 291 E−03
Area/Mass (m²/kg)	cm²/g	m²/kg		1.0* E−01
Volume, Capacity (m³)	mi³	km³		4.168 182 E+00
	acre·ft	m³	ha·m	1.233 489 E+03 / 1.233 489 E−01
	m³	m³		1
	yd³	m³		7.645 549 E−01
	bbl (42 U.S. gal)	m³		1.589 873 E−01
	ft³	m³	dm³ (L)	2.831 685 E−02 / 2.831 685 E+01
	Can. gal	m³	dm³ (L)	4.546 09* E−03 / 4.546 09* E+00
	U.K. gal	m³	dm³ (L)	4.546 092 E−03 / 4.546 092 E+00
	U.S. gal	m³	dm³ (L)	3.785 412 E−03 / 3.785 412 E+00
	L	dm³ (L)		1
	U.K. qt	dm³ (L)		1.136 523 E+00
	U.S. qt	dm³ (L)		9.463 529 E−01
	U.K. pt	dm³ (L)		5.682 615 E−01
	U.S. pt	dm³ (L)		4.731 765 E−01

Table C.2, continued

Quantity	SI Unit	Customary Unit	Metric Unit API Preferred	Metric Unit Other Allowable	Conversion Factor (multiply quantity expressed in customary units by factor to get metric equivalent)
Space, Time (continued)					
Volume, Capacity	m^3	U.K. fl oz	cm^3		2.841 308 E+01
		U.S. fl oz	cm^3		2.957 353 E+01
		in.^{3}	cm^3		1.638 706 E+01
		mL	cm^3		1
Volume/Length (linear displacement)	m^3/m	bbl/in.	m^3/m		6.259 342 E+00
		bbl/ft	m^3/m		5.216 119 E−01
		ft^3/ft	m^3/m		9.290 304* E−02
		U.S. gal/ft	dm^3/m	L/m	1.241 933 E+01
Volume/Mass	m^3/kg	see Density, Specific Volume, Concentration, Dosage			
Plane Angle	rad	rad	rad		1
		deg (°)	rad	°	1.745 329 E−02
					1
		min (')	rad	'	2.908 882 E−04
					1
		sec ('')	rad	''	4.848 137 E−06
					1

Quantity	Unit	Value		Unit symbol	Alt symbol
Solid Angle	sr	1		sr	
Time					
	million years (MY)	1		Ma	
	yr	1		a	
	wk	7.0*	E+00	d	
	d	1		d	
	h	1		h	
		6.0*	E+01		min
	min	6.0*	E+01	s	
		1.666 667	E−02		h
		1			min
	s	1		s	
	millimicrosecond	1		ns	

Mass, Amount of Substance

Quantity	Unit	Value		Unit symbol	Alt symbol
Mass					
	U.K. ton (long ton)	1.016 047	E+00	Mg	t
	U.S. ton (short ton)	9.071 847	E−01	Mg	t
	U.K. cwt	5.080 235	E+01	kg	
	U.S. cwt	4.535 924	E+01	kg	
	kg	1		kg	
	lb	4.535 924	E−01	kg	
	oz (troy)	3.110 348	E+01	g	
	oz (avdp)	2.834 952	E+01	g	
	g	1		g	
	grain	6.479 891	E+01	mg	
	mg	1		mg	
	μg	1		μg	

Table C.2, continued

Quantity	SI Unit	Customary Unit	Metric Unit		Conversion Factor (multiply quantity expressed in customary units by factor to get metric equivalent)
			API Preferred	Other Allowable	
Mass, Amount of Substance (continued)					
Mass/Length	kg/m	see Mechanics			
Mass/Area	kg/m^2	see Mechanics			
Mass/Volume	kg/m^3	see Density, Specific Volume, Concentration, Dosage			
Mass/Mass	kg/kg	see Density, Specific Volume, Concentration, Dosage			
Amount of Substance	mol	ft^3 (60°F, 1 atm)	kmol		1.195 29 E−03
		ft^3 (60°F, 14.73 lbf/in.^{2})	kmol		1.198 06 E−03
		m^3 (0°C, 1 atm)	kmol		4.461 53 E−02
		m^3 (15°C, 1 atm)	kmol		4.229 28 E−02
		m^3 (20°C, 1 atm)	kmol		4.157 15 E−02
		m^3 (25°C, 1 atm)	kmol		4.087 43 E−02
Heating Value, Entropy, Heat Capacity					
Heating Value (mass basis)	J/kg	Btu/lb	MJ/kg	J/g	2.326 000 E−03
			kJ/kg		2.326 000 E+00
				kW·h/kg	6.461 112 E−04
		cal/g	kJ/kg		4.184* E+00
		cal/lb		J/g	9.224 141 E+00

Quantity	SI unit	To convert from	To	To	Multiply by
Heating Value (mole basis)	J/mol	kcal/g mol	kJ/kmol		4.184* E+03
		Btu/lb mol	MJ/kmol		2.326 000 E−03
			kJ/kmol		2.326 000 E+00
Heating Value (volume basis- solids and liquids)	J/m³	therm/U.S. gal	MJ/m³	kJ/dm³	2.787 163 E+04
			kJ/m³		2.787 163 E+07
				kW·h/dm³	7.742 119 E+00
		therm/U.K. gal	MJ/m³	kJ/dm³	2.320 798 E+04
			kJ/m³		2.320 798 E+07
				kW·h/dm³	6.446 660 E+00
		therm/Can. gal	MJ/m³	kJ/dm³	2.320 799 E+04
			kJ/m³		2.320 799 E+07
				kW·h/dm³	6.446 663 E+00
		Btu/U.S. gal	MJ/m³	kJ/dm³	2.787 163 E−01
			kJ/m³		2.787 163 E+02
				kW·h/m³	7.742 119 E−02
		Btu/U.K. gal	MJ/m³	kJ/dm³	2.320 800 E−01
			kJ/m³		2.320 800 E+02
				kW·h/m³	6.446 660 E−02
		Btu/Can. gal	MJ/m³	kJ/dm³	2.320 799 E−01
			kJ/m³		2.320 799 E+02
				kW·h/m³	6.446 663 E−02
		Btu/ft³	MJ/m³	kJ/dm³	3.725 895 E−02
			kJ/m³		3.725 895 E+01
				kW·h/m³	1.034 971 E−02
		kcal/m³	MJ/m³		4.184* E−03
			kJ/m³		4.184* E+00
		cal/mL	MJ/m³	kJ/dm³	4.184* E+00
		ft·lbf/U.S. gal	kJ/m³		3.581 692 E−01

Table C.2, continued

Quantity	SI Unit	Customary Unit	Metric Unit		Conversion Factor (multiply quantity expressed in customary units by factor to get metric equivalent)
			API Preferred	Other Allowable	
Heating Value, Entropy, Heat Capacity (continued)					
Heating Value (volume basis—gases)	J/m^3	cal/mL^3	kJ/m^3	J/dm^3	4.184* E+03
		$kcal/m^3$	kJ/m^3	J/dm^3	4.184* E+00
		Btu/ft^3	kJ/m^3	J/dm^3	3.725 895 E+01
Specific Entropy	$J/(kg \cdot K)$	$Btu/(lb \cdot °R)$	$kJ/(kg \cdot K)$	$J/(g \cdot K)$	4.186 8* E+00
		$cal/(g \cdot K)$	$kJ/(kg \cdot K)$	$J/(g \cdot K)$	4.184* E+00
		$kcal/(kg \cdot °C)$	$kJ/(kg \cdot K)$	$J/(g \cdot K)$	4.184* E+00
Specific Heat Capacity (mass basis)	$J/(kg \cdot K)$	$kW \cdot h/(kg \cdot °C)$	$kJ/(kg \cdot K)$	$J/(g \cdot °C)$	3.6* E+03
		$Btu/(lb \cdot °F)$	$kJ/(kg \cdot K)$	$J/(g \cdot °C)$	4.186 8* E+00
		$kcal/(kg \cdot °C)$	$kJ/(kg \cdot K)$	$J/(g \cdot °C)$	4.184* E+00
Molar Heat Capacity	$J/(mol \cdot K)$	$Btu/(lb\ mol \cdot °F)$	$kJ/(kmol \cdot K)$	$J/(g \cdot °C)$	4.186 8* E+00
		$cal/(g\ mol \cdot °C)$	$kJ/(kmol \cdot K)$	$J/(g \cdot °C)$	4.184* E+00
Temperature, Pressure, Vacuum					
Temperature (absolute)	K	$°R$	K		5/9
		K	K		1
Temperature (traditional)	K	$°F$	$°C$		$(°F - 32)/1.8$
		$°C$	$°C$		1

Quantity	SI unit	To convert from	to	Multiply by
Temperature (difference)	K	°F	°C	5/9
		°C	°C	1
Temperature/Length (geothermal gradient)	K/m	°F/100 ft	mK/m	1.822 689 E+01
Length/Temperature (geothermal step)	m/K	ft/°F	m/K	5.486 4* E-01
Pressure	Pa	atm (14.696 lbf/in.^{2} or 760 mm Hg at 0°C)	MPa	1.013 250* E-01
			kPa	1.013 250* E+02
			bar	1.013 250* E+00
		bar	MPa	1.0* E-01
			kPa	1.0* E+02
			bar	1
		at (kgf/cm^2) (technical atmosphere)	MPa	9.806 650* E-02
			kPa	9.806 650* E+01
			bar	9.806 650* E-01
		lbf/in.^{2} (psi)	MPa	6.894 757 E-03
			kPa	6.894 757 E+00
			bar	6.894 757 E-02
		in Hg at 60°F	kPa	3.376 85 E+00
		in Hg at 32°F	kPa	3.386 38 E+00
		in H_2O at 39.2°F	kPa	2.490 82 E-01
		in H_2O at 60°F	kPa	2.488 4 E-01
		mm Hg at 0°C (torr)	kPa	1.333 22 E-01
		cm H_2O at 4°C	kPa	9.806 38 E-02
		lbf/ft^2(psf)	kPa	4.788 026 E-02
		μm Hg at 0°C	Pa	1.333 22 E-01
		μbar	Pa	1.0* E-01
		dyn/cm^2	Pa	1.0* E-01

Table C.2, continued

Quantity	SI Unit	Customary Unit	Metric Unit — API Preferred	Metric Unit — Other Allowable	Conversion Factor (multiply quantity expressed in customary units by factor to get metric equivalent)
Temperature, Pressure, Vacuum (continued)					
Vacuum, Draft	Pa	in Hg at 60°F	kPa		3.376 85 E+00
		in H$_2$O at 39.2°F	kPa		2.490 82 E−01
		in H$_2$O at 60°F	kPa		2.488 4 E−01
		mm Hg at 0°C (torr)	kPa		1.333 22 E−01
		cm H$_2$O at 4°C	kPa		9.806 38 E−02
Liquid Head	m	ft	m		3.048* E−01
		in.	mm		2.54* E+01
Pressure Drop/Length	Pa/m	psi/ft	kPa/m		2.262 059 E+01
		psi/100 ft	kPa/m		2.262 059 E−01
		psi/mi	kPa/km		4.284 203 E+00
Density, Specific Volume, Concentration, Dosage					
Density (gases)	kg/m^3	lb/ft^3	kg/m^3		1.601 846 E+01
Density (liquids)	kg/m^3	lb/U.S. gal	kg/m^3		1.198 264 E+02
				kg/dm^3	1.198 264 E−01
		lb/U.K. gal	kg/m^3		9.977 633 E+01
				kg/dm^3	9.977 633 E−02
		lb/ft^3	kg/m^3		1.601 846 E+01
				kg/dm^3	1.601 846 E−02

Quantity	Conventional unit	SI unit	Conversion factor
Density (solids) (kg/m³)	g/cm³	kg/m³	1.0* E+03
		kg/dm³	1 E+03
	kg/L	kg/m³	1.0* E+03
	°API	kg/m³	Use tables
	lb/ft³	kg/m³	1.601 846 E+01
		kg/dm³	1.601 846 E−02
Specific Volume (gases) (m³/kg)	ft³/lb	m³/kg	6.242 796 E−02
		dm³/kg	6.242 796 E+01
Specific Volume (liquids) (m³/kg)	ft³/lb	m³/kg	6.242 796 E−02
		dm³/kg	6.242 796 E+01
	U.K. gal/lb	m³/kg	1.022 242 E−02
		dm³/kg	1.022 242 E+01
	U.S. gal/lb	m³/kg	8.345 404 E−03
		dm³/kg	8.345 404 E+00
Molar Volume (m³/mol)	L/g mol	m³/kmol	1
	ft³/lb mol	m³/kmol	6.242 796 E−02
Specific Volume (clay yield) (m³/kg)	bbl/U.S. ton	m³/Mg (m³/t)	1.752 535 E−01
	bbl/U.K. ton	m³/Mg (m³/t)	1.564 763 E−01
Yield (shale distillation) (m³/kg)	bbl/U.S. ton	dm³/Mg (dm³/t)	1.752 535 E+02
	bbl/U.K. ton	dm³/Mg (dm³/t)	1.564 763 E+02
	U.S. gal/U.S. ton	dm³/Mg (dm³/t)	4.172 702 E+00
	U.S. gal/U.K. ton	dm³/Mg (dm³/t)	3.725 627 E+00
Concentration (mass/mass) (kg/kg)	wt %	kg/kg	1.0* E−02
		g/kg	1.0* E+01
	wt ppm	mg/kg	1

Table C.2, continued

Quantity	SI Unit	Customary Unit	Metric Unit API Preferred	Metric Unit Other Allowable	Conversion Factor (multiply quantity expressed in customary units by factor to get metric equivalent)
Density, Specific Volume, Concentration, Dosage (continued)					
Concentration (mass/volume)	kg/m³	lb/bbl	kg/m³	g/dm³	2.853 010 E+00
		g/U.S. gal	kg/m³		2.641 720 E−01
		g/U.K. gal	kg/m³		2.199 692 E−01
		lb/1000 U.S. gal	g/m³	mg/dm³	1.198 264 E+02
		lb/1000 U.K. gal	g/m³	mg/dm³	9.977 633 E+01
		gr/U.S. gal	g/m³	mg/dm³	1.711 806 E+01
		lb/1000 bbl	g/m³	mg/dm³	2.853 010 E+00
		mg/U.S. gal	g/m³	mg/dm³	2.641 720 E−01
		gr/100 ft³	mg/m³		2.288 352 E+01
		gr/ft³	mg/m³		2.288 352 E+03
Concentration (volume/volume)	m³/m³	bbl/bbl	m³/m³		1
		ft³/ft³	m³/m³		1
		bbl/(acre·ft)	dm³/m³	L/m³	1.288 923 E−01
		U.K. gal/ft³	dm³/m³	L/m³	1.605 437 E+02
		U.S. gal/ft³	dm³/m³	L/m³	1.336 806 E+02
		mL/U.S. gal	dm³/m³	L/m³	2.641 720 E−01
		mL/U.K. gal	dm³/m³	L/m³	2.199 692 E−01
		Vol %	m³/m³		1.0* E−02
			cm³/m³		1
		Vol ppm	dm³/m³	L/m³	1.0* E−03

Quantity	SI unit	To convert from	To	To (alt)	Multiply by
Concentration (mole/volume)	mol/m³	U.K. gal/1000 bbl	cm³/m³		2.859 406 E+01
		U.S. gal/1000 bbl	cm³/m³		2.380 952 E+01
		lb mol/U.S. gal	kmol/m³		1.198 264 E+02
		lb mol/U.K. gal	kmol/m³		9.977 633 E+01
		lb mol/ft³	kmol/m³		1.601 846 E+01
		std ft³(60°F, 1 atm)/bbl	kmol/m³		7.518 18 E−03
Concentration (volume/mole)	m³/mol	U.S. gal/1000 std ft³ (60°F/60°F)	dm³/kmol	L/kmol	3.166 93 E+00
		bbl/million std ft³ (60°F/60°F)	dm³/kmol	L/kmol	1.330 11 E−01

Facility Throughput, Capacity

Quantity	SI unit	To convert from	To	To (alt)	Multiply by
Throughput (mass basis)	kg/s	million lb/yr	Mg/a	t/a	4.535 924 E+02
		U.K. ton/yr	Mg/a	t/a	1.016 047 E+00
		U.S. ton/yr	Mg/a	t/a	9.071 847 E−01
		U.K. ton/d	Mg/d	t/d	1.016 047 E+00
				t/h, Mg/h	4.233 529 E−02
		U.S. ton/d	Mg/d	t/d	9.071 847 E−01
				t/h, Mg/h	3.779 936 E−02
		U.K. ton/h	Mg/h	t/h	1.016 047 E+00
		U.S. ton/h	Mg/h	t/h	9.071 847 E−01
		lb/h	kg/h		4.535 924 E−01
Throughput (volume basis)	m³/s	bbl/d	m³/a		5.803 036 E+01
			m³/d		1.589 873 E−01
			m³/h		6.624 471 E−03
		ft³/d	m³/h		1.179 869 E−03
			m³/d		2.831 685 E−02

Table C.2, continued

Quantity	SI Unit	Customary Unit	Metric Unit API Preferred	Metric Unit Other Allowable	Conversion Factor (multiply quantity expressed in customary units by factor to get metric equivalent)
Facility Throughput, Capacity (continued)					
		bbl/h	m³/h		1.589 873 E−01
		ft³/h	m³/h		2.831 685 E−02
		U.K. gal/h	m³/h		4.546 092 E−03
				L/h	4.546 092 E+00
		U.S. gal/h	m³/h		3.785 412 E−03
				L/h	3.785 412 E+00
		U.K. gal/min	m³/h		2.727 655 E−01
				L/min	4.546 092 E+00
		U.S. gal/min	m³/h		2.271 247 E−01
				L/min	3.785 412 E+00
Throughput (mole basis)	mol/s	lb mol/h	kmol/h		4.535 924 E−01
				kmol/s	1.259 979 E−04
Pipeline Capacity	m³/m	bbl/mi	m³/km		9.879 013 E−02
Flowrate					
Flowrate (mass basis)	kg/s	U.K. ton/min	kg/s		1.693 412 E+01
		U.S. ton/min	kg/s		1.511 975 E+01
		U.K. ton/h	kg/s		2.822 353 E−01
		U.S. ton/h	kg/s		2.519 958 E−01

Quantity	From	To	Factor
	million lb/d	kg/s	5.249 912 E+00
	U.K. ton/d	kg/s	1.175 980 E−02
	U.S. ton/d	kg/s	1.049 982 E−02
	million lb/yr	kg/s	1.438 332 E−02
	U.K. ton/yr	kg/s	3.221 864 E−05
	U.S. ton/yr	kg/s	2.876 664 E−05
	lb/s	kg/s	4.535 924 E−01
	lb/min	kg/s	7.559 873 E−03
	lb/h	kg/s	1.259 979 E−04
Flowrate (volume basis) m^3/s	bbl/d	dm^3/s	1.840 131 E−03
	ft^3/d	dm^3/s	3.277 413 E−04
	bbl/h	dm^3/s	4.416 314 E−02
	ft^3/h	dm^3/s	7.865 791 E−03
	U.K. gal/h	dm^3/s	1.262 803 E−03
	U.S. gal/h	dm^3/s	1.051 503 E−03
	U.K. gal/min	dm^3/s	7.576 820 E−02
	U.S. gal/min	dm^3/s	6.309 020 E−02
	ft^3/min	dm^3/s	4.719 474 E−01
	ft^3/s	dm^3/s	2.831 685 E+01
Flowrate (mole basis) mol/s	lb mol/s	kmol/s	4.535 924 E−01
	lb mol/h	kmol/s	1.259 979 E−04
	million scf/sd	kmol/s	1.383 449 E−02
Flowrate/Length (mass basis) $kg/(s·m)$	lb/(s·ft)	$kg/(s·m)$	1.488 164 E+00
	lb/(h·ft)	$kg/(s·m)$	4.133 789 E−04
Flowrate/Length (volume basis) m^2/s	U.K. gal/(min·ft)	$m^3/(s·m)$	2.485 833 E−04
	U.S. gal/(min·ft)	$m^3/(s·m)$	2.069 888 E−04

Table C.2, continued

Quantity	SI Unit	Customary Unit	Metric Unit		Conversion Factor (multiply quantity expressed in customary units by factor to get metric equivalent)
			API Preferred	Other Allowable	
Flowrate (continued)					
		U.K. gal/(h·in.)	m^2/s	$m^3/(s \cdot m)$	4.971 667 E-05
		U.S. gal/(h·in.)	m^2/s	$m^3/(s \cdot m)$	4.139 776 E-05
		U.K. gal/(h·ft)	m^2/s	$m^3/(s \cdot m)$	4.143 055 E-06
		U.S. gal/(h·ft)	m^2/s	$m^3/(s \cdot m)$	3.449 814 E-06
Flowrate/Area (mass basis)	$kg/(s \cdot m^2)$	lb/(s·ft^2)	$kg/(s \cdot m^2)$		4.882 428 E+00
		lb/(h·ft^2)	$kg/(s \cdot m^2)$		1.356 230 E-03
Flowrate/Area (volume basis)	m/s	ft^3/(s·ft^2)	m/s	$m^3/(s \cdot m^2)$	3.048* E-01
		ft^3/(min·ft^2)	m/s	$m^3/(s \cdot m^2)$	5.08* E-03
		U.K. gal/(h·in.^{2})	m/s	$m^3/(s \cdot m^2)$	1.957 349 E-03
		U.S. gal/(h·in.^{2})	m/s	$m^3/(s \cdot m^2)$	1.629 833 E-03
		U.K. gal/(min·ft^2)	m/s	$m^3/(s \cdot m^2)$	8.155 621 E-04
		U.S. gal/(min·ft^2)	m/s	$m^3/(s \cdot m^2)$	6.790 972 E-04
		U.K. gal/(h·ft^2)	m/s	$m^3/(s \cdot m^2)$	1.359 270 E-05
		U.S. gal/(h·ft^2)	m/s	$m^3/(s \cdot m^2)$	1.131 829 E-05
Flowrate/Pressure Drop (productivity index)	$m^3/(s \cdot Pa)$	bbl/(d·psi)	$m^3/(d \cdot kPa)$		2.305 916 E-02

Energy, Work, Quantity of Heat, Power

Energy, Work, Quantity of Heat	J				
	quad	EJ	1.055 056 E+00		
		TW·h	2.930 711 E+02		
	therm	MJ	1.055 056 E+02		
		kJ	1.055 056 E+05	kW·h	2.930 711 E+01
	U.S. tonf·mi	MJ	1.431 744 E+01		
	hp·h	MJ	2.684 520 E+00		
		kJ	2.684 520 E+03	kW·h	7.456 999 E-01
	ch·h or CV·h	MJ	2.647 796 E+00		
		kJ	2.647 796 E+03	kW·h	7.354 99 E-01
	kW·h	MJ	3.6* E+00		
		kJ	3.6* E+03		
	Chu	kJ	1.899 101 E+00	kW·h	5.275 280 E-04
	Btu	kJ	1.055 056 E+00	kW·h	2.930 711 E-04
	kcal	kJ	4.184* E+00		
	cal	kJ	4.184* E-03		
	ft·lbf	kJ	1.355 818 E-03		
	J	kJ	1.0* E-03		
	lb·ft²/s²(ft·pdl)	kJ	4.214 011 E-05		
	erg	J	1.0* E-07		

Table C.2, continued

Quantity	SI Unit	Customary Unit	Metric Unit		Conversion Factor (multiply quantity expressed in customary units by factor to get metric equivalent)
			API Preferred	Other Allowable	
Energy, Work, Quantity of Heat Power (continued)					
Impact Energy	J	kgf·m	J		9.806 650* E+00
		ft·lbf	J		1.355 818 E+00
Work/Length	J/m	U.S. tonf·mi/ft	MJ/m		4.697 322 E+01
Surface Energy	J/m²	erg/cm²	mJ/m²		1.0* E+00
Power	W	quad/yr	EJ/a	GW	1.055 056 E+00
					3.345 561 E+01
		erg/a	TW		3.170 979 E-27
			GW		3.170 979 E-24
		million Btu/h	MW		2.930 711 E-01
		ton of refrigeration	kW		3.516 853 E+00
		Btu/s	kW		1.055 056 E+00
		kW	kW		1
		hydraulic horse-power-hhp	kW		7.460 43 E-01
		hp (electric)	kW		7.46* E-01
		hp (550 ft·lbf/s)	kW		7.456 999 E-01
		ch or CV	kW		7.354 99 E-01
		Btu/min	kW		1.758 427 E-02
		ft·lbf/s	kW		1.355 818 E-03

Quantity	(SI)	From	To	Factor	Exp
Power/Area	W/m³	kcal/h	W	1.162 222	E+00
		Btu/h	W	2.930 711	E−01
		ft·lbf/min	W	2.259 697	E−02
Heat Flow Unit-hfu (geothermics)		Btu/(s·ft²)	kW/m²	1.135 653	E+01
		cal/(h·cm²)	kW/m²	1.162 222	E−02
		Btu/(h·ft²)	kW/m³	3.154 591	E−03
		μcal/(s·cm²)	mW/m²	4.184*	E+01
Heat Release Rate, Mixing Power	W/m³	hp/ft³	kW/m³	2.633 414	E+01
		cal/(h·cm³)	kW/m³	1.162 222	E+00
		Btu/(s·ft³)	kW/m³	3.725 895	E+01
		Btu/(h·ft³)	kW/m³	1.034 971	E−02
Heat Generation Unit-hgu (radioactive rocks)		cal/(s·cm³)	μW/m³	4.184*	E+12
Cooling Duty (machinery)	W/W	Btu/(bhp·h)	W/kW	3.930 148	E−01
Mass Fuel Consumption	kg/J	lb/(hp·h)	kg/MJ	1.689 659	E−01
		lb/(hp·h)	kg/(kW·h)	6.082 774	E−01
Volume Fuel Consumption	m³/J	m³/(kW·h)	dm³/MJ	2.777 778	E+02
		m³/(kW·h)	dm³/(kW·h)	1.0*	E+03
		U.S. gal/(hp·h)	dm³/MJ	1.410 089	E+00
		U.S. gal/(hp·h)	dm³/(kW·h)	5.076 321	E+00
		U.K. pt/(hp·h)	dm³/MJ	2.116 809	E−01
		U.K. pt/(hp·h)	dm³/(kW·h)	7.620 512	E−01

Table C.2, continued

Quantity	SI Unit	Customary Unit	Metric Unit		Conversion Factor (multiply quantity expressed in customary units by factor to get metric equivalent)
			API Preferred	Other Allowable	
Energy, Work, Quantity of Heat Power (continued)					
Fuel Consumption (automotive)	m^3/m	U.K. gal/mi	$dm^3/100$ km	L/100 km	2.824 811 E+02
		U.S. gal/mi	$dm^3/100$ km	L/100 km	2.352 146 E+02
		mi/U.S. gal	km/dm^3	km/L	4.251 437 E-01
		mi/U.K. gal	km/dm^3	km/L	3.540 060 E-01
Mechanics					
Velocity (linear), Speed	m/s	knot	km/h		1.852 E+00
				knot	1
		mi/h	km/h		1.609 344* E+00
		m/s	m/s		1
		ft/s	m/s		3.048* E-01
				cm/s	3.048* E+01
				m/ms	3.048* E-04
		ft/min	m/s		5.08* E-03
				cm/s	5.08* E-01
		ft/h	mm/s		8.466 667 E-02
				cm/s	8.466 667 E-03
		ft/d	mm/s		3.527 778 E-03
				m/d	3.048* E-01

Quantity	SI unit	To convert from	to	Multiply by
Velocity (angular)	rad/s	in./s	mm/s	2.54* E+01
		in./s	cm/s	2.54* E+00
		in./min	mm/s	4.233 333 E−01
		in./min	cm/s	4.233 333 E−02
		r/min	rad/s	1.047 198 E−01
		r/min	r/min	1
		r/s	rad/s	6.283 185 E+00
		r/s	r/s	1
		deg/min	rad/s	2.908 882 E−04
		deg/s	rad/s	1.745 329 E−02
Reciprocal Velocity	μs/m	μs/ft	μs/m	3.280 840 E+00
Acceleration (linear)	m/s²	ft/s²	m/s²	3.048* E−01
		gal (cm/s²)	m/s²	1.0* E−02
Acceleration (angular)	rad/s²	rad/s²	rad/s²	1
		rpm	rad/s²	1.047 198 E−01
Corrosion Rate	mm/a	in./yr (ipy)	mm/a	2.54* E+01
Momentum	kg·m/s	lb·ft/s	kg·m/s	1.382 550 E−01
Force	N	U.K. tonf	kN	9.964 016 E+00
		U.S. tonf	kN	8.896 443 E+00
		kgf (kp)	N	9.806 650* E+00
		lbf	N	4.448 222 E+00
		N	N	1
		pdl	mN	1.382 550 E+02
		dyn	mN	1.0* E−02

Table C.2, continued

Quantity	SI Unit	Customary Unit	Metric Unit API Preferred	Metric Unit Other Allowable	Conversion Factor (multiply quantity expressed in customary units by factor to get metric equivalent)
Mechanics (continued)					
Bending Moment, Torque	$N \cdot m$	U.S. tonf·m	$kN \cdot m$		2.711 636 E+00
		kgf·m	$N \cdot m$		9.806 650* E+00
		lbf·ft	$N \cdot m$		1.355 818 E+00
		lbf·in.	$N \cdot m$		1.129 848 E−01
		pdl·ft	$N \cdot m$		4.214 011 E−02
Bending Movement, Length	$N \cdot m/m$	lbf·ft/in.	$N \cdot m/m$		5.337 866 E+01
		kgf·m/m	$N \cdot m/m$		9.806 650* E+00
		lbf·in./in.	$N \cdot m/m$		4.448 222 E+00
Moment of Inertia	$kg \cdot m^2$	$lb \cdot ft^2$	$kg \cdot m^2$		4.214 011 E−02
		$in.^4$	cm^4		4.162 314 E+01
Stress	Pa	U.S. $tonf/in.^2$	MPa	N/mm^2	1.378 951 E+00
		kgf/mm^2	MPa	N/mm^2	9.806 650* E+00
		U.S. $tonf/ft^2$	MPa	N/mm^2	9.576 052 E−02
		$lbf/in.^2$ (psi)	MPa	N/mm^2	6.894 757 E−03
		lbf/ft^2 (psf)	kPa		4.788 026 E−02
		dyn/cm^2	Pa		1.0* E−01
Yield Point, Gel Strength (drilling fluid)		$lbf/100 \ ft^2$	Pa		4.788 026 E+01

Quantity	SI Unit	Convert from	To Unit (alternate)	Multiply by
Mass/Length	kg/m	lb/ft	kg/m	1.488 164 E+00
Mass/Area Structural Loading, Bearing Capacity (mass basis)	kg/m²	U.S. ton/ft²	Mg/m² (t/m²)	9.764 855 E+00
		lb/ft²	kg/m²	4.882 428 E+00
Modulus of Elasticity	Pa	lbf/in.² (psi)	MPa (N/mm²)	6.894 757 E-03
Section Modulus	m³	in.³	cm³	1.638 706 E+01
Coefficient of Thermal Expansion	m/(m·K)	in./(in.°F)	mm/(mm·°C)	5.555 556 E-01

Transport Properties

Quantity	SI Unit	Convert from	To Unit (alternate)	Multiply by
Diffusivity	m²/s	ft²/s	mm²/s	9.290 304* E+04
		cm²/s	mm²/s	1.0* E+02
		ft²/h	mm²/s	2.580 64* E+01
Thermal Resistance	K·m²/W	°C·m²·h/kcal	K·m²/kW	8.604 208 E+02
		°F·ft²·h/Btu	K·m²/kW	1.761 102 E+02
Heat Flux	W/m²	Btu/(h·ft²)	kW/m²	3.154 591 E-03
Thermal Conductivity	W/(m·K)	cal/(s·cm²·°C/cm)	W/(m²·°C/m)	4.184* E+02
		Btu/(h·ft²·°F/ft)	W/(m²·°C/m)	1.730 735 E+00
		kcal/(h·m²·°C/m)	W/(m²·°C/m)	1.162 222 E+00
		Btu/(h·ft²·°F/in.)	W/(m²·°C/m)	1.422 279 E-01
		cal/(h·cm²·°C/cm)	W/(m²·°C/m)	1.162 222 E-01
Heat Transfer Coefficient	W/(m²·K)	cal/(s·cm²·°C)	kW/(m²·K)	4.184* E+01
		Btu/(s·ft²·°F)	kW/(m²·K)	2.044 175 E+01
		cal/(h·cm²·°C)	kW/(m²·K)	1.162 222 E-02
		Btu/(h·ft²·°F)	kW/(m²·K)	5.678 263 E-03
		Btu/(h·ft²·°R)	kW/(m²·K)	5.678 263 E-03
		kcal/(h·m²·°C)	kW/(m²·K)	1.162 222 E-03

Table C.2, continued

Quantity	SI Unit	Customary Unit	Metric Unit — API Preferred	Metric Unit — Other Allowable	Conversion Factor (multiply quantity expressed in customary units by factor to get metric equivalent)
Transport Properties (continued)					
Volumetric Heat Transfer Coefficient	W/(m³·K)	Btu/(s·ft³·°F)	kW/(m³·K)		6.706 611 E+01
		Btu/(h·ft³·°F)	kW/(m³·K)		1.862 947 E-02
Surface Tension	N/m	dyn/cm	mN/m		1.0* E+00
Viscosity (dynamic)	Pa·s	lbf·s/in.²	mPa·s	cP	6.894 757 E+06
		lbf·s/ft2	mPa·s	cP	4.788 026 E+04
		kgf·s/m²	mPa·s		9.806 650* E+03
		dyn·s/cm²	Pa·s		1.0* E-01
			mPa·s	cP	1.0* E+02
		P	Pa·s		1.0* E-01
			mPa·s	cP	1.0* E+02
		cP	mPa·s	cP	1.0* E+00
Viscosity (kinematic)	m²/s	ft²/s	mm²/s	cSt	9.290 304* E+04
		in.²/s	mm²/s	cSt	6.451 6* E+02
		ft²/h	mm²/s	cSt	2.580 64* E+01
		m²/h	mm²/s		2.777 778 E+02
		cm²/s	mm²/s		1.0* E+02
		St	mm²/s	cSt	1.0* E+02
		cSt	mm²/s	cSt	1.0* E+00

Quantity	SI Unit	To convert from	to	Multiply by
Permeability	m^2	D	μm^2	1.0* E+00
		mD	μm^2	1.0* E−03
Electricity, Magnetism				
Admittance	S	S	S	1
Capacitance	F	μF	μF	1
Capacity, Storage Battery	C	A·h	kC	3.6* E+00
Charge Density	C/m^3	C/mm^3	C/mm^3	1
Conductance	S	S	S	1
		℧	S	1
Conductivity	S/m	S/m	S/m	1
		$m℧/m$	mS/m	
Current Density	A/m^2	A/mm^2	A/mm^2	1
		$A/in.^2$	A/mm^2	1.550 003 E−03
Displacement	C/m^2	C/cm^2	C/cm^2	1
Electric Charge	C	C	C	1
Electric Current	A	A	A	1
Electric Dipole Moment	C·m	C·m	C·m	1
Electric Field Strength	V/m	V/m	V/m	1
Electric Flux	C	C	C	1
Electric Polarization	C/m^2	C/cm^2	C/cm^2	1

Table C.2, continued

Quantity	SI Unit	Customary Unit	Metric Unit		Conversion Factor (multiply quantity expressed in customary units by factor to get metric equivalent)
			API Preferred	Other Allowable	
Electricity, Magnetism (continued)					
Electric Potential	V	V mV	V mV		1 1
Electromagnetic Moment	A·m²	A·m²	A·m²		1
Electromotive Force	V	V	V		1
Flux of Displacement	C	C	C		1
Frequency	Hz	cycles/s	Hz		1
Impedance	Ω	Ω	Ω		1
Linear Current Density	A/m	A/mm	A/mm		1
Magnetic Dipole Moment	Wb·m	Wb·m	Wb·m		1
Magnetic Field Strength	A/m	A/mm oersted	A/mm A/m		1 7.957 747 E+01
Magnetic Flux	Wb	mWb maxwell	mWb nWb		1 1.0* E+01
Magnetic Flux Density	T	mT gauss gamma	mT T nT		1 1.0* E-04 1

Magnetic Induction	T	mT	mT	1
Magnetic Moment	$A \cdot m^2$	$A \cdot m^2$	$A \cdot m^2$	1
Magnetic Polarization	T	mT	mT	1
Magnetic Potential Difference	A	A	A	1
Magnetic Vector Potential	Wb/m	Wb/mm	Wb/mm	1
Magnetization	A/m	A/mm	A/mm	1
Modulus of Admittance	S	S	S	1
Modulus of Impedance	Ω	Ω	Ω	1
Mutual Inductance	H	H	H	1
Permeability	H/m	μH/m	μH/m	1
Permeance	H	H	H	1
Permittivity	F/m	μF/m	μF/m	1
Potential Difference	V	V	V	1
Quantity of Electricity	C	C	C	1
Reactance	Ω	Ω	Ω	1
Reluctance	H^{-1}	H^{-1}	H^{-1}	1
Resistance	Ω	Ω	Ω	1
Resistivity	$\Omega \cdot m$	$\Omega \cdot cm$ $\Omega \cdot m$	$\Omega \cdot cm$ $\Omega \cdot m$	1 1

Table C.2, continued

Quantity	SI Unit	Customary Unit	Metric Unit		Conversion Factor (multiply quantity expressed in customary units by factor to get metric equivalent)
			API Preferred	Other Allowable	
Electricity, Magnetism (continued)					
Self Inductance	H	mH	mH		1
Surface Density of Charge	C/m^2	mC/m^2	mC/m^2		1
Susceptance	S	S	S		1
Volume Density of Charge	C/mm^3	C/mm^3	C/mm^3		1
Acoustics, Light, Radiation					
Acoustical Energy	J	J	J		1
Acoustical Intensity	W/m^2	W/cm^2	W/m^2		1.0* E+04
Acoustical Power	W	W	W		1
Activity (of a radionuclide)	Bq	curie	Bq		3.7* E+10
Illuminance, Illumination	lx	footcandle (fc) lm/ft^2	lx lx		1.076 391 E+01 1.076 391 E+01
Irradiance	W/m^2	W/m^2	W/m^2		1
Light Exposure	$lx \cdot s$	$fc \cdot s$	$lx \cdot s$		1.076 391 E+01

Luminance	cd/m²	cd/m²	1
	footlambert (fL)	cd/m²	3.426 259 E+00
Luminous Efficacy	lm/W	lm/W	1
Luminous Exitance	lm/m²	lm/m²	1
	lm/ft²	lm/m²	1.076 391 E+01
Luminous Flux	lm	lm	1
Luminous Intensity	cd	cd	1
Quantity of Light	lm·s	lm·s	1
	talbot		
Radiance	W/(m²·sr)	W/(m²·sr)	1
Radiant Energy	J	J	1
Radiant Flux	W	W	1
Radiant Intensity	W/sr	W/sr	1
Radiant Power	W	W	1
Sound Pressure	Pa	Pa	1
Wavelength	m	nm	1.0* E−01
	Å		